LONDON MATHEMATICAL SOCIETY LECTURE NOTE SERIES

Managing Editor: Professor N. J. Hitchin, Mathematical Institute, University of Oxford, 24-29 St Giles, Oxford OX1 3LB,
United Kingdom

The titles below are available from booksellers, or from Cambridge University Press at
www.cambridge.org/mathematics

159 Groups St Andrews 1989 volume 1, C.M. CAMPBELL & E.F. ROBERTSON (eds)
160 Groups St Andrews 1989 volume 2, C.M. CAMPBELL & E.F. ROBERTSON (eds)
161 Lectures on block theory, B. KÜLSHAMMER
163 Topics in varieties of group representations, S.M. VOVSI
164 Quasi-symmetric designs, M.S. SHRIKANDE & S.S. SANE
166 Surveys in combinatorics, 1991, A.D. KEEDWELL (ed)
168 Representations of algebras, H. TACHIKAWA & S. BRENNER (eds)
169 Boolean function complexity, M.S. PATERSON (ed)
170 Manifolds with singularities and the Adams-Novikov spectral sequence, B. BOTVINNIK
171 Squares, A.R. RAJWADE
172 Algebraic varieties, G.R. KEMPF
173 Discrete groups and geometry, W.J. HARVEY & C. MACLACHLAN (eds)
174 Lectures on mechanics, J.E. MARSDEN
175 Adams memorial symposium on algebraic topology 1, N. RAY & G. WALKER (eds)
176 Adams memorial symposium on algebraic topology 2, N. RAY & G. WALKER (eds)
177 Applications of categories in computer science, M. FOURMAN, P. JOHNSTONE & A. PITTS (eds)
178 Lower K- and L-theory, A. RANICKI
179 Complex projective geometry, G. ELLINGSRUD *et al*
180 Lectures on ergodic theory and Pesin theory on compact manifolds, M. POLLICOTT
181 Geometric group theory I, G.A. NIBLO & M.A. ROLLER (eds)
182 Geometric group theory II, G.A. NIBLO & M.A. ROLLER (eds)
183 Shintani zeta functions, A. YUKIE
184 Arithmetical functions, W. SCHWARZ & J. SPILKER
185 Representations of solvable groups, O. MANZ & T.R. WOLF
186 Complexity: knots, colourings and counting, D.J.A. WELSH
187 Surveys in combinatorics, 1993, K. WALKER (ed)
188 Local analysis for the odd order theorem, H. BENDER & G. GLAUBERMAN
189 Locally presentable and accessible categories, J. ADAMEK & J. ROSICKY
190 Polynomial invariants of finite groups, D.J. BENSON
191 Finite geometry and combinatorics, F. DE CLERCK *et al*
192 Symplectic geometry, D. SALAMON (ed)
194 Independent random variables and rearrangement invariant spaces, M. BRAVERMAN
195 Arithmetic of blowup algebras, W. VASCONCELOS
196 Microlocal analysis for differential operators, A. GRIGIS & J. SJÖSTRAND
197 Two-dimensional homotopy and combinatorial group theory, C. HOG-ANGELONI *et al*
198 The algebraic characterization of geometric 4-manifolds, J.A. HILLMAN
199 Invariant potential theory in the unit ball of C", M. STOLL
200 The Grothendieck theory of dessins d'enfant, L. SCHNEPS (ed)
201 Singularities, J.-P. BRASSELET (ed)
202 The technique of pseudodifferential operators, H.O. CORDES
203 Hochschild cohomology of von Neumann algebras, A. SINCLAIR & R. SMITH
204 Combinatorial and geometric group theory, A.J. DUNCAN, N.D. GILBERT & J. HOWIE (eds)
205 Ergodic theory and its connections with harmonic analysis, K. PETERSEN & I. SALAMA (eds)
207 Groups of Lie type and their geometries, W.M. KANTOR & L. DI MARTINO (eds)
208 Vector bundles in algebraic geometry, N.J. HITCHIN, P. NEWSTEAD & W.M. OXBURY (eds)
209 Arithmetic of diagonal hypersurfaces over finite fields, F.Q. GOUVÉA & N. YUI
210 Hilbert C*-modules, E.C. LANCE
211 Groups 93 Galway / St Andrews I, C.M. CAMPBELL *et al* (eds)
212 Groups 93 Galway / St Andrews II, C.M. CAMPBELL *et al* (eds)
214 Generalised Euler-Jacobi inversion formula and asymptotics beyond all orders, V. KOWALENKO *et al*
215 Number theory 1992–93, S. DAVID (ed)
216 Stochastic partial differential equations, A. ETHERIDGE (ed)
217 Quadratic forms with applications to algebraic geometry and topology, A. PFISTER
218 Surveys in combinatorics, 1995, P. ROWLINSON (ed)
220 Algebraic set theory, A. JOYAL & I. MOERDIJK
221 Harmonic approximation., S.J. GARDINER

222 Advances in linear logic, J.-Y. GIRARD, Y. LAFONT & L. REGNIER (eds)
223 Analytic semigroups and semilinear initial boundary value problems, KAZUAKI TAIRA
224 Computability, enumerability, unsolvability, S.B. COOPER, T.A. SLAMAN & S.S. WAINER (eds)
225 A mathematical introduction to string theory, S. ALBEVERIO, *et al*
226 Novikov conjectures, index theorems and rigidity I, S. FERRY, A. RANICKI & J. ROSENBERG (eds)
227 Novikov conjectures, index theorems and rigidity II, S. FERRY, A. RANICKI & J. ROSENBERG (eds)
228 Ergodic theory of Z^d actions, M. POLLICOTT & K. SCHMIDT (eds)
229 Ergodicity for infinite dimensional systems, G. DA PRATO & J. ZABCZYK
230 Prolegomena to a middlebrow arithmetic of curves of genus 2, J.W.S. CASSELS & E.V. FLYNN
231 Semigroup theory and its applications, K.H. HOFMANN & M.W. MISLOVE (eds)
232 The descriptive set theory of Polish group actions, H. BECKER & A.S. KECHRIS
233 Finite fields and applications, S. COHEN & H. NIEDERREITER (eds)
234 Introduction to subfactors, V. JONES & V.S. SUNDER
235 Number theory 1993–94, S. DAVID (ed)
236 The James forest, H. FETTER & B. G. DE BUEN
237 Sieve methods, exponential sums, and their applications in number theory, G.R.H. GREAVES *et al*
238 Representation theory and algebraic geometry, A. MARTSINKOVSKY & G. TODOROV (eds)
240 Stable groups, F.O. WAGNER
241 Surveys in combinatorics, 1997, R.A. BAILEY (ed)
242 Geometric Galois actions I, L. SCHNEPS & P. LOCHAK (eds)
243 Geometric Galois actions II, L. SCHNEPS & P. LOCHAK (eds)
244 Model theory of groups and automorphism groups, D. EVANS (ed)
245 Geometry, combinatorial designs and related structures, J.W.P. HIRSCHFELD *et al*
246 p-Automorphisms of finite p-groups, E.I. KHUKHRO
247 Analytic number theory, Y. MOTOHASHI (ed)
248 Tame topology and o-minimal structures, L. VAN DEN DRIES
249 The atlas of finite groups: ten years on, R. CURTIS & R. WILSON (eds)
250 Characters and blocks of finite groups, G. NAVARRO
251 Gröbner bases and applications, B. BUCHBERGER & F. WINKLER (eds)
252 Geometry and cohomology in group theory, P. KROPHOLLER, G. NIBLO & R. STÖHR (eds)
253 The q-Schur algebra, S. DONKIN
254 Galois representations in arithmetic algebraic geometry, A.J. SCHOLL & R.L. TAYLOR (eds)
255 Symmetries and integrability of difference equations, P.A. CLARKSON & F.W. NIJHOFF (eds)
256 Aspects of Galois theory, H. VÖLKLEIN *et al*
257 An introduction to noncommutative differential geometry and its physical applications 2ed, J. MADORE
258 Sets and proofs, S.B. COOPER & J. TRUSS (eds)
259 Models and computability, S.B. COOPER & J. TRUSS (eds)
260 Groups St Andrews 1997 in Bath, I, C.M. CAMPBELL *et al*
261 Groups St Andrews 1997 in Bath, II, C.M. CAMPBELL *et al*
262 Analysis and logic, C.W. HENSON, J. IOVINO, A.S. KECHRIS & E. ODELL
263 Singularity theory, B. BRUCE & D. MOND (eds)
264 New trends in algebraic geometry, K. HULEK, F. CATANESE, C. PETERS & M. REID (eds)
265 Elliptic curves in cryptography, I. BLAKE, G. SEROUSSI & N. SMART
267 Surveys in combinatorics, 1999, J.D. LAMB & D.A. PREECE (eds)
268 Spectral asymptotics in the semi-classical limit, M. DIMASSI & J. SJÖSTRAND
269 Ergodic theory and topological dynamics, M.B. BEKKA & M. MAYER
270 Analysis on Lie groups, N.T. VAROPOULOS & S. MUSTAPHA
271 Singular perturbations of differential operators, S. ALBEVERIO & P. KURASOV
272 Character theory for the odd order theorem, T. PETERFALVI
273 Spectral theory and geometry, E.B. DAVIES & Y. SAFAROV (eds)
274 The Mandlebrot set, theme and variations, TAN LEI (ed)
275 Descriptive set theory and dynamical systems, M. FOREMAN *et al*
276 Singularities of plane curves, E. CASAS-ALVERO
277 Computational and geometric aspects of modern algebra, M.D. ATKINSON *et al*
278 Global attractors in abstract parabolic problems, J.W. CHOLEWA & T. DLOTKO
279 Topics in symbolic dynamics and applications, F. BLANCHARD, A. MAASS & A. NOGUEIRA (eds)
280 Characters and automorphism groups of compact Riemann surfaces, T. BREUER
281 Explicit birational geometry of 3-folds, A. CORTI & M. REID (eds)
282 Auslander-Buchweitz approximations of equivariant modules, M. HASHIMOTO
283 Nonlinear elasticity, Y. FU & R.W. OGDEN (eds)
284 Foundations of computational mathematics, R. DEVORE, A. ISERLES & E. SÜLI (eds)
285 Rational points on curves over finite, fields, H. NIEDERREITER & C. XING
286 Clifford algebras and spinors 2ed, P. LOUNESTO
287 Topics on Riemann surfaces and Fuchsian groups, E. BUJALANCE *et al*

288 Surveys in combinatorics, 2001, J. HIRSCHFELD (ed)
289 Aspects of Sobolev-type inequalities, L. SALOFF-COSTE
290 Quantum groups and Lie theory, A. PRESSLEY (ed)
291 Tits buildings and the model theory of groups, K. TENT (ed)
292 A quantum groups primer, S. MAJID
293 Second order partial differential equations in Hilbert spaces, G. DA PRATO & J. ZABCZYK
294 Introduction to the theory of operator spaces, G. PISIER
295 Geometry and Integrability, L. MASON & YAVUZ NUTKU (eds)
296 Lectures on invariant theory, I. DOLGACHEV
297 The homotopy category of simply connected 4-manifolds, H.-J. BAUES
298 Higher operads, higher categories, T. LEINSTER
299 Kleinian Groups and Hyperbolic 3-Manifolds Y. KOMORI, V. MARKOVIC & C. SERIES (eds)
300 Introduction to Möbius Differential Geometry, U. HERTRICH-JEROMIN
301 Stable Modules and the D(2)-Problem, F.E.A. JOHNSON
302 Discrete and Continuous Nonlinear Schrödinger Systems, M. J. ABLORWITZ, B. PRINARI & A. D. TRUBATCH
303 Number Theory and Algebraic Geometry, M. REID & A. SKOROBOGATOV (eds)
304 Groups St Andrews 2001 in Oxford Vol. 1, C.M. CAMPBELL, E.F. ROBERTSON & G.C. SMITH (eds)
305 Groups St Andrews 2001 in Oxford Vol. 2, C.M. CAMPBELL, E.F. ROBERTSON & G.C. SMITH (eds)
306 Peyresq lectures on geometric mechanics and symmetry, J. MONTALDI & T. RATIU (eds)
307 Surveys in Combinatorics 2003, C. D. WENSLEY (ed.)
308 Topology, geometry and quantum field theory, U. L. TILLMANN (ed)
309 Corings and Comodules, T. BRZEZINSKI & R. WISBAUER
310 Topics in Dynamics and Ergodic Theory, S. BEZUGLYI & S. KOLYADA (eds)
311 Groups: topological, combinatorial and arithmetic aspects, T. W. MÜLLER (ed)
312 Foundations of Computational Mathematics, Minneapolis 2002, FELIPE CUCKER *et al* (eds)
313 Transcendental aspects of algebraic cycles, S. MÜLLER-STACH & C. PETERS (eds)
314 Spectral generalizations of line graphs, D. CVETKOVIC, P. ROWLINSON & S. SIMIC
315 Structured ring spectra, A. BAKER & B. RICHTER (eds)
316 Linear Logic in Computer Science, T. EHRHARD *et al* (eds)
317 Advances in elliptic curve cryptography, I. F. BLAKE, G. SEROUSSI & N. SMART
318 Perturbation of the boundary in boundary-value problems of Partial Differential Equations, D. HENRY
319 Double Affine Hecke Algebras, I. CHEREDNIK
320 L-Functions and Galois Representations, D. BURNS, K. BUZZARD & J. NEKOVÁR (eds)
321 Surveys in Modern Mathematics, V. PRASOLOV & Y. ILYASHENKO (eds)
322 Recent perspectives in random matrix theory and number theory, F. MEZZADRI & N. C. SNAITH (eds)
323 Poisson geometry, deformation quantisation and group representations, S. GUTT *et al* (eds)
324 Singularities and Computer Algebra, C. LOSSEN & G. PFISTER (eds)
325 Lectures on the Ricci Flow, P. TOPPING
326 Modular Representations of Finite Groups of Lie Type, J. E. HUMPHREYS
328 Fundamentals of Hyperbolic Manifolds, R. D. CANARY, A. MARDEN & D. B. A. EPSTEIN (eds)
329 Spaces of Kleinian Groups, Y. MINSKY, M. SAKUMA & C. SERIES (eds)
330 Noncommutative Localization in Algebra and Topology, A. RANICKI (ed)
331 Foundations of Computational Mathematics, Santander 2005, L. PARDO, A. PINKUS, E. SULI & M. TODD (eds)
332 Handbook of Tilting Theory, L. ANGELERI HÜGEL, D. HAPPEL & H. KRAUSE (eds)
333 Synthetic Differential Geometry 2ed, A. KOCK
334 The Navier-Stokes Equations, P. G. DRAZIN & N. RILEY
335 Lectures on the Combinatorics of Free Probability, A. NICA & R. SPEICHER
336 Integral Closure of Ideals, Rings, and Modules, I. SWANSON & C. HUNEKE
337 Methods in Banach Space Theory, J. M. F. CASTILLO & W. B. JOHNSON (eds)
338 Surveys in Geometry and Number Theory, N. YOUNG (ed)
339 Groups St Andrews 2005 Vol. 1, C.M. CAMPBELL, M. R. QUICK, E.F. ROBERTSON & G.C. SMITH (eds)
340 Groups St Andrews 2005 Vol. 2, C.M. CAMPBELL, M. R. QUICK, E.F. ROBERTSON & G.C. SMITH (eds)
341 Ranks of Elliptic Curves and Random Matrix Theory, J. B. CONREY, D. W. FARMER, F. MEZZADRI & N. C. SNAITH (eds)
342 Elliptic Cohomology, H. R. MILLER & D. C. RAVENEL (eds)
343 Algebraic Cycles and Motives Vol. 1, J. NAGEL & C. PETERS (eds)
344 Algebraic Cycles and Motives Vol. 2, J. NAGEL & C. PETERS (eds)
345 Algebraic and Analytic Geometry, A. NEEMAN
346 Surveys in Combinatorics, 2007, A. HILTON & J. TALBOT (eds)
347 Surveys in Contemporary Mathematics, N. YOUNG & Y. CHOI (eds)

L-functions and Galois Representations

Edited by

DAVID BURNS,

KEVIN BUZZARD

AND

JAN NEKOVÁŘ

CAMBRIDGE
UNIVERSITY PRESS

CAMBRIDGE
UNIVERSITY PRESS

University Printing House, Cambridge CB2 8BS, United Kingdom

One Liberty Plaza, 20th Floor, New York, NY 10006, USA

477 Williamstown Road, Port Melbourne, VIC 3207, Australia

314-321, 3rd Floor, Plot 3, Splendor Forum, Jasola District Centre, New Delhi - 110025, India

103 Penang Road, #05-06/07, Visioncrest Commercial, Singapore 238467

Cambridge University Press is part of the University of Cambridge.

It furthers the University's mission by disseminating knowledge in the pursuit of education, learning and research at the highest international levels of excellence.

www.cambridge.org
Information on this title: www.cambridge.org/9780521694155

First published 2007

A catalogue record for this publication is available from the British Library

ISBN 978-0-521-69415-5 Paperback

Contents

Preface

The London Mathematical Society symposium on L-functions and Galois representations took place at the University of Durham from the 19th to the 30th of July, 2004; this book is a collection of research articles in the areas covered by the conference, in many cases written by the speakers or audience members. There were series of lectures in each of the following subject areas:

- Local Langlands programme
- Local p-adic Galois representations
- Modularity of Galois representations
- Automorphic forms and Selmer groups
- p-adic modular forms and eigenvarieties
- The André-Oort conjecture

In practice it is becoming harder to distinguish some of these areas from others, because of major recent progress, much of which is documented in this volume. As well as these courses, there were 19 individual lectures. The organisers would like to thank the lecturers, and especially those whom we persuaded to contribute to this volume.

The symposium received generous financial support from both the EPSRC and the London Mathematical Society. These symposia now command a certain reputation in the number theory community and the organisers found it easy to attract many leading researchers to Durham; this would not have been possible without the financial support given to us, and we would like to heartily thank both organisations.

The conference could not possibly have taken place if it had not been for the efforts of John Bolton, James Blowey and Rachel Duke of the Department of Mathematics at the University of Durham, and for the hospitality of Grey College. We are grateful to both these institutions for their help in making the operation run so smoothly.

The feedback from the participants to the organisers seemed to indicate that many participants found the symposium mathematically stimulating; and the organisers can only hope that this volume serves a similar purpose.

David Burns
Kevin Buzzard
Jan Nekovář

List of participants

Viktor Abrashkin (Durham)
Amod Agashe (Missouri)
Adebisi Agboola (UCSB)
Konstantin Ardakov (Cambridge)
Julian Arndts (Cambridge)
Joel Bellaïche (IPDE - Roma I)
Denis Benois (Bordeaux I)
Laurent Berger (IHES)
Massimo Bertolini (Milano)
Amnon Besser (Beér Sheba)
Bryan Birch (Oxford)
Thanasis Bouganis (Cambridge)
Christophe Breuil (IHES)
Manuel Breuning (King's College London)
David Burns (King's College London)
Colin Bushnell (King's College London)
Kevin Buzzard (Imperial College)
Nigel Byott (Exeter)
Frank Calegari (Harvard)
Laurent Clozel (Orsay, Paris-Sud)
Pierre Colmez (Jussieu, Paris 6)
James Cooper (Oxford)
Christophe Cornut (Jussieu, Paris 6)
Anton Deitmar (Exeter)
Rob de Jeu (Durham)
Daniel Delbourgo (Nottingham)
Ehud de Shalit (Hebrew)
Fred Diamond (Brandeis)
Vladimir Dokchitser (Cambridge)
Neil Dummigan (Sheffield)
Matthew Emerton (Northwestern)
Ivan Fesenko (Nottingham)
Tom Fisher (Cambridge)
Jean-Marc Fontaine (Orsay, Paris-Sud)
Kazuhiro Fujiwara (Nagoya)
Toby Gee (Imperial College)
Sasha Goncharov (MPI (Bonn), Brown)
Ralph Greenberg (Washington)
Hannu Harkonen (Cambridge)
Guy Henniart (Orsay, Paris-Sud)
Haruzo Hida (UCLA)
Richard Hill (University College London)
Ben Howard (Harvard)
Susan Howson (Oxford)
Annette Huber-Klawitter (Leipzig)
Frazer Jarvis (Sheffield)
Adam Joyce (Imperial College)
Chandrashekhar Khare (Utah)
Mark Kisin (Chicago)
Bruno Klingler (Chicago)
Bernhard Koeck (Southampton)
Stephen Kudla (Maryland)
Masato Kurihara (Tokyo Metropolitan)
Mathias Lederer (Bielefeld)
Stephen Lichtenbaum (Brown)
Ron Livné (Hebrew)
Jayanta Manoharmayum (Sheffield)
Ariane Mézard (Orsay, Paris-Sud)
Jan Nekovář (Jussieu, Paris 6)
Wieslawa Niziol (Utah)
Rachel Ollivier (Jussieu, Paris 6)
Louisa Orton (Paris 13)
Pierre Parent (Bordeaux I)
Vytautas Paskunas (Bielefeld)
Mark Pavey (Exeter)
Karl Rubin (Stanford)
Mohamed Saidi (MPI (Bonn))
Takeshi Saito (Tokyo)
Kanetomo Sato (Nottingham)
Michael Schein (Harvard)
Peter Schneider (Munster)
Tony Scholl (Cambridge)
Alexei Skorobogatov (Imperial College)
Paul Smith (Nottingham)
Victor Snaith (Southampton)
David Solomon (King's College London)
Noam Solomon (Beér Sheba)
Michael Spiess (Bielefeld)
Nelson Stephens (Royal Holloway)
Shaun Stevens (East Anglia)
Peter Swinnerton-Dyer (Cambridge)
Martin Taylor (UMIST)
Richard Taylor (Harvard)
Jacques Tilouine (Paris 13)
Douglas Ulmer (Arizona)
Eric Urban (Columbia)
Otmar Venjakob (Heidelberg)
Marie-France Vignéras (Jussieu, Paris 7)
Stephen Wilson (Durham)
Christian Wuthrich (Cambridge)
Andrei Yafaev (University College London)
Atsushi Yamagami (Kyoto)
Sarah Zerbes (Cambridge)
Shou-Wu Zhang (Columbia)

Stark–Heegner points and special values of L-series

Massimo Bertolini

Dipartimento di Matematica
Universita' degli Studi di Milano
Via Saldini 50 20133
Milano, Italy
Massimo.Bertolini@mat.unimi.it

Henri Darmon

McGill University Mathematics Department,
805 Sherbrooke Street West Montreal,
QC H3A-2K6 CANADA
darmon@math.mcgill.ca

Samit Dasgupta

Department of Mathematics,
Harvard University,
1 Oxford St,
Cambridge, MA 02138, U.S.A.,
dasgupta@math.harvard.edu

Introduction

Let E be an elliptic curve over $\mathbb{Q}$ attached to a newform f of weight two on $\Gamma_0(N)$. Let K be a real quadratic field, and let $p\|N$ be a prime of multiplicative reduction for E which is inert in K, so that the p-adic completion K_p of K is the quadratic unramified extension of $\mathbb{Q}_p$.

Subject to the condition that all the primes dividing $M := N/p$ are split in K, the article [Dar] proposes an analytic construction of "Stark–Heegner points" in $E(K_p)$, and conjectures that these points are defined over specific class fields of K. More precisely, let

$$R := \left\{ \begin{pmatrix} a & b \\ c & d \end{pmatrix} \in M_2(\mathbb{Z}[1/p]) \text{ such that } M \text{ divides } c \right\}$$

be an Eichler $\mathbb{Z}[1/p]$-order of level M in $M_2(\mathbb{Q})$, and let $\Gamma := R_1^\times$ denote the group of elements in R of determinant 1. This group acts by Möbius transformations on the K_p-points of the p-adic upper half-plane

$$\mathcal{H}_p := \mathbb{P}^1(K_p) - \mathbb{P}^1(\mathbb{Q}_p),$$

and preserves the non-empty subset $\mathcal{H}_p \cap K$. In [Dar], modular symbols attached to f are used to define a map

$$\Phi : \Gamma\backslash(\mathcal{H}_p \cap K) \longrightarrow E(K_p), \tag{0.1}$$

whose image is conjectured to consist of points defined over ring class fields of K. Underlying this conjecture is a more precise one, analogous to the classical Shimura reciprocity law, which we now recall.

Given $\tau \in \mathcal{H}_p \cap K$, the collection $\mathcal{O}_\tau$ of matrices $g \in R$ satisfying

$$g\begin{pmatrix} \tau \\ 1 \end{pmatrix} = \lambda_g \begin{pmatrix} \tau \\ 1 \end{pmatrix} \quad \text{for some } \lambda_g \in K, \tag{0.2}$$

is isomorphic to a $\mathbb{Z}[1/p]$-order in K, via the map $g \mapsto \lambda_g$. This order is also equipped with the attendant ring homomorphism $\eta : \mathcal{O}_\tau \longrightarrow \mathbb{Z}/M\mathbb{Z}$ sending g to its upper left-hand entry (taken modulo M). The map η is sometimes referred to as the *orientation* at M attached to τ. Conversely, given any $\mathbb{Z}[1/p]$-order $\mathcal{O}$ of discriminant prime to M equipped with an orientation η, the set $\mathcal{H}_p^{\mathcal{O}}$ of $\tau \in \mathcal{H}_p$ with associated oriented order equal to $\mathcal{O}$ is preserved under the action of Γ, and the set of orbits $\Gamma\backslash\mathcal{H}_p^{\mathcal{O}}$ is equipped with a natural simply transitive action of the group $G = \mathrm{Pic}^+(\mathcal{O})$, where $\mathrm{Pic}^+(\mathcal{O})$ denotes the narrow Picard group of oriented projective $\mathcal{O}$-modules of rank one. Denote this action by $(\sigma, \tau) \mapsto \tau^\sigma$, for $\sigma \in G$ and $\tau \in \Gamma\backslash\mathcal{H}_p^{\mathcal{O}}$. Class field theory identifies G with the Galois group of the *narrow ring class field* of K attached to $\mathcal{O}$, denoted H_K. It is conjectured in [Dar] that the points $\Phi(\tau)$ belong to $E(H_K)$ for all $\tau \in \mathcal{H}_p^{\mathcal{O}}$, and that

$$\Phi(\tau)^\sigma = \Phi(\tau^\sigma), \qquad \text{for all } \sigma \in \mathrm{Gal}(H_K/K) = \mathrm{Pic}^+(\mathcal{O}). \tag{0.3}$$

In particular it is expected that the point

$$P_K := \Phi(\tau_1) + \cdots + \Phi(\tau_h)$$

should belong to $E(K)$, where $\tau_1, \ldots, \tau_h$ denote representatives for the distinct orbits in $\Gamma\backslash\mathcal{H}_p^{\mathcal{O}}$. The article [BD3] shows that the image of P_K in $E(K_p) \otimes \mathbb{Q}$ is of the form $t \cdot \mathbf{P}_K$, where

(i) t belongs to $\mathbb{Q}^\times$;
(ii) $\mathbf{P}_K \in E(K)$ is of infinite order precisely when $L'(E/K, 1) \neq 0$;

provided the following ostensibly extraneous assumptions are satisfied

(i) $\bar{P}_K = a_p P_K$, where $\bar{P}_K$ is the Galois conjugate of P_K over K_p, and a_p is the pth Fourier coefficient of f.
(ii) The elliptic curve E has at least two primes of multiplicative reduction.

The main result of [BD3] falls short of being definitive because of these two assumptions, and also because it only treats the image of P_K modulo the torsion subgroup of $E(K_p)$.

The main goal of this article is to examine certain "finer" invariants associated to P_K and to relate these to special values of L-series, guided by the analogy between the point P_K and classical Heegner points attached to imaginary quadratic fields.

In setting the stage for the main formula, let $E/\mathbb{Q}$ be an elliptic curve of conductor M; it is essential to assume that all the primes dividing M are *split* in K. This hypothesis is very similar to the one imposed in [GZ] when K is imaginary quadratic, where it implies that $L(E/K,1)$ vanishes systematically because the sign in its functional equation is -1. In the case where K is real quadratic the "Gross-Zagier hypothesis" implies that the sign in the functional equation for $L(E/K,s)$ is 1 so that $L(E/K,s)$ vanishes to even order and is expected to be frequently non-zero at $s=1$. Consistent with this expectation is the fact that the Stark–Heegner construction is now unavailable, in the absence of a prime $p\|M$ which is inert in K.

The main idea is to bring such a prime into the picture by "raising the level at p" to produce a newform g of level $N = Mp$ which is *congruent* to f. The congruence is modulo an appropriate ideal λ of the ring $\mathcal{O}_g$ generated by the Fourier coefficients of g. Let A_g denote the abelian variety quotient of $J_0(N)$ attached to g by the Eichler-Shimura construction. The main objective, which can now be stated more precisely, is to relate the *local behaviour at* p of the Stark–Heegner points in $A_g(K_p)$ to the algebraic part of the special value of $L(E/K,1)$, taken modulo λ.

The first key ingredient in establishing such a relationship is an extension of the map Φ of (0.1) to arbitrary eigenforms of weight 2 on $\Gamma_0(Mp)$ such as g, and not just eigenforms with rational Fourier coefficients attached to elliptic curves, in a precise enough form so that phenomena related to congruences between modular forms can be analyzed. Let $\mathbb{T}$ be the full algebra of Hecke operators acting on the space of forms of weight two on $\Gamma_0(Mp)$. The theory presented in Section 1, based on the work of the third author [Das], produces a torus T over K_p equipped with a natural $\mathbb{T}$-action, whose character group (tensored with $\mathbb{C}$) is isomorphic as a $\mathbb{T}\otimes\mathbb{C}$-module to the space of weight 2 modular forms on $\Gamma_0(Mp)$ which are new at p. It also builds a Hecke-stable lattice $L \subset T(K_p)$, and a map Φ generalising (0.1)

$$\Phi : \Gamma\backslash(\mathcal{H}_p \cap K) \longrightarrow T(K_p)/L. \tag{0.4}$$

It is conjectured in Section 1 that the quotient T/L is isomorphic to the rigid analytic space associated to an abelian variety J defined over $\mathbb{Q}$. A strong partial result in this direction is proven in [Das], where it is shown that T/L is isogenous over K_p to the rigid analytic space associated to the p-new quotient $J_0(N)^{p\text{-new}}$ of the jacobian $J_0(N)$. In Section 1, it is further conjectured that the points $\Phi(\tau) \in J(K_p)$ satisfy the same algebraicity properties as were stated for the map Φ of (0.1).

Letting Φ_p denote the group of connected components in the Néron model of J over the maximal unramified extension of $\mathbb{Q}_p$, one has a natural Hecke-equivariant projection

$$\partial_p : J(\mathbb{C}_p) \longrightarrow \Phi_p. \tag{0.5}$$

The group Φ_p is described explicitly in Section 1, yielding a concrete description of the Hecke action on Φ_p and a description of the primes dividing the cardinality of Φ_p in terms of "primes of fusion" between forms on $\Gamma_0(M)$ and forms on $\Gamma_0(Mp)$ which are new at p.

This description also makes it possible to attach to E and K an explicit element

$$\mathcal{L}(E/K,1)_{(p)} \in \bar{\Phi}_p,$$

where $\bar{\Phi}_p$ is a suitable f-isotypic quotient of Φ_p. Thanks to a theorem of Popa [Po], this element is closely related to the special value $L(E/K,1)$, and, in particular, one has the equivalence

$$L(E/K,1) = 0 \iff \mathcal{L}(E/K,1)_{(p)} = 0 \text{ for all } p.$$

Section 2 contains an exposition of Popa's formula.

Section 3 is devoted to a discussion of $\mathcal{L}(E/K,1)_{(p)}$; furthermore, by combining the results of Sections 1 and 2, it proves the main theorem of this article, an avatar of the Gross-Zagier formula which relates Stark–Heegner points to special values of L-series.

Main Theorem. *For all primes p which are inert in K,*

$$\partial_p(P_K) = \mathcal{L}(E/K,1)_{(p)}.$$

Potential arithmetic applications of this theorem (conditional on the validity of the deep conjectures of Section 1) are briefly discussed in Section 4.

Aknowledgements. It is a pleasure to thank the anonymous referee, for some comments which led us to improve our exposition.

1 Stark–Heegner points on $J_0(Mp)^{p\text{-new}}$

Heegner points on an elliptic curve E defined over $\mathbb{Q}$ can be defined analytically by certain complex line integrals involving the modular form

$$f := \sum_{n=1}^{\infty} a_n(E) e^{2\pi i n z}$$

corresponding to E, and the Weierstrass parametrization of E. To be precise, let τ be any point of the complex upper half plane $\mathcal{H} := \{z \in \mathbb{C} | \Im z > 0\}$. The complex number

$$J_\tau := \int_\infty^\tau 2\pi i f(z) dz \in \mathbb{C}$$

gives rise to an element of $\mathbb{C}/\Lambda_E \cong E(\mathbb{C})$, where Λ_E is the Néron lattice of E, and hence to a complex point $P_\tau \in E(\mathbb{C})$. If τ also lies in an imaginary quadratic subfield K of $\mathbb{C}$, then P_τ is a *Heegner point* on E. The theory of complex multiplication shows that this analytically defined point is actually defined over an abelian extension of K, and it furthermore prescribes the action of the Galois group of K on this point.

The Stark–Heegner points of [Dar], defined on elliptic curves over $\mathbb{Q}$ with multiplicative reduction at p, are obtained by replacing complex integration on $\mathcal{H}$ with a double integral on the product of a p-adic and a complex upper half plane $\mathcal{H}_p \times \mathcal{H}$.

We now very briefly describe this construction. Let E be an elliptic curve over $\mathbb{Q}$ of conductor $N = Mp$, with $p \nmid M$. The differential $\omega := 2\pi i f(z) dz$ and its anti-holomorphic counterpart $\bar{\omega} = -2\pi i f(\bar{z}) d\bar{z}$ give rise to two elements in the DeRham cohomology of $X_0(N)(\mathbb{C})$:

$$\omega^\pm := \omega \pm \bar{\omega}.$$

To each of these differential forms is attached a *modular symbol*

$$m_E^\pm\{x \to y\} := (\Omega_E^\pm)^{-1} \int_x^y \omega^\pm, \qquad \text{for } x, y \in \mathbb{P}^1(\mathbb{Q}).$$

Here $\Omega_E^\pm$ is an appropriate complex period chosen so that $m_E^\pm$ takes values in $\mathbb{Z}$ and in no proper subgroup of $\mathbb{Z}$.

The group Γ defined in the Introduction acts on $\mathbb{P}^1(\mathbb{Q}_p)$ by Möbius transformations. For each pair of cusps $x, y \in \mathbb{P}^1(\mathbb{Q})$ and choice of sign $\pm$, a $\mathbb{Z}$-valued additive measure $\mu^\pm\{x \to y\}$ on $\mathbb{P}^1(\mathbb{Q}_p)$ can be defined by

$$\mu^\pm\{x \to y\}(\gamma \mathbb{Z}_p) = m_E^\pm\{\gamma^{-1} x \to \gamma^{-1} y\}, \tag{1.1}$$

where γ is an element of Γ. Since the stabilizer of $\mathbb{Z}_p$ in Γ is $\Gamma_0(N)$, equation (1.1) is independent of the choice of γ by the $\Gamma_0(N)$-invariance of $m_E^\pm$. The motivation for this definition, and a proof that it extends to an additive measure on $\mathbb{P}^1(\mathbb{Q}_p)$, comes from "spreading out" the modular symbol $m_E^{\pm}$ along the Bruhat-Tits tree of $\mathrm{PGL}_2(\mathbb{Q}_p)$ (see [Dar], [Das], and Section 1.2 below). For any $\tau_1, \tau_2 \in \mathcal{H}_p$ and $x, y \in \mathbb{P}^1(\mathbb{Q}_p)$, a multiplicative double integral on $\mathcal{H}_p \times \mathcal{H}$ is then defined by (multiplicatively) integrating the function $(t-\tau_1)/(t-\tau_2)$ over $\mathbb{P}^1(\mathbb{Q}_p)$ with respect to the measure $\mu^{\pm}\{x \to y\}$:

$$\begin{aligned} ⨰_{\tau_1}^{\tau_2} \int_x^y \omega_{\pm} &:= ⨰_{\mathbb{P}^1(\mathbb{Q}_p)} \left(\frac{t-\tau_2}{t-\tau_1}\right) d\mu^{\pm}\{x \to y\}(t) \\ &= \lim_{||\mathcal{U}|| \to 0} \prod_{U \in \mathcal{U}} \left(\frac{t_U-\tau_2}{t_U-\tau_1}\right)^{\mu^{\pm}\{x \to y\}(U)} \in \mathbb{C}_p^{\times}. \end{aligned} \tag{1.2}$$

Here the limit is taken over uniformly finer disjoint covers $\mathcal{U}$ of $\mathbb{P}^1(\mathbb{Q}_p)$ by open compact subsets U, and t_U is an arbitrarily chosen point of U. Choosing special values for the limits of integration, in a manner motivated by the classical Heegner construction described above, one produces special elements in $\mathbb{C}_p^{\times}$. These elements are transferred to E using Tate's p-adic uniformization $\mathbb{C}_p^{\times}/q_E \cong E(\mathbb{C}_p)$ to define Stark–Heegner points.

In order to lift the Stark–Heegner points on E to the Jacobian $J_0(N)^{p\text{-new}}$, one can replace the modular symbols attached to E with the universal modular symbol for $\Gamma_0(N)$. In this section, we review this construction of Stark–Heegner points on $J_0(N)^{p\text{-new}}$, as described in fuller detail in [Das].

1.1 The universal modular symbol for $\Gamma_0(N)$

The first step is to generalize the measures $\mu^{\pm}\{x \to y\}$ on $\mathbb{P}^1(\mathbb{Q}_p)$. As we will see, the new measure naturally takes values in the p-new quotient of the homology group $H_1(X_0(N), \mathbb{Z})$. Once this measure is defined, the construction of Stark–Heegner points on $J_0(N)^{p\text{-new}}$ can proceed as the construction of Stark–Heegner points on E given in [Dar]. The Stark–Heegner points on $J_0(N)^{p\text{-new}}$ will map to those on E under the modular parametrization $J_0(N)^{p\text{-new}} \to E$.

We begin by recalling the universal modular symbol for $\Gamma_0(N)$. Let $\mathcal{M} := \mathrm{Div}_0\, \mathbb{P}^1(\mathbb{Q})$ be the group of degree zero divisors on the set of cusps of the complex upper half plane, defined by the exact sequence

$$0 \to \mathcal{M} \to \mathrm{Div}\, \mathbb{P}^1(\mathbb{Q}) \to \mathbb{Z} \to 0. \tag{1.3}$$

The group Γ acts on $\mathcal{M}$ via its action on $\mathbb{P}^1(\mathbb{Q})$ by Möbius transformations.

For any abelian group G, a *G-valued modular symbol* is a homomorphism $m : \mathcal{M} \longrightarrow G$; we write $m\{x \to y\}$ for $m([x]-[y])$. Let $\mathcal{M}(G)$ denote the left Γ-module of G-valued modular symbols, where the action of Γ is defined by the rule

$$(\gamma m)\{x \to y\} = m\{\gamma^{-1}x \to \gamma^{-1}y\}.$$

Note that the natural projection onto the group of coinvariants

$$\mathcal{M} \longrightarrow \mathcal{M}_{\Gamma_0(N)} = H_0(\Gamma_0(N), \mathcal{M})$$

is a $\Gamma_0(N)$-invariant modular symbol. Furthermore, this modular symbol is universal, in the sense that any other $\Gamma_0(N)$-invariant modular symbol factors through this one.

One can interpret $H_0(\Gamma_0(N), \mathcal{M})$ geometrically as follows. Given a divisor $[x]-[y] \in \mathcal{M}$, consider any path from x to y in the completed upper half plane $\mathcal{H} \cup \mathbb{P}^1(\mathbb{Q})$. Identifying the quotient $\Gamma_0(N)\backslash(\mathcal{H} \cup \mathbb{P}^1(\mathbb{Q}))$ with $X_0(N)(\mathbb{C})$, this path gives a well-defined element of $H_1(X_0(N), \text{cusps}, \mathbb{Z})$, the singular homology of the Riemann surface $X_0(N)(\mathbb{C})$ relative to the cusps. Manin [Man] proves that this map induces an isomorphism between the maximal torsion-free quotient $H_0(\Gamma_0(N), \mathcal{M})_T$ and $H_1(X_0(N), \text{cusps}, \mathbb{Z})$. Furthermore, the torsion of $H_0(\Gamma_0(N), \mathcal{M})$ is finite and supported at 2 and 3. The projection

$$\mathcal{M} \to \mathcal{M}_{\Gamma_0(N)} \to H_1(X_0(N), \text{cusps}, \mathbb{Z})$$

is called the *universal modular symbol for* $\Gamma_0(N)$.

The points of $X_0(N)$ over $\mathbb{C}$ correspond to isomorphism classes of pairs (E, C_N) of (generalized) elliptic curves $E/\mathbb{C}$ equipped with a cyclic subgroup $C_N \subset E$ of order N. To such a pair we can associate two points of $X_0(M)$, namely the points corresponding to the pairs (E, C_M) and $(E/C_p, C_N/C_p)$, where C_p and C_M are the subgroups of C_N of size p and M, respectively. This defines two morphisms of curves

$$f_1 : X_0(N) \to X_0(M) \text{ and } f_2 : X_0(N) \to X_0(M), \tag{1.4}$$

each of which is defined over $\mathbb{Q}$. The map f_2 is the composition of f_1 with the Atkin-Lehner involution W_p on $X_0(N)$. Write $f_* = f_{1*} \oplus f_{2*}$ and $f^* = f_1^* \oplus f_2^*$ (resp. $\overline{f_*}$ and $\overline{f^*}$) for the induced maps on singular homology (resp. relative singular homology):

$$\begin{aligned}
f_* :&\quad H_1(X_0(N), \mathbb{Z}) \to H_1(X_0(M), \mathbb{Z})^2 \\
\overline{f_*} :&\quad H_1(X_0(N), \text{cusps}, \mathbb{Z}) \to H_1(X_0(M), \text{cusps}, \mathbb{Z})^2 \\
f^* :&\quad H_1(X_0(M), \mathbb{Z})^2 \to H_1(X_0(N), \mathbb{Z}) \\
\overline{f^*} :&\quad H_1(X_0(M), \text{cusps}, \mathbb{Z})^2 \to H_1(X_0(N), \text{cusps}, \mathbb{Z}).
\end{aligned}$$

The abelian variety $J_0(N)^{p\text{-new}}$ is defined to be the quotient of $J_0(N)$ by the images of the Picard maps on Jacobians associated to f_1 and f_2. Define $\overline{H}$ and H to be the maximal torsion-free quotients of the cokernels of $\overline{f^*}$ and f^*, respectively:

$$\overline{H} := (\mathrm{Coker}\overline{f^*})_T \text{ and } H := (\mathrm{Coker} f^*)_T.$$

If we write g for the dimension of $J_0(N)^{p\text{-new}}$, the free abelian groups $\overline{H}$ and H have ranks $2g+1$ and $2g$, respectively, and the natural map $H \to \overline{H}$ is an injection ([Das, Prop. 3.2]).

The groups H and $\overline{H}$ have Hecke actions generated by T_ℓ for $\ell \nmid N$, U_ℓ for $\ell | N$, and W_p. We omit the proof of the following proposition.

Proposition 1.1 *The group $(\overline{H}/H)_T \cong \mathbb{Z}$ is Eisenstein; that is, T_ℓ acts as $\ell+1$ for $\ell \nmid N$, U_ℓ acts as ℓ for $\ell | M$, and W_p acts as -1.*

Proposition 1.1 implies that it is possible to choose a Hecke equivariant map $\psi : \overline{H} \to H$ such that the composition

$$H \longrightarrow \overline{H} \stackrel{\psi}{\longrightarrow} H \tag{1.5}$$

has finite cokernel. For example, we may take ψ to be the Hecke operator $(p^2-1)(T_r - (r+1))$ for any prime $r \nmid N$. We fix a choice of ψ for the remainder of the paper.

1.2 A p-adic uniformization of $J_0(N)^{p\text{-new}}$

For any free abelian group G, let $\mathrm{Meas}(\mathbb{P}^1(\mathbb{Q}_p), G)$ denote the Γ-module of G-valued measures on $\mathbb{P}^1(\mathbb{Q}_p)$ with total measure zero, where Γ acts by $(\gamma\mu)(U) := \mu(\gamma^{-1}U)$.

In order to construct a Γ-invariant $\mathrm{Meas}(\mathbb{P}^1(\mathbb{Q}_p), H)$-valued modular symbol, we recall the Bruhat-Tits tree $\mathcal{T}$ of $\mathrm{PGL}_2(\mathbb{Q}_p)$. The set of vertices $\mathcal{V}(\mathcal{T})$ of $\mathcal{T}$ is identified with the set of homothety classes of $\mathbb{Z}_p$-lattices in $\mathbb{Q}_p^2$. Two vertices v and v' are said to be adjacent if they can be represented by lattices L and L' such that L contains L' with index p. Let $\mathcal{E}(\mathcal{T})$ denote the set of oriented edges of $\mathcal{T}$, that is, the set of ordered pairs of adjacent vertices of $\mathcal{T}$. Given $e = (v_1, v_2)$ in $\mathcal{E}(\mathcal{T})$, call $v_1 = s(e)$ the source of e, and $v_2 = t(e)$ the target of e. Define the standard vertex v^o to be the class of $\mathbb{Z}_p^2$, and the standard oriented edge $e^o = (v^o, v)$ to be the edge whose source is v^o and whose stabilizer in Γ is equal to $\Gamma_0(N)$. Note that $\mathcal{E}(\mathcal{T})$ is equal to the disjoint union of the Γ-orbits of e^o and $\bar{e}^o$, where $\bar{e}^o = (v, v^o)$ is the opposite edge of e^o. A *half line* of $\mathcal{T}$ is a sequence (e_n) of oriented edges such that $t(e_n) = s(e_{n+1})$. Two half lines are said to be equivalent if they have in common all but a finite number of edges. It is known that the boundary $\mathbb{P}^1(\mathbb{Q}_p)$ of the p-adic upper half plane bijects onto the set of equivalence classes of half lines. For an oriented edge e, write U_e for the subset of $\mathbb{P}^1(\mathbb{Q}_p)$ whose elements correspond to classes of half lines passing through e. The sets U_e are determined by the rules: (1) $U_{\bar{e}^o} = \mathbb{Z}_p$, (2) $U_{\bar{e}} = \mathbb{P}^1(\mathbb{Q}_p) - U_e$, and (3) $U_{\gamma e} = \gamma U_e$ for all $\gamma \in \Gamma$. The U_e give a covering of $\mathbb{P}^1(\mathbb{Q}_p)$ by compact open sets. Finally, recall the existence of a Γ-equivariant reduction map

$$r : (K_p - \mathbb{Q}_p) \longrightarrow \mathcal{V}(\mathcal{T}),$$

defined on the K_p-points of $\mathcal{H}_p$. (As before, K_p is an unramified extension of $\mathbb{Q}_p$.) See [GvdP] for more details.

Define a function

$$\kappa\{x \to y\} : \mathcal{E}(\mathcal{T}) \longrightarrow H$$

as follows. When e belongs to the Γ-orbit of e^o and $\gamma \in \Gamma$ is chosen so that $\gamma e = e^o$, let $\kappa\{x \to y\}(e)$ be ψ applied to the image of $\gamma^{-1}([x] - [y])$ in $\overline{H}$ under the universal modular symbol for $\Gamma_0(N)$. Let $\kappa\{x \to y\}(e)$ be the negative of this value when the relation $\gamma e = \bar{e}^o$ holds.

The function $\kappa\{x \to y\}$ is a *harmonic cocycle on $\mathcal{T}$*, that is, it obeys the rules

(i) $\kappa\{x \to y\}(\bar{e}) = -\kappa\{x \to y\}(e)$ for all $e \in \mathcal{E}(\mathcal{T})$, and
(ii) $\sum_{s(e)=v} \kappa\{x \to y\}(e) = 0$ for all $v \in \mathcal{V}(\mathcal{T})$, where the sum is taken over the $p+1$ oriented edges e whose source $s(e)$ is v.

Furthermore, we have the Γ-invariance property

$$\kappa\{\gamma x \to \gamma y\}(\gamma e) = \kappa\{x \to y\}(e)$$

for all $\gamma \in \Gamma$.

The natural bijection between $\mathrm{Meas}(\mathbb{P}^1(\mathbb{Q}_p), H)$ and the group of harmonic cocycles on $\mathcal{T}$ valued in H shows that the definition

$$\mu\{x \to y\}(U_e) := \kappa\{x \to y\}(e)$$

yields a Γ-invariant $\mathrm{Meas}(\mathbb{P}^1(\mathbb{Q}_p), H)$-valued modular symbol μ ([Das, Prop. 3.1]). When $m = [x] - [y] \in \mathcal{M}$, we write μ_m for $\mu\{x \to y\}$.

We can now define, for $\tau_1, \tau_2 \in \mathcal{H}_p$ and $m \in \mathcal{M}$, a multiplicative double integral attached to the universal modular symbol for $\Gamma_0(N)$:

$$\begin{aligned} \times\!\!\!\!\!\!\int_{\tau_1}^{\tau_2}\int_m \omega &:= \times\!\!\!\!\!\!\int_{\mathbb{P}^1(\mathbb{Q}_p)} \left(\frac{t-\tau_2}{t-\tau_1}\right) d\mu_m(t) \\ &= \lim_{||\mathcal{U}|| \to 0} \prod_{U \in \mathcal{U}} \left(\frac{t_U - \tau_2}{t_U - \tau_1}\right) \otimes \mu_m(U) \in \mathbb{C}_p^\times \otimes_{\mathbb{Z}} H, \end{aligned}$$

with notations as in (1.2). One shows that this integral is Γ-invariant:

$$\times\!\!\!\!\!\!\int_{\gamma\tau_1}^{\gamma\tau_2}\int_{\gamma m}\omega = \times\!\!\!\!\!\!\int_{\tau_1}^{\tau_2}\int_{m}\omega \quad \text{for } \gamma\in\Gamma.$$

Letting T denote the torus $T=\mathbb{G}_m\otimes_{\mathbb{Z}} H$, we thus obtain a homomorphism

$$\begin{aligned} ((\mathrm{Div}_0\,\mathcal{H}_p)\otimes\mathcal{M})_\Gamma &\rightarrow T \\ ([\tau_1]-[\tau_2])\otimes m &\mapsto \times\!\!\!\!\!\!\int_{\tau_1}^{\tau_2}\int_{m}\omega. \end{aligned} \tag{1.6}$$

Consider the short exact sequence of Γ-modules defining $\mathrm{Div}_0\,\mathcal{H}_p$:

$$0\rightarrow \mathrm{Div}_0\,\mathcal{H}_p \rightarrow \mathrm{Div}\,\mathcal{H}_p\rightarrow\mathbb{Z}\rightarrow 0.$$

After tensoring with $\mathcal{M}$, the long exact sequence in homology gives a boundary map

$$\delta_1: H_1(\Gamma,\mathcal{M})\rightarrow((\mathrm{Div}_0\,\mathcal{H}_p)\otimes\mathcal{M})_\Gamma. \tag{1.7}$$

The long exact sequence in homology associated to the sequence (1.3) defining $\mathcal{M}$ gives a boundary map

$$\delta_2: H_2(\Gamma,\mathbb{Z})\rightarrow H_1(\Gamma,\mathcal{M}). \tag{1.8}$$

Define L to be the image of $H_2(\Gamma,\mathbb{Z})$ under the composed homomorphisms in (1.6), (1.7), and (1.8): $H_2(\Gamma,\mathbb{Z})\rightarrow T(\mathbb{Q}_p)$. Note that the Hecke algebra $\mathbb{T}$ of H acts on T.

Theorem 1.2 ([Das], Thm. 3.3) *Let K_p denote the quadratic unramified extension of $\mathbb{Q}_p$. The group L is a discrete, Hecke stable subgroup of $T(\mathbb{Q}_p)$ of rank $2g$. The quotient T/L admits a Hecke-equivariant isogeny over K_p to the rigid analytic space associated to the product of two copies of $J_0(N)^{p\text{-new}}$.*

Remark 1.3 If one lets the nontrivial element of $\mathrm{Gal}(K_p/\mathbb{Q}_p)$ act on T/L by the Hecke operator U_p, the isogeny of Theorem 1.2 is defined over $\mathbb{Q}_p$.

Remark 1.4 As described in [Das, §5.1], Theorem 1.2 is a generalization of a conjecture of Mazur, Tate, and Teitelbaum [MTT, Conjecture II.13.1] which was proven by Greenberg and Stevens [GS].

Theorem 1.2 implies that T/L is isomorphic to the rigid analytic space associated to an abelian variety J defined over a number field (which can be embedded in $\mathbb{Q}_p$). We now state a conjectural refinement of Theorem 1.2.

Conjecture 1.5 *The quotient T/L is isomorphic over K_p to the rigid analytic space associated to an abelian variety J defined over $\mathbb{Q}$.*

Presumably, the abelian variety J will have a natural Hecke action, and the isomorphism of Conjecture 1.5 will be Hecke equivariant; furthermore we expect that if one lets the nontrivial element of $\mathrm{Gal}(K_p/\mathbb{Q}_p)$ act on T/L by the Hecke operator U_p, the isomorphism will be defined over $\mathbb{Q}_p$.

The abelian variety J breaks up (after perhaps an isogeny of 2-power degree) into a product $J^+ \times J^-$, where the signs represent the eigenvalues of complex conjugation on H, and Theorem 1.2 (or rather its proof) implies that each of $J^{\pm}$ admits an isogeny denoted $\nu_{\pm}$ to $J_0(N)^{p\text{-new}}$.

Throughout this article, we will need to avoid a certain set of bad primes. Let S denote a finite set of primes containing those dividing $6\varphi(M)(p^2-1)$ or the size of the cokernel of the composite map (1.5). We say that two abelian varieties (or two analytic spaces) are S-isogenous if there is an isogeny between them whose degree is divisible only by primes in S. We expect that $\nu_{\pm}$ may be chosen to be S-isogenies defined over $\mathbb{Q}$, but as we will not need this result in the current article, we refrain from stating it as a formal conjecture.

1.3 Stark–Heegner points on J and $J_0(N)^{p\text{-new}}$

Fix $\tau \in \mathcal{H}_p$ and $x \in \mathbb{P}^1(\mathbb{Q})$. The significance of the subgroup L is that it is the smallest subgroup of T such that the cohomology class in $H^2(\Gamma, T/L)$ given by the 2-cocycle

$$d_{\tau,x}(\gamma_1,\gamma_2) := \times\!\!\!\!\!\!\int_{\tau}^{\gamma_1^{-1}\tau} \int_{x}^{\gamma_2 x} \omega \pmod{L}$$

vanishes (the cohomology class of this cocycle, and hence the smallest trivializing subgroup L, is independent of τ and x). Thus there exists a map $\beta_{\tau,x} : \Gamma \to T/L$ such that

$$\beta_{\tau,x}(\gamma_1\gamma_2) - \beta_{\tau,x}(\gamma_1) - \beta_{\tau,x}(\gamma_2) = \times\!\!\!\!\!\!\int_{\tau}^{\gamma_1^{-1}\tau} \int_{x}^{\gamma_2 x} \omega \pmod{L}. \tag{1.9}$$

The 1-cochain $\beta_{\tau,x}$ is defined uniquely up to an element of $\mathrm{Hom}(\Gamma, T/L)$. The following proposition, which follows from the work of Ihara and whose proof is reproduced in [Das, Prop. 3.7], allows us to deal with this ambiguity.

Proposition 1.6 *The abelianization of Γ is finite, and any prime dividing its size divides $6\varphi(M)(p^2-1)$.*

We may now define Stark–Heegner points on J and $J_0(N)^{p\text{-new}}$. Let K be a real quadratic field such that p is inert in K, and choose a real embedding σ of K. For each $\tau \in \mathcal{H}_p \cap K$, consider its associated order $\mathcal{O}_\tau$ as defined in (0.2). Let γ_τ be the generator of the group of units in $\mathcal{O}_\tau^\times$ of norm 1 whose associated λ_γ (see (0.2)) is greater than 1 under σ. Finally, choose any $x \in \mathbb{P}^1(\mathbb{Q})$, and let t denote the exponent of the abelianization of Γ. We then define the *Stark–Heegner point associated to* τ by

$$\Phi(\tau) := t \cdot \beta_{\tau,x}(\gamma_\tau) \in T(K_p)/L.$$

The multiplication by t ensures that this definition is independent of the choice of $\beta_{\tau,x}$, and one also checks that $\Phi(\tau)$ is independent of x. Furthermore, the point $\Phi(\tau)$ depends only on the Γ-orbit of τ, so we obtain a map

$$\Phi : \Gamma\backslash(\mathcal{H}_p \cap K) \to T(K_p)/L = J(K_p). \tag{1.10}$$

Following [Das], we conjecture that the images of Φ satisfy explicit algebraicity properties analogous to those mentioned in the Introduction. Fix a $\mathbb{Z}[1/p]$-order $\mathcal{O}$ in K, and let H_K be the narrow ring class field of K attached to $\mathcal{O}$, whose Galois group is canonically identified by class field theory with $\mathrm{Pic}^+(\mathcal{O})$. If h is the size of this Galois group, there are precisely h distinct Γ-orbits of points in $\mathcal{H}_p \cap K$ whose associated order is $\mathcal{O}$. Let $\tau_1, \ldots, \tau_h$ be representatives for these orbits.

Conjecture 1.7 *The points $\Phi(\tau_i)$ are global points defined over H_K:*

$$\Phi(\tau_i) \in J(H_K). \tag{1.11}$$

They are permuted simply transitively by $\mathrm{Gal}(H_K/K)$*, so the point*

$$P_K := \Phi(\tau_1) + \cdots + \Phi(\tau_h) \tag{1.12}$$

lies in $J(K)$.

While a proof of this conjecture (particularly, of equation (1.11)) seems far from the methods we have currently developed, one may still hope to glean some information from the p-adic invariants of Stark–Heegner points, and it seems of independent interest to relate such invariants to special values of Rankin L-series. Let $P_K = J(K_p)$ be as in (1.12). The goal of Section 3 is to relate P_K to a certain *algebraic part* of $L(E/K, 1)$, the latter being defined in terms of a formula of Popa that is explained in Section 2. This approach lends itself to generalisations to linear combinations of the points $\Phi(\tau_i)$ associated to the complex characters of $\mathrm{Gal}(H_K/K)$ (see Section 4 for more details).

We now conclude this section by remarking that Stark–Heegner points on $J_0(N)^{p\text{-new}}$ are defined by composing the map Φ of (1.10) with the maps $\nu_\pm$

resulting from Theorem 1.2. In [Das] it is conjectured that Stark–Heegner points on $J_0(N)^{p\text{-new}}$ are defined over H_K; Conjecture 1.7 may thus be viewed as a refinement.

2 Popa's formula

Let D denote the discriminant of K, and fix an orientation $\eta : \mathcal{O}_K \longrightarrow \mathbb{Z}/M\mathbb{Z}$ of the ring of integers $\mathcal{O} := \mathcal{O}_K$ of K. With notations as in the Introduction, there are exactly $h = \#G$ different $R_0(M)^\times$ conjugacy classes of oriented optimal embeddings of $\mathcal{O}$ into the order $R_0(M)$ of matrices in $M_2(\mathbb{Z})$ which are upper triangular modulo M. Let $\Psi_1, \ldots, \Psi_h$ denote representatives for these classes of embeddings. After fixing a fundamental unit ϵ_K of K of norm one, normalised so that $\epsilon_K > 1$ with respect to the fixed real embedding of K, set

$$\gamma_j := \Psi_j(\epsilon_K) \in \Gamma_0(M). \tag{2.1}$$

Let f be the normalised eigenform attached to E. Then we have

Proposition 2.1 (Popa) *The equality*

$$L(E/K, 1) \cdot (D^{1/2}/4\pi^2) = \left(\sum_{j=1}^{h} \int_{z_0}^{\gamma_j z_0} f(z)dz \right)^2$$

holds, for any choice of z_0 in the extended complex upper half plane.

Proof See Theorem 6.3.1 of [Po]. □

Remark 2.2 The result of Popa, which is stated here for simplicity in the case of the trivial character, deals more generally with twists of the L-series of E/K by complex characters of $\mathrm{Pic}^+(\mathcal{O})$ (and even with twists by complex characters attached to more general orders of K). In order to formulate the result in this more general form, one needs to define an action of $\mathrm{Pic}(\mathcal{O}_K)$ on the set of conjugacy classes of oriented optimal embeddings of $\mathcal{O}$ into the order $R_0(M)$. See [Po] for more details.

The eigenform f determines an algebra homomorphism $\varphi_f : \mathbb{T} \longrightarrow \mathbb{Z}$ satisfying

$$\varphi_f(T_n) = a_n(f), \quad \text{for } (n, N) = 1, \quad \varphi_f(U_\ell) = a_\ell(f), \quad \text{for } \ell | N.$$

Write I_f for the kernel of φ_f. For a $\mathbb{T}$-module A, let $A_f := A/I_f A$ be the largest quotient of A on which $\mathbb{T}$ acts via φ_f. Note that $H_1(X_0(M), \mathbb{Z})_f$ is a $\mathbb{Z}$-module of rank 2. Given a finite set of primes S, let $\mathbb{Z}_S$ denote the localization of $\mathbb{Z}$ in which the primes of S are inverted. By possibly enlarging S, we may assume that $H_1(X_0(M), \mathbb{Z}_S)_f$ is torsion-free, and hence a free $\mathbb{Z}_S$-module of rank 2.

For any such S, denote by $[\gamma_j] \in H_1(X_0(M), \mathbb{Z}_S)$ the homology class corresponding to γ_j, and set

$$[\gamma_K] := \sum_{j=1}^{h} [\gamma_j].$$

Define the *algebraic part* of $L(E/K, 1)$ by the formula

$$\begin{aligned} \mathcal{L}(E/K,1) &= \mathcal{L}(E/K,1)_S \\ &:= \text{the natural image of } [\gamma_K] \text{ in } H_1(X_0(M), \mathbb{Z}_S)_f. \end{aligned}$$

Proposition 2.1 directly implies the following

Corollary 2.3 *$L(E/K, 1) \neq 0$ if and only if $\mathcal{L}(E/K, 1) \neq 0$.*

3 The main formula

The goal of this section is to compute the image of the Stark-Heegner point P_K in the group of connected components at p of the abelian variety J introduced in Section 1, and relate it to $\mathcal{L}(E/K, 1)$.

3.1 The p-adic valuation

The image of the multiplicative double integral under the p-adic valuation map has a simple combinatorial description.

Proposition 3.1 ([BDG], Lemma 2.5 or [Das], Lemma 4.2) *For $\tau_1, \tau_2 \in K_p - \mathbb{Q}_p$, and $x, y \in \mathbb{P}^1(\mathbb{Q})$, the equality*

$$\mathrm{ord}_p \times\!\!\!\!\!\!\int_{\tau_1}^{\tau_2} \int_x^y \omega = \sum_{e: v_1 \to v_2} \kappa\{x \to y\}(e)$$

holds in H, where $v_i \in \mathcal{V}(\mathcal{T})$, $i = 1, 2$ is the image of τ_i by the reduction map, and the sum is taken over the edges in the path joining v_1 to v_2.

This proposition implies:

Proposition 3.2 ([Das], Props. 4.1, 4.9) *The image of L under*

$$\partial_p = \mathrm{ord}_p \otimes \mathrm{Id} : T(\mathbb{Q}_p) = \mathbb{Q}_p^\times \otimes H \to \mathbb{Z} \otimes H = H$$

is equal to the image of $\ker \bar{f}_*$ *by the composition of* ψ *with the natural projection* $H_1(X_0(N), \mathrm{cusps}, \mathbb{Z}) \longrightarrow \bar{H}$.

3.2 Connected components and primes of fusion

Let Φ_p denote the quotient

$$\mathrm{Coker} f^* / \ker f_*.$$

Let S be a finite set of primes chosen as at the end of section 1.2, that is, S contains the primes dividing $6\varphi(M)(p^2-1)$ or the size of the cokernel of the composite map (1.5).

The group Φ_p is finite, and the primes dividing the cardinality of $\Phi_p \otimes \mathbb{Z}_S$ are "congruence primes." This will be discussed further below.

Let $\Phi_{p,S}$ denote the $\mathbb{Z}_S$-module $\Phi_p \otimes \mathbb{Z}_S$. By Proposition 3.2, combined with the results of [Das], pp. 438-441, for any unramified extension K_p of $\mathbb{Q}_p$, the p-adic valuation gives a well-defined homomorphism

$$\partial_{p,S} : T(K_p)/L \to \Phi_{p,S}.$$

By the theory of p-adic uniformisation of abelian varieties, the group of connected components of the Néron model of J over the maximal unramified extension of $\mathbb{Q}_p$, tensored with $\mathbb{Z}_S$, is equal to $\Phi_{p,S}$.

Let $\tilde{\mathbb{T}}$, resp. $\mathbb{T}$ denote the Hecke algebra acting faithfully on

$$H_1(X_0(N), \mathbb{Z}), \text{ resp. } H_1(X_0(M), \mathbb{Z}).$$

This algebra is generated by the Hecke operators $\tilde{T}_q$ for $q \nmid N$ and $\tilde{U}_q$ for $q \mid N$, resp. T_q for $q \nmid M$ and U_q for $q \mid M$.

Identify

$$H_1(X_0(M), \mathbb{Z})^2$$

with a submodule of $H_1(X_0(N), \mathbb{Z})$ via f^*. Note that $H_1(X_0(M), \mathbb{Z})^2$ is stable for the action of $\tilde{\mathbb{T}}$. For $(n, p) = 1$, the action of the operator $\tilde{T}_n \in \tilde{\mathbb{T}}$ on $H_1(X_0(M), \mathbb{Z})^2$ is equal to the diagonal action of $T_n \in \mathbb{T}$; moreover, the action of $\tilde{U}_p \in \tilde{\mathbb{T}}$ is equal to that of the operator $U_p := \begin{pmatrix} T_p & -1 \\ p & 0 \end{pmatrix}$. Note that U_p and T_p (with T_p acting diagonally) satisfy the relation

$$U_p^2 - T_p U_p + p = 0.$$

The maximal quotient of $\tilde{\mathbb{T}}$ acting on $H_1(X_0(M), \mathbb{Z})^2$ is called the p-old quotient of $\tilde{\mathbb{T}}$, and is denoted $\tilde{\mathbb{T}}^{p\text{-old}}$.

Proposition 3.3 *There is a $\tilde{\mathbb{T}}$-equivariant isomorphism*

$$\Phi_{p,S} \simeq H_1(X_0(M), \mathbb{Z}_S)^2/\mathrm{Im}(f_* \circ f^*). \tag{3.1}$$

Proof The module $\Phi_{p,S}$ is isomorphic to the quotient of $H_1(X_0(N), \mathbb{Z}_S)$ by the $\mathbb{Z}_S$-submodule generated by the image of f^* and the kernel of f_*. It follows from a result of Ribet that the size of the cokernel of f_* divides $\varphi(M)$ (see [Rib1, Thm 4.3]). Thus, having tensored with $\mathbb{Z}_S$, we find that $\Phi_{p,S}$ is isomorphic to the cokernel of the endomorphism $f_* \circ f^*$ of $H_1(X_0(M), \mathbb{Z}_S)^2$. □

Corollary 3.4 *There is an isomorphism*

$$\Phi_{p,S} \cong H_1(X_0(M), \mathbb{Z}_S)/(T_p^2 - (p+1)^2), \tag{3.2}$$

which is compatible for the action of the Hecke operators $\tilde{T}_n \in \tilde{\mathbb{T}}$, resp. $T_n \in \mathbb{T}$, for $(n, p) = 1$, on the left-, resp. right-hand side.

Proof The endomorphism $f_* \circ f^*$ is given explicitly by the matrix

$$f_* \circ f^* = \begin{pmatrix} p+1 & T_p \\ T_p & p+1 \end{pmatrix}.$$

Since $p + 1$ is invertible in $\mathbb{Z}_S$, the cokernel of this map is isomorphic to

$$H_1(X_0(M), \mathbb{Z}_S)/(T_p^2 - (p+1)^2).$$

□

Remark 3.5 In this remark, assume as in Section 2 that f is the normalised eigenform attached to E, and write $a_n(f) \in \mathbb{Z}$ for the n-th Fourier coefficient of f. Let $S = S_f$ be a finite set of primes containing those dividing $6\varphi(M)(p^2 - 1)(a_r(f) - (r+1))$, for a prime r not dividing N. An argument similar to the proof of Corollary 3.4 shows that there is an isomorphism

$$(\Phi_{p,S})_f = \Phi_{p,S}/I_f\Phi_{p,S} \cong H_1(X_0(M), \mathbb{Z}_S)_f/(a_p(f)^2 - (p+1)^2). \tag{3.3}$$

Let λ be a maximal ideal of $\mathbb{T}$ belonging to the support of the module $H_1(X_0(M), \mathbb{Z}_S)/(T_p^2 - (p+1)^2)$, and let ℓ be the characteristic of the finite field $\mathbb{T}/\lambda\mathbb{T}$. The algebra homomorphism

$$\pi : \mathbb{T} \longrightarrow \mathbb{T}/\lambda\mathbb{T}$$

is identified with the reduction in characteristic ℓ of a modular form f on $\Gamma_0(M)$. (The normalised eigenform attached to an elliptic curve of conductor M has Fourier coefficients in $\mathbb{Z}$, and therefore arises in this way.)

Let $\tilde{\lambda}$ be a maximal ideal of $\tilde{\mathbb{T}}$ compatible with λ, in the sense that $\tilde{\lambda}$ arises from a maximal ideal $\bar{\lambda}$ of $\tilde{\mathbb{T}}^{p\text{-old}}$, and both λ and $\bar{\lambda}$ are contained in a maximal ideal of the Hecke ring $\mathbb{T}[U_p]$. (Note that the existence of $\tilde{\lambda}$ is guaranteed by the going-up theorem of Cohen-Seidenberg.) The isomorphism (3.1) shows that $\tilde{\lambda}$ is p-new (besides being p-old), since it appears in the support of the component group $\Phi_{p,S}$, on which $\tilde{\mathbb{T}}$ acts via its maximal p-new quotient. Therefore, $\tilde{\lambda} = \tilde{\lambda}_g$ corresponds to the reduction in characteristic ℓ of a p-new modular form g on $\Gamma_0(N)$.

In the terminology of Mazur, $\tilde{\lambda}$ is an *ideal of fusion* between the p-old and the p-new subspaces of modular forms on $\Gamma_0(N)$. The forms f and g are called *congruent* modular forms. For more details on these concepts, see [Rib2].

3.3 Specialisation of Stark-Heegner points

This section computes the image $\partial_{p,S}(\Phi(\tau))$ of a Stark-Heegner point $\Phi(\tau)$ in the group of connected components $\Phi_{p,S}$.

Assume that S contains the primes dividing $6\varphi(M)(p^2-1)(a_r(f)-(r+1))$, for a prime r not dividing N, and is such that $H_1(X_0(M), \mathbb{Z}_S)_f$ is a free $\mathbb{Z}_S$-module of rank 2.

We begin by imitating the definition of the map κ, with the group $\Gamma_0(M)$ replacing $\Gamma_0(N)$. Let $\tilde{\Gamma} = R^\times$, where R is the order appearing in the Introduction. Define a function

$$\bar{\kappa}\{x \to y\} : \mathcal{V}(\mathcal{T}) \longrightarrow H_1(X_0(M), \text{cusps}, \mathbb{Z}_S)$$

by setting $\bar{\kappa}\{x \to y\}(v) =$ image of $([\gamma x] - [\gamma y])$, where $\gamma \in \tilde{\Gamma}$ is chosen so that $\gamma v = v^o$. Since the stabilizer of v^o in $\tilde{\Gamma}$ is $\Gamma_0(M)$, and the natural homomorphism from $\mathcal{M}$ to $H_1(X_0(M), \text{cusps}, \mathbb{Z}_S)$ factors through $\Gamma_0(M)$, it follows that the map $\bar{\kappa}$ is well defined.

Recall the compatible ideals $\lambda_f \subset \mathbb{T}$ and $\tilde{\lambda}_g \subset \tilde{\mathbb{T}}$ introduced in Section 3.2. Assume in the sequel that f is the eigenform with rational coefficients attached to E, and that the p-th Hecke operator $\tilde{U}_p \in \tilde{\mathbb{T}}$ maps to 1 in the quotient ring $\tilde{\mathbb{T}}/\tilde{\lambda}_g$. (Since $\tilde{\lambda}_g$ is p-new, $\tilde{U}_p$ maps to ± 1 in $\tilde{\mathbb{T}}/\tilde{\lambda}_g$; the condition we are imposing is equivalent to requiring that λ_f belongs to the support of the module $H_1(X_0(M), \mathbb{Z}_S)/(T_p - (p+1))$.)

The identification

$$H_1(X_0(M), \text{cusps}, \mathbb{Z}_S)/\lambda_f H_1(X_0(M), \text{cusps}, \mathbb{Z}_S) \\ = H/\tilde{\lambda}_g H = \Phi_{p,S}/\tilde{\lambda}_g \Phi_{p,S},$$

which follows from Remark 3.5, implies that the reduction modulo λ_f of $\bar{\kappa}$ can also be viewed as a $\Phi_{p,S}/\tilde{\lambda}_g\Phi_{p,S}$-valued function.

Lemma 3.6 *The relation*

$$\bar{\kappa}\{x \to y\}(v') - \bar{\kappa}\{x \to y\}(v) = \kappa\{x \to y\}(e) \pmod{\tilde{\lambda}_g \Phi_{p,S}}$$

holds for all the oriented edges $e = (v, v')$.

Proof By our choice of S, the reduction modulo λ_f of $\bar{\kappa}$ yields a map

$$\eta_f : \mathcal{V}(\mathcal{T}) \longrightarrow H_1(X_0(M), \mathbb{Z}_S)/\lambda_f H_1(X_0(M), \mathbb{Z}_S) \simeq \mathbb{F}^2,$$

where $\mathbb{F}$ is the finite field with ℓ elements. The map

$$\eta_f^\sharp : \mathcal{E}(\mathcal{T}) \longrightarrow \mathbb{F}^2$$

given by the rule $\eta_f^\sharp(e) = \eta_f(v') - \eta_f(v)$, for $e = (v, v')$, defines the p-stabilised eigenform associated to η_f. It satisfies the relation $\tilde{U}_p \eta_f^\sharp = \eta_f^\sharp$, where the operator $\tilde{U}_p$ acts by sending an oriented edge e to the formal sum of the oriented edges originating from e. Since $\tilde{\lambda}_g$ is a prime of fusion, it follows that $\eta_f^\sharp$ coincides with the reduction of κ modulo $\tilde{\lambda}_g$. □

Let $v \in \mathcal{V}(\mathcal{T})$ denote the reduction of τ. Define a 1-cochain

$$\bar{\beta}_{\tau,x} : \Gamma \to \Phi_{p,S}/\tilde{\lambda}_g \Phi_{p,S}$$

by the rule

$$\bar{\beta}_{\tau,x}(\gamma) = \bar{\kappa}\{x \to \gamma x\}(v).$$

A direct calculation using the equation

$$\bar{\kappa}\{x \to y\}(\gamma v) = \bar{\kappa}\{\gamma^{-1}x \to \gamma^{-1}y\}(v)$$

along with Lemma 3.6 and Proposition 3.1 shows that

$$\bar{\beta}_{\tau,x}(\gamma_1\gamma_2) - \bar{\beta}_{\tau,x}(\gamma_1) - \bar{\beta}_{\tau,x}(\gamma_2) = \text{ord}_p \left(\times\!\!\!\!\!\!\int_{\tau}^{\gamma_1^{-1}\tau} \int_x^{\gamma_2 x} \omega \right) \tag{3.4}$$

in $\Phi_{p,S}/\tilde{\lambda}_g \Phi_{p,S}$. From equation (1.9) defining $\beta_{\tau,x}$ and the fact that $\beta_{\tau,x}$ is unique up to translation by a homomorphism from Γ, it follows that

$$t \cdot \mathrm{ord}_p\left(\beta_{\tau,x}(\gamma)\right) = t \cdot \bar{\beta}_{\tau,x}(\gamma)$$

in $\Phi_{p,S}/\tilde{\lambda}_g \Phi_{p,S}$. In particular, we have

Proposition 3.7 *The equality*

$$\partial_{p,S}(\Phi(\tau)) = t \cdot \bar{\kappa}\{x \to \gamma_\tau x\}(v)$$

holds in $\Phi_{p,S}/\tilde{\lambda}_g \Phi_{p,S}$.

With notations as in the Introduction, let $\tau = \tau_1, \ldots, \tau_h$ be representatives for the distinct Γ-orbits of points in $\mathcal{H}_p \cap K$ corresponding to a real quadratic order $\mathcal{O}$. Define the Stark-Heegner point

$$P_K := \Phi(\tau_1) + \ldots + \Phi(\tau_h).$$

Write v_j, $j = 1, \ldots, h$ for the image of τ_j by the reduction map, and $\gamma_j \in \mathcal{O}_1^\times$ for the element appearing in the definition of $\Phi(\tau_j)$. Normalise the τ_j so that the associated γ_j are defined as in equation (2.1). This implies that the vertices v_j all coincide with the standard vertex v^o.

One finds

Corollary 3.8 *The equality*

$$\partial_{p,S}(P_K) = \sum_{i=1}^{h} t \cdot \bar{\kappa}\{x \to \gamma_i x\}(v^o)$$

holds in $\Phi_{p,S}/\tilde{\lambda}_g \Phi_{p,S}$.

Recall the homology element $\mathcal{L}(E/K,1) \in H_1(X_0(M), \mathbb{Z}_S)$ defined in Section 2. In light of Proposition 3.3, let $\mathcal{L}(E/K,1)_{(p)}$ denote the natural image of $t \cdot \mathcal{L}(E/K,1)$ in $\Phi_{p,S}/\tilde{\lambda}_g \Phi_{p,S}$. (Note that t is a unit in $\mathbb{Z}_S$ by Proposition 1.6, so that $\mathcal{L}(E/K,1)_{(p)}$ is non-zero if and only if the image of $\mathcal{L}(E/K,1)$ in $\Phi_{p,S}/\tilde{\lambda}_g \Phi_{p,S}$ is non-zero.) Then, combining Corollary 3.8 with Proposition 2.1 yields the main theorem of the Introduction:

Theorem 3.9 *For all primes p which are inert in K,*

$$\partial_{p,S}(P_K) = \mathcal{L}(E/K,1)_{(p)}.$$

Remark 3.10 The element $\mathcal{L}(E/K,1)_{(p)}$ depends on the choice of primes p and ℓ (the residue characteristic of λ_f), and of the set S. Given a rational prime p which is inert in K, it is certainly possible that the module $H_1(X_0(M),\mathbb{Z}_S)_f/\ (a_p(f)\ -\ (p\ +\ 1))$ be zero, and that no modular form g, congruent to f, be available for which the quotient $\Phi_{p,S}/\tilde{\lambda}_g\Phi_{p,S}$ is non-zero. For such a choice of p, the statement of Theorem 3.9 amounts to a trivial equality. However, the Chebotarev density theorem can be used to produce infinitely many p (for a fixed ℓ) for which the equality of Theorem 3.9 is non-trivial. Assume that $L(E/K,1)$ is non-zero, or equivalently by Corollary 2.3, that $\mathcal{L}(E/K,1) = \mathcal{L}(E/K,1)_S$ is non-zero, where S is such that $H_1(X_0(M),\mathbb{Z}_S)_f$ is a free $\mathbb{Z}_S$-module of rank 2. By a theorem of Serre, for all but finitely many primes ℓ, the element $\mathcal{L}(E/K,1)$ is non-zero modulo ℓ and the Galois representation $\rho_{E,\ell}$ attached to $E[\ell]$—the ℓ-torsion of E—is surjective. A standard application of the Chebotarev density theorem (see for example [BD2]) shows that there exist infinitely many primes p which are inert in K and such that the following conditions are satisfied:

(i) ℓ divides the integer $a_p(f)-(p+1)$,
(ii) ℓ does not divide $6\varphi(M)(p^2-1)(a_r(f)-(r+1))$, for a prime $r\nmid Mp$.

Enlarge the set S above by including all the primes dividing the quantity $6\varphi(M)(p^2-1)(a_r(f)-(r+1))$. For such an S, the results of Section 3.2—see in particular Remark 3.5—show that there exists a congruent form g for which $\mathcal{L}(E/K,1)_{(p)}$ is a non-zero element of $\Phi_{p,S}/\tilde{\lambda}_g\Phi_{p,S}$. (Note that in this case, the latter quotient is identified with $\Phi_p/\tilde{\lambda}_g\Phi_p$, since all the primes in S are units modulo ℓ. Thus, the formula of Theorem 3.9 can be written by omitting a reference to S.)

4 Arithmetic applications

The Shimura reciprocity law implies that the points $\Phi(\tau)$, as τ varies over $\mathcal{H}_p\cap K$, satisfy the same norm-compatibility properties as classical Heegner points attached to an imaginary quadratic K, and it is expected that they should yield an "Euler system" in the sense of Kolyvagin. (Cf. [BD1], Prop. 6.18). Theorem 3.9 gives a relationship between Stark–Heegner points and special values of related Rankin L-series, and one might ask whether this result could have applications to the arithmetic of elliptic curves analogous to those of the Gross-Zagier theorem.

For example, assume Conjecture 1.7 that P_K belongs to $J(K)$. Theorem 3.9 then shows that, when $L(E/K,1)\neq 0$, the points P_K are of infinite order for infinitely many p (in fact, for precisely those p for which $\mathcal{L}(E/K,1)_{(p)}\neq 0$).

The P_K can then be used to construct a large and well-behaved supply of cohomology classes in $H^1(K, E_p)$. Following the methods of Kolyvagin, such classes could be used to prove the following theorem:

Theorem 4.1 *Assume Conjecture 1.7. If* $L(E/K, 1) \neq 0$, *the Mordell-Weil group and Shafarevich-Tate group of* E *over* K *are finite.*

We omit the details of the proof, but point out that such a proof would follow that same strategy as in [BD2], but with Stark-Heegner points replacing the classical Heegner points that are used in [BD2]. (See also [Lo] where a similar strategy is used to prove the Birch and Swinnerton-Dyer conjecture for elliptic curves of analytic rank 0 defined over totally real fields which do *not* necessarily arise as quotients of modular or Shimura curves.)

Theorem 4.1 has the drawback of being conditional on Conjecture 1.7—a limitation that appears all the more flagrant when one notes that the conclusion of this theorem already follows, unconditionally, from earlier results of Kolyvagin (or of Kato) applied in turn to $E/\mathbb{Q}$ and to the twist of E by the even Dirichlet character associated to K.

However, greater generality could be achieved by introducing a ring class character

$$\chi : \mathrm{Gal}(H_K/K) \longrightarrow \mathbb{C}^\times$$

and considering twisted special values of $L(E/K, \chi, 1)$ along with related eigencomponents of the Mordell-Weil group $E(H_K)$ and of the Shafarevich-Tate group $Ш := Ш(E/H_K)$:

$$E(H_K)^\chi := \{P \in E(H_K) \otimes \mathbb{C} \text{ such that } \sigma P = \chi(\sigma) P, \ \forall \sigma \in \mathrm{Gal}(H_K/K)\}$$

$$Ш^\chi := \{x \in Ш \otimes \mathbb{Z}[\chi] \text{ such that } \sigma x = \chi(\sigma) x, \ \ \forall \, \sigma \in \mathrm{Gal}(H_K/K)\}.$$

In light of Proposition 2.1 and Remark 2.2, Theorem 3.9 generalises directly to a relation between the special value $L(E/K, \chi, 1)$ and the images of the χ-parts of Stark-Heegner points in connected components. Furthermore, when χ is not a quadratic character an unconditional proof of the following theorem would appear to lie beyond the scope of the known Euler systems discovered by Kolyvagin and Kato, and would yield a genuinely new arithmetic application of the conjectural Euler system made from Stark–Heegner points:

Theorem 4.2 *Assume conjecture 1.7 of Section 1. If* $L(E/K, \chi, 1) \neq 0$, *then the Mordell-Weil group* $E(H_K)^\chi$ *is trivial and the Shafarevich-Tate group* $Ш^\chi$ *is finite.*

Part of the inspiration for Theorem 3.9 is the strong analogy between it and the "first explicit reciprocity law" of chapter 4 of [BD2] in the setting where K is imaginary. (See in particular the displayed formula in lemma 8.1 of [BD2].) In [BD2] this first explicit reciprocity law was used in conjunction with a "second explicit reciprocity law" to prove one divisibility in the anticyclotomic Main Conjecture of Iwasawa Theory (for K imaginary quadratic). Such a main conjecture has no counterpart when K is real quadratic (since K has no "anticyclotomic $\mathbb{Z}_p$-extension") but versions of this statement over ring class fields of K (of finite degree) remain non-trivial and meaningful. Note in this connection that it may be interesting to formulate a convincing substitute for the "second explicit reciprocity law" of [BD2] describing the local behaviour of the point P_K at primes different from p.

Bibliography

[BD1] M. Bertolini, H.Darmon, *The p-adic L-functions of modular elliptic curves.* in Mathematics unlimited—2001 and beyond, 109–170, Springer, Berlin, 2001.

[BD2] M. Bertolini, H. Darmon, *The main conjecture of Iwasawa theory for elliptic curves over anticyclotomic $\mathbb{Z}_p$-extensions.* Ann. of Math. (2) 162 (2005), no. 1, 1–64.

[BD3] M. Bertolini, H. Darmon, *The rationality of Stark-Heegner points over genus fields of real quadratic fields.* Ann. of Math., to appear.

[BDG] M. Bertolini, H. Darmon, P. Green, *Periods and points attached to quadratic algebras.* MSRI Publ. 49, Cambridge Univ. Press, 323-367, 2004.

[Dar] H. Darmon, *Integration on $\mathcal{H}_p \times \mathcal{H}$ and arithmetic applications.* Ann. of Math. (2) **154** (2001), no. 3, 589–639.

[Das] S. Dasgupta, *Stark–Heegner points on modular Jacobians.* Ann. Scient. Éc. Norm. Sup., 4e sér., 38 (2005), 427-469.

[DG] H. Darmon, P. Green, *Elliptic curves and class fields of real quadratic fields: algorithms and verifications.* Experimental Mathematics, 11:1, 2002, 37-55.

[GvdP] L. Gerritzen, M. van der Put, *Shottky Groups and Mumford Curves.* Lecture Notes in Mathematics, **817**. Springer, Berlin, 1980.

[GS] R. Greenberg, G. Stevens, *p-adic L-functions and p-adic periods of modular forms.* Invent. Math. **111** (1993), no. 2, 407–447.

[GZ] B. H. Gross, D. B. Zagier, *Heegner points and derivatives of L-series.* Invent. Math. **84** (1986), no. 2, 225–320.

[Lo] M. Longo, *On the Birch and Swinnerton-Dyer for modular elliptic curves over totally real fields,* Ann. Inst. Fourier (Grenoble) **56** (2006) no. 3, 689–733.

[Man] J.I. Manin, *Parabolic Points and Zeta Functions of Modular Curves.* Izv. Akad. Nauk SSSR Ser. Mat. **36** (1972), no. 1, 19–66.

[MTT] B. Mazur, J. Tate, J. Teitelbaum, *On p-adic analogues of the conjectures of Birch and Swinnerton-Dyer.* Invent. Math. **84** (1986), no. 1, 1–48.

[Po] A. Popa, *Central values of Rankin L-series over real quadratic fields.* Compos. Math. **142** (2006) no. 4, 811–866.

[Rib1] K. Ribet, *Congruence Relations Between Modular Forms.* Proceedings of the International Congress of Mathematicians, Warsaw, August 16-24, 1983.

[Rib2] K. Ribet, *On modular representation of* $\mathrm{Gal}(\bar{\mathbb{Q}}/\mathbb{Q})$ *arising from modular forms,* Invent. Math. **100** (1990) 431-476.

[Se] J.P. Serre, *Trees.* Springer, Berlin, 1980.

[W] A. Wiles, *Modular elliptic curves and Fermat's Last Theorem,* Ann. of Math. **141** (1995) 443-551.

Presentations of universal deformation rings

Gebhard Böckle
Fachbereich Mathematik
Universität Duisburg-Essen
Campus Essen
45117 Essen
Germany
boeckle@iem.uni-due.de

Abstract

[1] Let $\mathbb{F}$ be a finite field of characteristic $\ell > 0$, F a number field, G_F the absolute Galois group of F and let $\bar{\rho} : G_F \to \mathrm{GL}_N(\mathbb{F})$ be an absolutely irreducible continuous representation. Suppose S is a finite set of places containing all places above ℓ and above ∞ and all those at which $\bar{\rho}$ ramifies. Let $\mathcal{O}$ be a complete discrete valuation ring of characteristic zero with residue field $\mathbb{F}$. In such a situation one may consider all deformations of $\bar{\rho}$ to $\mathcal{O}$-algebras which are unramified outside S and satisfy certain local deformation conditions at the places in S. This was first studied by Mazur, [16], and under rather general hypotheses, the existence of a universal deformation ring was proven.

In [2] I studied, among other things, the number of generators needed for an ideal I in a presentation of such a universal deformation ring as a quotient of a power series ring over $\mathcal{O}$ by I. The present manuscript is an update of this part of [2]. The proofs have been simplified, the results slightly generalized. We also treat $\ell = 2$, more general groups than GL_N, and cases where not all relations are local. The results in [2] and hence also in the present manuscript are one of the (many) tools used in the recent attacks on Serre's conjecture by C. Khare and others.

Introduction

Let us consider the following simple lemma from commutative algebra:

Lemma 0.1 *Suppose that a ring R has a presentation of the form $R = W(\mathbb{F})[[T_1, \ldots, T_n]]/(f_1, \ldots, f_m)$. If $R/(\ell)$ is finite, and if $n \geq m$, then $n = m$ and R is a complete intersection that is finite flat over $W(\mathbb{F})$.*

1 *1991 Mathematics Subject Classification.* Primary 11F34, 11F70, Secondary 14B12.

If the ring R in the above lemma is a universal deformation ring for certain deformation types of a given residual representation, then the conclusion of the lemma would provide one with a lift to characteristic zero of this deformation type. This observation was first made by A.J. de Jong, [9], *(3.14)*, in 1996. Using obstruction theory and Galois cohomology, in [2] we investigated the existence of presentations of universal deformation rings of the type required in the lemma. In many cases such a presentation was found. However the finiteness of the ring $R/(\ell)$ seemed to be out of reach.

This was changed enormously by the ground breaking work [22] of R. Taylor where a potential version of Serre's conjecture was proved. The results of Taylor do allow one in many cases to prove the finiteness of $R/(\ell)$. I first learned about this from C. Khare soon after [22] was available. This gives a powerful tool to construct ℓ-adic Galois representations that are (potentially) semistable or ordinary at ℓ and have prescribed ramification properties at primes away from ℓ. Besides the deep modularity results for such representations provided by Wiles, Taylor, Skinner et al., the results in [2] were one of the ingredients of the recent proof of Serre's conjecture for conductor $N = 1$ and arbitrary weight by Khare in [10] (cf. also [11]). This is based on previous joint work between Khare and Wintenberger [12], and a result by Dieulefait [7]. Dieulefait also has some partial results on Serre's conjecture [8].

The present manuscript is an update of those parts of [2] which study the number of generators needed for an ideal I in a presentations of a given universal deformation ring as a quotient of a power series ring over $\mathcal{O}$ by I. The proofs have been simplified and the results generalized. We also treat $\ell = 2$, more general groups than GL_N and cases where not all relations are local. A key improvement is that the use of auxiliary primes has been avoided entirely. We hope that this will be useful for the interested reader.

We now give a summary of the individual sections. In Section 1, we start by briefly recalling Mazur's fundamental results on universal deformations with the main emphasis on presentations of universal deformation rings. In Section 2 we give a first link between the ideals of presentations of local and of global deformation rings in the setting of Mazur adapted to global number fields. The discrepancy is measured by Ш^2_S of the adjoint representation of the given residual representation.

It is natural to put further local restrictions on the initial deformation problem studied by Mazur. To obtain again a representable functor the local conditions need to be relatively representable. In Section 3 we present a possibly useful variant of this notion.

The core of the present article is Section 4, cf. Corollary 4.3. Here we study presentations of (uni)versal deformation rings for deformations in the sense of Mazur that moreover satisfy a number of local conditions that follow the axiomatics in Section 3. The obstruction module Ш^2_S is replaced by a naturally occurring dual Selmer group. The main novelty of the present paper is that unlike in [2] we do not require that this dual Selmer group vanishes. Instead we incorporate it into the presentation of the corresponding ring.

The final three sections investigate consequences of our results. In Section 5 we make some general comments and study the case of GL_2 in detail. In particular, we present the numerology for local ordinary deformation rings over arbitrary local fields (of any characteristic). In Section 6 we compare our results to those in [15, 23] of Mauger and Tilouine. The last section, Section 7, is dedicated to deriving a presentation of a global universal deformation rings as the quotient of a power series ring over the completed tensor product over all local versal deformation rings. The main result here is due to M. Kisin [13]. We show how to derive it using the results of Section 4.

Notation: For the rest of this article, we fix the following notation: $\mathbb{F}$ is a finite field of characteristic ℓ. The ring of Witt vectors of $\mathbb{F}$ is denoted $W(\mathbb{F})$. For a local ring R its maximal ideal is denoted by $\mathfrak{m}_R$. By $\mathcal{O}$ we denote a complete discrete valuation ring of characteristic zero with residue field $\mathbb{F}$, so that in particular $\mathcal{O}$ is finite over $W(\mathbb{F})$. The category of complete noetherian local $\mathcal{O}$-algebras R with a fixed isomorphism $R/\mathfrak{m}_R \cong \mathbb{F}$ will be $\mathcal{C}_\mathcal{O}$. Here and in the following $\mathbb{F}[\varepsilon]/(\varepsilon^2)$ is an $\mathcal{O}$-algebra via $\mathcal{O} \to \mathbb{F} \to \mathbb{F}[\varepsilon]/(\varepsilon^2)$.

For a ring R of $\mathcal{C}_\mathcal{O}$ its mod $\mathfrak{m}_\mathcal{O}$ tangent space is defined as

$$\mathfrak{t}_R := \mathrm{Hom}_\mathcal{O}(R, \mathbb{F}[\varepsilon]/(\varepsilon^2)) \cong \mathrm{Hom}_\mathbb{F}(\mathfrak{m}_R/(\mathfrak{m}_R^2 + \mathfrak{m}_\mathcal{O} R), \mathbb{F}).$$

For J an ideal of a ring R in $\mathcal{C}_\mathcal{O}$, we define $\mathrm{gen}(J) := \dim_\mathbb{F} J/(\mathfrak{m}_R J)$. By Nakayama's lemma, $\mathrm{gen}(J)$ is the minimal number of generators of J as an ideal in R.

By F we denote a number field and by S a finite set of places of F. We always assume that S contains all places of F above ℓ and ∞. The maximal outside S unramified extension of F inside a fixed algebraic closure F^{alg} of F is denoted F_S. It is a Galois extension of F whose corresponding Galois group is $G_{F,S} := \mathrm{Gal}(F_S/F)$.

For each place ν of F let F_ν be the completion of F at ν, let G_ν be the absolute Galois group of F_ν, and $I_\nu \subset G_\nu$ the inertia subgroup. Choosing for each such ν a field homomorphism $F_S \hookrightarrow F_\nu^{\mathrm{alg}}$, we obtain induced group homomorphisms $G_\nu \to G_{F,S}$.

Acknowledgments: This article owes many ideas and much inspiration to the work of Mazur, Wiles, Taylor, de Jong, and many others. Many thanks go to C. Khare for constantly reminding me to write un 'update' of the article [2] and for many comments. Many thanks also to Mark Kisin for having made available [13] and for some interesting related discussions.

1 A simple deformation problem

In this section we recall various basic notions and concepts from [16]. In terms of generality, we follow [23], and so we fix a smooth linear algebraic group $\mathcal{G}$ over $\mathcal{O}$. By $\mathcal{Z}_\mathcal{G}$ we denote the center of $\mathcal{G}$, by $\mathcal{T}$ we denote a smooth affine algebraic group over $\mathcal{O}$ that is a quotient of $\mathcal{G}$ via some surjective homomorphism $d\colon \mathcal{G} \to \mathcal{T}$ of algebraic groups over $\mathcal{O}$. The kernel of d is denoted $\mathcal{G}_0$. The Lie algebras over $\mathcal{O}$ corresponding to $\mathcal{G}$ and $\mathcal{G}_0$ will be $\mathfrak{g}$ and $\mathfrak{g}_0$, respectively.

Example 1.1

(i) $d := \det\colon \mathcal{G} := \mathrm{GL}_N \to \mathcal{T} := \mathrm{GL}_1$. Then $\mathfrak{g} = M_N(\mathbb{F})$ and $\mathfrak{g}_0 \subset \mathfrak{g}$ is the subset of trace zero matrices.

(ii) $\mathcal{G}$ is the Borel subgroup of GL_N formed by the set of upper triangular matrices, $\mathcal{T} := \mathrm{GL}_1^N$, and $d : \mathcal{G} \to \mathrm{GL}_1^N$ is the the quotient homomorphism of $\mathcal{G}$ modulo its unipotent radical. The corresponding Lie algebras are the obvious ones.

Throughout this section let Π be a profinite group such that the pro-ℓ completion of every open subgroup is topologically finitely generated. (This is the finiteness condition Φ_ℓ of [16], Def. 1.1.) Let us fix a continuous (residual) representation

$$\bar{\rho} : \Pi \to \mathcal{G}(\mathbb{F}).$$

The **adjoint representation of** Π **on** $\mathfrak{g}(\mathbb{F})$ is denoted by $\mathrm{ad}_{\bar{\rho}}$, **its subrepresentation on** $\mathfrak{g}_0(\mathbb{F}) \subset \mathfrak{g}(\mathbb{F})$ by $\mathrm{ad}^0_{\bar{\rho}}$. For M an $\mathbb{F}[\Pi]$-module, we define its **dimension** as $h^i(\Pi, M) := \dim_\mathbb{F} H^i(\Pi, M)$.

Following Mazur we first consider the following simple deformation problem: A **lifting** of $\bar{\rho}$ to $R \in \mathcal{C}_\mathcal{O}$ is a continuous representation $\rho : \Pi \to \mathcal{G}(R)$, such that $\rho \pmod{\mathfrak{m}_R} = \bar{\rho}$. A **deformation** of $\bar{\rho}$ to R is a strict equivalence class $[\rho]$ of liftings ρ of $\bar{\rho}$ to R, where two liftings ρ_1 and ρ_2 from Π to $\mathcal{G}(R)$ are strictly equivalent, if there exists an element in the kernel of $\mathcal{G}(R) \to \mathcal{G}(\mathbb{F})$ which conjugates one into the other.

We consider the functor

$$\mathrm{Def}_{\mathcal{O},\Pi} : \mathcal{C}_{\mathcal{O}} \to \mathbf{Sets} : R \mapsto \{[\rho] \mid [\rho] \text{ is a deformation of } \bar{\rho} \text{ to } R\}.$$

Theorem 1.2 ([16]) *Suppose* Π *and* $\bar{\rho}$ *are as above. Then*

(i) *The functor* $\mathrm{Def}_{\mathcal{O},\Pi}$ *has a versal hull, which we denote by* $\rho_{\bar{\rho},\mathcal{O}} : \Pi \to \mathcal{G}(R_{\bar{\rho},\mathcal{O}})$.
(ii) *If furthermore the centralizer of* $\mathrm{Im}(\bar{\rho})$ *in* $\mathcal{G}(\mathbb{F})$ *is contained in* $\mathcal{Z}_{\mathcal{G}}(\mathbb{F})$, *then* $\mathrm{Def}_{\mathcal{O},\Pi}$ *is representable by the above pair* $(R_{\bar{\rho},\mathcal{O}}, \rho_{\bar{\rho},\mathcal{O}})$.
(iii) $\mathfrak{t}_{R_{\bar{\rho},\mathcal{O}}} \cong H^1(\Pi, \mathrm{ad}_{\bar{\rho}})$.
(iv) $R_{\bar{\rho},\mathcal{O}}$ *has a presentation* $R_{\bar{\rho},\mathcal{O}} \cong \mathcal{O}[[T_1, \ldots, T_h]]/J$ *for some ideal* $J \subset \mathcal{O}[[T_1, \ldots, T_h]]$, *where* $h = h^1(\Pi, \mathrm{ad}_{\bar{\rho}})$ *and* $\mathrm{gen}(J) \leq h^2(\Pi, \mathrm{ad}_{\bar{\rho}})$.

Proof The proof is essentially contained in [16] §1.2, §1.6, where a criterion of Schlessinger is verified. For (ii) Mazur originally assumed that $\bar{\rho}$ was absolutely irreducible. It was later observed by Ramakrishna, [20], that this could be weakened to the condition given.

A proof for GL_N instead of a general group $\mathcal{G}$ in the precise form above can be found in [2], Thm. 2.4. The adaption to general $\mathcal{G}$ is obvious, and so we omit details.

Since this will be of importance later, we remark that the proofs in [16] or [2] show that there is a canonical surjective homomorphism

$$H^2(\Pi, \mathrm{ad}_{\bar{\rho}})^* \twoheadrightarrow J/\mathfrak{m}_{\mathcal{O}[[T_1,\ldots,T_h]]}J. \tag{1.1}$$

of vector spaces over $\mathbb{F}$ □

Remark 1.3 If $R_{\bar{\rho},\mathcal{O}}/(\ell)$ is known to have Krull dimension $h^1(\Pi, \mathrm{ad}_{\bar{\rho}}) - h^2(\Pi, \mathrm{ad}_{\bar{\rho}})$, then Theorem 1.2, which is obtained entirely by the use of obstruction theory, implies that $R_{\bar{\rho},\mathcal{O}}$ is flat over $\mathcal{O}$ of relative dimension $h^1(\Pi, \mathrm{ad}_{\bar{\rho}}) - h^2(\Pi, \mathrm{ad}_{\bar{\rho}})$, and a complete intersection. So in this situation Theorem 1.2 has some strong ring-theoretic consequences for $R_{\bar{\rho},\mathcal{O}}$.

In the generality of the present section, it can not be expected that the Krull dimension of $R_{\bar{\rho},\mathcal{O}}/(\ell)$ is always equal to $h^1(\Pi, \mathrm{ad}_{\bar{\rho}}) - h^2(\Pi, \mathrm{ad}_{\bar{\rho}}) + 1$. Recent work [1] by Bleher and Chinburg shows that this fails for finite groups Π and also for Galois groups $\Pi = G_{F,S}$ in case S does not contain all the primes above ℓ.

If $\mathcal{G} = \mathrm{GL}_n$ and $\bar{\rho}$ is absolutely irreducible, and if $\Pi = G_{F,S}$ and S contains all primes above ℓ, or if $\Pi = G_{F_\nu}$ (cf. Remark 5.2), all evidence suggests that $R_{\bar{\rho},\mathcal{O}}$ is a complete intersection, flat over $\mathcal{O}$ and of relative dimension

$h^1(\Pi, \mathrm{ad}_{\bar{\rho}}) - h^2(\Pi, \mathrm{ad}_{\bar{\rho}})$. But the amount of evidence is small. On the one hand, there is Leopoldt's conjecture, cf. [16], 1.10. On the other, if one is bootstrapping the content of [4] in light of the recent modularity results of Taylor et al., e.g. [22], there is some evidence for odd two-dimensional representations $\bar{\rho}$ of $G_{F,S}$ where F is a totally real field.

One often considers the following subfunctor of $\mathrm{Def}_{\Pi,\mathcal{O}}$: Let $\eta\colon \Pi \to \mathcal{T}(\mathcal{O})$ be a fixed lift of the residual representation $d \circ \bar{\rho} : \Pi \to \mathcal{T}(\mathbb{F})$. Then one defines the subfunctor $\mathrm{Def}^{\eta}_{\Pi,\mathcal{O}}$ of $\mathrm{Def}_{\Pi,\mathcal{O}}$ as the functor which to $R \in \mathcal{C}_{\mathcal{O}}$ assigns the set of all deformations $[\rho]$ of $\bar{\rho}$ to R for which the composite $d \circ \rho\colon \Pi \to \mathcal{T}(R)$ and the composite $\tau \circ \eta$ of η with the canonical homomorphism $\tau : \mathcal{T}(\mathcal{O}) \to \mathcal{T}(R)$ only differ by conjugation inside $\mathcal{T}(R)$. One obtains the following analog of Theorem 1.2:

Theorem 1.4 ([16]) *Suppose Π and $\bar{\rho}$ are as above. Then:*

(i) *The functor $\mathrm{Def}^{\eta}_{\Pi,\mathcal{O}}$ has a versal hull, which we denote by $\rho^{\eta}_{\bar{\rho},\mathcal{O}} : \Pi \to \mathcal{G}(R^{\eta}_{\bar{\rho},\mathcal{O}})$.*

(ii) *If furthermore the centralizer of $\mathrm{Im}(\bar{\rho})$ in $\mathcal{G}(\mathbb{F})$ is contained in $\mathcal{Z}_{\mathcal{G}}(\mathbb{F})$, then $\mathrm{Def}^{\eta}_{\mathcal{O},\Pi}$ is representable by the above pair $(R^{\eta}_{\bar{\rho},\mathcal{O}}, \rho_{\bar{\rho},\mathcal{O}})$.*

(iii) $\mathrm{t}_{R^{\eta}_{\bar{\rho},\mathcal{O}}} \cong H^1(\Pi, \mathrm{ad}_{\bar{\rho}})^{\eta} := \mathrm{Im}(H^1(\Pi, \mathrm{ad}^0_{\bar{\rho}}) \to H^1(\Pi, \mathrm{ad}_{\bar{\rho}}))$.

(iv) *$R^{\eta}_{\bar{\rho},\mathcal{O}}$ has a presentation $R^{\eta}_{\bar{\rho},\mathcal{O}} \cong \mathcal{O}[[T_1, \ldots, T_{h^{\eta}}]]/J^{\eta}$ for some ideal $J^{\eta} \subset \mathcal{O}[[T_1, \ldots, T_{h^{\eta}}]]$, where $h^{\eta} = \dim_{\mathbb{F}} H^1(\Pi, \mathrm{ad}^0_{\bar{\rho}})^{\eta}$ and $\mathrm{gen}(J^{\eta}) \le \dim_{\mathbb{F}} H^2(\Pi, \mathrm{ad}^0_{\bar{\rho}})$.*

Remark 1.5 From the long exact cohomology sequence for

$$0 \longrightarrow \mathfrak{g}_0 \longrightarrow \mathfrak{g} \longrightarrow \mathfrak{g}/\mathfrak{g}_0 \longrightarrow 0$$

it follows that $H^1(\Pi, \mathrm{ad}_{\bar{\rho}})^{\eta} \cong H^1(\Pi, \mathrm{ad}^0_{\bar{\rho}})$ is an isomorphism if and only if

$$H^0(\Pi, \mathrm{ad}_{\bar{\rho}}) \twoheadrightarrow H^0(\Pi, \mathrm{ad}_{\bar{\rho}}/\mathrm{ad}^0_{\bar{\rho}})$$

is surjective. To measure the discrepancy between h^{η} and $h^1(\Pi, \mathrm{ad}^0_{\bar{\rho}})$, we define $\delta(\Pi, \mathrm{ad}_{\bar{\rho}}) = 0$ and

$$\begin{aligned}\delta(\Pi, \mathrm{ad}_{\bar{\rho}})^{\eta} &:= \dim_{\mathbb{F}} \mathrm{Coker}(H^0(\Pi, \mathrm{ad}_{\bar{\rho}}) \longrightarrow H^0(\Pi, \mathrm{ad}_{\bar{\rho}}/\mathrm{ad}^0_{\bar{\rho}})) \\ &= h^1(\Pi, \mathrm{ad}^0_{\bar{\rho}}) - h^{\eta}. \quad (1.2)\end{aligned}$$

As an example consider the case $d = \det : \mathcal{G} = \mathrm{GL}_N \to \mathcal{T} = \mathrm{GL}_1$. If ℓ does not divide N, then $\mathrm{ad}_{\bar{\rho}} = \mathrm{ad}^0_{\bar{\rho}} \oplus \mathbb{F}$, where here $\mathbb{F}$ denotes the trivial representation of Π, and so $\delta(\Pi, \mathrm{ad}_{\bar{\rho}})^{\eta} = 0$. However for $\ell|N$ and absolutely irreducible $\bar{\rho}$, one finds that $\delta(\Pi, \mathrm{ad}_{\bar{\rho}})^{\eta} = 1$.

If $\eta = 1$, one can in fact consider two deformation functors: (i) the functor that arises from considering deformations into $\mathcal{G}_0$ instead of $\mathcal{G}$, and (ii) the functor $\mathrm{Def}^{\eta}_{\mathcal{O},\Pi}$ considered above. If $\delta(\Pi, \mathrm{ad}_{\bar{\rho}})^{\eta} = 0$, the two agree. Otherwise, the functor for $\mathcal{G}_0$ is less rigid, and in fact its mod $\mathfrak{m}_{\mathcal{O}}$ tangent space has a larger dimension (the difference being given by $\delta(\Pi, \mathrm{ad}_{\bar{\rho}})^{\eta}$).

Note also that the bound for $\mathrm{gen}(J^{\eta})$ in part (iv) is solely described in terms of $\mathrm{ad}^0_{\bar{\rho}}$.

2 A first local to global principle

For the remainder of this article we fix a residual representation

$$\bar{\rho}\colon G_{F,S} \to \mathcal{G}(\mathbb{F}).$$

Whenever it makes sense, we fix a lift $\eta\colon G_{F,S} \to \mathcal{T}(\mathcal{O})$ of $d \circ \bar{\rho}$. (If $\mathcal{T}(\mathbb{F})$ is of order prime to ℓ, such a lift always exists.) As in the previous section, the adjoint representation of $G_{F,S}$ on $\mathfrak{g}(\mathbb{F})$ is denoted by $\mathrm{ad}_{\bar{\rho}}$, its subrepresentation on $\mathfrak{g}_0(\mathbb{F}) \subset \mathfrak{g}(\mathbb{F})$ by $\mathrm{ad}^0_{\bar{\rho}}$.

To $\bar{\rho}$ we can attach the following canonical deformation functors: First, we define

$$\mathrm{Def}_{S,\mathcal{O}} := \mathrm{Def}_{G_{F,S},\mathcal{O}}, \quad \text{and} \quad \mathrm{Def}^{\eta}_{S,\mathcal{O}} := \mathrm{Def}^{\eta}_{G_{F,S},\mathcal{O}}.$$

The first functor parameterizes all deformations of $\bar{\rho}$ which are unramified outside S, the second (sub)functor moreover fixes the chosen determinant η.

Let ν be any place of F. The restriction of ρ to G_{ν} defines a residual representation $G_{\nu} \to \mathcal{G}(\mathbb{F})$, the restriction of η to G_{ν} a lift $G_{\nu} \to \mathcal{T}(\mathcal{O})$ of $d \circ \bar{\rho}$ restricted to G_{ν}. Thus we may define local deformation functors by

$$\mathrm{Def}_{\nu,\mathcal{O}} := \mathrm{Def}_{G_{\nu},\mathcal{O}}, \quad \text{and} \quad \mathrm{Def}^{\eta}_{\nu,\mathcal{O}} := \mathrm{Def}^{\eta}_{G_{\nu},\mathcal{O}}.$$

Notational convention: In the sequel we often write $?^{(\eta)}$ in formulas. This expresses two assertions at once: First, the formula is true if the round brackets are missing throughout. Second, the formula is also true if (η) is entirely omitted throughout the formula. Corresponding to the above cases the usage of $\mathrm{ad}^{(0)}_{\bar{\rho}}$ has to be interpreted as follows: If brackets around (η) are omitted, then they are to be omitted around (0) in $\mathrm{ad}^{(0)}_{\bar{\rho}}$, too; if (η) is omitted, then (0) in $\mathrm{ad}^{(0)}_{\bar{\rho}}$ is to be omitted, as well.

By global, respectively local class field theory, the groups $G_{F,S}$ and G_{ν} satisfy the conditions imposed on the abstract profinite group Π in Section 1. Therefore Theorems 1.2 and 1.4 are applicable to $\bar{\rho}$ and its restriction to the groups G_ν. The resulting (uni)versal global deformations are denoted by $\rho^{(\eta)}_{S,\mathcal{O}}: G_{F,S} \to R^{(\eta)}_{S,\mathcal{O}}$, and the local ones by $\rho^{(\eta)}_{\nu,\mathcal{O}} : G_\nu \to R^{(\eta)}_{\nu,\mathcal{O}}$. We also set

$$h := h^1(G_{F,S}, \mathrm{ad}_{\bar\rho}), h^\eta := h^1(G_{F,S}, \mathrm{ad}_{\bar\rho})^\eta,$$
$$h_\nu := h^1(G_\nu, \mathrm{ad}_{\bar\rho}), h^\eta_\nu := h^1(G_\nu, \mathrm{ad}_{\bar\rho})^\eta.$$

With the above notation, Theorem 1.2 shows that there exist presentations

$$0 \longrightarrow J^{(\eta)}_\nu \longrightarrow \mathcal{O}[[T_{\nu,1}, \ldots, T_{\nu,h^{(\eta)}_\nu}]] \longrightarrow R^{(\eta)}_{\nu,\mathcal{O}} \longrightarrow 0, \qquad (2.1)$$

$$0 \longrightarrow J^{(\eta)} \longrightarrow \mathcal{O}[[T_1, \ldots, T_{h^{(\eta)}}]] \longrightarrow R^{(\eta)}_{S,\mathcal{O}} \longrightarrow 0. \qquad (2.2)$$

The restriction $G_\nu \to G_{F,S}$ applied to deformations, induces a natural transformation of functors

$$\mathrm{Def}_{S,\mathcal{O}} \to \prod_{\nu \in S} \mathrm{Def}_{\nu,\mathcal{O}} .$$

This yields a ring homomorphism

$$\widehat{\bigotimes_{\nu \in S}} R^{(\eta)}_\nu \longrightarrow R^{(\eta)}_S ,$$

where by $\widehat{\otimes}$, we denote the completed tensor product over the ring $\mathcal{O}$. Using the smoothness of $\mathcal{O}[[T_1, \ldots, T_{h^{(\eta)}}]]$, and the above presentations, we obtain a commutative diagram with inserted dashed arrows, where α is a product of local maps $\alpha_\nu : \mathcal{O}[[T_{\nu,1}, \ldots, T_{\nu,h^{(\eta)}_\nu}]] \to \mathcal{O}[[T_1, \ldots, T_{h^{(\eta)}}]]$ and where $\langle J^{(\eta)}_\nu \rangle$ denotes the ideal generated by the $J^{(\eta)}_\nu$ in

$$\begin{array}{ccccc} \langle J^{(\eta)}_\nu : \nu \in S\rangle & \hookrightarrow & \mathcal{O}[[T_{\nu,1}, \ldots, T_{\nu,h^{(\eta)}_\nu} \mid \nu \in S]] & \twoheadrightarrow & \hat{\otimes}_{\nu \in S} R^{(\eta)}_\nu \\ \downarrow & & \downarrow \alpha = \prod \alpha_\nu & & \downarrow \\ J^{(\eta)} & \hookrightarrow & \mathcal{O}[[T_1, \ldots, T_{h^{(\eta)}}]] & \twoheadrightarrow & R^{(\eta)}_{S,\mathcal{O}}. \end{array} \qquad (2.3)$$

For any $\mathbb{F}[G_{F,S}]$ module M, define $\text{Ш}^2_S(M) = \mathrm{Ker}(H^2(G_{F,S}, M) \longrightarrow \oplus_{\nu \in S} H^2(G_\nu, M))$. Our first result on a local to global relation is the following simple consequence of Theorems 1.2 and 1.4:

Theorem 2.1 *The ideal $J^{(\eta)}$ is generated by the images of the ideals $J^{(\eta)}_\nu$, $\nu \in S$, together with at most $\dim_{\mathbb{F}} \text{Ш}^2_S(\mathrm{ad}^{(0)}_{\bar\rho})$ further elements.*

In particular, if the corresponding $\text{Ш}^2_S(\ldots)$ vanishes, then all relations in $J^{(\eta)}$ are local.

Proof By (1.1), there is a surjection

$$H^2(G_{F,S}, \mathrm{ad}^{(0)}_{\bar\rho})^* \twoheadrightarrow J^{(\eta)}/\mathfrak{m}_{\mathcal{O}[[T_1,\ldots,T_{h(\eta)}]]}J^{(\eta)},$$

and similarly for the local terms. Comparing local and global terms yields the commutative diagram

$$\begin{array}{ccccc} \bigoplus_{\nu\in S} H^2(G_\nu, \mathrm{ad}^{(0)}_{\bar\rho})^* & \longrightarrow & H^2(G_{F,S}, \mathrm{ad}^{(0)}_{\bar\rho})^* & \twoheadrightarrow & Ш^2_S(\mathrm{ad}^{(0)}_{\bar\rho})^* \\ \downarrow & & \downarrow & & \\ \bigoplus_{\nu\in S} J^{(\eta)}_\nu/\mathfrak{m}_{\mathcal{O}[[T_1,\ldots,T_{h^{(\eta)}_\nu}]]}J^{(\eta)}_\nu & \longrightarrow & J^{(\eta)}/\mathfrak{m}_{\mathcal{O}[[T_1,\ldots,T_{h(\eta)}]]}J^{(\eta)} & & \end{array}$$

where the lower horizontal homomorphism is induced from the ring homomorphism α of the previous diagram, and where the vertical homomorphisms are surjective.

By Nakayama's Lemma, any subset of $J^{(\eta)}$ whose image generates $J^{(\eta)}/\mathfrak{m}_{\mathcal{O}[[T_1,\ldots,T_{h(\eta)}]]}J^{(\eta)}$ forms a generating system for $J^{(\eta)}$. Therefore the assertion of the theorem follows immediately from the above diagram. □

Remark 2.2 An obvious consequence of Theorem 2.1 is the inequality

$$\mathrm{gen}(J^{(\eta)}) \leq \dim_{\mathbb{F}} Ш^2_S(\mathrm{ad}^{(0)}_{\bar\rho}) + \sum_{\nu\in S} \mathrm{gen}(J^{(\eta)}_\nu).$$

In general, this inequality is not best possible, since one has the exact sequence

$$0 \longrightarrow Ш^2_S(\mathrm{ad}^{(0)}_{\bar\rho}) \longrightarrow H^2(G_{F,S}, \mathrm{ad}^{(0)}_{\bar\rho}) \longrightarrow \oplus_{\nu\in S} H^2(G_\nu, \mathrm{ad}^{(0)}_{\bar\rho}) \longrightarrow H^0(G_{F,S}, (\mathrm{ad}^{(0)}_{\bar\rho})^\vee)^* \longrightarrow 0.$$

3 Local conditions

For the applications to modularity questions, the functors considered in the previous section are too general. At places ν above the prime ℓ modular Galois representations are potentially semistable; at places ν away from ℓ, one often wants to prescribe a certain behavior of the local Galois representations in question. This leads one to consider subfunctors $\widetilde{\mathrm{Def}}^{(\eta)}_{\nu,\mathcal{O}}$ of the functors $\mathrm{Def}^{(\eta)}_{\nu,\mathcal{O}}$ that describe a certain type of local deformation.

An important requirement on these subfunctors is that the resulting global deformation problems should have a versal hull. There are various approaches to achieve this. We find it most convenient to work with the notion of relative representability, which is basically described in [17], § 19.

Let us recall from [2], § 2, the relevant notion of relative representability: Following Schlessinger a homomorphism $\pi : A \to C$ of Artin rings in $\mathcal{C}_{\mathcal{O}}$ is called a **small extension** if π is surjective and if the kernel of π is isomorphic to the A-module $\mathbb{F}$.

In [17], p. 277, in the definition of *small*, the requirement of surjectivity is left out. Therefore the statement of Schlessinger's Theorem as given there is weaker than that given in [21]. The statement in [17], p. 277, is also true if small morphisms are assumed to be surjective.

A covariant functor $F\colon \mathcal{C}_{\mathcal{O}} \to$ **Sets** is called **continuous**, if for any directed inverse system $(A_i)_{i\in I}$ of Artin rings in $\mathcal{C}_{\mathcal{O}}$ with limit $A := \varprojlim A_i$ in $\mathcal{C}_{\mathcal{O}}$, one has

$$F(A) = \varprojlim F(A_i).$$

Definition 3.1 *Given two covariant continuous functors $F, G : \mathcal{C}_{\mathcal{O}} \to$* **Sets** *such that G is a subfunctor of F, we say that G is* **relatively representable** *if*

(i) *$G(k) \neq \varnothing$, and*

(ii) *for all small surjections $f_1 : A_1 \to A_0$ and maps $f_2 : A_2 \to A_0$ of artinian rings $\mathcal{C}_{\mathcal{O}}$, the following is a pullback diagram:*

$$\begin{array}{ccc} G(A_1 \times_{A_0} A_2) & \longrightarrow & G(A_1) \times_{G(A_0)} G(A_2) \\ \downarrow & & \downarrow \\ F(A_1 \times_{A_0} A_2) & \longrightarrow & F(A_1) \times_{F(A_0)} F(A_2). \end{array}$$

Remark 3.2 The definition of relative representability given in [17] seems at the outset more restrictive. However, by a reduction procedure similar to that of Schlessinger in [21], our definition might be equivalent to the the one given in [17].

The property from [17] is the one that is satisfied for essentially all subfunctors $\widetilde{\mathrm{Def}}^{(\eta)}_{\nu,\mathcal{O}} \subset \mathrm{Def}^{(\eta)}_{\nu,\mathcal{O}}$ that have been considered in deformation problems for Galois representations. Hence in all of these cases, the local deformation problems are relatively representable in the above sense.

Proposition 3.3 *Suppose $F, F_i, G_i : \mathcal{C}_{\mathcal{O}} \to$* **Sets**, *$i \in I$, I a finite set, are covariant continuous functors. Suppose for each $i \in I$ that G_i is a relatively representable subfunctor of F_i. Then the following holds:*

(i) *If F_i has a hull, i.e., F_i satisfies conditions $(\mathbf{H_1})$, $(\mathbf{H_2})$ and $(\mathbf{H_3})$ of Schlessinger, [21], Thm. 2.11, or [17], § 18, then so does G_i. If F_i is representable, then so is G_i.*

(ii) *The product* $\prod_{i\in I} G_i$ *is a continuous subfunctor of* $\prod_i F_i$ *which is relatively representable.*

(iii) *Suppose the* F_i *have a versal hull. Let* $\alpha : F \to \prod_i F_i$ *be a natural transformation, and let* G *be defined as the pullback of*

$$\begin{array}{ccc} G & \dashrightarrow & \prod G_i \\ \downarrow & & \downarrow \\ F & \longrightarrow & \prod F_i. \end{array}$$

Then, if F *has a versal hull, then so does* G*, and if* F *is representable, then so is* G*.*

The proof exploits the representability criterion of Schlessinger. It is a simple exercise in diagram chasing, and left to the reader.

After the above detour on general representability criteria, let us come back to the deformation functors we introduced in the previous section. The functors $\mathrm{Def}^{(\eta)}_{S,\mathcal{O}}$ and $\mathrm{Def}^{(\eta)}_{\nu,\mathcal{O}}$ are continuous. To work with finer local conditions, for each place ν in S we fix relatively representable subfunctors

$$\widetilde{\mathrm{Def}}^{(\eta)}_{\nu,\mathcal{O}} \subset \mathrm{Def}^{(\eta)}_{\nu,\mathcal{O}}.$$

We also define $\widetilde{\mathrm{Def}}^{(\eta)}_{S,\mathcal{O}}$ as the pullback of functors in the diagram

$$\begin{array}{ccc} \widetilde{\mathrm{Def}}^{(\eta)}_{S,\mathcal{O}} & \dashrightarrow & \prod_{\nu\in S} \widetilde{\mathrm{Def}}^{(\eta)}_{\nu,\mathcal{O}} \\ \downarrow & & \downarrow \\ \mathrm{Def}^{(\eta)}_{S,\mathcal{O}} & \longrightarrow & \prod \mathrm{Def}^{(\eta)}_{\nu,\mathcal{O}}. \end{array}$$

By Proposition 3.3, we obtain:

Proposition 3.4 *The functors* $\widetilde{\mathrm{Def}}^{(\eta)}_{\nu,\mathcal{O}}$ *have a versal hull* $\widetilde{\rho}^{(\eta)}_{\nu,\mathcal{O}} : G_\nu \to \mathcal{G}(\widetilde{R}^{(\eta)}_{\nu,\mathcal{O}})$*. The functor* $\widetilde{\mathrm{Def}}^{(\eta)}_{S,\mathcal{O}}$ *is representable, say by,* $\widetilde{\rho}^{(\eta)}_{S,\mathcal{O}} : G_\nu \to \mathcal{G}(\widetilde{R}^{(\eta)}_{S,\mathcal{O}})$*. The induced ring homomorphisms* $R^{(\eta)}_{\nu,\mathcal{O}} \to \widetilde{R}^{(\eta)}_{\nu,\mathcal{O}}$ *and* $R^{(\eta)}_{S,\mathcal{O}} \to \widetilde{R}^{(\eta)}_{S,\mathcal{O}}$ *are surjective.*

4 A refined local to global principle

We keep the hypotheses of the previous sections that the subfunctors $\widetilde{\mathrm{Def}}^{(\eta)}_{\nu,\mathcal{O}} \subset \mathrm{Def}^{(\eta)}_{\nu,\mathcal{O}}$ are relatively representable. In this section, we want to derive an analog of Theorem 2.1, i.e., some kind of local to global principle for the refined deformation problem $\widetilde{\mathrm{Def}}^{(\eta)}_{S,\mathcal{O}}$. The necessary substitute for $Ш^2_S(\mathrm{ad}_{\bar\rho})$ is a certain dual Selmer group. In our exposition of generalized Selmer groups, we follow Wiles, cf. also [18], (8.6.19) and (8.6.20).

Let us consider a place ν of S. Since $R^{(\eta)}_{\nu,\mathcal{O}} \to \widetilde{R}^{(\eta)}_{\nu,\mathcal{O}}$ is an epimorphism, there is an inclusion of mod ℓ tangent spaces $\mathfrak{t}_{\widetilde{R}^{(\eta)}_{\nu,\mathcal{O}}} \hookrightarrow \mathfrak{t}_{R^{(\eta)}_{\nu,\mathcal{O}}}$. Via the isomorphism $H^1(G_\nu, \mathrm{ad}_{\bar\rho})^{(\eta)} \cong \mathfrak{t}_{R^{(\eta)}_{\nu,\mathcal{O}}}$ this yields a subspace $L^{(\eta)}_\nu \subset H^1(G_\nu, \mathrm{ad}_{\bar\rho})^{(\eta)}$ canonically attached to $\widetilde{\mathrm{Def}}^{(\eta)}_{\nu,\mathcal{O}}$. Its dimension will be denoted $\widetilde{h}^{(\eta)}_\nu$. From the interpretation of $L^{(\eta)}_\nu$ as a mod $\mathfrak{m}_\mathcal{O}$ tangent space, we deduce the existence of a presentation

$$0 \longrightarrow \widetilde{J}^{(\eta)}_\nu \longrightarrow \mathcal{O}[[T_{\nu,1}, \ldots, T_{\nu,\widetilde{h}^{(\eta)}_\nu}]] \longrightarrow \widetilde{R}^{(\eta)}_{\nu,\mathcal{O}} \longrightarrow 0. \tag{4.1}$$

The collection $(L^{(\eta)}_\nu)_{\nu\in S}$ is often abbreviated by $\mathcal{L}^{(\eta)}$. Let us also denote by $L^0_\nu \subset H^1(G_\nu, \mathrm{ad}^0_{\bar\rho})$ the inverse image of $L^\eta_\nu \subset H^1(G_\nu, \mathrm{ad}_{\bar\rho})^\eta$ under the surjection $H^1(G_\nu, \mathrm{ad}^0_{\bar\rho}) \twoheadrightarrow H^1(G_\nu, \mathrm{ad}_{\bar\rho})^\eta$.

Convention on notation: For the refined deformation problems, the universal ring and the ideals in a presentation, and the dimensions of the mod $\mathfrak{m}_\mathcal{O}$ tangent spaces are given a tilde. For the corresponding subspaces of $H^1(\ldots)^{(\eta)}$ we stick to the commonly used notation $L^?_\nu$.

We denote by $\bar\chi_{\mathrm{cyc}}$ the mod ℓ cyclotomic character. For any finite $\mathbb{F}[G_{F,S}]$-module M, we define $M(i) := M \otimes \bar\chi^i_{\mathrm{cyc}}$ and denote by $M^\vee$ the Cartier dual of M as an $\mathbb{F}[G_{F,S}]$-module, i.e., $M^\vee = \mathrm{Hom}_\mathbb{F}(M, \mathbb{F})(1)$.

Example 4.1 Any simple Lie algebra is self-dual via the Killing form. This often proves $\mathrm{ad}^{(0)}_{\bar\rho} \cong (\mathrm{ad}^{(0)}_{\bar\rho})^\vee$. For instance consider $d = \det : \mathcal{G} = \mathrm{GL}_N \to \mathrm{GL}_1$. If $\ell \not| N$, then $\mathfrak{g}^0$ is simple, and so $(\mathrm{ad}^0_{\bar\rho})^\vee \cong \mathrm{ad}^0_{\bar\rho}$. This self-duality can be realized quite explicitly by the perfect trace pairing $(A, B) \mapsto \mathrm{Tr}(AB)$ on $M_N(\mathbb{F})$ (which also shows that $\mathrm{ad}_{\bar\rho}$ is self-dual for $\mathcal{G} = \mathrm{GL}_N$). If $\ell \not| N$ this pairing restricts to a non-degenerate pairing on $\mathfrak{g}^0(\mathbb{F})$. For $\ell | N$, the pairing is degenerate on the traceless matrices $M^0_N(\mathbb{F})$, but induces a non-degenerate pairing on $M^0_N(\mathbb{F})$ modulo the subrepresentation of scalar matrices.

The obvious pairing $M \times M^\vee \to \mathbb{F}(1)$ yields the perfect Tate duality pairing

$$H^{2-i}(G_\nu, M) \times H^i(G_\nu, M^\vee) \to H^2(G_\nu, \mathbb{F}(1)) \cong \mathbb{F},$$

$i \in \{0, 1, 2\}$. Applied to $M = \mathrm{ad}_{\bar\rho}$, one defines $L^\perp_\nu \subset H^1(G_\nu, \mathrm{ad}^\vee_{\bar\rho})$ as the annihilator of $L_\nu \subset H^1(G_\nu, \mathrm{ad}_{\bar\rho})$ under this pairing for $M = \mathrm{ad}_{\bar\rho}$, and one sets $\mathcal{L}^\perp := (L_\nu^\perp)_{\nu\in S}$. For $M = \mathrm{ad}^0_{\bar\rho}$, one defines $L_\nu^{\eta,\perp}$ as the annihilator of L_ν^0 under this pairing, and one sets $\mathcal{L}^{\eta,\perp} := (L_\nu^{\eta,\perp})_{\nu\in S}$.

It is now standard to define the Selmer group $H^1_{\mathcal{L}}(G_{F,S}, \mathrm{ad}_{\bar\rho})$ as the pullback of the diagram

$$\begin{array}{ccc} H^1_{\mathcal{L}}(G_{F,S}, \mathrm{ad}_{\bar\rho}) & \dashrightarrow & \bigoplus_{\nu\in S} L_\nu \\ \downarrow & & \downarrow \\ H^1(G_{F,S}, \mathrm{ad}_{\bar\rho}) & \xrightarrow{\mathrm{res}} & \bigoplus_{\nu\in S} H^1(G_\nu, \mathrm{ad}_{\bar\rho}), \end{array}$$

where the lower horizontal map is the restriction on cohomology. The analogous diagram with $\mathrm{ad}^\vee_{\bar\rho}$ in place of $\mathrm{ad}_{\bar\rho}$ and $L_\nu^\perp$ in place of L_ν defines the dual Selmer group $H^1_{\mathcal{L}^\perp}(G_{F,S}, \mathrm{ad}^\vee_{\bar\rho})$.

By analogy, we define $H^1_{\mathcal{L}}(G_{F,S}, \mathrm{ad}_{\bar\rho})^\eta$ as the pullback of the diagram

$$\begin{array}{ccc} H^1_{\mathcal{L}}(G_{F,S}, \mathrm{ad}_{\bar\rho})^\eta & \dashrightarrow & \bigoplus_{\nu\in S} L_\nu^\eta \\ \downarrow & & \downarrow \\ H^1(G_{F,S}, \mathrm{ad}_{\bar\rho})^\eta & \xrightarrow{\mathrm{res}} & \bigoplus_{\nu\in S} H^1(G_\nu, \mathrm{ad}_{\bar\rho})^\eta. \end{array}$$

The space $H^1_{\mathcal{L}}(G_{F,S}, \mathrm{ad}_{\bar\rho})^{(\eta)}$ is readily identified with the tangent space of $\widetilde{R}^{(\eta)}_{S,\mathcal{O}}$. For its dimension we write $\widetilde{h}^{(\eta)}$. Thus we have presentations:

$$0 \longrightarrow \widetilde{J}^{(\eta)} \longrightarrow \mathcal{O}[[T_1, \ldots, T_{\widetilde{h}^{(\eta)}}]] \longrightarrow \widetilde{R}^{(\eta)}_{S,\mathcal{O}} \longrightarrow 0. \tag{4.2}$$

Note that $\mathrm{Im}(H^0(G_{F,S}, \mathrm{ad}_{\bar\rho}/\mathrm{ad}^0_{\bar\rho}) \to H^1(G_{F,S}, \mathrm{ad}^0_{\bar\rho}))$ injects under the canonical restriction homomorphism into each of the $H^1(G_\nu, \mathrm{ad}^0_{\bar\rho})$. From this and our definition of the L_ν^0, one deduces that there is a short exact sequence

$$\begin{aligned} 0 \to \mathrm{Im}(H^0(G_{F,S}, \mathrm{ad}_{\bar\rho}/\mathrm{ad}^0_{\bar\rho}) \to H^1(G_{F,S}, \mathrm{ad}^0_{\bar\rho})) \\ \to H^1_{\mathcal{L}^0}(G_{F,S}, \mathrm{ad}^0_{\bar\rho}) \to H^1_{\mathcal{L}}(G_{F,S}, \mathrm{ad}_{\bar\rho})^\eta \to 0. \end{aligned} \tag{4.3}$$

For the proof of Theorem 4.2 below, we recall the following consequence of Poitou-Tate global duality, [18], (8.6.20): For $M \in \{\mathrm{ad}_{\bar\rho}, \mathrm{ad}^0_{\bar\rho}\}$ and $\mathcal{L}$ the usual $\mathcal{L}$ or $\mathcal{L}^0$, respectively, there is a five term exact sequence

$$\begin{aligned} 0 \longrightarrow H^1_{\mathcal{L}}(G_{F,S}, M) \longrightarrow H^1(G_{F,S}, M) \longrightarrow \bigoplus_{\nu\in S} H^1(G_\nu, M)/L_\nu \\ \longrightarrow H^1_{\mathcal{L}^\perp}(G_{F,S}, M^\vee)^* \longrightarrow Ш^2_S(M) \longrightarrow 0. \end{aligned}$$

By our definition of L_ν^0, we have $H^1(G_\nu, \mathrm{ad}^0_{\bar\rho})/L_\nu^0 \cong H^1(G_\nu, \mathrm{ad}_{\bar\rho})^\eta/L_\nu^\eta$. From the exact sequence (4.3) and the above 5-term sequence we thus obtain the 5-term exact sequence

$$0 \to H^1_{\mathcal{L}}(G_{F,S}, \mathrm{ad}_{\bar\rho})^{(\eta)} \to H^1(G_{F,S}, \mathrm{ad}_{\bar\rho})^{(\eta)} \to \bigoplus_{\nu \in S} H^1(G_\nu, \mathrm{ad}_{\bar\rho})^{(\eta)}/L_\nu^{(\eta)}$$
$$\to H^1_{\mathcal{L}^{(\eta)},\perp}(G_{F,S}, (\mathrm{ad}_{\bar\rho}^{(0)})^\vee)^* \to Ш^2_S(\mathrm{ad}_{\bar\rho}^{(0)}) \to 0. \quad (4.4)$$

As in Section 2, one can compare local and global presentations of deformation rings also for the more restricted deformation problems.

$$\begin{array}{ccccc} \langle \widetilde{J}_\nu^{(\eta)} : \nu \in S\rangle & \hookrightarrow & \mathcal{O}[[T_{\nu,1}, \dots, T_{\nu,\widetilde{h}_\nu^{(\eta)}} \mid \nu \in S]] & \twoheadrightarrow & \hat\otimes_{\nu \in S} \widetilde{R}_{\nu,\mathcal{O}}^{(\eta)} \\ \downarrow & & \downarrow & & \downarrow \\ \widetilde{J}^{(\eta)} & \hookrightarrow & \mathcal{O}[[T_1, \dots, T_{\widetilde{h}^{(\eta)}}]] & \twoheadrightarrow & \widetilde{R}_{S,\mathcal{O}}^{(\eta)}. \end{array}$$

Theorem 4.2 *As an ideal, $\widetilde{J}^{(\eta)}$ is generated by the images of the ideals $\widetilde{J}_\nu^{(\eta)}$, $\nu \in S$ together with at most $\dim_{\mathbb{F}} H^1_{\mathcal{L}^{(\eta)},\perp}(G_{F,S}, \mathrm{ad}_{\bar\rho}^{(0)})$ other elements. In particular*

$$\mathrm{gen}(\widetilde{J}^{(\eta)}) \le \sum_{\nu \in S} \mathrm{gen}(\widetilde{J}_\nu^{(\eta)}) + \dim_{\mathbb{F}} H^1_{\mathcal{L}^{(\eta)},\perp}(G_{F,S}, \mathrm{ad}_{\bar\rho}^{(0)}).$$

Proof Let us first consider the local situation. The following diagram compares the local presentations (2.1) and (4.1) for the functors $\mathrm{Def}_{\nu,\mathcal{O}}^{(\eta)}$ and $\widetilde{\mathrm{Def}}_{\nu,\mathcal{O}}^{(\eta)}$, respectively:

$$\begin{array}{ccccc} J_\nu^{(\eta)} & \longrightarrow & \mathcal{O}[[T_{\nu,1}, \dots, T_{\nu,h_\nu^{(\eta)}}]] & \twoheadrightarrow & R_{\nu,\mathcal{O}}^{(\eta)} \\ \downarrow & & \| & & \downarrow \\ \widetilde{\mathcal{J}}_\nu^{(\eta)} & \longrightarrow & \mathcal{O}[[T_{\nu,1}, \dots, T_{\nu,h_\nu^{(\eta)}}]] & \twoheadrightarrow & \widetilde{R}_{\nu,\mathcal{O}}^{(\eta)} \\ \downarrow & & \downarrow \pi_\nu & & \| \\ \widetilde{J}_\nu^{(\eta)} & \longrightarrow & \mathcal{O}[[T_{\nu,1}, \dots, T_{\nu,\widetilde{h}_\nu^{(\eta)}}]] & \twoheadrightarrow & \widetilde{R}_{\nu,\mathcal{O}}^{(\eta)}. \end{array}$$

The ideal $\widetilde{\mathcal{J}}_\nu^{(\eta)}$ is the kernel of the composite $\mathcal{O}[[T_{\nu,1}, \dots, T_{\nu,h_\nu^{(\eta)}}]] \to R_{\nu,\mathcal{O}}^{(\eta)} \to \widetilde{R}_{\nu,\mathcal{O}}^{(\eta)}$. The epimorphism π_ν is chosen so that the lower right square commutes. We may rearrange the variables in such a way that π_ν is concretely given by mapping $T_{\nu,i}$ to $T_{\nu,i}$, for $i \le \widetilde{h}_\nu^{(\eta)}$, and by mapping $T_{\nu,i}$ to zero for $i > \widetilde{h}_\nu^{(\eta)}$. Let moreover $f_{\nu,1}, \dots, f_{\nu,r_\nu}$ denote a minimal set of elements in $\mathcal{O}[[T_{\nu,1},\ldots,T_{\nu,\widetilde{h}_\nu^{(\eta)}}]]$ whose images in $\mathcal{O}[[T_{\nu,1},\ldots,T_{\nu,h_\nu^{(\eta)}}]]$ generate $\widetilde{J}_\nu^{(\eta)}$. Then a set of generators of $\widetilde{\mathcal{J}}_\nu^{(\eta)}$ is formed by the elements

$$f_{\nu,1},\ldots,f_{\nu,r_\nu},T_{\nu,\widetilde{h}_\nu^{(\eta)}+1},\ldots,T_{\nu,h_\nu^{(\eta)}}.$$

Now we turn to the global situation. By Theorem 2.1 the relation ideal in the presentation (2.2) of $R_{S,\mathcal{O}}^{(\eta)}$ is generated by local relations together with at most $r := \dim_{\mathbb{F}} Ш^2_S(M)$ further elements $f_1,\ldots,f_r$. Let the α_ν and $\alpha = \prod_\nu \alpha_\nu$ be homomorphisms as in diagram (2.3). For the ring $\widetilde{R}_{S,\mathcal{O}}^{(\eta)}$ we have the following two presentations. First, since $\widetilde{\mathrm{Def}}_{S,\mathcal{O}}^{(\eta)} \subset \mathrm{Def}_{S,\mathcal{O}}^{(\eta)}$ is defined by imposing local conditions, we may take the presentation of $R_{S,\mathcal{O}}^{(\eta)}$ and consider its quotient by further local relations. Second, we have the presentation (4.2). We obtain the following commutative diagram with exact rows

$$\begin{array}{ccccc}
\langle\{\alpha_\nu(\widetilde{\mathcal{J}}_\nu^{(\eta)});\nu\in S\}\cup\{f_1,\ldots,f_r\}\rangle & \hookrightarrow & \mathcal{O}[[T_1,\ldots,T_{h^{(\eta)}}]] & \twoheadrightarrow & \widetilde{R}_{S,\mathcal{O}}^{(\eta)} \\
\downarrow & & \downarrow\pi & & \| \\
\widetilde{\mathcal{J}}^{(\eta)} & \hookrightarrow & \mathcal{O}[[T_1,\ldots,T_{\widetilde{h}^{(\eta)}}]] & \twoheadrightarrow & \widetilde{R}_{S,\mathcal{O}}^{(\eta)}.
\end{array}$$

Since $\widetilde{h}^{(\eta)} = \dim \mathfrak{t}_{\widetilde{R}_{S,\mathcal{O}}^{(\eta)}}$, the homomorphism π is surjective. By properly choosing the coordinate functions T_i, we may thus assume that π is given as $T_i \mapsto T_i$ for $i = 1\ldots,\widetilde{h}^{(\eta)}$ and $T_i \mapsto 0$ for $i > \widetilde{h}^{(\eta)}$.

To further understand π, we interprete the $\mathbb{F}$-dual of sequence (4.4) as an assertion on the variables of our local and global presentations. Defining Δ via

$$0 \longrightarrow Ш^2_S(\mathrm{ad}_{\bar\rho}^{(0)})^* \longrightarrow H^1_{\mathcal{L}^{(\eta)},\perp}(G_{F,S},(\mathrm{ad}_{\bar\rho}^{(0)})^\vee) \longrightarrow \Delta \longrightarrow 0,$$

we have

$$0 \longrightarrow \Delta \longrightarrow \bigoplus_{\nu\in S}(H^1(G_\nu,M)^{(\eta)}/L_\nu^{(\eta)})^* \longrightarrow (H^1(G_{F,S},M)^{(\eta)})^*$$
$$\longrightarrow (H^1_{\mathcal{L}}(G_{F,S},M)^{(\eta)})^* \longrightarrow 0.$$

For $R \in \mathcal{C}_\mathcal{O}$ we have $\mathfrak{t}_R^* = \mathfrak{m}_R/(\mathfrak{m}_\mathcal{O} + \mathfrak{m}_R^2)$. This gives an interpretation for the $H^1(\ldots)^*$-terms:

- The (images of the) elements $T_1,\ldots,T_{\widetilde{h}^{(\eta)}}$ form an $\mathbb{F}$-basis of $(H^1_{\mathcal{L}}(G_{F,S},M)^{(\eta)})^*$.
- The (images of the) elements $T_1,\ldots,T_{h^{(\eta)}}$ form an $\mathbb{F}$-basis of $(H^1(G_{F,S},M)^{(\eta)})^*$.

- The (images of the) elements $T_{\nu,l_\nu^{(\eta)}+1},\ldots,T_{\nu,h_\nu^{(\eta)}}$ form an $\mathbb{F}$-basis of $(H^1(G_\nu,M)^{(\eta)}/L_\nu^{(\eta)})^*$.

Thus in the set $V := \bigcup_{\nu\in S}\alpha_\nu(\{T_{\nu,\widetilde{h}_\nu^{(\eta)}+1},\ldots,T_{\nu,h_\nu^{(\eta)}}\})$ we may choose $h^{(\eta)}-\widetilde{h}^{(\eta)}$ elements which form a basis of the $\mathbb{F}$-span of $\{T_{\widetilde{h}^{(\eta)}+1},\ldots,T_{h^{(\eta)}}\}$. Using the freedom we have in choosing the variables $T_{\widetilde{h}^{(\eta)}+1},\ldots,T_{h^{(\eta)}}$, we may assume that these are precisely the chosen ones from V. Hence under π, these chosen variables all map to zero.

We may therefore conclude the following: The ideal $\widetilde{J}^{(\eta)}$ is spanned by the images of the relations $f_{\nu,j}$, $\nu\in S$, $j=1,\ldots,l_\nu^{(\eta)}$, i.e., the local relations in a minimal presentation of $\widetilde{R}_{\nu,\mathcal{O}}^{(\eta)}$, together with the images of the elements f_j, $j=1,\ldots,r$, and together with the

$$d := \sum_{\nu\in S}(h_\nu^{(\eta)}-\widetilde{h}_\nu^{(\eta)}) - (h^{(\eta)}-\widetilde{h}^{(\eta)})$$

further elements in V which may or may not map to zero under π. Since $d = \dim_{\mathbb{F}}\Delta$, and $d+r = \dim_{\mathbb{F}} H^1_{\mathcal{L}^{(\eta)},\perp}(G_{F,S},\mathrm{ad}_{\bar\rho}^{(0)})$, the assertion of the theorem is shown. □

Corollary 4.3 *For the presentation*

$$0 \longrightarrow \widetilde{J}^{(\eta)} \longrightarrow \mathcal{O}[[T_1,\ldots,T_{\widetilde{h}^{(\eta)}}]] \longrightarrow \widetilde{R}_{S,\mathcal{O}}^{(\eta)} \longrightarrow 0$$

one has

$$\begin{aligned}\widetilde{h}^{(\eta)} - \mathrm{gen}(\widetilde{J}^{(\eta)}) \geq\\ & h^0(G_{F,S},\mathrm{ad}_{\bar\rho}^{(0)}) - h^0(G_{F,S},(\mathrm{ad}_{\bar\rho}^{(0)})^\vee) - \delta(G_{F,S},\mathrm{ad}_{\bar\rho})^{(\eta)}\\ & +\sum_{\nu\in S}\Big(\widetilde{h}_\nu^{(\eta)} + \delta(G_\nu,\mathrm{ad}_{\bar\rho})^{(\eta)} - h^0(G_\nu,\mathrm{ad}_{\bar\rho}^{(0)}) - \mathrm{gen}(\widetilde{J}_\nu^{(\eta)})\Big). \quad (4.5)\end{aligned}$$

Proof Following Wiles, cf. [18] (8.6.20), and using (4.3) we have

$$\begin{aligned}&\widetilde{h}^{(\eta)} + \delta(G_{F,S},\mathrm{ad}_{\bar\rho})^{(\eta)} - \dim_{\mathbb{F}} H^1_{\mathcal{L}^{(\eta)},\perp}(G_{F,S},(\mathrm{ad}_{\bar\rho}^{(0)})^\vee)\\ &= h^0(G_{F,S},\mathrm{ad}_{\bar\rho}^{(0)}) - h^0(G_{F,S},(\mathrm{ad}_{\bar\rho}^{(0)})^\vee) + \sum_{\nu\in S}(\widetilde{h}_\nu^{(0)} - h^0(G_\nu,\mathrm{ad}_{\bar\rho}^{(0)})).\end{aligned}$$

By our definition of $\widetilde{h}_\nu^{(0)}$ we have $\widetilde{h}_\nu^{(0)} = \widetilde{h}_\nu^{(\eta)} + \delta(G_\nu,\mathrm{ad}_{\bar\rho})^{(\eta)}$. Subtracting the bound for $\mathrm{gen}(J^{(\eta)})$ from Theorem 4.2 from the quantity $\widetilde{h}^{(\eta)}$ yields the desired estimate. □

Remark 4.4 Because of Remark 2.2, we expect the above estimate to be optimal in the case that $h^0(G_{F,S},(\mathrm{ad}_{\bar\rho}^{(0)})^\vee)=0$. If F contains ℓ-th roots of unity, the same remark shows that for $\mathrm{ad}_{\bar\rho}$ the above estimate will not be optimal. (For $\mathcal{G} = \mathrm{GL}_1$, i.e, for class field theory, the reader may easily verify this.) If $\mathrm{ad}_{\bar\rho} = \mathrm{ad}^0_{\bar\rho} \oplus \mathbb{F}$ this problem can be remedied since then the universal ring $\widetilde{R}_{S,\mathcal{O}}$ is the completed tensor product of $\widetilde{R}^{\eta}_{S,\mathcal{O}}$ with the deformation ring for one-dimensional representations. By class field theory (and Leopoldt's conjecture) the latter is well-understood.

5 General remarks and the case $\mathcal{G} = \mathrm{GL}_2$

The aim of this section is to analyze the terms occurring in estimate (4.5) given in Corollary 4.3 for the number of variables minus the number of relations in a presentation of $\widetilde{R}^{(\eta)}_{S,\mathcal{O}}$. After some initial general remarks we shall soon focus on the case $\mathcal{G} = \mathrm{GL}_2$. The main result is Theorem 5.8.

For many naturally defined subfunctors $\widetilde{\mathrm{Def}}^{(\eta)}_{\nu,\mathcal{O}} \subset \mathrm{Def}^{(\eta)}_{\nu,\mathcal{O}}$ (for $\nu \in S$) (for instance for the examples presented below) one has the following:

(i) If $\nu \not| \ell$, then $\widetilde{h}^{(0)}_\nu - h^0(G_\nu, \mathrm{ad}^{(0)}_{\bar\rho}) - \mathrm{gen}(\widetilde{J}^{(\eta)}_\nu) \geq 0$.

(ii) If one imposes a suitable semistability condition on deformations at places $\nu|\ell$, and a suitable parity condition at places above ∞, then

$$\sum_{\nu|\ell \text{ or } \nu|\infty} \Big(\widetilde{h}^{(0)}_\nu - h^0(G_\nu, \mathrm{ad}^{(0)}_{\bar\rho}) - \mathrm{gen}(\widetilde{J}^{(\eta)}_\nu)\Big) \geq 0.$$

The estimate in (i) is typically easy to achieve, and without any requirements on the restriction of $\bar\rho$ to G_ν. This is presently not so for (ii) at places $\nu|\ell$: If $\bar\rho$ satisfies some ordinariness condition at ν, then the ring parameterizing deformations satisfying a similar ordinariness conditions is relatively well understood. If on the other hand $\bar\rho$ is flat at ν, then suitable deformation rings are only well understood and well-behaved if the order of ramification of $\bar\rho$ at ν is relatively small.

If (i), resp. (ii) are satisfied, then in all known cases the corresponding local deformation ring is a complete intersection, finite flat over $\mathcal{O}$ thus of Krull dimension $\widetilde{h}^{(\eta)}_\nu - \mathrm{gen}(\widetilde{J}^{(\eta)}_\nu)$.

One may ask what consequences known properties of the global universal ring will have on the local universal rings. To discuss this, suppose (i) and (ii) are satisfied and that the dimension of $\widetilde{R}^{(\eta)}_{S,\mathcal{O}}$ is equal to $\widetilde{h}^{(\eta)} - \mathrm{gen}(\widetilde{J}^{(\eta)})$. Since typically the term (ii) arises from precise estimates for all $\nu|\ell, \infty$, having equality in (4.5) yields explicit expressions for all the terms $\mathrm{gen}(\widetilde{J}^{(\eta)}_\nu)$. Hence each of them will be equal to some "expected" expression. If furthermore the term

$\mathrm{gen}(\widetilde{J}_\nu^{(\eta)})$ is at most 1, then one can even conclude that $\widetilde{R}^{(\eta)}_{S,\mathcal{O}}$ is a complete intersection.

We now turn to some examples, first for the local situation:

Example 5.1 *$\nu \not| \ell, \infty$:*

(i) At such places one has

$$h^1(G_\nu, M) - h^0(G_\nu, M) - h^2(G_\nu, M) = 0$$

for the local Euler-Poincaré characteristic for any finite $\mathbb{F}[G_\nu]$-module M. Thus for $\mathrm{Def}^{(\eta)}_{\nu,\mathcal{O}}$ one obtains $\widetilde{h}_\nu^{(0)} - h^0(G_\nu, \mathrm{ad}_{\bar\rho}^{(0)}) - \mathrm{gen}(J_\nu^{(\eta)}) \geq 0$.

(ii) For the local deformation problems defined by Ramakrishna in [19], Prop. 1, p. 122, the ring $\widetilde{R}^\eta_{\nu,\mathcal{O}}$ is smooth over $\mathcal{O}$ of relative dimension $h^0(G_\nu, \mathrm{ad}^0_{\bar\rho}) = h^1(G_\nu, \mathrm{ad}^0_{\bar\rho}) - h^2(G_\nu, \mathrm{ad}^0_{\bar\rho})$ over $\mathcal{O}$; cf. the remark in [19], p. 124. Here $\widetilde{h}^0_\nu - h^0(G_\nu, \mathrm{ad}^0_{\bar\rho}) - \mathrm{gen}(J^\eta_\nu) = 0$.

(iii) The local deformation problem defined in [10], Prop. 3.4, is smooth of relative dimension 1 over $\mathcal{O}$ and again one has $\widetilde{h}^0_\nu - h^0(G_\nu, \mathrm{ad}^0_{\bar\rho}) - \mathrm{gen}(J^\eta_\nu) = 0$.

(iv) The local deformation problem in [6], p. 141, in the definition of $R^\flat$ at places $\nu \in P$, i.e., at prime number p with $p \equiv -1 \pmod{\widetilde{h}}$ again defines a local deformation problem with versal representing ring smooth of relative dimension 1 over $\mathcal{O}$. As in the previous cases one has $\widetilde{h}^0_\nu - h^0(G_\nu, \mathrm{ad}^0_{\bar\rho}) - \mathrm{gen}(J^\eta_\nu) = 0$.

Remark 5.2 Let $\bar\rho\colon G_{F_\nu} \to \mathrm{GL}_2(\mathbb{F})$ be arbitrary. Building on previous work of Boston, Mazur and Taylor-Wiles it is shown for $\nu \not| \ell$ in [3] that the Krull dimension of $R^\eta_{\nu,\mathcal{O}}$ is for any choice of $\bar\rho$ equal to $h^1(G_\nu, \mathrm{ad}^0_{\bar\rho}) - h^2(G_\nu, \mathrm{ad}^0_{\bar\rho})$. In [3] the same is shown for $\nu | \ell$ for all possible $\bar\rho$. Thus by Remark 1.3 in these cases the rings $R^\eta_{\nu,\mathcal{O}}$ are known to be complete intersections of the expected dimensions. So the estimate in part (i) above is optimal for $d = \det : \mathcal{G} = \mathrm{GL}_2 \to \mathcal{T} = \mathrm{GL}_1$. For parts (ii), (iii) and (iv) the same can be shown (e.g. by explicit calculation).

Example 5.3 *$\nu|\infty$, $\ell > 2$:* If ν is real, then G_ν is generated by a complex conjugation c_ν (of order 2). For $\ell > 2$ and $R \in \mathcal{C}_\mathcal{O}$, the group ring $R[\mathbb{Z}/(2)]$ has idempotents for the two R-projective irreducible representations of c_ν. Hence for any deformation $[\rho]$ of $\bar\rho$ to R, the lift of $\bar\rho_{G_\nu}$ is unique up to isomorphism. Therefore $\widetilde{h}_\nu^{(\eta)} = 0 = \mathrm{gen}(J_\nu^{(\eta)})$, and $h^0(G_\nu, \mathrm{ad}_{\bar\rho}^{(0)})$ depends on the action of c_ν on $\mathrm{ad}_{\bar\rho}^{(0)}$, more precisely on the conjugacy class of $\bar\rho(c_\nu)$. For $\ell = 2$ the problem is more subtle, cf. Example 5.4.

If ν is complex, then G_ν acts trivially, and again $\widetilde{h}_\nu^{(\eta)} = 0 = \mathrm{gen}(J_\nu^{(\eta)})$ (this also holds for $\ell = 2$). Clearly one has $h^0(G_\nu, \mathrm{ad}_{\bar\rho}^{(0)}) = \dim_{\mathbb{F}} \mathrm{ad}_{\bar\rho}^{(0)}$.

For cases with $\nu|\ell$ we refer to Examples 5.5 and 6.1. For a case with $\nu|\infty$ and $\ell = 2$, we refer to Example 5.4.

For the remainder of this section, we assume that $d = \det\colon \mathcal{G} = \mathrm{GL}_2 \to \mathcal{T} = \mathrm{GL}_1$.

One calls a residual representation $\bar\rho$ **odd**, if for any real place ν of F one has $\det\bar\rho(c_\nu) = -1$. Note that for $\ell = 2$, the condition $\det\bar\rho(c_\nu) = -1$ is vacuous.

Example 5.4 Suppose ν is real and $\bar\rho$ is odd. If $\ell \neq 2$, then $h^0(G_\nu, \mathrm{ad}_{\bar\rho}^0) = 1$, and so from the remarks in Example 5.3 it is clear that

$$\widetilde{h}_\nu^0 - h^0(G_\nu, \mathrm{ad}_{\bar\rho}^0) - \mathrm{gen}(J_\nu^\eta) = -1.$$

If $\ell = 2$, the main interest lies in deformations which are odd, i.e., for which the image of c_ν is non-trivial. The following two cases are the important ones for $\mathcal{G} = \mathrm{GL}_2$, and say with η fixed:

Case I: $\bar\rho(c_\nu)$ is conjugate to $\left(\begin{smallmatrix} 0 & 1 \\ 1 & 0 \end{smallmatrix}\right)$. The versal hull for this problem is then given by

$$\rho_{\nu,\mathcal{O}}^\eta\colon \mathbb{Z}/(2) = \langle c_\nu\rangle \longrightarrow \mathrm{GL}_2(\mathcal{O}) : c_\nu \mapsto \left(\begin{smallmatrix} 0 & 1 \\ 1 & 0 \end{smallmatrix}\right).$$

We have $\widetilde{h}_\nu^\eta = 0$, $h^0(G_\nu, \mathrm{ad}_{\bar\rho}^0) = 2$ and $\mathrm{gen}(J_\nu^\eta) = 0$, so that

$$\delta(G_\nu, \mathrm{ad}_{\bar\rho})^\eta + \widetilde{h}_\nu^\eta - h^0(G_\nu, \mathrm{ad}_{\bar\rho}^0) - \mathrm{gen}(J_\nu^\eta) = 1 + 0 - 2 - 0 = -1.$$

Case II: $\bar\rho(c_\nu)$ is conjugate to $\left(\begin{smallmatrix} 1 & 0 \\ 0 & 1 \end{smallmatrix}\right)$. Then the versal hull of a good deformation problem at ν (so that the deformations are odd whenever this is reasonable) is given by

$$\rho_{\nu,\mathcal{O}}^\eta\colon \mathbb{Z}/(2) = \langle c_\nu\rangle \longrightarrow \mathrm{GL}_2(\mathcal{O}[[a,b,c]]/(a^2 + 2a + bc))$$
$$c_\nu \mapsto \left(\begin{smallmatrix} 1+a & b \\ c & -1-a \end{smallmatrix}\right).$$

We have $\widetilde{h}_\nu^\eta = 3$, $h^0(G_\nu, \mathrm{ad}_{\bar\rho}^0) = 3$, $\mathrm{gen}(J_\nu^\eta) = 1$ and (!) $\delta(G_\nu, \mathrm{ad}_{\bar\rho})^\eta = 0$, so that

$$\delta(G_\nu, \mathrm{ad}_{\bar\rho})^\eta + \widetilde{h}_\nu^\eta - h^0(G_\nu, \mathrm{ad}_{\bar\rho}^0) - \mathrm{gen}(J_\nu^\eta) = 0 + 3 - 3 - 1 = -1.$$

Example 5.5 We now turn to the case $\nu|\ell$.
Case I: $F_\nu = \mathbb{Q}_\ell$, $h^0(G_\nu, \mathrm{ad}^0_{\bar\rho}) = 0$, and $\bar\rho_{G_K}$ is flat at ν for some finite extension K of $\mathbb{Q}_\ell$ of ramification degree at most $\ell - 1$ so that the corresponding group scheme and its Cartier dual are both connected. Then by [5] and [20], one has $\widetilde{R}^\eta_{\nu,\mathcal{O}} \cong \mathcal{O}[[T]]$ for a suitable flat deformation functor $\widetilde{\mathrm{Def}}^\eta_{\nu,\mathcal{O}}$. Because $\ell \neq 2$ by Remark 1.5 we have $\delta(G_\nu, \mathrm{ad}_{\bar\rho})^\eta = 0$. Hence

$$\delta(G_\nu, \mathrm{ad}_{\bar\rho})^\eta + \widetilde{h}^\eta_\nu - h^0(G_\nu, \mathrm{ad}^0_{\bar\rho}) - \mathrm{gen}(J^\eta_\nu) = 1 = [F_\nu : \mathbb{Q}_\ell].$$

Case II: $\bar\rho$ is ordinary at ν. We recall the computation of the obstruction theoretic invariants for $\widetilde{\mathrm{Def}}^\eta_{\nu,\mathcal{O}} \subset \mathrm{Def}^\eta_{\nu,\mathcal{O}}$ (much of these computations is contained in work of Mazur and Wiles): Suppose we are given a residual representation

$$\bar\rho\colon G_\nu \to \mathrm{GL}_2(\mathbb{F}) : \sigma \mapsto \begin{pmatrix} \bar\chi & \bar b \\ 0 & \bar\chi^{-1}\bar\eta_\nu \end{pmatrix}, \tag{5.1}$$

where $\bar\chi$ is unramified and $\bar\eta_\nu$ denotes the mod $\mathfrak{m}_\mathcal{O}$ reduction of $\eta_\nu = \eta_{|G_\nu}$.

We make the following (standard) hypotheses: The image $\mathrm{Im}(\bar\rho)$ is not contained in the set of scalar matrices, and, if $\bar\chi = \bar\chi^{-1}\bar\eta_\nu$, then (after possibly twisting by a character) we assume that $\bar\chi = \bar\chi^{-1}\bar\eta_\nu$ is the trivial character. In particular this means that if $\bar b = 0$, then $\bar\chi^2 \neq \bar\eta_\nu$.

By an ordinary lift of fixed determinant we mean a lift of the form

$$\rho\colon G_\nu \to \mathrm{GL}_2(R) : \sigma \mapsto \begin{pmatrix} \chi & b \\ 0 & \chi^{-1}\bar\eta_\nu \end{pmatrix},$$

where χ is unramified. Since $\mathrm{Im}(\bar\rho)$ is not contained in the set of scalar matrices, passing to strict equivalence classes of such lifts defines a relatively representable subfunctor $\widetilde{\mathrm{Def}}^\eta_{\nu,\mathcal{O}} \subset \mathrm{Def}^\eta_{\nu,\mathcal{O}}$ in the sense of Definition 3.1.

To compute the mod $\mathfrak{m}_\mathcal{O}$ tangent space of the corresponding ring $\widetilde{R}^\eta_{\nu,\mathcal{O}}$, and a bound on the number of relations in a minimal presentation we distinguish several subcases: (i) $\ell = 2$ (ii) $\ell \neq 2$ and $\bar\chi^2 \neq \bar\eta_\nu$, (iii) $\ell \neq 2$ and $\bar\chi^2 = \bar\eta_\nu$ (and so by our assumptions on $\bar\rho$, we have $\bar\chi = \bar\eta_\nu = 1$.).

Subcase (i), $\ell = 2$: Let us denote by ρ a lift to $\mathbb{F}[\varepsilon]/(\varepsilon^2)$, and use $\bar\rho$ also to denote the trivial lift. Then

$$\begin{aligned} &\sigma \mapsto \rho(\sigma)\bar\rho(\sigma)^{-1} \\ &= \begin{pmatrix} \chi(\sigma)\bar\chi^{-1}(\sigma) & \bar\eta_\nu^{-1}(\sigma)(b(\sigma)\bar\chi(\sigma) - \bar\chi(\sigma)b(\sigma)) \\ 0 & \chi^{-1}(\sigma)\bar\chi(\sigma) \end{pmatrix} =: \begin{pmatrix} 1+\varepsilon c_1(\sigma) & \varepsilon c_2(\sigma) \\ 0 & 1-\varepsilon c_1(\sigma) \end{pmatrix} \end{aligned} \tag{5.2}$$

defines a 1-cocycle into the upper triangular matrices in $\mathrm{ad}^0_{\bar\rho}$. Because we assume $\ell = 2$, one may in fact verify that the matrix entries c_1 and c_2 are also 1-cocycles for a suitable module. In fact they yield classes $[c_1] \in H^1(G_\nu/I_\nu, \mathbb{F})$ and $[c_2] \in H^1(G_\nu, \bar\chi^2\bar\eta_\nu^{-1})$. Since $G_\nu/I_\nu \cong \hat{\mathbb{Z}}$, one has $H^1(G_\nu/I_\nu, \mathbb{F}) \cong \mathbb{F}$.

If ρ and ρ' are lifts to $\mathbb{F}[\varepsilon]/(\varepsilon^2)$ of the required form, such that ρ' and ρ are conjugate by $1+\varepsilon a$ for some $a \in \mathrm{ad}^0_{\bar\rho}$ which is upper triangular, then 1-cocycles for ρ and ρ' give rise to the same cohomology classes. Conversely, if to a given pair of classes, one chooses different 1-cocycles, the resulting lifts ρ', ρ differ by conjugation by a $1+\varepsilon a$ for some $a \in \mathrm{ad}^0_{\bar\rho}$ which is upper triangular.

Regarding strict equivalence one has the following easy if tedious result: For arbitrary $a \in \mathrm{ad}^0_{\bar\rho}$ the conjugate of any lift ρ to $\mathbb{F}[\varepsilon]/(\varepsilon^2)$ under $1+\varepsilon a$ is again a lift of the required form if and only if one of the following happens:

(i) $a \in \mathrm{ad}^0_{\bar\rho}$ is arbitrary if $\bar\chi^2 = \bar\eta_\nu$ and $\bar b(I_\nu) = \{0\}$
(ii) $a \in \mathrm{ad}^0_{\bar\rho}$ is upper triangular otherwise.

Case (i) means that the image of $\bar\rho$ is an ℓ-group and that $\bar\rho$ is unramified. We define $\delta_\nu^{\ell,\mathrm{unr}}$ to be 1 in case (i) and 0 in case (ii). Then we have

$$\widetilde{h}^0_\nu = \dim_{\mathbb{F}} \mathrm{t}_{\widetilde{R}^\eta_{\nu,\mathcal{O}}} = 1 + h^1(G_\nu, \bar\chi^2\bar\eta_\nu^{-1}) - \delta_\nu^{\ell,\mathrm{unr}}.$$

Similarly, one can compute the obstruction to further lift a representation

$$\rho\colon G_\nu \to \mathrm{GL}_2(R) : \sigma \mapsto \begin{pmatrix} \chi & b \\ 0 & \chi^{-1}\eta_\nu \end{pmatrix}$$

to a representation ρ' given by $\begin{pmatrix} \chi' & b \\ 0 & (\chi')^{-1}\eta_\nu \end{pmatrix}$ for a small surjection $R' \to R$. Letting χ' be an unramified character which lifts χ (and always exists since $G_\nu/I_\nu \cong \hat{\mathbb{Z}}$ is of cohomological dimension one) and b' a set-theoretic continuous lift, as is standard, one shows that

$$(s,t) \mapsto \rho'(st)\rho'(t)^{-1}\rho'(s)^{-1} =: \begin{pmatrix} 1 & c_{s,t} \\ 0 & 1 \end{pmatrix}$$

defines a 2-cocycle of G_ν with values in $\bar\chi^2\bar\eta_\nu^{-1}$, and so we obtain a class in $H^2(G_\nu, \bar\chi^2\bar\eta_\nu^{-1})$. This gives the bound $\mathrm{gen}(\widetilde{J}^\eta_{\nu,\mathcal{O}}) \le h^2(G_\nu, \bar\chi^2\bar\eta_\nu^{-1})$.

As a last ingredient, we compute $h^0(G_\nu, \mathrm{ad}^0_{\bar\rho})$. This leads to the identity

$$\bar\rho\begin{pmatrix} \alpha & \beta \\ \gamma & -\alpha \end{pmatrix}\bar\rho^{-1} = \begin{pmatrix} \alpha+\gamma\bar b\bar\chi^{-1} & -2\alpha\bar\chi\bar b\bar\eta_\nu^{-1}+\beta\bar\chi^2\bar\eta_\nu^{-1}-\gamma\bar b^2\bar\eta_\nu^{-1} \\ \gamma\bar\chi^{-2}\bar\eta_\nu & -\alpha-\gamma\bar b\bar\chi^{-1} \end{pmatrix} \stackrel{!}{=} \begin{pmatrix} \alpha & \beta \\ \gamma & -\alpha \end{pmatrix},$$

and so gives the conditions

$$\gamma\bar b = 0, \gamma(1-\bar\eta_\nu\bar\chi^{-2}) = 0, \beta(1-\bar\eta_\nu\bar\chi^{-2}) = 2\alpha\bar\chi^{-1}\bar b = 0.$$

Since under our hypotheses we cannot have $\bar b = 0$ and $\bar\eta_\nu = \bar\chi^2$ simultaneously, we obtain $\gamma = 0$. From the last condition we see that the vanishing of β depends on $\bar\eta_\nu = \bar\chi^2$ or not. So we find $h^0(G_\nu, \mathrm{ad}^0_{\bar\rho}) = 1 + h^0(G_\nu, \bar\chi^2\bar\eta_\nu^{-1})$.

Using the formula for the local Euler-Poincaré characteristic at a place $\nu|\ell$ one obtains $\sum_{i=0}^{2} h^i(G_\nu, \bar{\chi}^2\bar{\eta}_\nu^{-1}) = -[F_\nu : \mathbb{Q}_\ell]$. Hence

$$\begin{aligned} & \widetilde{h}_\nu^0 - h^0(G_\nu, \mathrm{ad}_{\bar{\rho}}^0) - \mathrm{gen}(J_\nu^\eta) \\ \geq\ & h^0(G_\nu, \bar{\chi}^2\bar{\eta}_\nu^{-1}) - h^0(G_\nu, \mathrm{ad}_{\bar{\rho}}^0) + [F_\nu : \mathbb{Q}_\ell] + 1 - \delta_\nu^{\ell,\mathrm{unr}} \\ =\ & [F_\nu : \mathbb{Q}_\ell] - \delta_\nu^{\ell,\mathrm{unr}} \end{aligned}$$

for the corresponding ring $\widetilde{R}_{\nu,\mathcal{O}}^\eta$.

From now on, we assume $\ell \neq 2$. In this case $\delta(G_\nu, \mathrm{ad}_{\bar{\rho}})^\eta = 0$ by Remark 1.5, and so $\widetilde{h}_\nu^0 = \widetilde{h}_\nu^\eta$. Now for $\ell \neq 2$, the 1-cocycle defined in (5.2) cannot be decomposed in two independent 1-cocycles, and so one proceeds differently: Let $(\mathfrak{n} \subset)\mathfrak{b} \subset \mathrm{ad}_{\bar{\rho}}^0$ denote the subrepresentations on (strictly) upper triangular matrices of $\mathrm{ad}_{\bar{\rho}}^0$. Following Wiles, we see that the cocycle defines a cohomology class in

$$H_{\mathrm{str}}^1 := \mathrm{Ker}(H^1(G_\nu, \mathfrak{b}) \longrightarrow H^1(I_\nu, \mathfrak{b}/\mathfrak{n})).$$

One observes that $H^1(G_\nu, \mathfrak{b}) \longrightarrow H^1(I_\nu, \mathfrak{b}/\mathfrak{n})$ factors via $H^1(G_\nu, \mathfrak{b}/\mathfrak{n})$, and that the action of G_ν on $\mathfrak{b}/\mathfrak{n} \cong \mathbb{F}$ is trivial. Using the left exact inflation-restriction sequence one finds that H_{str}^1 is the pullback of the diagram

$$\begin{array}{ccc} & & H^1(G_\nu/I_\nu, \mathfrak{b}/\mathfrak{n}) \\ & & \downarrow \\ H^1(G_\nu, \mathfrak{b}) & \longrightarrow & H^1(G_\nu, \mathfrak{b}/\mathfrak{n}). \end{array}$$

Case (ii), $\ell \neq 2$ and $\bar{\chi}^2 \neq \bar{\eta}_\nu$. We claim that $H^1(G_\nu, \mathfrak{b}) \to H^1(G_\nu, \mathfrak{b}/\mathfrak{n})$ is surjective: Using the long exact sequence of cohomology it suffices to show that

$$H^2(G_\nu, \mathfrak{n}) \longrightarrow H^2(G_\nu, \mathfrak{b}) \longrightarrow H^2(G_\nu, \mathfrak{b}/\mathfrak{n}) \longrightarrow 0$$

is also exact on the left. Using Tate local duality, one has $h^2(G_\nu, \mathrm{ad}_{\bar{\rho}}^{(0)}) = h^0(G_\nu, (\mathrm{ad}_{\bar{\rho}}^{(0)})^\vee)$. The formulas

$$h^2(G_\nu, \mathfrak{n}) = \begin{cases} 1 & \text{if } \bar{\chi}^2 = \bar{\eta}_\nu \\ 0 & \text{else} \end{cases} \qquad h^2(G_\nu, \mathfrak{b}/\mathfrak{n}) = \begin{cases} 1 & \text{if } \bar{\chi}_{\mathrm{cyc}} \text{ is trivial} \\ 0 & \text{else} \end{cases}$$

follow readily. Representing matrices $\left(\begin{smallmatrix} \alpha & \beta \\ 0 & -\alpha \end{smallmatrix}\right)$ in $\mathfrak{b}$ by column vectors $\left(\begin{smallmatrix} \alpha \\ \beta \end{smallmatrix}\right)$, the representation of $\sigma \in G_\nu$ on $\mathfrak{b}$ is given by

$$\begin{pmatrix} \alpha \\ \beta \end{pmatrix} \mapsto \begin{pmatrix} 1 & 0 \\ -2\bar{\chi}(\sigma)\bar{b}(\sigma) & \bar{\eta}_\nu^{-1}(\sigma)\bar{\chi}^2(\sigma) \end{pmatrix} \begin{pmatrix} \alpha \\ \beta \end{pmatrix}.$$

The invariants of the $\vee$-dual of $\mathfrak{b}$ are the solutions (α, β) in $\mathbb{F}^2$ to the equations

$$\begin{pmatrix} 1 & 2\bar{\chi}^{-1}(\sigma)\bar{b}(\sigma)\bar{\eta}_\nu(\sigma) \\ 0 & \bar{\eta}_\nu(\sigma)\bar{\chi}^{-2}(\sigma) \end{pmatrix}\begin{pmatrix} \alpha \\ \beta \end{pmatrix} = \bar{\chi}_{\text{cyc}}^{-1}(\sigma)\begin{pmatrix} \alpha \\ \beta \end{pmatrix},$$

where σ ranges over all elements of G_ν. For fixed σ, the dimension of the solution space is 2 minus the rank of the matrix

$$\begin{pmatrix} \bar{\chi}_{\text{cyc}}(\sigma)-1 & 2\bar{\chi}^{-1}(\sigma)\bar{\chi}_{\text{cyc}}(\sigma)\bar{b}(\sigma)\bar{\eta}_\nu(\sigma) \\ 0 & \bar{\eta}_\nu(\sigma)\bar{\chi}^{-2}(\sigma)\bar{\chi}_{\text{cyc}}(\sigma)-1 \end{pmatrix}.$$

If $\bar{\chi}_{\text{cyc}}(\sigma)$ is non-trivial and different from $\bar{\eta}_\nu^{-1}(\sigma)\bar{\chi}^2(\sigma)$, it follows that $H^0(G_\nu, (\text{ad}^0_{\bar{\rho}})^\vee) = 0$. For varying σ, the maximal rank has to be non-zero, since otherwise we would have $1 = \bar{\chi}_{\text{cyc}} = \bar{\eta}_\nu^{-1}\bar{\chi}^2$, contradicting our hypotheses. One concludes that if one of the identities $1 = \bar{\chi}_{\text{cyc}}$ or $1 = \bar{\eta}_\nu^{-1}\bar{\chi}^2$ holds, then the rank of $H^0(G_\nu, (\text{ad}^0_{\bar{\rho}})^\vee)$ is 1, and otherwise, it is 2. It follows that

$$h^2(G_\nu, \mathfrak{b}) - h^2(G_\nu, \mathfrak{n}) - h^2(G_\nu, \mathfrak{b}/\mathfrak{n}) = 0,$$

and so the claim is shown.

By the claim the horizontal homomorphism in the above pullback diagram is surjective. By the inflation restriction sequence the vertical homomorphism $H^1(G_\nu/I_\nu, \mathfrak{b}/\mathfrak{n}) \to H^1(G_\nu, \mathfrak{b}/\mathfrak{n})$ is injective. As $H^1(G_\nu/I_\nu, \mathfrak{b}/\mathfrak{n}) \cong H^1(\hat{\mathbb{Z}}, \mathbb{F}) \cong \mathbb{F}$, we therefore deduce that

$$\dim_{\mathbb{F}} H^1_{\text{str}} = h^1(G_\nu, \mathfrak{b}) - h^1(G_\nu, \mathfrak{b}/\mathfrak{n}) + 1.$$

Using the local Euler-Poincaré formula and the above results, we find that

$$\begin{aligned} & \dim_{\mathbb{F}} H^1_{\text{str}} \\ = \ & [F_\nu : \mathbb{Q}_\ell] + 1 + h^2(G_\nu, \mathfrak{b}) - h^2(G_\nu, \mathbb{F}) + h^0(G_\nu, \mathfrak{b}) - h^0(G_\nu, \mathbb{F}) \\ = \ & [F_\nu : \mathbb{Q}_\ell] + h^2(G_\nu, \mathfrak{n}) + h^0(G_\nu, \mathfrak{b}). \end{aligned}$$

One easily shows that $h^0(G_\nu, \mathfrak{b}) = h^0(G_\nu, \text{ad}^0_{\bar{\rho}})$.

To compute $\widetilde{h}^\eta_\nu$ there is as in case (i) the question about strict equivalence. The definition of H^1_{str} only takes conjugation by upper triangular elements into account. However, from $\chi^2 \neq \bar{\eta}_\nu$, one may easily deduce (as in case (i)) that conjugating by a matrix of the form $1 + \varepsilon a$, $a \in \text{ad}^0_{\bar{\rho}}$, preserves the upper diagonal form of a lift to $\mathbb{F}[\varepsilon]/(\varepsilon^2)$ only if a lies in $\mathfrak{b}$. Hence in fact H^1_{str} does describe the mod $\mathfrak{m}_{\mathcal{O}}$ tangent space of the versal hull of $\widetilde{\text{Def}}^\eta_{\nu,\mathcal{O}} \subset \text{Def}^\eta_{\nu,\mathcal{O}}$. Thus $\widetilde{h}^\eta_\nu = \dim_{\mathbb{F}} H^1_{\text{str}}$.

The computation of possible obstructions proceeds as in case (i). The analysis given there does not depend on $\ell = 2$. Again one finds $\text{gen}(\widetilde{J}^\eta_{\nu,\mathcal{O}}) \leq h^2(G_\nu, \mathfrak{n})$. Combining the above results, we find that

$$\widetilde{h}^\eta_\nu - \text{gen}(\widetilde{J}^\eta_{\nu,\mathcal{O}}) - h^0(G_\nu, \text{ad}^0_{\bar{\rho}}) \geq [F_\nu : \mathbb{Q}_\ell].$$

Case (iii), $\ell \neq 2$ and $\bar{\chi}^2 = \bar{\eta}_\nu$. In this case the image of $\bar{\rho}$ is an elementary abelian ℓ-group. Therefore all the lifts factor via the pro-ℓ quotient $\widehat{G}_\nu^\ell$ of G_ν. This group is known to a sufficient degree, as to yield a precise estimate for $\widetilde{h}_\nu^\eta - \mathrm{gen}(\widetilde{J}_{\nu,\mathcal{O}}^\eta) - h^0(G_\nu, \mathrm{ad}_{\bar{\rho}}^0)$ by direct computation. This could be deduced from [3]. But for completeness we chose to give a simple direct argument.

There are two cases. Suppose first that F_ν does not contain a primitive ℓ-th root of unity. Then by [14], Thm. 10.5, the group $\widehat{G}_\nu^\ell$ is a free pro-ℓ group on (topological) generators $s, t_1, \ldots, t_n$, $n = [F_\nu : \mathbb{Q}_\ell]$, such the normal closure of the t_i form the inertia subgroup of $\widehat{G}_\nu^\ell$. By our hypothesis on $\bar{\rho}$, for any lift to some ring R in $\mathcal{C}_\mathcal{O}$ one has

$$s \mapsto \begin{pmatrix} 1+\alpha & \beta \\ 0 & (1+\alpha)^{-1}\eta_\nu(s) \end{pmatrix}, t_1 \mapsto \begin{pmatrix} 1 & \tau_1 \\ 0 & \eta_\nu(t_1) \end{pmatrix}, \ldots, t_n \mapsto \begin{pmatrix} 1 & \tau_n \\ 0 & \eta_\nu(t_n) \end{pmatrix}.$$

Obviously there are no obstructions to lifting. The element α lies in $\mathfrak{m}_R$. The elements τ_i lie in $\mathfrak{m}_R$ precisely if $\bar{\rho}(t_i)$ is trivial. Not having taken strict equivalence into account we have therefore $n + 2$ independent variables.

The analysis of the effect of strict equivalence proceeds as in case (i) and leads to the same cases (a) and (b) as described there. Hence one finds

$$\widetilde{h}_\nu^\eta = n + 2 - (1 + \delta_\nu^{\ell,unr}) = [F_\nu : \mathbb{Q}_\ell] + 1 - \delta_\nu^{\ell,unr}.$$

Moreover $h^0(\mathrm{ad}_{\bar{\rho}}^0) = 1$, and since there are no obstructions to lifting, we find

$$\widetilde{h}_\nu^\eta - \mathrm{gen}(\widetilde{J}_{\nu,\mathcal{O}}^\eta) - h^0(G_\nu, \mathrm{ad}_{\bar{\rho}}^0) = [F_\nu : \mathbb{Q}_\ell] - \delta_\nu^{\ell,unr}.$$

Let us now assume that F_ν does contain a primitive ℓ-th root of unity, and let q be the largest ℓ-power so that F_ν contains a primitive q-th root of unity. Let F be the free pro-ℓ group on (topological) generators $s, t_0, \ldots, t_n$, $n = [F_\nu : \mathbb{Q}_\ell]$. For closed subgroups H, K of F, let $[H, K]$ denote the closed subgroup of F generated by commutators $[h, k] = h^{-1}k^{-1}hk$, $h \in H$, $k \in K$, and denote by N the closed normal subgroup $[F, F] \cap F^{q^2}[F^q, F][F^q[F, F], F]$. Then by [14], Thm. 10.9, the group $\widehat{G}_\nu^\ell$ is the quotient of F by the closed normal subgroup generated by the element

$$r := t_0^{q_0} t_1^{q_1} [t_0, s][t_1, t_2] \ldots [t_{n-1}, t_n] r'$$

for some ℓ-powers q_0, q_1 which are divisible by q, and for some element $r' \in N$. The isomorphism may be chosen, so that the closed normal subgroup generated by the t_i maps to the inertia subgroup of $\widehat{G}_\nu^\ell$. By our hypothesis on $\bar{\rho}$, for any lift to some ring R in $\mathcal{C}_\mathcal{O}$ one has

$$s \mapsto \begin{pmatrix} 1+\alpha & \beta \\ 0 & (1+\alpha)^{-1}\eta_\nu(s) \end{pmatrix}, t_0 \mapsto \begin{pmatrix} 1 & \tau_1 \\ 0 & \eta_\nu(t_0) \end{pmatrix}, \ldots, t_n \mapsto \begin{pmatrix} 1 & \tau_n \\ 0 & \eta_\nu(t_n) \end{pmatrix}.$$

If the variables α, β and τ_i are chosen arbitrarily, the image of r in $\mathrm{GL}_2(R)$ is of the form $\begin{pmatrix} 1 & x \\ 0 & 1 \end{pmatrix}$ for some $x \in \mathfrak{m}_R$. The reason is as follows: The image of such a ρ is upper triangular. Passing to the quotient modulo the unipotent upper triangular normal subgroup gives a representation into $R^* \times R^*$, which by our hypothesis factors via G_ν/I_ν. Now r lies in the inertia subgroup of $\widehat{G}_\nu^\ell$, and so its image in $R^* \times R^*$ is zero.

In fact the expression x is computable in terms of α, β and the τ_i up to some error coming from r'. It is useful to write each variable ξ as $\xi_0 + \tilde{\xi}$, where ξ_0 is the Teichmüller lift of the reduction mod $\mathfrak{m}_R$ of ξ, and $\tilde{\xi} \in \mathfrak{m}_R$ for $R = \mathcal{O}[[\tilde{\alpha}, \tilde{\beta}, \tilde{\tau}_0, \ldots, \tilde{\tau}_n]]$. Working modulo $\mathfrak{m}_R^3$, one can evaluate x and show that it does not vanish. It may happen that x lies in $\mathfrak{m}_R \setminus (\mathfrak{m}_R^2, \mathfrak{m}_\mathcal{O})$. Therefore in a minimal presentation (4.1) of $\widetilde{R}_{\nu,\mathcal{O}}^\eta$ one has $\mathrm{gen}(\widetilde{R}_{\nu,\mathcal{O}}^\eta) \leq 1$, and moreover $\widetilde{h}_\nu^\eta - \mathrm{gen}(\widetilde{R}_{\nu,\mathcal{O}}^\eta)$ is $n + 3 - 1 = n + 2$ minus the number of variables that disappear when passing from lifts to deformations, i.e., to strict equivalence classes of lifts.

The analysis of the effect of strict equivalence is as in the case where no primitive ℓ-th root of unity lies in F_ν, and so we need to subtract $(1 + \delta_\nu^{\ell,\mathrm{unr}})$ from $n + 2$. This yields

$$\widetilde{h}_\nu^\eta - \mathrm{gen}(\widetilde{R}_{\nu,\mathcal{O}}^\eta) = n + 1 - \delta_\nu^{\ell,\mathrm{unr}}.$$

Since $h^0(G_\nu, \mathrm{ad}_{\bar{\rho}}^0) = 1$ in case (iii), we found, independently of a primitive ℓ-th root of unity being in F_ν or not, that

$$\widetilde{h}_\nu^\eta - \mathrm{gen}(\widetilde{J}_{\nu,\mathcal{O}}^\eta) - h^0(G_\nu, \mathrm{ad}_{\bar{\rho}}^0) = n - \delta_\nu^{\ell,\mathrm{unr}} = [F_\nu : \mathbb{Q}_\ell] - \delta_\nu^{\ell,\mathrm{unr}}.$$

Let us summarize our results:

Proposition 5.6 *Suppose $\bar{\rho}$ is ordinary at ν, i.e., of the form (5.1). Let $\widetilde{\mathrm{Def}}_{\nu,\mathcal{O}}^\eta \subset \mathrm{Def}_{\nu,\mathcal{O}}^\eta$ denote the subfunctor of ordinary deformations with fixed determinant. Define $\delta_\nu^{\ell,\mathrm{unr}}$ to be 1 if at the same time $\bar{\rho}$ is unramified and $\mathrm{Im}(\bar{\rho})$ is an ℓ-group, and to be zero otherwise. Then*

$$\widetilde{h}_\nu^\eta - \mathrm{gen}(\widetilde{J}_{\nu,\mathcal{O}}^\eta) - h^0(G_\nu, \mathrm{ad}_{\bar{\rho}}^0) \geq [F_\nu : \mathbb{Q}_\ell] - \delta_\nu^{\ell,\mathrm{unr}}.$$

Example 5.7 Lastly, we need to discuss the global terms in the estimate (4.5). Let $\overline{G}$ denote the quotient of $\mathrm{Im}(\bar{\rho})$ modulo its intersection with the center of $\mathrm{GL}_2(\mathbb{F})$, and assume that $\overline{G}$ is non-trivial.

We first give the results for $\ell = 2$. There, independently of F, by explicit computation one finds:

$$h^0(G_{F,S}, \mathrm{ad}_{\bar{\rho}}) = h^0(G_{F,S}, (\mathrm{ad}_{\bar{\rho}})^\vee) = \begin{cases} 2 & \text{if } \overline{G} \text{ is abelian,} \\ 1 & \text{otherwise.} \end{cases}$$

$$h^0(G_{F,S}, \mathrm{ad}^0_{\bar\rho}) = \begin{cases} 2 & \text{if } \overline{G} \text{ is a 2-group (and hence abelian),} \\ 1 & \text{otherwise.} \end{cases}$$

$$\delta(G_{F,S}, \mathrm{ad}_{\bar\rho})^\eta = \begin{cases} 0 & \text{if } \overline{G} \text{ is of order prime to 2,} \\ 1 & \text{otherwise.} \end{cases}$$

$$h^0(G_{F,S}, (\mathrm{ad}^0_{\bar\rho})^\vee) = \begin{cases} 2 & \text{if } \overline{G} \text{ is a 2-group} \\ 1 & \text{if } \overline{G} \text{ is dihedral or of Borel type and not a 2-group} \\ 0 & \text{otherwise.} \end{cases}$$

For $\ell \neq 2$, the result depends on F, because $\bar\chi_{\mathrm{cyc}}$ will in general be non-trivial. If $\bar\rho$ when considered over $\mathbb{F}^{\mathrm{alg}}$ is reducible, let $\bar\chi_2$ denote the character of $G_{F,S}$ on a one-dimensional quotient and $\bar\chi_1$ on the corresponding 1-dimensional subspace, and define $\bar\chi = \bar\chi_1\bar\chi_2^{-1}$. One finds:

$$h^0(G_{F,S}, \mathrm{ad}_{\bar\rho}) - 1 = h^0(G_{F,S}, \mathrm{ad}^0_{\bar\rho}) = \begin{cases} 1 & \text{if } \overline{G} \text{ is abelian,} \\ 0 & \text{otherwise.} \end{cases}$$

$h^0(G_{F,S}, (\mathrm{ad}_{\bar\rho})^\vee) = h^0(G_{F,S}, (\mathrm{ad}^0_{\bar\rho})^\vee) + h^0(G_{F,S}, \mathbb{F}(1))$, and

$$h^0(G_{F,S}, (\mathrm{ad}^0_{\bar\rho})^\vee) = \begin{cases} 2 & \text{if } \overline{G} \cong \mathbb{Z}/(2) \text{ and } \bar\chi_{\mathrm{cyc}} = \bar\chi \\ 1 & \text{if } \overline{G} \cong \mathbb{Z}/(2) \text{ and } \bar\chi_{\mathrm{cyc}} = 1 \\ 1 & \text{if } \overline{G} \not\cong \mathbb{Z}/(2) \text{ is abelian and } \bar\chi_{\mathrm{cyc}} \in \{\bar\chi, \bar\chi^{-1}, 1\} \\ 1 & \text{if } \overline{G} \text{ is non-abelian of Borel type and } \bar\chi_{\mathrm{cyc}} = \bar\chi \\ 1 & \text{if } \overline{G} \text{ is dihedral, } [F(\zeta_\ell) : F] = 2 \text{ and } \bar\rho_{|G_{F(\zeta_\ell)}} \text{ is reducible} \\ 0 & \text{otherwise.} \end{cases} \tag{5.3}$$

Combining the above results, we obtain the following general theorem in the case $\mathcal{G} = \mathrm{GL}_2$:

Theorem 5.8 *Suppose F is totally real and $\bar\rho$ is odd. Suppose further that*

(i) *At $\nu \nmid \ell, \infty$ the local deformation problem satisfies $\tilde{h}^\eta_\nu - h^0(G_\nu, \mathrm{ad}^0_{\bar\rho}) - \mathrm{gen}(J^\eta_\nu) \geq 0$.*

(ii) *At $\nu | \infty$ we choose either of the versal hulls in Example 5.4 depending on whether $\bar\rho(c_\nu)$ is trivial or not.*

(iii) *At $\nu | \ell$, either (i) $F_\nu = \mathbb{Q}_\ell$, and $\bar\rho$ satisfies the requirements in 5.5 case I, and $\widetilde{\mathrm{Def}}^\eta_{\nu,\mathcal{O}}$ is the functor of "flat deformations", or (ii) $\bar\rho$ is ordinary and $\delta^{\ell,\mathrm{unr}}_\nu = 0$, and $\widetilde{\mathrm{Def}}^\eta_{\nu,\mathcal{O}}$ is the functor of ordinary deformations with fixed determinant.*

(iv) $h^0(G_{F,S}, (\mathrm{ad}^0_{\bar\rho})^\vee) = 0$. *(cf. Example 5.7 for explicit conditions.)*

Then $\widetilde{R}^{\eta}_{S,\mathcal{O}}$ *has a presentation* $\mathcal{O}[[T_1,\dots,T_n]]/(f_1,\dots,f_n)$ *for suitable* $f_i \in \mathcal{O}[[T_1,\dots,T_n]]$.

Proof Example 5.7 and (iv) imply that

$$h^0(G_{F,S},\mathrm{ad}^0_{\bar\rho}) - \delta(G_{F,S},\mathrm{ad}^0_{\bar\rho})^{\eta} = 0.$$

Hence, in view of (i) it suffices to show that the joint contribution in (4.5) from the places above ℓ and ∞ under the stated hypotheses is zero. But using Proposition 5.6 and Example 5.4 yields

$$\sum_{\nu|\ell \text{ or } \nu|\infty} \Big(\widetilde{h}^0_\nu - h^0(G_\nu,\mathrm{ad}^0_{\bar\rho}) - \mathrm{gen}(J^{\eta}_\nu)\Big) = \sum_{\nu|\ell}[F_\nu:\mathbb{Q}_\ell] - \sum_{\nu|\infty} 1$$
$$= [F:\mathbb{Q}] - [F:\mathbb{Q}] = 0.$$

□

6 Comparison to the results of Tilouine and Mauger

In this section we will apply the estimate from Corollary 4.3 to obtain another approach to the results by Tilouine and Mauger in [23, 15] on presentations for universal deformations. While their main interest was in representations into symplectic groups, their results are rather general. If $h^0(G_{F,S},(\mathrm{ad}^{(0)}_{\bar\rho})^\vee) = 0$, we completely recover their results with fewer hypothesis. If not, a comparison is less clear. It seems however, that in most cases where their results are applicable, the term $h^0(G_{F,S},(\mathrm{ad}^{(0)}_{\bar\rho})^\vee)$ will be zero. Our main result is Theorem 6.6.

Example 6.1 Let $d\colon \mathcal{G} \to \mathcal{T}$ be arbitrary and let $S_{\mathrm{ord}} \subset S$ be a set of places of F which contains all places above ℓ and none above ∞. For each $\nu \in S_{\mathrm{ord}}$, we fix a smooth closed $\mathcal{O}$-subgroup scheme $\mathcal{P}_\nu \subset \mathcal{G}$.

For each place ν in S_{ord}, we consider the subfunctor $\mathrm{Def}^{\nu\text{-n.o.}}_{\nu,\mathcal{O}} \subset \mathrm{Def}_{\nu,\mathcal{O}}$ of deformations $[\rho_\nu\colon G_\nu \to \mathcal{G}(R)]$ such that there exists some $g_\nu \in \mathcal{G}(R)$, whose reduction mod $\mathfrak{m}_R$ is the identity, such that $g_\nu\rho_\nu g_\nu^{-1}(G_\nu) \subset \mathcal{P}_\nu(R)$. For this subfunctor to make sense, one obviously requires that $\bar\rho(G_\nu) \subset \mathcal{P}_\nu(\mathbb{F})$.

Following [23], a deformation $[\rho]$ is called **$\mathcal{P}$-nearly ordinary** (at S_{ord}) (where $\mathcal{P}$ stands for the family $(\mathcal{P}_\nu)_{\nu\in S_{\mathrm{ord}}}$) if for each $\nu \in S_{\mathrm{ord}}$ the restriction $[\rho_{|G_\nu}]$ satisfies the above condition. By $\widetilde{\mathrm{Def}}^{S_0\text{-n.o.}}_{S,\mathcal{O}} \subset \mathrm{Def}_{S,\mathcal{O}}$ we denote the global deformation functor of deformations which are $\mathcal{P}$-nearly ordinary at

$S_{\text{ord}} \subset S$, and are described by some other relatively representable functors $\widetilde{\mathrm{Def}}^{\eta}_{\nu,\mathcal{O}} \subset \mathrm{Def}^{\eta}_{\nu,\mathcal{O}}$ at places $\nu \in S \setminus S_{\text{ord}}$. If one furthermore fixes a lift

$$\eta\colon G_{F,S} \longrightarrow \mathcal{T}(\mathcal{O})$$

of $d \circ \bar{\rho}\colon G_{F,S} \to \mathcal{G}(\mathbb{F}) \to \mathcal{T}(\mathbb{F})$, the corresponding subfunctor is

$$\widetilde{\mathrm{Def}}^{S_0\text{-n.o.},\eta}_{S,\mathcal{O}} := \widetilde{\mathrm{Def}}^{S_0\text{-n.o.}}_{S,\mathcal{O}} \cap \mathrm{Def}^{\eta}_{S,\mathcal{O}}.$$

For each $\nu \in S_{\text{ord}}$, let $\mathfrak{p}_\nu \subset \mathfrak{g}$ denote the Lie-subalgebra of $\mathfrak{g}$ which corresponds to $\mathcal{P}_\nu \subset \mathcal{G}$. It carries a natural $\mathcal{P}_\nu$-action, so that $\mathfrak{g}/\mathfrak{p}_\nu(\mathbb{F})$ is a finite $\mathcal{P}_\nu$-module. Again following [23], we define the condition

$$(\mathbf{Reg}):\ \text{For all } \nu \in S_{\text{ord}} \text{ one has } h^0(G_\nu, \mathfrak{g}/\mathfrak{p}_\nu(\mathbb{F})) = 0.$$

One has the following simple result whose proof we omit:

Lemma 6.2 *If the condition* (**Reg**) *holds, for all* $\nu \in S_{\text{ord}}$, *the subfunctor* $\mathrm{Def}^{\nu\text{-n.o.}}_{\nu,\mathcal{O}} \subset \mathrm{Def}_{\nu,\mathcal{O}}$ *is relatively representable. Hence in this case* $\widetilde{\mathrm{Def}}^{S_0\text{-n.o.},(\eta)}_{S,\mathcal{O}}$ *has a versal hull*

$$\rho^{S_0\text{-n.o.},(\eta)}_{S,\mathcal{O}}\colon G_{F,S} \to \mathcal{G}(R^{S_0\text{-n.o.},(\eta)}_{S,\mathcal{O}}).$$

Locally at $\nu \in S_{\text{ord}}$ denote by $\mathrm{Def}^{(\eta)}_{\mathcal{P}_\nu,\mathcal{O}}$ the functor of deformations for representations of G_ν into $\mathcal{P}_\nu$ (possibly with the additional condition that the deformations are compatible with the chosen η.) Let

$$\rho^{(\eta)}_{\mathcal{P}_\nu,\mathcal{O}}\colon G_{F,S} \to \mathcal{P}_\nu(R^{(\eta)}_{\mathcal{P}_\nu,\mathcal{O}})$$

denote a corresponding versal hull and define $\mathfrak{p}^0_\nu$ to be the Lie-Algebra of the kernel of the composite $\mathcal{P}_\nu \hookrightarrow \mathcal{G} \xrightarrow{d} \mathcal{T}$. By Theorems 1.2 and 1.4 we find:

Proposition 6.3 *The mod* $\mathfrak{m}_{\mathcal{O}}$ *tangent space of* $R^{(\eta)}_{\mathcal{P}_\nu,\mathcal{O}}$ *is isomorphic to*

$$H^1(G_\nu, \mathfrak{p}_\nu)^{(\eta)} := \mathrm{Im}(H^1(G_\nu, \mathfrak{p}^{(0)}_\nu) \to H^1(G_\nu, \mathfrak{p}_\nu)).$$

Let $h^{(\eta)}_{\mathcal{P}_\nu} := \dim_{\mathbb{F}} H^1(G_\nu, \mathfrak{p}_\nu)^{(\eta)}$. *Then there exists a presentation*

$$0 \longrightarrow J^{(\eta)}_{\mathcal{P}_\nu} \longrightarrow \mathcal{O}[[T_1, \ldots, T_{h^{(\eta)}_{\mathcal{P}_\nu}}]] \longrightarrow R^{(\eta)}_{\mathcal{P}_\nu,\mathcal{O}} \longrightarrow 0$$

for some ideal $J^{(\eta)}_{\mathcal{P}_\nu} \subset \mathcal{O}[[T_1, \ldots, T_{h^{(\eta)}_{\mathcal{P}_\nu}}]]$ *with* $\mathrm{gen}(J^{(\eta)}_{\mathcal{P}_\nu}) \le \dim_{\mathbb{F}} H^2(G_\nu, \mathfrak{p}^{(0)}_\nu)$.

The two functors $\mathrm{Def}^{\nu\text{-n.o.},(\eta)}_{\nu,\mathcal{O}}$ and $\mathrm{Def}^{(\eta)}_{\mathcal{P}_\nu,\mathcal{O}}$ essentially describe the same deformation problem, except that a priori they work with a different notion of strict equivalence.

Lemma 6.4 *The obvious surjection*

$$\mathrm{Def}^{(\eta)}_{\mathcal{P}_\nu,\mathcal{O}}(\mathbb{F}[\varepsilon]/(\varepsilon^2)) \twoheadrightarrow \mathrm{Def}^{\nu\text{-n.o.},(\eta)}_{\nu,\mathcal{O}}(\mathbb{F}[\varepsilon]/(\varepsilon^2))$$

is a bijection provided that $(\mathbf{Reg})$ *holds.*

Proof Clearly, by definition, every lift ρ of $\bar{\rho}$ to $\mathbb{F}[\varepsilon]/(\varepsilon^2)$ whose class lies in $\mathrm{Def}^{\nu\text{-n.o.},(\eta)}_{\nu,\mathcal{O}}(\mathbb{F}[\varepsilon]/(\varepsilon^2))$ can be conjugated to take its image inside $\mathcal{P}_\nu(\mathbb{F}[\varepsilon]/(\varepsilon^2))$. Moreover the notion of strict equivalence for $\mathrm{Def}^{(\eta)}_{\mathcal{P}_\nu,\mathcal{O}}$ is an a priori weaker one than for $\mathrm{Def}^{\nu\text{-n.o.},(\eta)}_{\nu,\mathcal{O}}$, so that the orbits under the second notion of strict equivalence may be larger. This shows that the map in the lemma is well-defined and surjective. Let us now show injectivity, i.e., that the orbits under both notions of strict equivalence agree.
Let $\rho = (1+\varepsilon a)\bar{\rho}$ be a lift of $\bar{\rho}$ to $\mathbb{F}[\varepsilon]/(\varepsilon^2)$ with image inside $\mathcal{P}_\nu(\mathbb{F}[\varepsilon]/(\varepsilon^2))$, so that $a : G_\nu \to \mathfrak{p}_\nu$ is a 1-cocycle. Let $g = 1+\varepsilon b$ be arbitrary with $b \in \mathfrak{g}$. We need to show that the set of those b for which $g\rho g^{-1}$ lies in $\mathcal{P}_\nu(\mathbb{F}[\varepsilon]/(\varepsilon^2))$ (for all a as above) is exactly the set $\mathfrak{p}_\nu$: One computes explicitly

$$g\rho g^{-1} = (1+\varepsilon(a+gbg^{-1}-b))\bar{\rho}.$$

So independently of a, the element $gbg^{-1}-g$ must lie in $\mathfrak{p}_\nu$ for all $g \in \bar{\rho}(G_\nu)$. Equivalently, the image of b under the surjection $\mathfrak{g} \twoheadrightarrow \mathfrak{g}/\mathfrak{p}_\nu$ must lie in $H^0(G_\nu, \mathfrak{g}/\mathfrak{p}_\nu)$. By $(\mathbf{Reg})$ the latter set is zero, and so b lies indeed in $\mathfrak{p}_\nu = \mathrm{Ker}(\mathfrak{g} \twoheadrightarrow \mathfrak{g}/\mathfrak{p}_\nu)$. □

Recall that $0 \le \delta(G_\nu, \mathfrak{p}_\nu)^\eta = h^1(G_\nu, \mathfrak{p}^0_\nu) - \dim H^1(G_\nu, \mathfrak{p})^\eta$. The formula for the local Euler-Poincaré characteristic yields:

Proposition 6.5 *For* $\nu \in S_0$ *and the functor* $\mathrm{Def}^{(\eta)}_{\nu,\mathcal{O}} = \mathrm{Def}^{\nu\text{-n.o.},(\eta)}_{\nu,\mathcal{O}}$ *one has*

$$\widetilde{h}^{(\eta)}_\nu + \delta(G_\nu, \mathfrak{p}_\nu)^{(\eta)} - h^0(G_\nu, \mathrm{ad}^{(0)}_{\bar{\rho}}) - \mathrm{gen}(J^{(\eta)}_\nu) = \begin{cases} 0, & \text{if } \nu \nmid \ell, \\ [F_\nu : \mathbb{Q}_\ell] \dim_{\mathbb{F}} \mathfrak{p}_\nu, & \text{if } \nu \mid \ell. \end{cases}$$

In [23, 15] a lift of $d \circ \bar{\rho}$ is never chosen. Hence the term $\delta(G_\nu, \mathfrak{p}_\nu)^{(\eta)}$ is not present in their formulas.

Combining the above with Corollary 4.3 shows:

Theorem 6.6 *Fix* $\mathcal{P} = (\mathcal{P}_\nu)_{\nu \in S_0}$ *as above, and assume that:*

(i) $\bar{\rho} \in \widetilde{\mathrm{Def}}^{S_0\text{-n.o.},(\eta)}_{S,\mathcal{O}}(\mathbb{F})$.

(ii) *At* $\nu \in S \smallsetminus (S_{\mathrm{ord}} \cup \{\nu : \nu|\infty\})$ *we have* $\widetilde{h}_\nu^{(0)} - h^0(G_\nu, \mathrm{ad}_{\bar\rho}^{(0)}) - \mathrm{gen}(J_\nu^{(\eta)}) \geq 0$.

(iii) *The condition* $(\mathbf{Reg})$ *is satisfied.*

Then for the presentation

$$0 \longrightarrow \widetilde{J}^{(\eta)} \longrightarrow \mathcal{O}[[T_1, \ldots, T_{\widetilde{h}^{(\eta)}}]] \longrightarrow \widetilde{R}_{S,\mathcal{O}}^{S_0\text{-n.o.},(\eta)} \longrightarrow 0$$

one has

$$\begin{aligned}\widetilde{h}^{(\eta)} - \mathrm{gen}(J^{(\eta)}) &\geq h^0(G_{F,S}, \mathrm{ad}_{\bar\rho}^{(0)}) \\ &- h^0(G_{F,S}, (\mathrm{ad}_{\bar\rho}^{(0)})^\vee) - \delta(G_{F,S}, \mathrm{ad}_{\bar\rho})^{(\eta)} + \sum_{\nu|\ell} [F_\nu : \mathbb{Q}_\ell] \dim_{\mathbb{F}} \mathfrak{p}_\nu^{(0)} \\ &+ \sum_{\nu|\infty} \Big(\widetilde{h}_\nu^{(\eta)} + \delta(G_\nu, \mathrm{ad}_{\bar\rho})^{(\eta)} - h^0(G_\nu, \mathrm{ad}_{\bar\rho}^{(0)}) - \mathrm{gen}(\widetilde{J}_\nu^{(\eta)})\Big).\end{aligned}$$

If $\ell \neq 2$, *or if no constraints are imposed for the deformation at the infinite places, then their contribution in the above formula simplifies to give* $-\sum_{\nu|\infty} h^0(G_\nu, \mathrm{ad}_{\bar\rho}^{(0)})$.

Remark 6.7 Since in [23] or [15] no homomorphism η is fixed, and there are no conditions at ∞, the above is (philosophically) the same formula as that in [23], Prop. 7.3 or [15], Prop. 3.9, except for the term $-h^0(G_{F,S}, (\mathrm{ad}_{\bar\rho}^{(0)})^\vee)$. As noted in Remark 4.4, we expect that usually this term is not present in the formula – but that technically we are not able to remove it.

By 'philosophically' we mean that their formula was used primarily to bound the Krull dimension of some deformation ring. Our formula can obviously serve the same purpose.

Our hypotheses and those in [23, 15] are however different. If $h^0(G_{F,S}, (\mathrm{ad}_{\bar\rho}^{(0)})^\vee) = 0$ our result holds under much weaker hypotheses, namely without the hypothesis $(\mathbf{Reg}')$ in [15], Prop. 3.9. The latter seems to be rather hard to verify in practice.

If $h^0(G_{F,S}, (\mathrm{ad}_{\bar\rho}^{(0)})^\vee)$ is non-zero the comparison is less clear. The non-vanishing either means that we are in the case $\mathrm{ad}_{\bar\rho}$ and F contains a primitive ℓ-th root of unity, or that $\mathrm{ad}_{\bar\rho}^0$ surjects onto a one-dimensional quotient representation on which $G_{F,S}$ acts by the inverse of the mod ℓ-cyclotomic character. In the former case we'd expect that the $\mathfrak{p}_\nu$ typically also contain a trivial subrepresentation, and then the terms $h^0(G_\nu, (\mathfrak{p}_\nu^{(0)})^\vee) = h^2(G_\nu, \mathfrak{p}_\nu^{(0)})$ would be non-zero, so that the hypothesis $(\mathbf{Reg}')$ in [15], Prop. 3.9, would not be satisfied. In the latter case it is not clear to us whether this one-dimensional quotient will typically also occur as a quotient of one of the $\mathfrak{p}_\nu^{(0)}$. In any case, if $\bar\rho$ is 'highly irreducible' which is the generic case, the second case is unlikely to occur.

7 Relative presentations

In this last section we deduce some results on presentations of global deformation rings as quotients of power series rings over the completed tensor product of the corresponding local versal deformation rings from the results in Section 4. This is inspired by M. Kisin's theory of framed deformations. The results below are due to Kisin [13], who has given a different more direct approach.

This section makes no reference to Sections 5 and 6. *We let the notation be as in Section 4.*

Lemma 7.1 *There exists a natural number s and a presentation*

$$0 \longrightarrow \widetilde{J} \longrightarrow \Big(\widehat{\bigotimes}_{\nu\in S} \widetilde{R}^{(\eta)}_{\nu,\mathcal{O}}\Big)[[U_1,\ldots,U_s]] \longrightarrow \widetilde{R}^{(\eta)}_{S,\mathcal{O}} \longrightarrow 0$$

with $\mathrm{gen}(\widetilde{J})$ *being bounded from above by*

$$s+\delta(G_{F,S},\mathrm{ad}_{\bar\rho})^{(\eta)} - h^0(G_{F,S},\mathrm{ad}^{(0)}_{\bar\rho}) + h^0(G_{F,S},(\mathrm{ad}^{(0)}_{\bar\rho})^\vee) + \sum_{\nu\in S} h^0(G_\nu,\mathrm{ad}^{(0)}_{\bar\rho}) - \delta(G_\nu,\mathrm{ad}_{\bar\rho})^{(\eta)}.$$

Proof The proof of Theorem 4.2 yields the following commutative diagram

$$\begin{array}{ccccc} \widetilde{J}_{\mathrm{loc}} & \hookrightarrow & \mathcal{O}[[T_{\nu,j} : \nu\in S, j=1,\ldots,\widetilde{h}^{(\eta)}_\nu]] & \twoheadrightarrow & \mathcal{R}_{\mathrm{loc}} \\ \downarrow & & \downarrow{\scriptstyle\pi} & & \downarrow \\ \langle \pi(\widetilde{J}_{\mathrm{loc}})\cup\{g_1,\ldots,g_r\}\rangle & \hookrightarrow & \mathcal{O}[[T_1,\ldots,T_{\widetilde{h}^{(\eta)}}]] & \twoheadrightarrow & \widetilde{R}^{(\eta)}_{S,\mathcal{O}} \end{array}$$

where we set $\widetilde{J}_{\mathrm{loc}} := \langle\{\widetilde{J}^{(\eta)}_\nu : \nu\in S\}\rangle$, $\mathcal{R}_{\mathrm{loc}} := \Big(\widehat{\bigotimes}_{\nu\in S}\widetilde{R}^{(\eta)}_{\nu,\mathcal{O}}\Big)$, $r = \dim_{\mathbb{F}} H^1_{\mathcal{L}^{(\eta)},\perp}(G_{F,S},(\mathrm{ad}^{(0)}_{\bar\rho})^\vee)$ and the g_j are suitable elements of $\mathcal{O}[[T_1,\ldots,T_{\widetilde{h}^{(\eta)}}]]$. The failure of the surjectivity of $\mathcal{R}_{\mathrm{loc}} \to \widetilde{R}^{(\eta)}_{S,\mathcal{O}}$ can be measured by considering the induced homomorphism on mod $\mathfrak{m}_{\mathcal{O}}$ tangent spaces. Let s denote the dimension of the cokernel of $\mathfrak{t}_{\mathcal{R}_{\mathrm{loc}}} \to \mathfrak{t}_{\widetilde{R}^{(\eta)}_{S,\mathcal{O}}}$. (It is not difficult to show that $s = \dim Ш^1(\mathrm{ad}^{(0)}_{\bar\rho})$, but we do not need this.) Then there is a surjective homomorphism $\mathcal{R}_{\mathrm{loc}}[[U_1,\ldots,U_s]] \to \widetilde{R}^{(\eta)}_{S,\mathcal{O}}$ for variables U_i. Abbreviating $\mathcal{S}_{\text{loc}} := \mathcal{O}[[T_{\nu,j} : \nu \in S, j = 1, \ldots, \widetilde{h}_\nu^{(\eta)}]]$, there is a commutative diagram

$$\begin{array}{ccccc} \langle \widetilde{J}_{\text{loc}} \rangle & \hookrightarrow & \mathcal{S}_{\text{loc}}[[U_1, \ldots, U_s]] & \longrightarrow & \mathcal{R}_{\text{loc}}[[U_1, \ldots, U_s]] \\ \downarrow & & \downarrow \tilde{\pi} & & \downarrow \\ \langle \{\tilde{\pi}(\widetilde{J}_{\text{loc}}) \cup \{g_1, \ldots, g_r\} \rangle & \hookrightarrow & \mathcal{O}[[T_1, \ldots, T_{\widetilde{h}^{(\eta)}}]] & \twoheadrightarrow & \widetilde{R}_{S,\mathcal{O}}^{(\eta)} \end{array}$$

with surjective middle and right vertical homomorphisms. Since $\mathcal{S}_{\text{loc}}$ is a power series ring over $\mathcal{O}$, the kernel of $\tilde{\pi}$ is generated by

$$u := s + \sum_{\nu \in S} \widetilde{h}_\nu^{(\eta)} - \widetilde{h}^{(\eta)}$$

elements $H_1, \ldots, H_u$. Because $\tilde{\pi}$ is smooth, we may choose elements $G_1, \ldots,$ G_r in the ring $\mathcal{S}_{\text{loc}}[[U_1, \ldots, U_s]]$ whose images in $\mathcal{O}[[T_1, \ldots, T_{\widetilde{h}^{(\eta)}}]]$ agree with $g_1, \ldots, g_r$. Thus $\widetilde{R}_{S,\mathcal{O}}^{(\eta)}$ is the quotient of $\mathcal{R}_{\text{loc}}[[U_1, \ldots, U_s]]$ by the ideal $\widetilde{J}$ generated by the images of the elements $G_1, \ldots, G_r, H_1, \ldots, H_u$. Using the first formula in the proof of Corollary 4.3, we have

$$\begin{aligned} \text{gen}(\widetilde{J}) - s &= \sum_{\nu \in S} \widetilde{h}_\nu^{(\eta)} - \widetilde{h}^{(\eta)} + \dim_{\mathbb{F}} H^1_{\mathcal{L}^{(\eta)},\perp}(G_{F,S}, (\text{ad}_{\bar{\rho}}^{(0)})^\vee) \\ &= \delta(G_{F,S}, \text{ad}_{\bar{\rho}})^{(\eta)} - h^0(G_{F,S}, \text{ad}_{\bar{\rho}}^{(0)}) + h^0(G_{F,S}, (\text{ad}_{\bar{\rho}}^{(0)})^\vee) \\ &\qquad + \sum_{\nu \in S} h^0(G_\nu, \text{ad}_{\bar{\rho}}^{(0)}) - \delta(G_\nu, \text{ad}_{\bar{\rho}})^{(\eta)}. \end{aligned}$$

□

If R is flat over $\mathcal{O}$, its relative Krull dimension over $\mathcal{O}$ is denoted by $\dim_{\text{Krull}/\mathcal{O}} R$.

Corollary 7.2 *Suppose that*

(i) $\widetilde{R}_{S,\mathcal{O}}^{(\eta)}/(\ell)$ *is finite.*

(ii) *The rings* $\widetilde{R}_{\nu,\mathcal{O}}^{(\eta)}$, $\nu \in S$, *are flat over* $\mathcal{O}$.

(iii) $\dim_{\text{Krull}/\mathcal{O}} \widetilde{R}_{\nu,\mathcal{O}}^{(\eta)} \geq h^0(G_\nu, \text{ad}_{\bar{\rho}}^{(0)}) - \delta(G_\nu, \text{ad}_{\bar{\rho}})^{(\eta)}$ *for* $\nu \nmid \ell, \infty$.

(iv) *One has*

$$\sum_{\nu | \ell \text{ or } \nu | \infty} \dim_{\text{Krull}/\mathcal{O}} \widetilde{R}_{\nu,\mathcal{O}}^{(\eta)} \geq \sum_{\nu | \ell \text{ or } \nu | \infty} (h^0(G_\nu, \text{ad}_{\bar{\rho}}^{(0)}) - \delta(G_\nu, \text{ad}_{\bar{\rho}})^{(\eta)}).$$

(v) $\delta(G_{F,S}, \text{ad}_{\bar{\rho}})^{(\eta)} - h^0(G_{F,S}, \text{ad}_{\bar{\rho}}^{(0)}) + h^0(G_{F,S}, (\text{ad}_{\bar{\rho}}^{(0)})^\vee) = 0.$

Then the ℓ*-torsion of* $\widetilde{R}_{S,\mathcal{O}}^{(\eta)}$ *is finite, and* $\widetilde{R}_{S,\mathcal{O}}^{(\eta)}$ *modulo its* ℓ*-torsion is non-zero over* $\mathcal{O}$. *Hence this quotient is non-zero and finite flat over* $\mathcal{O}$.

Proof Since $\widetilde{R}^{(\eta)}_{S,\mathcal{O}}$ is noetherian the ℓ-torsion submodule of $\widetilde{R}^{(\eta)}_{S,\mathcal{O}}$ is finitely generated. Therefore there exists some $m \geq 0$ such that the ℓ-torsion submodule injects into $\widetilde{R}^{(\eta)}_{S,\mathcal{O}}/(\ell^m)$. By condition (i) (and noetherianess of $\widetilde{R}^{(\eta)}_{S,\mathcal{O}}$) the latter is finite. To complete the proof of the corollary, it suffices to show that the Krull dimension of $\widetilde{R}^{(\eta)}_{S,\mathcal{O}}$ is at least one.

We first compute the relative Krull dimension of the middle term in the presentation of Lemma 7.1:

$$\dim_{\mathrm{Krull}/\mathcal{O}}\Big(\widehat{\bigotimes}_{\nu\in S}\widetilde{R}^{(\eta)}_{\nu,\mathcal{O}}\Big) \overset{\text{(ii)}}{=} \sum_{\nu\in S}\dim_{\mathrm{Krull}/\mathcal{O}}\widetilde{R}^{(\eta)}_{\nu,\mathcal{O}}$$
$$\overset{\text{(iii),(iv)}}{\geq} \sum_{\nu\in S}(h^0(G_\nu,\mathrm{ad}^{(0)}_{\bar\rho})-\delta(G_\nu,\mathrm{ad}_{\bar\rho})^{(\eta)}).$$

Using (v), the relative Krull dimension of $\widehat{\bigotimes}_{\nu\in S}\widetilde{R}^{(\eta)}_{\nu,\mathcal{O}}[[U_1,\ldots,U_s]]$ over $\mathcal{O}$ is therefore at least

$$s+\delta(G_{F,S},\mathrm{ad}_{\bar\rho})^{(\eta)}-h^0(G_{F,S},\mathrm{ad}^{(0)}_{\bar\rho})+h^0(G_{F,S},(\mathrm{ad}^{(0)}_{\bar\rho})^\vee)$$
$$+\sum_{\nu\in S}h^0(G_\nu,\mathrm{ad}^{(0)}_{\bar\rho})-\delta(G_\nu,\mathrm{ad}_{\bar\rho})^{(\eta)}.$$

This is also the bound on $\mathrm{gen}(\widetilde{J})$ in the presentation of Lemma 7.1. Now the quotient of a local ring by a number of relations decreases the Krull dimension of the ring by at most this number (unless the quotient is zero). Since the Krull dimension is one more than the relative Krull dimension over $\mathcal{O}$, it follows that the Krull dimension of

$$\widetilde{R}^{(\eta)}_{S,\mathcal{O}} \cong \Big(\widehat{\bigotimes}_{\nu\in S}\widetilde{R}^{(\eta)}_{\nu,\mathcal{O}}\Big)[[U_1,\ldots,U_s]]/\widetilde{J}$$

is at least one, as was to be shown. □

Remark 7.3 The following is a ring theoretic example which shows that without any further hypotheses, one cannot rule out the possibility of ℓ-torsion in the ring $\widetilde{R}^{(\eta)}_{S,\mathcal{O}}$:

Suppose $F=\mathbb{Q}$, $S=\{\ell,\infty\}$, $\widetilde{R}^{(\eta)}_{\ell,\mathcal{O}} \cong \mathbb{Z}_\ell[[S,T]]/((\ell+S)T,T^2)$, that (iv) and (v) of the corollary are satisfied, and that $\widetilde{R}^{(\eta)}_{S,\mathcal{O}}$ is the quotient of $\widetilde{R}^{(\eta)}_{\ell,\mathcal{O}}$ by the ideal (S). Then $\widetilde{R}^{(\eta)}_{\ell,\mathcal{O}} \cong \mathbb{Z}_\ell[[T]]/(\ell T,T^2)$ has ℓ-torsion, although the remaining assertions (i) and (ii) of the corollary hold ((iii) holds trivially).

However, if in addtion to (i)–(v) one imposes the further condition that $\widehat{\bigotimes}_{\nu\in S}\widetilde{R}^{(\eta)}_{\nu,\mathcal{O}}$ is Cohen-Macaulay, then from standard results in commutative algebra one may indeed deduce that $\widetilde{R}^{(\eta)}_{S,\mathcal{O}}$ is flat over $\mathcal{O}$. This was pointed out by M. Kisin.

We now apply the previous corollary to the situation of Theorem 5.8, where however we relax the condition at the places above ℓ:

Theorem 7.4 *Suppose* $d = \det : \mathcal{G} = \mathrm{GL}_2 \to \mathcal{T} = \mathrm{GL}_1$, *$F$ is totally real and $\bar{\rho}$ is odd. Suppose further that*

(i) $\widetilde{R}^{\eta}_{S,\mathcal{O}}/(\ell)$ *is finite.*
(ii) *The rings* $\widetilde{R}^{\eta}_{\nu,\mathcal{O}}$, $\nu \in S$, *are flat over* $\mathcal{O}$.
(iii) *At* $\nu \not| \ell, \infty$ *the local deformation problem satisfies* $\widetilde{h}^0_\nu - h^0(G_\nu, \mathrm{ad}^0_{\bar{\rho}}) - \mathrm{gen}(J^{\eta}_\nu) \geq 0$.
(iv) *At* $\nu|\infty$ *we choose either of the versal hulls in Example 5.4 depending on whether* $\bar{\rho}(c_\nu)$ *is trivial or not.*
(v) *At* $\nu|\ell$, *one has* $\dim_{\mathrm{Krull}/\mathcal{O}} \widetilde{R}^{\eta}_{\nu,\mathcal{O}} = [F_\nu : \mathbb{Q}_\ell] + h^0(G_\nu, \mathrm{ad}^0_{\bar{\rho}}) - \delta(G_\nu, \mathrm{ad}_{\bar{\rho}})^{\eta}$.
(vi) $h^0(G_{F,S}, (\mathrm{ad}^0_{\bar{\rho}})^{\vee}) = 0$. *(cf. Example 5.7 for explicit conditions.)*

Then $\widetilde{R}^{\eta}_{S,\mathcal{O}}$ *has finite ℓ-torsion, and its quotient modulo ℓ-torsion is non-zero and finite flat over* $\mathcal{O}$.

Proof It suffices to verify the hypothesis of Corollary 7.2. Conditions (i), (ii), (iii) and (vi) imply conditions (i), (v), (ii) and (iv) of Corollary 7.2, respectively. At the infinite places ν, condition (iv) implies $\dim_{\mathrm{Krull}/\mathcal{O}} \widetilde{R}^{(\eta)}_{\nu,\mathcal{O}} = h^0(G_\nu, \mathrm{ad}^0_{\bar{\rho}}) - \delta(G_\nu, \mathrm{ad}_{\bar{\rho}})^{\eta} - 1$. Because of the identity $\sum_{\nu|\ell}[F_\nu : \mathbb{Q}_\ell] = [F : \mathbb{Q}] = \sum_{\nu|\infty} 1$, the latter observation combined with condition (v) implies condition (iii) of Corollary 7.2. □

Bibliography

[1] F. Bleher, T. Chinburg, *Universal deformation rings need not be complete intersection rings*, C. R. Math. Acad. Sci. Paris **342** (2006), no. 4, 229–232.

[2] G. Böckle, *A local-to-global principle for deformations of Galois representations*, J. Reine Angew. Math. **509** (1999), 199–236.

[3] G. Böckle, *Demuškin groups with group actions and applications to deformations of Galois representations*, Compositio Math. **121** (2000), no. 2, 109–154.

[4] G. Böckle, *On the density of modular points in universal deformation spaces*, Amer. J. Math. **123** (2001), no. 5, 985–1007.

[5] B. Conrad, *Ramified deformation problems*, Duke Math. J. **97** (1999), no. 3, 439–513.

[6] F. Diamond, *The Taylor-Wiles construction and multiplicity one*, Invent. Math. **128** (1997), no. 2, 379–391.

[7] L. Dieulefait, *Existence of families of Galois representations and new cases of the Fontaine-Mazur conjecture*, J. Reine Angew. Math. **577** (2004), 147–151.

[8] L. Dieulefait, *The level 1 weight 2 case of Serre's conjecture*, preprint available at http://arxiv.org/abs/math/0412099.

[9] A. J. de Jong, *A conjecture on arithmetic fundamental groups*, Israel J. Math. **121** (2001), 61–84.

[10] C. Khare, *Serre's Modularity Conjecture: The Level One Case*, Duke Math. J. **134** (2006) no. 3, 557–589.

[11] C. Khare, *Serre's modularity conjecture: a survey of the level one case*, this volume, pp. 270-299.

[12] C. Khare, J.-P. Wintenberger, *On Serre's reciprocity conjecture for 2-dimensional mod p representations of the Galois group of $\mathbb{Q}$*, preprint 2004, http://arXiv.org/abs/math/0412076

[13] M. Kisin, *Modularity of Potentially Barsotti-Tate Galois Representations*, to appear in 'Current Developments in Mathematics, 2005', International Press, Somerville, U.S.A.

[14] H. Koch, *Galois Theory of p-Extensions*, Springer Monographs in Mathematics, Springer Verlag, Berlin 2002.

[15] D. Mauger, *Algèbra de Hecke quasi-ordinaire universelle d'un group réductive*, thesis, 2000.

[16] B. Mazur, Deforming Galois Representations, in Galois groups over $\mathbb{Q}$, ed. Ihara et al., Springer-Verlag 1987.

[17] B. Mazur, *An introduction to the deformation theory of Galois representations*, in "Modular forms and Fermat's last theorem" (Boston, MA, 1995), pp. 243–311, Springer, New York, 1997.

[18] J. Neukirch, A. Schmidt, K. Wingberg, *Cohomology of number fields*, Grundlehren der Math. Wiss. **323**, Springer, Berlin, 2000.

[19] R. Ramakrishna, *Deforming Galois representations and the conjectures of Serre and Fontaine-Mazur*, Ann. of Math. (2) **156** (2002), no. 1, 115–154.

[20] R. Ramakrishna, *On a variation of Mazur's deformation functor*, Compositio Math. 87 (1993), 269–286.

[21] M. Schlessinger, *Functors of Artin rings*, Trans. A. M. S. 130 (1968), 208–222.

[22] R. Taylor, *Remarks on a conjecture of Fontaine and Mazur*, J. Inst. Math. Jussieu **1** (2002), no. 1, 125–143.

[23] J. Tilouine, *Deformations of Galois representations and Hecke algebras*, Publ. Mehta Research Institute, Narosa Publishing House, New Delhi, 1996.

Eigenvarieties

Kevin Buzzard
Department of Mathematics
Imperial College
London U.K.
buzzard@ic.ac.uk

Abstract

We axiomatise and generalise the "Hecke algebra" construction of the Coleman-Mazur Eigencurve. In particular we extend the construction to general primes and levels. Furthermore we show how to use these ideas to construct "eigenvarieties" parametrising automorphic forms on totally definite quaternion algebras over totally real fields.

1 Introduction

In a series of papers in the 1980s, Hida showed that classical ordinary eigenforms form p-adic families as the weight of the form varies. In the non-ordinary finite slope case, the same turns out to be true, as was established by Coleman in 1995. Extending this work, Coleman and Mazur construct a geometric object, the eigencurve, parametrising such modular forms (at least for forms of level 1 and in the case $p > 2$). On the other hand, Hida has gone on to extend his work in the ordinary case to automorphic forms on a wide class of reductive groups. One might optimistically expect the existence of non-ordinary families, and even an "eigenvariety", in some of these more general cases.

Anticipating this, we present in Part I of this paper (sections 2–5) an axiomatisation and generalisation of the Coleman-Mazur construction. In his original work on families of modular forms, Coleman in [10] developed Riesz theory for orthonormalizable Banach modules over a large class of base rings, and, in the case where the base ring was 1-dimensional, constructed the local pieces of a parameter space for normalised eigenforms. There are two places where we have extended Coleman's work. Firstly, we set up Coleman's Fredholm theory and Riesz theory (in sections 2 and 3 respectively) in a slightly more general situation, so that they can be applied to spaces such as direct summands of orthonormalizable Banach modules; the motivation for this is that at times in the theory we meet Banach modules which are invariants of orthonormalizable Banach modules under the action of a finite group; such modules are not necessarily orthonormalizable, but we want to use Fredholm theory anyway. And secondly we show in sections 4–5 that given a projective Banach module and a collection of commuting operators, one of which is compact, one can glue the local pieces constructed by Coleman to form an eigenvariety, in the case where the base ring is an arbitrary reduced affinoid. At one stage we are forced to use Raynaud's theory of formal models; in particular this generalisation is not an elementary extension of Coleman's ideas.

The resulting machine can be viewed as a construction of a geometric object from a family of Banach spaces equipped with certain commuting linear maps. Once one has this machine, one can attempt to feed in Banach spaces of "overconvergent automorphic forms" into the machine, and get "eigenvarieties" out. We extend the results of [9] in Part II of this paper (sections 6 and 7), constructing an eigencurve using families of overconvergent modular forms, and hence removing some of the assumptions on p and N in the main theorems of [9]. Note that here we do not need the results of section 4, as weight space is 1-dimensional and Coleman's constructions are enough.

There are still technical geometric problems to be resolved before one can give a definition of an overconvergent automorphic form on a general reductive group, but one could certainly hope for an elementary definition if the group in question is compact mod centre at infinity, as the geometry then becomes essentially non-existent. As a concrete example of this, we propose in Part III (sections 8–13) a definition of an overconvergent automorphic form in the case when the reductive group is a compact form of GL_2 over a totally real field, and apply our theory to this situation to construct higher-dimensional eigenvarieties.

Chenevier has constructed Banach spaces of overconvergent automorphic forms for compact forms of GL_n over $\mathbf{Q}$ and one can feed his spaces into the machine also to get eigenvarieties for these unitary groups.

This work began in 2001 during a visit to Paris-Nord, and the author would like to thank Jacques Tilouine for the invitation and Ahmed Abbes for several useful conversations. In fact the author believes that he was the first to coin the phrase "eigenvarieties", in 2001. Part I of this paper was written at that time, as well as some of Part III. The paper then remained in this state for three years, and the author most sincerely thanks Gaetan Chenevier for encouraging him to finish it off. In fact Theorem 4.6 of this paper is assumed both by Chenevier in [8], and Yamagami in [17], who independently announced results very similar to those in Part III of this paper, the main difference being that Yamagami works with the U operator at only one prime above p and fixes weights at the other places, hence his eigenvarieties can have smaller dimension than ours, but they see more forms (they are only assumed to have finite slope at one place above p). My apologies to both Chenevier and Yamagami for the delay in writing up this construction; I would also like to thank both of them for several helpful comments.

A lot has happened in this subject since 2001. Matthew Emerton has recently developed a general theory of eigenvarieties which in many cases produces cohomological eigenvarieties associated to a large class of reductive algebraic groups. As well as Coleman and Mazur, many other people (including Emerton, Ash and Stevens, Skinner and Urban, Mazur and Calegari, Kassaei, Kisin and Lai, Chenevier, and Yamagami), have made contributions to the area, all developing constructions of eigenvarieties in other situations. We finish this introduction with an explanation of the relationship between Emerton's work and ours. Emerton's approach to eigenvarieties is more automorphic and more conceptual than ours. His machine currently needs a certain spectral sequence to degenerate, but this degeneration occurs in the case of the Coleman-Mazur eigencurve and hence Emerton has independently given a construction of this eigencurve for arbitrary N and p as in Part II of this paper. However, Emerton's construction is less "concrete" and in particular the results in [6] and [7] rely on the construction of the 2-adic eigencurve presented in this paper. On the other hand Emerton's ideas give essentially the same construction of the eigenvariety associated to a totally definite quaternion algebra over a totally real field, in the sense that one can check that his more conceptual approach, when translated down, actually becomes equivalent to ours.

We would like to thank Peter Schneider for pointing out an error in an earlier version of Lemma 5.6, Elmar Grosse-Kloenne for pointing out a simplification in the definition of our admissible cover, and the referee for several helpful remarks, in particular for pointing out that flatness was necessary in Lemma 5.5.

PART I: The eigenvariety machine.

2 Compact operators on K-Banach modules

In this section we collect together the results we need from the theory of commutative Banach algebras. A comprehensive source for the terminology we use is [1]. Throughout this section, K will be a field complete with respect to a non-trivial non-archimedean valuation $|.|_K$, and A will be a commutative Noetherian K-Banach algebra. That is, A is a commutative Noetherian K-algebra equipped with a function $|.| : A \to \mathbf{R}_{\geq 0}$, and satisfying

- $|1| \leq 1$, and $|a| = 0$ iff $a = 0$,
- $|a + b| \leq \max\{|a|, |b|\}$,
- $|ab| \leq |a||b|$,
- $|\lambda a| = |\lambda|_K |a|$ for $\lambda \in K$,

and such that A is complete with respect to the metric induced by $|.|$. For elementary properties of such algebras we refer the reader to [1], §3.7 and thereafter. Such algebras are Banach algebras in the sense of Coleman [10]. Later on we shall assume (mostly for simplicity) that A is a reduced K-affinoid algebra with its supremum norm, but this stronger assumption does not make the arguments of this section or the next any easier.

From the axioms one sees that either $|1| = 0$ and hence $A = 0$, or $|1| = 1$, in which case the map $K \to A$ is injective and the norm on A extends the norm on K. Fix once and for all $\rho \in K^\times$ with $|\rho|_K < 1$. Such ρ exists as we are assuming the valuation on K is non-trivial. We use ρ to "normalise" vectors in several proofs. If A^0 denotes the subring $\{a \in A : |a| \leq 1\}$ then one easily checks that the ideals of A^0 generated by ρ^n, $n = 1, 2, \ldots$, form a basis of open neighbourhoods of zero in A. Note that A^0 may not be Noetherian (for example if $A = K = \mathbf{C}_p$).

Let A be a commutative Noetherian K-Banach algebra. A *Banach A-module* is an A-module M equipped with $|.| : M \to \mathbf{R}_{\geq 0}$ satisfying

- $|m| = 0$ iff $m = 0$,
- $|m + n| \leq \max\{|m|, |n|\}$,
- $|am| \leq |a||m|$ for $a \in A$ and $m \in M$,

and such that M is complete with respect to the metric induced by $|.|$. Note that A itself is naturally a Banach A-module, as is any closed ideal of A. In fact all ideals of A are closed, by Proposition 3.7.2/2 of [1].

If M and N are Banach A-modules, then we define a norm on $M \oplus N$ by $|m \oplus n| = \mathrm{Max}\{|m|, |n|\}$. This way $M \oplus N$ becomes a Banach A-module. In particular we can give A^r the structure of a Banach A-module in a natural way.

By a finite Banach A-module we mean a Banach A-module which is finitely-generated as an abstract A-module. We use the following facts several times in what follows:

Proposition 2.1

(a) (Open Mapping Theorem) A continuous surjective K-linear map between Banach K-modules is open.

(b) The category of finite Banach A-modules, with continuous A-linear maps as morphisms, is equivalent to the category of finite A-modules. In particular, any A-module homomorphism between finite Banach A-modules is automatically continuous, and if M is any finite A-module then there is a unique (up to equivalence) complete norm on M making it into a Banach A-module.

Proof (a) is Théorème 1 in Chapter I, §3.3 of [3] (but note that "homomorphisme" here has the meaning assigned to it in §2.7 of Chapter III of [2], and in particular is translated as "strict morphism" rather than "homomorphism").

(b) is proved in Propositions 3.7.3/2 and 3.7.3/3 of [1]. □

Note that by (b), a finite A-module M has a canonical topology, induced by any norm that makes M into a Banach A-module. We call this topology the *Banach topology* on M.

As an application of these results, we prove the following useful lemma:

Lemma 2.2 *If M is a Banach A-module, and P is a finite Banach A-module, then any abstract A-module homomorphism $\phi : P \to M$ is continuous.*

Proof Let $\pi : A^r \to P$ be a surjection of A-modules, and give A^r its usual Banach A-module norm. Then π is open by the Open Mapping Theorem, and $\phi\pi$ is bounded and hence continuous. So ϕ is also continuous. □

If I is a set, and for every $i \in I$ we have a_i, an element of A, then by the statement $\lim_{i\to\infty} a_i = 0$, we simply mean that for all $\epsilon > 0$ there are only finitely many $i \in I$ with $|a_i| > \epsilon$. This is no condition if I is finite, and is the usual condition if $I = \mathbf{Z}_{\geq 0}$. For general I, if $\lim_{i\to\infty} a_i = 0$ then only countably many of the a_i can be non-zero. We also mention here a useful convention: occasionally we will take a max or a supremum over a set (typically a set of norms) which can be empty in degenerate cases (e.g., if a certain module or ring is zero). In these cases we will define the max or the supremum to be zero. In other words, throughout the paper we are implicitly taking suprema in the set of non-negative reals rather than the set of all reals.

Let A be a non-zero commutative Noetherian K-Banach algebra, let M be a Banach A-module, and consider a subset $\{e_i : i \in I\}$ of M such that $|e_i| = 1$ for all $i \in I$. Then for any sequence $(a_i)_{i\in I}$ of elements of A with $\lim_{i\to\infty} a_i = 0$, the sum $\sum_i a_i e_i$ converges. We say that a Banach A-module

M is *orthonormalizable*, or *ONable* for short, if there exists such a subset $\{e_i : i \in I\}$ of M with the following two properties:

- Every element m of M can be written uniquely as $\sum_{i \in I} a_i e_i$ with $\lim_{i \to \infty} a_i = 0$, and
- If $m = \sum_i a_i e_i$ then $|m| = \max_{i \in I} |a_i|$.

Such a set of elements $\{e_i\}$ is called an *orthonormal basis*, or an ON basis, for M. Note that the second condition implies that $|e_i| = 1$ for all $i \in I$.

Again assume $A \neq 0$. If I is a set, we define $c_A(I)$ to be A-module of functions $f : I \to A$ such that $\lim_{i \to \infty} f(i) = 0$. Addition and the A-action are defined pointwise. We define $|f|$ to be $\mathrm{Max}\{|a_i| : i \in I\}$. With respect to this norm, $c_A(I)$ becomes a Banach A-module. If $i \in I$ and we define e_i to be the function sending $j \in I$ to 0 if $i \neq j$, and to 1 if $i = j$, and if furthermore $A \neq 0$, then it is easily checked that the e_i are an ON basis for $c_A(I)$, and we call $\{e_i : i \in I\}$ the *canonical ON basis* for $c_A(I)$. If M is any ONable Banach A-module, then to give an ON basis $\{e_i : i \in I\}$ for M is to give an isometric (that is, metric-preserving) isomorphism $M \cong c_A(I)$. Note also that $c_A(I)$ has the following universal property: if M is any Banach A-module then there is a natural bijection between $\mathrm{Hom}_A(c_A(I), M)$ and the set of bounded maps $I \to M$, given by sending $\phi : c_A(I) \to M$ to the map $i \mapsto \phi(e_i)$.

If $A = 0$ then the only A-module is $M = 0$, and we regard this module as being ONable of arbitrary rank. We have chosen to ignore this case in the definitions above because if we had included it then we should have to define an ONable Banach module as being a collection of e_i as above but with $|e_i| = |1|$ and so on; however this just clutters notation. There is no other problem with the zero ring in this situation. We will occasionally assume $A \neq 0$ in proofs, and leave the interested reader to fill in the trivial details in the case $A = 0$.

We recall some basic results on "matrices" associated to endomorphisms of Banach A-modules. The proofs are elementary exercises in analysis. Let M and N be Banach A-modules, and let $\phi : M \to N$ be an A-module homomorphism. Then a standard argument (see Corollary 2.1.8/3 of [1]) using the fact that one can use ρ to renormalise elements of M, shows that ϕ is continuous iff it is bounded, and in this case we define $|\phi| = \sup_{0 \neq m \in M} \frac{|\phi(m)|}{|m|}$ (this set of reals is bounded above if ϕ is continuous). Now assume that M is ONable, with ON basis $\{e_i : i \in I\}$. One easily checks that if ϕ is continuous and $\phi(e_i) = n_i$, then the n_i are a bounded collection of elements of N which uniquely determine ϕ. Furthermore, if n_i are an arbitrary bounded collection of elements of N there is a unique continuous map $\phi : M \to N$ such that $\phi(e_i) = n_i$ for all i, and $|\phi| = \sup_{i \in I} |n_i|$.

Now assume that N is also ONable, with basis $\{f_j : j \in J\}$. If $\phi : M \to N$ is a continuous A-module homomorphism, we can define its associated matrix coefficients $(a_{i,j})_{i \in I, j \in J}$ by[1]

$$\phi(e_i) = \sum_{j \in J} a_{i,j} f_j.$$

One checks easily from the arguments above that the collection $(a_{i,j})$ has the following two properties:

- For all i, $\lim_{j \to \infty} a_{i,j} = 0$.
- There exists a constant $C \in \mathbf{R}$ such that $|a_{i,j}| \leq C$ for all i, j.

In fact C can be taken to be $|\phi|$, and furthermore we have $|\phi| = \sup_{i,j} |a_{i,j}|$.

Conversely, given a collection $(a_{i,j})_{i \in I, j \in J}$ of elements of A, satisfying the two conditions above, there is a unique continuous $\phi : M \to N$ with norm $\sup_{i,j} |a_{i,j}|$ whose associated matrix is $(a_{i,j})$. As a useful consequence of this, we see that if ϕ and $\psi : M \to N$ are continuous, with associated matrices $(a_{i,j})$ and $(b_{i,j})$, then $|\phi - \psi| \leq \epsilon$ iff $|a_{i,j} - b_{i,j}| \leq \epsilon$ for all i and j.

Let A be a commutative Noetherian K-Banach algebra and let M, N be Banach A-modules. The A-module $\mathrm{Hom}(M, N)$ of continuous A-linear homomorphisms from M to N is then also a Banach A-module: completeness follows because if ϕ_n is a Cauchy sequence in $\mathrm{Hom}(M, N)$ then for all $m \in M$, $\phi_n(m)$ is a Cauchy sequence in N, and one can define $\phi(m)$ as its limit; then ϕ is the limit of the ϕ_n.

A continuous A-module homomorphism $M \to N$ is said to be of *finite rank* if its image is contained in a finitely-generated A-submodule of N. The closure in $\mathrm{Hom}(M, N)$ of the finite rank homomorphisms is the set of *compact* homomorphisms (many authors use the term "completely continuous"). Let M and N be ONable Banach modules, and let $\phi : M \to N$ be a continuous homomorphism, with associated matrix $(a_{i,j})$. We wish to give a simple condition which is expressible only in terms of the $a_{i,j}$, and which is equivalent to compactness. Such a result is announced in Lemma A1.6 of [10] for a more general class of rings A, but the proof seems to be incomplete. This is not a problem with the theory however, as the proof can be completed in all cases of interest without too much trouble. We complete the proof here in the case of commutative Noetherian K-Banach algebras. We start with some preliminary results. As ever, A is a Noetherian K-Banach algebra. If M is an ONable Banach A-module, with ON basis $\{e_i : i \in I\}$, and if $S \subseteq I$ is a finite subset, then we define A^S to be the submodule $\oplus_{i \in S} Ae_i$, and we define the projection

1 Here we follow Serre's conventions in [15], rather than writing $a_{j,i}$ for $a_{i,j}$.

$\pi_S : M \to A^S$ to be the map sending $\sum_{i \in I} a_i e_i$ to $\sum_{i \in S} a_i e_i$. Note that this projection is norm-decreasing onto a closed subspace of M.

Lemma 2.3 *Let M be an ONable Banach A-module, with basis $\{e_i : i \in I\}$, and let P a finite submodule of M.*

(a) There is a finite set $S \subseteq I$ such that $\pi_S : M \to A^S$ is injective on P.

(b) P is a closed subset of M, and hence is complete.

(c) For all $\epsilon > 0$, there is a finite set $T \subseteq I$ such that for all $p \in P$ we have $|\pi_T(p) - p| \leq \epsilon |p|$.

Proof Say P is generated by $m_1, \ldots, m_r$ and for $1 \leq \alpha \leq r$ we have $m_\alpha = \sum_i a_{\alpha,i} e_i$.

(a) For $i \in I$ let v_i be the element $(a_{\alpha,i})_{1 \leq \alpha \leq r}$ of A^r. The A-submodule of A^r generated by the v_i is finitely-generated, as A is Noetherian, and hence there is a finite set $S \subseteq I$ such that this module is generated by $\{v_i : i \in S\}$. It is now an easy exercise to check that this S works, because if $\pi_S(\sum_\alpha b_\alpha m_\alpha)$ is zero, then $\sum_\alpha b_\alpha a_{\alpha,i}$ is zero for all $i \in S$ and hence for all $i \in I$.

(b) P is a finite A-module, and hence there is, up to equivalence, a unique complete A-module norm on P. Let Q denote P equipped with this norm. The algebraic isomorphism $Q \to P$ induces a map $Q \to M$ which is continuous by Lemma 2.2, and hence the algebraic isomorphism $Q \to P$ is continuous. On the other hand, if S is chosen as in part (a), then the injection $P \to A^S$ induces a continuous injection from P onto a submodule of A^S which is closed by Proposition 3.7.3/1 of [1], and this submodule is algebraically isomorphic to Q, and hence isomorphic to Q as a Banach A-module. We hence have continuous maps $Q \to P \to Q$, which are algebraic isomorphisms, and hence the norms on P and Q are equivalent. So the maps are also homeomorphisms, and P is complete with respect to the metric induced from M, and is hence a closed submodule of M.

(c) By (b), P is complete and hence a K-Banach space. The map $A^r \to P$ sending (a_α) to $\sum_\alpha a_\alpha m_\alpha$ is thus a continuous surjection between K-Banach spaces, and hence by the open mapping theorem there exists $\delta > 0$ such that if $p \in P$ with $|p| \leq \delta$ then $p = \sum_{\alpha=1}^r a_\alpha m_\alpha$ with $|a_\alpha| \leq 1$ for all α. Choose T such that for all m_α we have $|\pi_T(m_\alpha) - m_\alpha| \leq \epsilon\delta|\rho|$. This T works: if $p \in P$ is arbitrary, then either $p = 0$ and hence the condition we are checking is automatic, or $p \neq 0$. In this case, there exists some $n \in \mathbf{Z}$ such that $|\rho|\delta < |\rho|^n |p| \leq \delta$, and then $\rho^n p = \sum_\alpha a_\alpha m_\alpha$ with $|a_\alpha| \leq 1$ for all α.

Then

$$\begin{aligned}|\pi_T(p) - p| &= \left|\sum_\alpha \rho^{-n} a_\alpha(\pi_T(m_\alpha) - m_\alpha)\right| \\ &\leq |\rho|^{-n}\epsilon\delta|\rho| \\ &\leq \epsilon|p|\end{aligned}$$

and we are done. □

Proposition 2.4 *Let M, N be ONable Banach A-modules, with ON bases $\{e_i : i \in I\}$ and $\{f_j : j \in J\}$. Let $\phi : M \to N$ be a continuous A-module homomorphism, with basis $(a_{i,j})$. Then ϕ is compact if and only if $\lim_{j\to\infty} \sup_{i\in I} |a_{i,j}| = 0$.*

Proof If the matrix of ϕ satisfies $\lim_{j\to\infty} \sup_{i\in I} |a_{i,j}| = 0$, then ϕ is easily seen to be compact: for any $\epsilon > 0$ there is a finite subset $S \subseteq J$ such that $|\phi - \pi_S\phi| \leq \epsilon$.

The other implication is somewhat more delicate. It suffices to prove the result when ϕ has finite rank. If $\phi = 0$ then the result is trivial, so assume $0 \neq \phi$ and $\phi(M) \subseteq P$, where $P \subseteq N$ is finite. By part (c) of Lemma 2.3, for any $\epsilon > 0$ we may choose T such that $|\pi_T(p) - p| \leq \epsilon|p|/|\phi|$, and hence $|\pi_T\phi - \phi| \leq \epsilon$. Hence $|a_{i,j}| \leq \epsilon$ if $j \notin T$, and we are home because ϵ was arbitrary. □

Remark If we allow A to be non-Noetherian then we do not know whether the preceding proposition remains true.

From this result, it easily follows that a compact operator $\phi : M \to M$, where M is an ONable Banach A-module, has a characteristic power series $\det(1 - X\phi) = \sum_{n\geq 0} c_n X^n \in A[[X]]$, defined in terms of the matrix coefficients of ϕ using the usual formulae, which we recall from §5 of [15] for convenience: firstly choose an ON basis $\{e_i : i \in I\}$ for M, and say ϕ has matrix $(a_{i,j})$ with respect to this basis. If S is any finite subset of I, then define $c_S = \sum_{\sigma:S\to S} \mathrm{sgn}(\sigma) \prod_{\mathrm{i}\in \mathrm{S}} \mathrm{a}_{\mathrm{i},\sigma(\mathrm{i})}$, where the sum ranges over all permutations of S, and for $n \geq 0$ define $c_n = (-1)^n \sum_S c_S$, where the sum is over all finite subsets of I of size n. One easily checks that this sum converges, using Proposition 2.4. Furthermore, again using Proposition 2.4 and following Proposition 7 of [15], one sees that the resulting power series $\det(1 - X\phi) = \sum_n c_n X^n$ converges for all $X \in A$. However, from our definition it is not clear to what extent the power series depends on the choice of

ON basis for M. We now investigate to what extent this is the case. We begin with some observations in the finite-dimensional case.

If P is any finite free ONable Banach A-module, with ON basis (e_i), and $\phi : P \to P$ is any A-module homomorphism, then $\det(1 - X\phi)$, defined as above with respect to the e_i, is the usual algebraically-defined $\det(1 - X\phi)$, because the definition above coincides with the usual classical definition, which is independent of choice of basis. Next we recall the well-known fact that if P and Q are both free A-modules of finite rank, and $u : P \to Q$ and $v : Q \to P$ are A-module homomorphisms, then $\det(1 - Xuv) = \det(1 - Xvu)$. Now let M be an ONable Banach A-module, with ON basis $\{e_i : i \in I\}$. We use this fixed basis for computing characteristic power series in the lemma below.

Lemma 2.5

(a) If $\phi_n : M \to M$, $n = 1, 2, \ldots$ are a sequence of compact operators that tend to a compact operator ϕ, then $\lim_n \det(1 - X\phi_n) = \det(1 - X\phi)$, *uniformly in the coefficients.*

(b) If $\phi : M \to M$ is compact, and furthermore if the image of ϕ is contained in $P := \oplus_{i \in S} Ae_i$, for S a finite subset of I, then $\det(1 - X\phi) = \det(1 - X\phi|_P)$*, the right hand determinant being the usual algebraically-defined one.*

(c) (strengthening of (b)) If $\phi : M \to M$ is compact, and if the image of ϕ is contained within an arbitrary submodule Q of M which is free of finite rank, then $\det(1 - X\phi) = \det(1 - X\phi|_Q)$*, where again the right hand side is the usual algebraically-defined determinant.*

Proof (a) This follows *mutatis mutandis* from [15], Proposition 8.

(b) If $(a_{i,j})$ is the matrix of ϕ then $a_{i,j} = 0$ for $j \notin S$ and the result follows immediately from the definition of the characteristic power series.

(c) Choose $\epsilon > 0$. By Lemma 2.3(c), there is a finite set $T \subseteq I$ such that $\pi_T : Q \to P := A^T$ has the property that $|\pi_T - i| \leq \epsilon$, where $i : Q \to M$ is the inclusion. Define $\phi^T = \pi_T \phi : M \to P \subseteq M$. By (b) we see that $\det(1 - X\phi^T)$ equals the algebraically-defined polynomial $\det(1 - X\phi^T|_P)$. Furthermore, by consideration of the maps $\phi : P \to Q$ and $\pi_T : Q \to P$, we see that this polynomial also equals the algebraically-defined $\det(1 - X\phi_T)$, where $\phi_T = \phi\pi_T : Q \to Q$. One can compute this latter determinant with respect to an arbitrary algebraic A-basis of Q. By Lemma 2.3(b), Q with its subspace topology is complete, and hence the topology on Q is the Banach topology. Now as ϵ tends to zero, $\phi_T : Q \to Q$ tends to $\phi : Q \to Q$ and $\phi^T : M \to M$ tends to $\phi : M \to M$, and the result follows by part (a). □

Corollary 2.6 *Let M be an A-module, and let $|.|_1$ and $|.|_2$ be norms on M both making M into an ONable Banach A-module, and both inducing the same topology on M. Then an A-linear map $\phi : M \to M$ is compact with respect to $|.|_1$ iff it is compact with respect to $|.|_2$, and furthermore if $\{e_i : i \in I\}$ and $\{f_j : j \in J\}$ are ON bases for $(M, |.|_1)$ and $(M, |.|_2)$ respectively, then the definitions of* $\det(1 - X\phi)$ *with respect to these bases coincide.*

Proof All one has to do is to check that ϕ can be written as the limit as maps ϕ_n which have image contained in free modules of finite rank, and then the result follows from parts (a) and (c) of the Lemma. To do this, one can simply use Lemma 2.3 to construct ϕ as a limit of $\pi_{T_n}\phi$, for T_n running through appropriate finite subsets of I. Note that ϕ_n will then tend to ϕ with respect to both norms (recall that two norms on M are equivalent iff they induce the same topology, because the valuation on K is non-trivial) and the result follows. □

The corollary enables us to conclude that the notion of a characteristic power series only depends on the topology on M, when A is a commutative Noetherian K-Banach algebra. In particular it does not depend on the choice of an orthonormal basis for M. Coleman proves in corollary A2.6.1 of [10] that the definition of the characteristic power series only depends on the topology on M when A is semi-simple; on the other hand neither of these conditions on A implies the other.

Next we show that the analogue of Corollaire 2 to Proposition 7 of [15] is true in this setting. Coleman announces such an analogue in Proposition A2.3 of [10] but again we have not been able to complete the proof in the generality in which Coleman is working. We write down a complete proof when A is a commutative Noetherian K-Banach algebra and remark that it is actually slightly delicate. We remark also that in the case where A is a reduced affinoid, which will be true in the applications, one can give an easier proof by using Corollary 2.10 to reduce to the case treated by Serre.

Lemma 2.7 *If M and N are ONable Banach A-modules, if $u : M \to N$ is compact and $v : N \to M$ is continuous, then uv and vu are compact, and* $\det(1 - Xuv) = \det(1 - Xvu)$.

Proof If there exist finite *free* sub-A-modules F of M and G of N such that $u(M)$ is contained in G and $v(G)$ is contained in F, then $u : F \to G$ and $v : G \to F$, and by Lemma 2.5(c) it suffices to check that the algebraically-defined characteristic polynomials of $uv : G \to G$ and $vu : F \to F$ are the same, which is a standard result. We reduce the general case to this case by several applications of Lemma 2.3, the catch being that it is not clear (to the author at least) whether any finite submodule of an ONable Banach module is contained within a finite free submodule.

We return to the general case. Write u as a limit of finite rank operators u_n. Then $u_n v$ and vu_n are finite rank, so uv and vu are both the limit of finite rank operators and are hence compact. By Lemma 2.5(a), it suffices to prove that $\det(1 - Xu_n v) = \det(1 - Xvu_n)$ for all n, and hence we may assume that u is finite rank. Let $Q \subseteq N$ denote a finite A-module containing the image of u.

Choose ON bases $\{e_i : i \in I\}$ for M and $\{f_j : j \in J\}$ for N. Now for any positive integer n we may, by Lemma 2.3(c), choose a finite subset $T_n \subseteq J$ such that $|\pi_{T_n} q - q| \leq \frac{1}{n}|q|$ for all $q \in Q$. It follows easily that $|\pi_{T_n} u - u| \leq |u|/n$ and hence $\lim_{n\to\infty} \pi_{T_n} u = u$. Hence $v\pi_{T_n} u \to vu$ and $\pi_{T_n} uv \to uv$ and again by Lemma 2.5(a) we may replace u by $\pi_{T_n} u$ and in particular we may assume that the image of u is contained in a finite *free* A-submodule of N. Let G denote this submodule. Now $P = v(G)$ is a finite submodule of M, and for any positive integer n we may, as above, choose a finite subset $S_n \subseteq I$ such that $|\pi_{S_n} p - p| \leq \frac{1}{n}|p|$ for all $p \in P$.

It is unfortunately not the case that $\pi_{S_n} v \to v$ as $n \to \infty$, as v is not in general compact. However we do have $\pi_{S_n} vu \to vu$ and hence the characteristic power series of $\pi_{S_n} vu$ tends (uniformly in the coefficients) to the characteristic power series of vu. Also, the image of $uv : N \to N$ and $u\pi_{S_n} v$ are both contained within G and hence the characteristic power series of uv (resp. $u\pi_{S_n} v$) is equal to the algebraically-defined characteristic power series of $uv : G \to G$ (resp. $u\pi_{S_n} v : G \to G$). Once one has restricted to G, one *does* have $u\pi_{S_n} v \to uv$, and hence the characteristic power series of $u\pi_{S_n} v$ tends to the characteristic power series of uv. We may hence replace v by $\pi_{S_n} v$ and in particular may assume that the image of v is contained within a finite free A-submodule F of M. We have now reduced to the algebraic case dealt with at the beginning of the proof. □

Corollary 2.6 also enables us to slightly extend the domain of definition of a characteristic power series: if M is a Banach A-module, then we say that M is *potentially ONable* if there exists a norm on M equivalent to the given norm, for which M becomes an ONable Banach A-module. Equivalently, M is potentially ONable if there is a bounded collection $\{e_i : i \in I\}$ of elements of M with the following two properties: firstly, every element m of M can be uniquely written as $\sum_i a_i e_i$ with $\lim_{i\to\infty} a_i = 0$, and secondly there exist positive constants c_1 and c_2 such that for all $m = \sum_i a_i e_i$ in M, we have $c_1 \sup_i |a_i| \leq |m| \leq c_2 \sup_i |a_i|$. We call the collection $\{e_i : i \in I\}$ a *potentially ON basis* for M. Being potentially ONable is probably a more natural notion than being ONable, because it is useful to be able to work with norms only up to equivalence, whereas ONability of a module really depends on the precise norm on the module. Note that to say a module is ONable is equivalent to saying that it is isometric to some $c_A(I)$, and to say that it is potentially ONable is just to say that it is isomorphic to some $c_A(I)$ (in the category of Banach modules, with continuous maps as morphisms).

If M is potentially ONable then one still has the notion of the characteristic power series of a compact operator on M, defined by choosing an equivalent ONable norm and using this norm to define the characteristic power series. By Corollary 2.6, this is independent of all choices. We note that certainly there can exist Banach A-modules which are potentially ONable but not ONable, for example if $A = K = \mathbf{Q}_p$ and $M = \mathbf{Q}_p(\sqrt{p})$ with its usual norm, then $|M| \neq |A|$ and so M is not ONable, but is potentially ONable.

A useful result is

Lemma 2.8 *If $h : A \to B$ is a continuous morphism of Noetherian K-Banach algebras, and M is a potentially ONable Banach A-module, then $M\widehat{\otimes}_A B$ is a potentially ONable Banach B-module, and furthermore if $\{e_i : i \in I\}$ is a potentially ON basis for M, then $\{e_i \otimes 1 : i \in I\}$ is a potentially ON basis for $M\widehat{\otimes}_A B$.*

Proof Set $N = c_B(I)$, and let $\{f_i : i \in I\}$ be its canonical ON basis. Then there is a natural A-bilinear bounded map $M \times B \to N$ sending $(\sum_i a_i e_i, b)$ to $\sum_i bh(a_i) f_i$, which induces a continuous map $M\widehat{\otimes}_A B \to N$. On the other hand, if $n \in N$, one can write n as a limit of elements of the form $\sum_{i \in S} b_i f_i$, where S is a finite subset of I. The element $\sum_{i \in S} e_i \otimes b_i$ of $M \otimes_A B$ has norm bounded above by a constant multiple of $\max_{i \in S} |b_i|$ and hence as S increases, the resulting sequence $\sum_{i \in S} e_i \otimes b_i$ is Cauchy and so its image in $M\widehat{\otimes}_A B$ tends to a limit. This construction gives a well-defined continuous A-module homomorphism $N \to M\widehat{\otimes}_A B$ which is easily checked to be an inverse to the natural map $M\widehat{\otimes}_A B \to N$, and now everything follows. □

Note that because we are only working in the "potential" world, we do not need to assume the map $A \to B$ is contractive, although in the applications we have in mind it usually will be.

Corollary 2.9 *If $h : A \to B$ is a continuous morphism of commutative Noetherian K-Banach algebras, M and N are potentially ONable Banach A-modules with potentially ON bases (e_i) and (f_j), and $\phi : M \to N$ is compact, with matrix $(a_{i,j})$, then $\phi \otimes 1 : M\widehat{\otimes}_A B \to N\widehat{\otimes}_A B$ is also compact and* *if* $(b_{i,j})$ *is the matrix of* $\phi \otimes 1$ *with respect to the bases* $(e_i \otimes 1)$ *and* $(f_j \otimes 1)$ *then* $b_{i,j} = h(a_{i,j})$.

Proof Compactness of $\phi \otimes 1$ follows from Proposition 2.4 and the rest is easy. □

Corollary 2.10 *With notation as above, if* $\det(1 - X\phi) = \sum_n c_n X^n$ *then* $\det(1 - X(\phi \otimes 1)) = \sum_n h(c_n) X^n$.

Proof Immediate. □

In practice we need to extend the notion of the characteristic power series of a compact operator still further, to the natural analogue of projective modules in this setting. Let us say that a Banach A-module P *satisfies property* (Pr) if there is a Banach A-module Q such that $P \oplus Q$, equipped with its usual norm, is potentially ONable. I am grateful to the referee for pointing out the following universal property for such modules: P has property (Pr) if and only if for every surjection $f : M \to N$ of Banach A-modules and for every continuous map $\alpha : P \to N$, α lifts to a map $\beta : P \to M$ such that $f\beta = \alpha$. The proof is an elementary application of the Open Mapping Theorem; the key point is that if $P = c_A(I)$ for some set I, then to give $\alpha : P \to N$ is to give a bounded map $I \to N$, and such a map lifts to a bounded map $I \to M$ by the Open Mapping Theorem. Note however that it would be perhaps slightly disingenuous to call such modules "projective", as there are epimorphisms in the category of Banach A-modules whose underlying module map is not surjective.

One can easily check that if P is a finite Banach A-module which is projective as an A-module, then P has property (Pr). The converse is also true:

Lemma 2.11 *If* P *is a finite Banach* A*-module with property* (Pr) *then* P *is projective as an* A*-module.*

Proof Choose a surjection $A^n \to P$ for some n and then use the universal property above. □

Note that potentially ONable Banach A-modules have property (Pr), but in general the converse is false—for example if there are finite A-modules which are projective but not free then such modules, equipped with any complete Banach A-module norm, will satisfy (Pr) but will not be potentially ONable.

Say P satisfies property (Pr) and $\phi : P \to P$ is a compact morphism. Define $\det(1 - X\phi)$ thus: firstly choose Q such that $P \oplus Q$ is potentially

ONable, and define $\det(1 - X\phi) = \det(1 - X(\phi \oplus 0))$; note that $\phi \oplus 0 : P \oplus Q \to P \oplus Q$ is easily seen to be compact. This definition may *a priori* depend on the choice of Q, but if R is another Banach A-module such that $P \oplus R$ is also potentially ONable, then so is $P \oplus Q \oplus P \oplus R$, and the maps $\phi \oplus 0 \oplus 0 \oplus 0$ and $0 \oplus 0 \oplus \phi \oplus 0$ are conjugate via an isometric A-module isomorphism, and hence have the same characteristic power series. Now the fact that $\det(1-X\phi)$ is well-defined independent of choice of Q follows easily from the fact that if M and N are ONable A-modules, and $\phi : M \to M$ is compact, then the characteristic power series of ϕ and $\phi \oplus 0 : M \oplus N \to M \oplus N$ coincide.

Many results that we have already proved for potentially ONable Banach A-modules are also true for modules with property (Pr), and the proofs are typically easy, because one can reduce to the potentially ONable case without too much difficulty. Indeed the trick used in the example above is typically the only idea one needs. One sometimes has to also use the following standard ingredients: Firstly, if R is any commutative ring, P is a finite projective R-module, and $\phi : P \to P$ is an R-module homomorphism, then there is an algebraically-defined $\det(1 - X\phi)$, defined either by localising and reducing to the free case, or by choosing a finite projective R-module Q such that $P \oplus Q$ is free, and defining $\det(1 - X\phi)$ to be $\det(1 - X(\phi \oplus 0))$. And secondly, if M and N both have property (Pr) and $\phi : M \to M$ and $\psi : N \to N$ are compact, then $\det(1 - X(\phi \oplus \psi)) = \det(1 - X\phi)\det(1 - X\psi)$. Finally, we leave it as an exercise for the reader to check the following generalisations of Lemma 2.7 and Lemma 2.8–Corollary 2.10.

Lemma 2.12 *If M and N are Banach A-modules with property (Pr), if $u : M \to N$ is compact and $v : N \to M$ is continuous, then uv and vu are compact, and* $\det(1 - Xuv) = \det(1 - Xvu)$.

Lemma 2.13 *If M is a Banach A-module with property (Pr), $\phi : M \to M$ is compact, and $h : A \to B$ is a continuous morphism of commutative Noetherian K-Banach algebras, then $M\widehat{\otimes}_A B$ has property (Pr) as a B-module, $\phi \otimes 1$ is compact, and* $\det(1 - X(\phi \otimes 1))$ *is the image of* $\det(1 - X\phi)$ *under h.*

3 Resultants and Riesz theory

We wish now to mildly extend the results in sections A3 and A4 of [10] to the case where A is a Noetherian K-Banach algebra and M is a Banach A-module satisfying property (Pr). Fortunately much of what Coleman proves already applies to our situation, or can easily be modified to do so. We make what are hopefully some helpful comments in case the reader wants to check the details. This section is not self-contained, and anyone wishing to check the details should read it in conjunction with §A3 and §A4 of [10].

Section A3 of [10] applies to commutative Noetherian K-Banach algebras already (apart from the comments relating to semi-simple algebras, because in general a commutative Noetherian K-Banach algebra may contain nilpotents). We give some hints for following the proofs in this section of [10]. We define the ring $A\{\{T\}\}$ to be the subring of $A[[T]]$ consisting of power series $\sum_{n\geq 0} c_n T^n$ with the property that for all $R \in \mathbf{R}_{>0}$, we have $|c_n|R^n \to 0$ as $n \to \infty$. One could put a norm on $A\{\{T\}\}$, for example $|\sum_n c_n T^n| = \mathrm{Max}_n |c_n|$, but $A\{\{T\}\}$ is not in general complete with respect to this norm. One very useful result about this ring is that if $H(T) \in A\{\{T\}\}$ and $D(T)$ is a monic polynomial of degree $d \geq 0$ then $H(T) = Q(T)D(T) + R(T)$ with $Q(T) \in A\{\{T\}\}$ and $R(T)$ a polynomial of degree less than d. Furthermore, $Q(T)$ and $R(T)$ are uniquely determined. A word on the proof: uniqueness uses the kind of trick in Lemma A3.1 of [10]. For existence one reduces to the case where all the coefficients of D have norm at most 1 and proves the result for polynomials first, and then takes a limit.

If $Q \in A[T]$ is a monic polynomial, and $P \in A\{\{T\}\}$, then Coleman defines the *resultant* $\mathrm{Res}(Q, P)$ on the top of p434 of [10]. Many of the formulae that Coleman needs are classical when P is a polynomial, and can be extended to the power series case using the following trick: straight from the definition it follows that if $u \in A^\times$ then $\mathrm{Res}(Q, P) = \mathrm{Res}(u^{-n}Q(uT), P(uT))$. This normalisation can be used to renormalise either Q or P into $A^0\langle T\rangle$, where $A^0 := \{a \in A : |a| \leq 1\}$. If S_n denotes the symmetric group acting naturally on $A^0\langle T_1, \ldots, T_n\rangle$ then the subring left invariant by the action is $A^0\langle e_1, \ldots, e_n\rangle$, where the e_i are the elementary symmetric functions of the T_i. Hence if $P, Q \in A^0\langle T\rangle$ then $\mathrm{Res}(Q, P) \in A^0$. If $Q \in A^0[T]$ is monic then one can check that the definition of a resultant makes sense for $P \in A\langle T\rangle$, and furthermore that $\mathrm{Res}(Q, -)$ is locally uniformly continuous in the second variable (in the sense that for all $M \in \mathbf{R}$, $\mathrm{Res}(Q, -)$ is a uniformly continuous function from $\{P \in A\langle T\rangle : |P| \leq M\}$ to A).

Coleman defines a function D sending a pair $B, P \in A[X]$ to an element $D(B, P) \in A[T]$. In fact if P has degree n and we define $P^*(X) = X^n P(X^{-1})$, then $D(B, P) = \mathrm{Res}(P^*(X), 1 - TB(X))$ where the resultant is computed in $A\langle T\rangle\{\{X\}\}$. One can check that $D(B(uX), P(u^{-1}X)) = D(B, P)$ if $u \in A^\times$, and that D is locally uniformly continuous in the B variable. It is also locally uniformly continuous in the P variable, because $\mathrm{Res}(X, C(X)) = 1$ if $C(0) = 1$. This is enough to check that Coleman's definition of $D(B, P)$ makes sense when B and P are in $A\{\{T\}\}$. In fact we shall only need it when B is a polynomial. Another useful formula is that $D(uB, P)(T) = D(B, P)(uT)$ for $u \in A^\times$.

In §A4 of [10] Coleman assumes his hypothesis (M), which tends not to be true for affinoids over K if K is not algebraically closed. Coleman also assumes that he is working with an ONable Banach A-module. We work in our more general situation. Hence let A denote a commutative Noetherian K-Banach algebra, let M be a Banach A-module satisfying property (Pr) and let $\phi : M \to M$ be a compact morphism, with characteristic power series $P(X) = \det(1 - X\phi)$. We define the *Fredholm resolvant* of ϕ to be $P(X)/(1 - X\phi) \in A[\phi][[X]]$. Exactly as in Proposition 10 of [15], one can prove that if $F(X) = \sum_{n\geq 0} v_m X^m$ then for all $R \in \mathbf{R}_{>0}$ the sequence $|v_m|R^m$ tends to zero, where v_m is thought of as being an element of $\mathrm{Hom}(M, M)$. Lemma A4.1 of [10] goes through unchanged, and we recall it here (Notation: if $Q(X)$ is a polynomial of degree n then $Q^*(X)$ denotes $X^n Q(X^{-1})$):

Lemma 3.1 *With A, M, ϕ and P as above, if $Q(X) \in A[X]$ is monic then Q and P generate the unit ideal in $A\{\{X\}\}$ if and only if $Q^*(\phi)$ is an invertible operator on M.* □

Before we continue, let us make some remarks on zeroes of power series. If $f = \sum_{n\geq 0} a_n T^n$ is in $A[[T]]$ and $s \in \mathbf{Z}_{\geq 0}$ then we define $\Delta^s f = \sum_{n\geq 0} \binom{n+s}{s} a_{n+s} T^n \in A[[T]]$. If $f, g \in A[[T]]$ then it is possible to check that $\Delta^s(fg) = \sum_{i=0}^{s} \Delta^i(f)\Delta^{s-i}(g)$. One also easily checks that if A is a Noetherian K-Banach algebra then Δ^s sends $A\{\{T\}\}$ to itself. We say that $a \in A$ is a *zero of order* h of $H \in A\{\{T\}\}$ if $(\Delta^s H)(a) = 0$ for $s < h$ and $(\Delta^h H)(a)$ is a unit. If $h \geq 1$ and $H = 1 + a_1 T + \ldots$ then this implies that $-1 = a(a_1 + a_2 a + \ldots)$ and hence that a is a unit. One now checks by induction on h that $H(T) = (1 - a^{-1}T)^h G(T)$, where $G \in A\{\{T\}\}$, and then that $G(a)$ is a unit.

Again let M be a Banach A-module with property (Pr) and let $\phi : M \to M$ be a compact morphism, with characteristic power series $P(T)$. Say $a \in A$ is a zero of $P(T)$ of order h.

Proposition 3.2 *There is a unique decomposition $M = N \oplus F$ into closed ϕ-stable submodules such that $1 - a\phi$ is invertible on F and $(1 - a\phi)^h = 0$ on N. The submodules N and F are defined as the kernel and the image of a projector which is in the closure in* $\mathrm{Hom}(M, M)$ *of $A[\phi]$. Moreover, N is projective of rank h, and assuming $h > 0$ then a is a unit and the characteristic power series of ϕ on N is $(1 - a^{-1}T)^h$.*

Proof We start by following Proposition 12 of [15], much of which goes through unchanged in our setting. We find that there are maps p and q in $\mathrm{Hom}(M, M)$, both in the closure of $A[\phi]$, such that $p^2 = p$, $q^2 = q$ and $p + q = 1$, and if we consider the decomposition $M = N \oplus F$ corresponding to these projections, $N = \ker(p)$, then $(1 - a\phi)^h = 0$ on N, and $(1 - a\phi)$ is invertible on F. The decomposition is visibly unique, as if $\psi = (1 - a\phi)^h$ then $N = \ker(\psi)$ and $F = \mathrm{Im}(\psi)$. We now diverge from Proposition 12 of [15].

It is clear that N satisfies (Pr), but furthermore we have $(1 - a\phi)^h = 0$ on N which implies that the identity is compact on N. An elementary argument (change the metric on N to an equivalent one if necessary and reduce to a computation of matrices) shows that if $\beta \in \mathrm{Hom}(N, N)$ has sufficiently small norm, then $|\beta^n| \to 0$ and hence $1 - \beta$ is invertible. Because 1 is compact, we can choose $\alpha : N \to N$ of finite rank such that $1 - \alpha$ is sufficiently small, and hence α is invertible and so N is finitely-generated. By Lemma 2.11, N is projective.

If $h = 0$ then $N = 0$ and $F = M$, as can be seen from Lemma 3.1, and we are home. So assume for the rest of the proof that $h > 0$. Then $P(a) = 0$ and this implies that a must be a unit. If P_N and P_F denote the characteristic power series of ϕ on N and F respectively, then $P = P_N P_F$ and by Lemma 3.1 we see that $(T - a)^h$ and P_F generate the unit ideal in $A\{\{T\}\}$. Hence $(1 - a^{-1}T)^h$ divides P_N in $A\{\{T\}\}$. Moreover, P_N is a polynomial because N is finitely-generated, and hence $(1 - a^{-1}T)^h$ divides P_N in $A[T]$. Moreover, $(1 - a^{-1}T)^h$ has constant term 1 and is hence not a zero-divisor in $A\{\{T\}\}$, hence if $P_N(T) = D(T)(1 - a^{-1}T)^h$ and $P(T) = (1 - a^{-1}T)^h G(T)$ then $D(T)$ divides $G(T)$ in $A\{\{T\}\}$ and so $D(a)$ is a unit.

We know that $D(T)$ is a polynomial. Furthermore, because $(1 - a\phi)^h = 0$ on N we see that ϕ has an inverse on N and hence that the determinant of ϕ is in $A^\times$. Hence the leading term of D is a unit. Reducing the situation modulo a maximal ideal of A we see that the reduction of P_N must be a power of the reduction of $(1 - a^{-1}T)$ and this is enough to conclude that $D = 1$. Hence the characteristic power series of ϕ on N is $(1 - a^{-1}T)^h$. Finally, the fact that ϕ is invertible on N means that the rank of N at any maximal ideal must equal the degree of P_N modulo this ideal, and hence the rank is h everywhere. □

Keep the notation: M has (Pr) and $\phi : M \to M$ is compact, with characteristic power series $P(T)$.

Theorem 3.3 *Suppose $P(T) = Q(T)S(T)$, where $S = 1 + \ldots \in A\{\{T\}\}$ and $Q = 1 + \ldots$ is a polynomial of degree n whose leading coefficient is a unit, and which is relatively prime to S. Then there is a unique direct sum decomposition*

$M = N \oplus F$ of M into closed ϕ-invariant submodules such that $Q^*(\phi)$ is zero on N and invertible on F. The projectors $M \to N$ and $M \to F$ are elements of $\mathrm{Hom}(M, M)$ *which are in the closure of $A[\phi]$. Furthermore, N is projective of rank n and the characteristic power series of ϕ on N is $Q(T)$.*

Proof We follow Theorem A4.3 of [10]. The operator $v = 1 - Q^*(\phi)/Q^*(0)$ has a characteristic power series which has a zero at $T = 1$ of order n. Applying the previous proposition to v, we see $M = N \oplus F$, where N and F are defined as the kernel and the image of a projector in the closure of $A[v]$ and hence in the closure of $A[\phi]$. Hence both N and F are ϕ-stable. Unfortunately, by the end of the proof of Theorem A4.3 of [10] one can only deduce that $Q^*(\phi)^n$ is zero on N and invertible on F, so we are not quite home yet. However, by Proposition 3.2, N is projective of rank n. Moreover, the characteristic power series of ϕ on F is coprime to Q, by Lemma 3.1. Hence if $G(T)$ is the characteristic power series of ϕ on N, we see that Q divides G. But G and Q have degree n and the same constant term, and furthermore the leading coefficient of Q is a unit. This is enough to prove that $G = Q$. □

4 An admissible covering

The key aim in this section is to generalise some of the results of section A5 of [10] (especially Proposition A5.8) to the case where the base is an arbitrary reduced affinoid. In fact almost all of Coleman's results go through unchanged, but there are some differences, which we summarise here. Firstly it is not true in general that the image of an affinoid under a quasi-finite map is still affinoid. However if one works with finite unions of affinoids then one can deal with the problems that this causes. Secondly Coleman uses the notion of a strict neighbourhood of a subspace of the unit disc. We slightly modify this notion to one which suits our purpose. Lastly we need some kind of criterion for when a quasi-finite map of rigid spaces of constant degree is finite. We use a theorem of Conrad whose proof invokes Raynaud's theory of admissible formal models of rigid spaces.

We set up some notation. Let K be a field with a complete non-trivial non-Archimedean valuation. Let R denote a reduced K-affinoid algebra, and let $B = \mathrm{Max}(R)$ be the associated affinoid variety. We equip R with its supremum semi-norm, which is a norm in this case. Let $R\{\{T\}\}$ denote the ring of power series $\sum_{n \geq 0} a_n T^n$ with $a_n \in R$ such that for all real $r > 0$ we have $|a_n| r^n \to 0$ as $n \to \infty$. Then $R\{\{T\}\}$ is just the ring of functions on $B \times_K \mathbf{A}^{1,\mathrm{an}}$, where here $\mathbf{A}^{1,\mathrm{an}}$ denotes the analytification of affine 1-space over K.

Let $P(T) = \sum_{n \geq 0} r_n T^n \in R\{\{T\}\}$ be a function with $r_0 = 1$. Our main object of study is the rigid space cut out by $P(T)$, that is, the space $Z \subseteq B \times_K \mathbf{A}^{1,\mathrm{an}}$ defined by the zero locus of $P(T)$. In practice, $P(T)$ will be the characteristic power series of a compact endomorphism of a Banach R-module.

Certainly Z is a rigid analytic variety, equipped with projection maps $f : Z \to B$ and $g : Z \to \mathbf{A}^{1,\mathrm{an}}$. We frequently make use of the following cover of Z: If $r \in \sqrt{|K^\times|}$ (that is, some power of r is the norm of a non-zero element of K) then let $B[0, r]$ denote the closed affinoid disc over K of radius r, considered as an admissible open subspace of $\mathbf{A}^{1,\mathrm{an}}$. Let Z_r denote the zero locus of $P(T)$ on the space $B \times_K B[0, r]$. Then Z_r is an affinoid, and the Z_r admissibly cover Z. Let $f_r : Z_r \to B$ denote the canonical projection. Note that any affinoid subdomain of Z will be admissibly covered by its intersections with the Z_r, which are affinoids, and hence will be contained within some Z_r.

Now let $\mathcal{C}$ denote the set of affinoid subdomains Y of Z with the following property: there is an affinoid subdomain X of B (depending on Y) with the property that $Y \subseteq Z_X := f^{-1}(X)$, the induced map $f : Y \to X$ is finite and surjective, and Y is disconnected from its complement[2] in Z_X, that is, there is a function $e \in \mathcal{O}(Z_X)$ such that $e^2 = e$ and Y is the locus of Z_X defined by $e = 1$.

Our goal is (c.f. Proposition A5.8 of [10])

Theorem *$\mathcal{C}$ is an admissible cover of Z.*

The reason we want this result is that in later applications Z will be a "spectral variety", and the $Y \in \mathcal{C}$ are exactly the affinoid subdomains of Z over which one can construct a Hecke algebra, and hence an eigenvariety, without any technical difficulties.

We prove the theorem after establishing some preliminary results.

Lemma 4.1 *$f_r : Z_r \to B$ is quasi-finite and flat.*

Proof By increasing r if necessary, we may assume $r \in |K^\times|$ and hence, by rescaling, that $r = 1$. The situation we are now in is as follows: R is an affinoid and $P(T) = \sum_{n \geq 0} r_n T^n \in R\langle T\rangle$ with $r_0 = 1$, and we must show that $R \to R\langle T\rangle/(P(T))$ is quasi-finite and flat. Quasi-finiteness is immediate from the Weierstrass preparation theorem. For flatness observe that $R\langle T\rangle$ is

2 Elmar Grosse-Kloenne has pointed out that this condition in fact follows from the others; one can use 9.6.3/3 and 9.5.3/5 of [1] to check that $Y \to Z_X$ is both an open and a closed immersion.

flat[3] over R and that if P is any maximal ideal of $R\langle T\rangle$ then $\mathcal{P} := P \cap R$ is a maximal ideal of R. Hence $R\langle T\rangle/(\mathcal{P}R\langle T\rangle) = (R/\mathcal{P}R)\langle T\rangle$ is an integral domain and the image of $P(T)$ in $(R/\mathcal{P}R)\langle T\rangle$ is non-zero, as its constant term is non-zero. Hence the image of $P(T)$ is not a zero-divisor and flatness now follows from Theorem 22.6 of [14]. □

Corollary 4.2 *If $Y \subseteq Z$ is an affinoid then $f : Y \to B$ is quasi-finite and flat.*

Proof Y is affinoid and hence $Y \subseteq Z_r$ for some $r \in \sqrt{|K^\times|}$, so the result follows from the previous lemma. □

Corollary 4.3 *If $Y \subseteq Z$ is an affinoid, and $X \subseteq B$ is an admissible open such that $Y \subseteq f^{-1}(X)$, and if there is an integer $d \geq 0$ such that all fibres of the induced map $f : Y \to X$ have degree d, then $f : Y \to X$ is finite and flat.*

Proof $f : Y \to X$ is flat by Corollary 4.2. It is also quasi-compact and separated, so finiteness follows from Theorem A.1.2 of [11]. □

Note that this latter result uses the full force of Raynaud's theory of formal models.

Lemma 4.4 *If $r \in \sqrt{|K^\times|}$ and $f_r : Z_r \to B$, then for $i \geq 0$ define $U_i := \{x \in B : \deg(f_r^{-1}(x)) \geq i\}$. Then each U_i is a finite union of affinoid subdomains of B, and U_i is empty for i sufficiently large.*

Proof The sequence $|r_n|r^n$ tends to zero as $n \to \infty$, and hence for any $x \in B$, the set $\{|r_n(x)|r^n : n \geq 0\}$ has a maximum, denoted M_x, which is attained. Note that $|r_0(x)| = 1$ and hence $M_x \geq 1$, and in particular if N is an integer such that $|r_n|r^n < 1$ for all $n \geq N$ then $M_x = \mathrm{Max}\{|r_n(x)|r^n : 0 \leq n < N\}$ and $M_x > |r_n(x)|r^n$ for all $n \geq N$. For $i \geq 0$ let S_i denote the affinoid subdomain of B defined by $\{x \in B : |r_i(x)|r^i = M_x\}$. Then S_i is empty for $i \geq N$. A calculation on the Newton polygon shows that

$$U_i = \cup_{j \geq i} S_j$$

and the result follows. □

Definition *If S and T are admissible open subsets of the affinoid B, such that both S and T are finite unions of affinoid subdomains of B, then we say that T*

3 One can prove flatness by using the Open Mapping Theorem and mimicking the proof of the result stated in Exercise 7.4 of [14], noting that the solution to the exercise is on p289 of loc. cit.

is a strict neighbourhood *of S (in B) if $S \subseteq T$ and there is an admissible open subset U of B with the following properties:*

- *U is a finite union of affinoid subdomains of B,*
- *$U \cap S$ is empty,*
- *$U \cup T = B$.*

The intersection of two affinoid subdomains of B is an affinoid subdomain of B. Hence if U and V are admissible open subsets of B which are both the union of finitely many affinoid subdomains, then so is $U \cap V$. As a consequence, we see that if T_α is a strict neighbourhood of S_α for $1 \leq \alpha \leq n$ then $\cup_\alpha T_\alpha$ is a strict neighbourhood of $\cup_\alpha S_\alpha$. We now prove the key technical lemma that we need.

Lemma 4.5 *Suppose $r \in \sqrt{|K^\times|}$ and $V \subseteq B$ is an affinoid subdomain with the property that $f_r : f_r^{-1}(V) \to V$ is finite of constant degree $d > 0$. Then there is an affinoid subdomain X of B which is a strict neighbourhood of V in B, and $s \in \sqrt{|K^\times|}$ with $s > r$ such that the affinoid $Y = f_s^{-1}(X)$ contains $f_r^{-1}(V)$, lies in $\mathcal{C}$, and is finite flat of degree d over X.*

Proof (c.f. Lemma A5.9 of [10]). If $x \in V$ then let $P_x(T) = \sum_{n \geq 0} r_n(x) T^n$ denote the specialisation of $P(T)$ to $K(x)\{\{T\}\}$. The statement that the degree of $f_r^{-1}(x)$ is d translates by the theory of the Newton polygon to the statement that for all $x \in V$ we have $|r_d(x)| \neq 0$ (so r_d is a unit in $\mathcal{O}^{\mathrm{an}}(V)$) and furthermore that for all integers $n \geq 0$ we have $-\log(|r_n(x)|) \geq (n-d)\log(r) - \log(|r_d(x)|)$, with strict inequality when $n > d$. Here log is the usual logarithm, with the usual convention that $-\log(0) = +\infty$. Because $P(T)$ is entire, there exists an integer $N > d$ such that for $n \geq N$ we have $-\log|r_n(x)| > n\log(r+1)$ for all $x \in B$ and hence $-\log(|r_n(x)|) \geq (n-d)\log(r+1) - \log(|r_d(x)|)$ for all $n \geq N$. For $d < n < N$ we have $-\log(|r_n(x)|) > (n-d)\log(r) - \log(|r_d(x)|)$ and hence $|r_n(x)/r_d(x)| < r^{d-n}$ for all $x \in V$. Because functions on affinoids attain their bounds, there is some $t \in \sqrt{|K^\times|}$ with $r < t < r+1$ and $|r_n(x)/r_d(x)| < t^{d-n}$ for $d < n < N$, and hence for all $x \in V$ we have $-\log(|r_n(x)|) \geq (n-d)\log(t) - \log(|r_d(x)|)$ for all $n \geq 0$, with equality iff $n = d$. Now choose $\gamma_1, \gamma_2, \delta_1, \delta_2 \in \log\left(\sqrt{|K^\times|}\right)$ such that $\delta_2 > -\log|r_d(x)| > \delta_1$ for all $x \in V$ and $\log(r) < \gamma_1 < \gamma_2 < \log(t)$. Let X be the affinoid subdomain of B defined by the N equations

$$\delta_1 \leq -\log|r_d(x)| \leq \delta_2,$$

$$-\log|r_n(x)| + \log|r_d(x)| \geq (n-d)\gamma_1 \text{ for } 0 < n < d,$$

and

$$-\log|r_n(x)| + \log|r_d(x)| \geq (n-d)t \text{ for } d < n < N.$$

These equations define X as a Laurent subdomain of B, and if $x \in V$ then not only are these equations satisfied, but strict inequality holds in every case. Hence $V \subset X$ and moreover if we consider the N affinoids, each defined by one of the N equations

$$\delta_1 \geq -\log|r_d(x)|,$$

$$-\log|r_d(x)| \geq \delta_2,$$

$$-\log|r_n(x)| + \log|r_d(x)| \leq (n-d)\gamma_1,\ 0 < n < d,$$

$$-\log|r_n(x)| + \log|r_d(x)| \leq (n-d)t,\ d < n < N,$$

and let W be the union of these N affinoids, then $X \cap W$ is empty and $X \cup V = B$. Hence X is a strict neighbourhood of V in B, in the sense we defined above. Let $s = \exp(\gamma_2)$ and set $Y = f_s^{-1}(X)$. Then Y is an affinoid in Z and by the previous corollary and the way we have arranged the Newton polygons, $f : Y \to X$ is finite of degree d (note that by our choice of t we have

$$-\log|r_n(x)| + \log|r_d(x)| \geq (n-d)t \text{ for all } n \geq N,$$

with strict inequality for $x \in V$). Furthermore if $x \in X$ then no slope of the Newton polygon of $P_x(T)$ can equal s, and hence the projection from $f^{-1}(X)$ to $\mathbf{A}^{1,\mathrm{an}}$ contains no elements of norm s. Hence Y is disconnected from its complement in $f^{-1}(X)$, and in particular is an affinoid subdomain of Z, so $Y \in \mathcal{C}$. □

We are now ready to prove the theorem.

Theorem 4.6 *$\mathcal{C}$ is an admissible cover of Z.*

Proof Again we follow Coleman. We know that Z is admissibly covered by the Z_r, $r \in \sqrt{|K^\times|}$, and hence it suffices to prove that for every Z_r, there is a finite collection of affinoids in $\mathcal{C}$ whose union contains Z_r. Recall that for $i \geq 0$, U_i is the subset of points in B such that $\deg(f_r^{-1}(a)) \geq i$, and that U_i is a finite union of affinoids. Furthermore, clearly $U_{i+1} \subseteq U_i$. If U_1 is empty there is nothing to prove, so let us assume that it is not. Let d denote the largest i such that U_i is non-empty. For $1 \leq i \leq d$ let $H(i)$ denote the following statement:

$H(i)$: "There is a finite set $Y_1, Y_2 \dots Y_{n(i)}$ of affinoid subdomains of Z, and a finite set $X_1, X_2 \dots X_{n(i)}$ of affinoid subdomains of B, such that for $1 \leq \alpha \leq n(i)$ we have $f : Y_\alpha \to X_\alpha$ is finite flat and surjective, $Y_\alpha \in \mathcal{C}$, $f_r^{-1}(U_i) \subseteq \cup_{\alpha=1}^{n(i)} Y_\alpha$, and $\cup_{\alpha=1}^{n(i)} X_\alpha$ is a strict neighbourhood of U_i in B."

If we can establish $H(1)$ then we are home because $f_r^{-1}(U_1) = Z_r$. We firstly establish $H(d)$, and then show that $H(i)$ implies $H(i-1)$ for $i \geq 2$, and this will be enough. For $H(d)$ we cover U_d by finitely many affinoid subdomains $V_1, V_2, \dots V_{n(1)}$ of B (in fact it is not difficult to show that U_d is itself an affinoid, but we shall not need this). By Corollary 4.3 we know that $f_r^{-1}(V_\alpha) \to V_\alpha$ is finite and flat. Now applying Lemma 4.5 to V_α we get a strict affinoid neighbourhood X_α of V_α, and if $Y_\alpha = f_s^{-1}(X_\alpha)$ (s as in the Lemma) then $H(d)$ follows immediately.

Now let us assume $H(i)$, $i \geq 2$. Then choose a finite union of affinoid subdomains $W \subset B$ such that $W \cap U_i$ is empty and $W \cup \bigcup_{\alpha=1}^{n(i)} X_\alpha = B$. Then $W \cap U_{i-1}$ is a finite union $V_{n(i)+1}, V_{n(i)+2} \dots V_{n(i)+m}$ of affinoid subdomains. Set $n(i-1) = n(i) + m$. Note that $f_r : f_r^{-1}(V_\alpha) \to V_\alpha$ is finite of degree $i-1$ for $n(i) < \alpha \leq n(i-1)$, hence one is in a position to apply Lemma 4.5 to get $Y_\alpha \to X_\alpha$ finite flat of degree $i-1$, $Y_\alpha \in \mathcal{C}$, $f_r^{-1}(V_\alpha) \subseteq Y_\alpha$, and X_α a strict neighbourhood of V_α, for $n(i) < \alpha \leq n(i-1)$. We now show that $\bigcup_{\alpha=1}^{n(i-1)} X_\alpha$ is a strict neighbourhood of U_{i-1}. We know that $\bigcup_{\alpha=n(i)+1}^{n(i-1)} X_\alpha$ is a strict neighbourhood of $\bigcup_{\alpha=n(i)+1}^{n(i-1)} V_\alpha$, so choose a finite union of affinoid subdomains W' such that $W' \cap (\bigcup_{\alpha=n(i)+1}^{n(i-1)} V_\alpha)$ is empty and $W' \cup (\bigcup_{\alpha=n(i)+1}^{n(i-1)} X_\alpha) = B$. Now set $W'' = W \cap W'$. Then W'' is a finite union of affinoids, $W'' \cap U_{i-1} = W' \cap W \cap U_{i-1} = W' \cap (\bigcup_{\alpha=n(i)+1}^{n(i-1)} V_\alpha)$ is empty, and $W'' \cup (\bigcup_{\alpha=1}^{n(i-1)} X_\alpha) = B$, and we are done. □

5 Spectral varieties and eigenvarieties

Let R be a reduced affinoid K-algebra equipped with its supremum norm, let M be a Banach R-module satisfying (Pr), and let $\mathbf{T}$ be a commutative R-algebra equipped with an R-algebra homomorphism to $\mathrm{End}_R(M)$, the continuous R-module endomorphisms of M. In practice $\mathbf{T}$ will be a polynomial R-algebra generated by (typically infinitely many) Hecke operators. We frequently identify $t \in \mathbf{T}$ with the endomorphism of M associated to it. Fix once and for all an element $\phi \in \mathbf{T}$, and assume that the induced endomorphism $\phi : M \to M$ is compact. Let $F(T) = 1 + \sum_{n \geq 1} c_n T^n$ be the characteristic power series of ϕ. We define the *spectral variety* Z_ϕ associated to ϕ to be the closed subspace of the rigid space $\mathrm{Max}(R) \times \mathbf{A}^1$ cut out by F. The spectral variety is a geometric object parametrising, in some sense, the reciprocals of the non-zero eigenvalues of ϕ. Its formulation is compatible with base change, by Lemma 2.13. Our main goal in this section is to write down a finite cover of this spectral variety, the *eigenvariety* associated to the data $(R, M, \mathbf{T}, \phi)$. Points on the eigenvariety will correspond to systems of eigenvalues for all the operators in $\mathbf{T}$, such that the eigenvalue for ϕ is non-zero. The construction is just an axiomatisation of Chapter 7 of [9] and is really not deep (in fact by far the deepest part of the entire construction is the fact that the cover $\mathcal{C}$ of Section 4 is admissible, as this appealed to the theory of formal models at one point). Unfortunately the construction does involve a lot of bookkeeping.

We begin with a finite-dimensional example, where ϕ is invertible and hence where we may avoid the technicalities of §4. Let R be a reduced affinoid K-algebra, and let M denote a finitely-generated projective R-module of rank d. Let $\mathbf{T}$ be an arbitrary R-algebra equipped with an R-algebra homomorphism to $\mathrm{End}_R(M)$, and let ϕ be an element of $\mathbf{T}$. Assume furthermore that $\phi : M \to M$ has an inverse, that is, there is an R-linear $\phi^{-1} : M \to M$ such that $\phi \circ \phi^{-1} = \phi^{-1} \circ \phi$ is the identity on M. Define $P(T) = \det(1 - T\phi) = 1 + \ldots \in R[T]$; then the leading term of P, that is, the coefficient of T^d, is a unit. Let Z_ϕ denote the zero locus of $P(T)$ regarded as a function on $\mathrm{Max}(R) \times \mathbf{A}^1$. Then $R[T]/(P(T))$ is a finite R-algebra and hence an affinoid algebra, and Z_ϕ is the affinoid rigid space associated to this affinoid algebra. Let $\mathbf{T}(Z_\phi)$ denote the image of $\mathbf{T}$ in $\mathrm{End}_R(M)$; then $\mathbf{T}(Z_\phi)$ is a finite R-algebra and hence an affinoid algebra. By the Cayley-Hamilton theorem we have $\phi^{-1} \in \mathbf{T}(Z_\phi)$, and furthermore there is a natural map $R[T]/P(T) \to \mathbf{T}(Z_\phi)$ sending T to ϕ^{-1}. Set $D_\phi = \mathrm{Max}(\mathbf{T}(Z_\phi))$. Then the maps $R \to R[T]/P(T) \to \mathbf{T}(Z_\phi)$ of affinoids give maps $D_\phi \to Z_\phi \to \mathrm{Max}(R)$. We call Z_ϕ the *spectral variety* and D_ϕ the *eigenvariety* associated to this data. As a concrete example, consider the case where $R = K\langle X, Y\rangle$, $M = R^2$, $\mathbf{T} = R[\phi, t]$, where ϕ acts on M as the matrix $\left(\begin{smallmatrix} 1 & X \\ 0 & 1 \end{smallmatrix}\right)$, and t acts as $\left(\begin{smallmatrix} 0 & Y \\ 0 & 0 \end{smallmatrix}\right)$. Then $\det(1 - T\phi) = (1 - T)^2$ so Z_ϕ is non-reduced, and $\mathbf{T}(Z_\phi)$ is the ring $R \oplus I\epsilon$, with $I = (X, Y)$ and $\epsilon^2 = 0$. Note that in this case the maps $D_\phi \to Z_\phi$ and $D_\phi \to \mathrm{Max}(R)$ are not flat, and D_ϕ is not reduced either.

It would not be unreasonable to say that what follows in this section is just a natural generalisation of this set-up, the main complication being that M is not necessarily finitely-generated, the purpose of the admissible cover $\mathcal{C}$ being to remedy this. See also Chenevier's thesis, where he develops essentially the same theory in essentially the same way (assuming Theorem 4.6 of this paper). The example above shows that in this generality one cannot expect D_ϕ and Z_ϕ to have too many "good" geometric properties; however one can hope that the examples of spectral and eigenvarieties arising "in nature" are better behaved.

Let us now go back to our more general situation, where M is a Banach R-module satisfying property (Pr) and $\mathbf{T}$ is a commutative R-algebra equipped with an R-algebra map $\mathbf{T} \to \mathrm{End}_R(M)$, such that the endomorphism of M induced by $\phi \in \mathbf{T}$ is compact. Let Z_ϕ be the closed subspace of $\mathrm{Max}(R) \times \mathbf{A}^1$ defined by the zero locus of the characteristic power series of ϕ. Let $\mathcal{C}$ be the admissible cover of Z_ϕ constructed in section 4. Let Y be an element of this admissible cover, with image $X \subseteq \mathrm{Max}(R)$. By definition, X is an affinoid subdomain of $\mathrm{Max}(R)$, so set $A = \mathcal{O}(X)$. Then A is reduced by Corollary 7.3.2/10 of [1]. Define $M_A := M\widehat{\otimes}_R A$ and, for $t \in \mathbf{T}$, let t_A denote the A-linear continuous endomorphism of M_A induced by $t : M \to M$. Note that $\phi_A : M_A \to M_A$ is still compact, by Lemma 2.13. Let $F_A(T)$ be the characteristic power series of ϕ_A on M_A. Again by Lemma 2.13, F_A is just the image of F in $A\{\{T\}\}$.

Let us assume first that X is connected. Then we wish to associate to Y a factor of $F_A(T)$ so that we are in a position to apply Theorem 3.3. We do this as follows. We know that $\mathcal{O}(Y)$ is a finite flat A-module, and hence it is projective of some rank d. The element T of $\mathcal{O}(Y)$ is a root of its characteristic polynomial Q', which is monic of degree d, and hence gives us a map $A[T]/(Q'(T)) \to \mathcal{O}(Y)$. In fact, Y is a closed subspace of $X \times B[0,r]$ for some r, and hence if S is some appropriate K-multiple of T then the natural map $A\langle S\rangle \to \mathcal{O}(Y)$ is surjective. By Proposition 3.7.4/1 of [1] and its proof, any residue norm on $\mathcal{O}(Y)$ will be equivalent to any of the Banach norms that $\mathcal{O}(Y)$ inherits from being a finite complete A-module. One can deduce from this that the map $A[T]/(Q'(T)) \to \mathcal{O}(Y)$ is surjective. Hence $A[T]/(Q'(T)) \to \mathcal{O}(Y)$ is an isomorphism, because both sides are locally free A-modules of rank d. This means that the image of $F_A(T)$ in $A\{\{T\}\}/(Q'(T))$ is zero, and hence that Q' divides F_A in $A\{\{T\}\}$. Comparing constant terms, we see that $Q' = a_0 + a_1 T + \ldots$ with a_0 a unit, and hence we can define $Q = a_0^{-1}Q'$ and we are in a position to invoke Theorem 3.3 to give a decomposition $M_A = N \oplus F$ where N is projective of rank d over A. Note that in general N will not be free. Because the projector $M_A \to N$ is in the closure of $A[\phi]$, it commutes with all the endomorphisms of M_A induced by elements of $\mathbf{T}$, and hence N is t-invariant for all $t \in \mathbf{T}$. Define $\mathbf{T}(Y)$ to be the A-sub-algebra of $\mathrm{End}_A(N)$ generated by all the elements of $\mathbf{T}$. Now $\mathrm{End}_A(N)$ is a finite A-module, and hence $\mathbf{T}(Y)$ is a finite A-algebra and hence an affinoid. Let $D(Y)$ denote the associated affinoid variety. We know that $Q^*(\phi)$ is zero on N, and hence $\mathbf{T}(Y)$ is naturally a finite $A[S]/(Q^*(S))$-algebra, via the map sending S to ϕ. Because the constant term of Q^* is a unit, there is a canonical isomorphism $A[S]/(Q^*(S)) = \mathcal{O}(Y)$ sending S to T^{-1}.

Hence $\mathbf{T}(Y)$ is a finite $\mathcal{O}(Y)$-algebra, and thus there is a natural finite map $D(Y) \to Y$.

For general $Y \in \mathcal{C}$, the image of Y in $\mathrm{Max}(R)$ may not be connected, but Y can be written as a finite disjoint union $Y = \cup Y_i$ corresponding to the connected components of the image of Y in $\mathrm{Max}(R)$. We define $D(Y)$ as the disjoint union of the $D(Y_i)$. This construction gives us, for each $Y \in \mathcal{C}$, a finite cover $D(Y)$ of Y. We wish to glue together the $D(Y)$, as Y ranges through all elements of $\mathcal{C}$, and the resulting curve D, which will be a finite cover of Z_ϕ, will be the *eigenvariety* associated to the data $(R, M, \mathbf{T}, \phi)$. We firstly establish a few lemmas.

Lemma 5.1 *If $Y \in \mathcal{C}$ with image $X \subseteq \mathrm{Max}(R)$, and X' is an affinoid subdomain of X then Y', the pre-image of X' under the map $Y \to X$, is in $\mathcal{C}$, and is an affinoid subdomain of Y. Furthermore, $D(Y')$ is canonically isomorphic to the pre-image of Y' under the map $D(Y) \to Y$.*

Proof Y' is the pre-image of X' under the map $Y \to X$ and is hence an affinoid subdomain of Y by Proposition 7.2.2/4 of [1]. The map $Y' \to X'$ is finite and surjective, and if e is the idempotent in $\mathcal{O}(Z_X)$ showing that Y is disconnected from its complement (that is, $e|_Y = 1$ and $e|_{Z_X \backslash Y} = 0$), then the restriction of e to $\mathcal{O}(Z_{X'})$ will do the same for Y'. Hence $Y' \in \mathcal{C}$. It is now elementary to check that $\mathbf{T}(Y') = \mathbf{T}(Y)\widehat{\otimes}_{\mathcal{O}(X)}\mathcal{O}(X')$ and hence $D(Y')$ is the pre-image of Y' under the map $D(Y) \to Y$. □

Lemma 5.2 *If $Y_1, Y_2 \in \mathcal{C}$ then $Y := Y_1 \cap Y_2 \in \mathcal{C}$. Furthermore for $1 \leq i \leq 2$, Y is an affinoid subdomain of Y_i, and $D(Y)$ is canonically isomorphic to the pre-image of Y under the map $D(Y_i) \to Y_i$.*

Proof Let X_i denote the image of Y_i in $\mathrm{Max}(R)$. Then the X_i are affinoid subdomains of $\mathrm{Max}(R)$, and hence so is their intersection. Let X denote a component of $X_1 \cap X_2$. It suffices to prove the assertions of the lemma with Y replaced by $Y \cap Z_X$, so let us re-define Y to be $Y \cap Z_X$.

Let Y_i' denote the pre-image of X under the map $Y_i \to X_i$. Then Y_i' is an affinoid subdomain of Y_i containing Y and by Lemma 5.1 we have $Y_i' \in \mathcal{C}$ with $D(Y_i')$ the pre-image of Y_i' under the map $D(Y_i) \to Y_i$. Now $Y = Y_1' \cap Y_2'$ is finite and flat over Y_1' and hence finite and flat over X. Let $e_i \in \mathcal{O}(Z_X)$ be the idempotent associated to Y_i', and set $e = e_1 e_2$. Then Y is the subset of Z_X defined by $e = 1$, and hence Y is locally free of finite rank over X. One easily checks that Y is a union of components of Y_i' for $1 \leq i \leq 2$, in fact. If Y is empty then the rest of the lemma is clear. If not then the map

$Y \to X$ is surjective, and $Y \in \mathcal{C}$. Finally, for $1 \leq i \leq 2$, the idempotents e and $e_i(1 - e_{3-i})$ sum to 1 on Y_i' showing that $D(Y)$ is actually a union of connected components of $D(Y_i')$, pulling back the inclusion $Y \subseteq Y_i'$. □

Now by Proposition 9.3.2/1 of [1] we can glue the $D(Y)$ for $Y \in \mathcal{C}$ to get a rigid space D_ϕ (the cocycle conditions are satisfied because they are satisfied for the cover $\mathcal{C}$ of Z_ϕ), and by Proposition 9.3.3/1 of [1] we can glue the maps $D(Y) \to Y$ to get a map $D_\phi \to Z_\phi$. We say that the rigid space D_ϕ is the *eigenvariety* associated to the data $(R, M, \mathbf{T}, \phi)$. We have already seen in the finite-dimensional case that the map $D_\phi \to Z_\phi$ might not be flat, and that Z_ϕ and D_ϕ may be non-reduced. We summarise the obvious positive results about Z_ϕ and D_ϕ that come out of their construction:

Lemma 5.3 *D_ϕ and Z_ϕ are separated, and the map $D_\phi \to Z_\phi$ is finite.*

Proof Z_ϕ is separated by, for example, Proposition 9.6/7 of [1] (applied to the admissible covering $\{Z_r\}$ of Z_ϕ defined in the previous section). The construction of D_ϕ over Z_ϕ shows that $D_\phi \to Z_\phi$ is finite; hence $D_\phi \to Z_\phi$ is separated, which implies that D_ϕ is separated. □

We need to establish further functorial properties of this construction. As before, let R be a reduced K-affinoid algebra equipped with its supremum norm, let M be a Banach R-module satisfying (Pr), and let $\mathbf{T}$ be a commutative R-algebra equipped with a distinguished element ϕ and an R-algebra homomorphism $\mathbf{T} \to \mathrm{Hom}_R(M, M)$, such that the image of ϕ is a compact endomorphism. We now consider what happens when we change R. More specifically, let R' denote another reduced K-affinoid algebra equipped with a map $R \to R'$, and let $M', \mathbf{T}', \phi'$ denote the obvious base extensions. The constructions above give us maps $D_\phi \to Z_\phi \to \mathrm{Max}(R)$ and $D_{\phi'} \to Z_{\phi'} \to \mathrm{Max}(R')$. The map $R \to R'$ gives us a map $\mathrm{Max}(R') \to \mathrm{Max}(R)$.

Lemma 5.4 *$Z_{\phi'} \to \mathrm{Max}(R')$ is canonically isomorphic to the pullback of $Z_\phi \to \mathrm{Max}(R)$ to $\mathrm{Max}(R')$.*

Proof This is an immediate consequence of Lemma 2.13. □

In particular there is a natural map $Z_{\phi'} \to Z_\phi$.

Lemma 5.5 *If $R \to R'$ is flat then $D_{\phi'} \to Z_{\phi'}$ is canonically isomorphic to the pullback of $D_\phi \to Z_\phi$ under the map $Z_{\phi'} \to Z_\phi$.*

Proof Let $\mathcal{C}$ and $\mathcal{C}'$ denote usual the admissible covers of Z_ϕ and $Z_{\phi'}$. If $Y \in \mathcal{C}$ and Y' is the pullback of Y to $Z_{\phi'}$ then one checks without too much difficulty that $Y' \in \mathcal{C}'$. It is not immediately clear whether every element of $\mathcal{C}'$ arises in this way (although this is the case if $\mathrm{Max}(R') \subseteq \mathrm{Max}(R)$ is an affinoid subdomain, which will be the only case we are interested in in practice). However, this does not matter because the elements of $\mathcal{C}'$ which arise in this way still form an admissible covering of $Z_{\phi'}$, as they cover the separated space $Z_{\phi'}$, and all the elements of $\mathcal{C}'$ are affinoids so one can use Proposition 9.1.4/2 of [1]. Hence we may construct $D_{\phi'}$ by gluing the $D(Y')$ for all Y' which arise in this way. One checks that $D(Y')$ is the pullback of $D(Y)$ under the map $Y' \to Y$ (this is where we use flatness, the point being that without flatness one cannot deduce that the natural map from $D(Y')$ to the pullback of $D(Y)$ is an isomorphism) and that everything is compatible with gluing, and after this somewhat tedious procedure one deduces that the maps $D(Y') \to D(Y)$ identify $D_{\phi'}$ with the pullback of D_ϕ as indicated. □

We will only be applying the above lemma in the case where $\mathrm{Max}(R')$ is an affinoid subdomain of $\mathrm{Max}(R)$ and in particular $R \to R'$ is flat in this case. I am grateful to the referee for pointing out that flatness is necessary for this lemma to be true. Indeed, one checks that in the example at the beginning of this section (with $R = K\langle X, Y\rangle$), if $R' = K$ and the map $R \to R'$ sends X and Y to zero, then the pullback of the eigenvariety is not isomorphic to the eigenvariety associated to the pullback. On the other hand, one can use the q-expansion principle to check that the construction of the cuspidal Coleman-Mazur eigencurve (see part II of this paper) will commute with any base change of reduced affinoids. Had we set up the theory for non-reduced bases one could no doubt even check that the construction commutes with arbitrary base change. The same arguments do not work for the eigenvarieties associated to totally definite quaternion algebras over totally real fields (see part III of this paper) and for Chenevier's unitary group eigenvarieties [8]. One does not have a q-expansion principle in these cases, and whether construction of these eigenvarieties commutes with all base changes (even those coming from the inclusion of a point into weight space) seems to be an open question, related to multiplicity one issues for overconvergent automorphic eigenforms in these settings. See Lemma 5.9 for a partial result. Another way of setting up the foundations of the theory of eigenvarieties might be to construct the eigenvariety as a limit of spectral varieties such as those in sublemma 6.2.3 of [9]; such a construction might well commute with all base changes, but would not see subtleties such as non-semisimplicity of eigenspaces at a fixed weight.

We now analyse how eigenvarieties change under a specific type of change of module. As always, let R be a reduced affinoid, let M and M' denote Banach modules satisfying property (Pr), let $\mathbf{T}$ be a commutative R-algebra equipped with maps $\mathbf{T} \to \mathrm{End}_R(M)$ and $\mathbf{T} \to \mathrm{End}_R(M')$, such that our chosen element $\phi \in \mathbf{T}$ acts compactly on both M and M'.

In practice we are interested only in modules M and M' which are related in a specific way, which we now axiomatise. We say that a continuous R-module and $\mathbf{T}$-module homomorphism $\alpha : M' \to M$ is a "primitive link" if there is a compact R-linear and $\mathbf{T}$-linear map $c : M \to M'$ such that $\phi : M \to M$ is $\alpha \circ c$ and $\phi : M' \to M'$ is $c \circ \alpha$. Note that these assumptions force the characteristic power series of ϕ on M and M' to coincide, by Lemma 2.12. Note also that the identity map $M \to M$ is a primitive link (take $c = \phi$). We say that a continuous R-module and $\mathbf{T}$-module homomorphism $\alpha : M' \to M$ is a "link" if one can find a sequence $M' = M_0, M_1, M_2, \ldots, M_n = M$ of Banach R-modules satisfying property (Pr) with $\mathbf{T}$-actions, and continuous R-module and $\mathbf{T}$-module maps $\alpha_i : M_i \to M_{i+1}$ such that each α_i is a primitive link, and α is the compositum of the α_i. We apologise for this terrible notation but the underlying notion is what occurs in applications; our motivation is the study of r-overconvergent modular forms as r changes. More precisely, with notation as in Part II of this manuscript, if $0 < r \leq r' < p/(p+1)$ and α is the inclusion from r'-overconvergent forms to r-overconvergent forms, then α will be a primitive link if $r' \leq pr$, but if $r' > pr$ then α may only be a link. Perhaps all of this can be avoided if one sets up the theory with a slightly more general class of topological modules.

Lemma 5.6 *Let R, M, M', $\mathbf{T}$, ϕ be as above, and assume that we are given a link $\alpha : M' \to M$ in the sense above. Let D_ϕ denote the eigenvariety associated to $(R, M, \mathbf{T}, \phi)$ and let D'_ϕ denote the eigenvariety associated to $(R, M', \mathbf{T}, \phi)$. Then D_ϕ and D'_ϕ are isomorphic.*

Proof This is clear if α is an isomorphism, so we may assume that α is a primitive link, and thus that there is a compact $c : M \to M'$ such that αc and $c\alpha$ are equal to the endomorphisms of M and M' induced by ϕ. We use a dash to indicate the analogue of one of our standard constructions, applied to M' (for example Z'_ϕ, $\mathcal{C}'$ and so on). By Lemma 2.13, Z_ϕ and Z'_ϕ are equal, as are $\mathcal{C}$ and $\mathcal{C}'$ (as their construction does not depend on the underlying Banach module). Choose $Y \in \mathcal{C}$ with connected image $X \subseteq \mathrm{Max}(R)$. It will suffice to prove that α induces an isomorphism $D'(Y) = D(Y)$ that commutes with all the glueing data on both sides, and this will follow if we can show that, after base extension to $A = \mathcal{O}(X)$, α induces an isomorphism between the finite flat sub-R-modules N' and N of M' and M corresponding to Y, and hence that α (which recall is $\mathbf{T}$-linear) induces an isomorphism $\mathbf{T}(Y) = \mathbf{T}'(Y)$.

Recall that there is a polynomial $Q = 1 + \ldots$ associated to Y as in the definition of $D(Y)$, such that the leading term of Q is a unit, and such that N' and N are the kernels of $Q^*(\phi)$ on M' and M respectively. From this one can conclude that α maps N' to N, that c maps N to N', and that there is an element ψ of $R[\phi] \subseteq \mathbf{T}$, such that ψ is an inverse to ϕ on both N and N'. We now see that $\alpha : N' \to N$ must be an R-module isomorphism, because it is elementary to check that $\psi \circ c$ is a two-sided inverse (recall that ψ is a polynomial in ϕ and hence commutes with c). Now everything else follows without too much trouble. □

I thank Peter Schneider for pointing out a problem with the proof of the above lemma in the initial version of this manuscript.

We now have enough for our eigenvariety machine. The data we are given is the following: we have a reduced rigid space $\mathcal{W}$, a commutative R-algebra $\mathbf{T}$, and an element $\phi \in \mathbf{T}$. For any admissible affinoid open $X \subseteq \mathcal{W}$, with $\mathcal{O}(X) = R_X$ (equipped with its supremum norm), we have a Banach R_X-module M_X satisfying (Pr), and an R-module homomorphism $\mathbf{T} \to \mathrm{Hom}_{R_X}(M_X, M_X)$, denoted $t \mapsto t_X$, such that ϕ_X is compact. Finally, if $Y \subseteq X \subseteq \mathcal{W}$ are two admissible affinoid opens, then we have a continuous $\mathcal{O}(Y)$-module homomorphism $\alpha : M_Y \to M_X \widehat{\otimes}_{R_X} R_Y$ which is a "link" in the above sense, and such that if $X_1 \subseteq X_2 \subseteq X_3 \subseteq \mathcal{W}$ are all affinoid subdomains then $\alpha_{13} = \alpha_{23}\alpha_{12}$ where α_{ij} denotes the map $M_{X_i} \to M_{X_i} \widehat{\otimes}_{\mathcal{O}(X_i)} \mathcal{O}(X_j)$.

Construction 5.7 (eigenvariety machine) *To the above data we may canonically associate the* eigenvariety D_ϕ, *a rigid space equipped with a map to* $\mathcal{W}$, *with the property that for any affinoid open* $X \subseteq \mathcal{W}$, *the pullback of* D_ϕ *to* X *is canonically isomorphic to the eigenvariety associated to the data* $(R_X, M_X, \mathbf{T}, \phi_X)$.

There is very little left to check in this construction. If $Y \subseteq X$ are affinoid subdomains of $\mathcal{W}$ then by Lemma 5.6 the eigenvarieties associated to $(R_Y, M_Y, \mathbf{T}, \phi_Y)$ and $(R_Y, M_X \widehat{\otimes}_{R_X} R_Y, \mathbf{T}, \phi_X)$ are isomorphic. By Lemma 5.5 the eigenvariety associated to $(R_Y, M_X \widehat{\otimes}_{R_X} R_Y, \mathbf{T}, \phi_X)$ is isomorphic to the pullback to Y of the eigenvariety that is associated to $(R_X, M_X, \mathbf{T}, \phi_X)$. The assumption on compatibility of the α ensures that the cocycle condition is satisfied, and hence the D_{ϕ_i} glue together to give an eigenvariety D_ϕ over $\mathcal{W}$ whose restriction to X_i is D_{ϕ_i}.

As we have seen earlier, one cannot expect D_ϕ to be reduced or flat over $\mathcal{W}$ in this generality. However, here are some positive results.

Lemma 5.8 *Assume $\mathcal{W}$ is equidimensional of dimension n. Then D_ϕ is also equidimensional of dimension n. The finite map $D_\phi \to Z_\phi$ has the property that each irreducible component of D_ϕ maps surjectively to an irreducible component of Z_ϕ. Moreover, the image in $\mathcal{W}$ of each irreducible component of D_ϕ is Zariski-dense in a component of $\mathcal{W}$.*

Proof This is Proposition 6.4.2 of [8]. □

We now explain how the points of D_ϕ are in bijection with systems of eigenvalues of $\mathbf{T}$. For L a complete extension of K, say that a map $\lambda : \mathbf{T} \to L$ is an *L-valued system of eigenvalues* if there is an affinoid $X = \mathrm{Max}(R_X) \subseteq \mathcal{W}$, a point in $X(L)$ (giving a map $R_X \to L$) and $0 \neq m \in M_X \widehat{\otimes}_{R_X} L$ such that $tm = \lambda(t)m$ for all $t \in \mathbf{T}$. Say that an L-valued system of eigenvalues is *ϕ-finite* if $\lambda(\phi) \neq 0$.

Lemma 5.9 *There is a natural bijection between ϕ-finite systems of eigenvalues and L-points of D_ϕ.*

Proof Because D_ϕ is separated, no pathologies occur when base extending to L and hence we may assume $L = K$. Recall that D_ϕ is covered by the $D(Y)$ for $Y \in \mathcal{C}$; choose $Y \in \mathcal{C}$ and let $X \subseteq \mathcal{W}$ be its image in $\mathcal{W}$. Choose a K-point P of X. This K-point corresponds to a map $R_X \to K$ and it suffices to construct a bijection between the K-points of $D(Y)$ lying above P and the ϕ-finite systems of eigenvalues coming from eigenvectors in $N \otimes_{R_X} K$, where $N \subseteq M_X$ is the subspace corresponding to Y. The result then follows from the following purely algebraic lemma. □

Lemma 5.10 *Let R be a commutative Noetherian ring and let N be a projective module of finite rank over R. Let T be a commutative subring of $\mathrm{End}_R(N)$. Let $\mathfrak{m}$ denote a maximal ideal of R, and let S denote the image of the natural map $T/\mathfrak{m}T \to \mathrm{End}_{R/\mathfrak{m}}(N/\mathfrak{m}N)$. Then the natural map $T/\mathfrak{m}T \to S$ induces a bijection between the prime ideals of $T/\mathfrak{m}T$ and the prime ideals of S.*

Proof It suffices to show that the kernel of the map $T/\mathfrak{m}T \to S$ is nilpotent. After localising at $\mathfrak{m}$ we may assume that N is free; choose a basis for N. Let t be an element of T whose image in $\mathrm{End}_{R/\mathfrak{m}}(N/\mathfrak{m}N)$ is zero. Then all the matrix coefficients of t with respect to this basis are in $\mathfrak{m}$. Thinking of t as a matrix with coefficients in R, we see that t is a root of its characteristic polynomial, which is monic and all of whose coefficients other than its leading term are in $\mathfrak{m}R$. Hence t is nilpotent in $T/\mathfrak{m}T$ and we are home. □

Part II: The Coleman-Mazur eigencurve.

6 Overconvergent modular forms.

One can say much more about the eigenvariety D_ϕ in the specific case for which all this machinery was originally invented, namely the Coleman-Mazur eigencurve. Again we shall not give a complete treatment of this topic, but will refer to [9] for many of the basic definitions and results we need. The paper [9] gives constructions of two objects, called C and D, both in the case of level 1 and $p > 2$. The results in sections 2–3 of this paper are enough for us to be able to extend the construction of D to the case of an arbitrary level and an arbitrary prime p, and we shall give details of the construction here. Note that we do not need the results in sections 4–5 of this paper here, because the eigenvarieties constructed are over a 1-dimensional base, and the rigid analytic results that Coleman develops in section A5 of [10] are sufficient.

Fix a prime p, set $K = \mathbf{Q}_p$, and let $\mathcal{W}$ be weight space, that is the rigid space whose $\mathbf{C}_p$-points are naturally the continuous group homomorphisms $\mathbf{Z}_p^\times \to \mathbf{C}_p^\times$ (see section 2 of [5] for more details on representability of such functors). Then $\mathcal{W}$ is the disjoint union of finitely many open discs, and there is a natural affinoid covering of $\mathcal{W}$ which on each component is a cover of the open disc by countably many closed discs. Coleman and Mazur restrict to the case $p > 2$, and for an affinoid Y in weight space define M_Y to be the space of r-overconvergent modular forms of level 1, for some appropriate real number r. As Y gets bigger one has to consider forms which overconverge less and less; this is why we must include cases where (in the notation of section 5) the "links" α are not the identity. Finally the map ϕ is chosen to be the Hecke operator U_p, which is compact. See [9] for rigorous definitions of the above objects, and verification that they satisfy the necessary criteria for the machine to work. In [9] it is proved that (for $N = 1$ and $p > 2$) the resulting eigencurve D_ϕ is reduced, and flat over Z_ϕ (see [9], Proposition 7.4.5 and the remarks before Theorem 7.1.1 respectively). Our Lemma 5.9 is just the statement that points on the eigencurve are overconvergent systems of finite slope eigenvalues, and the existence of q-expansions assures us that, at least in the cuspidal case, points on the eigencurve correspond bijectively with normalised overconvergent eigenforms.

In fact much more is proved in [9], where two rigid spaces are constructed for each odd prime p: a curve C, constructed via deformation theory and the theory of pseudorepresentations, and a curve D constructed via glueing Hecke algebras as above. In Theorem 7.5.1 of [9] it is proven that D is isomorphic to the space C^{red}. Since the paper [9] appeared, various authors have assumed that the constructions in it would generalise to the cases $N > 1$ and $p = 2$. In fact, it seems to us that the following are the main reasons that $N = 1$ and $p > 2$ are assumed in [9]. Firstly, the theory of pseudorepresentations does not work quite so well in the case $p = 2$. Secondly, there are some issues to be resolved when writing down the local conditions at primes dividing N on the deformation theory side (we remark that Kisin tells us that both of these issues can be resolved without too much trouble). And thirdly one sometimes has to deal with eigenspaces for the action of $(\mathbf{Z}/4\mathbf{Z})^\times$ on a 2-adic Banach module when $p = 2$ on the Hecke algebra side, causing problems when looking for orthonormal bases. The first two issues will not concern us in this paper, as we do not talk about the construction of C, and the results in sections 2 and 3 of this paper are enough to deal with the third issue. In fact Chenevier has pointed out to us that one can also avoid the troubles caused by the third issue when constructing eigencurves for $p = 2$ by appealing to the corollary of lemma 1 in [15]. We do not construct a generalisation of C here, but we do show how to construct an eigencurve D for a general prime p and level N prime to p.

We first establish some generalities and notation. All our rigid spaces will be over $K = \mathbf{Q}_p$ in this section. Recall that $\mathcal{W}$ is the rigid space over $\mathbf{Q}_p$ representing maps $\mathbf{Z}_p^\times \to \mathbf{G}_m$. Define $q = p$ if $p > 2$, and $q = 4$ if $p = 2$. Define $D = (\mathbf{Z}/q\mathbf{Z})^\times$, regarded as a quotient of $\mathbf{Z}_p^\times$ in the natural way. Define $\widehat{D}$ to be the set of group homomorphisms $D \to \mathbf{C}_p^\times$. Set $\gamma = 1 + q \in \mathbf{Z}_p^\times$. The natural surjection $\mathbf{Z}_p^\times \to D$ has kernel $1 + q\mathbf{Z}_p$, which is topologically isomorphic to $\mathbf{Z}_p$, and is topologically generated by γ. The map $\mathbf{Z}_p^\times \to D$ induces an isomorphism between D and the roots of unity in $\mathbf{Z}_p^\times$, and hence the surjection splits and we have an isomorphism $\mathbf{Z}_p^\times \cong D \times \mathbf{Z}_p$; we shall thus identify $\mathbf{Z}_p^\times$ with $D \times \mathbf{Z}_p$. If $\chi \in \widehat{D}$ then the composite of χ with the natural projection $\mathbf{Z}_p^\times \to D$ is an element of $\mathcal{W}$, and one easily checks that distinct elements of $\widehat{D}$ are in different components of $\mathcal{W}$. Hence this construction establishes a bijection between the components of $\mathcal{W}$ and the group $\widehat{D}$. Let $\mathcal{W}_\chi$ denote the component of $\mathcal{W}$ corresponding to the character $\chi \in \widehat{D}$. Let $\mathbf{1}$ denote the trivial character of D (sending everything to 1) and let $\mathcal{B}$ denote the component $\mathcal{W}_\mathbf{1}$ of $\mathcal{W}$. Note that if $\chi \in \widehat{D}$ then multiplication by χ gives an isomorphism $\mathcal{W}_\mathbf{1} \to \mathcal{W}_\chi$.

For $n \geq 1$ let X_n denote the affinoid subdomain of $\mathcal{W}$ corresponding to group homomorphisms $\psi : \mathbf{Z}_p^\times \to \mathbf{C}_p^\times$ such that $|\psi(1+q)^{p^{n-1}} - 1| \leq |q|$. It is easily checked that for any $\chi \in \widehat{D}$, $X_n \cap \mathcal{W}_\chi$ is an affinoid disc, that

$X_1 \subseteq X_2 \subseteq \ldots$, and that the X_i give an admissible cover of $\mathcal{W}$. The inclusion $X_i \subset \mathcal{W}$ induces a bijection of the connected components of X_i with the connected components of $\mathcal{W}$; if $\chi \in \widehat{D}$ then write $X_{i,\chi}$ for the closed disc $X_i \cap \mathcal{W}_\chi$. We remark here that if $k \in \mathbf{Z}$ and $\chi : (\mathbf{Z}/qp^{n-1}\mathbf{Z})^\times \to \mathbf{C}_p^\times$ then $\chi(1+q)^{p^{n-1}} = 1$ and hence the map $\psi \in \mathcal{W}$ defined by $\psi(x) = x^k\chi(x)$ is in $X_n(\mathbf{C}_p)$.

Define $R_i = \mathcal{O}(X_i)$. Then R_i is an affinoid for each i, and $R_i = \oplus_\chi R_{i,\chi}$, where χ runs through $\widehat{D}$ and $R_{i,\chi} = \mathcal{O}(X_{i,\chi})$.

In preparation for the application of our eigenvariety machine, we have to choose a family of radii of overconvergence. Fortunately Coleman and Mazur have done enough for us here, even if $p = 2$. We give a brief description of the modular curves and affinoids that we shall use. Let $\mathbf{A}_f$ denote the finite adeles. For a compact open subgroup $\Gamma \subset \mathrm{GL}_2(\mathbf{A}_f)$ that contains the principal congruence subgroup Γ_N for some N prime to p, we define the compact modular curve $X(\Gamma)$ over $\mathbf{Q}_p$ in the usual way. Let us firstly assume that Γ is sufficiently small to ensure that the associated moduli problem on generalised elliptic curves has no non-trivial automorphisms (we will remove this assumption below). Now recall from section 3 of [4], for example, that for an elliptic curve E over a finite extension of $\mathbf{Q}_p$, there is a measure $v(E)$ of its supersingularity, and that $v(E) < p^{2-m}/(p+1)$ implies that E possesses a canonical subgroup of order p^m. So for $r \in \mathbf{Q}$ with $0 \le r < p/(p+1)$ we define $X(\Gamma)_{\ge p^{-r}}$ to be the affinoid subdomain of the rigid space over $\mathbf{Q}_p$ associated to $X(\Gamma)$ whose non-cuspidal points parametrise elliptic curves E with a level Γ structure and such that $v(E) \le r$. For example if $r = 0$ then $X(\Gamma)_{\ge p^{-r}}$ is the ordinary locus of $X(\Gamma)$.

For $m \ge 1$ there is a fine moduli space $X(\Gamma, \Gamma_1(p^m))$ (resp. $X(\Gamma, \Gamma_0(p^m))$) over $\mathbf{Q}_p$ whose non-cuspidal points parametrise elliptic curves equipped with a level Γ structure and a point (resp. cyclic subgroup) of order p^m over $\mathbf{Q}_p$-schemes. There are natural forgetful functors

$$X(\Gamma, \Gamma_1(p^m)) \to X(\Gamma, \Gamma_0(p^m)) \to X(\Gamma).$$

If $0 \le r < p^{2-m}/(p+1)$ and E is an elliptic curve over a finite extension of $\mathbf{Q}_p$ with $v(E) \le r$ then, as mentioned above, E has a canonical subgroup of order p^m. For r in this range we define $X(\Gamma, \Gamma_0(p^m))_{\ge p^{-r}}$ to be the components of the pre-image of $X(\Gamma)_{\ge p^{-r}}$ in $X(\Gamma, \Gamma_0(p^m))$ whose non-cuspidal points parametrise elliptic curves with the property that their given cyclic subgroup of order p^m equals their canonical subgroup, and we define the rigid space $X(\Gamma, \Gamma_1(p^m))_{\ge p^{-r}}$ to be the pre-image of $X(\Gamma, \Gamma_0(p^m))_{\ge p^{-r}}$ in $X(\Gamma, \Gamma_1(p^m))$.

All these spaces are affinoids; this follows from the fact that $X(\Gamma)_{\geq p^{-r}}$ is an affinoid, being the complement of a non-zero finite number of open discs in a complete curve. There is a natural action of the finite group $(\mathbf{Z}/p^m\mathbf{Z})^\times$ on $X(\Gamma,\Gamma_1(p^m))$ and on $X(\Gamma,\Gamma_1(p^m))_{\geq p^{-r}}$ via the (weight 0) Diamond operators.

Finally if Γ is a compact open subgroup of $\mathrm{GL}_2(\mathbf{A}_f)$ containing Γ_N for some N prime to p, but which is not "sufficiently small", then choose some prime $l \nmid 2Np$ and define $\Gamma' := \Gamma \cap \Gamma_l$; then Γ' is a normal subgroup of Γ and Γ' is sufficiently small. Hence one may apply all the constructions above to Γ' and then define $X(\Gamma)_{\geq p^{-r}}$, $X(\Gamma,\Gamma_1(p^m))_{\geq p^{-r}}$ and so on by taking Γ/Γ'-invariants. The resulting objects are only coarse moduli spaces but this will not trouble us. A standard argument shows that this construction is independent of l. We define $X_1(p^m)_{\geq p^{-r}} := X(\mathrm{GL}_2(\widehat{\mathbf{Z}}),\Gamma_1(p^m))_{\geq p^{-r}}$ and $X_0(p^m)_{\geq p^{-r}} := X(\mathrm{GL}_2(\widehat{\mathbf{Z}}),\Gamma_0(p^m))_{\geq p^{-r}}$. Similarly if $N \geq 1$ is prime to p then we define $X_0(Np^m)_{\geq p^{-r}} := X(\Gamma_0(N),\Gamma_0(p^m))_{\geq p^{-r}}$, where $\Gamma_0(N)$ is as usual the matrices in $\mathrm{GL}_2(\widehat{\mathbf{Z}})$ which are upper triangular mod N. Note that $X_1(q)_{\geq 1}$ is the curve that Coleman and Mazur refer to as $Z_1(q)$, and that the quotient of $X_1(p^m)_{\geq p^{-r}}$ by the action of $(\mathbf{Z}/p^m\mathbf{Z})^\times$ is $X_0(p^m)_{\geq p^{-r}}$. Note also that $X_0(p^m)_{\geq p^{-r}}$ is "independent of m", in the sense that the natural (forgetful) map $X_0(p^m)_{\geq p^{-r}} \to X_0(p)_{\geq p^{-r}}$ is an isomorphism (as both rigid spaces represent the same functor).

We now come to the definition of the radii of overconvergence r_i. Note that these numbers depend only on p and not on any level structure. We let $\mathbf{E}_p$ denote the function on $X_1(q)_{\geq 1} \times \mathcal{B}$ defined in Proposition 2.2.7 of [9] (briefly, $\mathbf{E}_p$ is the function which, when restricted to a classical even weight $k \geq 4$ in $\mathcal{B}$, corresponds to the function $E_k(q)/E_k(q^p)$, where $E_k(q)$ is the p-deprived ordinary old Eisenstein series of weight k and level p). In Proposition 2.2.7 of [9] it is proved that $\mathbf{E}_p$ is overconvergent over $\mathcal{B}$. The specialisations to a classical weight of $\mathbf{E}_p$ are fixed by the weight 0 Diamond operators, and hence $\mathbf{E}_p$ descends to an overconvergent function on $X_0(q)_{\geq 1} \times \mathcal{B} = X_0(p)_{\geq 1} \times \mathcal{B}$. Furthermore, the assertions about the q-expansion coefficients of $\mathbf{E}_p$ made in Proposition 2.2.7 of [9], and the q-expansion principle, are enough to ensure that $\mathbf{E}_p$ has no zeroes on $X_0(p)_{\geq 1} \times \mathcal{W}_1$. Hence the inverse of $\mathbf{E}_p$ is a function on $X_0(p)_{\geq 1} \times \mathcal{B}$ and it is elementary to check that it is also overconvergent over $\mathcal{B}$. In particular, for any $i \geq 1$ there exists a rational $0 < r_i < 1/(p+1)$ such that the restrictions of both $\mathbf{E}_p$ and $\mathbf{E}_p^{-1}$ to $X_0(p)_{\geq 1} \times X_{i,1}$ extend to functions on $X_0(p)_{\geq p^{-r_i}} \times X_{i,1}$. We choose a sequence of rationals $r_1 \geq r_2 \geq r_3 \geq \ldots \geq 0$ such that each r_i has the aforementioned property. We may furthermore assume (for technical convenience) that $r_i < p^{2-i}/q(p+1)$ (although this may be implied by the other assumptions).

Remark It is not important for us to establish concrete values for the r_i, so we do not. However for other applications where one wants to have explicit knowledge about how far one can extend overconvergent modular forms, it may in future be important to understand exactly how the r_i behave. For example, machine computations for $p \in \{3, 5, 7, 13\}$ and $i = 1$ show that it is *not* the case that r_1 can be taken to be an arbitrary rational number less than $1/(p+1)$, because there are classical Eisenstein eigenforms of level p which have zeroes in $\cup_{r<1/(p+1)} X_1(p)_{\geq p^{-r}}$. Can one take r_1 to be any positive rational less than $(p-1)/p(p+1)$? There are other interesting questions here, which we shall not attempt to answer here.[4]

We are now ready to begin our definitions of the Banach modules of overconvergent forms. These modules depend on an auxiliary level structure, which we now choose. Fix a compact open subgroup $\Gamma \subseteq \mathrm{GL}_2(\mathbf{A}_f)$ that contains the principal congruence subgroup Γ_N for some N prime to p. For each $n \geq 1$ we wish now to define a Banach module $M_n = M_{\Gamma,n}$ over $R_n := \mathcal{O}(X_n)$. Recall that $R_n = \oplus_{\chi \in \widehat{D}} R_{n,\chi}$. If $\chi \in \widehat{D}$ then we define the Banach module $B_{n,\chi}$ to be

$$B_{n,\chi} := R_{n,\chi} \widehat{\otimes}_{\mathbf{Q}_p} \mathcal{O}(X(\Gamma, \Gamma_1(q))_{\geq p^{-r_n}})$$

where r_n is the radius of convergence defined previously. Note that $B_{n,\chi}$ is naturally a potentially ONable Banach $R_{n,\chi}$-module, because the ring $\mathcal{O}(X(\Gamma, \Gamma_1(q)))$ can be viewed as a Banach space over a discretely-valued field and is hence potentially ONable by Proposition 1 of [15] and the remarks before it.

We give the module $B_{n,\chi}$ an $R_{n,\chi}$-linear action of the group $(\mathbf{Z}/q\mathbf{Z})^\times$ by letting it act trivially on $R_{n,\chi}$ and via the (weight 0) Diamond operators on $\mathcal{O}(X(\Gamma, \Gamma_1(q)))$. We define $M_{n,\chi}$ to be the direct summand of $B_{n,\chi}$ where $(\mathbf{Z}/q\mathbf{Z})^\times$ acts via χ. Note that by definition $M_{n,\chi}$ satisfies property (Pr) (one can check that it is even potentially ONable, but we shall not need this because of our extension of Coleman's theory). We remark in passing that if $\Gamma = \mathrm{GL}_2(\widehat{\mathbf{Z}})$ then $M_{n,\chi} = 0$ if $\chi(-1) \neq 1$. Finally we define $M_n = M_{\Gamma,n}$ to be the module over $R_n = \oplus_{\chi \in \widehat{D}} R_{n,\chi}$ whose $R_{n,\chi}$-part is $M_{n,\chi}$. We give M_n an R_n-linear action of $(\mathbf{Z}/q\mathbf{Z})^\times$ by letting it act via χ on $M_{n,\chi}$. We will shortly see that the fibre of M_n at a point $\kappa \in X_n$ can be naturally identified with the

4 See however [7] for explicit calculations in the case $p = 2$, and note that these calculations may well generalise to odd primes p.

r_n-overconvergent forms of weight κ (although there are caveats regarding compatibility of this map with Diamond operators; see below).

The modules M_n we have described have the following functorial property: if Γ_1 and Γ_2 both satisfy the conditions imposed on Γ above (that is, they are of level prime to p), if $\gamma \in \mathrm{GL}_2(\mathbf{A}_f)$ has determinant which is a unit at p, and if $\gamma\Gamma_1\gamma^{-1} \subseteq \Gamma_2$, then there is a natural induced finite flat map $X(\Gamma_1) \to X(\Gamma_2)$ and the assumption on the determinant of γ means that if $0 \leq r \leq r_1$ then $X(\Gamma_1, \Gamma_1(q))_{\geq p^{-r}}$ is the pre-image of $X(\Gamma_2, \Gamma_1(q))_{\geq p^{-r}}$. Hence there is a natural inclusion $M_{\Gamma_2,n} \to M_{\Gamma_1,n}$. If furthermore $\gamma\Gamma_1\gamma^{-1} = \Gamma_2$ then this inclusion has an inverse and is hence an isomorphism. One can check using these ideas that if Γ_1 is a normal subgroup of Γ_2 then $M_{\Gamma_2,n}$ is the Γ_2/Γ_1-invariants of $M_{\Gamma_1,n}$. Note also that for $n \geq 1$ there is a natural map $M_n \to M_{n+1}\widehat{\otimes}_{R_{n+1}} R_n$, induced by restriction, which we shall later see is a link, in the sense of Part I.

We now make explicit the relation between these spaces and Katz' spaces of overconvergent modular forms. Firstly we recall the definitions (in the form due to Coleman). If $\Gamma \subset \mathrm{GL}_2(\mathbf{A}_f)$ is a sufficiently small compact open subgroup then there is a sheaf commonly denoted ω on $X(\Gamma)$ which, on non-cuspidal points, is the pushforward of the differentials on the universal elliptic curve. A weight k modular form of level Γ, defined over $\mathbf{Q}_p$, is a global section of $\omega^{\otimes k}$. If L is a field extension of $\mathbf{Q}_p$ then a weight k modular form of level Γ defined over L is a global section of $\omega^{\otimes k}$ on the base change of $X(\Gamma)$ to L. If Γ contains Γ_N for some positive integer N prime to p and $0 \leq r < p/(p+1)$ then a p^{-r}-overconvergent modular form of level Γ and weight k defined over $\mathbf{Q}_p$ is a section of (the analytic sheaf associated to) $\omega^{\otimes k}$ on the rigid space $X(\Gamma)_{\geq p^{-r}}$, and similarly for L a complete extension of $\mathbf{Q}_p$. If Γ is not sufficiently small then one can still make sense of these definitions by replacing Γ by a sufficiently small normal subgroup Γ' and then taking Γ/Γ'-invariants, as Γ/Γ' will act on everything. Finally, if $0 \leq r < p^{2-n}/(p+1)$ then one can define p^{-r}-overconvergent modular forms of weight k and level $\Gamma \cap \Gamma_1(p^n)$ as sections of $\omega^{\otimes k}$ on $X(\Gamma, \Gamma_1(p^n))_{\geq p^{-r}}$ (again using the standard tricks if Γ is not sufficiently small). There is a natural action of $(\mathbf{Z}/p^n\mathbf{Z})^\times$ on these spaces via the (weight k) Diamond operators.

Fix $n \geq 1$ and let L be the field $\mathbf{Q}_p(\zeta_{p^{n-1}})$ generated by a primitive p^{n-1}th root of unity $\zeta_{p^{n-1}}$. Fix $k \in \mathbf{Z}$, and a character $\varepsilon : (\mathbf{Z}/qp^{n-1}\mathbf{Z})^\times \to L^\times$. Define $\psi : \mathbf{Z}_p^\times \to L^\times$ by $\psi(x) = x^k\varepsilon(x)$. Then ψ is an L-valued point of weight space and in fact $\psi \in X_n(L)$. Choose $\chi \in \widehat{D}$ such that $\psi \in \mathcal{W}_\chi$. Define $\kappa = \psi/\chi \in \mathcal{W}_1$ and let $E_\kappa = 1 + \cdots$ be the associated Eisenstein series (see p39 of [9] and note that $\kappa \in \mathcal{W}_1 = \mathcal{B}$ so there are no problems with zeros of p-adic L-functions).

Proposition 6.1 *The q-expansion E_κ is the q-expansion of a p^{-r_n}- overconvergent weight k modular form of level $\Gamma_1(qp^{n-1})$ defined over L, which is non-vanishing on $X_1(qp^{n-1})_{\geq p^{-r_n}}$. Furthermore E_κ is in the ε/χ-eigenspace for the Diamond operators.*

Proof The fact that the q-expansion E_κ is the q-expansion of a section of $\omega^{\otimes k}$ on $X_1(qp^{n-1})_{\geq p^{-r}}$ for some $r > 0$ is Corollary 2.2.6 of [9]. Now the fact that E_κ is an eigenvector for U_p, which increases overconvergence, implies that E_κ extends to a section of $\omega^{\otimes k}$ on of $X_1(qp^{n-1})_{\geq p^{-r}}$ for any rational r with $0 \leq r < p^{3-n}/q(p+1)$. For r in this range, there is a map Frob : $X_1(qp^{n-1})_{\geq p^{-r/p}} \to X_1(qp^{n-1})_{\geq p^{-r}}$ which is finite and flat of degree p and which induces, by pullback, the map $F(q) \mapsto F(q^p)$ on modular forms (on non-cuspidal points the map sends (E, P) to $(E/C, \overline{Q})$ where C is the canonical subgroup of order p of E and $\overline{Q}$ is the image in E/C of any generator Q of the canonical subgroup of order qp^n of E such that $pQ = P$). By the q-expansion principle, E_κ has no zeroes on the ordinary locus $X_1(qp^{n-1})_{\geq 1}$. Let S denote the set of zeroes of E_κ on the non-ordinary locus of $X_1(qp^{n-1})_{\geq p^{-r_n}}$. It suffices to show that S is empty. We know that S is finite, because $X_1(qp^{n-1})_{\geq p^{-r_n}}$ is a connected affinoid curve and $E_\kappa \neq 0$ (as its q-expansion is non-zero). We also know that $E_\kappa(q)/E_\kappa(q^p)$ has no zeroes on $X_1(qp^{n-1})_{\geq p^{-r_n}}$, by definition of r_n, and that $r_n < p^{2-n}/q(p+1)$. Hence any zero of E_κ is also a zero of $E_\kappa(q^p) = \text{Frob}^* E_\kappa$. But if S is non-empty then let P denote a point of S which is "nearest to the ordinary locus" (that is, such that $v'(P)$ is minimal, where v' is the composite of the natural projection $X_1(qp^{n-1}) \to X_0(p)$ and the function denoted v' in section 4 of [4]). Then P is also a zero of $E_\kappa(q^p)$ and hence if $P = \text{Frob}(Q)$ for some point $Q \in X_1(qp^{n-1})_{\geq p^{-r_n/p}}$ then Q is also a zero of $E_\kappa(q)$, and furthermore Q is closer to the ordinary locus than P (in fact $v'(Q) = \frac{1}{p}v'(P) < v'(P)$ by Theorem 3.3(ii) of [4]), a contradiction.

The assertion about the Diamond operators is classical if $k \geq 2$ (see for example Proposition 7.1.1 of [13]). For general k it can be deduced as follows: by Theorem B4.1 of [10] applied with $i = 0$, the function on $X_1(q)_{\geq 1} \times \mathcal{B} \times \mathcal{B}$ denoted $E_\alpha(q)E_\beta(q)/E_{\alpha\beta}(q)$ in Theorem 2.2.2 of [9], when restricted to $X_1(q)_{\geq 1} \times \mathcal{B}^* \times \mathcal{B}^*$ (in the notation of the proof of Theorem 2.2.2 of [9]) is invariant under the natural action of $(\mathbf{Z}/q\mathbf{Z})^\times$ (acting trivially on $\mathcal{B}^\times$ and via Diamond operators on $X_1(q)$). But this action is continuous, so $(\mathbf{Z}/q\mathbf{Z})^\times$ acts trivially on $E_\alpha(q)E_\beta(q)/E_{\alpha\beta}(q)$. Hence, the function $E_\alpha(q)E_\beta(q)/E_{\alpha\beta}(q)$

descends to a function on $X_0(q)_{\geq 1} \times \mathcal{B} \times \mathcal{B} = X_0(qp^{n-1})_{\geq 1} \times \mathcal{B} \times \mathcal{B}$. The result for general k now follows from the result for $k \geq 2$. □

Corollary 6.2 *Let $k, \varepsilon, \psi, \chi, \kappa$ be as above, and let $\psi : R_n \to L$ also denote the homomorphism corresponding to the L-point of X_n induced by ψ. Then multiplication by E_κ induces an isomorphism between $M_{\Gamma,n} \otimes_{R_n} L$ (the tensor product formed via the homomorphism $\psi : R_n \to L$) and the space of p^{-r_n}-overconvergent modular forms of level Γ, weight k and character ε defined over L.*

Proof Say $\psi \in \mathcal{W}_\chi$. Then unravelling the definitions gives that $M_{\Gamma,n} \otimes_{R_n} L = M_{n,\chi} \otimes_{R_{n,\chi}} L$ is equal to the χ-eigenspace of $\mathcal{O}(X(\Gamma, \Gamma_1(q))_{\geq p^{-r_n}}) \otimes_{\mathbf{Q}_p} L$, and hence the χ-eigenspace of $\mathcal{O}(X(\Gamma, \Gamma_1(qp^{n-1}))_{\geq p^{-r_n}}) \otimes_{\mathbf{Q}_p} L$, where χ here is regarded as a character of $(\mathbf{Z}/qp^{n-1}\mathbf{Z})^\times$ (note that the forgetful functor

$$X(\Gamma, \Gamma_1(qp^{n-1}))_{\geq p^{-r_n}} \to X(\Gamma, \Gamma_1(q))_{\geq p^{-r_n}}$$

is the map induced by quotienting $X(\Gamma, \Gamma_1(qp^{n-1}))_{\geq p^{-r_n}}$ out by the group $(1 + q\mathbf{Z}_p/1 + qp^{n-1}\mathbf{Z}_p)$). The result now follows from the fact that E_κ has weight k, character ε/χ and is non-vanishing on $X(\Gamma, \Gamma_1(qp^{n-1}))_{\geq p^{-r_n}}$. □

Motivated by this Corollary, we define the q-expansion of an element of M_n to be the following element of $R_n[[q]]$, as follows: it suffices to attach a q-expansion in $R_{n,\chi}[[q]]$ to an element of $M_{n,\chi}$, and hence it suffices to attach a q-expansion in $R_{n,\chi}[[q]]$ to an element of $B_{n,\chi}$. Now an element of $\mathcal{O}(X(\Gamma, \Gamma_1(q))_{\geq p^{-r_n}})$ has a q-expansion in $\mathbf{Q}_p[[q]]$ in the usual way, and hence an element of $B_{n,\chi}$ has a q-expansion in $R_{n,\chi}[[q]]$. This is not the q-expansion that we are interested in however—this q-expansion corresponds to a family of weight 0 overconvergent forms. we twist this q-expansion by multiplying it by $\mathbf{E}$, the q-expansion of the restricted Eisenstein family defined in Section 2.2 of [9]. Note that the restricted Eisenstein family is a family over $\mathcal{B}$ and so we have to explain how to regard it as a family over $X_{n,\chi}$; we do this by pulling back via the composite of the natural inclusion $X_{n,\chi} \to \mathcal{W}_\chi$ and the natural isomorphism $\mathcal{W}_\chi \to \mathcal{W}_1$. The resulting power series is defined to be the q-expansion of m, and with this normalisation, the isomorphism of the previous corollary preserves q-expansions.

As we shall see in the next section, we will be defining Hecke operators on $M_{\Gamma,n}$ (at least for certain choices of Γ) so that they agree with the standard Hecke operators on overconvergent modular forms, via the isomorphism of the previous Corollary. We finish this section by remarking that the isomorphism of the previous corollary is however not compatible with Diamond operators in general, for two reasons—firstly, overconvergent forms of classical weight-character ψ naturally have an action of the group $(\mathbf{Z}/qp^{n-1}\mathbf{Z})^\times$, but the full space M_n only has an action of $(\mathbf{Z}/q\mathbf{Z})^\times$, and secondly even the actions of $(\mathbf{Z}/q\mathbf{Z})^\times$ do not in general coincide, as one is defined in weight 0 and the other in weight k so the two actions differ (at least for $p > 2$) by the kth power of the Teichmüller character.

7 Hecke operators and classical eigencurves

We now restrict to the case where Γ is either the congruence subgroup $\Gamma_0(N)$ (the subgroup of $\mathrm{GL}_2(\widehat{\mathbf{Z}})$ consisting of matrices which are upper triangular mod N) or $\Gamma_1(N)$ (the subgroup of $\Gamma_0(N)$ consisting of matrices of the form $\left(\begin{smallmatrix} * & * \\ 0 & 1 \end{smallmatrix}\right)$ mod N) of $\mathrm{GL}_2(\mathbf{A}_f)$, for some N prime to p. The associated moduli problems are then just the usual problems of representing cyclic subgroups or points of order N. We now define Hecke operators T_m for m prime to p, and a compact operator U_p, on the spaces $M_{\Gamma,n}$ for $n \geq 1$. Almost all of the work has been done for us, in section B5 of [10] (where Hecke operators are defined on overconvergent forms over $\mathcal{B}^*$) and in section 3.4 of [9] (where this work is extended to $\mathcal{B}$). Note that the arguments in these references do not assume $p > 2$ or $N = 1$. We do not reproduce the arguments here, we just mention that the key point is that because the argument is *not* a geometric one, the construction of T_m is done at the level of q-expansions and the resulting definitions initially go from forms of level N to forms of level Nm. However one can prove that the resulting maps do in fact send forms of level N to forms of level N by noting that the result is true for forms of classical weight, where the Hecke operator can be defined via a correspondence, and deducing the general case by considering a trace map and noting that a family of forms that vanishes at infinitely many places must be zero.

The one lacuna in the arguments in the references above is that in both cases the operators are defined as endomorphisms of overconvergent forms, rather than r-overconvergent forms for some fixed r. What we need to do is to prove that the Hecke operators defined by Coleman and Mazur send r-overconvergent forms to r-overconvergent forms, for some appropriate choice of r. Let $\mathbf{E}$ denote the restricted Eisenstein family (that is, the usual family of Eisenstein series over $\mathcal{B}$), and for $\ell \neq p$ a prime number, let $\mathbf{E}_\ell(q)$ denote the ratio $\mathbf{E}(q)/\mathbf{E}(q^\ell)$ as in Proposition 2.2.7 of [9], thought of as an overconvergent function on $X_0(\ell p)_{\geq 1} \times \mathcal{B}$ (note that Coleman and Mazur only assert that this function lives on $X_1(q,\ell)_{\geq 1} \times \mathcal{B}$ but it is easily checked to be invariant under the Diamond operators at p).

Lemma 7.1 *The restriction of* $\mathbf{E}_\ell(q)$ *to* $X_0(\ell p)_{\geq 1} \times X_{n,1}$ *extends to a non-vanishing function on* $X_0(\ell p)_{\geq p^{-r_n}} \times X_{n,1}$.

Proof For simplicity in this proof, we say that a function on $X_0(p)_{\geq 1} \times X_{n,1}$ is r-overconvergent if it extends to a function on $X_0(p)_{\geq p^{-r}} \times X_{n,1}$, and similarly we say that such a function is r-overconvergent and non-vanishing if it extends to a non-vanishing function on $X_0(p)_{\geq p^{-r}} \times X_{n,1}$.

We first prove that $\mathbf{E}_\ell(q)$ is r_n-overconvergent. Observe that Proposition 2.2.7 of [9] tells us that $\mathbf{E}_\ell(q)$ is r-overconvergent for some $r > 0$. Let us assume $r < r_n$ and explain how to analytically continue $\mathbf{E}_\ell(q)$ a little further. By definition of r_n, we know that $\mathbf{E}_p(q)$ is r_n-overconvergent and non-vanishing. Furthermore the non-trivial degeneracy map $X_0(\ell p) \to X_0(p)$ which on q-expansions sends $F(q)$ to $F(q^\ell)$ induces a morphism of rigid spaces $X_0(\ell p)_{\geq p^{-r_n}} \to X_0(p)_{\geq p^{-r_n}}$ and hence $\mathbf{E}_p(q^\ell)$ is also an r_n-overconvergent non-vanishing function. The ratio $\mathbf{E}_p(q^\ell)/\mathbf{E}_p(q)$ is therefore also an r_n-overconvergent non-vanishing function. But

$$\mathbf{E}_p(q^\ell)/\mathbf{E}_p(q) = \mathbf{E}_\ell(q^p)/\mathbf{E}_\ell(q)$$

and $\mathbf{E}_\ell(q)$ is r-overconvergent, and thus $\mathbf{E}_\ell(q^p)$ is also r-overconvergent. Let $r' = \min\{pr, r_n\}$. We claim that $\mathbf{E}_\ell(q)$ is r'-overconvergent and this clearly is enough because repeated applications of this idea will analytically continue $\mathbf{E}_\ell(q)$ until it is r_n-overconvergent, which is what we want. But it is a standard fact that the U operator increases overconvergence by a factor of p, and hence if $\mathbf{E}_\ell(q^p)$ is r-overconvergent, then $\mathbf{E}_\ell(q) = U(\mathbf{E}_\ell(q^p))$ is r'-overconvergent.

Finally we show that $\mathbf{E}_\ell(q)$ is non-vanishing on $X_0(\ell p)_{\geq p^{-r_n}} \times X_{n,1}$ and this follows from an argument similar to the non-vanishing statement proved in Proposition 6.1—if $\mathbf{E}_\ell(q)$ had a zero then choose a zero (x, κ) and then specialise to weight κ; we may assume that x is a zero closest to the ordinary locus in weight κ, and then $\mathbf{E}_\ell(q^p)$ has a zero closer to the ordinary locus in weight κ and hence $\mathbf{E}_\ell(q)/\mathbf{E}_\ell(q^p)$ would have a pole in weight κ, contradicting the fact that $\mathbf{E}_\ell(q)/\mathbf{E}_\ell(q^p) = \mathbf{E}_p(q)/\mathbf{E}_p(q^\ell)$ is r_n-overconvergent. □

We now have essentially everything we need to apply our eigenvariety machine. The preceding lemma and the arguments in section 3.4 of [9] can be used to define Hecke operators T_m (m prime to p) and U_p on r_n-overconvergent forms over X_n. If X is any admissible affinoid open subdomain of $\mathcal{W}$ then there exists some $n \geq 1$ such that $X \subseteq X_n$; we choose the smallest such n and define the Banach module M_X to be the pullback to $\mathcal{O}(X)$ of the $\mathcal{O}(X_n)$-module M_n. We let $\mathbf{T}$ denote the abstract polynomial algebra over R generated by the Hecke operators T_m for m prime to p, the operator $\phi = U_p$, and the Diamond operators at N if $\Gamma = \Gamma_1(N)$. These Hecke operators are well-known to commute, as can be seen by checking on classical points. We need to check that the natural restriction maps α between spaces of overconvergent forms of different radii are all links, but this follows easily from the technique used in the standard proof that the characteristic power series of U_p on r-overconvergent forms is independent of the choices of $r > 0$. The key point is that the endomorphism U_p of r-overconvergent forms can be checked to factor as a continuous map from r-overconvergent forms to s-overconvergent forms, for some $s > r$, followed by the (compact) restriction map from s-overconvergent forms to r-overconvergent forms. In fact one can take s to be anything less than both pr and $p/(p+1)$. One deduces that for any $0 < r < r' < p/(p+1)$, the natural map from r'-overconvergent forms to r-overconvergent forms is a "link". Our conclusion is that the construction of the "D" eigencurve in [9] can be generalised to all p and $N \geq 1$ prime to p.

PART III: Eigenvarieties for Hilbert modular forms.

8 Thickenings of K-points and weight spaces

Let K be a non-archimedean local field (that is, a field either isomorphic to a finite extension of $\mathbf{Q}_p$ or to the field of fractions of $k[[T]]$ with k finite). Let $\mathcal{O}$ denote the integers of K, and let V denote the closed affinoid unit disc over K. As an example of a construction which will be used many times in the sequel, we firstly show that is not difficult to construct a sequence $U_1 \supset U_2 \supset \ldots$ of affinoid subdomains of V, defined over K, with the property that

$$\bigcap_{t \geq 1} (U_t(L)) = V(K) = \mathcal{O} \qquad (*)$$

for all complete extensions L of K. Note that $(*)$ implies that $U_t(K) = V(K)$ for all t (set $L = K$), but also that no non-empty K-affinoid subdomain of V can be contained in all of the U_t. The U_t should be thought of as a system of affinoid neighbourhoods of $V(K) = \mathcal{O}$ in V. The construction of the U_t is simple: Let $\pi \in K$ be a uniformiser and define $U_t = \bigcup_{\alpha \in X_t} B(\alpha, |\pi|^t)$, where X_t is a set of representatives in $\mathcal{O}$ for $\mathcal{O}/(\pi)^t$, and $B(\alpha, |\pi|^t)$ is the closed affinoid disc with centre α and radius $|\pi|^t$. Note that X_t is finite because K is a local field, and hence U_t is an affinoid subdomain of V: it is a finite union of affinoid subdiscs of V of radius $|\pi|^t$. It is easy to check moreover that the U_t satisfy $(*)$ (use the fact that $\mathcal{O}$ is compact to get the harder inclusion).

We in fact need a "twisted" n-dimensional version of this construction, which is more technical to state but which requires essentially no new ideas.

Before we explain this generalisation, we make an observation about the possible radii of discs defined over non-archimedean fields. Let K be an arbitrary field complete with respect to a non-trivial non-archimedean norm. If L is a finite extension of K then there is a unique way to extend the norm on K to a norm on L, and hence there is a unique way to extend the norm on K to a norm on an algebraic closure $\overline{K}$ of K. Let $\left|\overline{K}^\times\right|$ denote the set $\left\{|x| : x \in \overline{K}^\times\right\}$. It is easily checked that this set is just $\{|y|^{1/d} : y \in K^\times, d \in \mathbf{Z}_{\geq 1}\}$. Note that for $r \in \left|\overline{K}^\times\right|$ the closed disc $B(0,r)$ with centre zero and radius r is a rigid space *defined over* K: if $r \leq 1$ then $r = |y|^{1/d}$ for some $y \in K^\times$ and $d \in \mathbf{Z}_{\geq 1}$, and one can construct $B(0,r)$ as the space associated to the affinoid algebra $K\langle T,S\rangle/(T^d - yS)$; the general case can be reduced to this by scaling. By using products of these discs, one sees that one can construct polydiscs with "fractional" radii over K.

Let M/M_0 be a fixed finite extension of non-archimedean local fields, and assume that the restriction of the norm on M is the norm on M_0. Let K denote any complete extension of M_0, again with the norm on K assumed to extend the norm on M_0. Assume moreover that K has the property that the image of any M_0-algebra homomorphism $M \to \overline{K}$ (an algebraic closure of K) lands in K. The U_t above will correspond to the case $M_0 = M = K$ of the construction below. Later on M_0 will be $\mathbf{Q}_p$ but we do not need to assume that we are in mixed characteristic yet.

Let I denote the set of M_0-algebra homomorphisms $M \to K$. We will use $|.|$ to denote the norms on both M and K, and there is of course no ambiguity here because any M_0-algebra map $i : M \to K$ will be norm-preserving. Let $\mathcal{O}$ now denote the integers of M (in particular $\mathcal{O}$ is no longer the integers of K), and let π be a uniformiser of M. Let V be the unit polydisc over K of dimension $|I|$, the number of elements of I, and think of the coordinates of V as being indexed by elements of I. Let $\mathcal{N}_K$ denote the set $\{|x| : x \in \overline{K}^\times, |x| \leq 1\}$ and let $\mathcal{N}_K^\times$ denote $\mathcal{N}_K \backslash \{1\} = \{|x| : x \in \overline{K}^\times, |x| < 1\}$.

If $\alpha = (\alpha_1, \alpha_2, \ldots,) \in K^I$ and $r \in \mathcal{N}_K$ then we define $B(\alpha, r)$ to be the K-polydisc whose L-points, for L any complete extension of K, are

$$B(\alpha,r)(L) = \{(x_1, x_2, \ldots) \in L^I : |x_i - \alpha_i| \leq r \text{ for all } i\}.$$

For example we have $V = B(0,1)$. Note that $B(\alpha, r)$ is defined over K by the comments above.

There is a natural map $M \to K^I$ which on the ith component sends m to $i(m)$, and we frequently write m_i for $i(m)$. Furthermore, we implicitly identify $m \in M$ with its natural image $(m_i)_{i \in I}$ in K^I. In particular, if $\alpha \in \mathcal{O}$ then α can be thought of as a K-point of V.

Definition *If $r \in \mathcal{N}_K$ then define $\mathbf{B}_r := \bigcup_{\alpha\in\mathcal{O}} B(\alpha, r)$, and if furthermore $r \in \mathcal{N}_K^\times$ then define $\mathbf{B}_r^\times := \bigcup_{\alpha\in\mathcal{O}^\times} B(\alpha, r)$.*

We see that $\mathbf{B}_r$ and $\mathbf{B}_r^\times$ are finite unions of polydiscs, because $B(\alpha, r) = B(\beta, r)$ if $|\alpha - \beta| \le r$, and $\mathcal{O}$ is compact. In particular $\mathbf{B}_r$ and $\mathbf{B}_r^\times$ are K-affinoid subdomains of V. The space $\mathbf{B}_r$ should be thought of as a thickening of $\mathcal{O}$ in V; similarly $\mathbf{B}_r^\times$ should be thought of as a thickening of $\mathcal{O}^\times$. One can check that for all complete extensions L of K, we have

$$\bigcap_{r\in\mathcal{N}_K} \mathbf{B}_r(L) = \mathcal{O}$$

and

$$\bigcap_{r\in\mathcal{N}_K^\times} \mathbf{B}_r^\times(L) = \mathcal{O}^\times.$$

The proof follows without too much difficulty from the fact that $\mathcal{O}$ and $\mathcal{O}^\times$ are compact subsets of the metric space K^I. One can also check that for any $r \in \mathcal{N}_K$, the space $\mathbf{B}_r$ is an affinoid subgroup of $(\mathbf{A}^1)^I$, the product of I copies of the additive group, and that if $r \in \mathcal{N}_K^\times$ then $\mathbf{B}_r^\times$ is an affinoid subgroup of $\mathbf{G}_m^I$, the product of I copies of the multiplicative group. An example of the idea continually used in the argument is that if $(y_i) \in \mathbf{B}_r^\times(L)$ for L some complete extension of K, then (y_i) is close to some element α of $\mathcal{O}^\times$, and hence (y_i^{-1}) is close to α^{-1}. See the lemma below for other examples of this type of argument.

We record some elementary properties of $\mathbf{B}_r$ and $\mathbf{B}_r^\times$ that we shall use later. Let $\gamma = \left(\begin{smallmatrix} a & b \\ c & d \end{smallmatrix}\right)$ be an element of $\mathrm{M}_2(\mathcal{O})$ with $|c| < 1$, $|d| = 1$ and $\det(\gamma) \ne 0$. Define $m \in \mathcal{N}_K$ by $|\det(\gamma)| = m$. Choose $r \in \mathcal{N}_K$, and $t \in \mathbf{Z}_{>0}$ such that $|c| \le |\pi^t|$.

Lemma 8.1 *(a) There is a map of rigid spaces $\mathbf{B}_r \to \mathbf{B}_{r|\pi^t|}^\times$ which on points sends (z_i) to $(c_i z_i + d_i)$ (where here as usual c_i denotes the image of $c \in M$ in K via the map i and so on).*

(b) There is a map of rigid spaces $\mathbf{B}_r \to \mathbf{B}_{rm}$ which on points sends (z_i) to $\left(\frac{a_i z_i + b_i}{c_i z_i + d_i}\right)$.

Proof (a) Clearly there is a map $V \to V$ sending (z_i) to $(c_i z_i + d_i)$; we must check that the image of $\mathbf{B}_r$ is contained within $\mathbf{B}_{r|\pi^t|}^\times$. Because all the rigid spaces in question are finite unions of affinoid polydiscs, it suffices to check this on L-points, for L any complete extension of K. So let (z_i) be an L-point of $\mathbf{B}_r$. Then there exists $\alpha \in \mathcal{O}$ such that $|z_i - \alpha_i| \le r \le 1$ for all $i \in I$. In particular $|z_i| \le 1$ so $|c_i z_i + d_i| = 1$, and so $c_i z_i + d_i$

is invertible. Note now that $\beta := c\alpha + d \in \mathcal{O}^\times$ and for $i \in I$ we have $|(c_i z_i + d_i) - \beta_i| = |c_i(z_i - \alpha_i)| \le |c|r \le |\pi^t|r$, which is what we wanted.

(b) A similar argument works, the key point being that if $(z_i) \in \mathbf{B}_r(L)$ and $\alpha \in \mathcal{O}$ is chosen such that $|z_i - \alpha_i| \le r$ for all i, then defining $\beta = \frac{a\alpha+b}{c\alpha+d} \in \mathcal{O}$, we see that $\left|\frac{a_i z_i + b_i}{c_i z_i + d_i} - \beta_i\right| = |\det(\gamma)_i(z_i - \alpha_i)| \le |\det(\gamma)|r = mr$ (as $|c_i z_i + d_i| = 1 = |c\alpha + d|$ as in (a)) and this is enough. □

Now let M_0, M, K be as before, and let $p > 0$ denote the residue characteristic of K. Let Γ be a profinite abelian group containing an open subgroup topologically isomorphic to $\mathbf{Z}_p^d$ for some d. If U is a rigid space over K, let $\mathcal{O}(U)$ denote the ring of rigid functions on U. Note that $\mathcal{O}$ is still the integer ring of M but this should not cause confusion. We say that a group homomorphism $\Gamma \to \mathcal{O}(U)$ (resp. $\Gamma \to \mathcal{O}(U)^\times$) is *continuous* if, for all affinoid subdomains X of U, the induced map $\Gamma \to \mathcal{O}(X)$ (resp. $\Gamma \to \mathcal{O}(X)^\times$) is continuous. We recall some results on representability of certain group functors. By a K-group we mean a group object in the category of rigid spaces over K.

Lemma 8.2 *(a) The functor from K-rigid spaces to groups, sending a space X to the group $\mathcal{O}(X)$ under addition (resp. the group $\mathcal{O}(X)^\times$ under multiplication) is represented by the K-group $\mathbf{A}^1$ (resp. $\mathbf{G}_m$), the analytification of the affine line (resp. the affine line with zero removed).*

(b) If Γ is as above, then there is a separated K-group $\mathcal{X}_\Gamma$ representing the functor which sends a K-rigid space U to the group of continuous group homomorphisms $\Gamma \to \mathcal{O}(U)^\times$. Moreover $\mathcal{X}_\Gamma$ is isomorphic to the product of an open unit polydisc and a finite rigid space over K.

(c) If L is a complete extension of K then the base change of $\mathcal{X}_\Gamma$ to L represents the functor on L-rigid spaces sending U to the group of continuous group homomorphisms $\Gamma \to \mathcal{O}(U)^\times$.

Proof

(a) If X is a rigid space and $f \in \mathcal{O}(X)$ then for all affinoid subdomains $U \subseteq X$, f induces a unique map $U \to \mathbf{A}^1$ because there exists $0 \ne \lambda \in K$ such that λf is power-bounded on U, and then there is a unique map $K\langle T\rangle \to \mathcal{O}(U)$ sending T to λf by Proposition 1.4.3/1 of [1]; everything is compatible and glues, and the result for $\mathbf{A}^1$ follows easily. Moreover, if f is in $\mathcal{O}(X)^\times$ then for every affinoid subdomain U of X, it is possible to find an affinoid annulus in $\mathbf{G}_m$ containing $f(U)$, as both $f(U)$ and $(1/f)(U)$ lie in an affinoid subdisc of $\mathbf{A}^1$. Hence $f \in \mathcal{O}(X)^\times$ gives a map $X \to \mathbf{G}_m$, and the result in the multiplicative case now follows without too much difficulty.

(b) The existence of $\mathcal{X}_\Gamma$ is Lemma 2(i) of [5]. We recall the idea of the proof: the structure theorem for topologically finitely-generated profinite abelian groups shows that Γ is topologically isomorphic to a product of groups which are either finite and cyclic, or copies of $\mathbf{Z}_p$. Now by functorial properties of products in the rigid category, it suffices to show representability in the cases $\Gamma = \mathbf{Z}_p$ and Γ a finite cyclic group. The case $\Gamma = \mathbf{Z}_p$ is treated in Lemma 1 of [5] and the remarks after it, which show that the functor is represented by the open unit disc centre 1, and the case of Γ cyclic of order n is represented by the analytification of μ_n over K. That χ_Γ has the stated structure is now clear.

(c) By functoriality it is enough to verify these base change properties in the two cases $\Gamma = \mathbf{Z}_p$ and Γ finite cyclic; but in both of these cases the result is clear. □

We now assume that $M_0 = \mathbf{Q}_p$, and hence that M is a finite extension of $\mathbf{Q}_p$. We assume (merely for notational ease) that the norms on M_0, M and K are all normalised such that $|p| = p^{-1}$. We remind the reader that if t is an element of an affinoid K-algebra and $|t| < 1$ then the power series for $\log(1+t)$ converges, and if $|t| < p^{-1/(p-1)}$ then the power series for $\exp(t)$ converges; furthermore log and exp give isomorphisms of rigid spaces from the open disc with centre 1 and radius $p^{-1/(p-1)}$ to the open disc with centre 0 and radius $p^{-1/(p-1)}$. We would like to use logs to analyse $\mathbf{B}_r^\times$ and hence are particularly interested in the spaces $\mathbf{B}_r^\times$ for $r < p^{-1/(p-1)}$; we call such r "sufficiently small". For these r, we see that the component of $\mathbf{B}_r^\times$ containing 1 is isomorphic, via the logarithm on each coordinate, to the component of $\mathbf{B}_r$ containing 0.

Recall that $\mathcal{O}$ is the integers of M, and hence $\Gamma = \mathcal{O}^\times$ satisfies the conditions just before Lemma 8.2. If $n \in \mathbf{Z}^I$ then there is a group homomorphism $\mathcal{O}^\times \to K^\times = \mathbf{G}_m(K)$, which sends α to $\prod_i \alpha_i^{n_i}$. It is easily checked (via exp and log) that if r is sufficiently small then this map is the K-points of a map of K-rigid spaces $\mathbf{B}_r^\times \to \mathbf{G}_m$. For more arithmetically complicated continuous maps from $\mathcal{O}^\times$ to invertible functions on affinoids, we might have to make r smaller still before such an analytic extension exists, but the proposition below shows that we can always do this. An important special case of this proposition is the case of an arbitrary continuous homomorphism $\mathcal{O}^\times \to K^\times$, but the proof is essentially no more difficult if $K^\times$ is replaced by the invertible functions on an arbitrary affinoid, so we work in this generality.

Proposition 8.3 *If X is a K-affinoid space and $n : \mathcal{O}^\times \to \mathcal{O}(X)^\times$ is a continuous group homomorphism, then there is at least one $r \in \mathcal{N}_K^\times$ and, for* *any such r, a unique map of K-rigid spaces $\beta_r : \mathbf{B}_r^\times \times X \to \mathbf{G}_m$, such that for all $\alpha \in \mathcal{O}^\times$, $n(\alpha)$ is the element of $\mathcal{O}(X)^\times$ corresponding (via Lemma 8.2(a)) to the map $X \to \mathbf{G}_m$ obtained by evaluating β_r at the K-valued point α of $\mathbf{B}_r^\times$.*

Definition *We call β_r a* thickening *of n.*

Proof of Proposition. It suffices to prove that there exists at least one sufficiently small r such that if B is the component of $\mathbf{B}_r^\times$ containing the K-point 1, then there is a unique $\beta : B \times X \to \mathbf{G}_m$ with a property analogous to that of β_r above for all $\alpha \in \Delta := \{\alpha \in \mathcal{O}^\times : |\alpha - 1| \leq r\}$ (or even in some subgroup of finite index). Via the logarithm map one sees that Δ is isomorphic to $\mathcal{O}$. It hence suffices to prove that for any continuous group homomorphism $\chi : \mathcal{O} \to \mathcal{O}(X)^\times$ there exists $N \in \mathbf{Z}_{\geq 0}$ such that the induced homomorphism $p^N\mathcal{O} \to \mathcal{O}(X)^\times$ is induced by a unique map of K-rigid spaces $B(0, p^{-N}) \times X \to \mathbf{G}_m$. Here $B(0, p^{-N})$ denotes the polydisc in V with radius p^{-N}, and $p^N\mathcal{O}$ is embedded as a subset of the K-points of $B(0, p^{-N})$ in the usual way. Now observe that if $d = [M : \mathbf{Q}_p]$ then as a topological group, $\mathcal{O}$ is isomorphic to $\mathbf{Z}_p^d$. Hence, if one fixes a K-Banach algebra norm on $\mathcal{O}(X)$ and a $\mathbf{Z}_p$-basis $e_1, e_2, \ldots, e_d$ of $\mathcal{O}$, one sees using Lemma 1 of [5] that there exists a positive integer N such that $|\chi(p^N e_j) - 1| < p^{-1/(p-1)}$ for all j. Observe now that $\log(\chi)$ is a continuous group homomorphism $p^N\mathcal{O} \to \mathcal{O}(X)$, and $\mathcal{O}(X)$ is a K-vector space; furthermore, the image of $p^N\mathcal{O}$ will land in a finite-dimensional K-subspace of $\mathcal{O}(X)$. It is a standard fact (linear independence of distinct field embeddings) that the continuous group homomorphisms $\mathcal{O} \to K$ form a finite-dimensional K-vector space with basis the set I, now regarded as the ring homomorphisms $\mathcal{O} \to K$, and hence there exists $f_1, f_2, \ldots, f_d \in \mathcal{O}(X)^\times$ such that for all $\alpha \in p^N\mathcal{O}$ we have $\chi(\alpha) = \exp\left(\sum_i \alpha_i f_i\right)$, where α_i denotes $i(\alpha) \in K$. By increasing N if necessary, we may assume that $\left|p^N f_i\right| < p^{-1/(p-1)}$ for all i, and we claim that this N will work. To construct $\beta : B(0, p^{-N}) \times X \to \mathbf{G}_m$ it suffices, by Lemma 8.2(a), to construct a unit in the affinoid $\mathcal{O}(X)\langle T_1, T_2, \ldots, T_d\rangle = \mathcal{O}\left(B(0, p^{-N}) \times X\right)$ which specialises to $\chi(\alpha)$ via the map sending T_i to α_i/p^N, for all $\alpha \in p^N\mathcal{O}$. The unit $\exp\left(\sum_i p^N T_i f_i\right)$ is easily seen to do the trick.

For uniqueness it suffices (again via exp and log) to prove that a map of rigid spaces $f : \mathbf{B}_1 \to \mathbf{A}^1$ which sends every element of $\mathcal{O}$ to 0 must be identically 0, that is, that $\mathcal{O}$ is Zariski-dense in $\mathbf{B}_1$. It suffices to show that f vanishes on a small polydisc centre 0, and one can check this on points. Again choose a $\mathbf{Z}_p$-basis $(e_1, e_2, \ldots, e_d)$ of $\mathcal{O}$ as a $\mathbf{Z}_p$-module. It suffices to prove that for all complete extensions L of K, f is zero on all L-points of $\mathbf{B}_1$ of the form $z_1e_1 + z_2e_2 + \ldots + z_de_d$ with $z_i \in \mathcal{O}_L$, as this contains all the L-points of a small polydisc in $\mathbf{B}_1$ by a determinant calculation. Note that if all the z_β are in $\mathbf{Z}_p$ then certainly $f(z_1e_1 + z_2e_2 + \ldots) = 0$. Now fix $z_\beta \in \mathbf{Z}_p$ for $\beta \geq 2$ and consider the function on the affinoid unit disc over L sending z_1 to $f(z_1e_1 + z_2e_2 + \ldots)$. This is a function on a closed 1-ball that vanishes at infinitely many points, and hence it is identically zero. Now fix $z_1 \in \mathcal{O}_L$ and $z_\beta \in \mathbf{Z}_p$ for $\beta \geq 3$, and let z_2 vary, and so on, to deduce that f is identically 0. □

For applications, we want to consider products of the $\mathbf{B}_r$ and $\mathbf{B}_r^\times$ constructed above. Let F denote a number field, with integers $\mathcal{O}_F$, and let p be a prime. Let $\mathcal{O}_p$ denote $\mathcal{O}_F \otimes \mathbf{Z}_p$, the product of the integer rings in the completions of F at all the primes above p, and let K_0 be the closure in $\overline{\mathbf{Q}}_p$ of the compositum of the images of all the field homomorphisms $F \to \overline{\mathbf{Q}}_p$. Then K_0 is a finite Galois extension of $\mathbf{Q}_p$. Then K_0 contains the image of any field homomorphism $F_v \to \overline{\mathbf{Q}}_p$, where v is any place of F above p, so we are in a position to apply the previous constructions with $M_0 = \mathbf{Q}_p$, $M = F_v$, and K any complete extension of K_0.

Let J denote the set of places of F above p, and let I denote the set of field homomorphisms $F \to K$. Note that each $i \in I$ extends naturally to a map $i : F_p \to K$ where $F_p = F \otimes \mathbf{Q}_p = \oplus_{j \in J} F_j$. For $j \in J$, let I_j denote the subset of I consisting of $i : F_p \to K$ which factor through the completion $F \to F_j$. Then I is the disjoint union of the I_j. For $r \in (\mathcal{N}_K)^J$ and $j \in J$ write r_j for the component of r at j. Let $\mathbf{B}_r$ (resp. $\mathbf{B}_r^\times$ if $r_j < 1$ for all j) denote the rigid space over K which is the product over $j \in J$ of the rigid spaces $\mathbf{B}_{r_j}$ (resp. $\mathbf{B}_{r_j}^\times$) defined above. Then $\mathbf{B}_r$ (resp. $\mathbf{B}_r^\times$) is a thickening of $\mathcal{O}_p$ (resp. $\mathcal{O}_p^\times$) in the unit g-polydisc over K, where now $g = [F : \mathbf{Q}]$. Indeed, it is easily checked that for all complete extensions L of K we have

$$\mathbf{B}_r(L) = \{z \in L^I \ : \text{there is } \alpha \in \mathcal{O}_p \text{ with } |z_i - \alpha_i| \leq r_i\}$$

and, when $r_i < 1$ for all i,

$$\mathbf{B}_r^\times(L) = \{z \in L^I \ : \text{there is } \alpha \in \mathcal{O}_p^\times \text{ with } |z_i - \alpha_i| \leq r_i\}$$

just as before, where, for $\alpha \in \mathcal{O}_p$, α_i denotes $i(\alpha) \in K$.

Now assume that F is totally real. Let G denote a subgroup of $\mathcal{O}_F^\times$ of finite index, and let Γ_G be the quotient of $\mathcal{O}_p^\times \times \mathcal{O}_p^\times$ by the closure of the image of G via the map $\gamma \mapsto (\gamma, \gamma^2)$. Then Γ_G is topologically isomorphic to $(\mathcal{O}_p^\times/\overline{G}) \times \mathcal{O}_p^\times$, so its dimension is related to the defect of Leopoldt's conjecture for the pair (F, p) (in particular, the dimension is at least $g+1$ and conjecturally equal to $g+1$). Let $\mathcal{X}_{\Gamma_G}$ be the rigid space associated to Γ_G in Lemma 8.2(b), and let $\mathcal{W}$ to be the direct limit $\lim \mathcal{X}_{\Gamma_G}$ as G varies over the set of subgroups of finite index of $\mathcal{O}_F^\times$, partially ordered by inclusion. The fact that $\mathcal{W}$ exists is an easy consequence of Lemma 2(iii) of [5], which shows that the transition morphisms are closed and open immersions: if $G_1 \subseteq G_2 \subseteq \mathcal{O}_F^\times$ are subgroups of $\mathcal{O}_F^\times$ of finite index and $\Gamma_i = \Gamma_{G_i}$, then there is a surjection $\Gamma_1 \to \Gamma_2$ with finite kernel, and the corresponding map $\mathcal{X}_{\Gamma_2} \to \mathcal{X}_{\Gamma_1}$ is a closed immersion which geometrically identifies $\mathcal{X}_{\Gamma_2}$ with a union of components of $\mathcal{X}_{\Gamma_1}$. In particular we see that the $\mathcal{X}_{\Gamma_G}$, as G varies through the subgroups of $\mathcal{O}_F^\times$ of finite index, form an admissible cover of $\mathcal{W}$. A K-point of $\mathcal{W}$ corresponds to a continuous group homomorphism $\mathcal{O}_p^\times \times \mathcal{O}_p^\times \to K^\times$ whose kernel contains a subgroup of $\mathcal{O}_F^\times$ of finite index (we always regard $\mathcal{O}_F^\times$ as being embedded in $\mathcal{O}_p^\times \times \mathcal{O}_p^\times$ via the map $\gamma \mapsto (\gamma, \gamma^2)$). More generally, we define a *weight* to be a continuous group homomorphism $\kappa : \mathcal{O}_p^\times \times \mathcal{O}_p^\times \to \mathcal{O}(X)^\times$, for X any affinoid, such that the kernel of κ contains a subgroup of $\mathcal{O}_F^\times$ of finite index.

If U is an affinoid K-space and $U \to \mathcal{W}$ is a map of rigid spaces, then (because the $\mathcal{X}_{\Gamma_G}$ cover $\mathcal{W}$ admissibly) there is a subgroup G of finite index of $\mathcal{O}_F^\times$ such that the image of U is contained within $\mathcal{X}_{\Gamma_G}$. In particular, by the universal property of $\mathcal{X}_{\Gamma_G}$ there is an induced continuous group homomorphism $\Gamma_G \to \mathcal{O}(U)^\times$, which induces a continuous group homomorphism $\kappa : \mathcal{O}_p^\times \times \mathcal{O}_p^\times \to \mathcal{O}(U)^\times$. By composing this map with the map $\mathcal{O}_p^\times \to \mathcal{O}_p^\times \times \mathcal{O}_p^\times$ sending γ to $(\gamma, 1)$, we get a continuous group homomorphism $n : \mathcal{O}_p^\times \to \mathcal{O}(U)^\times$, which can be written as a product over $j \in J$ of continuous group homomorphisms $n_j : \mathcal{O}_{F_j}^\times \to \mathcal{O}(U)^\times$. Hence by Proposition 8.3 there exists $r \in \left(\mathcal{N}_K^\times\right)^J$ and a map $\mathbf{B}_r^\times \times U \to \mathbf{G}_m$ giving rise to n. We call such a map a *thickening* of n. Because we have only set up our Fredholm theory on Banach modules, we will have to somehow single out one such thickening, which we do (rather arbitrarily) in the definition below. First we single out a discrete subset $\mathcal{N}_d^\times$ of $\left(\mathcal{N}_K^\times\right)^J$ as follows: let π_j denote a uniformiser of F_j and define $\mathcal{N}_d^\times \subset \left(\mathcal{N}_K^\times\right)^J$ to be the product over $j \in J$ of the sets $\{|\pi_j^t| : t \in \mathbf{Z}_{>0}\}$. We equip $\mathcal{N}_d^\times$ with the obvious partial ordering.

Definition *Let X be an affinoid and let $\kappa = (n, v) : \mathcal{O}_p^\times \times \mathcal{O}_p^\times \to \mathcal{O}(X)^\times$ be a weight. We define $r(\kappa)$ to be the largest element of $\mathcal{N}_d^\times$ such that the construction above works. Explicitly, we choose $r(\kappa)_j = |\pi_j^{t_j}|$ with $t_j \in \mathbf{Z}_{>0}$, and the t_j are chosen as small as possible such that the maps $n_j : \mathcal{O}_{F_j}^\times \to \mathcal{O}(X)^\times$ are induced by maps $\mathbf{B}_{r(\kappa)_j}^\times \times X \to \mathbf{G}_m$ and hence the map $n : \mathcal{O}_p^\times \to \mathcal{O}(X)^\times$ is induced by a map $\mathbf{B}_{r(\kappa)}^\times \times X \to \mathbf{G}_m$.*

This construction applies in particular when X is an affinoid subdomain of $\mathcal{W}$ (the inclusion $X \to \mathcal{W}$ induces a map κ as above). Note however that as the image of X in $\mathcal{W}$ gets larger, the $r(\kappa)_j$ will get smaller—there is in general no universal r and map $\mathbf{B}_r^\times \times \mathcal{W} \to \mathbf{G}_m$. Note also that the construction applies if X is a point, and in this case κ corresponds to a point of $\mathcal{W}$.

9 Classical automorphic forms.

Our exposition of the classical theory follows [12] for the most part. We recall the notation of the latter part of the previous section, and add a little more. Recall that F is a totally real field of degree g over $\mathbf{Q}$, and $\mathcal{O}_F$ is the integers of F. We fix an isomorphism $\mathbf{C} \xrightarrow{\sim} \overline{\mathbf{Q}}_p$; then we can think of I as the set of all infinite places of F, or as all the field embeddings $F \to \overline{\mathbf{Q}}_p$. Note that any such map i extends to a map $i : F_p := F \otimes \mathbf{Q}_p \to \overline{\mathbf{Q}}_p$. Recall that K is a complete extension of K_0, the compositum of the images $i(F)$ of F as i runs through I, and we may also think of I as the set of all field homomorphisms $F \to K$. We let J denote the set of primes of $\mathcal{O}_F$ dividing p. If $j \in J$ then let F_j denote the completion of F at j and let $\mathcal{O}_j$ denote the integers in F_j. We set $\mathcal{O}_p := \mathcal{O}_F \otimes \mathbf{Z}_p$; then $F_p = \oplus_{j \in J} F_j$ and $\mathcal{O}_p = \oplus_{j \in J} \mathcal{O}_j$. Choose once and for all uniformisers π_j of the local fields F_j for all j, and let $\pi \in F_p$ denote the element whose jth component is π_j. We will also use π to denote the ideal of $\mathcal{O}_F$ which is the product of the prime ideals above p. Note that some constructions (for example the Hecke operators U_{π_j} defined later) will depend to a certain extent on this choice, but others (for example the eigenvarieties we construct) will not.

Any $i \in I$ gives a map $F_p \to K$ and this map factors through the projection $F_p \to F_j$ for some $j := j(i) \in J$; hence we get a natural surjection $I \to J$. If S is any set then this surjection induces a natural injection $S^J \to S^I$, where as usual S^I denotes the set of maps $I \to S$. We continue to use the following very useful notation: if $(a_j) \in S^J$ and $i \in I$ then by a_i we mean a_j for $j = j(i)$.

Now let D be a quaternion algebra over F ramified at all infinite places. Let us assume that D is split at all places above p.[5] Let $\mathcal{O}_D$ denote a fixed maximal order of D, and fix an isomorphism $\mathcal{O}_D \otimes_{\mathcal{O}_F} \mathcal{O}_{F_v} = \mathrm{M}_2(\mathcal{O}_{F_v})$ for all finite places v of F where D splits (here F_v is the completion of F at v and $\mathcal{O}_{F_v}$ is the integers in this completion). In particular we fix an isomorphism $\mathcal{O}_D \otimes_{\mathcal{O}_F} \mathcal{O}_p = \mathrm{M}_2(\mathcal{O}_p)$, and this induces an isomorphism $D_p := D \otimes_F F_p = \mathrm{M}_2(F_p)$.

5 One can almost certainly develop some of the theory as long as at least one place above p is split, although one might have to fix the weights at the ramified places.

We recall the classical definitions of automorphic forms for D. If $n \in \mathbf{Z}^I_{\geq 0}$ then we define L_n to be the K-vector space with basis the monomials $\prod_{i\in I} Z_i^{m_i}$, where $m \in \mathbf{Z}^I_{\geq 0}$, $0 \leq m_i \leq n_i$, and where the Z_i are independent indeterminates. If $t \in \mathbf{Z}^J_{\geq 1}$ then define $\mathbf{M}_t$ to be the elements (γ_j) of $\mathrm{M}_2(\mathcal{O}_\mathrm{p}) = \prod_{\mathrm{j}\in \mathrm{J}} \mathrm{M}_2(\mathcal{O}_\mathrm{j})$ with the property that if $\gamma_j = \begin{pmatrix} a_j & b_j \\ c_j & d_j \end{pmatrix}$ then $\det(\gamma_j) \neq 0$, $\pi_j^{t_j}$ divides c_j, and π_j does not divide d_j. Then $\mathbf{M}_t$ is a monoid under multiplication. By $\mathbf{M}_1$ we mean the monoid $\mathbf{M}_t$ for $t = (1, 1, \ldots, 1)$. If $v \in \mathbf{Z}^I$ and $n \in \mathbf{Z}^I_{\geq 0}$ then define the right $\mathbf{M}_1$-module $L_{n,v}$ to be the K-vector space L_n equipped with the action of $\mathbf{M}_1$ defined by letting $(\gamma_j) = \left(\begin{pmatrix} a_j & b_j \\ c_j & d_j \end{pmatrix}\right)_{j\in J}$ send $\prod_i Z_i^{m_i}$ to $\prod_i (c_i Z_i + d_i)^{n_i} (a_i d_i - b_i c_i)^{v_i} \left(\frac{a_i Z_i + b_i}{c_i Z_i + d_i}\right)^{m_i}$ and extending K-linearly (note that here we are using the notation a_i for the image of $a_{j(i)}$ in K via the map i, as explained above). Note that in fact the same definition gives an action of $\mathrm{GL}_2(F_p)$ on $L_{n,v}$, but we never use this action.

The natural maps $\mathcal{O}_F \to \mathcal{O}_p \to \mathrm{M}_2(\mathcal{O}_\mathrm{p})$ (via the diagonal embedding) induce an embedding from $\mathcal{O}_F^\times$ into $\mathbf{M}_1$. An easy check shows that the totally positive units in $\mathcal{O}_F^\times$ act trivially on $L_{n,v}$ if $n + 2v \in \mathbf{Z}$.

Define $\mathbf{A}_{F,f}$ to be the finite adeles of F and $D_f := D \otimes_F \mathbf{A}_{F,f}$. If $x \in D_f$ then let $x_p \in D_p = \mathrm{M}_2(\mathrm{F_p})$ denote the projection onto the factor of D_f at p. If $t \in \mathbf{Z}^J_{\geq 1}$ then we say that a compact open subgroup $U \subset D_f^\times$ has *wild level* $\geq \pi^t$ if the projection $U \to D_p^\times$ is contained within $\mathbf{M}_t$. If $t = (1, 1, \ldots, 1)$ then we drop it from the notation and talk about compact open subgroups of wild level $\geq \pi$.

Say $D_f^\times = \coprod_{\lambda=1}^{\mu} D^\times \tau_\lambda U$. Then the groups $\Gamma_\lambda := \tau_\lambda^{-1} D^\times \tau_\lambda \cap U$ are finitely-generated and moreover $\tau_\lambda \Gamma_\lambda \tau_\lambda^{-1} \subset D^\times$ is commensurable with $\mathcal{O}_D^\times$ and hence with $\mathcal{O}_F^\times$. Hence Γ_λ is also commensurable with $\mathcal{O}_F^\times$. If $\mathfrak{n}$ is an ideal of $\mathcal{O}_F$ which is coprime to $\mathrm{Disc}(D)$ then we define $U_0(\mathfrak{n})$ (resp. $U_1(\mathfrak{n})$) in the usual way as being matrices in $(\mathcal{O}_D \otimes \widehat{\mathbf{Z}})^\times$ which are congruent to $\begin{pmatrix} * & * \\ 0 & * \end{pmatrix}$ (resp. $\begin{pmatrix} * & * \\ 0 & 1 \end{pmatrix}$) mod $\mathfrak{n}$. Note that for many such choices of U we see that the Γ_λ are all contained within $\mathcal{O}_F^\times$ (see, for example, Lemma 7.1 of [12] and the observation that, in Hida's notation, the groups $\overline{\Gamma}^i(U)$ are finite because D is totally definite). However we do not need to assume this because of our generalisation of Coleman's Fredholm theory.

Say $t \in \mathbf{Z}^J_{\geq 1}$, U is a compact open of wild level $\geq \pi^t$, and A is any right $\mathbf{M}_t$-module, with action written $(a, m) \mapsto a.m$. If $f : D_f^\times \to A$ and $u \in U$ then define $f|u : D_f^\times \to A$ by $(f|u)(g) := f(gu^{-1}).u_p$. Now set

$$\mathcal{L}(U, A) := \left\{ f : D^\times \backslash D_f^\times \to A \,:\, f|u = f \text{ for all } u \in U \right\}.$$

Note that $f \in \mathcal{L}(U, A)$ is determined by $f(\tau_\lambda)$ for $1 \leq \lambda \leq \mu$, and one checks easily that the map $f \mapsto (f(\tau_\lambda))_{1 \leq \lambda \leq \mu}$ induces an isomorphism

$$\mathcal{L}(U, A) \to \bigoplus_{\lambda=1}^{\mu} A^{\Gamma_\lambda}.$$

In particular, the functor $\mathcal{L}(U, -)$ is left exact. We remark that in the circumstances that will interest us later on, A will be an ONable Banach module over an affinoid in characteristic zero, the Γ_λ will all act via finite groups, and the invariants will hence be a Banach module with property (Pr). Indeed, this phenomenon was the main reason for extending Coleman's theory from ONable modules to modules with property (Pr).

If $\eta \in D_f^\times$ and $\eta_p \in \mathbf{M}_t$ then one can define an endomorphism $[U\eta U]$ of $\mathcal{L}(U, A)$ as follows: decompose $U\eta U = \coprod_i Ux_i$ (a finite union) and define

$$f|[U\eta U] := \sum_i f|x_i.$$

This operator is called the Hecke operator associated to η.

Now let $n \in \mathbf{Z}^I_{\geq 0}$ and $v \in \mathbf{Z}^I$ be such that $n + 2v \in \mathbf{Z}$. Set $k = n + 2$ and $w = v + n + 1$. Then $k - 2w \in \mathbf{Z}$ and $k \geq 2$ (that is, $k_i \geq 2$ for all i), and conversely given k and w with these properties one can of course recover n and v. We finish by recalling the definition of classical automorphic forms for D in this context. Let $U \subset D_f^\times$ be a compact open subgroup of wild level $\geq \pi$.

Definition *The space of classical automorphic forms $S^D_{k,w}(U)$ of weight (k, w) and level U for D is the space $\mathcal{L}(U, L_{n,v})$.*

This space is a finite-dimensional K-vector space. It is not, in the strict sense, a classical space of forms, because we have twisted the weight action from infinity to p. On the other hand if one chooses a field homomorphism $K \to \mathbf{C}$ then $S^D_{k,w}(U) \otimes_K \mathbf{C}$ is isomorphic to a classical space of Hilbert modular forms, as described in, for example, section 2 of [12]. We remark also that because the full group $\mathrm{GL}_2(F_p)$ acts naturally on $L_{n,v}$, our assumption that U has wild level $\geq \pi$ is unnecessary at this point. However, the forms that we shall p-adically interpolate will always have wild level $\geq \pi$, because of the standard phenomenon that to p-adically analytically interpolate forms on GL_2 one has to drop an Euler factor.

10 Overconvergent automorphic forms.

Let X be an affinoid over K, and let $\kappa = (n, v) : \mathcal{O}_p^\times \times \mathcal{O}_p^\times \to \mathcal{O}(X)^\times$ be a weight. In this section we will define $\mathcal{O}(X)$-modules of r-overconvergent automorphic forms of weight κ. In this generality, κ really is a family of weights; one important case to keep in mind is when X is a point, so n and v are continuous group homomorphisms $\mathcal{O}_p^\times \to K^\times$, the resulting spaces will then be Banach spaces over K and will be automorphic forms of a fixed weight. One important special case of this latter situation is when n and v are of the form $\alpha \mapsto \prod_i \alpha_i^{m_i}$ where the m_i are integers; the resulting spaces of automorphic forms will have a "classical weight" and there will be a natural finite-dimensional subspace corresponding to a space of classical automorphic forms as defined in the previous section.

The group homomorphism $\kappa : \mathcal{O}_p^\times \times \mathcal{O}_p^\times \to \mathcal{O}(X)^\times$ induces a map $f : X \to \mathcal{W}$ by Lemma 8.2(b) and (c). We extend the map $v : \mathcal{O}_p^\times \to \mathcal{O}(X)^\times$ to a group homomorphism $v : F_p^\times \to \mathcal{O}(X)^\times$ by defining $v(\pi_j) = 1$ for all $j \in J$. Note that this extension depends on our choice of π_j (it is analogous to Hida's choices of $\{x^v\}$ in [12]) but subsequent definitions will not depend seriously on this choice (in particular the eigenvariety we construct will not depend on this choice). Note also that the supremum semi-norm of every element in the image of n or v is 1.

Now for $r \in (\mathcal{N}_K)^J$ define $\mathcal{A}_{\kappa,r}$ to be the K-Banach algebra $\mathcal{O}(\mathbf{B}_r \times X)$. Note that $\mathcal{A}_{\kappa,r}$ does not yet depend on κ but we will define a monoid action below which does. Let us assume for simplicity that X is reduced (this is not really necessary, but will be true in practice and also gives us a canonical choice of norm on $\mathcal{O}(X)$, namely the supremum norm). Endow $\mathcal{A}_{\kappa,r}$ with the supremum norm. As usual write $\kappa = (n, v)$ and $n = \prod_{j \in J} n_j$ with $n_j : \mathcal{O}_{F_j}^\times \to \mathcal{O}(X)^\times$.

Definition *We say that $t = (t_j)_{j \in J} \in \mathbf{Z}_{>0}^J$ is* good *for the pair (κ, r) if for each $j \in J$ there is thickening of n_j to a map $\mathbf{B}^\times_{r_j|\pi_j^{t_j}|} \times X \to \mathbf{G}_m$.*

Equivalently, $t \in \mathbf{Z}_{>0}^J$ is good if $r|\pi^t| <= r(\kappa)$ in $(\mathcal{N}_K)^J$ with the obvious partial order. Given any (κ, r) as above, there will exist good $t \in \mathbf{Z}_{>0}^J$ by Proposition 8.3 (indeed, there will exist a unique minimal good t). The point of the definition is that if t is good for (κ, r) then we can define a right action of $\mathbf{M}_t$ on $\mathcal{A}_{\kappa,r}$ (we denote the action by a dot) by letting $\gamma = \left(\begin{smallmatrix} a & b \\ c & d \end{smallmatrix}\right) \in \mathbf{M}_t$ act as follows: if $h \in \mathcal{A}_{\kappa,r}$ and $(z, x) \in \mathbf{B}_r(L) \times X(L)$ for L any complete extension of K then

$$(h.\gamma)(z, x) := n(cz + d, x)\left(v\left(\det(\gamma)\right)(x)\right) h\left((az + b)/(cz + d), x\right).$$

This is really a definition "on points" but it is easily checked that $h.\gamma \in \mathcal{A}_{\kappa,r}$, using Lemma 8.1, and that the definition does give an action. It is elementary to check that for fixed $\gamma \in \mathbf{M}_t$, the map $\mathcal{A}_{\kappa,r} \to \mathcal{A}_{\kappa,r}$ defined by $h \mapsto h.\gamma$ is a continuous $\mathcal{O}(X)$-module homomorphism (but it is not in general a ring homomorphism if κ is non-trivial). The fact that n and v take values in elements of $\mathcal{O}(X)^\times$ with supremum norm 1 easily implies that $\gamma : \mathcal{A}_{\kappa,r} \to \mathcal{A}_{\kappa,r}$ is norm-decreasing. One also checks using Lemma 8.1(b) that if $|\det(\gamma_j)| = m_j$ then γ induces a continuous norm-decreasing $\mathcal{O}(X)$-module homomorphism from $\mathcal{A}_{\kappa,rm}$ to $\mathcal{A}_{\kappa,r}$.

We now have enough to define our Banach modules of overconvergent modular forms. This definition is ultimately inspired by [16], a preprint which sadly may well never see the light of day but which contained the crucial idea of beefing up a polynomial ring to a restricted power series ring in order to move from the classical to the overconvergent setting.

Definition *Let X be a reduced affinoid over K and let $X \to \mathcal{W}$ be a morphism of rigid spaces, inducing $\kappa : \mathcal{O}_p^\times \times \mathcal{O}_p^\times \to \mathcal{O}(X)^\times$. If $r \in (\mathcal{N}_K)^J$, if t is good for (κ, r), and if U is a compact open subgroup of $D_f^\times$ of wild level $\geq \pi^t$, then define the space of r-overconvergent automorphic forms of weight κ and level U to be the $\mathcal{O}(X)$-module*

$$\mathbf{S}_\kappa^D(U; r) := \mathcal{L}(U, \mathcal{A}_{\kappa,r}).$$

We remark that, just as in the case of "classical" overconvergent modular forms, the hypotheses of the definition imply that if κ is a weight near the boundary of weight space (that is, such that $r(\kappa)$ is small), then $r|\pi^t|$ must be small and hence for each j either r_j is small or there must be some large power of π_j in the level.

If $f \in \mathbf{S}_\kappa^D(U; r)$ then f is determined by $f(\tau_\lambda)$ for $\lambda = 1, \ldots, \mu$. Moreover, if $u \in U$ then so is u^{-1}, and hence both u_p and u_p^{-1} are in $\mathbf{M}_t$. In particular, both u_p and its inverse are norm-decreasing, and hence u_p is norm-preserving. We deduce that for $d \in D^\times$, $\tau \in D_f^\times$ and $u \in U$ we have $|f(d\tau u)| = |f(\tau).u_p| = |f(\tau)|$ and hence $|f(g)| \leq \max_\lambda f(\tau_\lambda)$ for all $g \in D_f^\times$. In particular we can define a norm on $\mathbf{S}_\kappa^D(U; r)$ by $|f| = \max_{g \in D_f^\times} |f(g)|$, and the isomorphism

$$\mathbf{S}_\kappa^D(U; r) \to \bigoplus_{\lambda=1}^{\mu} (\mathcal{A}_{\kappa,r})^{\Gamma_\lambda}$$

defined by $f \mapsto (f(t_\lambda))_\lambda$ is norm-preserving. Next observe that the group Γ_λ contains, with finite index, a subgroup of $\mathcal{O}_F^\times$ of finite index, and hence Γ_λ acts on $\mathcal{A}_{\kappa,r}$ via a finite quotient. Hence $\mathbf{S}_\kappa^D(U; r)$ is a direct summand of an ONable Banach $\mathcal{O}(X)$-module and our Fredholm theory applies.

11 Classical forms are overconvergent.

Fix $n \in \mathbf{Z}^I_{\geq 0}$ and $v \in \mathbf{Z}^I$ such that $n+2v \in \mathbf{Z}$. Set $k = n+2$ and $w = v+n+1$ as usual. Define $\kappa : \mathcal{O}_p^\times \times \mathcal{O}_p^\times \to K^\times$ by $\kappa(\alpha, \beta) = \prod_i \alpha_i^{n_i} \beta_i^{v_i}$. Note that κ is trivial on the totally positive units in $\mathcal{O}_F^\times$ (embedded via $\gamma \mapsto (\gamma, \gamma^2)$ as usual) and hence κ is a K-point of $\mathcal{W}$. With notation as above, we are taking $L = K$ and X a point. The map $n : \mathcal{O}_p^\times \to K^\times$ defined by $\alpha \mapsto \prod_i \alpha_i^{n_i}$ extends to a map of rigid spaces $\mathbf{B}_r^\times \to \mathbf{G}_m$ for any $r \in \left(\mathcal{N}_K^\times\right)^J$, so $r(\kappa)_j = |\pi_j|$ for all $j \in J$. If $r \in \left(\mathcal{N}_K\right)^J$ then there is a natural injection $L_{n,v} \to \mathcal{A}_{\kappa,r} = \mathcal{O}(\mathbf{B}_r)$ induced from the natural inclusion $\mathbf{B}_r \subset (\mathbf{A}^1)^I$ and one checks easily that this is an $\mathbf{M}_1$-equivariant inclusion. If $U \subset D_f^\times$ is a compact open subgroup of level $\geq \pi$ then we get an inclusion

$$S_{k,w}^D(U) = \mathcal{L}(U, L_{n,v}) \subseteq \mathcal{L}(U, \mathcal{A}_{\kappa,r}) = \mathbf{S}_\kappa^D(U; r)$$

between the finite-dimensional space of classical forms and the typically infinite-dimensional space of overconvergent ones.

This relationship between classical and overconvergent modular forms is however not quite the one that we want in general. When we construct our eigenvarieties in this setting, we will want the level structure at p to be $U_0(\pi)$, and hence we need to explain how to interpret finite slope classical forms of conductor π^2 and above, or forms with non-trivial character at p, as forms of level π and some appropriate weight. Briefly, the trick is that we firstly load the character at p into the weight of the overconvergent form, thus reducing us to level $U_0(\pi^n)$, and then decrease r and decrease the level to $U_0(\pi)$. Note that a variant of this trick is used to construct the classical eigencurve—although there the level structure is reduced only to $\Gamma_1(p)$ (or $\Gamma_1(4)$ if $p = 2$) because the p-adic zeta function may have zeroes on points of weight space which are not contained in the identity component.

Let us explain these steps in more detail. Let U_0 be a compact open subgroup of $D_f^\times$ of the form $U' \times \mathrm{GL}_2(\mathcal{O}_p)$, choose $t \in \mathbf{Z}^J_{\geq 1}$, and let U_1 denote the group $U_0 \cap U_1(\pi^t)$. Then U_1 is a normal subgroup of $U_0 \cap U_0(\pi^t)$; let Δ denote the quotient group. The map $\mathcal{O}_p^\times \to \mathrm{GL}_2(\mathcal{O}_p) \subset U_0$ sending d to $\left(\begin{smallmatrix} 1 & 0 \\ 0 & d \end{smallmatrix}\right)$ identifies Δ with the quotient $(\mathcal{O}_p/\pi^t)^\times$. If L is a complete extension of K, if $n \in \mathbf{Z}^I_{\geq 0}$ and $v \in \mathbf{Z}^I$ are chosen such that $n + 2v \in \mathbf{Z}$, and if $k = n + 2$ and $w = v + n + 1$ as usual, then for $f \in S_{k,w}^D(U_1) = \mathcal{L}(U_1, L_{n,v})$ and $u \in U_0 \cap U_0(\pi^t)$ we have $f|u^{-1} \in S_{k,w}^D(uU_1u^{-1}) = S_{k,w}^D(U_1)$ and hence there is a left action of $U_0 \cap U_0(\pi^t)$ on the finite-dimensional L-vector space $S_{k,w}^D(U_1)$ defined by letting u act by $f \mapsto f|u^{-1}$. This action is easily seen to factor through Δ, and is just the Diamond operators at primes above p in this setting. If L contains enough roots of unity then $S_{k,w}^D(U_1)$ is a direct sum of eigenspaces for this action. Choose a character $\varepsilon : \Delta \to L^\times$ and let ε also denote the induced character of $\mathcal{O}_p^\times$. Now define $\kappa : \mathcal{O}_p^\times \times \mathcal{O}_p^\times \to L^\times$ by $\kappa(\alpha,\beta) = \varepsilon(\alpha)\prod_i \alpha_i^{n_i}\beta_i^{v_i}$. The fact that $n+2v \in \mathbf{Z}$ means that κ vanishes on a subgroup of $\mathcal{O}_F^\times$ of finite index, and hence κ is a weight. One checks that $|\pi^t| \leq r(\kappa)$ and hence that if $r \in (\mathcal{N}_K)^J$ then t is good for (r,κ), so the spaces $\mathcal{L}(U_0, \mathcal{A}_{\kappa,r})$ and $\mathcal{L}(U_1, \mathcal{A}_{\kappa,r})$ are well-defined. Moreover, the natural map $L_{n,v} \to \mathcal{A}_{\kappa,r}$ is equivariant for the action of the submonoid of $\mathbf{M}_t$ consisting of matrices $\left(\begin{smallmatrix} a & b \\ c & d \end{smallmatrix}\right)$ with $\pi_j^{t_j}|(d_j - 1)$. Hence if $\mathcal{L}(U_1, L_{n,v})(\varepsilon)$ denotes the ε-eigenspace of $\mathcal{L}(U_1, L_{n,v})$ under the action of Δ, then we get an induced map $\mathcal{L}(U_1, L_{n,v})(\varepsilon) \to \mathcal{L}(U_1, \mathcal{A}_{\kappa,r})$ and unravelling the definitions one checks easily that the image of $\mathcal{L}(U_1, L_{n,v})(\varepsilon)$ is in fact contained in $\mathcal{L}(U_0, \mathcal{A}_{\kappa,r})$ (the point being that the map $L_{n,v} \to \mathcal{A}_{\kappa,r}$ is not $U_0(\pi^t)$-equivariant, and the two actions differ by ε). This construction embeds classical forms with non-trivial character at primes above p into overconvergent forms with $U_0(\pi^t)$ level structure at p, and should be thought of as the replacement in this setting of the construction of moving from a classical form of level p^n and character ε to an overconvergent function of level p^n and trivial character, by dividing by an appropriate Eisenstein series with character ϵ. Note the phenomenon, also present in the classical case, that forms in distinct eigenspaces of $S_{k,w}^D(U_1)$ for the Diamond operators above p actually become overconvergent eigenforms of distinct weights in this setting.

We now explain the relationship between forms of level $U_0(\pi)$ and forms of level $U_0(\pi^r)$ for any $r \geq 1$. Let $X \to \mathcal{W}$ be a map from a reduced affinoid to weight space, and let $\kappa = (n,v) : \mathcal{O}_p^\times \times \mathcal{O}_p^\times \to \mathcal{O}(X)$ be the induced weight. Let U be a compact open subgroup of $D_f^\times$ of the form $U' \times \mathrm{GL}_2(\mathcal{O}_p)$. Say $r \in (\mathcal{N}_K)^J$, $s \in \mathbf{Z}_{\geq 0}^J$ and $t \in \mathbf{Z}_{\geq 1}^J$ are chosen such that there is a thickening of n to $\mathbf{B}_{r|\pi^{s+t}|}^\times \times X \to \mathbf{G}_m$. Then we have defined spaces of $r|\pi^s|$-overconvergent weight κ automorphic forms of level $U \cap U_0(\pi^t)$ and also r-overconvergent weight κ automorphic forms of level $U \cap U_0(\pi^{t+s})$. We now show that these spaces are canonically isomorphic.

Proposition 11.1 *There is a canonical isomorphism*

$$\mathcal{L}(U \cap U_0(\pi^t), \mathcal{A}_{\kappa,r|\pi^s|}) \cong \mathcal{L}(U \cap U_0(\pi^{t+s}), \mathcal{A}_{\kappa,r}).$$

Remark We will see later that this isomorphism preserves the action of various Hecke operators when $U = U_1(\mathfrak{n})$ or $U_0(\mathfrak{n})$.

Proof The proof is an analogue of [5], Lemma 4, part 4, in this setting. We explain the construction of maps in both directions; it is then easy to check that these maps are well-defined and inverse to one another. As usual let π^s denote the element of $\mathcal{O}_p$ whose component at $j \in J$ is $\pi_j^{s_j}$. Then $\begin{pmatrix} \pi^s & 0 \\ 0 & 1 \end{pmatrix}$ is an element of $\mathrm{GL}_2(\mathcal{O}_p)$ and hence we can think of it as an element of $D_f^\times$.

If $f \in \mathcal{L}(U \cap U_0(\pi^t), \mathcal{A}_{\kappa,r|\pi^s|})$ then define $h : D_f^\times \to \mathcal{A}_{\kappa,r}$ by $h(g) = f\left(g\begin{pmatrix} \pi^{-s} & 0 \\ 0 & 1 \end{pmatrix}\right).\begin{pmatrix} \pi^s & 0 \\ 0 & 1 \end{pmatrix}$; note that if $\phi \in \mathcal{O}(\mathbf{B}_{r|\pi^s|} \times X)$ then $\phi.\begin{pmatrix} \pi^s & 0 \\ 0 & 1 \end{pmatrix}$ can be thought of as an element of $\mathcal{O}(\mathbf{B}_r \times X)$ as if $z \in \mathbf{B}_r(L)$ then $\pi^s z \in \mathbf{B}_{r|\pi^s|}(L)$. One checks that $h \in \mathcal{L}(U \cap U_0(\pi^{t+s}), \mathcal{A}_{\kappa,r})$.

Slightly harder work is the map the other way. First note that $\mathbf{B}_{r|\pi^s|}$ is the disjoint union of $\pi^s\mathbf{B}_r + \alpha$ as $\alpha \in \mathcal{O}_p$ runs through a set of coset representatives S for $\mathcal{O}_p/\pi^s$; hence $\mathcal{A}_{\kappa,r} = \bigoplus_{\alpha \in S} \mathcal{O}((\pi^s\mathbf{B}_r + \alpha) \times X)$. Now if $h \in \mathcal{L}(U \cap U_0(\pi^{t+s}), \mathcal{A}_{\kappa,r})$ then define $f : D_f^\times \to \mathcal{A}_{\kappa,r|\pi^s|}$ as follows: for $g \in D_f^\times$ we define $f(g)$ on $(\pi^s\mathbf{B}_r + \alpha) \times X$ by $f(g)(\pi^s z + \alpha, x) = h\left(g\begin{pmatrix} \pi^s & \alpha \\ 0 & 1 \end{pmatrix}\right)(z, x)$. One checks easily that this is well-defined (that is, independent of choice of coset representatives S). A little trickier is that $f \in \mathcal{L}(U \cap U_0(\pi^t), \mathcal{A}_{\kappa,r|\pi^s|})$, the hard part being to check that $f|u = f$ for $u \in U \cap U_0(\pi^t)$. We give a sketch of the idea, which is just algebra. Firstly one checks easily that $f|u = f$ if $u = \begin{pmatrix} 1 & \gamma \\ 0 & 1 \end{pmatrix}$ with $\gamma \in \mathcal{O}_p$. Now say $u \in U \cap U_0(\pi^t)$, and choose $\alpha \in \mathcal{O}_p$. We must check that $(f(g))(\pi^s z + \alpha, x) = ((f|u)(g))(\pi^s z + \alpha, x)$ for all $x \in X$ and $z \in \mathbf{B}_r$ (again we present the argument on points but of course this suffices). The trick is knowing how to unravel the right hand side. Because $u' := \begin{pmatrix} 1 & -\alpha \\ 0 & 1 \end{pmatrix} u \begin{pmatrix} 1 & \alpha \\ 0 & 1 \end{pmatrix} \in U \cap U_0(\pi^t)$, there exists $\beta \in \mathcal{O}_p$ such that $\begin{pmatrix} 1 & \beta \\ 0 & 1 \end{pmatrix} u' = v$ with $v_p = \begin{pmatrix} a & b \\ c & d \end{pmatrix}$ satisfying $\pi^s | b$. In particular $\begin{pmatrix} \pi^{-s} & 0 \\ 0 & 1 \end{pmatrix} v \begin{pmatrix} \pi^s & 0 \\ 0 & 1 \end{pmatrix} = v' \in U \cap U_0(\pi^{t+s})$. Hence

$$\begin{aligned} ((f|u)(g))(\pi^s z + \alpha, x) &= \left(f|v\begin{pmatrix} 1 & -\alpha \\ 0 & 1 \end{pmatrix}\right)(g)(\pi^s z + \alpha, x) \\ &= (f|v)\left(g\begin{pmatrix} 1 & \alpha \\ 0 & 1 \end{pmatrix}\right)(\pi^s z, x) \end{aligned}$$

and by expanding out the definition of $f|v$ and then using the definition of f in terms of h, one checks readily that this equals $(h|v')\left(g\begin{pmatrix} \pi^s & \alpha \\ 0 & 1 \end{pmatrix}\right)(z, x)$. We are now home, as $h|v' = h$.

Finally one checks easily that the above associations $f \mapsto h$ and $h \mapsto f$ are inverse to one another. □

No doubt one can now mimic the constructions of section 7 of [5] to deduce the existence of various canonical maps between spaces of overconvergent forms, and relate the kernels of these maps to spaces of classical forms; these maps, analogous to Coleman's θ^{k-1} operator, will not be considered here for reasons of space, as they are not necessary for the construction of eigenvarieties. The reader interested in these things might like, as an exercise, to verify that overconvergent forms of small slope are classical in this setting, following section 7 of [5].

12 Hecke operators.

Let X be a reduced affinoid over K and let $\kappa = (n, v) : X \to \mathcal{W}$ be a morphism of rigid spaces. If $r \in (\mathcal{N}_K)^J$, if $\rho = |\pi^t| \in \mathcal{N}_d^\times$ is such that n has a thickening to $\mathbf{B}_{r\rho}$, and if U is a compact open subgroup of $D_f^\times$ of wild level $\geq \pi^t$, then we have defined the r-overconvergent automorphic forms of weight κ and level U. If v is a finite place of F where D splits then we define $\eta_v \in D_f^\times$ to be the element which is the identity at all places away from v, and the matrix $\left(\begin{smallmatrix}\pi_v & 0\\ 0 & 1\end{smallmatrix}\right)$ at v, where $\pi_v \in F_v$ is a uniformiser. If v is prime to p then will not matter which uniformiser we choose, but if $v|p$ then for simplicity we use the uniformiser which we have already chosen earlier (this is really just for notational convenience though—a different choice would only change the operators we define by units). Let us assume in this section that U is a compact open of the form $U_0(\mathfrak{n}) \cap U_1(\mathfrak{r})$ for some integral ideals $\mathfrak{n}$ and $\mathfrak{r}$ of $\mathcal{O}_F$, both prime to $\mathrm{Disc}(D)$, with $\pi|\mathfrak{n}$ and π coprime to $\mathfrak{r}$. In this case, the resulting Hecke operator $T_v = [U\eta_v U]$, acting on $\mathcal{L}(U, \mathcal{A}_{\kappa,r})$, is easily checked to be independent of the choice of π_v, as long as v is prime to p. If furthermore v is prime to $\mathfrak{n}\mathfrak{r}$ then we may regard π_v as an element of the centre of $D_f^\times$ and we define S_v to be the resulting Hecke operator $[U\pi_v U]$.

A standard argument shows that the endomorphisms T_v and S_v all commute with one another. Furthermore, we have

Lemma 12.1 *The isomorphism of Proposition 11.1 is Hecke equivariant.*

Proof For the Hecke operators away from p this is essentially immediate. At primes above p things are slightly more delicate, because for $t_j \geq 1$ the natural left coset decomposition of the double coset $U_0(\pi_j^{t_j})\left(\begin{smallmatrix}\pi_j & 0\\ 0 & 1\end{smallmatrix}\right)U_0(\pi_j^{t_j})$ is $\coprod_{\alpha\in\mathcal{O}_j/\pi_j} U_0(\pi_j^{t_j})\left(\begin{smallmatrix}\pi_j & 0\\ \alpha\pi_j^{t_j} & 1\end{smallmatrix}\right)$ which depends on t_j. However, one checks easily that if (in the notation of Proposition 11.1) $f \in \mathcal{L}(U \cap U_0(\pi^t), \mathcal{A}_{r|\pi^s|})$ and h is the element of $\mathcal{L}(U \cap U_0(\pi^{t+s}), \mathcal{A}_r)$ associated to f in the proof, then $T_v h$ is indeed associated to $T_v f$, for all $v|p$, the calculation boiling down to the fact that

$$\begin{pmatrix}\pi_j & 0\\ \alpha\pi_j^{t_j} & 1\end{pmatrix}\begin{pmatrix}\pi^s & 0\\ 0 & 1\end{pmatrix} = \begin{pmatrix}\pi^s & 0\\ 0 & 1\end{pmatrix}\begin{pmatrix}\pi_j & 0\\ \alpha\pi_j^{t_j+s_j} & 1\end{pmatrix}.$$

□

If p factors in F as $\prod_j \mathfrak{p}_j^{e_j}$ then let U_j denote the Hecke operator $T_{\mathfrak{p}_j}$, let U_π denote $\prod_{j\in J} U_j$, and let η_j denote the matrix $\eta_{\mathfrak{p}_j}$.

Lemma 12.2 *The map $U_\pi : \mathbf{S}^D_\kappa(U;r) \to \mathbf{S}^D_\kappa(U;r)$ is the composite of the natural inclusion $\mathbf{S}^D_\kappa(U;r) \to \mathbf{S}^D_\kappa(U;r|\pi|)$ and a continuous norm-decreasing map $\mathbf{S}^D_\kappa(U;r|\pi|) \to \mathbf{S}^D_\kappa(U;r)$. The inclusion $\mathbf{S}^D_\kappa(U;r) \to \mathbf{S}^D_\kappa(U;r|\pi|)$ is norm-decreasing and compact, and hence U_π, considered as an endomorphism of $\mathbf{S}^D_\kappa(U;r)$, is also norm-decreasing and compact.*

Proof One checks easily that U_π is the Hecke operator $[U\eta U]$ associated to the matrix $\eta := \prod_j \eta_j$. If one decomposes $U\eta U$ into a finite disjoint union $\coprod_\delta Ux_\delta$ of cosets, then $\det((x_\delta)_p)/\det(\eta_p)$ is a unit at all places of F above p, and hence by Lemma 8.1(b) the endomorphism of $\mathcal{A}_{\kappa,r}$ induced by $(x_\delta)_p$ can be factored as the inclusion $\mathcal{A}_{\kappa,r} \subset \mathcal{A}_{\kappa,r|\pi|}$ followed by a norm-decreasing map $\mathcal{A}_{\kappa,r|\pi|} \to \mathcal{A}_{\kappa,r}$. The inclusion $\mathcal{A}_{\kappa,r} \subset \mathcal{A}_{\kappa,r|\pi|}$ is induced by the inner inclusion of affinoids $\mathbf{B}_{r|\pi|} \to \mathbf{B}_r$ and is hence compact and norm-decreasing; the result now follows easily. □

13 The characteristic power series of U_π.

We now have enough data to define the ingredients for our eigenvariety machine in this case. Let $\mathfrak{n}$ be an integral ideal of $\mathcal{O}_F$ prime to p and to $\mathrm{Disc}(D)$ (this latter hypothesis is not really necessary, but we enforce it for simplicity's sake), and set $U_0 = U_0(\mathfrak{n})$ or $U_1(\mathfrak{n})$. Define $U = U_0 \cap U_0(\pi)$; then U has wild level $\geq \pi$. If $X \subset \mathcal{W}$ is an affinoid subdomain, then set $R_X = \mathcal{O}(X)$, let $\kappa : \mathcal{O}_p^\times \times \mathcal{O}_p^\times \to \mathcal{O}(X)^\times$ denote the corresponding weights, and let $r = r(\kappa)$. Let $\mathbf{T}$ be the set of Hecke operators T_v (for v running through all the finite places of F where D splits) and S_v (for v running through all the finite places of F prime to $\mathfrak{n}p$ where D splits) defined above, and let ϕ denote the operator U_π. Define $M_X = \mathbf{S}^D_\kappa(U;r) = \mathcal{L}(U,\mathcal{A}_{\kappa,r})$. If $Y \subseteq X$ is an affinoid subdomain and κ' is the weight corresponding to Y then $r(\kappa) \leq r(\kappa')$ and hence there is an inclusion $\mathbf{B}_{r(\kappa)} \subseteq \mathbf{B}_{r(\kappa')}$. There is a canonical isomorphism $M_X \widehat{\otimes}_{R_X} R_Y = \mathcal{L}(U,\mathcal{A}_{\kappa',r(\kappa)})$, and the inclusion $\mathbf{B}_{r(\kappa)} \to \mathbf{B}_{r(\kappa')}$ induces an injection $\mathcal{A}_{\kappa',r(\kappa')} \to \mathcal{A}_{\kappa',r(\kappa)}$ and hence an injection $\alpha : M_Y \to M_X \widehat{\otimes}_{R_X} R_Y$. It is easy to check that this injection commutes with the action of all the Hecke operators T_v and S_v. We now check that α is a link; the argument is a slight variant on the usual one because we have allowed non-parallel radii of convergence in our definitions and hence have to make essential use of r-overconvergent forms with $r \notin \mathcal{N}_d^\times$.

Lemma 13.1 *If $U = U_0 \cap U_0(\pi)$ as above, if $Y \subseteq \mathcal{W}$ is a reduced affinoid with corresponding weight $\kappa : \mathcal{O}_p^\times \times \mathcal{O}_p^\times \to \mathcal{O}(Y)^\times$, and if $r, r' \in (\mathcal{N}_K)^J$*

with $r, r' \leq r(\kappa)$ and $r_j \leq r'_j$ for all j, then the natural map $\mathbf{S}^D_\kappa(U; r') \to \mathbf{S}^D_\kappa(U; r)$ is a link.

Proof It suffices to prove that α is a primitive link when $r'_j|\pi_j| < r_j \leq r'_j$ for all j. But this is not too hard: let c be the compositum of the (compact) restriction map $\mathbf{S}^D_\kappa(U; r) \to \mathbf{S}^D_\kappa(U; r'|\pi|)$ and the continuous norm-decreasing map $\beta : \mathbf{S}^D_\kappa(U; r'|\pi|) \to \mathbf{S}^D_\kappa(U; r')$ in the statement of Lemma 12.2; then it is not hard to check that αc and $c\alpha$ are both U_π as endomorphisms of their respective spaces. □

We may now apply our eigenvariety machine, and deduce the existence of an eigenvariety parametrising systems of Hecke eigenvalues on overconvergent automorphic forms, and in particular p-adically interpolating classical automorphic forms for D. The eigenvariety itself is a rigid space, the geometry of which we know very little about—indeed if we do not know Leopoldt's conjecture then we do not even know its dimension.

If one were to check (and it is no doubt not difficult, following the ideas of section 7 of [5]) that overconvergent forms of small slope were classical, then the existence of the eigenvariety implies results of Gouvêa-Mazur type for classical Hilbert modular forms over F, if $[F : \mathbf{Q}]$ is even (although there are probably more elementary ways of attacking analogues of the Gouvêa-Mazur conjectures in this setting—see for example the recent thesis of Aftab Pande).

Bibliography

[1] S. Bosch, U. Güntzer, and R. Remmert. *Non-Archimedean analysis*, volume 261 of *Grundlehren der Mathematischen Wissenschaften [Fundamental Principles of Mathematical Sciences]*. Springer-Verlag, Berlin, 1984.

[2] N. Bourbaki. *Éléments de mathématique. Part I. Les structures fondamentales de l'analyse. Livre III. Topologie générale. Chapitres III et IV*. Actual. Sci. Ind., no. 916. Hermann & Cie., Paris, 1942.

[3] N. Bourbaki. *Espaces vectoriels topologiques. Chapitres 1 à 5*. Masson, Paris, new edition, 1981. Éléments de mathématique. [Elements of mathematics].

[4] K. Buzzard. Analytic continuation of overconvergent eigenforms. *J. Amer. Math. Soc.*, 16(1):29–55 (electronic), 2003.

[5] K. Buzzard. On p-adic families of automorphic forms. *Progress in Mathematics*, 224:23–44, 2004.

[6] K. Buzzard and F. Calegari. The 2-adic eigencurve is proper. *Documenta Math.* Extra Volume: John H. Coates' Sixtieth Birthday (2006) 211–232.

[7] K. Buzzard and L. J. P. Kilford. The 2-adic eigencurve at the boundary of weight space. *Compos. Math.*, 141(3):605–619, 2005.

[8] G. Chenevier. Familles p-adiques de formes automorphes pour $\mathrm{GL}(n)$. *Journal für die reine und angewandte Mathematik*, 570:143–217, 2004.

[9] R. Coleman and B. Mazur. The eigencurve. In *Galois representations in arithmetic algebraic geometry (Durham, 1996)*, volume 254 of *London Math. Soc. Lecture Note Ser.*, pages 1–113. Cambridge Univ. Press, Cambridge, 1998.

[10] R. Coleman. p-adic Banach spaces and families of modular forms. *Invent. Math.*, 127(3):417–479, 1997.

[11] B. Conrad. Modular curves and rigid-analytic spaces. *Pure Appl. Math. Q.*, 2(1):29–110, 2006.

[12] H. Hida. On p-adic Hecke algebras for GL_2 over totally real fields. *Ann. of Math. (2)*, 128(2):295–384, 1988.

[13] H. Hida. *Elementary theory of L-functions and Eisenstein series*, volume 26 of *London Mathematical Society Student Texts*. Cambridge University Press, Cambridge, 1993.

[14] H. Matsumura. *Commutative ring theory*, volume 8 of *Cambridge Studies in Advanced Mathematics*. Cambridge University Press, Cambridge, second edition, 1989. Translated from the Japanese by M. Reid.

[15] J-P. Serre. Endomorphismes complètement continus des espaces de Banach p-adiques. *Inst. Hautes Études Sci. Publ. Math.*, (12):69–85, 1962.

[16] G. Stevens. Overconvergent modular symbols. Preprint.

[17] A. Yamagami. On p-adic families of Hilbert cusp forms of finite slope. Preprint.

Nontriviality of Rankin-Selberg L-functions and CM points

Christophe Cornut

Institut de Mathématiques de Jussieu
4 place Jussieu
F-75005 Paris
France
cornut@math.jussieu.fr

Vinayak Vatsal

University of British Columbia
Department of Mathematics
Vancouver,
BC V6T 1Z2
Canada
vatsal@math.ubc.ca

Contents

1 Introduction

1.1 Rankin-Selberg L-functions

Let π be an irreducible cuspidal automorphic representation of GL_2 over a totally real number field F. Let K be a totally imaginary quadratic extension of F. Given a quasi-character χ of $\mathbf{A}_K^\times/K^\times$, we denote by $L(\pi, \chi, s)$ the Rankin-Selberg L-function associated to π and $\pi(\chi)$, where $\pi(\chi)$ is the automorphic representation of GL_2 attached to χ – see [19] and [18] for the definitions. This L-function, which is first defined as a product of Euler factors over all places of F, is known to have a meromorphic extension to $\mathbf{C}$ with functional equation

$$L(\pi, \chi, s) = \epsilon(\pi, \chi, s) L(\tilde{\pi}, \chi^{-1}, 1 - s)$$

where $\tilde{\pi}$ is the contragredient of π and $\epsilon(\pi, \chi, s)$ is a certain ϵ-factor.

Let $\omega : \mathbf{A}_F^\times/F^\times \to \mathbf{C}^\times$ be the central quasi-character of π. The condition

$$\chi \cdot \omega = 1 \quad \text{on} \quad \mathbf{A}_F^\times \subset \mathbf{A}_K^\times \tag{1.1}$$

implies that $L(\pi, \chi, s)$ is entire and equal to $L(\tilde{\pi}, \chi^{-1}, s)$. The functional equation thus becomes

$$L(\pi, \chi, s) = \epsilon(\pi, \chi, s) L(\pi, \chi, 1 - s)$$

and the parity of the order of vanishing of $L(\pi, \chi, s)$ at $s = 1/2$ is determined by the value of

$$\epsilon(\pi, \chi) \stackrel{\text{def}}{=} \epsilon(\pi, \chi, 1/2) \in \{\pm 1\}.$$

We say that the pair (π, χ) is *even* or *odd*, depending upon whether $\epsilon(\pi, \chi)$ is $+1$ or -1. It is expected that the order of vanishing of $L(\pi, \chi, s)$ at $s = 1/2$ should 'usually' be minimal, meaning that either $L(\pi, \chi, 1/2)$ or $L'(\pi, \chi, 1/2)$ should be nonzero, depending upon whether (π, χ) is even or odd.

Calculation of sign

For the computation of $\epsilon(\pi, \chi)$, one first writes it as the product over all places v of F of the local signs $\epsilon(\pi_v, \chi_v)$ which are attached to the local components of π and χ, normalized as in [11, Section 9]. Let η be the quadratic Hecke character of F attached to K/F, and denote by η_v and ω_v the local components of η and ω. Then

$$\epsilon(\pi, \chi) = (-1)^{\#S(\chi)} \tag{1.2}$$

where

$$S(\chi) \stackrel{\text{def}}{=} \{v;\ \epsilon(\pi_v, \chi_v) \neq \eta_v \cdot \omega_v(-1)\}. \tag{1.3}$$

Indeed, $S(\chi)$ is finite because $\epsilon(\pi_v, \chi_v) = 1 = \eta_v \cdot \omega_v(-1)$ for all but finitely many v's, and (1.2) then follows from the product formula: $\eta \cdot \omega(-1) = 1 = \prod_v \eta_v \cdot \omega_v(-1)$.

The various formulae for the local ϵ-factors that are spread throughout [19] and [18] allow us to decide whether a given place v of F belongs to $S(\chi)$, provided that the local components π_v and $\pi(\chi_v)$ of π and $\pi(\chi)$ are not simultaneously supercuspidal. At the remaining places, one knows that χ ramifies and we may use a combination of [19, Proposition 3.8] and [18, Theorem 20.6] to conclude that, when χ is *sufficiently ramified* at v, v does not belong to $S(\chi)$. For our purposes, we just record the following facts.

For any finite place v of F in $S(\chi)$, K_v is a field and π_v is either special or supercuspidal. Conversely, if v is inert in K, χ is unramified at v and π_v is either special or supercuspidal, then v belongs to $S(\chi)$ if and only if the v-adic valuation of the conductor of π_v is odd. Finally an archimedean (real) place v of F belongs to $S(\chi)$ if $\chi_v = 1$ and π_v is the holomorphic discrete series of weight $k_v \geq 2$.

Ring class characters

In this paper, we regard the automorphic representation π and the field K as being fixed and let χ vary through the collection of ring class characters of P-power conductor, where P is a fixed maximal ideal in the ring of integers $\mathcal{O}_F \subset F$.

Here, we say that χ is a *ring class character* if there exists some $\mathcal{O}_F$-ideal $\mathcal{C}$ such that χ factors through the finite group

$$\mathbf{A}_K^\times / K^\times K_\infty^\times \widehat{\mathcal{O}}_{\mathcal{C}}^\times \simeq \mathrm{Pic}(\mathcal{O}_{\mathcal{C}})$$

where $K_\infty = K \otimes \mathbf{R}$ and $\mathcal{O}_{\mathcal{C}} \stackrel{\text{def}}{=} \mathcal{O}_F + \mathcal{C}\mathcal{O}_K$ is the $\mathcal{O}_F$-order of conductor $\mathcal{C}$ in K. The conductor $c(\chi)$ of χ is the largest such $\mathcal{C}$. Note that this definition differs from the classical one — the latter yields an ideal $c'(\chi)$ of $\mathcal{O}_K$ such that $c'(\chi) \mid c(\chi)\mathcal{O}_K$.

Equivalently, a ring class character is a finite order character whose restriction to $\mathbf{A}_F^\times$ is *everywhere unramified.* In view of (1.1), it thus make sense to require that ω is a finite order, everywhere unramified character of $\mathbf{A}_F^\times/F^\times$. Then, there are ring class characters of conductor P^n satisfying (1.1) for any sufficiently large n. Concerning our fixed representation π, we also require that

> π is cuspidal of parallel weight $(2, \cdots, 2)$ and level $\mathcal{N}$, and the prime-to-P part $\mathcal{N}'$ of $\mathcal{N}$ is relatively prime to the discriminant $\mathcal{D}$ of K/F.

In this situation, we can give a fairly complete description of $S(\chi)$.

Lemma 1.1 *For a ring class character χ of conductor P^n, $S(\chi) = S$ or $S \cup \{P\}$, where S is the union of all archimedean places of F, together with those finite places of F which do not divide P, are inert in K, and divide $\mathcal{N}$ to an odd power. Moreover, $S(\chi) = S$ if either P does not divide $\mathcal{N}$, or P splits in K, or n is sufficiently large.*

Remark 1.2 Note that π_v is indeed special or supercuspidal for any finite place v of F which divides $\mathcal{N}$ to an odd power, as the conductor of a principal series representation with unramified central character is necessarily a square.

It follows that the sign of the functional equation essentially does not depend upon χ, in the sense that for all but finitely many ring class characters of P-power conductor,

$$\epsilon(\pi, \chi) = (-1)^{|S|} = (-1)^{[F:\mathbf{Q}]}\eta(\mathcal{N}').$$

If $P \nmid \mathcal{N}$ or splits in K, this formula even holds for all χ's. We say that the triple (π, K, P) is *definite* or *indefinite* depending upon whether this generic sign $(-1)^{[F:\mathbf{Q}]}\eta(\mathcal{N}')$ equals $+1$ or -1.

Exceptional cases

In the definite case, it might be that the L-function $L(\pi, \chi, s)$ actually factors as the product of two odd L-functions, and therefore vanishes to order at least 2. This leads us to what Mazur calls the *exceptional case.*

Definition 1.3 We say that (π, K) is exceptional if $\pi \simeq \pi \otimes \eta$.

This occurs precisely when $\pi \simeq \pi(\alpha)$ for some quasi-character α of $\mathbf{A}_K^\times/K^\times$ in which case $L(\pi, \chi, s) = L(\alpha\chi, s) \cdot L(\alpha\chi', s)$ where χ' is the outer twist of χ by $\mathrm{Gal}(K/F)$. Moreover, both factors have a functional equations with sign ± 1 and it can and does happen that both signs are -1, in which case $L(\pi, \chi, s)$ has at least a double zero. It is then more natural to study the individual factors than the product; this is the point of view taken in [23]. In this paper, we will always assume that (π, K) is not exceptional when (π, K, P) is definite.

Mazur's conjectures

With this convention, we now do expect, in the spirit of Mazur's conjectures in [20], that the order of vanishing of $L(\pi, \chi, 1/2)$ should generically be 0 in the definite case and 1 in the indefinite case. Let us say that χ is *generic* if it follows this pattern. We will show that there are many generic χ's of conductor P^n for all sufficiently large n.

More precisely, let $K[P^n]/K$ be the abelian extension of K associated by class field theory to the subgroup $K^\times K_\infty^\times \widehat{\mathcal{O}}_{P^n}^\times$ of $\mathbf{A}_K^\times$, so that

$$G(n) \stackrel{\text{def}}{=} \mathrm{Gal}(K[P^n]/K) \simeq \mathbf{A}_K^\times / K^\times K_\infty^\times \widehat{\mathcal{O}}_{P^n}^\times \simeq \mathrm{Pic}(\mathcal{O}_{P^n}).$$

Put $K[P^\infty] = \cup K[P^n]$, $G(\infty) = \mathrm{Gal}(K[P^\infty]/K) = \varprojlim G(n)$ and let G_0 be the torsion subgroup of $G(\infty)$. It is shown in section 2 below that G_0 is a finite group and $G(\infty)/G_0$ is a free $\mathbf{Z}_p$-module of rank $[F_P : \mathbf{Q}_p]$, where p is the residue characteristic of P. Moreover, the reciprocity map of K maps $\mathbf{A}_F^\times \subset \mathbf{A}_K^\times$ onto a subgroup $G_2 \simeq \mathrm{Pic}(\mathcal{O}_F)$ of G_0 (the missing group G_1 will make an appearance latter). Using this reciprocity map to identify ring class characters of P-power conductor with finite order characters of $G(\infty)$, and ω with a character of G_2, we see that the condition (1.1) on χ is equivalent to the requirement that $\chi \cdot \omega = 1$ on G_2.

Conversely, a character χ_0 of G_0 induces a character on $\mathbf{A}_F^\times$, and it make sense therefore to require that $\chi_0 \cdot \omega = 1$ on $\mathbf{A}_F^\times$. Given such a character, we denote by $P(n, \chi_0)$ the set of characters of $G(n)$ which induce χ_0 on G_0 and do not factor through $G(n-1)$ – these are just the ring class characters of conductor P^n which, beyond (1.1), satisfy the stronger requirement that $\chi = \chi_0$ on G_0.

Theorem 1.4 *Let the data of (π, K, P) be given and* definite. *Let χ_0 be any character of G_0 with $\chi_0 \cdot \omega = 1$ on $\mathbf{A}_F^\times$. Then for all n sufficiently large, there exists a character $\chi \in P(n, \chi_0)$ for which $L(\pi, \chi, 1/2) \neq 0$.*

For the indefinite case, we obtain a slightly more restrictive result.

Theorem 1.5 *Let the data of (π, K, P) be given and* indefinite. *Suppose also that $\omega = 1$, and that $\mathcal{N}$, $\mathcal{D}$ and P are pairwise coprime. Let χ_0 be any character of G_0 with $\chi_0 = 1$ on $\mathbf{A}_F^\times$. Then for all n sufficiently large, there exists a character $\chi \in P(n, \chi_0)$ for which $L'(\pi, \chi, 1/2) \neq 0$.*

We prove these theorems using Gross-Zagier formulae to reduce the nonvanishing of L-functions and their derivatives to the nontriviality of certain CM points. The extra assumptions in the indefinite case are due to the fact that these formulae are not yet known in full generality, although great progress has been made by Zhang [32, 31, 33] in extending the original work of Gross and Zagier. We prove the relevant statements about CM points without these restrictions.

1.2 Gross-Zagier Formulae

Roughly speaking, the general framework of a Gross-Zagier formula yields a discrete set of *CM points* on which the Galois group of the maximal abelian extension of K acts continuously, together with a function ψ on this set with values in a complex vector space such that the following property holds: a character χ as above is generic if and only if

$$\mathbf{a}(x, \chi) \stackrel{\text{def}}{=} \int_{\mathrm{Gal}_K^{\mathrm{ab}}} \chi(\sigma)\psi(\sigma \cdot x) d\sigma \neq 0 \tag{1.4}$$

where x is any CM point whose conductor equals that of χ – we will see that CM points have conductors. Note that the above integral is just a finite sum.

In the indefinite case, the relevant set of CM points consists of those special points with complex multiplication by K in a certain Shimura curve M defined over F, and ψ takes its values in (the complexification of) the Mordell-Weil groups of a suitable quotient A of $J = \mathrm{Pic}^0_{M/F}$. In the definite case, a finite set M plays the role of the Shimura curve. The CM points project onto this M and the function ψ is the composite of this projection with a suitable complex valued function on M.

Quaternion algebras

In both cases, these objects are associated to a quaternion algebra B over F whose isomorphism class is uniquely determined by π, K and P. To describe this isomorphism class, we just need to specify the set $\mathrm{Ram}(B)$ of places of F where B ramifies. In the definite case, the set S of Lemma 1.1 has even order and we take $\mathrm{Ram}(B) = S$, so that B is totally definite. In the indefinite case, S is odd but it still contains all the archimedean (real) places of F. We fix arbitrarily a real place τ of F and take $\mathrm{Ram}(B) = S - \{\tau\}$.

We remark here that in both cases, B splits at P. Moreover, the Jacquet-Langlands correspondence implies that there is a unique cuspidal automorphic representation π' on B associated to $\pi = \mathrm{JL}(\pi')$, and π' occurs with multiplicity one in the space of automorphic cuspforms on B – this is the space denoted by $\mathcal{A}_0(G)$ in [30]. Finally, since K_v is a field for all v's in S, we may embed K into B as a maximal commutative F-subalgebra. We fix such an embedding.

Let $G \stackrel{\text{def}}{=} \mathrm{Res}_{F/\mathbf{Q}}(B^\times)$ be the algebraic group over $\mathbf{Q}$ whose set of points on a commutative $\mathbf{Q}$-algebra A is given by $G(A) = (B \otimes A)^\times$. Thus, G is a reductive group with center $Z \stackrel{\text{def}}{=} \mathrm{Res}_{F/\mathbf{Q}}(F^\times)$ and the reduced norm $\mathrm{nr} : B \to F$ induces a morphism $\mathrm{nr} : G \to Z$ which also identifies Z with the cocenter $G/[G,G]$ of G. Our chosen embedding $K \hookrightarrow B$ allows us to view $T \stackrel{\text{def}}{=} \mathrm{Res}_{F/\mathbf{Q}}(K^\times)$ as a maximal subtorus of G which is defined over $\mathbf{Q}$.

CM points

For any compact open subgroup H of $G(\mathbf{A}_f)$, we define a set of CM points by

$$\mathrm{CM}_H \stackrel{\text{def}}{=} T(\mathbf{Q})\backslash G(\mathbf{A}_f)/H.$$

There is an action of $T(\mathbf{A}_f)$ on CM_H, given by left multiplication in $G(\mathbf{A}_f)$. This action factors through the reciprocity map

$$\mathrm{rec}_K : T(\mathbf{A}_f) \twoheadrightarrow \mathrm{Gal}_K^{\mathrm{ab}}$$

and thus defines a Galois action on CM_H. For $x = [g]$ in CM_H (with g in $G(\mathbf{A}_f)$), the stabilizer of x in $T(\mathbf{A}_f)$ equals

$$U(x) \stackrel{\text{def}}{=} T(\mathbf{Q}) \cdot (T(\mathbf{A}_f) \cap gHg^{-1})$$

and we say that x is *defined* over the abelian extension of K which is fixed by $\mathrm{rec}_K(U(x))$. When $H = \widehat{R}^\times$ for some $\mathcal{O}_F$-order $R \subset B$, $T(\mathbf{A}_f) \cap gHg^{-1} = \widehat{\mathcal{O}(x)}^\times$ for some $\mathcal{O}_F$-order $\mathcal{O}(x) \subset K$, and we define the conductor of x to be that of $\mathcal{O}(x)$. In particular, a CM point of conductor P^n is defined over $K[P^n]$.

We shall also need a somewhat more technical notion, namely that of a *good* CM point.

Definition 1.6 Assume therefore that $H = \widehat{R}^\times$ as above, and that the P-component of R is an Eichler order of level P^δ in $B_P \simeq M_2(F_P)$. Then R is uniquely expressed as the (unordered) intersection of two $\mathcal{O}_F$-orders R_1 and R_2 in B, which are both maximal at P but agree with R outside P. We say that a CM point $x = [g] \in \mathrm{CM}_H$ is *good* if either $\delta = 0$ or $K_P \cap g_P R_1 g_P^{-1} \neq K_P \cap g_P R_2 g_P^{-1}$, and we say that x is *bad* otherwise.

It is relatively easy to check that if CM_H contains any CM point of P-power conductor, then it contains good CM points of conductor P^n for all sufficiently large n.

Automorphic forms

Let $\mathcal{S}$ denote the space of automorphic forms on B in the definite case, and the space of automorphic cuspforms on B in the indefinite case. As a first step towards the construction of the function ψ of (1.4), we shall now specify a certain line $\mathbf{C} \cdot \Phi$ in the realization $\mathcal{S}(\pi')$ of π' in $\mathcal{S}$. We will first define an admissible $G(\mathbf{A}_f)$-submodule $\mathcal{S}_2$ of $\mathcal{S}$, using the local behavior of π' at infinity. The line we seek then consists of those vectors in $\mathcal{S}_2(\pi') = \mathcal{S}_2 \cap \mathcal{S}(\pi')$ which are fixed by a suitable compact open subgroup H of $G(\mathbf{A}_f)$. We refer to [10] for a more comprehensive discussion of these issues.

Recall from [19] that $\mathcal{S}$ and π' are representations of $G(\mathbf{A}_f) \times \mathcal{H}_\infty$ where $\mathcal{H}_\infty$ is a certain sort of group algebra associated to $G(\mathbf{R})$. As a representation of $\mathcal{H}_\infty$, π' is the direct sum of copies of the irreducible representation $\pi'_\infty = \otimes_{v|\infty} \pi'_v$ of $\mathcal{H}_\infty$. Let V_∞ be the representation space of π'_∞. We claim that V_∞ is one dimensional in the definite case, while V_∞ has a "weight decomposition"

$$V_\infty = \oplus_{k \in 2\mathbf{Z} - \{0\}} V_{\infty,k} \tag{1.5}$$

into one dimensional subspaces in the indefinite case. Indeed, the compatibility of the global and local Jacquet-Langlands correspondence, together with our assumptions on $\pi = \mathrm{JL}(\pi')$, implies that for a real place v of F, π'_v is the trivial one dimensional representation of $B_v^\times$ if v ramifies in B, while for $v = \tau$ in the indefinite case, $\pi'_v \simeq \pi_v$ is the holomorphic discrete series of weight 2 which is denoted by σ_2 in [6, section 11.3], and the representation space of σ_2 is known to have a weight decomposition similar to (1.5).

Remark 1.7 In the indefinite case, the above decomposition is relative to the choice of an isomorphism between B_τ and $M_2(\mathbf{R})$. Given such an isomorphism, the subspace $V_{\infty,k}$ consists of those vectors in V_∞ on which $SO_2(\mathbf{R})$ acts by the character

$$\begin{pmatrix} \cos(\theta) & \sin(\theta) \\ -\sin(\theta) & \cos(\theta) \end{pmatrix} \mapsto e^{2ki\theta}.$$

Definition 1.8 We denote by $\mathcal{S}_2$ the admissible $G(\mathbf{A}_f)$-submodule of $\mathcal{S}$ which is the image of the $G(\mathbf{A}_f)$-equivariant morphism

$$\mathrm{Hom}_{\mathcal{H}_\infty}(V_\infty, \mathcal{S}) \hookrightarrow \mathcal{S} : \quad \varphi \mapsto \varphi(v_\infty)$$

where v_∞ is any nonzero element of V_∞ in the definite case, and any nonzero element of $V_{\infty,2}$ (a *lowest weight* vector) in the indefinite case. By construction, the $G(\mathbf{A}_f)$-submodule $\mathcal{S}_2(\pi') = \mathcal{S}_2 \cap \mathcal{S}(\pi')$ of $\mathcal{S}_2$ is isomorphic to $\mathrm{Hom}_{\mathcal{H}_\infty}(V_\infty, \mathcal{S}(\pi'))$. It is therefore irreducible.

Level subgroups

Turning now to the construction of H, let δ be the exponent of P in $\mathcal{N}$, so that $\mathcal{N} = P^\delta \mathcal{N}'$. Let $R_0 \subset B$ be an Eichler order of level P^δ such that the conductor of the $\mathcal{O}_F$-order $\mathcal{O} = \mathcal{O}_K \cap R_0$ is a power of P. The existence of R_0 is given by [29, II.3], and we may even require that $\mathcal{O} = \mathcal{O}_K$ if P does not divide $\mathcal{N}$ or splits in K. On the other hand, recall that the reduced discriminant of B/F is the squarefree product of those primes of F which are inert in K and divide $\mathcal{N}'$ to an odd power. We may thus find an ideal $\mathcal{M}$ in $\mathcal{O}_K$ such that

$$\mathrm{Norm}_{K/F}(\mathcal{M}) \cdot \mathrm{Disc}_{B/F} = \mathcal{N}'.$$

We then take

$$H \stackrel{\text{def}}{=} \widehat{R}^\times \quad \text{where} \quad R \stackrel{\text{def}}{=} \mathcal{O} + \mathcal{M} \cap \mathcal{O} \cdot R_0. \tag{1.6}$$

Note that R is an $\mathcal{O}_F$-order of reduced discriminant $\mathcal{N}$ in B. Since $R_P = R_{0,P}$ is an Eichler order (of level P^δ), we have the notion of good and bad CM points on CM_H. Since $x = [1]$ is a CM point of P-power conductor (with $\mathcal{O}(x) = \mathcal{O}$), there are good CM points of conductor P^n for all sufficiently large n.

We claim that $\mathcal{S}_2(\pi')^H$ is 1-dimensional. Indeed, for every finite place v of F, $(\pi'_v)^{R_v^\times}$ is 1-dimensional: this follows from [3, Theorem 1] when v does not divide $\mathcal{N}'$ (including $v = P$) and from [10, Proposition 6.4], or a mild generalization of [32, Theorem 3.2.2] in the remaining cases.

1.3 The indefinite case

Suppose first that (π, K, P) is indefinite, so that

$$B \otimes \mathbf{R} = \prod_{v|\infty} B_v \simeq M_2(\mathbf{R}) \times \mathbf{H}^{[F:\mathbf{Q}]-1}$$

where $\mathbf{H}$ is Hamilton's quaternion algebra and the $M_2(\mathbf{R})$ factor corresponds to $v = \tau$. We fix such an isomorphism, thus obtaining an action of

$$G(\mathbf{R}) \simeq GL_2(\mathbf{R}) \times (\mathbf{H}^\times)^{[F:\mathbf{Q}]-1}$$

on $X \stackrel{\text{def}}{=} \mathbf{C} - \mathbf{R}$ by combining the first projection with the usual action of $GL_2(\mathbf{R})$ on X.

For any compact open subgroup H of $G(\mathbf{A}_f)$, we then have a Shimura curve $\mathrm{Sh}_H(G, X)$ whose complex points are given by

$$\mathrm{Sh}_H(G, X)(\mathbf{C}) = G(\mathbf{Q})\backslash (G(\mathbf{A}_f)/H \times X).$$

The reflex field of this curve is the subfield $\tau(F)$ of $\mathbf{C}$, and its pull-back to F is a smooth curve M_H over F whose isomorphism class does not depend upon our choice of τ. When $S = \varnothing$, $F = \mathbf{Q}$, $G = GL_2$ and the M_H's are the classical (affine) modular curves over $\mathbf{Q}$. These curves can be compactified by adding finitely many cusps, and we denote by M_H^* the resulting proper curves. In all other cases, M_H is already proper over F and we put $M_H^* = M_H$. We denote by J_H the connected component of the relative Picard scheme of M_H^*/F.

Let x be the unique fixed point of $T(\mathbf{R})$ in the upper half plane $X^+ \subset X$. The map $g \mapsto (g, x)$ then defines a bijection between CM_H and the set of *special points with complex multiplication by* K in M_H. It follows from Shimura's theory that these points are defined over the maximal abelian extension K^{ab} of K, and that the above bijection is equivariant with respect to the Galois actions on both sides.

On the other hand, there is a natural $G(\mathbf{A}_f)$-equivariant isomorphism between the subspace $\mathcal{S}_2$ of $\mathcal{S}$ and the inductive limit (over H) of the spaces of holomorphic differentials on M_H^*. This is well-known in the classical case where $S = \varnothing$ – see for instance [6], section 11 and 12. For the general case, we sketch a proof in section 3.6 of this paper.

In particular, specializing now to the level structure H defined by (1.6), we obtain a line $\mathcal{S}_2(\pi')^H = \mathbf{C} \cdot \Phi$ in the space $\mathcal{S}_2^H$ of holomorphic differentials on M_H^*, a space isomorphic to the cotangent space of $J_H/\mathbf{C}$ at 0. By construction, this line is an eigenspace for the action of the universal Hecke algebra $\mathbf{T}_H$, with coefficients in $\mathbf{Z}$, which is associated to our H. Since the action of $\mathbf{T}_H$ on the cotangent space factors through $\mathrm{End}_F J_H$, the annihilator of $\mathbf{C} \cdot \Phi$ in $\mathbf{T}_H$ cuts out a quotient A of J_H:

$$A \stackrel{\text{def}}{=} J_H / \mathrm{Ann}_{\mathbf{T}_H}(\mathbf{C} \cdot \Phi) \cdot J_H.$$

The Zeta function of A is essentially the product of the L-function of π together with certain conjugates – see [32, Theorem B] for a special case.

The function ψ of (1.4) is now the composite of

- the natural inclusions $\mathrm{CM}_H \hookrightarrow M_H \hookrightarrow M_H^*$,
- a certain morphism $\iota_H \in \mathrm{Mor}(M_H^*, J_H) \otimes \mathbf{Q}$, and
- the quotient map $J_H \twoheadrightarrow A$.

In the classical case where $S = \varnothing$, ι_H is a genuine morphism $M_H^* \to J_H$ which is defined using the cusp at ∞ on M_H^*. In the general case, one has to use the so-called *Hodge class*. For a discussion of the Hodge class, we refer to [33, section 6], or [11, section 23]. A variant of this construction, adapted to our purposes, is given in section 3.5 below.

Statement of results

Now, let χ be a ring class character of conductor P^n such that $\chi \cdot \omega = 1$ on $\mathbf{A}_F^\times$. Suppose also that $\epsilon(\pi, \chi) = -1$: this holds true for any $n \geq 0$ if $P \nmid \mathcal{N}$ or P splits in K, but only for $n \gg 0$ in the general case. Then $L(\pi, \chi, 1/2) = 0$ and the Birch and Swinnerton-Dyer conjecture predicts that the χ^{-1}-component of $A(K[P^n]) \otimes \mathbf{C}$ should be non-trivial. If moreover $L'(\pi, \chi, 1/2) \neq 0$, the Gross-Zagier philosophy tells us more, namely that this non-triviality should be accounted for by the CM points of conductor P^n: if x is such a point, there should exists a formula relating $L'(\pi, \chi, 1/2)$ to the canonical height of

$$\mathbf{a}(x, \chi) \stackrel{\text{def}}{=} \frac{1}{|G(n)|} \sum_{\sigma \in G(n)} \chi(\sigma)\psi(\sigma x) \in A(K[P^n]) \otimes \mathbf{C},$$

thereby showing that $L'(\pi, \chi, 1/2)$ is nonzero precisely when $\mathbf{a}(x, \chi)$ is a nonzero element in the χ^{-1}-component of $A(K[P^n]) \otimes \mathbf{C}$.

Unfortunately, such a formula has not yet been proven in this degree of generality. For our purposes, the most general case of which we are aware is Theorem 6.1 of Zhang's paper [33], which gives a precise formula of this type under the hypotheses that the central character of π is trivial and that $\mathcal{N}$, $\mathcal{D}$ and P are pairwise prime to each other.

Remark 1.9 We point out that Zhang works with the Shimura curves attached to G/Z instead of G, and uses $\mathbf{a}(x, \chi^{-1})$ instead of $\mathbf{a}(x, \chi)$. The first distinction is not a real issue, and the second is irrelevant, as long as we are restricting our attention to the *anticyclotomic* situation where $\chi = \omega = 1$ on $\mathbf{A}_F^\times$. Indeed, χ^{-1} is then equal to the outer twist of χ by $\mathrm{Gal}(K/F)$, so that $L(\pi, \chi, s) = L(\pi, \chi^{-1}, s)$ and any lift of the non-trivial element of $\mathrm{Gal}(K/F)$ to $\mathrm{Gal}(K[P^n]/F)$ interchanges the eigenspaces for χ and χ^{-1} in $A(K[P^n]) \otimes \mathbf{C}$.

One has to be more careful when χ is non-trivial on $\mathbf{A}_F^\times$. To be consistent with the BSD conjecture, a Gross-Zagier formula should relate $L'(\pi, \chi, 1/2)$ to a point in the χ^{-1}-component of $A(K[P^n]) \otimes \mathbf{C}$.

In any case, Zhang's Gross-Zagier formula implies that Theorem 1.5 is now a consequence of the following result, which itself is a special case of Theorem 4.1 in the text.

Theorem 1.10 *Let χ_0 be any character of G_0 such that $\chi_0 \cdot \omega = 1$ on $\mathbf{A}_F^\times$. Then, for any good CM point x of conductor P^n with n sufficiently large, there exists a character $\chi \in P(n, \chi_0)$ such that $\mathbf{a}(x, \chi) \neq 0$.*

1.4 The definite case

Suppose now that the triple (π, K, P) is *definite*, so that π'_∞ is the trivial 1-dimensional representation of

$$G(\mathbf{R}) = \prod_{v|\infty} B_v^\times \simeq (\mathbf{H}^\times)^{[F:\mathbf{Q}]}.$$

Then $\mathcal{S}(\pi')$ is contained in $\mathcal{S}_2$, and the latter is simply the subspace of $\mathcal{S}$ on which $G(\mathbf{R})$ acts trivially; this is the space of all smooth functions

$$\phi : G(\mathbf{Q})\backslash G(\mathbf{A})/G(\mathbf{R}) = G(\mathbf{Q})\backslash G(\mathbf{A}_f) \longrightarrow \mathbf{C},$$

with $G(\mathbf{A}_f)$ acting by right translation. Note that the $G(\mathbf{A}_f)$-module underlying $\mathcal{S}(\pi') = \mathcal{S}_2(\pi')$ is admissible, infinite dimensional and irreducible; it contains no nonzero function which factors through the reduced norm, because any such function spans a finite dimensional $G(\mathbf{A}_f)$-invariant subspace.

For any compact open subgroup H of $G(\mathbf{A}_f)$, we may identify $\mathcal{S}_2^H$ with the set of complex valued functions on the finite set

$$M_H \stackrel{\text{def}}{=} G(\mathbf{Q})\backslash G(\mathbf{A}_f)/H,$$

and any such function may be evaluated on $\mathrm{CM}_H = T(\mathbf{Q})\backslash G(\mathbf{A}_f)/H$.

Specializing now to the H which is defined by (1.6), let ψ be the function induced on CM_H by some nonzero element Φ in the 1-dimensional space $\mathcal{S}_2(\pi')^H = \mathcal{S}(\pi')^H = \mathbf{C} \cdot \Phi$:

$$\psi : \mathrm{CM}_H \to M_H \xrightarrow{\Phi} \mathbf{C}.$$

For a ring class character χ of conductor P^n such that $\chi \cdot \omega = 1$ on $\mathbf{A}_F^\times$, the Gross-Zagier philosophy predicts that there should exist a formula relating $L(\pi, \chi, 1/2)$ to $|\mathbf{a}(x, \chi)|^2$, for some CM point $x \in \mathrm{CM}_H$ of conductor P^n, with

$$\mathbf{a}(x, \chi) \stackrel{\text{def}}{=} \frac{1}{|G(n)|} \sum_{\sigma \in G(n)} \chi(\sigma)\psi(\sigma \cdot x) \in \mathbf{C}.$$

Such a formula has indeed been proven by Zhang [33, Theorem 7.1], under the assumption that $\omega = 1$, and that $\mathcal{N}$, $\mathcal{D}$ and P are pairwise coprime. On the other hand, there is a more general theorem of Waldspurger which, although it does not give a precise *formula* for the central value of $L(\pi, \chi, s)$, still gives a criterion for its non-vanishing.

Statement of results

Thus, let χ be any character of $T(\mathbf{Q})\backslash T(\mathbf{A})$ such that $\chi \cdot \omega = 1$ on $\mathbf{A}_F^\times$. Such a character yields a linear form ℓ_χ on $\mathcal{S}(\pi')$, defined by

$$\ell_\chi(\phi) \stackrel{\text{def}}{=} \int_{Z(\mathbf{A})T(\mathbf{Q})\backslash T(\mathbf{A})} \chi(t)\phi(t)dt$$

where dt is any choice of Haar measure on $T(\mathbf{A})$. By a fundamental theorem of Waldspurger [30, Théorème 2], this linear form is nonzero on $\mathcal{S}(\pi')$ if and only if $L(\pi, \chi, 1/2) \neq 0$ and certain local conditions are satisfied. The results of Tunnell and Saito which are summarized in [11, Section 10] show that these local conditions are satisfied if and only if the set $S(\chi)$ of (1.3) is equal to the set S of places where B ramifies. For a ring class character χ of P-power conductor, Lemma 1.1 shows that $S(\chi) = S$ if and only if (π, χ) is even. We thus obtain the following simple criterion.

Theorem 1.11 *[Waldspurger] For a ring class character χ of P-power conductor such that $\chi \cdot \omega = 1$ on $\mathbf{A}_F^\times$,*

$$L(\pi, \chi, 1/2) \neq 0 \quad \Leftrightarrow \quad \exists \phi \in \mathcal{S}(\pi') : \ell_\chi(\phi) \neq 0.$$

Remark 1.12 Waldspurger's theorem does not give a precise formula for the *value* of $L(\pi, \chi, 1/2)$, and it does not specify a canonical choice of ϕ (a *test vector* in the language of [12]) on which to evaluate the linear functional ℓ_χ. The problem of finding such a test vector ϕ and a Gross-Zagier formula relating $\ell_\chi(\phi)$ to $L(\pi, \chi, 1/2)$ is described in great generality in [11], and explicit formulae are proven in [9] (for $F = \mathbf{Q}$) and for a general F in [31, 33], under various assumptions. A leisurely survey of this circle of ideas may be found in [28].

Recall that ψ is the function which is induced on CM_H by some nonzero Φ in $\mathcal{S}(\pi')^H$. For $\phi = g \cdot \Phi \in \mathcal{S}(\pi')$, with $g \in G(\mathbf{A}_f)$ corresponding to a CM point $x = [g] \in \mathrm{CM}_H$ whose conductor P^n equals that of χ, we find that, up to a nonzero constant,

$$\ell_\chi(\phi) \sim \mathbf{a}(x, \chi).$$

Theorem 1.4 therefore is a consequence of the following result, which itself is a special case of Theorem 5.10 in the text.

Theorem 1.13 *Let χ_0 be any character of G_0 such that $\chi_0 \cdot \omega = 1$ on $\mathbf{A}_F^\times$. Then, for any good CM point x of conductor P^n with n sufficiently large, there exists a character $\chi \in P(n, \chi_0)$ such that $\mathbf{a}(x, \chi) \neq 0$.*

1.5 Applications

As we have already explained, the present work was firstly motivated by a desire to prove non-vanishing of L-functions and their derivatives. However, the results we prove on general CM points have independent applications to Iwasawa theory, even when they are not yet known to be related to L-functions.

For instance, Theorem 1.10 implies directly that certain Euler systems of Bertolini-Darmon (when $F = \mathbf{Q}$) and Howard (for general F) are actually non-trivial. The non-triviality of these Euler systems is used by B. Howard in [17] to establish half of the relevant Main Conjecture, for the anti-cyclotomic Iwasawa theory of abelian varieties of $\mathrm{GL}(2)$-type. It is also used by J. Nekovář in [21] to prove new cases of parity in the Bloch-Kato conjecture, for Galois representations attached to Hilbert modular newforms over F with trivial central character and parallel weight $(2k, \cdots, 2k)$, $k \geq 1$.

1.6 Sketch of proof

We want to briefly sketch the proof of our nontriviality theorems for CM points. The basic ideas are drawn from our previous papers [4, 26, 27] with a few simplifications and generalizations.

Thus, let χ_0 be a character of G_0 such that $\chi_0 \cdot \omega = 1$ on $\mathbf{A}_F^\times$, let $H = \widehat{R}^\times$ be the compact open subgroup of $G(\mathbf{A}_f)$ defined by (1.6), and let x be a CM point of conductor P^n in CM_H. We have defined a function ψ on CM_H with values in a complex vector space, and we want to show that

$$\mathbf{a}(x, \chi) \stackrel{\mathrm{def}}{=} \frac{1}{|G(n)|} \sum_{\sigma \in G(n)} \chi(\sigma) \psi(\sigma \cdot x)$$

is nonzero for at least some $\chi \in P(n, \chi_0)$, provided that n is sufficiently large. The analysis of such sums proceeds in a series of reductions.

From $G(n)$ to G_0

To prove that $\mathbf{a}(x, \chi) \neq 0$ for some $\chi \in P(n, \chi_0)$, it suffices to show that the *sum* of these values is nonzero. A formal computation in the group algebra of

$G(n)$ shows that this sum,

$$\mathbf{b}(x, \chi_0) \stackrel{\text{def}}{=} \sum_{\chi \in P(n, \chi_0)} \mathbf{a}(x, \chi)$$

is given by

$$\mathbf{b}(x, \chi_0) = \frac{1}{q\,|G_0|} \sum_{\sigma \in G_0} \chi_0(\sigma) \psi_* (\sigma \cdot \tilde{x})$$

where ψ_* is the extension of ψ to $\mathbf{Z}[\mathrm{CM}_H]$ and

$$\tilde{x} = q \cdot x - \mathrm{Tr}_{Z(n)}(x) = \sum_{\sigma \in Z(n)} x - \sigma \cdot x.$$

Here, $Z(n) = \mathrm{Gal}(K[P^n]/K[P^{n-1}])$ and $q = |Z(n)| = |\mathcal{O}_F/P|$.

Distribution relations

To deal with $\tilde{x}$, we have to use distribution relations and Hecke correspondences, much as in the case of $F = \mathbf{Q}$ treated in our previous works. However, there are numerous technicalities to overcome, owing to the fact that we are now working over a more general field, with automorphic forms that may have a nontrivial central character, and with a prime P that may divide the level. Although the necessary arguments are ultimately quite simple, the details are somewhat tedious, and we request forgiveness for what might seem to be a rather opaque digression. To avoid obscuring the main lines of the argument, we have banished the discussion of distribution relations to the appendices.

Basically, these distribution relations will produce for us a level structure $H^+ \subset H$, a function ψ^+ on CM_{H^+} and a CM point $x^+ \in \mathrm{CM}_{H^+}$ of conductor P^n such that

$$\forall \sigma \in \mathrm{Gal}_K^{\mathrm{ab}} : \quad \psi^+(\sigma \cdot x^+) = \psi_*(\sigma \cdot \tilde{x}).$$

In fact, $H^+ = \widehat{R^+}^\times$ for some $\mathcal{O}_F$-order $R^+ \subset B$ which agrees with R outside P, and is an Eichler order of level $P^{\max(\delta,2)}$ at P. Here we remind the reader that δ is defined in Definition 1.6.

Note that this part of the proof is responsible for the goodness assumptions in our theorems. Indeed, the *bad* CM points simply do not seem to satisfy any distribution relations, and the above construction may therefore only be applied to a *good* CM point x. We also mention that this computation would not work with a more general function ψ: one needs ψ to be *new* at P, in some suitable sense.

From G_0 *to* G_0/G_2

We now have

$$\mathbf{b}(x,\chi_0) = \frac{1}{q\,|G_0|}\sum_{\sigma\in G_0}\chi_0(\sigma)\psi^+(\sigma\cdot x^+).$$

Using the fact that $\chi_0\cdot\omega = 1$ on G_2, we prove that

$$\mathbf{b}(x,\chi_0) = \frac{1}{q\,|G_0/G_2|}\sum_{\sigma\in G_0/G_2}\chi_0(\sigma)\psi^+(\sigma\cdot x^+).$$

Indeed, the map $\sigma\mapsto\chi_0(\sigma)\psi^+(\sigma\cdot x^+)$ factors through G_0/G_2.

From G_0/G_2 *to* G_0/G_1

We can reduce the above sum to something even simpler. Indeed, it turns out that there is a subgroup $G_1\subset G_0$, containing G_2, such that the *Galois* action of the elements in G_1 can be realized by *geometric* means. In fact, G_1 is the maximal such subgroup, and G_1/G_2 is generated by the classes of the frobeniuses in $G(\infty)$ of those primes of K which are ramified over F but do not divide P.

More precisely, we construct yet another level structure $H_1^+\subset H^+$, a function ψ_1^+ on $\mathrm{CM}_{H_1^+}$, and a CM point $x_1^+\in\mathrm{CM}_{H_1^+}$ of conductor P^n, such that

$$\forall\gamma\in\mathrm{Gal}_K^{\mathrm{ab}}:\quad \psi_1^+(\gamma\cdot x_1^+) = \sum_{\sigma\in G_1/G_2}\chi_0(\sigma)\psi^+(\sigma\gamma\cdot x^+).$$

This H_1^+ corresponds to an $\mathcal{O}_F$-order $R_1^+\subset B$ which only differs from R^+ at those finite places $v\neq P$ of F which ramify in K.

This part of the proof is responsible for our general assumption that $\mathcal{N}'$ and $\mathcal{D}$ are relatively prime. Indeed, to establish the above formula, we need to know that for all $v\neq P$ that ramify in K, the local component $H_v^+ = H_v = R_v^\times$ of H^+ is a *maximal* order in a *split* quaternion algebra. It seems likely that the case where R_v is an Eichler order of level v in a split algebra could still be handled by similar methods.

Dealing with G_0/G_1

We finally obtain

$$\mathbf{b}(x,\chi_0) = \frac{1}{q\,|G_0/G_2|}\mathbf{c}(x_1^+)\quad\text{with}\quad \mathbf{c}(y) = \sum_{\sigma\in G_0/G_1}\chi_0(\sigma)\psi_1^+(\sigma\cdot y).$$

We prove that $\mathbf{c}(y) \neq 0$ for sufficiently many y's in $\mathrm{CM}_{H_1^+}(P^n)$, $n \gg 0$, using a theorem of M. Ratner on uniform distribution of unipotent orbits on p-adic Lie groups. Just as in our previous work, we show that the elements of G_0/G_1 act *irrationally* on the relevant CM points. Slightly more precisely, we prove that for y as above, the images $\overline{\mathbf{r}}(z)$ of vectors of the form

$$\mathbf{r}(z) = (\sigma \cdot z)_{\sigma \in G_0/G_1} \in \mathrm{CM}_{H_1^+}(P^n)^{G_0/G_1}$$

are uniformly distributed in some appropriate space, as z runs through the Galois orbit of y, and n goes to infinity. In the indefinite case, $\overline{\mathbf{r}}(z)$ is the vector of supersingular points in characteristic ℓ which is obtained by reducing the coordinates of $\mathbf{r}(z)$ at some suitable place of $K[P^\infty]$. In the definite case, M_H itself plays the role of the supersingular locus. The uniform distribution theorem implies that the image of the Galois orbit of y tends to be large, and it easily follows that $\mathbf{c}(y)$ is nonzero.

We have chosen to present a more general variant of the uniform distribution property alluded to above in a separate paper [5], which is quoted here in propositions 4.17 and 5.6 (for respectively the indefinite and the definite case). Although it really is the kernel of our proof, one may consider [5] as a black box while reading this paper. On the other hand, we also provide a different proof for the indefinite case, based on a proven instance of the André-Oort conjecture, rather than Ratner's theorem.

Finally, the authors would like to thank Jan Nekovář and J. Milne for their help and encouragement.

1.7 Notations

For any place v of F, F_v is the completion of F at v and $\mathcal{O}_{F,v}$ is its ring of integers (if v is finite). If E is a vector field over F, such as K or B, we put $E_v = E \otimes_F F_v$. More generally, if R is a module over the ring of integers $\mathcal{O}_F$ of F, we put $R_v = R \otimes_{\mathcal{O}_F} \mathcal{O}_{F,v}$. We denote by $\mathbf{A}$ (resp. $\mathbf{A}_f$) the ring of adeles (resp. finite adeles) of $\mathbf{Q}$, so that $\mathbf{A} = \mathbf{A}_f \times \mathbf{R}$, and $\mathbf{A}_f$ is the restricted product of the $\mathbf{Q}_v$'s with respect to the $\mathbf{Z}_v$'s. We put $\mathbf{A}_F = \mathbf{A} \otimes_{\mathbf{Q}} F$ and $\mathbf{A}_K = \mathbf{A} \otimes_{\mathbf{Q}} K$. For the finite adeles, we write $\widehat{F} = \mathbf{A}_f \otimes_{\mathbf{Q}} F$ and $\widehat{K} = \mathbf{A}_f \otimes_{\mathbf{Q}} K$. Thus $\widehat{F} = \widehat{\mathcal{O}}_F \otimes \mathbf{Q}$ where $\widehat{M} = M \otimes \widehat{\mathbf{Z}}$ denotes the profinite completion of a finitely generated $\mathbf{Z}$-module M. For any affine algebraic group $G/\mathbf{Q}$, we topologize $G(\mathbf{A}) = G(\mathbf{A}_f) \times G(\mathbf{R})$ in the usual way. When G is the Weil restriction $G = \mathrm{Res}_{F/\mathbf{Q}} G'$ of an algebraic group G'/F, we denote by $g_v \in G'(F_v)$ the v-component of $g \in G(\mathbf{A}) = G'(\mathbf{A}_F)$ (or $G(\mathbf{A}_f) = G'(\widehat{F})$), and we identify $G'(F_v)$ with the subgroup $\{g \in G(\mathbf{A}); \forall w \neq v, \, g_w = 1\}$ of $G(\mathbf{A})$ (or $G(\mathbf{A}_f)$).

We put $\mathrm{Gal}_F = \mathrm{Gal}(\overline{F}/F)$ and $\mathrm{Gal}_K = \mathrm{Gal}(\overline{F}/K)$ where $\overline{F}$ is a fixed algebraic closure of F containing K. We denote by F^{ab} and K^{ab} the maximal abelian extensions of F and K inside $\overline{F}$, with Galois groups $\mathrm{Gal}_F^{\mathrm{ab}}$ and $\mathrm{Gal}_K^{\mathrm{ab}}$. We denote by Frob_v the *geometric* Frobenius at v (the inverse of $x \mapsto x^{N(v)}$) and normalize the reciprocity map

$$\mathrm{rec}_F : \mathbf{A}_F^\times \to \mathrm{Gal}_F^{\mathrm{ab}} \quad \text{and} \quad \mathrm{rec}_K : \mathbf{A}_K^\times \to \mathrm{Gal}_K^{\mathrm{ab}}$$

accordingly.

2 The Galois group of $K[P^\infty]/K$

Fix a prime P of F with residue field $\mathbf{F} = \mathcal{O}_F/P$ of characteristic p and order $q = |\mathbf{F}|$. We have assembled here the basic facts we need pertaining to the infinite abelian extension $K[P^\infty] = \cup_{n\geq 0} K[P^n]$ of K. Recall that

$$G(n) = \mathrm{Gal}(K[P^n]/K) \quad \text{and} \quad G(\infty) = \mathrm{Gal}(K[P^\infty]/K) = \varprojlim G(n).$$

The first section describes $G(\infty)$ as a topological group: it is an extension of a free $\mathbf{Z}_p$-module of rank $[F_P : \mathbf{Q}_p]$ by a finite group G_0, the torsion subgroup of $G(\infty)$. The second section defines a filtration

$$\{1\} \subset G_2 \subset G_1 \subset G_0$$

which plays a crucial role in the proof (and statement) of our main results. Finally, the third section gives an explicit formula for a certain idempotent in the group algebra of $G(n)$.

2.1 The structure of $G(\infty)$

Lemma 2.1 *The reciprocity map induces an isomorphism of topological groups between* $\widehat{K}^\times/K^\times U$ *and* $G(\infty)$ *where*

$$U = \cap \widehat{\mathcal{O}}_{P^n}^\times = \{\lambda \in \widehat{\mathcal{O}}_K^\times,\ \lambda_P \in \mathcal{O}_{F,P}^\times\}.$$

Proof We have to show that the natural continuous map

$$\phi : \widehat{K}^\times/K^\times U \to \varprojlim \widehat{K}^\times/K^\times \widehat{\mathcal{O}}_{P^n}^\times$$

is an isomorphism of topological groups. Put

$$X_n = K^\times \widehat{\mathcal{O}}_{P^n}^\times / K^\times U \simeq \mathcal{O}_{P^n,P}^\times / \mathcal{O}_{P^n}^\times \mathcal{O}_{F,P}^\times$$

so that $\ker(\phi) = \varprojlim X_n$ and $\mathrm{coker}(\phi) = \varprojlim^{(1)} X_n$. Note that $(\mathcal{O}_{P^n}^\times)_{n\geq 0}$ is a decreasing sequence of subgroups of $\mathcal{O}_K^\times$ with $\cap_{n\geq 0} \mathcal{O}_{P^n}^\times = \mathcal{O}_F^\times$: since

$\mathcal{O}_F^\times$ has finite index in $\mathcal{O}_K^\times$, $\mathcal{O}_{P^n}^\times = \mathcal{O}_F^\times$ and $X_n = \mathcal{O}_{P^n,P}^\times/\mathcal{O}_{F,P}^\times$ for all $n \gg 0$. It follows that $\varprojlim X_n = \varprojlim^{(1)} X_n = \{1\}$, so that ϕ is indeed a group isomorphism. This also shows that $K^\times U \cap \widehat{\mathcal{O}}_{P^n}^\times = U$ for $n \gg 0$. In particular, $K^\times U$ is a locally closed, hence closed subgroup of $\widehat{K}^\times$. Being a separated quotient of the compact group $\widehat{K}^\times/\overline{K^\times}$, $\widehat{K}^\times/K^\times U$ is also compact. Being a continuous bijection between compact spaces, ϕ is an homeomorphism. □

It easily follows that the open subgroup $\mathrm{Gal}(K[P^\infty]/K[1])$ of $G(\infty)$ is isomorphic to $\mathcal{O}_{K,P}^\times/\mathcal{O}_K^\times\mathcal{O}_{F,P}^\times$. Since $\mathcal{O}_K^\times/\mathcal{O}_F^\times$ is finite and $\mathcal{O}_{K,P}^\times/\mathcal{O}_{F,P}^\times$ contains an open subgroup topologically isomorphic to $\mathbf{Z}_p^{[F_P:\mathbf{Q}_p]}$, a classical result on profinite groups implies that

Corollary 2.2 *The torsion subgroup G_0 of $G(\infty)$ is finite and $G(\infty)/G_0$ is topologically isomorphic to $\mathbf{Z}_p^{[F_P:\mathbf{Q}_p]}$.*

2.2 *A filtration of* G_0

Let $G(\infty)'$ be the subgroup of $G(\infty)$ which is generated by the Frobeniuses of those primes of K which are not above P (these primes are unramified in $K[P^n]$ for all $n \geq 0$). In particular, $G(\infty)'$ is a *countable* but *dense* subgroup of $G(\infty)$.

Lemma 2.3 *The reciprocity map induces topological isomorphisms*

$$\begin{array}{rcl} (\widehat{K}^\times)^P/(S^{-1}\mathcal{O}_F)^\times(\widehat{\mathcal{O}}_K^\times)^P & \xrightarrow{\simeq} & G(\infty)' \\ \textit{and} \qquad K_P^\times/K^\times F_P^\times & \xrightarrow{\simeq} & G(\infty)/G(\infty)'. \end{array}$$

Here: $S = \mathcal{O}_F - P$ and $X^P = \{\lambda \in X, \lambda_P = 1\}$ for $X \subset \widehat{K}^\times$.

Proof Class field theory tells us that $G(\infty)'$ is the image of $\mathrm{rec}_K((\widehat{K}^\times)^P)$ in $G(\infty)$. Both statements thus follow from Lemma 2.1. □

The map $\lambda \mapsto \lambda\widehat{\mathcal{O}}_K \cap K^\times$ yields an isomorphism between $(\widehat{K}^\times)^P/(\widehat{\mathcal{O}}_K^\times)^P$ and the group $\mathcal{I}_K^P$ of all fractional ideals of K which are relatively prime to P. This bijection maps $(S^{-1}\mathcal{O}_F)^\times(\widehat{\mathcal{O}}_K^\times)^P/(\widehat{\mathcal{O}}_K^\times)^P$ to the group $\mathcal{P}_F^P$ of those ideals in $\mathcal{I}_K^P$ which are principal *and* generated by an element of $F^\times$ – which then necessarily belongs to $(S^{-1}\mathcal{O}_F)^\times$. We thus obtain a perhaps more enlightening description of $G(\infty)'$: it is isomorphic to $\mathcal{I}_K^P/\mathcal{P}_F^P$. The isomorphism sends the class of a prime $Q \nmid P$ of K to its Frobenius in $G(\infty)'$.

Definition 2.4 We denote by $G_1 \subset G_0$ the torsion subgroup of $G(\infty)'$.

There is an obvious finite subgroup in $\mathcal{I}_K^P/\mathcal{P}_F^P$. Indeed, let $\mathcal{I}_F^P$ be the group of all fractional ideals J in K for which $J = \mathcal{O}_K I$ for some fractional ideal I of F relatively prime to P. Then $\mathcal{P}_F^P \subset \mathcal{I}_F^P \subset \mathcal{I}_K^P$ and $\mathcal{I}_F^P/\mathcal{P}_F^P$ is finite. In fact,

$$\mathcal{I}_F^P/\mathcal{P}_F^P \simeq (\widehat{F}^\times)^P/(S^{-1}\mathcal{O}_F)^\times(\widehat{\mathcal{O}}_F^\times)^P \simeq \widehat{F}^\times/F^\times\widehat{\mathcal{O}}_F^\times \simeq \mathrm{Pic}(\mathcal{O}_F).$$

Definition 2.5 We denote by $G_2 \simeq \mathrm{Pic}(\mathcal{O}_F)$ the corresponding subgroup of G_1.

Note that G_2 is simply the image of $\mathrm{rec}_K(\widehat{F}^\times)$ in $G(\infty)$ and the isomorphism between G_2 and $\mathrm{Pic}(\mathcal{O}_F) \simeq \widehat{F}^\times/F^\times\widehat{\mathcal{O}}_F^\times$ is induced by the reciprocity map of K. By definition, G_1/G_2 is isomorphic to the torsion subgroup of $(\widehat{K}^\times)^P/(\widehat{F}^\times\widehat{\mathcal{O}}_K^\times)^P \simeq \mathcal{I}_K^P/\mathcal{I}_F^P$. We thus obtain:

Lemma 2.6 *G_1/G_2 is an $\mathbf{F}_2$-vector space with basis*

$$\{\sigma_Q \bmod G_2;\ Q \mid \mathcal{D}'\}$$

where $\mathcal{D}'$ is the squarefree product of those primes $Q \neq P$ of F which ramify in K, and $\sigma_Q = \mathrm{Frob}_{\mathcal{Q}} \in G_1$ with $\mathcal{Q}^2 = Q\mathcal{O}_K$. In particular,

$$G_1/G_2 = \{\sigma_D \bmod G_2;\ D \mid \mathcal{D}'\}$$

where $\sigma_D = \prod_{Q|D} \sigma_Q$ for $D \mid \mathcal{D}'$.

The following lemma is an easy consequence of the above discussion.

Lemma 2.7 *Let $Q \neq P$ be a prime of $\mathcal{O}_F$ which does not split in K and let $\mathcal{Q}$ be the unique prime of $\mathcal{O}_K$ above Q. Then the decomposition subgroup of $\mathcal{Q}$ in $G(\infty)$ is finite. More precisely, it is a subgroup of G_2 if $\mathcal{Q} = Q\mathcal{O}_K$ and a subgroup of G_1 not contained in G_2 if $\mathcal{Q}^2 = Q\mathcal{O}_K$.*

2.3 *A formula*

Let $\chi_0 : G_0 \to \mathbf{C}^\times$ be a fixed character of G_0. For $n > 0$, we say that a character $\chi : G(n) \to \mathbf{C}^\times$ is *primitive* if it does not factor through $G(n-1)$. We denote by $P(\chi_0, n)$ the set of primitive characters of $G(n)$ inducing χ_0 on G_0 and let $\mathbf{e}(\chi_0, n)$ be the sum of the orthogonal idempotents

$$\mathbf{e}_\chi \overset{\text{def}}{=} \frac{1}{|G(n)|} \sum_{\sigma \in G(n)} \overline{\chi}(\sigma) \cdot \sigma \in \mathbf{C}[G(n)], \quad \chi \in P(\chi_0, n).$$

Note that $\mathbf{e}(\chi_0, n)$ is yet another idempotent in $\mathbf{C}[G(n)]$.

Lemma 2.8 *For $n \gg 0$, we may identify G_0 with its image $G_0(n)$ in $G(n)$ and*

$$\mathbf{e}(\chi_0, n) = \frac{1}{q\,|G_0|} \cdot \left(q - \mathrm{Tr}_{Z(n)}\right) \cdot \sum_{\sigma \in G_0} \overline{\chi}_0(\sigma) \cdot \sigma \quad \text{in } \mathbf{C}[G(n)].$$

Here, $\mathrm{Tr}_{Z(n)} \stackrel{\mathrm{def}}{=} \sum_{\sigma \in Z(n)} \sigma$ *with* $Z(n) \stackrel{\mathrm{def}}{=} \mathrm{Gal}(K[P^n]/K[P^{n-1}])$.

Proof We denote by $G^\vee$ the group of characters of a given G. Write

$$\mathbf{e}(\chi_0, n) = \sum_{\sigma \in G(n)} \mathbf{e}_\sigma(\chi_0, n) \cdot \sigma \in \mathbf{C}[G(n)]$$

and put $H(n) = G(n)/G_0(n)$. If n is sufficiently large, (1) $G_0 \twoheadrightarrow G_0(n)$ is an isomorphism, (2) $G(n) \twoheadrightarrow H(n)$ induces an isomorphism from $Z(n) = \ker(G(n) \twoheadrightarrow G(n-1))$ to the kernel of $H(n) \twoheadrightarrow H(n-1)$, and (3) the kernel $X(n)$ of $G(n) \to H(n-1)$ is the direct sum of $G_0(n)$ and $Z(n)$ in $G(n)$. In particular, there exists an element $\chi_0' \in G(n)^\vee$ inducing χ_0 on $G_0(n) \simeq G_0$ and 1 on $Z(n)$, so that

$$P(\chi_0, n) = H(n)^\vee \chi_0' - H(n-1)^\vee \chi_0' = \left(H(n)^\vee - H(n-1)^\vee\right) \chi_0'.$$

For $\sigma \in G(n)$, we thus obtain

$$\begin{aligned} |G(n)| \cdot \mathbf{e}_\sigma(\chi_0, n) &= \left(\sum_{\chi \in H(n)^\vee} \overline{\chi}(\sigma) - \sum_{\chi \in H(n-1)^\vee} \overline{\chi}(\sigma) \right) \cdot \overline{\chi}_0'(\sigma) \\ &= \left\{ \begin{array}{lll} 0 & \text{if} & \sigma \notin X(n) \\ -|H(n-1)| & \text{if} & \sigma \in X(n) \smallsetminus G_0(n) \\ |H(n)| - |H(n-1)| & \text{if} & \sigma \in G_0(n) \end{array} \right\} \cdot \overline{\chi}_0'(\sigma). \end{aligned}$$

Since $X(n) = G_0(n) \oplus Z(n)$ with $\chi_0' = \chi_0$ on $G_0(n)$ and 1 on $Z(n)$,

$$\begin{aligned} |G(n)| \cdot \mathbf{e}(\chi_0, n) &= \sum_{\sigma \in G_0(n)} \sum_{\tau \in Z(n)} |G(n)| \cdot \mathbf{e}_{\sigma\tau}(\chi_0, n) \cdot \sigma\tau \\ &= \sum_{\sigma \in G_0} \left(|H(n)| - |H(n-1)| \cdot \mathrm{Tr}_{Z(n)}\right) \cdot \overline{\chi}_0(\sigma)\sigma. \end{aligned}$$

This is our formula. Indeed,

$$|G(n)| = |G_0|\,|H(n)|, \quad |H(n)| = |Z(n)|\,|H(n-1)|,$$

and $|Z(n)| = |\mathbf{F}| = q$ by Lemma 2.9 below. □

The reciprocity map induces an isomorphism between

$$K^\times \widehat{\mathcal{O}}_{P^{n-1}}^\times / K^\times \widehat{\mathcal{O}}_{P^n}^\times \simeq \mathcal{O}_{P^{n-1},P}^\times / \mathcal{O}_{P^{n-1}}^\times \mathcal{O}_{P^n,P}^\times$$

and $Z(n)$. For $n \gg 0$, $\mathcal{O}_{P^{n-1}}^\times = \mathcal{O}_F^\times$ is contained in $\mathcal{O}_{P^n,P}^\times$, so that $Z(n) \simeq \mathcal{O}_{P^{n-1},P}^\times / \mathcal{O}_{P^n,P}^\times$. On the other hand, for any $n \geq 1$, the $\mathbf{F}$-algebra $\mathcal{O}_{P^n}/P\mathcal{O}_{P^n}$ is isomorphic to $\mathbf{F}[\epsilon] = \mathbf{F}[X]/X^2\mathbf{F}[X]$, and the projection $\mathcal{O}_{P^n,P} \to \mathcal{O}_{P^n,P}/P\mathcal{O}_{P^n,P} \simeq \mathcal{O}_{P^n}/P\mathcal{O}_{P^n}$ induces an isomorphism between $\mathcal{O}_{P^n,P}^\times / \mathcal{O}_{P^{n+1},P}^\times$ and $\mathbf{F}[\epsilon]^\times / \mathbf{F}^\times \simeq \{1 + \alpha\epsilon;\ \alpha \in \mathbf{F}\}$. We thus obtain:

Lemma 2.9 *For $n \gg 0$, $Z(n) \simeq \mathbf{F}$ as a group.*

3 Shimura Curves

Let F be a totally real number field. To each finite set S of finite places of F such that $|S| + [F : \mathbf{Q}]$ is odd, we may attach a collection of Shimura curves over F. If K is a totally imaginary quadratic extension of F in which the primes of S do not split, these curves are provided with a systematic supply of CM points defined over the maximal abelian extension K^{ab} of K. As explained in the introduction, our aim in this paper (for the indefinite case) is to prove the non-triviality of certain cycles supported on these points.

This section provides some of the necessary background on Shimura curves, with [2] as our main reference. Further topics are discussed in Section 3.1 of [5]. To simplify the exposition, we require S to be nonempty if $F = \mathbf{Q}$. This rules out precisely the case where our Shimura curves are the classical modular curves over $\mathbf{Q}$. This assumption implies that our Shimura curves are complete – there are no cusps to be added, but then also no obvious way to embed the curves into their Jacobians. We note however that everything works with the obvious modifications in the non-compact case as well – see [4] and [27], as well as [5].

3.1 Shimura curves

Let $\{\tau_1, \cdots, \tau_d\} = \mathrm{Hom}_{\mathbf{Q}}(F, \mathbf{R})$ be the set of real embeddings of F. We shall always view F as a subfield of $\mathbf{R}$ (or $\mathbf{C}$) through τ_1. Let B be a quaternion algebra over F which ramifies precisely at $S \cup \{\tau_2, \cdots, \tau_d\}$, a finite set of even order. Let G be the reductive group over $\mathbf{Q}$ whose set of points on a commutative $\mathbf{Q}$-algebra A is given by $G(A) = (B \otimes A)^\times$. Let Z be the center of G.

In particular, $G_{\mathbf{R}} \simeq G_1 \times \cdots \times G_d$ where $B_{\tau_i} = B \otimes_{F,\tau_i} \mathbf{R}$ and G_i is the algebraic group over $\mathbf{R}$ whose set of points on a commutative $\mathbf{R}$-algebra A is given by $G_i(A) = (B_{\tau_i} \otimes_{\mathbf{R}} A)^\times$. Fix $\epsilon \in \{\pm 1\}$ and let X be the $G(\mathbf{R})$-conjugacy class of the morphism from $\mathbb{S} \stackrel{\mathrm{def}}{=} \mathrm{Res}_{\mathbf{C}/\mathbf{R}}(\mathbb{G}_{m,\mathbf{C}})$ to $G_{\mathbf{R}}$ which maps $z = x + iy \in \mathbb{S}(\mathbf{R}) = \mathbf{C}^\times$ to

$$\left[\begin{pmatrix} x & y \\ -y & x \end{pmatrix}^{\epsilon}, 1, \cdots, 1 \right] \in G_1(\mathbf{R}) \times \cdots \times G_d(\mathbf{R}) \simeq G(\mathbf{R}). \qquad (3.1)$$

We have used an isomorphism of $\mathbf{R}$-algebras $B_{\tau_1} \simeq M_2(\mathbf{R})$ to identify G_1 and $\mathrm{GL}_2/\mathbf{R}$; the resulting conjugacy class X does not depend upon this choice, but it does depend on ϵ, cf. Section 3.3.1 of [5] and Remark 3.1 below.

It is well-known that X carries a complex structure for which the left action of $G(\mathbf{R})$ is holomorphic. For every compact open subgroup H of $G(\mathbf{A}_f)$, the quotient of $G(\mathbf{A}_f)/H \times X$ by the diagonal left action of $G(\mathbf{Q})$ is a *compact* Riemann surface

$$M_H^{\mathrm{an}} \stackrel{\mathrm{def}}{=} G(\mathbf{Q}) \backslash (G(\mathbf{A}_f)/H \times X).$$

The *Shimura curve* M_H is Shimura's canonical model for M_H^{an}. It is a proper and smooth curve over F (the reflex field) whose underlying Riemann surface $M_H(\mathbf{C})$ equals M_H^{an}.

Remark 3.1 With notations as above, let $h : \mathbb{S} \to G_{\mathbf{R}}$ be the morphism defined by (3.1). There are $G(\mathbf{R})$-equivariant diffeomorphisms

$$\begin{array}{ccccc} X & \stackrel{\simeq}{\longleftarrow} & G(\mathbf{R})/H_\infty & \stackrel{\simeq}{\longrightarrow} & \mathbf{C} \smallsetminus \mathbf{R} \\ ghg^{-1} & \longleftarrow & g & \longmapsto & g \cdot \epsilon i \end{array}$$

where

$$\begin{aligned} H_\infty &= \mathrm{Stab}_{G(\mathbf{R})}(h) = \mathrm{Stab}_{G(\mathbf{R})}(\pm i) \\ &= \mathbf{R}^\times \mathrm{SO}_2(\mathbf{R}) \times G_2(\mathbf{R}) \times \cdots \times G_d(\mathbf{R}) \end{aligned}$$

with $G(\mathbf{R})$ acting on $\mathbf{C} \smallsetminus \mathbf{R}$ through its first component $G_1(\mathbf{R}) \simeq \mathrm{GL}_2(\mathbf{R})$ by $\left(\begin{smallmatrix} a & b \\ c & d \end{smallmatrix}\right) \cdot \lambda = \frac{a\lambda + b}{c\lambda + d}$. With these conventions, the derivative of $\lambda \mapsto gh(z)g^{-1} \cdot \lambda$ at $\lambda = g \cdot \epsilon i$ equals $z/\overline{z}$ (for $g \in G(\mathbf{R})$, $\lambda \in \mathbf{C} \smallsetminus \mathbf{R}$ and $z \in \mathbf{C}^\times = \mathbb{S}(\mathbf{R})$). In other words, the above bijection between X and $\mathbf{C} \smallsetminus \mathbf{R}$ is an *holomorphic* diffeomorphism.

This computation shows that the Shimura curves of the introduction, which are also those considered in [33] or [17], correspond to the case where $\epsilon = 1$. On the other hand, Carayol explicitly works with the $\epsilon = -1$ case in our main reference [2].

3.2 Connected components

We denote by

$$M_H \stackrel{c}{\longrightarrow} \mathcal{M}_H \to \mathrm{Spec}(F)$$

the *Stein factorization* of the structural morphism $M_H \to \mathrm{Spec}(F)$, so that $\mathcal{M}_H \stackrel{\text{def}}{=} \mathrm{Spec}\,\Gamma(M_H, \mathcal{O}_{M_H})$ is a finite étale F-scheme and the F-morphism $c : M_H \to \mathcal{M}_H$ is proper and smooth with geometrically connected fibers. In particular,

$$\mathcal{M}_H(\overline{F}) \simeq \pi_0(M_H \times_F \overline{F}) \simeq \pi_0(M_H(\mathbf{C})) \simeq \pi_0(M_H^{\mathrm{an}}). \tag{3.2}$$

Remark 3.2 (1) In the notations of Remark 3.1, let X^+ be the connected component of h in X, so that $X^+ = G(\mathbf{R})^+ \cdot h \simeq \mathcal{H}_\epsilon$ where $G(\mathbf{R})^+$ is the neutral component of $G(\mathbf{R})$ and $\mathcal{H}_\epsilon \stackrel{\text{def}}{=} \{\lambda \in \mathbf{C};\ \epsilon \cdot \Im(\lambda) > 0\}$. Put $G(\mathbf{Q})^+ \stackrel{\text{def}}{=} G(\mathbf{R})^+ \cap G(\mathbf{Q})$. Then $\pi_0(M_H^{\mathrm{an}}) \simeq G(\mathbf{Q})^+ \backslash G(\mathbf{A}_f)/H$, corresponding to the decomposition

$$\begin{array}{ccccc} \coprod_\alpha \overline{\Gamma}_\alpha \backslash \mathcal{H}_\epsilon & \stackrel{\simeq}{\longleftarrow} & \coprod_\alpha \Gamma_\alpha \backslash X^+ & \stackrel{\simeq}{\longrightarrow} & M_H^{\mathrm{an}} \\ [g \cdot \epsilon i] \in \overline{\Gamma}_\alpha \backslash \mathcal{H}_\epsilon & \longleftarrow & [x = ghg^{-1}] \in \Gamma_\alpha \backslash X^+ & \longmapsto & [(\alpha, x)] \end{array} \tag{3.3}$$

where $\alpha \in G(\mathbf{A}_f)$ runs through a set of representatives of the finite set $G(\mathbf{Q})^+ \backslash G(\mathbf{A}_f)/H$, Γ_α is the discrete subgroup $\alpha H \alpha^{-1} \cap G(\mathbf{Q})^+$ of $G(\mathbf{R})^+$ and $\overline{\Gamma}_\alpha \subset \mathrm{PGL}_2^+(\mathbf{R})$ is its image through the obvious map $G(\mathbf{R})^+ \to \mathrm{GL}_2^+(\mathbf{R}) \to \mathrm{PGL}_2^+(\mathbf{R})$.

(2) The strong approximation theorem [29, p. 81] and the norm theorem [29, p. 80] imply that the reduced norm $\mathrm{nr} : \widehat{B}^\times \to \widehat{F}^\times$ induces a bijection

$$\pi_0(M_H^{\mathrm{an}}) \simeq G(\mathbf{Q})^+ \backslash G(\mathbf{A}_f)/H \stackrel{\simeq}{\longrightarrow} Z(\mathbf{Q})^+ \backslash Z(\mathbf{A}_f)/\mathrm{nr}(H) \tag{3.4}$$

where $Z(\mathbf{Q})^+ = \mathrm{nr}(G(\mathbf{Q})^+)$ is the subgroup of totally positive elements in $Z(\mathbf{Q}) = F^\times$. Using also (3.2), we obtain a left action of Gal_F on the RHS of (3.4). The general theory of Shimura varieties implies that the latter action factors through $\mathrm{Gal}_F^{\mathrm{ab}}$, where it is given by the following *reciprocity law* (see Lemma 3.12 in [5]): for $\lambda \in \widehat{F}^\times$, the element $\sigma = \mathrm{rec}_F(\lambda)$ of $\mathrm{Gal}_F^{\mathrm{ab}}$ acts on the RHS of (3.4) as multiplication by λ^ϵ. In particular, this action is transitive and M_H is therefore a *connected* F-curve (although not a geometrically connected one).

3.3 Related group schemes

The *Jacobian* J_H of M_H is the identity component of the relative Picard scheme P_H of $M_H \to \mathrm{Spec}(F)$ and the *Néron-Severi* group NS_H of M_H is the quotient of P_H by J_H. By [15], J_H is an abelian scheme over F while NS_H is a "separable discrete" F-group scheme. The canonical isomorphism [14, V.6.1] of F-group schemes

$$P_H \stackrel{\simeq}{\longrightarrow} \mathrm{Res}_{\mathcal{M}_H/F}(\mathrm{Pic}_{M_H/\mathcal{M}_H})$$

induces an isomorphism between J_H and $\mathrm{Res}_{\mathcal{M}_H/F}(\mathrm{Pic}^0_{M_H/\mathcal{M}_H})$, so that

$$\mathrm{NS}_H \xrightarrow{\simeq} \mathrm{Res}_{\mathcal{M}_H/F}\left(\mathrm{Pic}_{M_H/\mathcal{M}_H}/\mathrm{Pic}^0_{M_H/\mathcal{M}_H}\right) \xrightarrow{\simeq} \mathrm{Res}_{\mathcal{M}_H/F}(\underline{\mathbf{Z}})$$

where we have identified $\mathrm{Pic}_{M_H/\mathcal{M}_H}/\mathrm{Pic}^0_{M_H/\mathcal{M}_H}$ with the constant $\mathcal{M}_H$-group scheme $\underline{\mathbf{Z}}$ using the degree map $\deg_H : \mathrm{Pic}_{M_H/\mathcal{M}_H} \to \underline{\mathbf{Z}}$. We denote by

$$\overline{\deg}_H : P_H \to \mathrm{NS}_H = P_H/J_H$$

the quotient map, so that $\overline{\deg}_H = \mathrm{Res}_{\mathcal{M}_H/F}(\deg_H)$ under the above identifications.

Remark 3.3 If $\mathcal{M}_H(\overline{F}) = \{s_\alpha\}$ with $s_\alpha : \mathrm{Spec}(\overline{F}) \to \mathcal{M}_H$,

$$\begin{array}{rclcrcl} M_H \times_F \mathrm{Spec}(\overline{F}) & = & \coprod_\alpha \mathcal{C}_\alpha, & & P_H \times_F \mathrm{Spec}(\overline{F}) & = & \prod_\alpha P_\alpha, \\ J_H \times_F \mathrm{Spec}(\overline{F}) & = & \prod_\alpha J_\alpha, & \text{and} & \mathrm{NS}_H \times_F \mathrm{Spec}(\overline{F}) & = & \prod_\alpha \underline{\mathbf{Z}}_\alpha \end{array}$$

where $\mathcal{C}_\alpha = c^{-1}(s_\alpha)$, $P_\alpha = \mathrm{Pic}(\mathcal{C}_\alpha)$, $J_\alpha = \mathrm{Pic}^0(\mathcal{C}_\alpha)$ and $\underline{\mathbf{Z}}_\alpha = s_\alpha^*(\underline{\mathbf{Z}})$ is isomorphic to $\underline{\mathbf{Z}}$ over $\overline{F}$. With these identifications, $\overline{\deg}_H$ maps $(p_\alpha) \in \prod_\alpha P_\alpha$ to $(\deg(p_\alpha)) \in \prod_\alpha \mathbf{Z}$.

3.4 Hecke operators

As H varies among the compact open subgroups of $G(\mathbf{A}_f)$, the Shimura curves $\{M_H\}_H$ form a projective system with finite flat transition maps which is equipped with a "continuous" right action of $G(\mathbf{A}_f)$. Specifically, for any element $g \in G(\mathbf{A}_f)$ and for any compact open subgroups H_1 and H_2 of $G(\mathbf{A}_f)$ such that $g^{-1}H_1 g \subset H_2$, multiplication on the right by g in $G(\mathbf{A}_f)$ defines a map $M^{\mathrm{an}}_{H_1} \to M^{\mathrm{an}}_{H_2}$ which descends to a finite flat F-morphism

$$[\cdot g] = [\cdot g]_{H_1,H_2} : M_{H_1} \to M_{H_2}.$$

We shall refer to such a map as *the degeneracy map* induced by g. Letting H_1 and H_2 vary, these degeneracy maps together define an automorphism $[\cdot g]$ of $\varprojlim\{M_H\}_H$.

There is a natural *left* action of the Hecke algebra

$$\mathbf{T}_H \stackrel{\mathrm{def}}{=} \mathrm{End}_{\mathbf{Z}[G(\mathbf{A}_f)]}\left(\mathbf{Z}[G(\mathbf{A}_f)/H]\right) \simeq \mathbf{Z}[H\backslash G(\mathbf{A}_f)/H]$$

on P_H, J_H and NS_H. With the ring structure induced by its representation as an endomorphism algebra, $\mathbf{T}_H$ is the opposite of the most frequently encountered Hecke algebra: for $\alpha \in G(\mathbf{A}_f)$, the Hecke operator $\mathcal{T}_H(\alpha) \in \mathbf{T}_H$

corresponding to the double class $H\alpha H$ acts by $\mathcal{T}_H(\alpha) = f'_* \circ [\cdot\alpha]_* \circ f^*$ where f and f' are the obvious transition maps in the following diagram

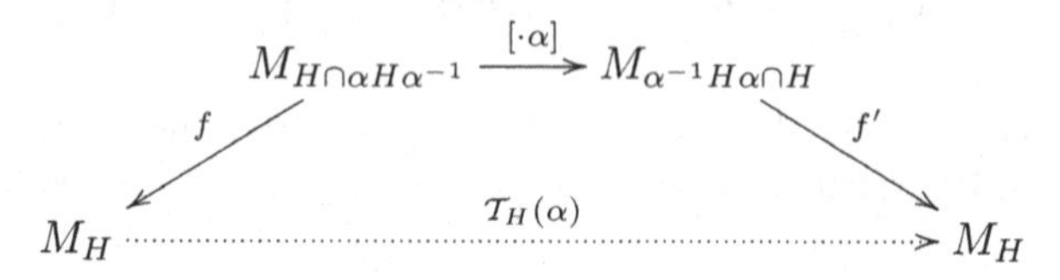

The *degree* of $\mathcal{T}_H(\alpha)$ is the degree of f, namely the index of $H \cap \alpha H \alpha^{-1}$ in H.[1] On the level of divisors, $\mathcal{T}_H(\alpha)$ maps $x = [g, h] \in M_H^{\mathrm{an}}$ to

$$\mathcal{T}_H(\alpha)(x) = \sum [g\alpha_i, h] \in \mathrm{Div} M_H^{\mathrm{an}}$$

where $H\alpha H = \coprod \alpha_i H$. If α belongs to the center of $G(\mathbf{A}_f)$, then $[\cdot\alpha] : M_H \to M_H$ is an automorphism of M_H/F and $\mathcal{T}_H(\alpha) = [\cdot\alpha]_*$.

Definition 3.4 We denote by θ_M and θ_J the induced left action of $Z(\mathbf{A}_f)$ on M_H and J_H: $\theta_M(\alpha) = [\cdot\alpha]$ and $\theta_J(\alpha) = [\cdot\alpha]_*$. These actions factor through $Z(\mathbf{Q})\backslash Z(\mathbf{A}_f)/Z(\mathbf{A}_f) \cap H$. When $H = \widehat{R}^\times$ for some $\mathcal{O}_F$-order $R \subset B$, $Z(\mathbf{A}_f) \cap H = \widehat{\mathcal{O}}_F^\times$ and we thus obtain left actions

$$\theta_M : \mathrm{Pic}(\mathcal{O}_F) \to \mathrm{Aut}_F M_H \quad \textit{and} \quad \theta_J : \mathrm{Pic}(\mathcal{O}_F) \to \mathrm{Aut}_F J_H.$$

3.5 The Hodge class and the Hodge embedding

Let S be a scheme. To any commutative group scheme $\mathcal{G}$ over S, and indeed to any presheaf of abelian groups $\mathcal{G}$ on the category of S-schemes, we may attach a presheaf of $\mathbf{Q}$-vector spaces $\mathcal{G} \otimes \mathbf{Q}$ by the following rule: for any S-scheme X, $\mathcal{G} \otimes \mathbf{Q}(X) \stackrel{\mathrm{def}}{=} \mathcal{G}(X) \otimes \mathbf{Q}$. To distinguish between the sections of $\mathcal{G}(X)$ and the sections of $\mathcal{G} \otimes \mathbf{Q}(X)$, we write $X \to \mathcal{G}$ for the former and $X \rightsquigarrow \mathcal{G}$ for the latter, but we will refer to both kind of sections as *morphisms*.

This construction is functorial in the sense that given two commutative group schemes $\mathcal{G}_1$ and $\mathcal{G}_2$ over S, any element α of

$$\mathrm{Hom}^0_S(\mathcal{G}_1, \mathcal{G}_2) \stackrel{\mathrm{def}}{=} \mathrm{Hom}_S(\mathcal{G}_1, \mathcal{G}_2) \otimes \mathbf{Q}$$

defines a morphism $\alpha : \mathcal{G}_1 \rightsquigarrow \mathcal{G}_2$: choose $n \geq 1$ such that $n\alpha = \alpha_0 \otimes 1$ for some $\alpha_0 \in \mathrm{Hom}_S(\mathcal{G}_1, \mathcal{G}_2)$, set $\alpha(f) = \alpha_0(f) \otimes \frac{1}{n} \in \mathcal{G}_2(X) \otimes \mathbf{Q}$ for $f \in \mathcal{G}_1(X)$ and extend by linearity to $\mathcal{G}_1(X) \otimes \mathbf{Q}$. The resulting morphism does

1 In general, the degree of $M_{H'} \to M_H$ equals

$$[HZ(\mathbf{Q}) : H'Z(\mathbf{Q})] = [H : H'] \cdot [H \cap Z(\mathbf{Q}) : H' \cap Z(\mathbf{Q})].$$

But here $H' = H \cap \alpha H \alpha^{-1}$, so that $H \cap Z(\mathbf{Q}) \simeq H' \cap Z(\mathbf{Q})$.

not depend upon the choice of n and α_0 and furthermore satisfies $\alpha(\lambda_1 f_1 + \lambda_2 f_2) = \lambda_1\alpha(f_1) + \lambda_2\alpha(f_2)$ in $\mathcal{G}_2 \otimes \mathbf{Q}(X)$ for any $\lambda_1, \lambda_2 \in \mathbf{Q}$ and $f_1, f_2 : X \rightsquigarrow \mathcal{G}_1$.

This said, the "Hodge embedding" is a morphism $\iota_H : M_H \rightsquigarrow J_H$ over F which we shall now define. To start with, consider the F-morphism $(\cdot)_H : M_H \to P_H$ which for any F-scheme X maps $x \in M_H(X)$ to the element $(x) \in P_H(X)$ which is represented by the effective relative Cartier divisor on $M_H \times_F X$ defined by x (viewing x as a section of $M_H \times_F X \to X$). Our morphism ι_H is the composite of this map with a retraction $P_H \rightsquigarrow J_H$ of $J_H \hookrightarrow P_H$. Defining the latter amounts to defining a section $\mathrm{NS}_H \rightsquigarrow P_H$ of $\overline{\deg}_H : P_H \twoheadrightarrow \mathrm{NS}_H$ and since $\overline{\deg}_H = \mathrm{Res}_{\mathcal{M}_H/F}(\deg_H)$, we may as well search for a section $\underline{\mathbf{Z}} \rightsquigarrow \mathrm{Pic}_{M_H/\mathcal{M}_H}$ of $\deg_H : \mathrm{Pic}_{M_H/\mathcal{M}_H} \twoheadrightarrow \underline{\mathbf{Z}}$. In other word, we now want to construct an element (the *Hodge class*)

$$\delta_H \in \mathrm{Pic}_{M_H/\mathcal{M}_H}(\mathcal{M}_H) \otimes \mathbf{Q} = \mathrm{Pic}(M_H) \otimes \mathbf{Q}$$

such that $\deg(\delta_H) = 1$ in $\underline{\mathbf{Z}}(\mathcal{M}_H) \otimes \mathbf{Q} = \mathbf{Q}$ (recall from Remark 3.2 that $\mathcal{M}_H$ is connected). The resulting morphism $\iota_H : M_H \rightsquigarrow J_H$ will thus be given by

$$\iota_H(x) = (x)_H - s_H \circ \overline{\deg}_H(x)_H \quad \text{in } J_H \otimes \mathbf{Q}(X)$$

for any F-scheme X and $x \in M_H(X)$, with $s_H : \mathrm{NS}_H \rightsquigarrow P_H$ defined by

$$s_H \stackrel{\text{def}}{=} \mathrm{Res}_{\mathcal{M}_H/F}\begin{pmatrix} \underline{\mathbf{Z}} & \rightsquigarrow & \mathrm{Pic}_{M_H/\mathcal{M}_H} \\ n & \mapsto & n \cdot \delta_H \end{pmatrix}.$$

We may now proceed to the definition of the Hodge class δ_H. For a compact open subgroup H' of H, the *ramification divisor* of the transition map $f = \mathcal{T}_{H',H} : M_{H'} \to M_H$ is defined by

$$R_{H'/H} \stackrel{\text{def}}{=} \sum_x \mathrm{length}_{\mathcal{O}_{M_{H'},x}} (\Omega_f)_x \cdot x \quad \text{in } \mathrm{Div}(M_{H'})$$

where x runs through the finitely many closed points of $M_{H'}$ in the support of the sheaf of relative differentials $\Omega_f = \Omega_{M_{H'}/M_H}$. The *branch divisor* is the flat push-out of $R_{H'/H}$:

$$B_{H'/H} \stackrel{\text{def}}{=} f_* R_{H'/H} \quad \text{in } \mathrm{Div}(M_H).$$

When H' is a normal subgroup of H, $M_H = M_{H'}/H$ and the branch divisor pulls-back to $f^* B_{H'/H} = \deg(f) \cdot R_{H'/H}$.

For $H'' \subset H' \subset H$, $f = \mathcal{T}_{H',H}$, $g = \mathcal{T}_{H'',H'}$ and $h = f \circ g = \mathcal{T}_{H'',H}$,

$$0 \to g^*\Omega_f \to \Omega_h \to \Omega_g \to 0$$

is an exact sequence of coherent sheaves on $M_{H''}$ (this may be proven using Proposition 2.1 of [16, Chapter IV], the flatness of g and the snake lemma). This exact sequence shows that

$$\begin{array}{rrcll} & R_{H''/H} & = & R_{H''/H'} + g^* R_{H'/H} & \text{in } \mathrm{Div}(M_{H''}) \\ \text{and} & B_{H''/H} & = & f_* B_{H''/H'} + \deg(g) \cdot B_{H'/H} & \text{in } \mathrm{Div}(M_H). \end{array} \tag{3.5}$$

If H' is sufficiently small, $R_{H''/H'} = 0$ and $B_{H''/H'} = 0$ for any $H'' \subset H'$ (see for instance [2, Corollaire 1.4.1.3]). In particular,

$$B_H \stackrel{\text{def}}{=} \frac{1}{\deg(f)} B_{H'/H} \in \mathrm{Div}(M_H) \otimes \mathbf{Q}$$

does not depend upon H', provided that H' is sufficiently small. When H itself is sufficiently small, $B_H = 0$. In general:

Lemma 3.5 $f^* B_H = B_{H'} + R_{H'/H}$ *in* $\mathrm{Div}(M_{H'}) \otimes \mathbf{Q}$.

Proof Let H'' be a sufficiently small normal subgroup of H contained in H'. With notations as above, $g^* : \mathrm{Div}(M_{H'}) \to \mathrm{Div}(M_{H''})$ is injective,

$$\begin{array}{rrcccl} & g^* f^* B_H & = & \frac{1}{\deg(h)} h^* h_* R_{H''/H} & = & R_{H''/H}, \\ \text{and} & g^* B_{H'} & = & \frac{1}{\deg(g)} g^* g_* R_{H''/H'} & = & R_{H''/H'}. \end{array}$$

The lemma thus follows from (3.5). □

On the other hand, Hurwitz's formula [16, IV, Prop. 2.3] tells us that

$$\mathcal{K}_{H'} = f^* \mathcal{K}_H + \text{class of } R_{H'/H} \quad \text{in } \mathrm{Pic}(M_{H'})$$

where $\mathcal{K}_H$ is the canonical class on M_H, namely the class of $\Omega_{M_H/F}$. It follows that

$$\begin{array}{rcll} f^*(\mathcal{K}_H + B_H) & = & \mathcal{K}_{H'} + B_{H'} & \text{in } \mathrm{Pic}(M_{H'}) \otimes \mathbf{Q}, \\ f_*(\mathcal{K}_{H'} + B_{H'}) & = & \deg(f) \cdot (\mathcal{K}_H + B_H) & \text{in } \mathrm{Pic}(M_H) \otimes \mathbf{Q}. \end{array} \tag{3.6}$$

If H' is sufficiently small, $B_{H'} = 0$ and $\deg(\mathcal{K}_{H'}) > 0$. The above formulae therefore imply that $\deg(\mathcal{K}_H + B_H) > 0$ for *any* H, and we may thus define

$$\delta_H \stackrel{\text{def}}{=} \frac{1}{\deg(\mathcal{K}_H + B_H)} \cdot (\mathcal{K}_H + B_H) \in \mathrm{Pic}(M_H) \otimes \mathbf{Q}.$$

By construction: $\deg(\delta_H) = 1$,

$$\begin{array}{rrcll} & f^* \delta_H & = & \deg(f) \cdot \delta_{H'} & \text{in } \mathrm{Pic}(M_{H'}) \otimes \mathbf{Q} \\ \text{and} & f_* \delta_{H'} & = & \delta_H & \text{in } \mathrm{Pic}(M_H) \otimes \mathbf{Q}. \end{array} \tag{3.7}$$

Lemma 3.6 *For any* $\alpha \in G(\mathbf{A}_f)$, $\mathcal{T}_H(\alpha)(\delta_H) = \deg(\mathcal{T}_H(\alpha)) \cdot \delta_H$.

Proof Given the definition of $\mathcal{T}_H(\alpha)$, this follows from (3.7) once we know that for any H and α, $[\cdot\alpha]_*\delta_H = \delta_{\alpha^{-1}H\alpha}$ in $\mathrm{Pic}(M_{\alpha^{-1}H\alpha}) \otimes \mathbf{Q}$. This is obvious if H is sufficiently small and the general case follows, using (3.7) again. □

Remark 3.7 Let d_H be the smallest positive integer such that $d_H B_H$ belongs to $\mathrm{Div}(M_H) \subset \mathrm{Div}(M_H) \otimes \mathbf{Q}$ and put $\delta_H^0 = d_H \cdot \mathcal{K}_H +$ class of $d_H B_H$ in $\mathrm{Pic}(M_H)$. The proof just given shows that the relation $\mathcal{T}_H(\alpha)(\delta_H^0) = \deg(\mathcal{T}_H(\alpha)) \cdot \delta_H^0$ already holds in $\mathrm{Pic}(M_H)$.

With notations as in Remark 3.2, the restriction of B_H to $\overline{\Gamma}_\alpha \backslash \mathcal{H}_\epsilon$ equals $\sum_x \left(1 - e_x^{-1}\right) \cdot x$ where x runs through a set of representatives of $\overline{\Gamma}_\alpha \backslash \mathcal{H}_\epsilon$ in $\mathcal{H}_\epsilon$ and e_x is the order of its stabilizer in $\overline{\Gamma}_\alpha \subset \mathrm{PGL}_2^+(\mathbf{R})$. Compare with [11, section 23]. Thus, d_H is the largest common multiple of the e_x's.

Remark 3.8 With the notations of Remark 3.3, let $\delta_\alpha \in \mathrm{Pic}(\mathcal{C}_\alpha) \otimes \mathbf{Q}$ be the restriction to $\mathcal{C}_\alpha$ of the pull-back of δ_H to

$$\mathrm{Pic}(M_H \times_F \mathrm{Spec}(\overline{F})) \otimes \mathbf{Q} = \prod_\alpha \mathrm{Pic}(\mathcal{C}_\alpha) \otimes \mathbf{Q}.$$

Then $\deg(\delta_\alpha) = 1$ and the restriction of ι_H to $\mathcal{C}_\alpha$ maps $x \in \mathcal{C}_\alpha(\overline{F})$ to $(x) - \delta_\alpha \in J_\alpha(\overline{F}) \otimes \mathbf{Q}$. In particular, the image of ι_H on $M_H(\overline{F})$ spans $J_H(\overline{F}) \otimes \mathbf{Q}$ over $\mathbf{Q}$.

The terminology *Hodge Class* is due to S. Zhang. In his generalization of the Gross-Zagier formulae to the case of Shimura curves, the morphism $\iota_H : M_H \rightsquigarrow J_H$ plays the role of the embedding $x \mapsto (x) - (\infty)$ of a classical modular curve into its Jacobian. This is why we refer to ι_H as the Hodge "embedding". It *is* a finite morphism, in the sense that some nonzero multiple $n\iota_H$ of ι_H is a genuine finite morphism from M_H to J_H. More generally:

Lemma 3.9 *Let $\pi : J_H \rightsquigarrow A$ be a nonzero morphism of abelian varieties over F. Then $\alpha = \pi \circ \iota_H : M_H \rightsquigarrow A$ is finite (in the above sense).*

Proof We may assume that $\pi : J_H \to A$ and $\alpha : M_H \to A$ are genuine morphisms. Since M_H is a connected complete curve over F, α is either finite or constant, and it can not be constant by Remark 3.8. □

3.6 Differentials and automorphic forms

Let $\Omega_H = \Omega_{M_H/F}$ be the sheaf of differentials on M_H and denote by Ω_H^{an} the pull-back of Ω_H to M_H^{an}, so that Ω_H^{an} is the sheaf of holomorphic 1-forms on

M_H^{an}. The *right* action of $G(\mathbf{A}_f)$ on the projective system $\{M_H\}_H$ induces a $\mathbf{C}$-linear *left* action of $G(\mathbf{A}_f)$ on the inductive system $\{\Gamma(\Omega_H^{\mathrm{an}})\}_H$ of global sections of these sheaves. We want to identify $\varinjlim \Gamma(\Omega_H^{\mathrm{an}})$, together with its $G(\mathbf{A}_f)$-action, with a suitable space $\mathcal{S}_2$ of automorphic forms on G.

Fix an isomorphism $G(\mathbf{R}) \simeq \mathrm{GL}_2(\mathbf{R}) \times G_2(\mathbf{R}) \times \cdots \times G_d(\mathbf{R})$ as in section 3.1 and let $\mathcal{S}_2$ be the complex vector space of all functions $\phi : G(\mathbf{A}) = G(\mathbf{A}_f) \times G(\mathbf{R}) \to \mathbf{C}$ with the following properties:

P1 ϕ is left $G(\mathbf{Q})$-invariant.
P2 ϕ is right invariant under $\mathbf{R}^* \times G_2(\mathbf{R}) \times \cdots \times G_d(\mathbf{R}) \subset G(\mathbf{R})$.
P3 ϕ is right invariant under some compact open subgroup of $G(\mathbf{A}_f)$.
P4 For every $g \in G(\mathbf{A})$ and $\theta \in \mathbf{R}$,

$$\phi\left(g\left(\begin{pmatrix} \cos(\theta) & \epsilon\sin(\theta) \\ -\epsilon\sin(\theta) & \cos(\theta) \end{pmatrix}, 1, \cdots, 1\right)\right) = \exp(2i\theta)\phi(g).$$

P5 For every $g \in G(\mathbf{A})$, the function

$$z = x + iy \mapsto \phi(g, z) \stackrel{\mathrm{def}}{=} \frac{1}{y}\phi\left(g \times \left(\begin{pmatrix} \epsilon y & x \\ 0 & 1 \end{pmatrix}, 1, \cdots, 1\right)\right)$$

is holomorphic on $\mathcal{H}_\epsilon$.

There is a left action of $G(\mathbf{A}_f)$ on $\mathcal{S}_2$ given by $(g \cdot \phi)(x) = \phi(xg)$.

Proposition 3.10 *There is a $G(\mathbf{A}_f)$-equivariant bijection*

$$\varinjlim \Gamma(\Omega_H^{\mathrm{an}}) \xrightarrow{\sim} \mathcal{S}_2$$

which identifies $\Gamma(\Omega_H^{\mathrm{an}})$ with $\mathcal{S}_2^H$.

Proof Recall from Remark 3.2 that there is a decomposition

$$M_H^{\mathrm{an}} = \textstyle\coprod_\alpha \overline{\Gamma}_\alpha \backslash \mathcal{H}_\epsilon$$

where α runs through a set of representatives of $G(\mathbf{Q})^+ \backslash G(\mathbf{A}_f)/H$ and $\overline{\Gamma}_\alpha$ is the image of $\Gamma_\alpha = \alpha H \alpha^{-1} \cap G(\mathbf{Q})^+$ in $\mathrm{PGL}_2^+(\mathbf{R})$.

Let $\omega \in \Gamma(\Omega_H^{\mathrm{an}})$ be a global holomorphic 1-form on M_H^{an}. The restriction of ω to the connected component $\overline{\Gamma}_\alpha \backslash \mathcal{H}_\epsilon$ of M_H^{an} pulls back to a $\overline{\Gamma}_\alpha$-invariant holomorphic form on $\mathcal{H}_\epsilon$. The latter equals $f_\alpha(z)dz$ for some holomorphic function f_α on $\mathcal{H}_\epsilon$ such that $f_\alpha|\gamma = f_\alpha$ for all $\gamma \in \overline{\Gamma}_\alpha$, where

$$(f|\gamma)(z) = \det(\gamma)(cz + d)^{-2} f\left(\frac{az + b}{cz + d}\right)$$

for a function f on $\mathcal{H}_\epsilon$ and $\gamma = \left(\begin{smallmatrix} a & b \\ c & d \end{smallmatrix}\right)$ in $\mathrm{GL}_2^+(\mathbf{R})$. Since

$$G(\mathbf{A}) = \coprod_\alpha G(\mathbf{Q}) \cdot \left(\alpha H \times G(\mathbf{R})^+\right),$$

we may write any element $g \in G(\mathbf{A})$ as a product $g = g_\mathbf{Q}(\alpha h \times g_\mathbf{R}^+)$ for some α with $g_\mathbf{Q} \in G(\mathbf{Q})$, $h \in H$ and

$$g_\mathbf{R}^+ \in G(\mathbf{R})^+ = \mathrm{GL}_2^+(\mathbf{R}) \times G_2(\mathbf{R}) \times \cdots \times G_d(\mathbf{R}).$$

We put $\phi_\omega(g) = \left(f_\alpha | g_{\mathbf{R},1}^+\right)(\epsilon i)$ where $g_{\mathbf{R},1}^+$ is the first component of $g_\mathbf{R}^+$. If g also equals $g'_\mathbf{Q}(\alpha' h' \times g'^+_\mathbf{R})$, then $\alpha = \alpha'$ and ${g'}_\mathbf{Q}^{-1} g_\mathbf{Q}$ belongs to $G^+(\mathbf{Q}) \cap \alpha H \alpha^{-1} = \Gamma_\alpha$, so that

$$f_{\alpha'} | {g'}^+_{\mathbf{R},1} = f_\alpha | {g'}_\mathbf{Q}^{-1} g_\mathbf{Q} g_{\mathbf{R},1}^+ = f_\alpha | g_{\mathbf{R},1}^+.$$

It follows that ϕ_ω is a well-defined complex valued function on $G(\mathbf{A})$. We leave it to the reader to check that ϕ_ω belongs to $\mathcal{S}_2^H$. Conversely, any H-invariant element ϕ in $\mathcal{S}_2$ defines an holomorphic differential 1-form ω_ϕ on M_H^{an}: with notations as in **P5**, the restriction of ω_ϕ to $\overline{\Gamma}_\alpha \backslash \mathcal{H}_\epsilon$ pulls back to $\phi(\alpha, z)dz$ on $\mathcal{H}_\epsilon$. □

Remark 3.11 The complex cotangent space of J_H at 0 is canonically isomorphic to $\Gamma(\Omega_H^{\mathrm{an}})$, and therefore also to $\mathcal{S}_2^H$. With these identifications, the right action of $\mathbf{T}_H = \mathrm{End}_{\mathbf{Z}[G(\mathbf{A}_f)]}(\mathbf{Z}[G(\mathbf{A}_f)/H])$ on $\mathcal{S}_2^H \simeq \mathrm{Hom}_{\mathbf{Z}[G(\mathbf{A}_f)]}(\mathbf{Z}[G(\mathbf{A}_f)/H], \mathcal{S}_2)$ coincides with the right action induced on the cotangent space by the left action of $\mathbf{T}_H$ on J_H.

Remark 3.12 When $\epsilon = 1$, $\mathcal{S}_2$ is exactly the subspace of $\mathcal{S}$ which is defined in the introduction, given our choice of an isomorphism between G_1 and $\mathrm{GL}_2(\mathbf{R})$ (cf. Remark 1.7 and Definition 1.8). This follows from the relevant properties of lowest weight vectors, much as in the classical case [6, Section 11.5].

3.7 The P-new quotient

Let P be a prime of F where B is split, and consider a compact open subgroup H of $G(\mathbf{A}_f)$ which decomposes as $H = H^P R_P^\times$, where R_P is an Eichler order in $B_P \simeq M_2(F_P)$ and H^P is a compact open subgroup of $G(\mathbf{A}_f)^P = \{g \in G(\mathbf{A}_f);\, g_P = 1\}$.

Definition 3.13 The *P-new quotient* of J_H is the largest quotient $\pi : J_H \twoheadrightarrow J_H^{P-\mathrm{new}}$ of J_H such that for any Eichler order $R'_P \subset B_P$ strictly containing R_P, $\pi \circ f^* = 0$ where $H' = H^P {R'}_P^\times$ and $f^* : J_{H'} \to J_H$ is the morphism induced by the degeneracy map $f : M_H \to M_{H'}$.

The quotient map $\pi : J_H \twoheadrightarrow J_H^{P-\mathrm{new}}$ induces an embedding from the complex cotangent space of $J_H^{P-\mathrm{new}}$ at 0 into the complex cotangent space of J_H at 0. Identifying the latter space first with $\Gamma(\Omega_H^{\mathrm{an}})$ and then with the space of H-invariant elements in $\mathcal{S}_2$ (Proposition 3.10), we obtain the P-new subspace $\mathcal{S}_{2,P-\mathrm{new}}^H$ of $\mathcal{S}_2^H$. By construction,

$$\mathcal{S}_2^H = \mathcal{S}_{2,P-\mathrm{new}}^H \oplus \mathcal{S}_{2,P-\mathrm{old}}^H$$

where $\mathcal{S}_{2,P-\mathrm{old}}^H$ is the subspace of $\mathcal{S}_2^H$ spanned by the elements fixed by $R'^{\times}_P$ for some Eichler order $R'_P \subset B_P$ strictly containing R_P.

3.8 CM Points

Let K be a totally imaginary quadratic extension of F and put $T = \mathrm{Res}_{K/\mathbf{Q}}(\mathbb{G}_{m,K})$. Any ring homomorphism $K \hookrightarrow B$ induces an embedding $T \hookrightarrow G$. A morphism $h : \mathbb{S} \to G_{\mathbf{R}}$ in X is said to have *complex multiplication* by K if it factors through the morphism $T_{\mathbf{R}} \hookrightarrow G_{\mathbf{R}}$ which is induced by an F-algebra homomorphism $K \hookrightarrow B$. For a compact open subgroup H of $G(\mathbf{A}_f)$, we say that $x \in M_H(\mathbf{C})$ is a *CM point* if $x = [g, h] \in M_H^{\mathrm{an}}$ for some $g \in G(\mathbf{A}_f)$ and $h \in X$ with complex multiplication by K.

We assume that K splits B, which amounts to requiring that K_v is a field for every finite place v of F where B ramifies. Then, there exists an F-algebra homomorphism $K \hookrightarrow B$, and any two such homomorphisms are conjugated by an element of $B^\times = G(\mathbf{Q})$. We fix such an homomorphism and let $T \hookrightarrow G$ be the induced morphism.

In each of the two connected components of X, there is exactly one morphism $\mathbb{S} \to G_{\mathbf{R}}$ which factors through $T_{\mathbf{R}} \hookrightarrow G_{\mathbf{R}}$. These two morphisms are permuted by the normalizer of $T(\mathbf{Q})$ in $G(\mathbf{Q})$, and $T(\mathbf{Q})$ is their common stabilizer in $G(\mathbf{Q})$. They correspond respectively to

$$z \in \mathbb{S} \mapsto (z \text{ or } \bar{z}, 1, \cdots, 1) \in T_1 \times \cdots \times T_d \simeq T_{\mathbf{R}}$$

where $K_i = K \otimes_{F,\tau_i} \mathbf{R}$, $T_i = \mathrm{Res}_{K_i/\mathbf{R}}(\mathbb{G}_{m,K_i})$ for $1 \leq i \leq d$, and where we have chosen an extension $\tau_1 : K \hookrightarrow \mathbf{C}$ of $\tau_1 : F \hookrightarrow \mathbf{R}$ to identify K_1 with $\mathbf{C}$ and T_1 with $\mathbb{S}$. We choose τ_1 in such a way that the morphism $h_K : \mathbb{S} \to G_{\mathbf{R}}$ corresponding to $z \mapsto (z, 1, \cdots 1)$ belongs to the connected component of the morphism $h : \mathbb{S} \to G_{\mathbf{R}}$ which is defined by (3.1). The map

$$g \in G(\mathbf{A}_f) \mapsto [g, h_K] \in M_H^{\mathrm{an}} = G(\mathbf{Q})\backslash\left(G(\mathbf{A}_f)/H \times X\right)$$

then induces a bijection between the set of K-CM points in M_H^{an} and the set $\mathrm{CM}_H = T(\mathbf{Q})\backslash G(\mathbf{A}_f)/H$ of the introduction. In the sequel, we will use this identification without any further reference. In particular, we denote by $[g] \in M_H^{\mathrm{an}} = M_H(\mathbf{C})$ the CM point corresponding to $[g, h_K]$.

By Shimura's theory, the CM points are algebraic and defined over the maximal abelian extension K^{ab} of K. Moreover, for $\lambda \in T(\mathbf{A}_f)$ and $g \in G(\mathbf{A}_f)$, the action of $\sigma = \mathrm{rec}_K(\lambda) \in \mathrm{Gal}_K^{\mathrm{ab}}$ on $x = [g] \in \mathrm{CM}_H$ is given by the following reciprocity law (viewing K as a subfield of $\mathbf{C}$ through τ_1):

$$\sigma \cdot x = [\lambda^{\epsilon} g] \in \mathrm{CM}_H.$$

We refer to Sections 3.1.2 and 3.2.1 of [5] for a more detailed discussion of CM points on Shimura curves, including a proof of the above facts.

4 The Indefinite Case

To the data of F, B, K (and ϵ), we have attached a collection of Shimura curves $\{M_H\}$ equipped with a systematic supply of CM points defined over the maximal abelian extension K^{ab} of K. We now also fix a prime P of F where B is split, and restrict our attention to the CM points of P-power conductor in a given Shimura curve $M = M_H$, with $H = \widehat{R}^{\times}$ for some $\mathcal{O}_F$-order $R \subset B$. We assume that

(H1) R_P is an Eichler order in $B_P \simeq M_2(F_P)$.
(H2) For any prime $Q \neq P$ of F which ramifies in K, B is split at Q and R_Q is a maximal order in $B_Q \simeq M_2(F_Q)$.

We put $J = J_H$, $\mathrm{CM} = \mathrm{CM}_H$, $\iota = \iota_H$ and so on... We denote by $\mathrm{CM}(P^n) \subset M(K[P^n])$ the set of CM points of conductor P^n and put

$$\mathrm{CM}(P^{\infty}) = \cup_{n \geq 0} \mathrm{CM}(P^n) \subset M(K[P^{\infty}]).$$

Thanks to **(H1)**, we have the notion of *good* CM points, as defined in the introduction. Recall that all CM points are good when R_P is maximal; otherwise, the good CM points are those which are of type I or II, in the terminology of section 6.

In this section, we study the contribution of these points to the growth of the Mordell-Weil groups of suitable quotients A of J, as one ascends the abelian extension $K[P^{\infty}]$ of K.

4.1 Statement of the main results

We say that $\pi : J \rightsquigarrow A$ is a *surjective morphism* if some nonzero multiple of π is a genuine surjective morphism $J \rightarrow A$ of abelian varieties over F. We say that π is *P-new* if it factors through the P-new quotient of J, cf. Definition 3.13. We say that π is $\mathrm{Pic}(\mathcal{O}_F)$*-equivariant* if A is endowed with an action θ_A of $\mathrm{Pic}(\mathcal{O}_F)$ such that

$$\forall \sigma \in \mathrm{Pic}(\mathcal{O}_F): \quad \theta_A(\sigma) \circ \pi = \pi \circ \theta_J(\sigma) \quad \text{in } \mathrm{Hom}^0(J, A),$$

cf. Definition 3.4.

We put $C \stackrel{\text{def}}{=} \mathbf{Z}[\mathrm{Pic}(\mathcal{O}_F)]$. For a character $\omega : \mathrm{Pic}(\mathcal{O}_F) \to \mathbf{C}^\times$, we let $\mathbf{Q}\{\omega\}$ be the image of the induced morphism $\omega_* : C \otimes \mathbf{Q} \to \mathbf{C}$. Then $\mathbf{Q}\{\omega\}$ only depends upon the $\mathrm{Aut}(\mathbf{C})$-conjugacy class $\{\omega\}$ of ω, and $C \otimes \mathbf{Q} \simeq \prod_{\{\omega\}} \mathbf{Q}\{\omega\}$. If A is endowed with an action of $\mathrm{Pic}(\mathcal{O}_F)$, then $\mathrm{End}^0(A)$ is a $C \otimes \mathbf{Q}$-algebra and we write

$$A \simeq \oplus_{\{\omega\}} A\{\omega\}$$

for the corresponding decomposition in the category $\mathbf{Ab}_F^0$ of abelian varieties over F up to isogenies. If $A = A\{\omega\}$ for some ω, then C acts on A through its quotient $\mathbf{Z}\{\omega\}$, the image of C in $\mathbf{Q}\{\omega\}$.

Recall from section 2 that the torsion subgroup G_0 of $G(\infty) = \mathrm{Gal}(K[P^\infty]/K)$ contains a subgroup G_2 which is canonically isomorphic to $\mathrm{Pic}(\mathcal{O}_F)$. For a character χ of $G(\infty)$ or G_0, we denote by $\mathrm{Res}(\chi)$ the induced character on $\mathrm{Pic}(\mathcal{O}_F)$. For a character χ of $G(n)$, we denote by $\mathbf{e}_\chi \in \mathbf{C}[G(n)]$ the idempotent of χ. We say that χ is *primitive* if it does not factor through $G(n-1)$.

Theorem 4.1 *Suppose that $\pi : J \rightsquigarrow A$ is a surjective,* $\mathrm{Pic}(\mathcal{O}_F)$*-equivariant and P-new morphism. Fix a character χ_0 of G_0 such that $A\{\omega\} \neq 0$ where $\omega^\epsilon = \mathrm{Res}(\chi_0)$. Then: for any $n \gg 0$ and any good CM point $x \in \mathrm{CM}(P^n)$, there exists a primitive character χ of $G(n)$ inducing χ_0 on G_0 such that $\mathbf{e}_\chi \alpha(x) \neq 0$ in $A \otimes \mathbf{C}$, where α is defined in Lemma 3.9.*

Replacing π by $\pi\{\omega\} : J \rightsquigarrow A \rightsquigarrow A\{\omega\}$ and using Lemma 4.6 below, one easily checks that Theorem 4.1 is in fact equivalent to the following variant, in which we use ω to embed $\mathbf{Z}\{\omega\}$ into $\mathbf{C}$.

Theorem 4.2 *Suppose that $\pi : J \rightsquigarrow A$ is a surjective,* $\mathrm{Pic}(\mathcal{O}_F)$*-equivariant and P-new morphism. Suppose also that $A = A\{\omega\} \neq 0$ for some character ω of* $\mathrm{Pic}(\mathcal{O}_F)$*. Fix a character χ_0 of G_0 inducing ω^ϵ on $G_2 \simeq \mathrm{Pic}(\mathcal{O}_F)$. Then: for any $n \gg 0$ and any good CM point $x \in \mathrm{CM}(P^n)$, there exists a primitive character χ of $G(n)$ inducing χ_0 on G_0 such that $\mathbf{e}_\chi \alpha(x) \neq 0$ in $A \otimes_{\mathbf{Z}\{\omega\}} \mathbf{C}$.*

Remark 4.3 In the situation of Theorem 1.10, $\epsilon = 1$ and

$$A = J/\mathrm{Ann}_{\mathbf{T}}(\mathbf{C} \cdot \Phi)J$$

where $\mathbf{C} \cdot \Phi = \mathcal{S}_2(\pi')^H$, π' is an automorphic representation of G with central character $\omega : Z(\mathbf{A}_f) \twoheadrightarrow \mathrm{Pic}(\mathcal{O}_F) \rightarrow \mathbf{C}^\times$ and $H = \widehat{R}^\times$ with R defined by (1.6) – see Remarks 3.1, 3.11 and 3.12. The assumptions **(H1)** and **(H2)** are then satisfied, and the projection $J \rightarrow A$ is surjective and $\mathrm{Pic}(\mathcal{O}_F)$-equivariant (with $A = A\{\omega\}$). By construction, there is no Eichler order in B_P strictly containing R_P whose group of invertible elements fixes Φ. It thus follows from the discussion after Definition 3.13 that $J \rightarrow A$ also factors through the P-new quotient of J. Since $\epsilon = 1$, Theorem 4.1 asserts that for any character χ_0 of G_0 such that $\chi_0 \cdot \omega = 1$ on $Z(\mathbf{A}_f)$, and any good CM point $x \in \mathrm{CM}(P^n)$ with n sufficiently large, there exists a character $\chi \in P(n, \chi_0)$ such that

$$\mathbf{a}(x, \chi) = \mathbf{e}_{\chi^{-1}}\alpha(x) \neq 0 \quad \text{in } A(K[P^n]) \otimes \mathbf{C}.$$

This is exactly the statement of Theorem 1.10.

4.2 An easy variant

As an introduction to this circle of ideas, we will first show that a weaker variant of Theorem 4.1 can be obtained by very elementary methods, in the spirit of [22]. Thus, let $\pi : J \rightsquigarrow A$ be a nonzero surjective morphism, and put $\alpha = \pi \circ \iota : M \rightsquigarrow A$ where $\iota : M \rightsquigarrow J$ is the "Hodge embedding" of section 3.5. Then:

Proposition 4.4 *For all $n \gg 0$ and all $x \in \mathrm{CM}(P^n)$,*

$$\alpha(x) \neq 0 \quad \textit{in } A(K[P^n]) \otimes \mathbf{Q}.$$

Proof Using Lemma 3.9, we may assume that $\alpha : M \rightarrow A$ is a finite morphism. In particular, there exists a positive integer d such that $\left|\alpha^{-1}(x)\right| \leq d$ for any $x \in A(\mathbf{C})$. On the other hand, it follows from Lemma 2.7 that the torsion subgroup of $A(K[P^\infty])$ is finite, say of order $t > 0$. Then α maps at most dt points in $\mathrm{CM}(P^\infty)$ to torsion points in A, and the proposition follows. □

Corollary 4.5 *There exists a character $\chi : G(n) \rightarrow \mathbf{C}^\times$ such that*

$$\mathbf{e}_\chi \alpha(x) \neq 0 \quad \textit{in } A(K[P^n]) \otimes \mathbf{C}.$$

Suppose moreover that π is a $\mathrm{Pic}(\mathcal{O}_F)$-equivariant morphism. On $A(K[P^n]) \otimes \mathbf{C}$, we then also have an action of $\mathrm{Pic}(\mathcal{O}_F)$. For a character $\omega : \mathrm{Pic}(\mathcal{O}_F) \to \mathbf{C}^\times$, let $\mathbf{e}_\chi^\omega \alpha(x)$ be the ω-component of $\mathbf{e}_\chi \alpha(x)$. For any $\sigma \in \mathrm{Aut}(\mathbf{C})$, the automorphism $1 \otimes \sigma$ of $A(K[P^n]) \otimes \mathbf{C}$ maps $\mathbf{e}_\chi^\omega \alpha(x)$ to $\mathbf{e}_{\sigma\circ\chi}^{\sigma\circ\omega} \alpha(x)$: if the former is nonzero, so is the latter.

Lemma 4.6 $\mathbf{e}_\chi^\omega \alpha(x) = 0$ *unless* $\mathrm{Res}(\chi) = \omega^\epsilon$ *on* $G_2 \simeq \mathrm{Pic}(\mathcal{O}_F)$.

Proof Write $x = [g]$ for some $g \in G(\mathbf{A}_f)$. For $\lambda \in Z(\mathbf{A}_f)$, put

$$\mathrm{Pic}(\mathcal{O}_F) \ni [\lambda] = \sigma = \mathrm{rec}_K(\lambda) \in G_2.$$

If ρ denotes the Galois action, we find that

$$\theta_M(\sigma)(x) = [g\lambda] = [\lambda g] = \rho(\sigma^\epsilon)(x) \quad \text{in } M(K[P^n]).$$

It follows that

$$\begin{array}{rcll} \theta_J(\sigma)(\iota x) & = & \rho(\sigma^\epsilon)(\iota x) & \text{in } J(K[P^n]) \otimes \mathbf{Q}, \\ \theta_A(\sigma)(\alpha(x)) & = & \rho(\sigma^\epsilon)(\alpha(x)) & \text{in } A(K[P^n]) \otimes \mathbf{Q}, \\ \text{and} \quad \omega(\sigma) \cdot \mathbf{e}_\chi^\omega \alpha(x) & = & \chi(\sigma^\epsilon) \cdot \mathbf{e}_\chi^\omega \alpha(x) & \text{in } A(K[P^n]) \otimes \mathbf{C}. \end{array} \tag{4.1}$$

In particular, $\mathbf{e}_\chi^\omega \alpha(x) = 0$ if $\omega(\sigma) \neq \chi(\sigma^\epsilon)$ for some $\sigma \in G_2$. □

We thus obtain the following refinement of Proposition 4.4.

Proposition 4.7 *Let* $\omega : \mathrm{Pic}(\mathcal{O}_F) \to \mathbf{C}^\times$ *be any character such that* $A\{\omega\} \neq 0$. *Then for all* $n \gg 0$ *and all* $x \in \mathrm{CM}(P^n)$, *there exists a character* $\chi : G(n) \to \mathbf{C}^\times$ *inducing* ω^ϵ *on* G_2 *such that*

$$\mathbf{e}_\chi \alpha(x) \neq 0 \quad \textit{in } A(K[P^n]) \otimes \mathbf{C}.$$

Proof Applying Proposition 4.4 to $\pi\{\omega\} : J \rightsquigarrow A \rightsquigarrow A\{\omega\}$, we find a character χ' on $G(n)$ such that $\mathbf{e}_{\chi'} \alpha(x) \neq 0$ in $A\{\omega\}(K[P^n]) \otimes \mathbf{C}$. Lemma 4.6 then implies that $\mathrm{Res}(\chi') = \sigma \cdot \omega^\epsilon$ for some $\sigma \in \mathrm{Aut}(\mathbf{C})$, and we take $\chi = \sigma^{-1} \circ \chi'$. □

Remark 4.8 In contrast to Theorem 4.1, this proposition does not require π to be P-new, nor x to be good. It holds true without the assumptions **(H1)** and **(H2)**. On the other hand, Theorem 4.1 yields a *primitive* character whose *tame part* χ_0 is fixed but arbitrary, provided it coincides with ω^ϵ on $\mathrm{Pic}(\mathcal{O}_F)$. This seems to entail a significantly deeper assertion on the growth of the Mordell-Weil groups of A along $K[P^\infty]/K$.

4.3 Proof of Theorem 4.2

To prove that $\mathbf{e}_\chi\alpha(x)$ is nonzero for *some* primitive character χ of $G(n)$ inducing χ_0 on G_0, it is certainly sufficient to show that the *sum* of these values is a nonzero element in $A(K[P^n]) \otimes_{\mathbf{Z}\{\omega\}} \mathbf{C}$. Provided that n is sufficiently large, Lemma 2.8 implies that this sum is equal to

$$\mathbf{e}(\chi_0, n) \cdot \alpha(x) = \frac{1}{q\,|G_0|} \cdot \pi \left(\sum_{\sigma \in G_0} \overline{\chi}_0(\sigma)\sigma \cdot d(x) \right) \tag{4.2}$$

where $d(x) = (q - \mathrm{Tr}_{Z(n)})(\iota x)$. When x is a *good* CM point, $d(x)$ may be computed using the distribution relations of section 6.

Lemma 4.9 *Let δ be the exponent of P in the level of the Eichler order $R_P \subset B_P$. If n is sufficiently large, the following relations hold in the P-new quotient $J^{P-new} \otimes \mathbf{Q}$ of $J \otimes \mathbf{Q}$.*

(i) *If $\delta = 0$, $d(x) = q \cdot \iota x - T_P^l \cdot \iota x' + \iota x''$ where $T_P^l \in \mathbf{T}$ is a certain Hecke operator, $x' = \mathrm{pr}_u(x)$ belongs to $\mathrm{CM}(P^{n-1})$ and $x'' = \mathrm{pr}_l(x')$ belongs to $\mathrm{CM}(P^{n-2})$.*

(ii) *If $\delta = 1$, $d(x) = q \cdot \iota x + \iota x'$ with $x' = \mathrm{pr}(x)$ in $\mathrm{CM}(P^{n-1})$.*

(iii) *If $\delta \geq 2$ and x is a* good *CM point, $d(x) = q \cdot \iota x$.*

Proof We refer the reader to section 6 for the notations and proofs. Strictly speaking, we do not show there that x' belongs to $\mathrm{CM}(P^{n-1})$ and x'' belongs to $\mathrm{CM}(P^{n-2})$. This however easily follows from the construction of these points. Also, Lemmas 6.6, 6.11 and 6.14 compute formulas involving the image of $\sum_{\lambda \in \mathcal{O}_{n-1}^\times/\mathcal{O}_n^\times} \mathrm{rec}_K(\lambda) \cdot x$ in the P-new quotient of the free abelian group $\mathbf{Z}[\mathrm{CM}]$. To retrieve the above formulas, use the discussion preceding Lemma 2.9 and the compatibility of ι with the formation of the P-new quotients of $\mathbf{Z}[\mathrm{CM}]$ and J. □

Since $\pi : J \rightsquigarrow A$ is P-new, these relations also hold in $A \otimes \mathbf{Q}$. In particular, for $\delta \geq 2$, part (iii) of the above lemma implies that Theorem 4.2 is now a consequence of the following theorem, whose proof will be given in sections 4.5-4.6.

Theorem 4.10 *Suppose that $\pi : J \rightsquigarrow A$ is a surjective, $\mathrm{Pic}(\mathcal{O}_F)$-equivariant morphism such that $A = A\{\omega\} \neq 0$ for some character ω of $\mathrm{Pic}(\mathcal{O}_F)$. Fix a character χ of G_0 inducing ω^ϵ on $G_2 \simeq \mathrm{Pic}(\mathcal{O}_F)$. Then for all but finitely* *many* $x \in \mathrm{CM}(P^\infty)$,

$$\mathbf{e}_\chi \cdot \alpha(x) = \frac{1}{|G_0|} \sum_{\sigma \in G_0} \overline{\chi}(\sigma)\sigma \cdot \alpha(x) \neq 0 \quad \text{in } A \otimes_{\mathbf{Z}\{\omega\}} \mathbf{C}.$$

Remark 4.11 This is the statement which is actually used in [17].

In the next subsection, we will show that Theorem 4.2 also follows from the above theorem when $\delta = 0$ or 1, provided that we change the original P-new parameterization $\pi : J \rightsquigarrow A$ of A to a non-optimal parameterization $\pi^+ : J^+ \rightsquigarrow A$. Although this new parameterization will still satisfy to the assumptions **(H1)** and **(H2)**, the proof of Theorem 4.10 only requires **(H2)** to hold.

4.4 Changing the level

Suppose first that $\delta = 0$. Let $R_P^+ \subset R_P$ be the Eichler order of level P^2 in B_P which is constructed in section 6.5. Put $H^+ = H^P R_P^+$, $M^+ = M_{H^+}$, $J^+ = J_{H^+}$ and so on. By Lemma 6.16, there exists degeneracy maps d_0, d_1 and $d_2 : M^+ \to M$ as well as an element $\vartheta \in C^\times$ with the property that for all $x \in \mathrm{CM}(P^n)$ with $n \geq 2$, there exists a CM point $x^+ \in \mathrm{CM}^+(P^n)$ such that

$$(d_0, d_1, d_2)(x^+) = (x, x', \vartheta^{-1} x'').$$

Combining this with part (1) of Lemma 4.9 (and using also the results of section 3.5, especially formula 3.7 and Lemma 3.6) we obtain:

$$d(x) = \left(q(d_0)_* - T_P^l (d_1)_* + \vartheta (d_2)_*\right)(\iota^+ x^+) \quad \text{in } J \otimes \mathbf{Q}$$

so that $\pi \circ d(x) = \alpha^+(x^+)$ in $A \otimes \mathbf{Q}$, where $\alpha^+ = \pi^+ \circ \iota^+$ with

$$\pi^+ = \pi \circ (q, -T_P^l, \vartheta) \circ \begin{pmatrix} d_0 \\ d_1 \\ d_2 \end{pmatrix}_* : J^+ \to J^3 \to J \rightsquigarrow A.$$

In particular, (4.2) becomes

$$\mathbf{e}(\chi_0, n) \cdot \alpha(x) = \frac{1}{q\,|G_0|} \cdot \sum_{\sigma \in G_0} \overline{\chi}_0(\sigma)\sigma \cdot \alpha^+(x^+) \quad \text{in } A \otimes_{\mathbf{Z}\{\omega\}} \mathbf{C}.$$

Theorem 4.2 for π thus follows from Theorem 4.10 for π^+, once we know that our new parameterization $\pi^+ : J^+ \rightsquigarrow A$ is surjective and $\mathrm{Pic}(\mathcal{O}_F)$-equivariant (these are the assumptions of Theorem 4.10). The $\mathrm{Pic}(\mathcal{O}_F)$-equivariance is straightforward. Since the second and third morphisms in the definition of π^+

are surjective, it remains to show that the first one is also surjective, which amounts to showing that the induced map on the (complex) cotangent spaces at 0 is an injection. In view of Proposition 3.10, this all boils down to the following lemma.

Lemma 4.12 *The kernel of* $(d_0^*, d_1^*, d_2^*) : (\mathcal{S}_2^H)^3 \to \mathcal{S}_2^{H^+}$ *is trivial.*

Proof With notations as in section 6.5, the above map is given by

$$(F_0, F_1, F_2) \mapsto F_0' + F_1' + F_2'$$

where $F_i'(g) = d_i^* F_i(g) = F_i(gb_i)$ with $b_i \in B_P^\times$ such that $b_i L(0) = L(2-i)$. Here, $L = (L(0), L(2))$ is a 2-lattice in some simple left $B_P \simeq M_2(F_P)$-module $V \simeq F_P^2$ such that $R_P = \{\alpha \in B_P; \alpha L(0) \subset L(0)\}$. Put

$$R_i = \{\alpha \in B_P;\ \alpha L(2-i) \subset L(2-i)\}$$

so that $R_2 = R_P$ and $R_i^\times = b_i R_2^\times b_i^{-1}$ fixes F_i'. One easily checks that $R_0^\times \cap R_1^\times$ and $R_1^\times \cap R_2^\times$ generate $R_1^\times$ inside $B_P^\times$. By [24, Chapter 2, section 1.4], $R_0^\times$ and $R_1^\times$ (resp. $R_1^\times$ and $R_2^\times$) generate the subgroup $(B_P^\times)^0$ of all elements in $B_P^\times \simeq \mathrm{GL}_2(F_P)$ whose reduced norm (=determinant) belongs to $\mathcal{O}_{F,P}^\times \subset F_P^\times$.

Suppose that $F_0' + F_1' + F_2' \equiv 0$ on $G(\mathbf{A})$. Then $F_2' = -F_0' - F_1'$ is fixed by $R_2^\times$ and $R_0^\times \cap R_1^\times$ and therefore also by $(B_P^\times)^0$. Being continuous, left invariant under $G(\mathbf{Q})$ and right invariant under $(B_P^\times)^0 H^P$, the function $F_2' : G(\mathbf{A}) \to \mathbf{C}$ is then also left (and right) invariant under the kernel $G^1(\mathbf{A})$ of the reduced norm $\mathrm{nr} : G(\mathbf{A}) = \widehat{B}^\times \to \widehat{F}^\times$ by the strong approximation theorem [29, p. 81]. For any $g \in G(\mathbf{A})$ and $\theta \in \mathbf{R}$, we thus obtain (using the property **P4** of section 3.6)

$$F_2'(g) = F_2'\left(g \times \left(\begin{pmatrix} \cos\theta & \epsilon\sin\theta \\ -\epsilon\sin\theta & \cos\theta \end{pmatrix}, 1, \cdots, 1\right)\right) = e^{2i\theta} F_2'(g),$$

so that $F_2' \equiv 0$ on $G(\mathbf{A})$. Similarly, $F_0' \equiv 0$ on $G(\mathbf{A})$. It follows that $F_1' \equiv 0$, hence $F_0 \equiv F_1 \equiv F_2 \equiv 0$ on $G(\mathbf{A})$ and (d_0^*, d_1^*, d_2^*) is indeed injective. □

Suppose next that $\delta = 1$. Using now Lemma 6.17, we find two degeneracy maps d_{01} and $d_{12} : M^+ \to M$, as well as an F-automorphism ϑ of M such that for all $x \in \mathrm{CM}(P^n)$ with $n \geq 2$, there exists a CM point $x^+ \in \mathrm{CM}^+(P^n)$ such that (x, x') equals

$$(d_{01}, \vartheta^{-1} d_{12})(x^+) \quad \text{or} \quad (d_{12}, \vartheta d_{01})(x^+).$$

Theorem 4.2 (with π) thus again follows from Theorem 4.10 with

$$\pi' = \pi \circ \left\{ \begin{array}{c} (q, \vartheta_*^{-1}) \\ \text{or} \\ (\vartheta_*, q) \end{array} \right\} \circ \left(\begin{array}{c} d_{01} \\ d_{12} \end{array} \right)_* : J^+ \to J^2 \to J \rightsquigarrow A$$

once we know that π' induces an injection on the complex cotangent spaces. Since π is P-new, this now amounts to the following lemma (see section 3.7).

Lemma 4.13 *The kernel of* $(d_{01}^*, d_{12}^*) : (\mathcal{S}_{2,P-new}^H)^2 \to \mathcal{S}_2^{H^+}$ *is trivial.*

Proof With notations as in section 6.5, the above map is now given by

$$(F_{01}, F_{12}) \mapsto F'_{01} + F'_{12}$$

where $F'_{01} = F_{01}$ and $F'_{12}(g) = F_{12}(gb_{12})$ with $b_{12} \in B_P^\times$ such that $b_{12}(L(0), L(1)) = (L(1), L(2))$ for some 2-lattice $L = (L(0), L(2))$ in V such that $R_P^\times = R_0^\times \cap R_1^\times$ with

$$R_i = \{\alpha \in B_P;\ \alpha L(i) \subset L(i)\}.$$

If $F'_{01} + F'_{12} = 0$, $F'_{01} = -F'_{12}$ is fixed by $R_0^\times \cap R_1^\times$ and $R_1^\times \cap R_2^\times$. It is therefore also fixed by $R_1^\times$ so that F_{01} and F_{12} both belong to the P-old subspace of $\mathcal{S}_2^H$. □

4.5 Geometric Galois action

We now turn to the proof of Theorem 4.10. Thus, let $\pi : J \rightsquigarrow A$ be a surjective and $\mathrm{Pic}(\mathcal{O}_F)$-equivariant morphism such that $A = A\{\omega\} \neq 0$ for some character ω of $\mathrm{Pic}(\mathcal{O}_F)$, and let χ be a fixed character of G_0 inducing ω^ϵ on $G_2 \simeq \mathrm{Pic}(\mathcal{O}_F)$.

For $0 \leq i \leq 2$, let C_i be the subring of $\mathbf{C}$ which is generated by the values of χ on G_i, so that $C_2 \subset C_1 \subset C_0$, C_1 is finite flat over C_2, and so is C_0 over C_1. Since χ induces ω^ϵ on G_2, the canonical factorization of $\omega_* : C \to \mathbf{C}$ yields an isomorphism between $\mathbf{Z}\{\omega\}$ and C_2. Let A_i be the (nonzero) abelian variety over F which is defined by

$$A_i(X) = A(X) \otimes_C C_i = A(X) \otimes_{\mathbf{Z}\{\omega\}} C_i$$

for any F-scheme X (note that $A_2 \simeq A$).

Upon multiplying π by a suitable integer, we may assume that π and α are genuine morphisms. For any $x \in \mathrm{CM}(P^n)$, we may then view

$$\mathbf{a}(x) \stackrel{\text{def}}{=} \sum_{\sigma \in G_0} \overline{\chi}(\sigma)\sigma \cdot \alpha(x)$$

as an element of

$$A_0(K[P^\infty]) = A_1(K[P^\infty]) \otimes_{C_1} C_0 = A(K[P^\infty]) \otimes_{\mathbf{Z}\{\omega\}} C_0,$$

and we now have to show that $\mathbf{a}(x)$ is a *nontorsion* element in this group, provided that n is sufficiently large.

Using formula (4.1), which applies thanks to the $\mathrm{Pic}(\mathcal{O}_F)$-equivariance of π, we immediately find that

$$\mathbf{a}(x) = |G_2| \sum_{\sigma \in \mathcal{R}'} \overline{\chi}(\sigma)\sigma \cdot \alpha(x)$$

where $\mathcal{R}' \subset G_0$ is the following set of representatives for G_0/G_2. We first choose a set of representatives $\mathcal{R} \subset G_0$ of G_0/G_1 containing 1, and then take

$$\mathcal{R}' = \{\tau\sigma_D^\epsilon;\ \tau \in \mathcal{R} \text{ and } D \mid \mathcal{D}'\}$$

where $\mathcal{D}' \subset \mathcal{O}_F$ and the σ_D's for $D \mid \mathcal{D}'$ were defined in Lemma 2.6. The next lemma will allow us to further simplify $\mathbf{a}(x)$.

Lemma 4.14 *There exists a Shimura curve M_1 and a collection of degeneracy maps $\{\mathbf{d}_D : M_1 \to M;\ D \mid \mathcal{D}'\}$ such that for all $n \geq 0$,*

$$\forall x \in \mathrm{CM}(P^n),\ \exists x_1 \in \mathrm{CM}_1(P^n) \text{ s.t. } \forall D \mid \mathcal{D}' : \quad \sigma_D^\epsilon x = \mathbf{d}_D(x_1).$$

Proof Our assumption **(H2)** asserts that for any $Q \mid \mathcal{D}'$, R_Q is a maximal order in $B_Q \simeq M_2(F_Q)$. Let Γ_Q be the set of elements in R_Q whose reduced norm is a uniformizer in $\mathcal{O}_{F,Q}$, and choose some α_Q in Γ_Q. Then $\Gamma_Q = R_Q^\times \alpha_Q R_Q^\times$ and

$$R_{1,Q} \stackrel{\text{def}}{=} R_Q \cap \alpha_Q R_Q \alpha_Q^{-1} \subset B_Q$$

is an Eichler order of level Q. Put $H_1 = \widehat{R}_1^\times$, where R_1 is the unique $\mathcal{O}_F$-order in B which agrees with R outside $\mathcal{D}'$, and equals $R_{1,Q}$ at $Q \mid \mathcal{D}'$. Put $M_1 = M_{H_1}$, $\mathrm{CM}_1 = \mathrm{CM}_{H_1}$ and so on. For $D \mid \mathcal{D}'$, put

$$\alpha_D \stackrel{\text{def}}{=} \prod_{Q \mid D} \alpha_Q \in G(\mathbf{A}_f). \tag{4.3}$$

Then $\alpha_D^{-1} H_1 \alpha_D \subset H$. Let $\mathbf{d}_D = [\cdot \alpha_D] : M_1 \to M$ be the corresponding degeneracy map.

Recall also that $\sigma_D = \prod_{Q \mid D} \sigma_Q$ for $D \mid \mathcal{D}'$, where $\sigma_Q \in G_1$ is the geometric Frobenius of the unique prime $\mathcal{Q}$ of K above Q (so that $\mathcal{Q}^2 = Q\mathcal{O}_K$). Let $\pi_Q \in \mathcal{O}_{K,Q}$ be a local uniformizer at $\mathcal{Q}$, and for $D \mid \mathcal{D}'$, put $\pi_D = \prod_{Q \mid D} \pi_Q$ in $\widehat{K}^\times$, so that σ_D is the restriction of $\mathrm{rec}_K(\pi_D)$ to $K[P^\infty]$.

Consider now some $x = [g] \in \mathrm{CM}(P^n)$, with $g \in G(\mathbf{A}_f)$ and $n \geq 0$. For each $Q \mid \mathcal{D}'$, $K_Q \cap g_Q R_Q g_Q^{-1} = \mathcal{O}_{K,Q}$. In particular, π_Q belongs to $g_Q R_Q g_Q^{-1}$ and $g_Q^{-1}\pi_Q g_Q$ belongs to Γ_Q: there exists $r_{1,Q}$ and $r_{2,Q}$ in $R_Q^\times$ such that $\pi_Q g_Q = g_Q r_{1,Q}\alpha_Q r_{2,Q}$. Put $r_i = \prod_{Q|\mathcal{D}'} r_{i,Q} \in H$ and $x_1 = [gr_1] \in \mathrm{CM}_1$. For $D \mid \mathcal{D}'$, we find that

$$\mathbf{d}_D(x_1) = [gr_1\alpha_D] = [gr_1\alpha_D r_2] = [\pi_D g] = \sigma_D^\epsilon x.$$

Finally, x_1 belongs to $\mathrm{CM}_1(P^n)$, because $\widehat{K}^\times \cap gr_1 H_1 (gr_1)^{-1}$ is obviously equal to $\widehat{K}^\times \cap gHg^{-1}$ away from $\mathcal{D}'$, and

$$\begin{aligned} &K_Q \cap g_Q r_{1,Q} R_{1,Q}^\times (g_Q r_{1,Q})^{-1} = \\ &= K_Q \cap g_Q r_{1,Q} R_Q^\times (g_Q r_{1,Q})^{-1} \cap g_Q r_{1,Q}\alpha_Q R_Q^\times \alpha_Q^{-1} (g_Q r_{1,Q})^{-1} \\ &= \left(K_Q \cap g_Q R_Q^\times g_Q^{-1}\right) \cap \left(K_Q \cap \pi_Q g_Q R_Q^\times g_Q^{-1} \pi_Q^{-1}\right) \\ &= K_Q \cap g_Q R_Q^\times g_Q^{-1} \end{aligned}$$

for $Q \mid \mathcal{D}'$. This finishes the proof of Lemma 4.14. □

Put $J_1 = \mathrm{Pic}^0 M_1$ and let $\iota_1 : M_1 \rightsquigarrow J_1$ be the corresponding "Hodge embedding". With notations as above, we find that $\mathbf{a}(x) = |G_2|\,\mathbf{b}(x_1)$, where for any CM point $y \in \mathrm{CM}_1(P^\infty)$,

$$\mathbf{b}(y) \stackrel{\text{def}}{=} \sum_{\tau \in \mathcal{R}} \overline{\chi}(\tau)\tau \cdot \alpha_1(y) \quad \text{in } A_1(K[P^\infty]) \otimes_{C_1} C_0.$$

Here, $\alpha_1 \stackrel{\text{def}}{=} \pi_1 \circ \iota_1$ with $\pi_1 : J_1 \to A_1$ defined by

$$\begin{array}{ccccccc} J_1 & \longrightarrow & J^{\{D|\mathcal{D}\}} & \stackrel{\pi}{\longrightarrow} & A^{\{D|\mathcal{D}\}} & \longrightarrow & A_1 = A \otimes_C C_1 \\ j & \longmapsto & ((\mathbf{d}_D)_*(j))_{D|\mathcal{D}} & & (a_D)_{D|\mathcal{D}} & \longmapsto & \sum_{D|\mathcal{D}} \overline{\chi}(\sigma_D^\epsilon) a_D \end{array}$$

Indeed, Lemma 4.14 (together with the formula (3.7)) implies that

$$\alpha_1(x_1) = \sum_{D|\mathcal{D}} \overline{\chi}(\sigma_D^\epsilon)\sigma_D^\epsilon \cdot \alpha(x) \quad \text{in } A(K[P^\infty]) \otimes_C C_1 = A_1(K[P^\infty]).$$

We now have to show that for all $x \in \mathrm{CM}_1(P^n)$ with $n \gg 0$,

$$\mathbf{b}(x) \neq \text{torsion} \quad \text{in } A_1(K[P^\infty]) \otimes_{C_1} C_0.$$

We will need to know that our new parameterization $\pi_1 : J_1 \to A_1$ is still surjective. As before (Lemmas 4.12 and 4.13), this amounts to the following lemma.

Lemma 4.15 *The kernel of* $\sum_{D|\mathcal{D}'} \mathbf{d}_D^* : (\mathcal{S}_2^H)^{\{D|\mathcal{D}'\}} \to \mathcal{S}_2^{H_1}$ *is trivial.*

Proof We retain the notations of the proof of Lemma 4.14. The map under consideration is given by

$$(F_D)_{D|\mathcal{D}'} \mapsto \sum_{D|\mathcal{D}'} \alpha_D \cdot F_D$$

where $(\alpha_D \cdot F)(g) = F(g\alpha_D)$ for any $F : G(\mathbf{A}) \to \mathbf{C}$ and $g \in G(\mathbf{A})$. We show that it is injective by induction on the number of prime divisors of $\mathcal{D}'$. There is nothing to prove if $\mathcal{D}' = \mathcal{O}_F$. Otherwise, let Q be a prime divisor of $\mathcal{D}'$. We put $\mathcal{D}'_Q = \mathcal{D}'/Q$ and $H'_1 = \widehat{R'_1}^{\times}$, where R'_1 is the unique $\mathcal{O}_F$-order in B which agrees with R_1 outside Q, and equals R_Q at Q. The functions

$$\mathcal{F}_0 = \sum_{D|\mathcal{D}'_Q} \alpha_D \cdot F_D \quad \text{and} \quad \mathcal{F}_1 = \sum_{D|\mathcal{D}'_Q} \alpha_D \cdot F_{DQ}$$

then belong to $\mathcal{S}_2^{H'_1}$, and

$$\sum_{D|\mathcal{D}'} \alpha_D \cdot F_D = \mathcal{F}_0 + \alpha_Q \cdot \mathcal{F}_1.$$

If this function is trivial on $G(\mathbf{A})$, $\mathcal{F}_0 = -\alpha_Q \cdot \mathcal{F}_1$ is fixed by $R_Q^\times$ and $\alpha_Q R_Q^\times \alpha_Q^{-1}$. Arguing as in the proof of Lemma 4.12, we obtain $\mathcal{F}_0 \equiv \mathcal{F}_1 \equiv 0$ on $G(\mathbf{A})$. By induction, $F_D \equiv 0$ on $G(\mathbf{A})$ for all $D \mid \mathcal{D}'$. □

4.6 Chaotic Galois action

We still have to show that for all but finitely many x in $\mathrm{CM}_1(P^\infty)$, $\mathbf{b}(x)$ is a nontorsion point in $A_0(K[P^\infty])$. Two proofs of this fact may be extracted from the results of [5]. These proofs are both based upon the following elementary observations:

- The torsion submodule of $A_0(K[P^\infty])$ is *finite*.

This easily follows from Lemma 2.7. We thus want $\mathbf{b}(x)$ to land away from a given finite set, provided that x belongs to $\mathrm{CM}_1(P^n)$ with $n \gg 0$.

- The map $x \mapsto \mathbf{b}(x)$ may be decomposed as follows:

$$\begin{array}{ccccccc} \mathrm{CM}_1(P^\infty) & \xrightarrow{\Delta} & M_1^{\mathcal{R}} & \xrightarrow{\alpha_1} & A_1^{\mathcal{R}} & \xrightarrow{\Sigma} & A_0 \\ x & \longmapsto & (\sigma x)_{\sigma \in \mathcal{R}} & & (a_\sigma)_{\sigma\in\mathcal{R}} & \mapsto & \sum \overline{\chi}(\sigma) a_\sigma \end{array}$$

In this decomposition, the second and third maps are algebraic morphisms defined over F. Moreover: Σ is surjective (this easily follows from the definitions) and α_1 is finite (by Lemmas 4.15 and 3.9). In some sense, this decomposition separates the geometrical and arithmetical aspects in the definition of $\mathbf{b}(x)$.

First proof (using a proven case of the André-Oort conjecture)

Suppose that $\mathbf{b}(x)$ is a torsion point in $A_0(K[P^\infty])$ for infinitely many $x \in \mathrm{CM}_1(P^\infty)$. We may then find some element a_0 in $A_0(\mathbf{C})$ such that $\mathcal{E} = \mathbf{b}^{-1}(a_0)$ is an infinite subset of $\mathrm{CM}_1(P^\infty)$. Since $\Sigma \circ \alpha_1 : M_1^{\mathcal{R}} \to A_0$ is continuous for the Zariski topology, we see that $(\Sigma \circ \alpha_1)^{-1}(a_0)$ contains the Zariski closure $\overline{\Delta(\mathcal{E})}^{\mathrm{Zar}}$ of $\Delta(\mathcal{E})$ in $M_1^{\mathcal{R}}(\mathbf{C})$. As explained in Remark 3.21 of [5], a proven case of the André-Oort conjecture implies that $\overline{\Delta(\mathcal{E})}^{\mathrm{Zar}}$ contains a connected component of $M_1^{\mathcal{R}}(\mathbf{C})$.

More precisely, let Z be any irreducible component of $\overline{\Delta(\mathcal{E})}^{\mathrm{Zar}}$ containing infinitely many points of $\Delta(\mathcal{E})$. The last remark of [8, Section 7.3] tells us that Z is a subvariety of Hodge type in $M_1^{\mathcal{R}}$, and the list of all such subvarieties is easy to compile, following the method of [7, Section 2], see the forthcoming addendum to [5]. We find that Z is a product of curves $Z_i \subset M_1^{\mathcal{R}_i}$ for a certain partition $\{\mathcal{R}_i\}$ of $\mathcal{R}$. Now Proposition 3.18 of [5] implies that the partition is trivial: each $\mathcal{R}_i$ is a singleton, and Z is indeed a connected component of $M_1^{\mathcal{R}}$.

Remark 4.16 This last reference requires $\mathcal{E}$ to be an infinite collection of *P-isogenous* CM points, where two CM points x and x' are said to be P-isogenous if they can be represented by g and $g' \in G(\mathbf{A}_f)$ with $g_v = g'_v$ for all $v \neq P$. Now, if a P-isogeny class contains a CM point of conductor P^n for some $n \geq 0$, it is actually contained in $\mathrm{CM}_1(P^\infty)$ and any other P-isogeny class in $\mathrm{CM}_1(P^\infty)$ also contains a point of conductor P^n. Since $\mathrm{CM}_1(P^n)$ is finite, we thus see that $\mathrm{CM}_1(P^\infty)$ is the disjoint union of *finitely* many P-isogeny classes, and one of them at least has infinite intersection with our infinite set $\mathcal{E}$.

We thus obtain a collection of connected components $(\mathcal{C}_\sigma)_{\sigma \in \mathcal{R}}$ of $M_1(\mathbf{C})$ with the property that for all $(x_\sigma)_{\sigma \in \mathcal{R}}$ in $\prod_{\sigma \in \mathcal{R}} \mathcal{C}_\sigma$,

$$\sum_{\sigma \in \mathcal{R}} \overline{\chi}(\sigma)\alpha_1(x_\sigma) = a_0.$$

It easily follows that α_1 should then be constant on $\mathcal{C}_1$. Being defined over F on the *connected* curve M_1 (cf. Remark 3.2), α_1 would then be constant on M_1, a contradiction.

Second proof (using a theorem of M. Ratner)

Let U be a nonempty open subscheme of $\mathrm{Spec}(\mathcal{O}_F)$ such that for every closed point $v \in U$, $v \neq P$, $B_v \simeq M_2(F_v)$ and $R_{1,v}$ is maximal. Shrinking U if necessary, we may assume that M_1 has a proper and smooth model $\mathbf{M}_1$ over U, which agrees locally with the models considered in [5], and $\alpha_1 : M_1 \to A_1$

extends uniquely to a *finite* morphism $\alpha_1 : \mathbf{M}_1 \to \mathbf{A}_1$, where $\mathbf{A}_1$ is the Néron model of A_1 over U. In the Stein factorization

$$\mathbf{M}_1 \xrightarrow{c} \mathcal{M}_{H_1} \to U$$

of the stuctural morphism $\mathbf{M}_1 \to U$, the *scheme of connected components* $\mathcal{M}_1$ is then a finite and étale cover of U, and the fibers of c are geometrically connected. For each closed point $v \in U$, we choose a place $\overline{v}$ of $\overline{F} \subset \mathbf{C}$ above v, with valuation ring $\mathcal{O}(\overline{v})$ and residue field $\mathbf{F}(\overline{v})$, an algebraic closure of the residue field $\mathbf{F}(v)$ of v. We thus obtain the following diagram of reduction maps

$$\begin{array}{lccccc} \mathrm{red}_v: & M_1(\overline{F}) & \xleftarrow{\simeq} & \mathbf{M}_1(\mathcal{O}(\overline{v})) & \longrightarrow & \mathbf{M}_1(\mathbf{F}(\overline{v})) \\ & \downarrow c & & \downarrow c & & \downarrow c \\ \mathrm{red}_v: & \mathcal{M}_1(\overline{F}) & \xleftarrow{\simeq} & \mathcal{M}_1(\mathcal{O}(\overline{v})) & \xrightarrow{\simeq} & \mathcal{M}_1(\mathbf{F}(\overline{v})). \end{array}$$

We denote by $\mathcal{C} \mapsto \mathcal{C}(v)$ the induced bijection between the sets of geometrical connected components in the generic and special fibers, and we denote by $\mathcal{C}_x$ the connected component of $x \in M_1(\overline{F})$. In particular, $\mathcal{C}_x(v) = c^{-1}(\mathrm{red}_v c(x))$ is the connected component of $\mathrm{red}_v(x)$. Inside $\mathbf{M}_1(\mathbf{F}(\overline{v}))$, there is a finite collection of distinguished points, namely the *supersingular points* as described in Section 3.1.3 of [5]. We denote by $\mathcal{C}^{\mathrm{ss}}(v)$ the set of supersingular points inside $\mathcal{C}(v)$.

We let $d > 0$ be a uniform upper bound on the number of geometrical points in the fibers of $\alpha_1 : \mathbf{M}_1 \to \mathbf{A}_1$ (such a bound does exist, thanks to the generic flatness theorem, see for instance [13, Corollaire 6.9.3]). We let $t > 0$ be the order of the torsion subgroup of $A_0(K[P^\infty])$. One easily checks that the order of $\mathcal{C}^{\mathrm{ss}}(v)$ goes to infinity with the order of the residue field $\mathbf{F}(v)$ of v. Shrinking U if necessary, we may therefore assume that for all $\mathcal{C}$ and v,

$$|\mathcal{C}^{\mathrm{ss}}(v)| > td.$$

Now, let v be a closed point of U which is *inert* in K (there are infinitely many such points). Then Lemma 3.1 of [5] states that any CM point $x \in \mathrm{CM}_1$ reduces to a supersingular point $\mathrm{red}_v(x) \in \mathcal{C}_x^{\mathrm{ss}}(v)$ and we have the following crucial result.

Proposition 4.17 *For all but finitely many x in $\mathrm{CM}_1(P^\infty)$, the following property holds. For any $(z_\sigma)_{\sigma\in\mathcal{R}}$ in $\prod_{\sigma\in\mathcal{R}} \mathcal{C}^{\mathrm{ss}}_{\sigma x}(v)$, there exists some $\gamma \in \mathrm{Gal}_K^{\mathrm{ab}}$ such that*

$$\forall \sigma \in \mathcal{R}: \quad \mathrm{red}_v(\gamma\sigma \cdot x) = z_\sigma \quad \textit{in } \mathbf{M}_1(\mathbf{F}(\overline{v})).$$

Proof This is a special case of Theorem 3.5 of [5], except that the latter deals with P-*isogeny classes* of CM points instead of the set of all CM points of P-power conductor that we consider here. However, we have already observed in Remark 4.16 that $\mathrm{CM}_1(P^\infty)$ is the disjoint union of finitely many such P-isogeny classes, and the proposition follows. □

Corollary 4.18 *For all but finitely many x in $\mathrm{CM}_1(P^\infty)$, the following property holds. For any z in $\mathcal{C}_x^{\mathrm{ss}}(v)$, there is a $\gamma \in \mathrm{Gal}_K^{\mathrm{ab}}$ such that*

$$\mathrm{red}_v(\gamma \cdot \mathbf{b}(x)) - \mathrm{red}_v(\mathbf{b}(x)) = \alpha_1(z) - \alpha_1(z_1)$$

in $\mathbf{A}_0(\mathbf{F}(\overline{v})) = \mathbf{A}_1(\mathbf{F}(\overline{v})) \otimes_{C_1} C_0$, with $z_1 = \mathrm{red}_v(x) \in \mathcal{C}_x^{\mathrm{ss}}(v)$.

Proof Take $z_\sigma = \mathrm{red}_v(\sigma x)$ for $\sigma \neq 1$ in $\mathcal{R}$. □

This finishes the proof of Theorem 4.10. Indeed, the Galois orbit of any torsion point in $A_0(K[P^\infty])$ has at most t elements, while the above corollary implies that for all but finitely many $x \in \mathrm{CM}_1(P^\infty)$,

$$\left|\mathrm{Gal}_K^{\mathrm{ab}} \cdot \mathbf{b}(x)\right| \geq \left|\mathrm{red}_v\left(\mathrm{Gal}_K^{\mathrm{ab}} \cdot \mathbf{b}(x)\right)\right| \geq \left|\alpha_1\left(\mathcal{C}_x^{\mathrm{ss}}(v)\right)\right| \geq \frac{1}{d}\left|\mathcal{C}_x^{\mathrm{ss}}(v)\right| > t.$$

5 The definite case

Suppose now that B is a *definite* quaternion algebra, so that $B \otimes \mathbf{R} \simeq \mathbf{H}^{[F:\mathbf{Q}]}$. Let K be a totally imaginary quadratic extension of F contained in B. We put $G = \mathrm{Res}_{F/\mathbf{Q}}(B^\times)$, $T = \mathrm{Res}_{F/\mathbf{Q}}(K^\times)$ and $Z = \mathrm{Res}_{F/\mathbf{Q}}(F^\times)$ as before.

5.1 Automorphic forms and representations

We denote by $\mathcal{S}_2$ the space of all *weight* 2 *automorphic forms* on G, namely the space of all smooth (=locally constant) functions

$$\theta : G(\mathbf{Q})\backslash G(\mathbf{A}_f) \to \mathbf{C}.$$

There is an admissible left action of $G(\mathbf{A}_f)$ on $\mathcal{S}_2$, given by right translations: for $g \in G(\mathbf{A}_f)$ and $x \in G(\mathbf{Q})\backslash G(\mathbf{A}_f)$,

$$(g \cdot \theta)(x) = \theta(xg).$$

This representation is semi-simple, and $\mathcal{S}_2$ is the algebraic direct sum of its irreducible subrepresentations. An irreducible representation π' of $G(\mathbf{A}_f)$ is *automorphic* if it occurs in $\mathcal{S}_2$. It then occurs with multiplicity one, and we denote by $\mathcal{S}_2(\pi')$ the corresponding subspace of $\mathcal{S}_2$, so that

$$\mathcal{S}_2 = \oplus_{\pi'} \mathcal{S}_2(\pi').$$

If π' is finite dimensional, it is of dimension 1 and corresponds to a smooth character χ of $G(\mathbf{A}_f)$. Then χ is trivial on $G(\mathbf{Q})$, $\mathcal{S}_2(\pi')$ equals $\mathbf{C} \cdot \chi$, and χ factors through the reduced norm $G(\mathbf{A}_f) \to Z(\mathbf{A}_f)$. A function $\theta \in \mathcal{S}_2$ is said to be *Eisenstein* if it belongs to the subspace spanned by these finite dimensional subrepresentations of $\mathcal{S}_2$. Equivalently, θ is an Eisenstein function if and only if it factors through the reduced norm (because any such function spans a finite dimensional $G(\mathbf{A}_f)$-invariant subspace of $\mathcal{S}_2$).

We say that π' is *cuspidal* if its representation space has infinite dimension. The space of (weight 2) *cuspforms* $\mathcal{S}_2^0$ is the $G(\mathbf{A}_f)$-invariant subspace of $\mathcal{S}_2$ which is spanned by its irreducible cuspidal subrepresentations. Thus, $\theta = 0$ is the only cuspform which is also Eisenstein.

5.2 The exceptional case

The Jacquet-Langlands correspondence assigns, to every cuspidal representation π' of $G(\mathbf{A}_f)$ as above, an irreducible automorphic representation $\pi = \mathrm{JL}(\pi')$ of GL_2/F, of weight $(2, \cdots, 2)$. We say that (π', K) is *exceptional* if (π, K) is exceptional. Thus, (π', K) is exceptional if and only if $\pi \simeq \pi \otimes \eta$, where η is the quadratic character attached to K/F. We want now to describe a simple characterization of these exceptional cases.

Write $\pi = \otimes \pi_v$, $\pi' = \otimes \pi'_v$ and let $\mathcal{N}$ be the conductor of π. For every finite place v of F not dividing $\mathcal{N}$,

$$\pi_v \simeq \pi'_v \simeq \pi(\mu_{1,v}, \mu_{2,v}) \simeq \pi(\mu_{2,v}, \mu_{1,v})$$

for some unramified characters $\mu_{i,v} : F_v^\times \to \mathbf{C}^\times$, $i = 1, 2$. These characters are uniquely determined by $\beta_{i,v} = \mu_{i,v}(\varpi_v)$, where ϖ_v is any local uniformizer at v, and by the strong multiplicity one theorem, the knowledge of all but finitely many of the unordered pairs $\{\beta_{1,v}, \beta_{2,v}\}$ uniquely determines π and π'. On the other hand, the representation space $\mathcal{S}(\pi_v)$ of π_v contains a unique line $\mathbf{C} \cdot \phi_v$ of vectors which are fixed by the maximal compact open subgroup $H_v = \mathrm{GL}_2(\mathcal{O}_{F,v})$ of $G_v = \mathrm{GL}_2(F_v) \simeq B_v^\times$. The spherical Hecke algebra

$$\mathrm{End}_{\mathbf{Z}[G_v]}(\mathbf{Z}[G_v/H_v]) \simeq \mathbf{Z}[H_v \backslash G_v / H_v]$$

acts on $\mathbf{C} \cdot \phi_v$, and the eigenvalues of the Hecke operators

$$T_v = \left[H_v \left(\begin{smallmatrix} \varpi_v & 0 \\ 0 & 1 \end{smallmatrix}\right) H_v\right] \quad \text{and} \quad S_v = \left[H_v \left(\begin{smallmatrix} \varpi_v & 0 \\ 0 & \varpi_v \end{smallmatrix}\right) H_v\right]$$

are respectively given by

$$a_v = (Nv)^{1/2}(\beta_{1,v} + \beta_{2,v}) \quad \text{and} \quad s_v = \beta_{1,v} \cdot \beta_{2,v}$$

where Nv is the order of the residue field of v. Note that $s_v = \omega_v(\varpi_v)$ where $\omega_v = \mu_{1,v}\mu_{2,v}$ is the local component of the central character ω of π and π'. Therefore, the knowledge of ω and all but finitely many of the a_v's uniquely determines π and π'.

Proposition 5.1 *(π, K) and (π', K) are exceptional if and only if $a_v = 0$ for all but finitely many of the v's which are inert in K.*

Proof With notations as above, the v-component of $\pi \otimes \eta$ is equal to $\pi(\mu_{1,v}\eta_v, \mu_{2,v}\eta_v)$ where η_v is the local component of η. Therefore, $\pi \simeq \pi \otimes \eta$ if and only if $\{\beta_{1,v}, \beta_{2,v}\} = \{\beta_{1,v}\eta(v), \beta_{2,v}\eta(v)\}$ for almost all v, where $\eta(v) = \eta_v(\varpi_v)$ equals 1 if v splits in K, and -1 if v is inert in K. The proposition easily follows. □

Remark 5.2 It is well-known that the field $E_\pi \subset \mathbf{C}$ generated by the a_v's and the values of ω is a number field. Moreover, for any finite place λ of E_π with residue characteristic ℓ, there exists a unique (up to isomorphism) continuous representation

$$\rho_{\pi,\lambda} : \mathrm{Gal}_F \to \mathrm{GL}_2(E_{\pi,\lambda})$$

such that for every finite place $v \nmid \ell N$, $\rho_{\pi,\lambda}$ is unramified at v and the characteristic polynomial of $\rho_{\pi,\lambda}(\mathrm{Frob}_v)$ equals

$$X^2 - a_v X + Nv \cdot s_v \in E_\pi[X] \subset E_{\pi,\lambda}[X].$$

See [25] and the reference therein.

Put $\mathcal{E} = E_{\pi,\lambda}$ and let $V = \mathcal{E}^2$ be the representation space of $\rho = \rho_{\pi,\lambda}$. If $\pi \simeq \pi \otimes \eta$, then $\rho \simeq \rho \otimes \eta$ (viewing now η as a Galois character). In particular, there exists $\theta \in \mathrm{GL}(V)$ such that

$$\theta \circ \rho = \eta \cdot \rho \circ \theta \quad \text{on } \mathrm{Gal}_F.$$

Since ρ is absolutely irreducible (this follows from the arguments in Section 2 of [25]), θ^2 is a scalar in $\mathcal{E}^\times$ but θ is not. Let $\mathcal{E}'$ be a quadratic extension of $\mathcal{E}$ containing a square root c of θ^2. Put $V' = V \otimes_{\mathcal{E}} \mathcal{E}'$. Then $V' = V'_+ \oplus V'_-$ where $\theta = \pm c$ on $V'_\pm$, and $\dim_{\mathcal{E}'} V'_\pm = 1$. Moreover,

$$\forall \sigma \in \mathrm{Gal}_F : \quad (\rho \otimes \mathbf{Id}_{\mathcal{E}'})(\sigma)(V'_\pm) = V'_{\pm\eta(\sigma)}.$$

It easily follows that $\rho \otimes \mathbf{Id}_{\mathcal{E}'} \simeq \mathrm{Ind}_{\mathrm{Gal}_K}^{\mathrm{Gal}_F}(\alpha)$, where $\alpha : \mathrm{Gal}_K \to \mathcal{E}'^{\times}$ is the continuous character giving the action of Gal_K on V'_+.

Conversely, suppose that the base change of ρ to an algebraic closure $\overline{\mathcal{E}}$ of $\mathcal{E}$ is isomorphic to $\mathrm{Ind}_{\mathrm{Gal}_K}^{\mathrm{Gal}_F}(\alpha)$ for some character $\alpha : \mathrm{Gal}_K \to \overline{\mathcal{E}}^{\times}$. Then $\pi \simeq \pi \otimes \eta$ by Proposition 5.1, since

$$a_v = \mathrm{Tr}(\rho(\mathrm{Frob}_v)) = 0$$

for almost all v's that are inert in K.

5.3 CM points and Galois actions

Given a compact open subgroup H of $G(\mathbf{A}_f)$, we say that $\theta \in \mathcal{S}_2$ has *level* H if it is fixed by H. The space $\mathcal{S}_2^H$ of all such functions may thus be identified with the finite dimensional space of all complex valued function on the *finite* set

$$M_H \stackrel{\mathrm{def}}{=} G(\mathbf{Q})\backslash G(\mathbf{A}_f)/H.$$

In particular, any such θ yields a function $\psi = \theta \circ \mathrm{red}$ on

$$\mathrm{CM}_H \stackrel{\mathrm{def}}{=} T(\mathbf{Q})\backslash G(\mathbf{A}_f)/H,$$

where $\mathrm{red} : \mathrm{CM}_H \to M_H$ is the obvious map.

Also, θ is an Eisenstein function if and only if it factors through the map $c : M_H \to N_H$ which is induced by the reduced norm, where

$$N_H \stackrel{\mathrm{def}}{=} Z(\mathbf{Q})^+\backslash Z(\mathbf{A}_f)/\mathrm{nr}(H)$$

and $Z(\mathbf{Q})^+ = \mathrm{nr}(G(\mathbf{Q}))$ is the subgroup of totally positive elements in $Z(\mathbf{Q}) = F^{\times}$. We will need to consider a somewhat weaker condition.

Recall from the introduction that the set CM_H of *CM points*, is endowed with the following *Galois action*: for $x = [g] \in \mathrm{CM}_H$ and $\sigma = \mathrm{rec}_K(\lambda) \in \mathrm{Gal}_K^{\mathrm{ab}}$ (with $g \in G(\mathbf{A}_f)$ and $\lambda \in T(\mathbf{A}_f)$),

$$\sigma \cdot x = [\lambda g] \in \mathrm{CM}_H.$$

The Galois group $\mathrm{Gal}_F^{\mathrm{ab}}$ similarly acts on N_H, and we thus obtain an action of $\mathrm{Gal}_K^{\mathrm{ab}}$ on N_H: for $x = [z] \in N_H$ and $\sigma = \mathrm{rec}_K(\lambda) \in \mathrm{Gal}_K^{\mathrm{ab}}$ (with $z \in Z(\mathbf{A}_f)$ and $\lambda \in T(\mathbf{A}_f)$),

$$\sigma \cdot x = [\mathrm{nr}(\lambda) z] \in N_H.$$

By construction, the composite map

$$\mathrm{CM}_H \xrightarrow{\mathrm{red}} M_H \xrightarrow{c} N_H$$

is $\mathrm{Gal}_K^{\mathrm{ab}}$-equivariant (and surjective).

Definition 5.3 We say that $\theta : M_H \to \mathbf{C}$ is *exceptional* (with respect to K) if there exists some $z_0 \in N_H$ with the property that θ is constant on $c^{-1}(z)$, for all z in $\mathrm{Gal}_K^{\mathrm{ab}} \cdot z_0$.

Remark 5.4 Since K is quadratic over F, there are at most two $\mathrm{Gal}_K^{\mathrm{ab}}$-orbits in N_H. If there is just one, $\theta : M_H \to \mathbf{C}$ is exceptional if and only if it is Eisenstein: this occurs for instance whenever $\mathrm{nr}(H)$ is the maximal compact open subgroup of $Z(\mathbf{A}_f)$, provided that K/F ramifies at some finite place. On the other hand, if there are two $\mathrm{Gal}_K^{\mathrm{ab}}$-orbits in N_H, there might be exceptional θ's which are not Eisenstein.

Lemma 5.5 *Let π' be any cuspidal representation of $G(\mathbf{A}_f)$. Suppose that $\mathcal{S}_2(\pi')$ contains a nonzero θ which is exceptional with respect to K. Then (π', K) is exceptional.*

Proof Let $H = \prod_v H_v$ be a compact open subgroup of $G(\mathbf{A}_f)$ such that θ is right invariant under H. Since π is cuspidal, θ is not Eisenstein and the above remark shows that there must be *exactly* two $\mathrm{Gal}_K^{\mathrm{ab}}$-orbits in N_H, say X and Y, with θ constant on $c^{-1}(z)$ for all z in X, but $\theta(x_1) \neq \theta(x_2)$ for some x_1 and x_2 in M_H with $c(x_1) = c(x_2) = y \in Y$.

For all but finitely many v's, $H_v = R_v^\times$ where $R_v \simeq M_2(\mathcal{O}_{F,v})$ is a maximal order in $B_v \simeq M_2(F_v)$. For any such v, we know that

$$\theta | T_v = a_v \theta$$

where for $x = [g] \in M_H$ with $g \in G(\mathbf{A}_f)$,

$$(\theta | T_v)(x) = \sum_{i \in I_v} \theta(x_{v,i}) \quad \text{with } x_{v,i} = [g\gamma_{v,i}] \in M_H.$$

Here, $H_v \begin{pmatrix} \varpi_v & 0 \\ 0 & 1 \end{pmatrix} H_v = \coprod_{i \in I_v} \gamma_{v,i} H_v$ with ϖ_v a local uniformizer in F_v. Note that for x in $c^{-1}(y)$, the $x_{v,i}$'s all belong to $c^{-1}(\mathrm{Frob}_v \cdot y)$. If v is inert in K, $\mathrm{Frob}_v \cdot y$ belongs to X and θ is constant on its fiber, say $\theta(x') = \theta(v, y)$ for all $x' \in c^{-1}(\mathrm{Frob}_v \cdot y)$. For such v's, we thus obtain

$$a_v \theta(x_1) = |I_v| \, \theta(v, y) = a_v \theta(x_2).$$

Since $\theta(x_1) \neq \theta(x_2)$ by construction, $a_v = 0$ whenever v is inert in K. The lemma now follows from Proposition 5.1. □

5.4 Main results

To prove our main theorems in the definite case, we shall proceed backwards, starting from the analog of Proposition 4.17, and ending with our target result, the analog of Theorem 4.1.

Thus, let $\mathcal{R}' \subset G_0$ be the set of representatives for G_0/G_2 which we considered in section 4.5. Recall that

$$\mathcal{R}' = \{\tau\sigma_D;\ \tau \in \mathcal{R} \text{ and } D \mid \mathcal{D}'\}$$

where $\mathcal{R} \subset G_0$ is a set of representatives for G_0/G_1 containing 1, while $\mathcal{D}'$ and the σ_D's for $D \mid \mathcal{D}'$ were defined in Lemma 2.6. Suppose that $H = \widehat{R}^\times$ for some $\mathcal{O}_F$-order $R \subset B$, and consider the following maps:

$$\begin{array}{ccc} \mathrm{CM}_H(P^\infty) \xrightarrow{\ \mathrm{RED}\ } M_H^{\mathcal{R}} & \qquad & M_H^{\mathcal{R}} \xrightarrow{\ C\ } N_H^{\mathcal{R}} \\ x \longmapsto (\mathrm{red}(\tau\cdot x))_{\tau\in\mathcal{R}} & \qquad & (a_\tau)_{\tau\in\mathcal{R}} \longmapsto (c(a_\tau))_{\tau\in\mathcal{R}} \end{array}$$

If we endow $N_H^{\mathcal{R}}$ with the diagonal Galois action, the composite

$$C \circ \mathrm{RED} : \mathrm{CM}_H(P^\infty) \to N_H^{\mathcal{R}}$$

becomes a $G(\infty)$-equivariant map. In particular, for any $x \in \mathrm{CM}_H(P^\infty)$,

$$\mathrm{RED}(G(\infty)\cdot x) \subset C^{-1}(G(\infty)\cdot C \circ \mathrm{RED}(x)).$$

The following key result is our initial input from [5].

Proposition 5.6 *For all but finitely many $x \in \mathrm{CM}_H(P^\infty)$,*

$$\mathrm{RED}(G(\infty)\cdot x) = C^{-1}(G(\infty)\cdot C \circ \mathrm{RED}(x)).$$

Proof Given the definition of G_1, this is just a special case of Corollary 2.10 of [5] except that the latter deals with P-isogeny classes of CM points instead of the set of all CM points of P-power conductor which we consider here. Nevertheless, $\mathrm{CM}_H(P^\infty)$ breaks up as the disjoint union of *finitely* many such P-isogeny classes, cf. Remark 4.16. The proposition follows. □

Corollary 5.7 *Let θ be any non-exceptional function on M_H, and let $\psi = \theta \circ \mathrm{red}$ be the induced function on CM_H. Let χ be any character of G_0. Then, for any CM point $x \in \mathrm{CM}_H(P^n)$ with n sufficiently large, there exists some $y \in G(\infty)\cdot x$ such that*

$$\sum_{\tau\in\mathcal{R}} \chi(\tau)\psi(\tau\cdot y) \neq 0.$$

Proof Replacing x by $\sigma \cdot x$ for some $\sigma \in G(\infty)$, we may assume that $\theta(p_1) \neq \theta(p_2)$ for some p_1 and p_2 in $c^{-1}(c \circ \mathrm{red}(x))$. If n is sufficiently large, the proposition then produces y_1 and y_2 in $G(\infty) \cdot x$ such that $\mathrm{red}(y_1) = p_1$, $\mathrm{red}(y_2) = p_2$ and

$$\mathrm{red}(\tau \cdot y_1) = \mathrm{red}(\tau \cdot x) = \mathrm{red}(\tau \cdot y_2)$$

for any $\tau \neq 1$ in $\mathcal{R}$. If $\mathbf{a}(y) = \sum_{\tau \in \mathcal{R}} \chi(\tau)\psi(\tau \cdot y)$, we thus obtain

$$\mathbf{a}(y_1) - \mathbf{a}(y_2) = \theta(p_1) - \theta(p_2) \neq 0,$$

and one at least of $\mathbf{a}(y_1)$ or $\mathbf{a}(y_2)$ is nonzero. □

Suppose now that we are given an irreducible cuspidal representation π' of $G(\mathbf{A}_f)$, with (unramified) central character ω. We still consider a level structure of the form $H = \widehat{R}^\times$ for some $\mathcal{O}_F$-order $R \subset B$, but we now also require the following condition.

(H2) For any prime $Q \neq P$ of F which ramifies in K, B is split at Q and R_Q is a maximal order in $B_Q \simeq M_2(F_Q)$.

Our next result is the analog of Theorem 4.10.

Proposition 5.8 *Suppose that θ is a nonzero function in $\mathcal{S}_2(\pi')^H$, and let ψ be the induced function on CM_H. Let χ be any character of G_0 such that $\chi \cdot \omega = 1$ on $\mathbf{A}_F^\times$. Then, for all $x \in \mathrm{CM}(P^n)$ with n sufficiently large, there exists some $y \in G(\infty) \cdot x$ such that*

$$\mathbf{a}(y) \stackrel{\text{def}}{=} \sum_{\sigma \in G_0} \chi(\sigma)\psi(\sigma \cdot y) \neq 0.$$

Proof Since $Z(\mathbf{A}_f)$ acts on $\mathcal{S}_2(\pi') \ni \theta$ through ω, we find that

$$\begin{aligned} \mathbf{a}(y) &= |G_2| \sum_{\sigma \in \mathcal{R}'} \chi(\sigma)\psi(\sigma \cdot y). \\ &= |G_2| \sum_{\tau \in \mathcal{R}} \chi(\tau) \sum_{D|\mathcal{D}'} \chi(\sigma_D)\psi(\sigma_D \tau \cdot y). \end{aligned}$$

Using Lemma 5.9 below, we obtain

$$\mathbf{a}(y) = |G_2| \sum_{\tau \in \mathcal{R}} \chi(\tau)\psi_1(\tau \cdot y_1)$$

where $\psi_1 : \mathrm{CM}_{H_1} \to \mathbf{C}$ is induced by a nonzero function θ_1 of level $H_1 = \widehat{R}_1^\times$ in $\mathcal{S}_2(\pi')$, and $y \mapsto y_1$ is a Galois equivariant map from $\mathrm{CM}_H(P^n)$ to

$\mathrm{CM}_{H_1}(P^n)$. Since θ_1 is non-exceptional by Lemma 5.5, we may apply Corollary 5.7 to θ_1 and $x_1 \in \mathrm{CM}_{H_1}(P^n)$, thus obtaining some $y_1 = \gamma \cdot x_1$ in $G(\infty) \cdot x_1$ such that

$$\sum_{\tau \in \mathcal{R}} \chi(\tau)\psi_1(\tau \cdot y_1) = |G_2|^{-1} \mathbf{a}(y) \neq 0,$$

with $y = \gamma \cdot x \in G(\infty) \cdot x$. □

Lemma 5.9 *There exists an $\mathcal{O}_F$-order $R_1 \subset R$, a nonzero function θ_1 of level $H_1 = \widehat{R}_1^\times$ in $\mathcal{S}_2(\pi')$, and for each $n \geq 0$, a Galois equivariant map $x \mapsto x_1$ from $\mathrm{CM}_H(P^n)$ to $\mathrm{CM}_{H_1}(P^n)$ such that*

$$\sum_{D|\mathcal{D}'} \chi(\sigma_D)\psi(\sigma_D \cdot x) = \psi_1(x_1), \tag{5.1}$$

where $\psi = \theta \circ \mathrm{red}$ and $\psi_1 = \theta_1 \circ \mathrm{red}$ as usual.

Proof The proof is very similar to that of Lemma 4.14. We put

$$R_1 = R \cap \alpha_{\mathcal{D}'} R \alpha_{\mathcal{D}'}^{-1} \subset B$$

where for any prime divisor Q of $\mathcal{D}'$, Γ_Q is the set of elements in $R_Q \simeq M_2(\mathcal{O}_{F,Q})$ whose reduced norm (=determinant) is a uniformizer in $\mathcal{O}_{F,Q}$, α_Q is a chosen element in Γ_Q, and $\alpha_D = \prod_{Q|D} \alpha_Q$ for any divisor D of $\mathcal{D}'$. We then define

$$\theta_1 = \sum_{D|\mathcal{D}'} \chi(\sigma_D)(\alpha_D \cdot \theta).$$

Thus, θ_1 is a function of level $H_1 = \widehat{R}_1^\times$ in $\mathcal{S}_2(\pi')$.

Consider now some $x = [g] \in \mathrm{CM}_H(P^n)$, with $g \in G(\mathbf{A}_f)$ and $n \geq 0$. For each $Q \mid \mathcal{D}'$, we know that $K_Q \cap g_Q R_Q g_Q^{-1} = \mathcal{O}_{K,Q}$. If π_Q denotes a fixed generator of the maximal ideal of $\mathcal{O}_{K,Q}$, we thus find that $g_Q^{-1}\pi_Q g_Q$ belongs to Γ_Q. Since $\Gamma_Q = R_Q^\times \alpha_Q R_Q^\times$, there exists $r_{1,Q}$ and $r_{2,Q}$ in $R_Q^\times$ such that $g_Q^{-1}\pi_Q g_Q = r_{1,Q}\alpha_Q r_{2,Q}$. For $i \in \{1,2\}$, we put $r_i = \prod_{Q|\mathcal{D}'} r_{i,Q}$ and view it as an element of $H \subset G(\mathbf{A}_f)$. One easily checks, as in the proof of Lemma 4.14, that the CM point $x_1 = [gr_1]$ in CM_{H_1} has conductor P^n, and we claim that

(i) the map $x \mapsto x_1$ is well-defined and Galois equivariant;
(ii) formula (5.1) holds for all $x \in \mathrm{CM}_H(P^\infty)$.

For (i), suppose that we replace g by $g' = \lambda g h$ for some $\lambda \in T(\mathbf{A}_f)$ and $h \in H$. For $Q \mid \mathcal{D}'$, let $r'_{1,Q}$ and $r'_{2,Q}$ be elements of $R_Q^\times$ such that

$$g'^{-1}_Q \pi_Q g'_Q = r'_{1,Q}\alpha_Q r'_{2,Q}.$$

Since $g'_Q = \lambda_Q g_Q h_Q$ and $\pi_Q \lambda_Q = \lambda_Q \pi_Q$ in $B_Q^\times$, we find that

$$r_{1,Q}\alpha_Q r_{2,Q} = g_Q^{-1}\pi_Q g_Q = h_Q r'_{1,Q}\alpha_Q r'_{2,Q} h_Q^{-1}.$$

In particular, $r_{1,Q}^{-1} h_Q r'_{1,Q}$ equals $\alpha_Q r_{2,Q} h_Q {r'}_{2,Q}^{-1}\alpha_Q^{-1}$, and thus belongs to $R_{1,Q}^\times = R_Q^\times \cap \alpha_Q R_Q^\times \alpha_Q^{-1}$. It follows that for $r'_1 = \prod_{Q|\mathcal{D}'} r'_{1,Q} \in H$, $[g' r'_1]$ equals $[\lambda g r_1]$ in $\mathrm{CM}_{H_1}(P^n)$, and this finishes the proof of (i).

For (ii), we simply have to observe that for any divisor D of $\mathcal{D}'$, if ψ_D denotes the function on CM_{H_1} which is induced by $\alpha_D \cdot \theta \in \mathcal{S}_2(\pi')$,

$$\psi_D(x_1) = \theta(g r_1 \alpha_D) = \theta(g r_1 \alpha_D r_2) = \theta(\pi_D g) = \psi(\sigma_D \cdot x)$$

where $\pi_D = \prod_{Q|D} \pi_Q$, so that $\sigma_D = \mathrm{rec}_K(\pi_D)$.

To complete the proof of the lemma, it remains to show that θ_1 is nonzero. This may be proved by induction, exactly as in Proposition 5.3 of [27], or Lemma 4.15 above in the indefinite case. The final step of the argument runs as follows: if $\vartheta + \rho(\pi_Q)(\vartheta') = 0$ for some ϑ and ϑ' in $\mathcal{S}_2(\pi')$ that are fixed by $R_Q^\times$, then $\vartheta = -\rho(\pi_Q)(\vartheta')$ is fixed by the group spanned by $R_Q^\times$ and $\pi_Q R_Q^\times \pi_Q^{-1}$. This group contains the kernel of the reduced norm $B_P^\times \to F_P^\times$, and the strong approximation theorem then implies that ϑ is Eisenstein, hence zero. □

Finally, suppose moreover that the following condition holds.

(H1) R_P is an Eichler order in $B_P \simeq M_2(F_P)$.

We then have the notion of *good* CM points. We say that $\theta \in \mathcal{S}_2(\pi')$ is *P-new* if it is fixed by $R_P^\times$, and π' contains no nonzero vectors which are fixed by ${R'_P}^\times$ for some Eichler order $R'_P \subset B_P$ strictly containing R_P. The following is the analog of Theorem 4.1.

Theorem 5.10 *Suppose that θ is a nonzero function in $\mathcal{S}_2(\pi')^H$. Suppose moreover that θ is P-new, and let ψ be the induced function on CM_H. Let χ_0 be any character of G_0 such that $\chi_0 \cdot \omega = 1$ on $\mathbf{A}_F^\times$. Then, for any good CM point $x \in \mathrm{CM}_H(P^n)$ with n sufficiently large, there exists a primitive character χ of $G(n)$ inducing χ_0 on G_0 such that*

$$\mathbf{a}(x,\chi) \stackrel{\mathrm{def}}{=} \sum_{\sigma \in G(n)} \chi(\sigma)\psi(\sigma \cdot x) \neq 0.$$

Proof Since $\mathbf{a}(\gamma \cdot x, \chi) = \chi^{-1}(\gamma)\mathbf{a}(x,\chi)$ for any $\gamma \in G(n)$, it suffices to show that for some y in the Galois orbit of x, the average of the $\mathbf{a}(y,\chi)$'s is nonzero (with χ running through the set $P(n,\chi_0)$ of primitive characters of

$G(n)$ inducing χ_0 on G_0). By Lemma 2.8, this amounts to showing that

$$\sum_{\sigma \in G_0} \chi_0(\sigma)\psi_*(\sigma \cdot d(y)) \neq 0$$

for some $y \in G(\infty) \cdot x$, where $\psi_* : \mathbf{Z}[\mathrm{CM}_H] \to \mathbf{C}$ is the natural extension of ψ and

$$d(y) \stackrel{\text{def}}{=} q \cdot y - \mathrm{Tr}_{Z(n)}(y) \in \mathbf{Z}[\mathrm{CM}_H].$$

Since θ is P-new, ψ_* factors through the P-new quotient $\mathbf{Z}[\mathrm{CM}_H]^{P-\text{new}}$ of $\mathbf{Z}[\mathrm{CM}_H]$. In the latter, the image of $d(y)$ may be computed using the distribution relations of the appendix, provided that y (or x) is a *good* CM point of conductor P^n with n sufficiently large. We find that

$$\psi_*(d(y)) = \psi^+(y^+)$$

where $H^+ = \widehat{R^+}^\times$ for some $\mathcal{O}_F$-order $R^+ \subset B$, θ^+ is a function of level H^+ in $\mathcal{S}_2(\pi')$, ψ^+ is the induced function on CM_{H^+}, and y^+ belongs to $\mathrm{CM}_{H^+}(P^n)$. Moreover, the map $y \mapsto y^+$ commutes with the action of $\mathrm{Gal}_K^{\mathrm{ab}}$, so that y^+ belongs to $G(\infty) \cdot x^+ \subset \mathrm{CM}_{H_1}(P^n)$ and

$$\psi_*(\sigma \cdot d(y)) = \psi_*(d(\sigma y)) = \psi^+((\sigma y)^+) = \psi^+(\sigma \cdot y^+)$$

for any $\sigma \in G_0$. We now have to show that

$$\mathbf{a}(y^+) \stackrel{\text{def}}{=} \sum_{\sigma \in G_0} \chi_0(\sigma)\psi^+(\sigma \cdot y^+) \neq 0$$

for some $y^+ \in G(\infty) \cdot x^+$, provided that n is sufficiently large.

When $\delta \geq 2$, $R^+ = R$, $x^+ = x$ and $\theta^+ = q \cdot \theta$ with $q = |\mathcal{O}_F/P|$. Otherwise, R^+ is the unique $\mathcal{O}_F$-order in B which agrees with R outside P, and whose localization R_P^+ at P is the Eichler order of level P^2 constructed in section 6.5. In particular, R^+ satisfies **(H2)**. Moreover:

$$\theta^+ = \begin{cases} b_0 \cdot \theta_0 + b_1 \cdot \theta_1 + b_2 \cdot \theta_2 & \text{if } \delta = 0 \\ b_{01} \cdot \theta_{01} + b_{12} \cdot \theta_{12} & \text{if } \delta = 1 \end{cases}$$

where the b_*'s are the elements of $B_P^\times$ defined in section 6.5, while the θ_*'s are the elements of $\mathcal{S}_2(\pi')^H$ which are respectively given by

$$\begin{array}{lrcll} & (\theta_0, \theta_1, \theta_2) & = & (q \cdot \theta, -T \cdot \theta, \gamma \cdot \theta) & \text{if } \delta = 0 \\ \text{and} & (\theta_{01}, \theta_{12}) & = & (q \cdot \theta, \gamma^{-1} \cdot \theta) & \\ \text{or} & (\theta_{01}, \theta_{12}) & = & (\gamma \cdot \theta, q \cdot \theta) & \text{if } \delta = 1. \end{array}$$

In the above formulas, T and γ are certain Hecke operators in $\mathbf{T}_H$, with $\gamma \in \mathbf{T}_H^\times$. The argument that we already used in Lemmas 4.12 and 4.13 shows that

$\theta^{+} \neq 0$ in all (four) cases, and we may therefore apply Proposition 5.8 to conclude the proof of our theorem. □

6 Appendix: Distribution Relations

Fix a number field F, a quadratic extension K of F and a quaternion algebra B over F containing K. Let $\mathcal{O}_F$ and $\mathcal{O}_K$ be the ring of integers in F and K. Let H be a compact open subgroup of $\widehat{B}^\times$ and put $\mathrm{CM}_H = K_+^\times \backslash \widehat{B}^\times / H$ where $K_+^\times$ is the subgroup of $K^\times$ which consists of those elements which are positive at every real place of K.

The Galois group $\mathrm{Gal}_K^{\mathrm{ab}} \simeq \overline{K_+^\times} \backslash \widehat{K}^\times$ acts on CM_H by $\sigma \cdot x = [\lambda^\epsilon b]$ for $\sigma = \mathrm{rec}_K(\lambda)$ and $x = [b]$ ($\lambda \in \widehat{K}^\times$, $b \in \widehat{B}^\times$). Here, ϵ is a fixed element in $\{\pm 1\}$. We extend this action by linearity to the free abelian group $\mathbf{Z}[\mathrm{CM}_H]$ generated by CM_H. On the latter, we also have a Galois equivariant left action of the Hecke algebra

$$\mathbf{T}_H \stackrel{\mathrm{def}}{=} \mathrm{End}_{\mathbf{Z}[\widehat{B}^\times]}(\mathbf{Z}[\widehat{B}^\times / H]) \simeq \mathbf{Z}[H \backslash \widehat{B}^\times / H].$$

An element $[\alpha] \in H \backslash \widehat{B}^\times / H$ acts on $\mathbf{Z}[\widehat{B}^\times / H]$ or $\mathbf{Z}[\mathrm{CM}_H]$ by

$$[b] \mapsto \sum_{i=1}^{n} [b\alpha_i] \quad \text{for } H\alpha H = \coprod_{i=1}^{n} \alpha_i H \text{ and } b \in \widehat{B}^\times.$$

A *distribution relation* is an expression relating these two actions. The aim of this section is to establish some of these relations when $H = H^P R_P^\times$ where P is a prime of F where B is split, H^P is any compact open subgroup of $(\widehat{B}^\times)^P = \{b \in \widehat{B};\ b_P = 1\}$ and $R_P \subset B_P$ is an *Eichler order* of level P^δ for some $\delta \geq 0$.

More precisely, we shall relate the action of the "decomposition group at P" to the action of the local Hecke algebra

$$\mathbf{T}(R_P^\times) = \mathrm{End}_{\mathbf{Z}[B_P^\times]}(\mathbf{Z}[B_P^\times / R_P^\times]) \simeq \mathbf{Z}[R_P^\times \backslash B_P^\times / R_P^\times] \subset \mathbf{T}_H$$

on CM_H. This naturally leads us to the study of the left action of $K_P^\times$ and $\mathbf{T}(R_P^\times)$ on $B_P^\times / R_P^\times$. For any $x = [b] \in B_P^\times / R_P^\times$, the stabilizer of x in $K_P^\times$ equals $\mathcal{O}(x)^\times$ where $\mathcal{O}(x) = K_P \cap bR_P b^{-1}$ is an $\mathcal{O}_{F,P}$-order in K_P. On the other hand, any $\mathcal{O}_{F,P}$-order $\mathcal{O} \subset K_P$ is equal to

$$\mathcal{O}_n \stackrel{\mathrm{def}}{=} \mathcal{O}_{F,P} + P^n \mathcal{O}_{K,P}$$

for a unique integer $n = \ell_P(\mathcal{O})$ (cf. section 6.1 below). For x as above, we put $\ell_P(x) \stackrel{\mathrm{def}}{=} \ell_P(\mathcal{O}(x))$. This function on $B_P^\times / R_P^\times$ obviously factors through

$K_P^\times \backslash B_P^\times / R_P^\times$. Using the decomposition

$$\mathrm{Gal}_K^{\mathrm{ab}} \backslash \mathrm{CM}_H \simeq \widehat{K}^\times \backslash \widehat{B}^\times / H \simeq (\widehat{K}^\times)^P \backslash (\widehat{B}^\times)^P / H^P \times K_P^\times \backslash B_P^\times / R_P^\times$$

we thus obtain a Galois invariant fibration $\ell_P : \mathrm{CM}_H \to \mathbf{N}$ with the property that for any $x \in \mathrm{CM}_H$ with $n = \ell_P(x)$, x is fixed by the closed subgroup $\mathrm{rec}_K(\mathcal{O}_n^\times)$ of $\mathrm{Gal}_K^{\mathrm{ab}}$. If $n \geq 1$, we put

$$\mathrm{Tr}(x) \stackrel{\mathrm{def}}{=} \sum_{\lambda \in \mathcal{O}_{n-1}^\times / \mathcal{O}_n^\times} \mathrm{rec}_K(\lambda) \cdot x \in \mathbf{Z}[\mathrm{CM}_H].$$

This is, on the Galois side, the expression that we will try to compute in terms of the action of the local Hecke algebra.

When $\delta \geq 1$ (so that R_P is not a maximal order), our formulas simplify in the P-new quotient $\mathbf{Z}[\mathrm{CM}_H]^{P-\mathrm{new}}$ of $\mathbf{Z}[\mathrm{CM}_H]$. The latter is the quotient of $\mathbf{Z}[\mathrm{CM}_H]$ by the $\mathbf{Z}$-submodule which is spanned by the elements of the form $\sum [b\alpha_i]$ where $b \in \widehat{B}^\times$ and $\{\alpha_i\} \subset B_P^\times$ is a set of representatives of $R'^\times_P / R_P^\times$ for some Eichler order $R_P \subset R'_P \subset B_P$ of level $P^{\delta'}$ with $\delta' < \delta$.

We start this section with a review on the arithmetic of $\mathcal{O}_n$. The next three sections establish the distribution relations for $\mathrm{Tr}(x)$ when $\delta = 0$, $\delta = 1$ and $\delta \geq 2$ respectively. The final section explains how the various points that are involved in the formulas for $\delta = 0$ or $\delta = 1$ may all be retrieved from a single CM point of higher level $\delta = 2$.

To fix the notation, we put $\mathbf{F} = \mathcal{O}_F / P \simeq \mathcal{O}_{F,P} / P\mathcal{O}_{F,P}$ and let $\mathbf{F}[\epsilon] = \mathbf{F}[X]/X^2\mathbf{F}[X]$ be the infinitesimal deformation $\mathbf{F}$-algebra. We choose a local uniformizer ϖ_P of F at P. We set $\varepsilon_P = -1$, 0 or 1 depending upon whether P is inert, ramifies or splits in K. We denote by $\mathcal{P}$ (resp. $\mathcal{P}$ and $\mathcal{P}^*$) the primes of K above P and let $\sigma_{\mathcal{P}}$ (resp. $\sigma_{\mathcal{P}}$ and $\sigma_{\mathcal{P}^*}$) be the corresponding *geometric* Frobeniuses.

6.1 Orders

Since $\mathcal{O}_{K,P}/\mathcal{O}_{F,P}$ is a torsionfree rank one $\mathcal{O}_{F,P}$-module, we may find an $\mathcal{O}_{F,P}$-basis $(1, \alpha_P)$ of $\mathcal{O}_{K,P}$. Let $\mathcal{O}$ be any $\mathcal{O}_{F,P}$-order in K_P. The projection of $\mathcal{O} \subset \mathcal{O}_{K,P} = \mathcal{O}_{F,P} \oplus \mathcal{O}_{F,P}\alpha_P$ to the second factor equals $P^n\mathcal{O}_{F,P}\alpha_P$ for a well-defined integer $n = \ell_P(\mathcal{O}) \geq 0$. Since $\mathcal{O}_{F,P} \subset \mathcal{O}$,

$$\mathcal{O} = \mathcal{O}_{F,P} \oplus P^n\mathcal{O}_{F,P}\alpha_P = \mathcal{O}_{F,P} + P^n\mathcal{O}_{K,P}.$$

Conversely, $\forall n \geq 0$, $\mathcal{O}_n \stackrel{\mathrm{def}}{=} \mathcal{O}_{F,P} + P^n\mathcal{O}_{K,P}$ is an $\mathcal{O}_{F,P}$-order in K_P.

Since any $\mathcal{O}_n$-ideal is generated by at most two elements (it is already generated by two elements as an $\mathcal{O}_{F,P}$-module), $\mathcal{O}_n$ is a Gorenstein ring for any

$n \geq 0$ [1]. For $n = 0$, $\mathcal{O}_0 = \mathcal{O}_{K,P}$ and the $\mathbf{F}$-algebra $\mathcal{O}_0/P\mathcal{O}_0$ is a degree 2 extension of $\mathbf{F}$ if $\varepsilon_P = -1$, is isomorphic to $\mathbf{F}[\epsilon]$ if $\varepsilon_P = 0$ and to $\mathbf{F}^2$ if $\varepsilon_P = 1$. For $n > 0$, $\mathcal{O}_n$ is a local ring with maximal ideal $P\mathcal{O}_{n-1}$ and $\mathcal{O}_n/P\mathcal{O}_n$ is again isomorphic to $\mathbf{F}[\epsilon]$.

Lemma 6.1 *For any $n \geq 0$, the left action of $\mathcal{O}_n^\times$ on $\mathbf{P}(\mathcal{O}_n/P\mathcal{O}_n)$ factors through $\mathcal{O}_n^\times/\mathcal{O}_{n+1}^\times$. Its set of fixed points is given by the following formula*

$$\mathbf{P}(\mathcal{O}_n/P\mathcal{O}_n)^{\mathcal{O}_n^\times} = \begin{cases} \varnothing & \text{if } n = 0 \text{ and } \varepsilon_P = -1, \\ \{\mathcal{P}\mathcal{O}_0/P\mathcal{O}_0\} & \text{if } n = 0 \text{ and } \varepsilon_P = 0, \\ \{\mathcal{P}\mathcal{O}_0/P\mathcal{O}_0, \mathcal{P}^*\mathcal{O}_0/P\mathcal{O}_0\} & \text{if } n = 0 \text{ and } \varepsilon_P = 1, \\ \{P\mathcal{O}_{n-1}/P\mathcal{O}_n\} & \text{if } n > 0. \end{cases}$$

The remaining points are permuted faithfully and transitively by $\mathcal{O}_n^\times/\mathcal{O}_{n+1}^\times$.

Proof This easily follows from the above discussion together with the observation that the quotient map $\mathcal{O}_n \to \mathcal{O}_n/P\mathcal{O}_n$ induces a bijection between $\mathcal{O}_n^\times/\mathcal{O}_{n+1}^\times$ and $(\mathcal{O}_n/P\mathcal{O}_n)^\times/\mathbf{F}^\times$. □

6.2 The $\delta = 0$ case

Let V be a simple left B_P-module, so that $V \simeq F_P^2$ as an F_P-vector space. The embedding $K_P \hookrightarrow B_P$ endows V with the structure of a (left) K_P-module for which V is free of rank one. Let $\mathcal{L}$ be the set of $\mathcal{O}_{F,P}$-lattices in V and pick $L_0 \in \mathcal{L}$ such that $\{\alpha \in B_P; \alpha L_0 \subset L_0\} = R_P$. Then $b \mapsto bL_0$ yields a bijection between $B_P^\times/R_P^\times$ and $\mathcal{L}$. The induced left actions of $K_P^\times$ and $\mathbf{T}(R_P^\times)$ on $\mathbf{Z}[\mathcal{L}]$ are respectively given by

$$(\lambda, L) \mapsto \lambda L \quad \text{and} \quad [R_P^\times \alpha R_P^\times](L) = \sum_{i=1}^{n} bL_i$$

for $\lambda \in K_P^\times$, $L = bL_0 \in \mathcal{L}$, $\alpha \in B_P^\times$, $R_P^\times \alpha R_P^\times = \coprod_{i=1}^n \alpha_i R_P^\times$ and $L_i = \alpha_i L_0$. The function ℓ_P on $\mathcal{L}$ maps a lattice L to the unique integer $n = \ell_P(L)$ such that $\{x \in K_P^\times; xL \subset L\}$ equals $\mathcal{O}_n$.

Lemma 6.2 *The function ℓ_P defines a bijection between $K_P^\times\backslash B_P^\times/R_P^\times \simeq K_P^\times\backslash\mathcal{L}$ and $\mathbf{N}$.*

Proof Fix a K_P-basis e of V. For any $n \geq 0$, $\ell_P(\mathcal{O}_n e) = n$ – this shows that ℓ_P is surjective. Conversely, let L be a lattice with $\ell_P(L) = n$. Then L is a free (rank one) $\mathcal{O}_n$-module by [1, Proposition 7.2]. In particular, there exists an element $\lambda \in K_P^\times$ such that $L = \mathcal{O}_n\lambda e = \lambda\mathcal{O}_n e$. This shows that $\ell_P : K_P^\times\backslash\mathcal{L} \to \mathbf{N}$ is also injective. $\square$

Definition 6.3 Let $L \subset V$ be a lattice.

(i) The *lower (resp. upper) neighbors* of L are the lattices $L' \subset L$ (resp. $L \subset L'$) such that $L/L' \simeq \mathbf{F}$ (resp. $L'/L \simeq \mathbf{F}$).
(ii) The *lower (resp. upper) Hecke operator* T_P^l (resp. T_P^u) on $\mathbf{Z}[\mathcal{L}]$ maps L to the sum of its lower (resp. upper) neighbors.
(iii) If $n = \ell_P(L) \geq 1$, the *lower (resp. upper) predecessor* of L is defined by

$$\mathrm{pr}_l(L) \stackrel{\mathrm{def}}{=} P\mathcal{O}_{n-1}L \quad (\text{resp. } \mathrm{pr}_u(L) \stackrel{\mathrm{def}}{=} \mathcal{O}_{n-1}L).$$

Remark 6.4 T_P^l and T_P^u are the local Hecke operators corresponding to respectively $R_P^\times \alpha R_P^\times$ and $R_P^\times \alpha^{-1} R_P^\times$ where α is any element of $R_P \simeq M_2(\mathcal{O}_{F,P})$ whose reduced norm (= determinant) is a uniformizer in F_P.

Lemma 6.5 *Let L be a lattice in V and put $n = \ell_P(L)$.*

(i) *If $n = 0$, there are exactly $1 + \varepsilon_P$ lower neighbors L' of L for which $\ell_P(L') = 0$, namely $L' = \mathcal{P}L$ if $\varepsilon_P = 0$ and $L' = \mathcal{P}L$ or $\mathcal{P}^*L$ if $\varepsilon_P = 1$.*
(ii) *If $n > 0$, there is a unique lower neighbor L' of L for which $\ell_P(L') \leq n$, namely $L' = \mathrm{pr}_l(L)$ for which $\ell_P(L') = n - 1$.*
(iii) *In both cases, the remaining lower neighbors have $\ell_P = n + 1$. They are permuted faithfully and transitively by $\mathcal{O}_n^\times/\mathcal{O}_{n+1}^\times$ and L is their common upper predecessor.*

Proof This is a straightforward consequence of Lemma 6.1, together with the fact already observed in the proof of Lemma 6.2 that any lattice L with $n = \ell_P(L)$ is free of rank one over $\mathcal{O}_n$. $\square$

We leave it to the reader to formulate and prove an "upper" variant of this lemma. The function $L \mapsto \mathrm{pr}_l(L)$ (resp. $\mathrm{pr}_u(L)$) commutes with the action of $K_P^\times$, and so does the induced function on $\{[b] \in B_P^\times/R_P^\times, \ell_P(b) \geq 1\}$. The latter function extends to a $\widehat{K}^\times$-equivariant function on $\{[b] \in \widehat{B}^\times/H, \ell_P(b_p) \geq 1\}$ with values in $\widehat{B}^\times/H$ (take the identity on $(\widehat{B}^\times)^P/H^P$). Dividing by $K_+^\times$, we finally obtain Galois equivariant functions pr_l and pr_u on $\{x \in \mathrm{CM}_H, \ell_P(x) \geq 1\}$ with values in CM_H. These functions do not depend upon the various choices that we made (V and L_0).

Corollary 6.6 *For $x \in \mathrm{CM}_H$ with $\ell_P(x) = n \geq 1$,*

$$\mathrm{Tr}(x) = T_P^l(x') - x''$$

where $x' = \mathrm{pr}_u(x)$, $x'' = \mathrm{pr}_l(x')$ if $n \geq 2$ and

$$x'' = \begin{cases} 0 & (\varepsilon_P = -1) \\ \sigma_{\mathcal{P}}^{\epsilon} x' & (\varepsilon_P = 0) \\ (\sigma_{\mathcal{P}}^{\epsilon} + \sigma_{\mathcal{P}^*}^{\epsilon}) x' & (\varepsilon_P = 1) \end{cases} \qquad \textit{if } n = 1.$$

Note that if $\ell_P(x) = 1$, $\ell_P(x') = 0$ and x' is indeed defined over an abelian extension of K which is unramified above P.

6.3 *The $\delta = 1$ case*

With V as above, $B_P^\times / R_P^\times$ may now be identified with the set $\mathcal{L}_1$ of all pairs of lattices $L = (L(0), L(1)) \in \mathcal{L}^2$ such that $L(1) \subset L(0)$ with $L(0)/L(1) \simeq \mathbf{F}$. Indeed, $B_P^\times \simeq \mathrm{GL}(V)$ acts transitively on $\mathcal{L}_1$ and there exists some $L_0 = (L_0(0), L_0(1)) \in \mathcal{L}_1$ whose stabilizer equals $R_P^\times$.

To each $L \in \mathcal{L}_1$, we may now attach two integers, namely

$$\ell_{P,0}(L) = \ell_P(L(0)) \quad \text{and} \quad \ell_{P,1}(L) = \ell_P(L(1)).$$

For $L \in \mathcal{L}_1$, $\ell_P(L) = \max(\ell_{P,0}(L), \ell_{P,1}(L))$ and exactly one of the following three situations occurs (see Lemma 6.5).

Definition 6.7 We say that

- L is of **type I** if $\ell_{P,0}(L) = n - 1 < \ell_{P,1}(L) = n$. The *leading vertex* of L equals $L(1)$ and if $n \geq 2$, we define the *predecessor* of L by
$$\mathrm{pr}(L) = (L(0), \mathrm{pr}_l L(0)) = (L(0), P\mathcal{O}_{n-2} L(0)).$$
- L is of **type II** if $\ell_{P,0}(L) = n > \ell_{P,1}(L) = n - 1$. Then $L(0)$ is the *leading* vertex and for $n \geq 2$, the *predecessor* of L is defined by
$$\mathrm{pr}(L) = (\mathrm{pr}_u L(1), L(1)) = (\mathcal{O}_{n-2} L(1), L(1)).$$
- L is of **type III** if $\ell_{P,0}(L) = n = \ell_{P,1}(L)$ (in which case $n = 0$, $\varepsilon_P = 0$ or 1 and $L(1) = \mathcal{P} L(0)$ or $L(1) = \mathcal{P}^* L(0)$). As a convention, we define the *leading* vertex of L to be $L(0)$.

Remark 6.8 The type of L together with the integer $n = \ell_P(L)$ almost determines the $K_P^\times$-homothety class of $\mathcal{L}$. Indeed, Lemma 6.2 implies that we can move the leading vertex of L to $\mathcal{O}_n e$ ($\{e\}$ is a K_P-basis of V). Then $L = (\mathcal{O}_{n-1} e, \mathcal{O}_n e)$ if L is of type I, $L = (\mathcal{O}_n e, P\mathcal{O}_{n-1} e)$ if L is of type II and $L = (\mathcal{O}_n e, \mathcal{P}\mathcal{O}_n e)$ or $(\mathcal{O}_n e, \mathcal{P}^*\mathcal{O}_n e)$ if L is of type III (in which case $n = 0$).

Definition 6.9 The lower (resp. upper) Hecke operator T_P^l (resp. T_P^u) on $\mathbf{Z}[\mathcal{L}_1]$ maps $L \in \mathcal{L}_1$ to the sum of all elements $L' \in \mathcal{L}_1$ such that $L'(0) = L(0)$ but $L'(1) \neq L(1)$ (resp. $L'(1) = L(1)$ but $L'(0) \neq L(0)$).

The P-new quotient $\mathbf{Z}[\mathcal{L}_1]^{P-\text{new}}$ of $\mathbf{Z}[\mathcal{L}_1]$ is the quotient of $\mathbf{Z}[\mathcal{L}_1]$ by the $\mathbf{Z}$-submodule which is spanned by the elements of the form $\sum_{L'(0)=M} L'$ or $\sum_{L'(1)=M} L'$ with M a lattice in V. By construction,

$$T_P^l \equiv T_P^u \equiv -1 \quad \text{on } \mathbf{Z}[\mathcal{L}_1]^{P-\text{new}}.$$

Remark 6.10 For $i \in \{0, 1\}$, put $R(i) = \{b \in B_P;\, bL_0(i) \subset L_0(i)\}$ so that $R_P = R(0) \cap R(1)$. Then T_P^l and T_P^u are the local Hecke operators corresponding to respectively $R_P^\times \alpha R_P^\times$ and $R_P^\times \beta R_P^\times$, for any α in $R(0)^\times - R(1)^\times$ and β in $R(1)^\times - R(0)^\times$. Also, $R(0)^\times = R_P^\times \coprod R_P^\times \beta R_P^\times$ and $R(1)^\times = R_P^\times \coprod R_P^\times \alpha R_P^\times$.

For $L \in \mathcal{L}_1$ and $\lambda \in K_P^\times$, L and λL have the same type and $\text{pr}(\lambda L) = \lambda \text{pr}(L)$ (if $\ell_P(L) \geq 2$). We thus obtain a Galois invariant notion of *type* on CM_H and a Galois equivariant map $x \mapsto \text{pr}(x)$ on $\{x \in \text{CM}_H;\, \ell_P(x) \geq 2\}$ with values in CM_H. The following is then an easy consequence of Lemma 6.5.

Lemma 6.11 *For $x \in \text{CM}_H$ with $\ell_P(x) \geq 2$,*

$$\text{Tr}(x) = \begin{cases} T_P^l(\text{pr}(x)), & \text{if } x \text{ is of type I} \\ T_P^u(\text{pr}(x)), & \text{if } x \text{ is of type II.} \end{cases}$$

In the P-new quotient of $\mathbf{Z}[\text{CM}_H]$, these relations simplify to:

$$\text{Tr}(x) = -\text{pr}(x).$$

Remark 6.12 In contrast to the $\delta = 0$ case, the above constructions do depend upon the choice of L_0. More precisely, our definition of types on CM_H are sensitive to the choice of an orientation on R_P: changing $L_0 = (L_0(0), L_0(1))$ to $L_0' = (L_0(1), PL_0(0))$ exchanges type I and type II points.

6.4 *The $\delta \geq 2$ case*

We now have $B_P^\times / R_P^\times \simeq \mathcal{L}_\delta$ where $\mathcal{L}_\delta$ is the set of all pairs of lattices $L = (L(0), L(\delta))$ in V such that $L(\delta) \subset L(0)$ with $L(0)/L(\delta) \simeq O_F/P^\delta$. We refer to such pairs as δ-lattices. To each $L \in \mathcal{L}_\delta$, we may attach the sequence of intermediate lattices

$$L(\delta) \subsetneq L(\delta-1) \subsetneq \cdots \subsetneq L(1) \subsetneq L(0)$$

and the sequence of integers $\ell_{P,i}(L) \stackrel{\text{def}}{=} \ell_P(L(i))$, for $0 \leq i \leq \delta$. The function ℓ_P corresponds to

$$\ell_P(L) = \max(\ell_{P,i}(L)) = \max(\ell_{P,0}(L), \ell_{P,\delta}(L)).$$

Using Lemma 6.5, one easily checks that the sequence $\ell_{P,i}(L)$ satisfies the following property: there exists integers $0 \leq i_1 \leq i_2 \leq \delta$ such that $\ell_{P,i+1}(L) - \ell_{P,i}(L)$ equals -1 for $0 \leq i < i_1$, 0 for $i_1 \leq i < i_2$ and 1 for $i_2 \leq i < \delta$. Moreover, $\ell_{P,i}(L) = 0$ for all $i_1 \leq i \leq i_2$ if $i_2 \neq i_1$ in which case $\varepsilon_P = 0$ or 1, and $i_2 - i_1 \leq 1$ if $\varepsilon_P = 0$. For our purposes, we only need to distinguish between three types of δ-lattices.

Definition 6.13 We say that $L \in \mathcal{L}_\delta$ is of **type I** if $\ell_{P,0}(L) < \ell_{P,\delta}(L)$, of **type II** if $\ell_{P,0}(L) > \ell_{P,\delta}(L)$ and of **type III** if $\ell_{P,0}(L) = \ell_{P,\delta}(L)$.

The P-new quotient $\mathbf{Z}[\mathcal{L}_\delta]^{P-\text{new}}$ of $\mathbf{Z}[\mathcal{L}_\delta]$ is the quotient of $\mathbf{Z}[\mathcal{L}_\delta]$ by the $\mathbf{Z}$-submodule which is spanned by the elements of the form

$$\textstyle\sum_{(L'(1),L'(\delta))=M} L' \quad \text{or} \quad \sum_{(L'(0),L'(\delta-1))=M} L'$$

with $M \in \mathcal{L}_{\delta-1}$. It easily follows from Lemma 6.5 that for any $L \in \mathcal{L}_\delta$ which is not of type III, $\mathrm{Tr}(L) = 0$ in $\mathbf{Z}[\mathcal{L}_\delta]^{P-\text{new}}$ where $\mathrm{Tr}(L) = \sum_{\lambda \in \mathcal{O}_{n-1}^\times/\mathcal{O}_n^\times} \lambda L$ for $n = \ell_P(L)$. Indeed,

$$\mathrm{Tr}(L) = \begin{cases} \sum_{(L'(0),L'(\delta-1))=(L(0),L(\delta-1))} L' & \text{if } L \text{ is of type I,} \\ \sum_{(L'(1),L'(\delta))=(L(1),L(\delta))} L' & \text{if } L \text{ is of type II.} \end{cases}$$

Extending the notion of types to CM_H as in the previous section, we obtain:

Lemma 6.14 *For any $x \in \mathrm{CM}_H$ which is not of type III,*

$$\mathrm{Tr}(x) = 0 \quad \textit{in } \mathbf{Z}[\mathrm{CM}_H]^{P-\textit{new}}.$$

Remark 6.15 If δ is odd, ℓ_P is bounded on the set of type III points in $\mathcal{L}_\delta$ or CM_H. If δ is even, there are type III points with $\ell_P = n$ for any $n \geq \delta/2$. In both cases, there are type I and type II points with $\ell_P = n$ for any $n > \delta/2$.

6.5 Predecessors and degeneracy maps

Suppose first that $\delta = 0$ and let $L_0(0)$ be a lattice in V such that $R_P = \{\alpha \in B_P;\, \alpha L_0(0) \subset L_0(0)\}$. Choose a lattice $L_0(2) \subset L_0(0)$ such that $L_0 = (L_0(0), L_0(2))$ is a 2-lattice and let

$$R_P^+ = \{\alpha \in B_P;\, \alpha L_0(0) \subset L_0(0) \text{ and } \alpha L_0(2) \subset L_0(2)\} \tag{6.1}$$

be the corresponding Eichler order (of level P^2). Put $H^+ = H^P(R_P^+)^\times$.

To a 2-lattice L, we may attach three lattices:

$$d_0(L) = L(2), \quad d_1(L) = L(1) \quad \text{and } d_2(L) = L(0).$$

Conversely, to each lattice L with $n = \ell_P(L) \geq 2$, we may attach a unique 2-lattice $L^+ = (\mathcal{O}_{n-2}L, L)$ with the property that

$$(d_0, d_1, d_2)(L^+) = (L, \mathrm{pr}_u L, \mathrm{pr}_u \circ \mathrm{pr}_u L) = (L, L', P^{-1}L'')$$

where $L' = \mathrm{pr}_u L$ and $L'' = \mathrm{pr}_l L'$. Being $K_P^\times$-equivariant, these constructions have Galois equivariant counterparts on suitable spaces of CM points. More precisely:

- Choose $b_i \in B_P^\times$ such that $b_i L_0(0) = L_0(2 - i)$.
- Define $d_i : \mathrm{CM}_{H^+} \to \mathrm{CM}_H$ by $d_i([b]) = [bb_i]$ for $b \in \widehat{B}^\times$.
- Define $\vartheta : \mathrm{CM}_H \to \mathrm{CM}_H$ by $\vartheta([b]) = [b\varpi_P]$ for $b \in \widehat{B}^\times$.
- Use the identifications $B_P^\times/R_P^\times \leftrightarrow \mathcal{L}$ and $B_P^\times/(R_P^+)^\times \leftrightarrow \mathcal{L}_2$ to define the $K_P^\times$-equivariant map $x \mapsto x^+$ on $\{[b] \in B_P^\times/R_P^\times;\, \ell_P(bL_0(0)) \geq 2\}$ with values in $B_P^\times/(R_P^+)^\times$ which corresponds to $L \mapsto L^+$ on the level of lattices.
- Using the decomposition $\widehat{B}^\times/H = (\widehat{B}^\times)^P/H^P \times B_P^\times/R_P^\times$ (and similarly for $\widehat{B}^\times/H^+$), extend $x \mapsto x^+$ to a $\widehat{K}^\times$-equivariant map defined on the suitable subset of $\widehat{B}^\times/H$ with values in $\widehat{B}^\times/H^+$ (take the identity on $(\widehat{B}^\times)^P$).
- Dividing out by $K_+^\times$, we thus obtain a Galois equivariant map $x \mapsto x^+$ on $\{x \in \mathrm{CM}_H;\, \ell_P(x) \geq 2\}$ with values in CM_{H^+}.

By construction:

Lemma 6.16 ($\delta = 0$) *For any $x \in \mathrm{CM}_H$ with $\ell_P(x) \geq 2$,*

$$(d_0, d_1, d_2)(x^+) = (x, x', \vartheta^{-1}x'') \quad \text{in } \mathrm{CM}_H^3$$

where $x' = \mathrm{pr}_u(x)$ and $x'' = \mathrm{pr}_l(x')$.

The $\delta = 1$ case is only slightly more difficult. Fix a 1-lattice $(L_0(0), L_0(1))$ whose stabilizer equals $R_P^\times$ and let $L_0(2)$ be a sublattice of $L_0(1)$ such that

$L_0 = (L_0(0), L_0(2))$ is a 2-lattice. Define R_P^+ by the same formula (6.1), so that R_P^+ is again an Eichler order of level P^2. Put $H^+ = H^P(R_P^+)^\times$.

To each 2-lattice L we may attach two 1-lattices, namely $d_{01}(L) = (L(0), L(1))$ and $d_{12}(L) = (L(1), L(2))$. Conversely suppose that L is a 1-lattice with $n = \ell_P(L) \geq 2$. If L is of type I, $L^+ = (\mathcal{O}_{n-2}L(0), L(1))$ is a 2-lattice and

$$(d_{01}, d_{12})(L^+) = (\vartheta^{-1}\mathrm{pr}(L), L)$$

where ϑ is now the permutation of $\mathcal{L}_1$ which maps $(L(0), L(1))$ to $(L(1), PL(0))$. If L is of type II, $L^+ = (L(0), P\mathcal{O}_{n-2}L(1))$ is a 2-lattice and

$$(d_{01}, d_{12})(L^+) = (L, \vartheta\mathrm{pr}(L)).$$

These constructions are again equivariant with respect to the action of $K_P^\times$, and may thus be extended to $\mathrm{Gal}_K^{\mathrm{ab}}$-equivariant constructions on CM points. More precisely,

- Choose $b_{01} = 1$ and $b_{12} \in B_P^\times$ such that $b_{12}(L_0(0), L_0(1)) = (L_0(1), L_0(2))$. Define d_{01} and $d_{12} : \mathrm{CM}_{H^+} \to \mathrm{CM}_H$ by $d_{01}([b]) = [bb_{01}]$, $d_{12}([b]) = [bb_{12}]$ for $b \in \widehat{B}^\times$.
- Choose ω in $B_P^\times$ such that $\omega(L_0(0), L_0(1)) = (L_0(1), PL_0(0))$ and define $\vartheta : \mathrm{CM}_H \to \mathrm{CM}_H$ by $\vartheta([b]) = [b\omega]$ for $b \in \widehat{B}^\times$.
- Proceeding as above in the $\delta = 0$ case, extend $L \mapsto L^+$ to a Galois equivariant function $x \mapsto x^+$ defined on $\{x \in \mathrm{CM}_H;\ \ell_P(x) \geq 2\}$ with values in CM_{H^+}.

With these notations, we obtain:

Lemma 6.17 *($\delta = 1$) For any $x \in \mathrm{CM}_H$ with $\ell_P(x) \geq 2$,*

$$(x, \mathrm{pr}(x)) = \begin{cases} (d_{12}, \vartheta d_{01})(x^+) & \text{if } x \text{ is of type I,} \\ (d_{01}, \vartheta^{-1} d_{12})(x^+) & \text{if } x \text{ is of type II.} \end{cases}$$

Bibliography

[1] H. Bass. On the ubiquity of Gorenstein rings. *Math. Z.*, 82:8–28, 1963.

[2] H. Carayol. Sur la mauvaise réduction des courbes de Shimura. *Compositio Math.*, 59(2):151–230, 1986.

[3] W. Casselman. On some results of Atkin and Lehner. *Math. Ann.*, 201:301–314, 1973.

[4] C. Cornut. Mazur's conjecture on higher Heegner points. *Invent. Math.*, 148(3):495–523, 2002.

[5] C. Cornut and V. Vatsal. CM points and quaternion algebras. *Doc. Math.*, 10:263–309 (electronic), 2005.

[6] F. Diamond and J. Im. Modular curves and modular forms. In K. Murty, editor, *Seminar on Fermat's last theorem, Toronto, 1993*, C.M.S. Conference Proceedings, 39–133. Amer. Math. Soc., 1995.

[7] B. Edixhoven. Special points on products of modular curves. *Duke Math. J.*, 126(2):325–348, 2005.

[8] B. Edixhoven and A. Yafaev. Subvarieties of Shimura varieties. *Ann. of Math. (2)*, 157(2):621–645, 2003.

[9] B. Gross. Heights and the special values of L-series. In H. Kisilevsky and J. Labute, editors, *Number Theory*, volume 7 of *CMS Conference Proceedings*, 115–189. Amer. Math. Soc., 1987.

[10] B. Gross. Local orders, root numbers, and modular curves. *Am. J. Math.*, 110:1153–1182, 1988.

[11] B. Gross. Heegner points and representation theory. In *Heegner points and Rankin L-series*, 37–66. MSRI Publications, 2004.

[12] B. Gross and D. Prasad. Test vectors for linear forms. *Math. Ann.*, 291:343–355, 1991.

[13] A. Grothendieck. Éléments de géométrie algébrique. IV. Étude locale des schémas et des morphismes de schémas. II. *Inst. Hautes Études Sci. Publ. Math.*, (24):231, 1965.

[14] A. Grothendieck. Technique de descente et théorèmes d'existence en géométrie algébrique. V. Les schémas de Picard: théorèmes d'existence. In *Séminaire Bourbaki, Vol. 7*, Exp. No. 232, 143–161. Soc. Math. France, Paris, 1995.

[15] A. Grothendieck. Technique de descente et théorèmes d'existence en géométrie algébrique. VI. Les schémas de Picard: propriétés générales. In *Séminaire Bourbaki, Vol. 7*, 221–243. Soc. Math. France, Paris, 1995.

[16] R. Hartshorne. *Algebraic geometry*. Springer-Verlag, New York, 1977. Graduate Texts in Mathematics, No. 52.

[17] B. Howard. Iwasawa theory of Heegner points on abelian varieties of $GL(2)$-type. *Duke Math. J.*, 124(1):1–45, 2004.

[18] H. Jacquet. *Automorphic forms on* $\mathrm{GL}(2)$. *Part II.* Lecture Notes in Mathematics, Vol. 278, Springer-Verlag, Berlin, 1972.

[19] H. Jacquet and R. P. Langlands. *Automorphic forms on* $\mathrm{GL}(2)$. Lecture Notes in Mathematics, Vol. 114, Springer-Verlag, Berlin, 1970.

[20] B. Mazur. Modular curves and arithmetic. In *Proceedings of the International Congress of Mathematicians, Vol. 1, 2 (Warsaw, 1983)*, 185–211. PWN, 1984.

[21] J. Nekovář. Selmer complexes. To appear in Astérisque. Available at `http://www.math.jussieu.fr/~nekovar/pu/`

[22] J. Nekovář and N. Schappacher. On the asymptotic behaviour of Heegner points. *Turkish J. Math.*, 23(4):549–556, 1999.

[23] D. E. Rohrlich. On L-functions of elliptic curves and anticyclotomic towers. *Invent. Math.*, 75(3):383–408, 1984.

[24] J.-P. Serre. *Arbres, amalgames,* SL_2. Société Mathématique de France, Paris, 1977. Avec un sommaire anglais, Rédigé avec la collaboration de Hyman Bass, Astérisque, No. 46.

[25] R. Taylor. On Galois representations associated to Hilbert modular forms. *Invent. Math.*, 98(2):265–280, 1989.
[26] V. Vatsal. Uniform distribution of Heegner points. *Invent. Math.*, 148:1–46, 2002.
[27] V. Vatsal. Special values of anticylotomic L-functions. *Duke Math J.*, 116(2):219–261, 2003.
[28] V. Vatsal. Special value formulae for Rankin L-functions. In *Heegner points and Rankin L-series*, volume 49 of *Math. Sci. Res. Inst. Publ.*, 165–190. Cambridge Univ. Press, Cambridge, 2004.
[29] M.-F. Vignéras. *Arithmétique des algèbres de quaternions*, Lecture Notes in Mathematics, Vol. 800, Springer-Verlag, 1980.
[30] J.-L. Waldspurger. Sur les valeurs de certaines fonctions L automorphes en leur centre de symmétrie. *Compos. Math.*, 54:174–242, 1985.
[31] S. Zhang. Gross-Zagier formula for GL_2. *Asian J. Math.*, 5(2):183–290, 2001.
[32] S. Zhang. Heights of Heegner points on Shimura curves. *Ann. of Math. (2)*, 153(1):27–147, 2001.
[33] S. Zhang. Gross-Zagier formula for GL_2 II. In *Heegner points and Rankin L-series*, 191–242. MSRI Publications, 2003.

A correspondence between representations of local Galois groups and Lie-type groups

Fred Diamond

Department of Mathematics,
Brandeis University,
Waltham,
MA 02454, USA.
Current address: Department of Mathematics,
King's College London,
WC2R 2LS, UK.
fred.diamond@kcl.ac.uk[a]

[a] *Research supported by NSF grants DMS-9996345, 0300434*

Introduction

Serre conjectured in [13] that every continuous, irreducible odd representation

$$\rho : G_{\mathbb{Q}} \to \mathrm{GL}_2(\overline{\mathbb{F}}_p)$$

arises from a modular form. Moreover he refines the conjecture by specifying an optimal weight and level for a Hecke eigenform giving rise to ρ. Viewing Serre's conjecture as a manifestation of Langlands' philosophy in characteristic p, this refinement can be viewed as a local-global compatibility principle, the weight of the form reflecting the behavior of ρ at p, the level reflecting the behavior at primes other than p. The equivalence between the "weak" conjecture and its refinement (for $\ell > 2$) follows from work of Ribet [11] and others (see [6]). Remarkable progress has recently been made on the conjecture itself by Khare and Wintenberger; see for example Khare's article in this volume.

Serre's conjecture is generalized in [5] to the context of Hilbert modular forms and two-dimensional representations of G_K where K is a totally real number field in which p is unramified. The difficulty in formulating the refinement lies in the specification of the weight. This is handled in [5] by giving a recipe for a set $W_{\mathfrak{p}}(\rho)$ of irreducible $\overline{\mathbb{F}}_p$-representations of $\mathrm{GL}_2(\mathcal{O}_K/\mathfrak{p})$ for each prime $\mathfrak{p}|p$ in terms of $\rho|_{I_{\mathfrak{p}}}$; the sets $W_{\mathfrak{p}}(\rho)$ then conjecturally characterize the types of local factors at primes over p of automorphic representations giving rise to ρ. We omit the subscript $\mathfrak{p}$ since we shall be concerned only with local behavior, so now K will denote a finite unramified extension of $\mathbb{Q}_p$ with residue field k.

The purpose of the paper is to prove that if the local Galois representation is semisimple, then $W(\rho)$ is essentially the set of Jordan-Hölder constituents of the reduction of an irreducible *characteristic zero* representation of $\mathrm{GL}_2(k)$. Moreover, denoting this representation by $\alpha(\rho)$ we obtain

Theorem 0.18 *There is a bijection*

$$\{\rho : G_K \longrightarrow \mathrm{GL}_2(\overline{\mathbb{F}}_p)\}/ \textit{ equivalence of } \rho|_{I_K}^{\mathrm{ss}}$$

$$\alpha \updownarrow$$

$$\{\textit{irreducible } \overline{\mathbb{Q}}_p\textit{-representations of } \mathrm{GL}_2(k) \textit{ not factoring through } \det\}/\sim$$

such that $W(\rho^{\mathrm{ss}})$ contains the set of Jordan-Hölder factors of the reduction of $\alpha(\rho)$.

Moreover, the last inclusion is typically an equality and one can explicitly describe the exceptional weights. We remark that the local Langlands correspondence also gives rise to a bijection between the sets in the theorem by taking the K-type corresponding to a tamely ramified lift of ρ. The bijection of the theorem however has a different flavor. Indeed if $[k : \mathbb{F}_p]$ is odd, then irreducible ρ correspond to principal series and special representations, while reducible ρ correspond to supercuspidal ones.

A generalization of Serre's Conjecture to the setting of GL_n was formulated by Ash and others in [1], [2], and Herzig's thesis [8] pursues the idea of relating the set of Serre weights of a semi-simple $\rho : G_{\mathbb{Q}_p} \to \mathrm{GL}_n(\overline{\mathbb{F}}_p)$ to the reduction of an irreducible characteristic zero representation of $\mathrm{GL}_n(\mathbb{F}_p)$. However Herzig shows that the phenomenon described in Theorem 0.18 does *not* persist for $n > 2$; instead he defines an operator $\mathcal{R}$ on the irreducible mod p representations of $\mathrm{GL}_n(\mathbb{F}_p)$ and shows that the regular (i.e., up to certain exceptions) Serre weights of ρ are given by applying $\mathcal{R}$ to the constituents of the reduction of a certain characteristic zero representation $V(\rho)$. Herzig also show that such a relationship holds in the context of $\mathrm{GL}_2(k)$. Moreover, the association $\rho \mapsto V(\rho)$ appears to be compatible with the local Langlands correspondence in the sense described above. In this light, Theorem 0.18 can be viewed as saying that Herzig's operator $\mathcal{R}$ typically sends the set of irreducible constituents of the reduction of one $\overline{\mathbb{Q}}_p$-representation of $\mathrm{GL}_2(k)$ to those of another.

One can also view Theorem 0.18 in the context of the theory of mod p and p-adic local Langlands correspondences being developed by Breuil and others (see [3], [4], [7]). In particular, one would like a mod p local Langlands correspondence to associate a mod p representation of $\mathrm{GL}_2(K)$ to ρ, and local-global compatibility considerations suggest that the set of Serre weights comprise the constituents of its $\mathrm{GL}_2(\mathcal{O}_K)$-socle. One would also like a p-adic local Langlands correspondence associating p-adic representations of $\mathrm{GL}_2(K)$ to suitable lifts of ρ, and satisfying some compatibility with the mod p correspondence with respect to reduction. One can thus speculate that the theorem reflects some property of the hypothetical p-adic correspondence for GL_2.

The paper is organized as follows: In Section 1, we compute the semisimplification of the reduction mod p of the irreducible characteristic zero representations of $\mathrm{GL}_2(k)$. The main theorem is proved in Section 2, and the exceptional weights are described in Section 3 for the sake of completeness.

The author is grateful to Florian Herzig, Richard Taylor and the referee for their feedback on an earlier draft.

1 A Brauer character computation

In this section we compute the Jordan-Hölder constituents of the reduction mod p of the irreducible characteristic zero representations of $\mathrm{GL}_2(k)$. For $\mathrm{SL}_2(k)$, this is essentially done in [14]. See also [8] for another method of doing this calculation based on work of Jantzen.

We first recall the irreducible $\overline{\mathbb{Q}}_p$-representations of $G = \mathrm{GL}_2(k)$ (see for example [9, Ch. 28] or [10, XVIII, §12]).

Let B denote the subgroup of upper-triangular matrices in G. For a pair of homomorphisms $\chi_1, \chi_2 : k^\times \to \overline{\mathbb{Q}}^\times$, we let $I(\chi_1, \chi_2)$ denote the $(q+1)$-dimensional representation of G induced from the character of B defined by

$$\begin{pmatrix} x & w \\ 0 & y \end{pmatrix} \mapsto \chi_1(x)\chi_2(y).$$

$I(\chi_1, \chi_2) \sim I(\chi_1', \chi_2')$ if and only if $\{\chi_1, \chi_2\} = \{\chi_1', \chi_2'\}$. If $\chi_1 \neq \chi_2$, then $I(\chi_1, \chi_2)$ is irreducible. $I(\chi, \chi) \sim \chi \circ \det \oplus \mathrm{sp}_\chi$ for an irreducible q-dimensional representation sp_χ.

The remaining irreducible $\overline{\mathbb{Q}}$-representations of G are parametrized as follows. Let k' be a quadratic extension of k, σ the non-trivial k-automorphism of k' and Nm the norm from k' to k. For each homomorphism $\xi : k'^\times \to \overline{\mathbb{Q}}^\times$ such that $\xi \neq \xi \circ \sigma$, there is an irreducible $(q-1)$-dimensional $\overline{\mathbb{Q}}$-representation $\Theta(\xi)$ of G, and $\Theta(\xi) \sim \Theta(\xi')$ if and only if $\xi' \in \{\xi, \xi \circ \sigma\}$. Moreover for any homomorphism $\chi : k^\times \to \overline{\mathbb{Q}}^\times$, we have $(\chi \circ \det)\Theta(\xi) \sim \Theta((\chi \circ \mathrm{Nm})\xi))$.

Letting i denote a k-algebra embedding $k' \to \mathrm{M}_2(k)$, the character table of G is as follows:

Conjugacy class of:	Representation			
	$\chi\circ\det$	sp_χ	$I(\chi_1,\chi_2)$	$\Theta(\xi)$
$\begin{pmatrix} x & 0 \\ 0 & x \end{pmatrix}$	$\chi(x)^2$	$q\chi(x)^2$	$(q+1)\chi_1(x)\chi_2(x)$	$(q-1)\xi(x)$
$\begin{pmatrix} x & 1 \\ 0 & x \end{pmatrix}$	$\chi(x)^2$	0	$\chi_1(x)\chi_2(x)$	$-\xi(x)$
$\begin{pmatrix} x & 0 \\ 0 & y \end{pmatrix} \notin k^\times$	$\chi(xy)$	$\chi(xy)$	$\chi_1(x)\chi_2(y)+\chi_1(y)\chi_2(x)$	0
$i(z)\notin k^\times$	$\chi(zz^\sigma)$	$-\chi(zz^\sigma)$	0	$-\xi(z)-\xi(z^\sigma)$

Next we recall the irreducible $\overline{\mathbb{F}}_p$-representations of $\mathrm{GL}_2(k)$. Let $S = k(\overline{\mathbb{F}}_p)$, the set of embeddings $k \to \overline{\mathbb{F}}_p$. For integers m_τ, n_τ with $n_\tau \geq 0$ for each $\tau \in S$, we have the representation

$$V_{\vec{m},\vec{n}} = \otimes_{\tau\in S} \det{}^{m_\tau} k^2 \otimes_k \mathrm{Sym}^{n_\tau-1} k^2 \otimes_{k,\tau} \overline{\mathbb{F}}_p.$$

We make the convention that $\mathrm{Sym}^{-1} = 0$, so that the dimension of $V_{\vec{m},\vec{n}}$ is equal to $\prod_{\tau\in S} n_\tau$. If $1 \leq n_\tau \leq p$ for all τ, then $V_{\vec{m},\vec{n}}$ is irreducible; assuming further that $0 \leq m_\tau \leq p-1$ for each τ and some $m_\tau < p-1$, then the $V_{\vec{m},\vec{n}}$ are inequivalent and form a complete list of the irreducible $\overline{\mathbb{F}}_p$-representations of $\mathrm{GL}_2(k)$.

Recall that the semisimplification of an $\overline{\mathbb{F}}_p$-representation of G is determined by its Brauer character, which is a $\overline{\mathbb{Q}}_p$-valued function on the p-regular conjugacy classes of G (see [12, 18.1, 18.2] for example). Letting $\tilde{}$ denote the Teichmüller lift, the Brauer character of $V_{\vec{m},\vec{n}}$, which we denote $\beta_{\vec{m},\vec{n}}$, is as follows:

$\begin{pmatrix} x & 0 \\ 0 & y \end{pmatrix}$	$\prod_{\tau\in S}\left(\tilde{\tau}(xy)^{m_\tau}\sum_{0\leq\nu\leq n_\tau-1}\tilde{\tau}(y)^\nu\tilde{\tau}(x)^{n_\tau-1-\nu}\right)$
$i(z)\notin k^\times$	$\prod_{\tau\in S}\left(\tilde{\tau}'(z)^{(q+1)m_\tau}\sum_{0\leq\nu\leq n_\tau-1}\tilde{\tau}'(z)^{n_\tau-1+(q-1)\nu}\right)$

where τ' denotes either extension of τ to k'.

If V is a finite-dimensional $\overline{\mathbb{Q}}_p$-representation of G, then there exists a $\overline{\mathbf{Z}}_p$-lattice $L \subset V$ stable under the action of G. Reducing L modulo the maximal ideal of $\overline{\mathbf{Z}}_p$ then yields an $\overline{\mathbb{F}}_p$-representation $\overline{L}$ of G whose Brauer character is the restriction of the character of V to the p-regular classes of G. In particular, the semisimplification of $\overline{L}$ is independent of the choice of lattice L, and we denote it $\overline{V}$ and call it the *reduction* of V.

We now compute $\overline{V}$ for all irreducible V (i.e., the decomposition matrix of G with respect to reduction mod p). First note that any homomorphism $\chi : k^\times \to \overline{\mathbb{Q}}_p^\times$ can be written in the form $\prod_\tau \tilde{\tau}^{a_\tau}$ for some integers a_τ with $0 \leq a_\tau \leq p-1$, in which case $\overline{\chi} = \prod_\tau \tau^{a_\tau}$. Moreover, if $V' \sim (\chi\circ\det)\otimes V$, then $\overline{V}' \sim (\overline{\chi}\circ\det)\otimes\overline{V}$, so we can replace V by such a twist in order to compute its reduction.

We first consider the representations $I(\chi_1, \chi_2)$. Twisting by $\chi_1^{-1} \circ \det$, we need only consider those of the form $I(1, \chi)$. The reduction is then given by the following proposition:

Proposition 1.1 *Let $V = I(1, \prod_\tau \tilde{\tau}^{a_\tau})$ with $0 \le a_\tau \le p-1$ for each $\tau \in S$. Then $\overline{V} \sim \oplus_{J \subset S} V_J$, where $V_J = V_{\vec{m}_J, \vec{n}_J}$ with $\vec{m}_J$ and $\vec{n}_J$ defined as follows:*

$$m_{J,\tau} = \begin{cases} 0, & \text{if } \tau \in J, \\ a_\tau + \delta_J(\tau), & \text{if } \tau \notin J, \end{cases}$$

$$\text{and} \quad n_{J,\tau} = \begin{cases} a_\tau + \delta_J(\tau), & \text{if } \tau \in J, \\ p - a_\tau - \delta_J(\tau), & \text{if } \tau \notin J, \end{cases}$$

where δ_J is the characteristic function of $J^{(p)} = \{\, \tau \circ \mathrm{Frob} \,|\, \tau \in J \,\}$. Moreover the non-zero V_J are inequivalent.

Proof. We need to show that the sum of the $\beta_{\vec{m}_J, \vec{n}_J}$ coincides with the character of V on p-regular conjugacy classes.

We first consider conjugacy classes of elements of the form $\begin{pmatrix} x & 0 \\ 0 & y \end{pmatrix}$ with $x, y \in k$. Let us choose an embedding $\tau_0 : k \to \overline{\mathbb{F}}_p$ and index the elements of S by setting $\tau_i = \tau \circ \mathrm{Frob}_p^i$ for $i \in \mathbf{Z}/f\mathbf{Z}$. We then have

$$\begin{aligned} & \beta_{\vec{m},\vec{n}}\left(\begin{pmatrix} x & 0 \\ 0 & y \end{pmatrix}\right) \\ = \; & \tilde{x}^{\sum_{i=0}^{f-1} m_i p^i} \tilde{y}^{\sum_{i=0}^{f-1} m_i p^i} \sum_{\vec{0} \le \vec{\nu} \le \vec{n} - \vec{1}} \tilde{y}^{\sum_{i=0}^{f-1} \nu_i p^i} \tilde{x}^{\sum_{i=0}^{f-1} (n_i - 1 - \nu_i) p^i} \\ = \; & \tilde{x}^{\sum_{i=0}^{f-1} (2m_i + n_i - 1) p^i} \sum_{\vec{m} \le \vec{b} \le \vec{m} + \vec{n} - \vec{1}} (\tilde{y}/\tilde{x})^{\sum_{i=0}^{f-1} b_i p^i}, \end{aligned}$$

where we have simply written m_i for m_{τ_i}, n_i for n_{τ_i} and $\tilde{w}$ for $\tilde{\tau}_0(w)$. (We also abuse notation in viewing i as an integer when it appears as an exponent of p and as a congruence class when it appears as an index.) Taking $(\vec{m}, \vec{n}) = (\vec{m}_J, \vec{n}_J)$ and viewing $J \subset \{0, 1, \ldots, f-1\}$, we have

$$\begin{aligned} & \textstyle\sum_{i=0}^{f-1} (2m_i + n_i - 1) p^i \\ = \; & \textstyle\sum_{i \in J} (a_i - 1 + \delta_J(i)) p^i + \sum_{i \notin J} (a_i - 1 + \delta_J(i) + p) p^i \\ = \; & (1 - \delta_J(0))(q-1) + \textstyle\sum_{i=0}^{f-1} a_i p^i, \end{aligned} \tag{1.1}$$

so that $\tilde{x}^{\sum_{i=0}^{f-1}(2m_i+n_i-1)p^i} = \tilde{x}^{\sum_{i=0}^{f-1} a_i p^i}$, giving

$$\beta_{\vec{m}_J,\vec{n}_J}\left(\begin{pmatrix} x & 0 \\ 0 & y \end{pmatrix}\right) = \tilde{x}^{\sum_{i=0}^{f-1} a_i p^i}\left(\sum_{d\in B_J}(\tilde{y}/\tilde{x})^d\right)$$

where

$$B_J = \left\{ d = \sum_{i=0}^{f-1} b_i p^i \;\middle|\; \begin{array}{ll} 0 \le b_i < a_i + \delta_J(i), & \text{if } i \in J, \\ a_i + \delta_J(i) \le b_i < p, & \text{if } i \notin J \end{array} \right\}. \tag{1.2}$$

Note that if $d \in B_J$, then $0 \le d \le q-1$. Since the only dependence relation among the functions $w \mapsto \tilde{w}^d$ on $k^\times$ for $0 \le d \le q-1$ is that $\tilde{w}^0 = \tilde{w}^{q-1}$, we see that $V_J = 0$ if and only if $B_J = \varnothing$. Moreover if $V_J \sim V_{J'}$, then either $B_J = B_{J'}$ or one is gotten from the other by replacing 0 by $q-1$. One sees easily that the first case implies that either $B_J = \varnothing$ or $J = J'$, and that the second is impossible. We thus conclude that the the non-zero $V_{\vec{m}_J,\vec{n}_J}$ are inequivalent.

To complete the proof of the proposition, we use another description of the B_J:

Lemma 1.2 *Suppose that* $0 \le d \le q-1$. *Write* $d = \sum_{i=0}^{f-1} b_i p^i$ *with* $0 \le b_i \le p-1$ *for each* $i \in \mathbf{Z}/f\mathbf{Z}$. *If* $d \ne \sum_{i=0}^{f-1} a_i p^i$, *then*

$$d \in B_J \iff J = \left\{ j \in \mathbf{Z}/f\mathbf{Z} \;\middle|\; \sum_{i=0}^{f-1} b_{i+j+1}p^i < \sum_{i=0}^{f-1} a_{i+j+1}p^i \right\}.$$

Furthermore, $d = \sum a_i p^i \in B_J$ *if and only if* $J = S$ *or* $J = \varnothing$.

Proof. First note that if $d = \sum_{i=0}^{f-1} b_i p^i \ne \sum_{i=0}^{f-1} a_i p^i$, then

$$\sum_{i=0}^{f-1} b_{i+j+1}p^i < \sum_{i=0}^{f-1} a_{i+j+1}p^i \qquad \text{if and only if} \qquad b_{j-r} < a_{j-r}$$

where $r \in \{0, \ldots, f-1\}$ is chosen so that $b_{j-r} \ne a_{j-r}$, $b_{j-r+1} = a_{j-r+1}$, $\ldots, b_j = a_j$. Indeed if $b_{j-r} < a_{j-r}$, then

$$\begin{aligned} & b_j p^{f-1} + b_{j-1}p^{f-2} + \cdots + b_{j+1} \\ < \;& b_j p^{f-1} + b_{j-1}p^{f-2} + \cdots + b_{j-r}p^{f-1-r} + p^{f-1-r} \\ \le \;& a_j p^{f-1} + a_{j-1}p^{f-2} + \cdots + a_{j-r}p^{f-1-r} \\ \le \;& a_j p^{f-1} + a_{j-1}p^{f-2} + \cdots + a_{j+1}. \end{aligned}$$

Now suppose that $d = \sum_{i=0}^{f-1} b_i p^i \in B_J$ and $j \in J$. Then $b_j \le a_j$ and if equality holds then $j-1 \in J$, so $b_{j-1} \le a_{j-1}$. Iterating, we find that either $d = \sum_{i=0}^{f-1} a_i p^i$ and $J = S$, or that

$$b_{j-r} < a_{j-r}, b_{j-r+1} = a_{j-r+1}, \ldots, b_j = a_j, \text{ for some } r \in \{0, \ldots, f-1\},$$

yielding the desired inequality. The case $j \notin J$ is similar.

Conversely, suppose that $d \ne \sum_{i=0}^{f-1} a_i p^i$ and J is given by the formula in the statement of the lemma. If $j \in J$, then we have $b_{j-r} < a_{j-r}, b_{j-r+1} = a_{j-r+1}, \ldots, b_j = a_j$, for some $r \in \{0, \ldots, f-1\}$, so either $b_j < a_j$, or $b_j = a_j$ and the inequality for $j-1$ gives $j-1 \in J$. In either case we have $b_j < a_j + \delta_J(j)$. Similarly we find that $j \notin J$ implies that $a_j + \delta_J(j) \le b_j$. Finally, it is clear that $\sum_{i=0}^{f-1} a_i p^i$ is in both $B_\varnothing$ and B_S. □

Returning to the proof of Proposition 1.1, the lemma gives

$$\begin{aligned} \sum_J \left(\sum_{d \in B_J} (\tilde{y}/\tilde{x})^d \right) &= (\tilde{y}/\tilde{x})^{\sum_{i=0}^{f-1} a_i p^i} + \sum_{d=0}^{q-1} (\tilde{y}/\tilde{x})^d \\ &= \begin{cases} 1+q & \text{if } \tilde{y} = \tilde{x}, \\ 1 + (\tilde{y}/\tilde{x})^{\sum_{i=0}^{f-1} a_i p^i} & \text{if } \tilde{y} \ne \tilde{x}, \end{cases} \end{aligned}$$

from which it follows that

$$\sum_J \beta_{\vec{m}_J, \vec{n}_J} \left(\begin{pmatrix} x & 0 \\ 0 & y \end{pmatrix} \right) = \begin{cases} (q+1) \prod_\tau \tilde{\tau}(x)^{a_\tau}, & \text{if } y = x, \\ \prod_\tau \tilde{\tau}(x)^{a_\tau} + \prod_\tau \tilde{\tau}(y)^{a_\tau}, & \text{if } y \ne x. \end{cases}$$

Now consider conjugacy classes of elements of the form $i(z)$ for $z \notin k^\times$. Choosing an embedding τ_0' of k extending τ_0 and writing $\tilde{z}$ for $\tilde{\tau}_0'(z)$, we have

$$\begin{aligned} \beta_{\vec{m}, \vec{n}}(i(z)) &= \tilde{z}^{\sum_{i=0}^{f-1} (q+1) m_i p^i} \sum_{\vec{0} \le \vec{\nu} \le \vec{n} - \vec{1}} \tilde{z}^{\sum_{i=0}^{f-1} (n_i - 1 + (q-1)\nu_i) p^i} \\ &= \tilde{z}^{\sum_{i=0}^{f-1} (2m_i + n_i - 1) p^i} \sum_{\vec{m} \le \vec{b} \le \vec{m} + \vec{n} - \vec{1}} \tilde{z}^{(q-1) \sum_{i=0}^{f-1} b_i p^i}. \end{aligned}$$

Summing over J and using (1.1) then gives

$$\begin{aligned} &\sum_J \beta_{\vec{m}_J, \vec{n}_J}(i(z)) \\ &= \tilde{z}^{\sum_{i=0}^{f-1} a_i p^i} \tilde{z}^{(q-1)(1-\delta_J(0))} \sum_J \left(\sum_{d \in B_J} \tilde{z}^{(q-1)d} \right) \\ &= \tilde{z}^{\sum_{i=0}^{f-1} a_i p^i} \left(\sum_{J \ni f-1} \left(\sum_{d \in B_J} \tilde{z}^{(q-1)d} \right) + \sum_{J \not\ni f-1} \left(\sum_{d \in B_J} \tilde{z}^{(q-1)(1+d)} \right) \right) \end{aligned}$$

where B_J is as in (1.2). According to Lemma 1.2, the values of $d \neq \sum_{i=0}^{f-1} a_i p^i$ contributing to the first sum are those with $0 \leq d < \sum_{i=0}^{f-1} a_i p^i$, the values contributing to the second are those with $\sum_{i=0}^{f-1} a_i p^i < d \leq q-1$, and there is one occurrence of $\sum_{i=0}^{f-1} a_i p^i$ in each. It follows that

$$\sum_J \beta_{\vec{m}_J,\vec{n}_J}(i(z)) = \tilde{z}^{\sum_{i=0}^{f-1} a_i p^i} \sum_{d=0}^{q} \tilde{z}^{(q-1)d} = 0,$$

since $\tilde{z}^{q-1} \neq 1$, but $\tilde{z}^{q^2-1} = 1$. This completes the proof of Proposition 1.1.

□

Note that when χ is trivial, so $V \sim \det \oplus \mathrm{sp}$, the proposition gives $\overline{V} \sim V_{\vec{0},\vec{1}} \oplus V_{\vec{0},\vec{p}}$, the first factor being the reduction of det and the second being that of sp. If χ is non-trivial, then $I(1,\chi)$ is irreducible and its reduction is given by the proposition.

Now we turn our attention to the $(q-1)$-dimensional representations $\Theta(\xi)$. Choosing $\tau_0' : k' \to \overline{\mathbb{F}}_p$ as in the proof of the theorem, we can write $\xi = (\tilde{\tau}_0')^n$ for some n, determined mod (q^2-1). Since $\xi \neq \xi \circ \sigma$, we have that n is not divisible by $q+1$ and can therefore be written in the form $\alpha + (q+1)\beta$ with $1 \leq \alpha \leq q$, $0 \leq \beta \leq q-2$. Twisting by $\tilde{\tau}_0{}^{-\beta} \circ \det$, we can assume $n = \alpha$ and write $\xi = \tilde{\tau}_0' \prod_{i=0}^{f-1} (\tilde{\tau}_i')^{a_{\tau_i}}$ where $\tau_i' = \tau_0' \circ \mathrm{Frob}_p^i$, $\tau_i = \tau_i'|k$ and $0 \leq a_{\tau_i} \leq p-1$ for $i = 0, \ldots, f-1$.

Proposition 1.3 *Let* $V = \Theta\left(\tilde{\tau}_0' \prod_{i=0}^{f-1} (\tilde{\tau}_i')^{a_{\tau_i}}\right)$ *with* $0 \leq a_\tau \leq p-1$ *for each* $\tau \in S$. *Then* $\overline{V} \sim \oplus_{J \subset S} V_J$, *where* $V_J = V_{\vec{m}_J,\vec{n}_J}$ *with* $\vec{m}_J$ *and* $\vec{n}_J$ *defined as follows:*

$$m_{J,\tau} = \begin{cases} \delta_J(\tau), & \text{if } \tau = \tau_0 \in J, \\ a_\tau + 1, & \text{if } \tau = \tau_0 \notin J, \\ 0, & \text{if } \tau \in J, \tau \neq \tau_0, \\ a_\tau + \delta_J(\tau), & \text{if } \tau \notin J, \tau \neq \tau_0, \end{cases}$$

$$\text{and} \quad n_{J,\tau} = \begin{cases} a_\tau + 1 - \delta_J(\tau), & \text{if } \tau = \tau_0 \in J, \\ p - a_\tau - 1 + \delta_J(\tau), & \text{if } \tau = \tau_0 \notin J, \\ a_\tau + \delta_J(\tau), & \text{if } \tau \in J, \tau \neq \tau_0, \\ p - a_\tau - \delta_J(\tau), & \text{if } \tau \notin J, \tau \neq \tau_0, \end{cases}$$

where δ_J *is the characteristic function of* $J^{(p)} = \{\, \tau \circ \mathrm{Frob} \mid \tau \in J \,\}$. *Moreover the non-zero* V_J *are inequivalent.*

Proof. Taking $(\vec{m}, \vec{n}) = (\vec{m}_J, \vec{n}_J)$ as in (1.1) now gives

$$\sum_{i=0}^{f-1}(2m_i + n_i - 1)p^i = 1 + (1 - \delta_J(0))(q-1) + \sum_{i=0}^{f-1} a_i p^i, \tag{1.3}$$

so that

$$\beta_{\vec{m}_J,\vec{n}_J}\left(\begin{pmatrix} x & 0 \\ 0 & y \end{pmatrix}\right) = \tilde{x}^{1+\sum_{i=0}^{f-1} a_i p^i}\left(\sum_{d\in B'_J} (\tilde{y}/\tilde{x})^d\right),$$

where

$$B'_J = \left\{ d = \sum_{i=0}^{f-1} b'_i p^i \;\middle|\; \begin{array}{ll} \delta_J(0) \le b'_0 < a_0 + 1, & \text{if } 0 \in J, \\ a_0 + 1 \le b'_0 < p + \delta_J(0), & \text{if } 0 \notin J, \\ 0 \le b'_i < a_i + \delta_J(i), & \text{if } i \in J, i \ne 0, \\ a_i + \delta_J(i) \le b'_i < p, & \text{if } i \notin J, i \ne 0 \end{array} \right\}. \tag{1.4}$$

Note that if $d \in B'_J$, then $1 \le d \le q-1$. Since there are no dependence relations among the functions $w \mapsto \tilde{w}^d$ on $k^\times$ for such d, we see as in the proof of Proposition 1.1 that the non-zero V_J are inequivalent.

Lemma 1.4 *Suppose that* $1 \le d \le q-1$. *Write* $d = \sum_{i=0}^{f-1} b_i p^i$ *with* $0 \le b_i \le p-1$ *for each* $i \in \mathbf{Z}/f\mathbf{Z}$. *If* $d \le \sum_{i=0}^{f-1} a_i p^i$, *then*

$$d \in B'_J \iff J = \left\{ j \in \{0, \ldots, f-1\} \;\middle|\; 0 < \sum_{i=0}^{j} b_i p^i \le \sum_{i=0}^{j} a_i p^i \right\}.$$

If $\sum_{i=0}^{f-1} a_i p^i < d$, *then*

$$d \in B'_J \iff J = \left\{ j \in \{0, \ldots, f-1\} \;\middle|\; \sum_{i=0}^{j} b_i p^i \le \sum_{i=0}^{j} a_i p^i \right\}.$$

Proof. Write $d = \sum_{i=0}^{f-1} b'_i p^i$ with $\delta_J(0) \le b'_0 < p + \delta_J(0)$, and $0 \le b'_i < p$ for $i = 1, \ldots, f-1$. We then have $(b'_0, b'_1, \ldots, b'_{f-1}) = (b_0, b_1, \ldots, b_{f-1})$ unless $f-1 \in J$ and $b_0 = 0$, in which case

$$(b'_0, b'_1, \ldots, b'_{f-1}) = (p, p-1, \ldots, p-1, b_r - 1, b_{r+1}, \ldots, b_{f-1})$$

where r is the least positive integer such that $b_r > 0$. It follows that if $f-1 \in J$, then

$$0 < \sum_{i=0}^{j} b_i p^i \le \sum_{i=0}^{j} a_i p^i \iff \sum_{i=0}^{j} b'_i p^i \le \sum_{i=0}^{j} a_i p^i.$$

If $d \in B'_J$, then we have

$$j \in J \Longleftrightarrow \sum_{i=0}^{j} b'_i p^i \leq \sum_{i=0}^{j} a_i p^i \tag{1.5}$$

for $j = 0, \ldots, f-1$ by induction on j. In particular, $f-1$ is in J if and only if $d \leq \sum_{i=0}^{f-1} a_i p^i$, and (1.5) translates into the desired formula for B'_J in either case.

Suppose conversely that J is as defined in the statement of the lemma. In particular, $f-1$ is in J if and only if $d \leq \sum_{i=0}^{f-1} a_i p^i$, so (1.5) holds for $j = 0, \ldots, f-1$. Therefore $b'_0 \leq a_0$ if and only if $0 \in J$, and we deduce that

$$j \in J \Longleftrightarrow b'_j < a_j + \delta_J(j)$$

for $j = 1, \ldots, f-1$ by induction. It follows that $d \in B'_J$. □

Returning to the proof of Proposition 1.3, the lemma gives

$$\sum_J \left(\sum_{d \in B'_J} (\tilde{y}/\tilde{x})^d \right) = \sum_{d=1}^{q-1} (\tilde{y}/\tilde{x})^d = \begin{cases} q-1 & \text{if } \tilde{y} = \tilde{x}, \\ 0 & \text{if } \tilde{y} \neq \tilde{x}, \end{cases}$$

since $(\tilde{y}/\tilde{x})^{q-1} = 1$. It follows that

$$\sum_J \beta_{\vec{m}_J, \vec{n}_J} \left(\begin{pmatrix} x & 0 \\ 0 & y \end{pmatrix} \right) = \begin{cases} (q-1)\xi(x), & \text{if } y = x, \\ 0, & \text{if } y \neq x. \end{cases}$$

where $\xi = \tilde{\tau}'_0 \prod_{i=0}^{f-1} (\tilde{\tau}'_i)^{a_{\tau_i}}$.

Now for conjugacy classes of elements of the form $i(z)$ for $z \notin k^\times$, we have

$$\begin{aligned} &\sum_J \beta_{\vec{m}_J, \vec{n}_J}(i(z)) \\ &= \tilde{z}^{1+\sum_{i=0}^{f-1} a_i p^i} \tilde{z}^{(q-1)(1-\delta_J(0))} \sum_J \left(\sum_{d \in B'_J} \tilde{z}^{(q-1)d} \right) \\ &= \tilde{z}^{1+\sum_{i=0}^{f-1} a_i p^i} \left(\sum_{J \ni f-1} \left(\sum_{d \in B'_J} \tilde{z}^{(q-1)d} \right) + \sum_{J \not\ni f-1} \left(\sum_{d \in B'_J} \tilde{z}^{(q-1)(1+d)} \right) \right) \end{aligned}$$

where B'_J is as in (1.4). According to Lemma 1.4, the values of d contributing to the first sum are those with $1 \leq d \leq \sum_{i=0}^{f-1} a_i p^i$, the values contributing to the second are those with $\sum_{i=0}^{f-1} a_i p^i < d \leq q-1$. It follows that

$$\begin{aligned}
&\sum_J \beta_{\vec{m}_J,\vec{n}_J}(i(z)) \\
&\quad = \tilde{z}^{1+\sum_{i=0}^{f-1} a_i p^i}\left(-1 - \tilde{z}^{(q-1)(1+\sum_{i=0}^{f-1} a_i p^i)} + \sum_{d=0}^{q} \tilde{z}^{(q-1)d}\right) \\
&\quad = -\tilde{z}^{1+\sum_{i=0}^{f-1} a_i p^i} - \tilde{z}^{q(1+\sum_{i=0}^{f-1} a_i p^i)} \\
&\quad = -\xi(z) - \xi(z^\sigma).
\end{aligned}$$

This completes the proof of Proposition 1.3.

□

2 The correspondence

In this section we construct the bijection of Theorem 0.18. Let ω_0 and ω_0' denote fundamental characters $I_K \to \overline{\mathbb{F}}_p^\times$ corresponding to embeddings $\tau_0 : k \to \overline{\mathbb{F}}_p$ and $\tau_0' : k' \to \overline{\mathbb{F}}_p$ chosen as in the preceding section. Thus ω_0' has order q^2-1, and $\omega_0 = (\omega_0')^{q+1}$. If $\rho : G_K \to \mathrm{GL}_2(\overline{\mathbb{F}}_p)$ is a continuous representation, then $\rho|_{I_K}^{\mathrm{ss}}$ is equivalent to one of the form

$$\begin{array}{rll}
& \omega_0^r \oplus \omega_0^s & \text{for some } r, s \in \mathbf{Z}, \\
\text{or} & (\omega_0')^t \oplus (\omega_0')^{qt} & \text{for some } t \in \mathbf{Z} \text{ not divisible by } q+1,
\end{array}$$

according to whether or not ρ is reducible.

In [5], the weight part of an analogue of Serre's conjecture is formulated over totally real fields by defining a set $W(\rho)$ of irreducible $\overline{\mathbb{F}}_p$-representations of $\mathrm{GL}_2(k)$. We recall the definition in the easiest case, when $\rho|_{I_K}$ is semi-simple. In the case $\rho \sim \omega_0^r \oplus \omega_0^s$, then we define $W(\rho)$ by the rule

$$V_{\vec{m},\vec{n}} \in W(\rho) \iff$$
$$\left\{\begin{array}{rcl} r & \equiv & \sum_{i=0}^{f-1} m_i p^i + \sum_{i\in J^*} n_i p^i \bmod (q-1), \\ s & \equiv & \sum_{i=0}^{f-1} m_i p^i + \sum_{i\notin J^*} n_i p^i \bmod (q-1), \end{array}\right. \quad \text{for some } J^* \subset S$$

(where as usual, $m_i = m_{\tau_i}$ with $\tau_i = \tau_0 \circ \mathrm{Frob}^i$). In the case $\rho \sim (\omega_0')^t \oplus (\omega_0')^{qt}$ for some $c \not\equiv 0\Lambda\text{-mod}(q+1)$, we let $S' = \{0, 1, \ldots, 2f-1\}$ and define $\pi : S' \to S$ by reduction mod f. We then define $W(\rho)$ by

$$V_{\vec{m},\vec{n}} \in W(\rho) \iff \left\{\begin{array}{c} t \equiv \sum_{i=0}^{f-1}(q+1)m_i p^i + \sum_{i\in J^*} n_i p^i \bmod (q^2-1) \\ \text{for some } J^* \subset S' \text{ such that } \pi : J^* \xrightarrow{\sim} S. \end{array}\right.$$

Let R_I denote the set of equivalence classes of $\overline{\mathbb{Q}}_p$-representations of I_K as above; note that there are $(q^2-q)/2$ of each type. Let R_G denote the set of equivalence classes of representations of $\mathrm{GL}_2(k)$ of the form $I(\chi_1, \chi_2)$ or $\Theta(\xi)$; note that there are $(q^2 - q)/2$ of each of these as well. Recall that if V is a $\overline{\mathbb{Q}}_p$ representation of $\mathrm{GL}_2(k)$, then $\overline{V}$ denotes the semi-simplification of its reduction modulo the maximal ideal of $\overline{Z}_p$.

Theorem 2.1 *There is a bijection $\beta : R_G \to R_I$ such that if $\rho : G_K \to \mathrm{GL}_2(\overline{\mathbb{F}}_p)$ restricts to $\beta(V)$ on I_K and $V_{\vec{m},\vec{n}}$ is an irreducible subrepresentation of $\overline{V}$, then $V_{\vec{m},\vec{n}} \in W(\rho)$.*

Note that Theorem 0.18 follows from Theorem 2.1 on replacing $I(\chi, \chi)$ by its q-dimensional irreducible subrepresentation and setting $\alpha(\rho) = \beta^{-1}(\rho|^{\mathrm{ss}}_{I_K})$.

Proof. We first define β for the representations considered in Propositions 1.1 and 1.3. Suppose that $b_0, \ldots, b_{f-1}$ are integers with $1 \le b_i \le p$. If f is odd, then we let

$$\beta(I(1, \tilde{\tau}_0^{\sum_{i=0}^{f-1}(b_i-1)p^i})) = (\omega_0')^{b_0+b_2p^2+\cdots+b_{f-1}p^{f-1}+b_1p^{f+1}+\cdots+b_{f-2}p^{2f-2}} \oplus (\omega_0')^{b_1p+b_3p^3+\cdots+b_{f-2}p^{f-2}+b_0p^f+b_2p^{f+2}+\cdots+b_{f-1}p^{2f-1}}$$

and

$$\beta(\Theta((\tilde{\tau}_0')^{1+\sum_{i=0}^{f-1}(b_i-1)p^i})) = \omega_0^{b_0+b_2p^2+\cdots+b_{f-1}p^{f-1}} \oplus \omega_0^{1+b_1p+b_3p^3+\cdots+b_{f-2}p^{f-2}}.$$

If f is even, then we let

$$\beta(I(1, \tilde{\tau}_0^{\sum_{i=0}^{f-1}(b_i-1)p^i})) = \omega_0^{b_0+b_2p^2+\cdots+b_{f-2}p^{f-2}} \oplus \omega_0^{b_1p+b_3p^3+\cdots+b_{f-1}p^{f-1}}$$

and

$$\beta(\Theta((\tilde{\tau}_0')^{1+\sum_{i=0}^{f-1}(b_i-1)p^i})) = (\omega_0')^{b_0+b_2p^2+\cdots+b_{f-2}p^{f-2}+p^f+b_1p^{f+1}+\cdots+b_{f-1}p^{2f-1}} \oplus (\omega_0')^{1+b_1p+b_3p^3+\cdots+b_{f-1}p^{f-1}+b_0p^f+b_2p^{f+2}+\cdots+b_{f-2}p^{2f-2}}.$$

There is no ambiguity in replacing $b_0 = \cdots = b_{f-1} = 1$ with $b_0 = \cdots = b_{f-1} = p$ in the formula for $\beta(I(1, \tilde{\tau}_0^{\sum_{i=0}^{f-1}(b_i-1)p^i}))$ as it just exchanges the two characters in the sum. We also need to check that the exponents of ω_0' are not divisible by $q+1$. This follows from the fact that if f is odd, then

$$\begin{aligned} &b_0 + b_2p^2 + \cdots + b_{f-1}p^{f-1} + b_1p^{f+1} + \cdots + b_{f-2}p^{2f-2} \\ &\quad \equiv b_0 - b_1p + b_2p^2 - \ldots - b_{f-2}p^{f-2} + b_{f-1}p^{f-1} \bmod (q+1) \end{aligned} \tag{2.1}$$

$$\text{and } 1 \le b_0 - b_1p + b_2p^2 - \ldots - b_{f-2}p^{f-2} + b_{f-1}p^{f-1} \le q,$$

and that if f is even, then

$$\begin{aligned} &b_0 + b_2p^2 + \cdots + b_{f-2}p^{f-2} + p^f + b_1p^{f+1} + \cdots + b_{f-1}p^{2f-1} \\ &\quad \equiv -1 + b_0 - b_1p + b_2p^2 - \ldots - b_{f-2}p^{f-2} - b_{f-1}p^{f-1} \bmod (q+1), \end{aligned}$$

$$\text{and } 1 - q \le b_0 - b_1p + b_2p^2 - \ldots - b_{f-2}p^{f-2} - b_{f-1}p^{f-1} \le 0. \tag{2.2}$$

We extend β to all of R_G by twisting. If $\chi : k^\times \to \overline{\mathbb{Q}}_p^\times$ is a character, then we let $\beta(\chi)$ denote the character $I_K \to \overline{\mathbb{F}}_p^\times$ corresponding to $\overline{\chi}$ by local class field theory, i.e., if $\chi = \tilde{\tau}_0^r$, then $\beta(\chi) = \omega_0^r$. Any representation in R_G can be written in the form $(\chi \circ \det) \otimes V$ for some χ and some V for which we have already defined $\beta(V)$. We then let $\beta((\chi \circ \det) \otimes V) = \beta(\chi) \otimes \beta(V)$. We need to check there is no ambiguity in the definition. If $\chi_1 \neq \chi_2$, then $I(\chi_1, \chi_2)$ has two expressions of the above form, namely $(\chi_2 \circ \det) \otimes I(1, \chi)$ and $(\chi_2\chi \circ \det) \otimes I(1, \chi^{-1})$, so it suffices to check that $\beta(I(1,\chi)) = \beta(\chi) \otimes \beta(I(1, \chi^{-1}))$. If $\chi = \tilde{\tau}_0^{\sum_{i=0}^{f-1}(b_i-1)p^i}$ with each $b_i \in \{1, \ldots, p\}$, then $\chi^{-1} = \tilde{\tau}_0^{\sum_{i=0}^{f-1}(b_i'-1)p^i}$ where $b_i' = p + 1 - b_i$. It is then straightforward to check that replacing $\beta(I(1,\chi))$ with $\beta(\chi)\beta(I(1,\chi^{-1}))$ simply interchanges the two characters of I_K. Similarly each $\Theta(\xi)$ has two expressions as above, given explicitly by twisting the identity $\Theta(\xi) \sim (\chi \circ \det) \otimes \Theta(\xi')$ where $\xi = (\tilde{\tau}_0')^{1+\sum_{i=0}^{f-1}(b_i-1)p^i}$, $\xi' = (\tilde{\tau}_0')^{1+\sum_{i=0}^{f-1}(b_i'-1)p^i}$ and $\chi = \tilde{\tau}_0^{\sum_{i=0}^{f-1}(b_i-1)p^i}$ with each $b_i \in \{1, \ldots, p\}$ and $b_i' = p + 1 - b_i$. We find that $\beta(\Theta(\xi)) = \beta(\chi) \otimes \beta(\Theta(\xi'))$, the characters of I_K again being interchanged.

Since R_I and R_G have the same cardinality, it suffices to show that β is surjective in order to conclude it is a bijection. Therefore it suffices to show that every representation in R_I is a twist of one of the form $\beta(V)$ for some V as in Proposition 1.1 or 1.3. For representations of the form $(\omega_0')^t \oplus (\omega_0')^{qt}$, this follows from (2.1) and (2.2). Indeed since the values of $b_0 - b_1p + \cdots \pm b_{f-1}p^{f-1}$ are distinct, we see that there is an exponent in every non-zero congruence class mod $(q+1)$. For representations of the form $\omega^r \oplus \omega^s$, it suffices to note similarly that $r - s \bmod (q-1)$ arises as the difference of exponents of ω_0 for some $\beta(V)$.

Suppose now that $\rho : G_K \to \mathrm{GL}_2(\overline{\mathbb{F}}_p)$ restricts to $\beta(V)$ on I_K. To prove the assertion about $W(\rho)$, we can twist V by $\chi \circ \det$ and ρ by a character restricting to $\beta(\chi)$ and so assume V is as in Proposition 1.1 or 1.3. We now need to show that each non-zero $V_{\vec{m}_J,\vec{n}_J}$ is in $W(\rho)$.

Suppose first that $V = I(1, \tilde{\tau}_0^{\sum_{i=0}^{f-1}(b_i-1)p^i})$ with each $b_i \in \{1,\ldots,p\}$ and f is odd. Given J, we let $J' = \{\, j \in S' \mid j \equiv i \bmod f \text{ for some } i \in J\,\}$, $J_0' = \{\, j \in S' \mid j \text{ is even}\,\}$, $J_1' = \{\, j \in S' \mid j \text{ is odd}\,\}$ and $J^* = (J_0' \cap J') \cup (J_1' \smallsetminus J')$. We then have $\pi : J^* \stackrel{\sim}{\to} S$ and $\sum_{i=0}^{f-1}(q+1)m_{J,i}p^i + \sum_{i\in J^*} n_{J,i}p^i$ is congruent mod (q^2-1) to

$$\begin{aligned}
&\sum_{i\notin J'}(b_i-1+\delta_J(i))p^i \\
&\qquad + \sum_{i\in J_0'\cap J'}(b_i-1+\delta_J(i))p^i + \sum_{i\in J_1'\smallsetminus J'}(p-b_i+1-\delta_J(i))p^i \\
&\qquad\qquad \equiv \sum_{i\in J_0'}(b_i-1+\delta_J(i))p^i + \sum_{i\in J_1'\smallsetminus J'} p^{i+1} \\
&\qquad\qquad \equiv \sum_{i\in J_0'} b_ip^i \bmod (q^2-1)
\end{aligned}$$

since $\sum_{i\in J_0'}(\delta_J(i)-1)p^i \equiv \sum_{i\in J_1'\smallsetminus J'} p^{i+1}$. It follows that $V_{\vec{m}_J,\vec{n}_J} \in W(\rho)$.

If $V = \Theta((\tilde{\tau}_0')^{1+\sum_{i=0}^{f-1}(b_i-1)p^i})$ and f is even, then we proceed exactly as above, but with $J_0' = \{0, 2, \ldots, f-2, f+1, f+3, \ldots, 2f-1\}$ and $J_1' = S' \smallsetminus J_0'$. The remaining cases are similar, but simpler. If $V = I(1, \tilde{\tau}_0^{\sum_{i=0}^{f-1}(b_i-1)p^i})$ (resp. $\Theta((\tilde{\tau}_0')^{1+\sum_{i=0}^{f-1}(b_i-1)p^i})$) and f is even (resp. odd), we let $J_0 = \{\, j \in S \mid j \text{ is even}\,\}$, $J_1 = \{\, j \in S \mid j \text{ is odd}\,\}$ and $J^* = (J_0 \cap J) \cup (J_1 \smallsetminus J)$. In each case a calculation similar to the one above shows that $V_{\vec{m}_J,\vec{n}_J} \in W(\rho)$. □

3 Exceptional weights

Let β be the bijection of Theorem 2.1, and suppose throughout this section that $\rho : G_K \to \mathrm{GL}_2(\overline{\mathbb{F}}_p)$ restricts to $\beta(V)$ on I_K. We say that $V_{\vec{m},\vec{n}}$ is an *exceptional weight* for ρ (or V) if it lies in the complement in $W(\rho)$ of the set of constituents of $\overline{V}$. In this section we characterize the exceptional weights. We first give a sufficient condition for there to be none.

Theorem 3.1 *Suppose* $V = I(\tilde{\tau}_0^c, \tilde{\tau}_0^{a+c})$ *with* $a = \sum_{i=0}^{f-1} a_ip^i$. *If f is odd and* $1 \le a_i \le p-2$ *for each i, then there are no exceptional weights for V. If f is even and* $a \equiv \pm\frac{q-1}{p+1} \bmod (q-1)$ *or* $1 \le a_i \le p-2$ *for each i, then there are* *no exceptional weights for V unless $a \equiv \pm 2\frac{q-1}{p+1} \bmod (q-1)$, in which case the only exceptional weights are those $V_{\vec{m},\vec{n}} \in W(\rho)$ with $\vec{n} = \vec{p}$.*

Suppose $V = \Theta((\tilde{\tau}_0')^{1+a+c(q+1)})$ with $a = \sum_{i=0}^{f-1} a_i p^i$. If f is even and $1 \le a_i \le p-2$ for each i, then there are no exceptional weights for V. If f is odd and $1 + a \equiv \pm\frac{q+1}{p+1} \bmod (q+1)$ or $1 \le a_i \le p-2$ for each i, then there are no exceptional weights for V unless $1+a \equiv \pm 2\frac{q+1}{p+1} \bmod (q+1)$, in which case the only exceptional weights are those $V_{\vec{m},\vec{n}} \in W(\rho)$ with $\vec{n} = \vec{p}$.

We remark that the special cases in the statement are precisely those where ρ is the sum of two characters whose ratio is trivial or cyclotomic on inertia.

Proof. We first treat the cases where $1 \le a_i \le p-2$ for each i. In this case the $V_{\vec{m}_J,\vec{n}_J}$ of Propositions 1.1 and 1.3 are all non-zero, so $\overline{V}$ has 2^f constituents. If $\beta(V)$ is irreducible, then Proposition 3.1 of [5] shows that $\#W(\rho) \le 2^f$, so it follows that equality holds and there are no exceptional weights in this case.

So suppose that $\beta(V)$ is reducible. Propositions 3.4 and 3.5 of [5] then show that $\#W(\rho) \le 2^f$ unless

$$\sum_{i=0}^{f-1}(-1)^i b_i p^i \equiv (p+1)\sum_{i\in J^*}(-1)^i p^i \bmod (q-1)$$

for some $J^* \subset S$, where each $b_i = a_i + 1 \in \{2, \dots, p-1\}$.

If f is even, then the left-hand side is strictly between $1-q$ and 0 while the right hand side is between $-2(q-1)$ and $q-1$. Setting

$$\sum_{i=0}^{f-1}(-1)^i b_i p^i = c(q-1) + (p+1)\sum_{i\in J}(-1)^i p^i$$

for $c = 0, \pm 1$ and solving p-adically, the restriction on b_i forces either $J = \{0, 2, \dots, f-2\}$ and $\vec{a} = (1, p-2, 1, \dots, p-2)$ or $J = \{1, 3, \dots, f-1\}$ and $\vec{a} = (p-2, 1, p-2, \dots, 1)$, which gives $a \equiv \pm 2\frac{q-1}{p+1} \bmod (q-1)$. In this case one has $\#W(\rho) \le 2^f + 1$ if $p > 3$, so there is at most one exceptional weight. Note also that there is an element of $W(\rho)$ of the form $V_{\vec{m},\vec{p}}$, but no such factor of $\overline{V}$ since $a \not\equiv 0 \bmod (q-1)$. If $p = 3$, one has $\#W(\rho) \le 2^f + 2$ and two elements of $W(\rho)$ of the form $V_{\vec{m},\vec{p}}$ accounting for all exceptional weights.

The argument in the case of odd f is similar, but the left-hand side is strictly between 0 and $q-1$ while the right-hand side is between $-(q-1)$ and $2(q-1)$ and we get that either $J = \{0, 2, \dots, f-1\}$ and $\vec{a} = (1, p-2, 1, \dots, p-2, 1)$ or $J = \{1, 3, \dots, f-2\}$ and $\vec{a} = (p-2, 1, p-2, \dots, 1, p-2)$, giving $1 + a \equiv \pm 2\frac{q+1}{p+1} \bmod (q+1)$.

We now turn our attention to the remaining cases. Suppose first that $p > 2$ and f is even. Twisting V and ρ, we can assume that

$$\vec{a} = (p-1, 0, p-1, 0, \ldots, p-1, 0) \quad \text{and}$$
$$\rho_{I_K} \sim \omega_0^{p+p^3+\cdots+p^{f-1}} \oplus \omega_0^{p+p^3+\cdots+p^{f-1}}.$$

For each $J^* \subset S$, we explicitly describe the $V_{\vec{m},\vec{n}}$ such that

$$\sum_{i=0}^{f/2} p^{1+2i} \equiv \sum_{i=0}^{f-1} m_i p^i + \sum_{i \in J^*} n_i p^i \equiv \sum_{i=0}^{f-1} m_i p^i + \sum_{i \notin J^*} n_i p^i \bmod (q-1).$$

Propositions 3.4 and 3.5 of [5] show that this holds for a unique $\vec{n}$ unless $J^* = \{0, 2, \ldots, f-2\}$ or $\{1, 3, \ldots, f-1\}$. For each of these two values of J^*, there are two possibilities for $\vec{n}$, namely $(p, 1, p, 1, \ldots, p, 1)$ and $(1, p, 1, p, \ldots, 1, p)$. Otherwise there is an i such that $\chi_{J^*}(i-1) = \chi_{J^*}(i)$ where χ_{J^*} is the characteristic function of J^*, and $\vec{n}$ is characterized as the unique f-tuple such that

- $n_i \in \{0, p-1, p\}$ for all i;
- if $n_{i-1} = 1$, then $n_i = p$;
- if $n_{i-1} = p-1$ or p, then $n_i = p-1$ if $\chi_{J^*}(i-1) = \chi_{J^*}(i)$;
- if $n_{i-1} = p-1$ or p, then $n_i = 1$ if $\chi_{J^*}(i-1) \neq \chi_{J^*}(i)$.

Note also that $\sum m_i p^i \bmod (q-1)$ is determined by $\vec{n}$ and J^*. It is then straightforward to check that each such $V_{\vec{m},\vec{n}}$ arises as $V_{\vec{m}_{J'},\vec{n}_{J'}}$ where

$$J' = \{j \in \{0, 2, \ldots, f-2\} \mid n_j = p-1 \text{ or } p\}$$
$$\cup \{j \in \{1, 3, \ldots, f-1\} \mid n_j = 1\},$$

so there are no exceptional weights. (Note that $(J')^*$ need not coincide with J^*.)

The case of odd f, $p > 2$ is similar. We assume

$$\vec{a} = (p-1, 0, p-1, 0, \ldots, p-1) \quad \text{and}$$
$$\rho_{I_K} \sim \omega_0^{1+p+p^3+\cdots+p^{f-2}} \oplus \omega_0^{1+p+p^3+\cdots+p^{f-2}}.$$

For each J^*, there is a unique possibility for $\vec{n}$ characterized exactly as in the case of f even, and we set

$$J' = \{j \in \{0, 2, \ldots, f-1\} \mid n_j = p-1 \text{ or } p\}$$
$$\cup \{j \in \{1, 3, \ldots, f-2\} \mid n_j = 1\}$$

to conclude there are no exceptional weights.

Finally if $p = 2$, then one argues as above using Proposition 3.6 of [5], but with two changes. Firstly, we find also that there are two possibilities for $\vec{n}$ if $J^* = \varnothing$ or S, namely $\vec{n} = (1, 1, \dots, 1)$ or $(2, 2, \dots, 2)$, the latter being exceptional. Secondly, to generalize the characterization of $\vec{n}$ one defines $n_i(x) \in \{0, x-1, x\}$ and then sets $\vec{n} = \vec{n}(p)$ with $p = 2$. □

We finish with a complete characterization of the exceptional weights.

Theorem 3.2 *Suppose that $V_{\vec{m},\vec{n}} \in W(\rho)$.*

If $\rho|_{I_K} \sim \omega_0^r \oplus \omega_0^s$, then $V_{\vec{m},\vec{n}}$ is exceptional if and only if for each $J^* \subset S$ such that

$$r \equiv \sum_{i=0}^{f-1} m_i p^i + \sum_{i \in J^*} n_i p^i, \quad s \equiv \sum_{i=0}^{f-1} m_i p^i + \sum_{i \notin J^*} n_i p^i \bmod (q-1), \quad (3.1)$$

we have $n_i = p$ and $\chi_{J^*}(i-1) = \chi_{J^*}(i)$ for some $i \in S$.

If $\rho|_{I_K} \sim (\omega_0')^t \oplus (\omega_0')^{qt}$, then $V_{\vec{m},\vec{n}}$ is exceptional if and only if for each $J^* \subset S'$ such that

$$t \equiv \sum_{i=0}^{f-1} (q+1) m_i p^i + \sum_{i \in J^*} n_i p^i \bmod (q^2 - 1) \quad \text{and} \quad \pi : J^* \xrightarrow{\sim} S, \quad (3.2)$$

we have $n_i = p$ and $\chi_{J^*}(i-1) = \chi_{J^*}(i)$ for some $i \in S'$.

Proof. We note first that every $V_{\vec{m},\vec{n}}$ as in the statement of the theorem is indeed exceptional, for if it is equivalent to $V_{\vec{m}_J,\vec{n}_J}$ for some $J \subset S$, then the proof of Theorem 2.1 provides a J^* such that (3.1) or (3.2) holds, but the explicit formula for $\vec{n}_J$ shows that $n_i < p$ whenever $\chi_{J^*}(i-1) = \chi_{J^*}(i)$.

Suppose on the other hand that (3.1) or (3.2) holds for some J^* such that $\chi_{J^*}(i-1) \neq \chi_{J^*}(i)$ whenever $n_i = p$. We then choose J so that J^* is as in the proof of Theorem 2.1 and verify that $V_{\vec{m},\vec{n}} = V_{\vec{m}_J,\vec{n}_J}$ (except possibly in the case of reducible ρ with $r = s$, where the result is already immediate from Theorem 3.1). We can twist V and ρ and so assume V is as in the statements of Proposition 1.1 or 1.3.

Suppose first that $V = I\left(1, \tilde{\tau}_0^{\sum_{i=0}^{f-1} a_i p^i}\right)$ with f even and $0 \leq a_i p - 1$ for each i. We then have

$$\sum_{i=0}^{f-1} m_i p^i + \sum_{i \in J^*} n_i p^i \equiv b_0 + b_2 p^2 + \cdots b_{f-2} p^{f-2} \bmod (q-1) \quad (3.3)$$

$$\sum_{i=0}^{f-1} m_i p^i + \sum_{i \notin J^*} n_i p^i \equiv b_1 p + b_3 p^3 + \cdots b_{f-1} p^{f-1} \bmod (q-1) \quad (3.4)$$

where each $b_i = a_i + 1$. Let $b'_i = n_i - \delta_J(i) + 1$ for $i \in J$ and $b'_i = p - n_i - \delta_J(i) + 1$ for $i \notin J$. The condition that $\chi_{J^*}(i-1) \neq \chi_{J^*}(i)$ whenever $n_i = p$ guarantees that $1 \leq b'_i \leq p$ for each i. It is then straightforward to check that

$$\sum_{i=0}^{f-1}(-1)^i b'_i p^i \equiv \sum_{i \in J^*} n_i p^i - \sum_{i \notin J^*} n_i p^i \bmod (q-1),$$

which by (3.3) is congruent to $\sum_{i=0}^{f-1} b_i p^i$. Since $1 \leq b_i, b'_i \leq p$ for each i and we have ruled out the case $\sum_{i=0}^{f-1} b_i p^i \equiv 0 \bmod (q-1)$, it follows that $b_i = b'_i$ for all i, and therefore that $\vec{n} = \vec{n}_J$. We then compute that

$$\begin{aligned} \sum_{i=0}^{f-1} m_i p^i &\equiv \sum_{\text{even } i} b_i p^i - \sum_{i \in J^*} n_i p^i \\ &\equiv \sum_{\text{even } i \notin J^*} (a_i + \delta_J(i))p^i + \sum_{\text{odd } i \in J^*} (a_i + \delta_J(i))p^i \\ &\equiv \sum_{i \in S} m_{J,i} p^i, \end{aligned}$$

hence $V_{\vec{m},\vec{n}} = V_{\vec{m}_J,\vec{n}_J}$.

Suppose next that $V = I\left(1, \tilde{\tau}_0^{\sum_{i=0}^{f-1} a_i p^i}\right)$ but f is odd. We then start with the congruence

$$\sum_{i=0}^{f-1} m_i(q+1)p^i + \sum_{i \in J^*} n_i p^i \equiv b_0 + b_2 p^2 + \cdots b_{2f-2}p^{2f-2} \bmod (q^2-1)$$

instead of (3.3). Defining b'_i and arguing as above then gives

$$\sum_{i=0}^{f-1}(-1)^i b'_i p^i \equiv \sum_{i=0}^{f-1}(-1)^i b_i p^i \bmod (q+1),$$

so $\vec{b} = \vec{b}'$ and $\vec{n} = \vec{n}_J$. Similarly one finds that

$$\begin{aligned} (q+1)\sum_{i=0}^{f-1} m_i p^i &\equiv \sum_{\text{even } i \in S'} b_i p^i - \sum_{i \in J^*} n_i p^i \\ &\equiv (q+1)\sum_{i \notin J}(a_i + \delta_J(i))p^i \bmod (q^2-1) \end{aligned}$$

giving $V_{\vec{m},\vec{n}} = V_{\vec{m}_J,\vec{n}_J}$.

Now suppose that $V = \Theta\left((\tilde{\tau}_0')^{1+\sum_{i=0}^{f-1} a_i p^1}\right)$ with f even. We then have

$$\sum_{i=0}^{f-1} m_i(q+1)p^i + \sum_{i\in J^*} n_i p^i$$
$$\equiv b_0 + b_2p^2 + \cdots + b_{f-2}p^{f-2} + p^f + b_{f+1}p^{f+1} + \cdots + b_{2f-1}p^{2f-1} \mod (q^2-1).$$

We define b_i' for $i > 0$ as above, but set $b_0' = n_0 + \delta_J(0)$ or $p - n_0 + \delta_J(0)$ according to whether $0 \in J^*$. Arguing as above, with special attention to the terms with $i = 0, f$, then gives

$$\sum_{i=0}^{f-1}(-1)^i b_i' p^i \equiv 1 + \sum_{i\in J^*, 0\le i<f} n_i p^i - \sum_{i\notin J^*, 0\le i<f} n_i p^i$$
$$\equiv \sum_{i=0}^{f-1}(-1)^i b_i p^i \mod (q+1),$$

so that $\vec{b} = \vec{b}'$ and $\vec{n} = \vec{n}_J$. Again computing mod (q^2-1), but with special attention to $i = 0, f$, gives

$$\sum_{i=0}^{f-1} m_i p^i \equiv \sum_{i=0}^{f-1} m_{J,i} p^i \mod (q-1)$$

so that $V_{\vec{m},\vec{n}} = V_{\vec{m}_J,\vec{n}_J}$.

Finally suppose that $V = \Theta\left((\tilde{\tau}_0')^{1+\sum_{i=0}^{f-1} a_i p^1}\right)$ with f odd. Starting with

$$\sum_{i=0}^{f-1} m_i p^i + \sum_{i\in J^*} n_i p^i \equiv b_0 + b_2p^2 + \cdots b_{f-1}p^{f-1} \mod (q-1)$$
$$\sum_{i=0}^{f-1} m_i p^i + \sum_{i\notin J^*} n_i p^i \equiv 1 + b_1 p + b_3 p^3 + \cdots b_{f-1}p^{f-1} \mod (q-1),$$

and defining b_i' as in the preceding case, we get

$$\sum_{i=0}^{f-1}(-1)^i b_i' p^i \equiv \sum_{i=0}^{f-1}(-1)^i b_i p^i \mod (q-1),$$

again giving $\vec{b} = \vec{b}'$ since $r \neq s$. Checking again that

$$\sum_{i=0}^{f-1} m_i p^i \equiv \sum_{i=0}^{f-1} m_{J,i} p^i \bmod (q-1)$$

yields $V_{\vec{m},\vec{n}} = V_{\vec{m}_J,\vec{n}_J}$. □

Bibliography

[1] A. Ash, W. Sinnott, *An analogue of Serre's conjecture for Galois representations and Hecke eigenclasses in the mod p cohomology of* $\mathrm{GL}(n, \mathbf{Z})$, Duke Math. J. **105** (2000), 1–24.

[2] A. Ash, D. Doud, D. Pollack, *Galois representations with conjectural connections to arithmetic cohomology.*, Duke Math. J. **112** (2002), 521–579.

[3] C. Breuil, *Sur quelques représentations modulaires et p-adiques de* $\mathrm{GL}_2(\mathbb{Q}_p)$*. I*, Comp. Math. **138** (2003), 165–188.

[4] C. Breuil, *Sur quelques représentations modulaires et p-adiques de* $\mathrm{GL}_2(\mathbb{Q}_p)$*. II*, J. Inst. Math. Jussieu **2** (2003), 23–58.

[5] K. Buzzard, F. Diamond, A.F. Jarvis, *On Serre's conjecture for mod* ℓ *Galois representations over totally real fields*, preprint.

[6] F. Diamond, *The refined conjecture of Serre*, in Elliptic Curves and Fermat's Last Theorem, J.Coates and S.-T.Yau (eds.), Hong Kong 1993, Intl. Press, 2nd ed. (1997), 172–186.

[7] M. Emerton, *A local-global compatibility conjecture in the p-adic Langlands programme for* $\mathrm{GL}_2/\mathbb{Q}$, Pure Appl. Math. Q. **2** (2006), 279-393.

[8] F. Herzig, *The weight in a Serre-type conjecture for tame* n*-dimensional Galois representations*, Ph.D. Thesis, Harvard University, 2006.

[9] G. James, M. Liebeck, *Representations and Characters of Groups*, 2nd ed. Cambridge Univ. Press, 2001.

[10] S. Lang, *Algebra*, 3rd ed. Addison-Wesley, 1993, GTM **211**, Springer, 2002.

[11] K. Ribet, *On modular representations of* $\mathrm{Gal}(\overline{\mathbb{Q}}/\mathbb{Q})$ *arising from modular forms*, Inv. Math. **100** (1990), 431–476.

[12] J.-P. Serre, *Linear Representations of Finite Groups*, Springer-Verlag, 1977.

[13] J.-P. Serre, *Sur les représentations modulaires de degré 2 de* $\mathrm{Gal}(\overline{\mathbb{Q}}/\mathbb{Q})$, Duke Math. J. **54** (1987), 179–230.

[14] B. S. Upadhyaya, *Composition factors of the principal indecomposable modules for the special linear groups* $\mathrm{SL}(2, q)$, Jour. London Math. Soc. **17** (1978), 437-445.

Non-vanishing modulo p of Hecke L–values and application

Haruzo Hida

Department of Mathematics
UCLA
Los Angeles, Ca 90095-1555
U.S.A.
hida@math.ucla.edu [a]

[a] *The author is partially supported by the NSF grants: DMS 0244401 and DMS 0456252*

Contents

1 Introduction

Let F be a totally real field and M/F be a totally imaginary quadratic extension (a CM field). We fix a prime $p > 2$ unramified in $M/\mathbb{Q}$ and suppose that all prime factors of p in F split in M (M is p–ordinary). Fixing two embeddings $i_\infty : \overline{\mathbb{Q}} \hookrightarrow \mathbb{C}$ and $i_p : \overline{\mathbb{Q}} \hookrightarrow \overline{\mathbb{Q}}_p$, we take a p–ordinary CM type Σ of M. Thus $\Sigma_p = \{i_p \circ \sigma\}$ is exactly a half of the p–adic places of M. Fix a Hecke character λ of infinity type $k\Sigma + \kappa(1-c)$ with $0 < k \in \mathbb{Z}$ and $\kappa = \sum_{\sigma\in\Sigma} \kappa_\sigma \sigma$ with $\kappa_\sigma \geq 0$. If the conductor $\mathfrak{C}$ of λ is a product of primes split in M/F, we call λ has *split* conductor. Throughout this paper, we assume that λ has split conductor. We fix a prime $\mathfrak{l} \nmid \mathfrak{C}p$ of F. As is well known ([K] and [Sh1]), for a finite order Hecke character χ and for a power Ω of the Néron period of an abelian scheme (over a p–adic valuation ring) of CM type Σ, the L–value $\frac{L^{(p)}(0,\lambda\chi)}{\Omega}$ is (p–adically) integral (where the superscript: "(p)" indicates removal of Euler factors at p). The purpose of this paper is three fold:

(1) To prove non-vanishing modulo p of Hecke L–values $\frac{L^{(p)}(0,\lambda\chi)}{\Omega}$ for "almost all" anticyclotomic characters χ of finite order with $\mathfrak{l}$–power conductor (under some mild assumptions; Theorem 4.3);

(2) To prove the divisibility: $L_p^-(\psi)|\mathcal{F}^-(\psi)$ in the anticyclotomic Iwasawa algebra Λ^- of M for an anticyclotomic character ψ of split conductor, where $L_p^-(\psi)$ is the anticyclotomic Katz p–adic L–function of the branch character ψ and $\mathcal{F}^-(\psi)$ is the corresponding Iwasawa power series (see Theorem 5.1).

(3) To prove the equality $L_p^-(\psi) = \mathcal{F}^-(\psi)$ up to units under some assumptions if $F/\mathbb{Q}$ is an abelian extension (Theorem 5.8) and $M = F[\sqrt{D}]$ for $0 > D \in \mathbb{Z}$.

Roughly speaking, $\mathcal{F}^-(\psi)$ is the characteristic power series of the ψ-branch of the Galois group of the only Σ_p–ramified p–abelian extension of the anticyclotomic tower over the class field of ψ.

The first topic is a generalization of the result of Washington [Wa] (see also [Si]) to Hecke L–values, and the case where λ has conductor 1 has been dealt with in [H04c] basically by the same technique. The phrase "almost all" is in the sense of [H04c] and means "Zariski densely populated characters". If $\mathfrak{l}$ has degree 1 over $\mathbb{Q}$, we can prove a stronger non-vanishing modulo p outside a (non-specified) finite set. In [HT1] and [HT2], we have shown the divisibility in item (2) in $\Lambda^- \otimes_{\mathbb{Z}} \mathbb{Q}$ and indicated that the full divisibility holds except for p outside an explicit finite set S of primes if one obtains the result claimed in (1). We will show that S is limited to ramified primes and even primes.

Though the result in (2) is a direct consequence of the vanishing of the μ–invariant of $L_p^-(\psi)$ proven in [H04c] by the divisibility in $\Lambda^- \otimes_{\mathbb{Z}} \mathbb{Q}$, we shall give another proof of this fact using the non-vanishing (1). We will actually show a stronger result (Corollary 5.6) asserting that the relative class number $h(M/F)$ times $L_p^-(\psi)$ divides the congruence power series of the CM component of the nearly ordinary Hecke algebra (which does not directly follow from the vanishing of μ). Our method to achieve (2) is a refinement of the work [HT1] and [HT2], and this subtle process explains the length of the paper.

Once the divisibility (2) is established, under the assumption of (3), if ψ descends to a character of $\mathrm{Gal}(\overline{\mathbb{Q}}/\mathbb{Q}[\sqrt{D}])$, we can restrict $L_p^-(\psi)$ and $\mathcal{F}^-(\psi)$ to a $\mathbb{Z}_p$-extension of an abelian extension of $\mathbb{Q}[\sqrt{D}]$, and applying Rubin's identity of the restricted power series ([R] and [R1]), we conclude the identity $L_p^-(\psi) = \mathcal{F}^-(\psi)$.

We should mention that the stronger divisibility of the congruence power series by $h(M/F)L_p^-(\psi)$ in this paper will be used to prove the equality of $L^-(\psi)$ and $\mathcal{F}^-(\psi)$ under some mild conditions on ψ for general base fields F in our forthcoming paper [H04d]. We shall keep the notation and the assumptions introduced in this introduction throughout the paper.

2 Hilbert Modular Forms

We shall recall algebro-geometric theory of Hilbert modular forms limiting ourselves to what we need later.

2.1 Abelian variety with real multiplication

Let O be the integer ring of F, and put $O^* = \{x \in F | \mathrm{Tr}(xO) \subset \mathbb{Z}\}$ (which is the inverse different $\mathfrak{d}^{-1}$). We fix an integral ideal $\mathfrak{N}$ and a fractional ideal $\mathfrak{c}$ of F prime to $\mathfrak{N}$. We write A for a fixed base algebra, in which $N(\mathfrak{N})$ and $N(\mathfrak{c})$ is invertible. The Hilbert modular variety $\mathfrak{M}(\mathfrak{c};\mathfrak{N})$ of level $\mathfrak{N}$ classifies triples $(X, \Lambda, i)_{/S}$ formed by

- An abelian scheme $\pi : X \to S$ for an A–scheme S with an embedding: $O \hookrightarrow \mathrm{End}(X_{/S})$ making $\pi_*(\Omega_{X/S})$ a locally free $O \otimes \mathcal{O}_S$–module of rank 1;
- An O–linear polarization $\Lambda : X^t = \mathrm{Pic}^0_{X/S} \cong X \otimes \mathfrak{c}$;
- A closed O–linear immersion $i = i_{\mathfrak{N}} : (\mathbb{G}_m \otimes O^*)[\mathfrak{N}] \hookrightarrow X$.

By Λ, we identify the O–module of symmetric O–linear homomorphisms with $\mathfrak{c}$. Then we require that the (multiplicative) monoid of symmetric O–linear isogenies induced locally by ample invertible sheaves be identified with the set of totally positive elements $\mathfrak{c}_+ \subset \mathfrak{c}$. Thus $\mathfrak{M}(\mathfrak{c};\mathfrak{N})_{/A}$ is the coarse moduli scheme of the following functor from the category of A–schemes into the category $SETS$:

$$\mathcal{P}(S) = \left[(X,\Lambda,i)_{/S}\right],$$

where $[\ ] = \{\ \}/\cong$ is the set of isomorphism classes of the objects inside the brackets, and we call $(X,\Lambda,i) \cong (X',\Lambda',i')$ if we have an O–linear isomorphism $\phi : X_{/S} \to X'_{/S}$ such that $\Lambda' = \phi \circ \Lambda \circ \phi^t$ and $\phi \circ i = i'$. The scheme $\mathfrak{M}$ is a fine moduli if $\mathfrak{N}$ is sufficiently deep. In [K] and [HT1], the moduli $\mathfrak{M}$ is described as an algebraic space, but it is actually a quasi-projective scheme (e.g. [C], [H04a] Lectures 5 and 6 and [PAF] Chapter 4).

2.2 Abelian varieties with complex multiplication

We write $|\cdot|_p$ for the p–adic absolute value of $\overline{\mathbb{Q}}_p$ and $\widehat{\mathbb{Q}}_p$ for the p–adic completion of $\overline{\mathbb{Q}}_p$ under $|\cdot|_p$. Recall the p–ordinary CM type (M,Σ), and let R be the integer ring of M. Thus $\Sigma \sqcup \Sigma c$ for the generator c of $\mathrm{Gal}(M/F)$ gives the set of all embeddings of M into $\overline{\mathbb{Q}}$. For each $\sigma \in (\Sigma \cup \Sigma c)$, $i_p\sigma$ induces a p–adic place $\mathfrak{p}_\sigma$ giving rise to the p–adic absolute value $|x|_{\mathfrak{p}_\sigma} = |i_p(\sigma(x))|_p$. We write $\Sigma_p = \{\mathfrak{p}_\sigma | \sigma \in \Sigma\}$ and $\Sigma_p c = \{\mathfrak{p}_{\sigma c} | \sigma \in \Sigma\}$. By ordinarity, we have $\Sigma_p \cap \Sigma_p c = \varnothing$.

For each O–lattice $\mathfrak{a} \subset M$ whose p–adic completion $\mathfrak{a}_p$ is identical to $R_p = R \otimes_{\mathbb{Z}} \mathbb{Z}_p$, we consider the complex torus $X(\mathfrak{a})(\mathbb{C}) = \mathbb{C}^\Sigma/\Sigma(\mathfrak{a})$, where $\Sigma(\mathfrak{a}) = \{(i_\infty(\sigma(a)))_{\sigma\in\Sigma} | a \in \mathfrak{a}\}$. By a theorem in [ACM] 12.4, this complex torus is algebraizable to an abelian variety $X(\mathfrak{a})$ of CM type (M,Σ) over a number field.

Let $\mathbb{F}$ be an algebraic closure of the finite field $\mathbb{F}_p$ of p–elements. We write $\widehat{W}$ for the p–adically closed discrete valuation ring inside $\widehat{\mathbb{Q}}_p$ unramified over $\mathbb{Z}_p$ with residue field $\mathbb{F}$. Thus $\widehat{W}$ is isomorphic to the ring of Witt vectors with coefficients in $\mathbb{F}$. Let $\mathcal{W} = i_p^{-1}(\widehat{W})$, which is a strict henselization of $\mathbb{Z}_{(p)} = \mathbb{Q} \cap \mathbb{Z}_p$. In general, we write W for a finite extension of $\widehat{W}$ in $\widehat{\mathbb{Q}}_p$, which is a complete discrete valuation ring. We suppose that p is unramified in $M/\mathbb{Q}$. Then the main theorem of complex multiplication ([ACM] 18.6) combined with the criterion of good reduction over $\widehat{W}$ [ST] tells us that $X(\mathfrak{a})$ is actually defined over the field of fractions $\mathcal{K}$ of $\mathcal{W}$ and extends to an abelian scheme over $\mathcal{W}$ (still written as $X(\mathfrak{a})_{/\mathcal{W}}$). All endomorphisms of $X(\mathfrak{a})_{/\mathcal{W}}$ are defined over $\mathcal{W}$. We write $\theta : M \hookrightarrow \mathrm{End}(X(\mathfrak{a})) \otimes_{\mathbb{Z}} \mathbb{Q}$ for the embedding of M taking $\alpha \in M$ to the complex multiplication by $\Sigma(\alpha)$ on $X(\mathfrak{a})(\mathbb{C}) = \mathbb{C}^\Sigma/\Sigma(\mathfrak{a})$.

Let $R(\mathfrak{a}) = \{\alpha \in R | \alpha\mathfrak{a} \subset \mathfrak{a}\}$. Then $R(\mathfrak{a})$ is an order of M over O. Recall the prime $\mathfrak{l} \nmid p$ of F in the introduction. The order $R(\mathfrak{a})$ is determined by its conductor ideal which we assume to be an $\mathfrak{l}$–power $\mathfrak{l}^e$. In other words, $R(\mathfrak{a}) = R_e := O + \mathfrak{l}^e R$. The following three conditions for a fractional R_e–ideal $\mathfrak{a}$ are equivalent (cf. [IAT] Proposition 4.11 and (5.4.2) and [CRT] Theorem 11.3):

(I1) $\mathfrak{a}$ is R_e–projective;
(I2) $\mathfrak{a}$ is locally principal;
(I3) $\mathfrak{a}$ is a proper R_e–ideal (that is, $R_e = R(\mathfrak{a})$).

Thus $Cl_e := \mathrm{Pic}(R_e)$ is the group of R_e–projective fractional ideals modulo principal ideals. The group Cl_e is finite and called the ring class group modulo $\mathfrak{l}^e$.

We choose and fix a differential $\omega = \omega(R)$ on $X(R)_{/\mathcal{W}}$ so that

$$H^0(X(R), \Omega_{X(R)/\mathcal{W}}) = (\mathcal{W} \otimes_{\mathbb{Z}} O)\omega.$$

If $\mathfrak{a}_p = R_p$, $X(R \cap \mathfrak{a})$ is an étale covering of both $X(\mathfrak{a})$ and $X(R)$; so, $\omega(R)$ induces a differential $\omega(\mathfrak{a})$ first by pull-back to $X(R\cap\mathfrak{a})$ and then by pull-back inverse from $X(R \cap \mathfrak{a})$ to $X(\mathfrak{a})$. As long as the projection $\pi : X(R \cap \mathfrak{a}) \twoheadrightarrow X(\mathfrak{a})$ is étale, the pull-back inverse $(\pi^*)^{-1} : \Omega_{X(R\cap\mathfrak{a})/\mathcal{W}} \to \Omega_{X(\mathfrak{a})/\mathcal{W}}$ is a surjective isomorphism. We thus have

$$H^0(X(\mathfrak{a}), \Omega_{X(R)/\mathcal{W}}) = (\mathcal{W} \otimes_{\mathbb{Z}} O)\omega(\mathfrak{a}).$$

We choose a totally imaginary $\delta \in M$ with $\mathrm{Im}(i_\infty(\sigma(\delta))) > 0$ for all $\sigma \in \Sigma$ such that $(a, b) \mapsto (c(a)b - ac(b))/2\delta$ gives the identification $R \wedge R \cong \mathfrak{d}^{-1}\mathfrak{c}^{-1}$. We assume that $\mathfrak{c}$ is prime to $p\ell$ ($(\ell) = \mathfrak{l} \cap \mathbb{Z}$). This Riemann form: $R \wedge R \cong \mathfrak{c}^* = \mathfrak{d}^{-1}\mathfrak{c}^{-1}$ gives rise to a $\mathfrak{c}$–polarization $\Lambda = \Lambda(R) : X(R)^t \cong X(R) \otimes \mathfrak{c}$, which is again defined over $\mathcal{W}$. Here $\mathfrak{d}$ is the different of $F/\mathbb{Q}$, and $\mathfrak{c}^* = \{x \in F | \mathrm{Tr}_{F/\mathbb{Q}}(x\mathfrak{c}) \subset \mathbb{Z}\}$. Since we have $R_e \wedge R_e = \mathfrak{l}^e(O \wedge R) + \mathfrak{l}^{2e}(R \wedge R)$, the pairing induces $R_e \wedge R_e \cong (\mathfrak{c}\mathfrak{l}^{-e})^*$, and this pairing induces a $\mathfrak{c}\mathfrak{l}^{-e}N_{M/F}(\mathfrak{a})^{-1}$–polarization $\Lambda(\mathfrak{a})$ on $X(\mathfrak{a})$ for a proper R_e–ideal $\mathfrak{a}$.

We choose a local generator a of $\mathfrak{a}_\mathfrak{l}$. Multiplication by a induces an isomorphism $R_{e,\mathfrak{l}} \cong \mathfrak{a}_\mathfrak{l}$. Since $X(R_e)_{/\mathcal{W}}$ has a subgroup $C(R_e) = R/(O + \mathfrak{l}^e R) \subset X(R_e)$ isomorphic étale-locally to $O/\mathfrak{l}^e$. This subgroup $C(R_e)$ is sent by multiplication by a to $C(\mathfrak{a}) \subset X(\mathfrak{a})_{/\mathcal{W}}$, giving rise to a $\Gamma_0(\mathfrak{l}^e)$–level structure $C(\mathfrak{a})$ on $X(\mathfrak{a})$.

For our later use, we choose ideals $\mathfrak{F}$ and $\mathfrak{F}_c$ of R prime to $\mathfrak{c}$ so that $\mathfrak{F} \subset \mathfrak{F}_c^c$ and $\mathfrak{F} + \mathfrak{F}_c = R$. The product $\mathfrak{C} = \mathfrak{F}\mathfrak{F}_c$ shall be later the conductor of the Hecke character we study. We put $\mathfrak{f} = \mathfrak{F} \cap O$ and $\mathfrak{f}' = \mathfrak{F}_c \cap O$; so, $\mathfrak{f} \subset \mathfrak{f}'$. We shall define a level $\mathfrak{f}^2$–structure on $X(\mathfrak{a})$: supposing that $\mathfrak{a}$ is prime to $\mathfrak{f}$, we have $\mathfrak{a}_\mathfrak{f} \cong R_\mathfrak{f} = R_\mathfrak{F} \times R_{\mathfrak{F}^c}$, which induces a canonical identification

$$i(\mathfrak{a}) : \mathfrak{f}^*/O^* = \mathfrak{f}^{-1}/O \cong \mathfrak{F}^{-1}/R \cong \mathfrak{F}^{-1}\mathfrak{a}_\mathfrak{F}/\mathfrak{a}_\mathfrak{F} \subset X(\mathfrak{a})[\mathfrak{f}]. \tag{2.1}$$

This level structure induces $i'(\mathfrak{a}) : X(\mathfrak{a})[\mathfrak{f}] \twoheadrightarrow \mathfrak{f}^{-1}/O$ by the duality under Λ. In this way, we get many sextuples:

$$(X(\mathfrak{a}), \Lambda(\mathfrak{a}), i(\mathfrak{a}), i'(\mathfrak{a}), C(\mathfrak{a})[\mathfrak{l}], \omega(\mathfrak{a})) \in \mathcal{M}(\mathfrak{c}\mathfrak{l}^{-e}(\mathfrak{a}\mathfrak{a}^c)^{-1}; \mathfrak{f}^2, \Gamma_0(\mathfrak{l}))(\mathcal{W}) \tag{2.2}$$

as long as $\mathfrak{l}^e$ is prime to p, where $C(\mathfrak{a})[\mathfrak{l}] = \{x \in C(\mathfrak{a}) | \mathfrak{l}x = 0\}$. A precise definition of the moduli scheme of Γ_0–type: $\mathcal{M}(\mathfrak{c}\mathfrak{l}^{-e}(\mathfrak{a}\mathfrak{a}^c)^{-1}; \mathfrak{f}^2, \Gamma_0(\mathfrak{l}))$ classifying such sextuples will be given in 2.7. The point $x(\mathfrak{a}) = (X(\mathfrak{a}), \Lambda(\mathfrak{a}), i(\mathfrak{a}), i'(\mathfrak{a}))$ of the moduli scheme $\mathfrak{M}(\mathfrak{c}(\mathfrak{a}\mathfrak{a}^c)^{-1}; \mathfrak{f}^2)$ is called a *CM point* associated to $X(\mathfrak{a})$.

2.3 Geometric Hilbert modular forms

We return to the functor $\mathcal{P}$ in 2.1. We could insist on freeness of the differentials $\pi_*(\Omega_{X/S})$, and for ω with $\pi_*(\Omega_{X/S}) = (\mathcal{O}_S \otimes_{\mathbb{Z}} O)\omega$, we consider the functor classifying quadruples (X, Λ, i, ω):

$$\mathcal{Q}(S) = \left[(X, \Lambda, i, \omega)_{/S}\right].$$

Let $T = \mathrm{Res}_{O/\mathbb{Z}}\mathbb{G}_m$. We let $a \in T(S) = H^0(S, (\mathcal{O}_S \otimes_{\mathbb{Z}} O)^\times)$ act on $\mathcal{Q}(S)$ by $(X, \Lambda, i, \omega) \mapsto (X, \Lambda, i, a\omega)$. By this action, $\mathcal{Q}$ is a T–torsor over $\mathcal{P}$; so, $\mathcal{Q}$ is representable by an A–scheme $\mathcal{M} = \mathcal{M}(\mathfrak{c}; \mathfrak{N})$ affine over $\mathfrak{M} = \mathfrak{M}(\mathfrak{c}; \mathfrak{N})_{/A}$. For each character $k \in X^*(T) = \mathrm{Hom}_{gp-sch}(T, \mathbb{G}_m)$, if $F \neq \mathbb{Q}$, the k^{-1}–eigenspace of $H^0(\mathcal{M}_{/A}, \mathcal{O}_{\mathcal{M}/A})$ is by definition the space of modular forms of weight k integral over A. We write $G_k(\mathfrak{c}, \mathfrak{N}; A)$ for this space of A–integral modular forms, which is an A–module of finite type. When $F = \mathbb{Q}$, as is well known, we need to take the subsheaf of sections with logarithmic growth towards cusps (the condition (G0) below). Thus $f \in G_k(\mathfrak{c}, \mathfrak{N}; A)$ is a functorial rule assigning a value in B to each isomorphism class of $(X, \Lambda, i, \omega)_{/B}$ (defined over an A–algebra B) satisfying the following three conditions:

- (G1) $f(X, \Lambda, i, \omega) \in B$ if (X, Λ, i, ω) is defined over B;
- (G2) $f((X, \Lambda, i, \omega) \otimes_B B') = \rho(f(X, \Lambda, i, \omega))$ for each morphism $\rho : B_{/A} \to B'_{/A}$;
- (G3) $f(X, \Lambda, i, a\omega) = k(a)^{-1} f(X, \Lambda, i, \omega)$ for $a \in T(B)$.

By abusing the language, we pretend f to be a function of isomorphism classes of test objects $(X, \Lambda, i, \omega)_{/B}$ hereafter. The sheaf of k^{-1}–eigenspace $\mathcal{O}_{\mathcal{M}}[k^{-1}]$ under the action of T is an invertible sheaf on $\mathfrak{M}_{/A}$. We write this sheaf as $\underline{\omega}^k$ (imposing (G0) when $F = \mathbb{Q}$). Then we have

$$G_k(\mathfrak{c}, \mathfrak{N}; A) = H^0(\mathfrak{M}(\mathfrak{c}; \mathfrak{N})_{/A}, \underline{\omega}^k_{/A})$$

as long as $\mathfrak{M}(\mathfrak{c};\mathfrak{N})$ is a fine moduli space. Writing $\underline{\mathbb{X}} = (\mathbb{X}, \boldsymbol{\lambda}, \mathbf{i}, \boldsymbol{\omega})$ for the universal abelian scheme over $\mathfrak{M}$, $s = f(\underline{\mathbb{X}})\boldsymbol{\omega}^k$ gives rise to the section of $\underline{\omega}^k$. Conversely, for any section $s \in H^0(\mathfrak{M}(\mathfrak{c};\mathfrak{N}), \underline{\omega}^k)$, taking a unique morphism $\phi : \mathrm{Spec}(B) \to \mathfrak{M}$ such that $\phi^*\underline{\mathbb{X}} = \underline{X}$ for $\underline{X} = (X, \Lambda, i, \omega)_{/B}$, we can define $f \in G_k$ by $\phi^* s = f(\underline{X})\omega^k$.

We suppose that the fractional ideal $\mathfrak{c}$ is prime to $\mathfrak{N}p$, and take two ideals $\mathfrak{a}$ and $\mathfrak{b}$ prime to $\mathfrak{N}p$ such that $\mathfrak{a}\mathfrak{b}^{-1} = \mathfrak{c}$. To this pair $(\mathfrak{a}, \mathfrak{b})$, we can attach the Tate AVRM $Tate_{\mathfrak{a},\mathfrak{b}}(q)$ defined over the completed group ring $\mathbb{Z}((\mathfrak{a}\mathfrak{b}))$ made of formal series $f(q) = \sum_{\xi \gg -\infty} a(\xi) q^\xi$ $(a(\xi) \in \mathbb{Z})$. Here ξ runs over all elements in $\mathfrak{a}\mathfrak{b}$, and there exists a positive integer n (dependent on f) such that $a(\xi) = 0$ if $\sigma(\xi) < -n$ for some $\sigma \in I$. We write $A[[(\mathfrak{a}\mathfrak{b})_{\geq 0}]]$ for the subring of $A[[\mathfrak{a}\mathfrak{b}]]$ made of formal series f with $a(\xi) = 0$ for all ξ with $\sigma(\xi) < 0$ for at least one embedding $\sigma : F \hookrightarrow \mathbb{R}$. Actually, we skipped a step of introducing the toroidal compactification of $\mathfrak{M}$ whose (completed) stalk at the cusp corresponding to $(\mathfrak{a}, \mathfrak{b})$ actually carries $Tate_{\mathfrak{a},\mathfrak{b}}(q)$. However to make exposition short, we ignore this technically important point, referring the reader to the treatment in [K] Chapter I, [C], [DiT], [Di], [HT1] Section 1 and [PAF] 4.1.4. The scheme $Tate(q)$ can be extended to a semi-abelian scheme over $\mathbb{Z}[[(\mathfrak{a}\mathfrak{b})_{\geq 0}]]$ adding the fiber $\mathbb{G}_m \otimes \mathfrak{a}^*$. Since $\mathfrak{a}$ is prime to p, $\mathfrak{a}_p = O_p$. Thus if A is a $\mathbb{Z}_p$–algebra, we have a canonical isomorphism:

$$Lie(Tate_{\mathfrak{a},\mathfrak{b}}(q) \bmod \mathfrak{A}) = Lie(\mathbb{G}_m \otimes \mathfrak{a}^*) \cong A \otimes_{\mathbb{Z}} \mathfrak{a}^* \cong A \otimes_{\mathbb{Z}} O^*.$$

By Grothendieck-Serre duality, we have $\Omega_{Tate_{\mathfrak{a},\mathfrak{b}}(q)/A[[(\mathfrak{a}\mathfrak{b})_{\geq 0}]]} \cong A[[(\mathfrak{a}\mathfrak{b})_{\geq 0}]]$. Indeed we have a canonical generator ω_{can} of $\Omega_{Tate(q)}$ induced by $\frac{dt}{t} \otimes 1$ on $\mathbb{G}_m \otimes \mathfrak{a}^*$. We have a canonical inclusion $(\mathbb{G}_m \otimes O^*)[\mathfrak{N}] = (\mathbb{G}_m \otimes \mathfrak{a}^*)[\mathfrak{N}]$ into $\mathbb{G}_m \otimes \mathfrak{a}^*$, which induces a canonical closed immersion $i_{can} : (\mathbb{G}_m \otimes O^*)[\mathfrak{N}] \hookrightarrow Tate(q)$. As described in [K] (1.1.14) and [HT1] page 204, $Tate_{\mathfrak{a},\mathfrak{b}}(q)$ has a canonical $\mathfrak{c}$–polarization Λ_{can}. Thus we can evaluate $f \in G_k(\mathfrak{c}, \mathfrak{N}; A)$ at $(Tate_{\mathfrak{a},\mathfrak{b}}(q), \Lambda_{can}, i_{can}, \omega_{can})$. The value $f(q) = f_{\mathfrak{a},\mathfrak{b}}(q)$ actually falls in $A[[(\mathfrak{a}\mathfrak{b})_{\geq 0}]]$ (if $F \neq \mathbb{Q}$: Koecher principle) and is called the q–expansion at the cusp $(\mathfrak{a}, \mathfrak{b})$. When $F = \mathbb{Q}$, we impose f to have values in the power series ring $A[[(\mathfrak{a}\mathfrak{b})_{\geq 0}]]$ when we define modular forms:

(G0) $f_{\mathfrak{a},\mathfrak{b}}(q) \in A[[(\mathfrak{a}\mathfrak{b})_{\geq 0}]]$ for all $(\mathfrak{a}, \mathfrak{b})$.

2.4 *p*–Adic Hilbert modular forms

Suppose that $A = \varprojlim_n A/p^n A$ and that $\mathfrak{N}$ is prime to p. We can think of a functor

$$\widehat{\mathcal{P}}(A) = \left[(X, \Lambda, i_p, i_{\mathfrak{N}})_{/S}\right]$$

similar to $\mathcal{P}$ that is defined over the category of p–adic A–algebras $B = \varprojlim_n B/p^n B$. An important point is that we consider an isomorphism of ind-group schemes $i_p : \mu_{p^\infty} \otimes_{\mathbb{Z}} O^* \hookrightarrow X[p^\infty]$ (in place of a differential ω), which induces $\widehat{\mathbb{G}}_m \otimes O^* \cong \widehat{X}$ for the formal completion $\widehat{V}$ at the characteristic p–fiber of a scheme V over A.

It is a theorem (due to Deligne-Ribet and Katz) that this functor is representable by the formal Igusa tower over the formal completion $\widehat{\mathfrak{M}}(\mathfrak{c};\mathfrak{N})$ of $\mathfrak{M}(\mathfrak{c};\mathfrak{N})$ along the ordinary locus of the modulo p fiber (e.g., [PAF] 4.1.9). A p–adic modular form $f_{/A}$ for a p–adic ring A is a function (strictly speaking, a functorial rule) of isomorphism classes of $(X,\Lambda,i_p,i_{\mathfrak{N}})_{/B}$ satisfying the following three conditions:

(P1) $f(X,\Lambda,i_p,i_{\mathfrak{N}}) \in B$ if $(X,\Lambda,i_p,i_{\mathfrak{N}})$ is defined over B;
(P2) $f((X,\Lambda,i_p,i_{\mathfrak{N}}) \otimes_B B') = \rho(f(X,\Lambda,i_p,i_{\mathfrak{N}}))$ for each continuous A–algebra homomorphism $\rho : B \to B'$;
(P3) $f_{\mathfrak{a},\mathfrak{b}}(q) \in A[[(\mathfrak{a}\mathfrak{b})_{\geq 0}]]$ for all $(\mathfrak{a},\mathfrak{b})$ prime to $\mathfrak{N}p$.

We write $V(\mathfrak{c},\mathfrak{N};A)$ for the space of p–adic modular forms satisfying (P1-3). This $V(\mathfrak{c},\mathfrak{N};A)$ is a p–adically complete A–algebra.

We have the q–expansion principle valid both for classical modular forms and p–adic modular forms f,

(q-exp) *f is uniquely determined by the q–expansion: $f \mapsto f_{\mathfrak{a},\mathfrak{b}}(q) \in A[[(\mathfrak{a}\mathfrak{b})_{\geq 0}]]$.*

This follows from the irreducibility of (the Hilbert modular version of) the Igusa tower proven in [DeR] (see also [PAF] 4.2.4).

Since $\widehat{\mathbb{G}}_m \otimes O^*$ has a canonical invariant differential $\frac{dt}{t}$, we have $\omega_p = i_{p,*}(\frac{dt}{t})$ on X. This allows us to regard $f \in G_k(\mathfrak{c},\mathfrak{N};A)$ a p–adic modular form by

$$f(X,\Lambda,i_p,i_{\mathfrak{N}}) := f(X,\Lambda,i_{\mathfrak{N}},\omega_p).$$

By (q-exp), this gives an injection of $G_k(\mathfrak{c},\mathfrak{N};A)$ into the space of p–adic modular forms $V(\mathfrak{c},\mathfrak{N};A)$ (for a p–adic ring A) preserving q–expansions.

2.5 Complex analytic Hilbert modular forms

Over $\mathbb{C}$, the category of test objects (X,Λ,i,ω) is equivalent to the category of triples $(\mathcal{L},\Lambda,i)$ made of the following data (by the theory of theta functions): $\mathcal{L}$ is an O–lattice in $O \otimes_{\mathbb{Z}} \mathbb{C} = \mathbb{C}^I$, an alternating pairing $\Lambda : \mathcal{L} \wedge_O \mathcal{L} \cong \mathfrak{c}^*$ and $i : \mathfrak{N}^*/O^* \hookrightarrow F\mathcal{L}/\mathcal{L}$. The alternating form Λ is supposed to be positive in the sense that $\Lambda(u,v)/\operatorname{Im}(uv^c)$ is totally positive definite. The differential ω can be recovered by $\iota : X(\mathbb{C}) = \mathbb{C}^I/\mathcal{L}$ so that $\omega = \iota^* du$ where $u = (u_\sigma)_{\sigma\in I}$ is the variable on $\mathbb{C}^I$. Conversely

$$\mathcal{L}_X = \left\{ \int_\gamma \omega \in O \otimes_{\mathbb{Z}} \mathbb{C} \middle| \gamma \in H_1(X(\mathbb{C}), \mathbb{Z}) \right\}$$

is a lattice in $\mathbb{C}^I$, and the polarization $\Lambda : X^t \cong X \otimes \mathfrak{c}$ induces $\mathcal{L} \wedge \mathcal{L} \cong \mathfrak{c}^*$.

Using this equivalence, we can relate our geometric definition of Hilbert modular forms with the classical analytic definition. Define $\mathfrak{Z}$ by the product of I copies of the upper half complex plane $\mathfrak{H}$. We regard $\mathfrak{Z} \subset F \otimes_{\mathbb{Q}} \mathbb{C} = \mathbb{C}^I$ made up of $z = (z_\sigma)_{\sigma\in I}$ with totally positive imaginary part. For each $z \in \mathfrak{Z}$, we define

$$\mathcal{L}_z = 2\pi\sqrt{-1}(\mathfrak{b}z + \mathfrak{a}^*),$$
$$\Lambda_z(2\pi\sqrt{-1}(az+b), 2\pi\sqrt{-1}(cz+d)) = -(ad-bc) \in \mathfrak{c}^*$$

with $i_z : \mathfrak{N}^*/O^* \to \mathbb{C}^I/\mathcal{L}_z$ given by $i_z(a \mod O^*) = (2\pi\sqrt{-1}a \mod \mathcal{L}_z)$.

Consider the following congruence subgroup $\Gamma_1^1(\mathfrak{N}; \mathfrak{a}, \mathfrak{b})$ given by

$$\left\{ \left(\begin{smallmatrix} a & b \\ c & d \end{smallmatrix}\right) \in SL_2(F) \middle| a, d \in O,\ b \in (\mathfrak{a}\mathfrak{b})^*,\ c \in \mathfrak{N}\mathfrak{a}\mathfrak{b}\mathfrak{d} \text{ and } d - 1 \in \mathfrak{N} \right\}.$$

We let $g = (g_\sigma) \in SL_2(F \otimes_{\mathbb{Q}} \mathbb{R}) = SL_2(\mathbb{R})^I$ act on $\mathfrak{Z}$ by linear fractional transformation of g_σ on each component z_σ. It is easy to verify

$$(\mathcal{L}_z, \Lambda_z, i_z) \cong (\mathcal{L}_w, \Lambda_w, i_w) \iff w = \gamma(z) \text{ for } \gamma \in \Gamma_1^1(\mathfrak{N}; \mathfrak{a}, \mathfrak{b}).$$

The set of pairs $(\mathfrak{a}, \mathfrak{b})$ with $\mathfrak{a}\mathfrak{b}^{-1} = \mathfrak{c}$ is in bijection with the set of cusps (unramified over ∞) of $\Gamma_1^1(\mathfrak{N}; \mathfrak{a}, \mathfrak{b})$. Two cusps are equivalent if they transform each other by an element in $\Gamma_1^1(\mathfrak{N}; \mathfrak{a}, \mathfrak{b})$. The standard choice of the cusp is $(O, \mathfrak{c}^{-1})$, which we call the infinity cusp of $\mathfrak{M}(\mathfrak{c}; \mathfrak{N})$. Write $\Gamma_1^1(\mathfrak{c}; \mathfrak{N}) = \Gamma_1^1(\mathfrak{N}; O, \mathfrak{c}^{-1})$. For each ideal $\mathfrak{t}$, $(\mathfrak{t}, \mathfrak{t}\mathfrak{c}^{-1})$ gives another cusp. The two cusps $(\mathfrak{t}, \mathfrak{t}\mathfrak{c}^{-1})$ and $(\mathfrak{s}, \mathfrak{s}\mathfrak{c}^{-1})$ are equivalent under $\Gamma_1^1(\mathfrak{c}; \mathfrak{N})$ if $\mathfrak{t} = \alpha\mathfrak{s}$ for an element $\alpha \in F^\times$ with $\alpha \equiv 1 \mod \mathfrak{N}$ in $F_{\mathfrak{N}}^\times$. We have

$$\mathfrak{M}(\mathfrak{c}; \mathfrak{N})(\mathbb{C}) \cong \Gamma_1^1(\mathfrak{c}; \mathfrak{N})\backslash\mathfrak{Z}, \text{ canonically.}$$

Let $G = \mathrm{Res}_{O/\mathbb{Z}} GL(2)$. Take the following open compact subgroup of $G(\mathbb{A}^{(\infty)})$:

$$U_1^1(\mathfrak{N}) = \left\{ \left(\begin{smallmatrix} a & b \\ c & d \end{smallmatrix}\right) \in G(\widehat{\mathbb{Z}}) \middle| c \in \mathfrak{N}\widehat{O} \text{ and } a \equiv d \equiv 1 \mod \mathfrak{N}\widehat{O} \right\},$$

and put $K = K_1^1(\mathfrak{N}) = \left(\begin{smallmatrix} d & 0 \\ 0 & 1 \end{smallmatrix}\right)^{-1} U_1^1(\mathfrak{N}) \left(\begin{smallmatrix} d & 0 \\ 0 & 1 \end{smallmatrix}\right)$ for an idele d with $dO = \mathfrak{d}$ and $d^{(\mathfrak{d})} = 1$. Then taking an idele c with $c\widehat{O} = \widehat{\mathfrak{c}}$ and $c^{(\mathfrak{c})} = 1$, we see that

$$\Gamma_1^1(\mathfrak{c}; \mathfrak{N}) \subset \left(\left(\begin{smallmatrix} c & 0 \\ 0 & 1 \end{smallmatrix}\right) K \left(\begin{smallmatrix} c & 0 \\ 0 & 1 \end{smallmatrix}\right)^{-1} \cap G(\mathbb{Q})_+ \right) \subset O^\times \Gamma_1^1(\mathfrak{c}; \mathfrak{N})$$

for $G(\mathbb{Q})_+$ made up of all elements in $G(\mathbb{Q})$ with totally positive determinant. Choosing a complete representative set $\{c\} \subset F_{\mathbb{A}}^{\times}$ for the strict ray class group $Cl_F^+(\mathfrak{N})$ modulo $\mathfrak{N}$, we find by the approximation theorem that

$$G(\mathbb{A}) = \bigsqcup_{c \in Cl_F^+(\mathfrak{N})} G(\mathbb{Q}) \left(\begin{smallmatrix} c & 0 \\ 0 & 1 \end{smallmatrix}\right) K \cdot G(\mathbb{R})^+$$

for the identity connected component $G(\mathbb{R})^+$ of the Lie group $G(\mathbb{R})$. This shows

$$G(\mathbb{Q})\backslash G(\mathbb{A})/KC_{\mathbf{i}} \cong G(\mathbb{Q})_+\backslash G(\mathbb{A})_+/KC_{\mathbf{i}} \cong \bigsqcup_{\mathfrak{c} \in Cl_F^+(\mathfrak{N})} \mathfrak{M}(\mathfrak{c};\mathfrak{N})(\mathbb{C}), \quad (2.3)$$

where $G(\mathbb{A})_+ = G(\mathbb{A}^{(\infty)})G(\mathbb{R})^+$ and $C_{\mathbf{i}}$ is the stabilizer in $G(\mathbb{R})^+$ of $\mathbf{i} = (\sqrt{-1}\ldots,\sqrt{-1}) \in \mathfrak{Z}$. By (2.3), a $Cl_F^+(\mathfrak{N})$–tuple $(f_\mathfrak{c})_\mathfrak{c}$ with $f_\mathfrak{c} \in G_k(\mathfrak{c},\mathfrak{N};\mathbb{C})$ can be viewed as a single automorphic form defined on $G(\mathbb{A})$.

Recall the identification of $X^*(T)$ with $\mathbb{Z}[I]$ so that $k(x) = \prod_\sigma \sigma(x)^{k_\sigma}$. Regarding $f \in G_k(\mathfrak{c},\mathfrak{N};\mathbb{C})$ as a holomorphic function of $z \in \mathfrak{Z}$ by $f(z) = f(\mathcal{L}_z,\Lambda_z,i_z)$, it satisfies the following automorphic property:

$$f(\gamma(z)) = f(z)\prod_\sigma (c^\sigma z_\sigma + d^\sigma)^{k_\sigma} \text{ for all } \gamma = \left(\begin{smallmatrix} a & b \\ c & d \end{smallmatrix}\right) \in \Gamma_1^1(\mathfrak{c};\mathfrak{N}). \quad (2.4)$$

The holomorphy of f is a consequence of the functoriality (G2). The function f has the Fourier expansion

$$f(z) = \sum_{\xi \in (\mathfrak{ab})_{\geq 0}} a(\xi)\mathbf{e}_F(\xi z)$$

at the cusp corresponding to $(\mathfrak{a},\mathfrak{b})$. Here $\mathbf{e}_F(\xi z) = \exp(2\pi\sqrt{-1}\sum_\sigma \xi^\sigma z_\sigma)$. This Fourier expansion gives the q–expansion $f_{\mathfrak{a},\mathfrak{b}}(q)$ substituting q^ξ for $\mathbf{e}_F(\xi z)$.

2.6 Differential operators

Shimura studied the effect on modular forms of the following differential operators on $\mathfrak{Z}$ indexed by $k \in \mathbb{Z}[I]$:

$$\delta_k^\sigma = \frac{1}{2\pi\sqrt{-1}}\left(\frac{\partial}{\partial z_\sigma} + \frac{k_\sigma}{2y_\sigma\sqrt{-1}}\right) \text{ and } \delta_k^r = \prod_\sigma \left(\delta_{k_\sigma+2r_\sigma-2}^\sigma \cdots \delta_{k_\sigma}^\sigma\right), \quad (2.5)$$

where $r \in \mathbb{Z}[I]$ with $r_\sigma \geq 0$. An important point is that the differential operator preserves rationality property at CM points of (arithmetic) modular forms, although it does not preserve holomorphy (see [AAF] III and [Sh1]). We shall describe the rationality. The complex uniformization $\iota : X(\mathfrak{a})(\mathbb{C}) \cong \mathbb{C}^\Sigma/\Sigma(\mathfrak{a})$

induces a canonical base $\omega_\infty = \iota^* du$ of $\Omega_{X(\mathfrak{a})/\mathbb{C}}$ over $R \otimes_{\mathbb{Z}} \mathbb{R}$, where $u = (u_\sigma)_{\sigma \in \Sigma}$ is the standard variable on $\mathbb{C}^\Sigma$. Define a period $\Omega_\infty \in \mathbb{C}^\Sigma = O \otimes_{\mathbb{Z}} \mathbb{C}$ by $\omega(R) = \omega(\mathfrak{a}) = \Omega_\infty \omega_\infty$. Here the first identity follows from the fact that $\omega(\mathfrak{a})$ is induced by $\omega(R)$ on $X(R)$. We suppose that $\mathfrak{a}$ is prime to p. Here is the rationality result of Shimura for $f \in G_k(\mathfrak{c}, \mathfrak{f}^2; \mathcal{W})$:

$$\frac{(\delta_k^r f)(x(\mathfrak{a}), \omega_\infty)}{\Omega_\infty^{k+2r}} = (\delta_k^r f)(x(\mathfrak{a}), \omega(\mathfrak{a})) \in \overline{\mathbb{Q}}. \tag{S}$$

Katz interpreted the differential operator in terms of the Gauss-Manin connection of the universal AVRM over $\mathfrak{M}$ and gave a purely algebro-geometric definition of the operator (see [K] Chapter II and [HT1] Section 1). Using this algebraization of δ_k^r, he extended the operator to geometric modular forms and p–adic modular forms. We write his operator corresponding to δ_*^k as $d^k : V(\mathfrak{c}, \mathfrak{N}; A) \to V(\mathfrak{c}, \mathfrak{N}; A)$. The level p–structure $i_p(\mathfrak{a}) : (\mathbb{G}_m \otimes O^*)[p^\infty] \cong M_\Sigma/\mathfrak{a}_\Sigma \hookrightarrow X(\mathfrak{a})[p^\infty]$ ($\mathfrak{a}_\Sigma = \prod_{\mathfrak{P} \in \Sigma_p} \mathfrak{a}_{\mathfrak{P}} = R_\Sigma$) induces an isomorphism $\iota_p : \widehat{\mathbb{G}}_m \otimes O^* \cong \widehat{X}(\mathfrak{a})$ for the p–adic formal group $\widehat{X}(\mathfrak{a})_{/W}$ at the origin. Then $\omega(R) = \omega(\mathfrak{a}) = \Omega_p \omega_p$ ($\Omega_p \in O \otimes_{\mathbb{Z}} W = W^\Sigma$) for $\omega_p = \iota_{p,*}\frac{dt}{t}$. An important formula given in [K] (2.6.7) is: for $f \in G_k(\mathfrak{c}, \mathfrak{f}^2; \mathcal{W})$,

$$\frac{(d^r f)(x(\mathfrak{a}), \omega_p)}{\Omega_p^{k+2r}} = (d^r f)(x(\mathfrak{a}), \omega(\mathfrak{a})) = (\delta_k^r f)(x(\mathfrak{a}), \omega(\mathfrak{a})) \in \mathcal{W}. \tag{K}$$

The effect of d^r on q–expansion of a modular form is given by

$$d^r \sum_\xi a(\xi) q^\xi = \sum_\xi a(\xi) \xi^r q^\xi. \tag{2.6}$$

See [K] (2.6.27) for this formula.

2.7 Γ_0–level structure and Hecke operators

We now assume that the base algebra A is a $\mathcal{W}$–algebra. Choose a prime $\mathfrak{q}$ of F. We are going to define Hecke operators $U(\mathfrak{q}^n)$ and $T(1, \mathfrak{q}^n)$ assuming for simplicity that $\mathfrak{q} \nmid p\mathfrak{N}$, though we may extend the definition for arbitrary $\mathfrak{q}$ (see [PAF] 4.1.10). Then $X[\mathfrak{q}^r]$ is an étale group over B if X is an abelian scheme over an A–algebra B. We call a subgroup $C \subset X$ cyclic of order $\mathfrak{q}^r$ if $C \cong O/\mathfrak{q}^r$ over an étale faithfully flat extension of B.

We can think of quintuples $(X, \Lambda, i, C, \omega)_{/S}$ adding an additional information C of a cyclic subgroup scheme $C \subset X$ cyclic of order $\mathfrak{q}^r$. We define the space of classical modular forms $G_k(\mathfrak{c}, \mathfrak{N}, \Gamma_0(\mathfrak{q}^r); A)$ (resp. the space $V(\mathfrak{c}, \mathfrak{N}, \Gamma_0(\mathfrak{q}^r); A)$ of p–adic modular forms) of level $(\mathfrak{N}, \Gamma_0(\mathfrak{q}^r))$ by (G1-4) (resp. (P1-3)) replacing test objects (X, Λ, i, ω) (resp. $(X, \Lambda, i_{\mathfrak{N}}, i_p)$) by $(X, \Lambda, i, C, \omega)$ (resp. $(X, \Lambda, i_{\mathfrak{N}}, C, i_p)$).

Our Hecke operators are defined on the space of level $(\mathfrak{N}, \Gamma_0(\mathfrak{q}^r))$. The operator $U(\mathfrak{q}^n)$ is defined only when $r > 0$ and $T(1, \mathfrak{q}^n)$ is defined only when $r = 0$. For a cyclic subgroup C' of $X_{/B}$ of order $\mathfrak{q}^n$, we can define the quotient abelian scheme X/C' with projection $\pi : X \to X/C'$. The polarization Λ and the differential ω induce a polarization $\pi_*\Lambda$ and a differential $(\pi^*)^{-1}\omega$ on X/C'. If $C' \cap C = \{0\}$ (in this case, we say that C' and C are *disjoint*), $\pi(C)$ gives rise to the level $\Gamma_0(\mathfrak{q}^r)$–structure on X/C'. Then we define for $f \in G_k(\mathfrak{c}\mathfrak{q}^n; \mathfrak{N}, \Gamma_0(\mathfrak{q}^r); A)$,

$$f|U(\mathfrak{q}^n)(X, \Lambda, C, i, \omega) = \frac{1}{N(\mathfrak{q}^n)} \sum_{C'} f(X/C', \pi_*\Lambda, \pi \circ i, \pi(C), (\pi^*)^{-1}\omega), \tag{2.7}$$

where C' runs over all étale cyclic subgroups of order $\mathfrak{q}^n$ disjoint from C. Since $\pi_*\Lambda = \pi \circ \Lambda \circ \pi^t$ is a $\mathfrak{c}\mathfrak{q}^n$–polarization, the modular form f has to be defined for abelian varieties with $\mathfrak{c}\mathfrak{q}^n$–polarization. Since $\mathfrak{q} \nmid \mathfrak{N}$, forgetting the $\Gamma_0(\mathfrak{q}^n)$–structure, we define for $f \in G_k(\mathfrak{c}\mathfrak{q}^n; \mathfrak{N}; A)$

$$f|T(1, \mathfrak{q}^n)(X, \Lambda, i, \omega) = \frac{1}{N(\mathfrak{q}^n)} \sum_{C'} f(X/C', \pi_*\Lambda, \pi \circ i, (\pi^*)^{-1}\omega), \tag{2.8}$$

where C' runs over all étale cyclic subgroups of order $\mathfrak{q}^n$. We can check that $f|U(\mathfrak{q}^n)$ and $f|T(1, \mathfrak{q}^n)$ belong to $V(\mathfrak{c}, \mathfrak{N}, \Gamma_0(\mathfrak{q}^r); A)$ and also stay in $G_k(\mathfrak{c}, \mathfrak{N}, \Gamma_0(\mathfrak{q}^r); A)$ if $f \in G_k(\mathfrak{c}\mathfrak{q}, \mathfrak{N}, \Gamma_0(\mathfrak{q}^r); A)$. We have

$$U(\mathfrak{q}^n) = U(\mathfrak{q})^n.$$

2.8 Hilbert modular Shimura varieties

We extend the level structure i limited to $\mathfrak{N}$–torsion points to far bigger structure $\eta^{(p)}$ including all prime-to–p torsion points. Since the prime-to–p torsion on an abelian scheme $X_{/S}$ is unramified at p (see [ACM] 11.1 and [ST]), the extended level structure $\eta^{(p)}$ is still defined over S if S is a $\mathcal{W}$–scheme. Triples $(X, \overline{\Lambda}, \eta^{(p)})_{/S}$ for $\mathcal{W}$–schemes S are classified by an integral model $Sh^{(p)}_{/\mathcal{W}}$ (cf. [Ko]) of the Shimura variety $Sh_{/\mathbb{Q}}$ associated to the algebraic $\mathbb{Q}$–group $G = \mathrm{Res}_{F/\mathbb{Q}} GL(2)$ (in the sense of Deligne [De] 4.22 interpreting Shimura's original definition in [Sh] as a moduli of abelian schemes up to isogenies). Here the classification is up to prime-to–p isogenies, and $\overline{\Lambda}$ is an equivalence class of polarizations up to prime-to–p O–linear isogenies.

To give a description of the functor represented by $Sh^{(p)}$, we introduce some more notations. We consider the fiber category $\mathcal{A}^{(p)}_F$ over schemes defined by

(Object) abelian schemes X with real multiplication by O;
(Morphism) $\mathrm{Hom}_{\mathcal{A}^{(p)}_F}(X, Y) = \mathrm{Hom}(X, Y) \otimes_{\mathbb{Z}} \mathbb{Z}_{(p)}$,

where $\mathbb{Z}_{(p)}$ is the localization of $\mathbb{Z}$ at the prime ideal (p), that is,

$$\mathbb{Z}_{(p)} = \left\{ \frac{a}{b} | b\mathbb{Z} + p\mathbb{Z} = \mathbb{Z},\ a, b \in \mathbb{Z} \right\}.$$

Isomorphisms in this category are isogenies with degree prime to p (called "prime-to-p isogenies"), and hence the degree of polarization Λ is supposed to be also prime to p. Two polarizations are equivalent if $\Lambda = c\Lambda' = \Lambda' \circ i(c)$ for a totally positive c prime to p. We fix an O–lattice $L \subset V = F^2$ with O–hermitian alternating pairing $\langle \cdot, \cdot \rangle$ inducing a self duality on $L_p = L \otimes_{\mathbb{Z}} \mathbb{Z}_p$. We consider the following condition on an AVRM $X_{/S}$ with $\theta : O \hookrightarrow \mathrm{End}(X_{/S})$:

(det) *the characteristic polynomial of $\theta(a)$ $(a \in O)$ on $Lie(X)$ over $\mathcal{O}_S$ is given by $\prod_{\sigma \in I}(T - \sigma(a))$, where I is the set of embeddings of F into $\overline{\mathbb{Q}}$.*

This condition is equivalent to the local freeness of $\pi_* \Omega_{X/S}$ over $\mathcal{O}_S \otimes_{\mathbb{Z}} O$ for $\pi : X \to S$.

For an open-compact subgroup K of $G(\mathbb{A}^{(\infty)})$ *maximal at* p (i.e. $K = GL_2(O_p) \times K^{(p)}$), we consider the following functor from $\mathbb{Z}_{(p)}$–schemes into $SETS$:

$$\mathcal{P}_K^{(p)}(S) = \left[(X, \overline{\Lambda}, \overline{\eta}^{(p)})_{/S} \text{ with (det)} \right]. \tag{2.9}$$

Here $\overline{\eta}^{(p)} : L \otimes_{\mathbb{Z}} \mathbb{A}^{(p\infty)} \cong V^{(p)}(X) = T(X) \otimes_{\mathbb{Z}} \mathbb{A}^{(p\infty)}$ is an equivalence class of $\eta^{(p)}$ modulo multiplication $\eta^{(p)} \mapsto \eta^{(p)} \circ k$ by $k \in K^{(p)}$ for the Tate module $T(X) = \varprojlim_{\mathfrak{N}} X[\mathfrak{N}]$ (in the sheafified sense that $\eta^{(p)} \equiv (\eta')^{(p)} \mod K$ étale-locally), and a $\Lambda \in \overline{\Lambda}$ induces the self-duality on L_p. As long as $K^{(p)}$ is sufficiently small (for K maximal at p), $\mathcal{P}_K^{(p)}$ is representable over any $\mathbb{Z}_{(p)}$–algebra A (e.g. [H04a], [H04b] Section 3.1 and [PAF] 4.2.1) by a scheme $Sh_{K/A} = Sh/K$, which is smooth by the unramifiedness of p in $F/\mathbb{Q}$. We let $g \in G(\mathbb{A}^{(p\infty)})$ act on $Sh^{(p)}_{/\mathbb{Z}_{(p)}}$ by

$$x = (X, \overline{\Lambda}, \eta) \mapsto g(x) = (X, \overline{\Lambda}, \eta \circ g),$$

which gives a right action of $G(\mathbb{A})$ on $Sh^{(p)}$ through the projection $G(\mathbb{A}) \twoheadrightarrow G(\mathbb{A}^{(p\infty)})$.

By the universality, we have a morphism $\mathfrak{M}(\mathfrak{c}; \mathfrak{N}) \to Sh^{(p)}/\widehat{\Gamma}_1^1(\mathfrak{c}; \mathfrak{N})$ for the open compact subgroup:

$$\widehat{\Gamma}_1^1(\mathfrak{c}; \mathfrak{N}) = \left(\begin{smallmatrix} c & 0 \\ 0 & 1 \end{smallmatrix}\right) K_1^1(\mathfrak{N}) \left(\begin{smallmatrix} c & 0 \\ 0 & 1 \end{smallmatrix}\right)^{-1} = \left(\begin{smallmatrix} cd^{-1} & 0 \\ 0 & 1 \end{smallmatrix}\right) U_1^1(\mathfrak{N}) \left(\begin{smallmatrix} cd^{-1} & 0 \\ 0 & 1 \end{smallmatrix}\right)^{-1}$$

maximal at p. The image of $\mathfrak{M}(\mathfrak{c}; \mathfrak{N})$ gives a geometrically irreducible component of $Sh^{(p)}/\widehat{\Gamma}_1^1(\mathfrak{c}; \mathfrak{N})$. If $\mathfrak{N}$ is sufficiently deep, by the universality of

$\mathfrak{M}(\mathfrak{c};\mathfrak{N})$, we can identify $\mathfrak{M}(\mathfrak{c};\mathfrak{N})$ with its image in $Sh^{(p)}/\widehat{\Gamma}_1^1(\mathfrak{c};\mathfrak{N})$. By the action on the polarization $\Lambda \mapsto \alpha\Lambda$ for a suitable totally positive $\alpha \in F$, we can bring $\mathfrak{M}(\mathfrak{c};\mathfrak{N})$ into $\mathfrak{M}(\alpha\mathfrak{c};\mathfrak{N})$; so, the image of $\varprojlim_{\mathfrak{N}} \mathfrak{M}(\mathfrak{c};\mathfrak{N})$ in $Sh^{(p)}$ only depends on the strict ideal class of $\mathfrak{c}$.

For each $x = (X, \Lambda, i) \in \mathfrak{M}(\mathfrak{c};\mathfrak{N})(S)$ for a $\mathcal{W}$–scheme S, choosing $\eta^{(p)}$ so that $\eta^{(p)} \mod \widehat{\Gamma}_1^1(\mathfrak{c};\mathfrak{N}) = i$, we get a point $x = (X, \overline{\Lambda}, \eta^{(p)}) \in Sh^{(p)}(S)$ projecting down to $x = (X, \Lambda, i)$. Each element $g \in G(\mathbb{A})$ with totally positive determinant in $F^\times$ acts on $x = (X, \Lambda, \eta^{(p)}) \in Sh^{(p)}$ by $x \mapsto g(x) = (X, \det(g)\Lambda, \eta^{(p)} \circ g)$. This action is geometric preserving the base scheme $\mathrm{Spec}(\mathcal{W})$ and is compatible with the action of $G(\mathbb{A}^{(p\infty)})$ given as above (see [PAF] 4.2.2), because $\overline{\Lambda} = \overline{\det(g)\Lambda}$. Then we can think of the projection of $g(x)$ in $\mathfrak{M}(\mathfrak{c};\mathfrak{N})$. By abusing the notation slightly, if the lift $\eta^{(p)}$ of i is clear in the context, we write $g(x) \in \mathfrak{M}(\mathfrak{c};\mathfrak{N})$ for the image of $g(x) \in Sh^{(p)}$. If the action of g is induced by a prime-to–p isogeny $\alpha : X \to g(X)$, we write $g(x,\omega) = (g(x), \alpha_*\omega)$ for $(x,\omega) \in \mathcal{M}(\mathfrak{c};\mathfrak{N})$ if there is no ambiguity of α. When $\det(g)$ is not rational, the action of g is often non-trivial on $\mathrm{Spec}(\mathcal{W})$; see [Sh] II, [Sh1] and [PAF] 4.2.2.

2.9 Level structure with "Neben" character

In order to make a good link between classical modular forms and adelic automorphic forms (which we will describe in the following subsection), we would like to introduce "Neben" characters. We fix two integral ideals $\mathfrak{N} \subset \mathfrak{n} \subset O$. We think of the following level structure on an AVRM X:

$$i : (\mathbb{G}_m \otimes O^*)[\mathfrak{N}] \hookrightarrow X[\mathfrak{N}] \text{ and } i' : X[\mathfrak{n}] \twoheadrightarrow O/\mathfrak{n} \tag{2.10}$$

with $\mathrm{Im}(i) \times_{X[\mathfrak{N}]} X[\mathfrak{n}] = \mathrm{Ker}(i')$, where the sequence $(\mathbb{G}_m \otimes O^*)[\mathfrak{N}] \xrightarrow{i} X[\mathfrak{N}] \xrightarrow{i'} O/\mathfrak{n}$ is required to induce an isomorphism

$$(\mathbb{G}_m \otimes O^*)[\mathfrak{N}] \otimes_O O/\mathfrak{n} \cong (\mathbb{G}_m \otimes O^*)[\mathfrak{n}]$$

under the polarization Λ. When $\mathfrak{N} = \mathfrak{n}$, this is exactly a $\Gamma_1^1(\mathfrak{N})$–level structure. We fix two characters $\epsilon_1 : (O/\mathfrak{n})^\times \to A^\times$ and $\epsilon_2 : (O/\mathfrak{N})^\times \to A^\times$, and we insist for $f \in G_k(\mathfrak{c},\mathfrak{N};A)$ on the version of (G0-3) for quintuples $(X, \Lambda, i \cdot d, a \cdot i', \omega)$ and the equivariancy:

$$f(X, \overline{\Lambda}, i \cdot d, a \cdot i', \omega) = \epsilon_1(a)\epsilon_2(d) f(X, \overline{\Lambda}, i, i', \omega) \text{ for } a, d \in (O/\mathfrak{N})^\times. \tag{Neben}$$

Here $\overline{\Lambda}$ is the polarization class modulo multiple of totally positive numbers in F prime to $\mathfrak{n}$. We write $G_k(\mathfrak{c}, \Gamma_0(\mathfrak{N}), \epsilon; A)$ ($\epsilon = (\epsilon_1, \epsilon_2)$) for the A–module of geometric modular forms satisfying these conditions.

2.10 Adelic Hilbert modular forms and Hecke algebras

Let us interpret what we have said so far in automorphic language and give a definition of the adelic Hilbert modular forms and their Hecke algebra of level $\mathfrak{N}$ (cf. [H96] Sections 2.2-4 and [PAF] Sections 4.2.8–4.2.12). We first recall formal Hecke rings of double cosets. For that, we fix a prime element $\varpi_\mathfrak{q}$ of $O_\mathfrak{q}$ for every prime ideal $\mathfrak{q}$ of O.

We consider the following open compact subgroup of $G(\mathbb{A}^{(\infty)})$:

$$\begin{aligned} U_0(\mathfrak{N}) &= \left\{ \left(\begin{smallmatrix} a & b \\ c & d \end{smallmatrix}\right) \in GL_2(\widehat{O}) \middle| c \equiv 0 \mod \mathfrak{N}\widehat{O} \right\}, \\ U_1^1(\mathfrak{N}) &= \left\{ \left(\begin{smallmatrix} a & b \\ c & d \end{smallmatrix}\right) \in U_0(\mathfrak{N}) \middle| a \equiv d \equiv 1 \mod \mathfrak{N}\widehat{O} \right\}, \end{aligned} \tag{2.11}$$

where $\widehat{O} = O \otimes_\mathbb{Z} \widehat{\mathbb{Z}}$ and $\widehat{\mathbb{Z}} = \prod_\ell \mathbb{Z}_\ell$. Then we introduce the following semi-group

$$\Delta_0(\mathfrak{N}) = \left\{ \left(\begin{smallmatrix} a & b \\ c & d \end{smallmatrix}\right) \in G(\mathbb{A}^{(\infty)}) \cap M_2(\widehat{O}) \middle| c \equiv 0 \mod \mathfrak{N}\widehat{O}, d_\mathfrak{N} \in O_\mathfrak{N}^\times \right\}, \tag{2.12}$$

where $d_\mathfrak{N}$ is the projection of $d \in \widehat{O}$ to $\prod_{\mathfrak{q}|\mathfrak{N}} O_\mathfrak{q}$ for prime ideals $\mathfrak{q}$. Writing T_0 for the maximal diagonal torus of $GL(2)_{/O}$ and putting

$$D_0 = \left\{ \mathrm{diag}[a,d] = \left(\begin{smallmatrix} a & 0 \\ 0 & d \end{smallmatrix}\right) \in T_0(F_{\mathbb{A}^{(\infty)}}) \cap M_2(\widehat{O}) \middle| d_\mathfrak{N} = 1 \right\}, \tag{2.13}$$

we have (e.g. [MFG] 3.1.6 and [PAF] Section 5.1)

$$\Delta_0(\mathfrak{N}) = U_0(\mathfrak{N}) D_0 U_0(\mathfrak{N}). \tag{2.14}$$

In this section, writing $\mathfrak{p}^\alpha = \prod_{\mathfrak{p}|p} \mathfrak{p}^{\alpha(\mathfrak{p})}$ with $\alpha = (\alpha(\mathfrak{p}))$, the group U is assumed to be a subgroup of $U_0(\mathfrak{N}\mathfrak{p}^\alpha)$ with $U \supset U_1^1(\mathfrak{N}\mathfrak{p}^\alpha)$ for some multi-exponent α (though we do not assume that $\mathfrak{N}$ is prime to p). Formal finite linear combinations $\sum_\delta c_\delta U\delta U$ of double cosets of U in $\Delta_0(\mathfrak{N}\mathfrak{p}^\alpha)$ form a ring $R(U, \Delta_0(\mathfrak{N}\mathfrak{p}^\alpha))$ under convolution product (see [IAT] Chapter 3 or [MFG] 3.1.6). The algebra is commutative and is isomorphic to the polynomial ring over the group algebra $\mathbb{Z}[U_0(\mathfrak{N}\mathfrak{p}^\alpha)/U]$ with variables $\{T(\mathfrak{q}), T(\mathfrak{q},\mathfrak{q})\}_\mathfrak{q}$ for primes $\mathfrak{q}$, $T(\mathfrak{q})$ corresponding to the double coset $U\left(\begin{smallmatrix} \varpi_\mathfrak{q} & 0 \\ 0 & 1 \end{smallmatrix}\right)U$ and $T(\mathfrak{q},\mathfrak{q})$ (for primes $\mathfrak{q} \nmid \mathfrak{N}\mathfrak{p}^\alpha$) corresponding to $U_0\varpi_\mathfrak{q}U$. Here we have chosen a prime element $\varpi_\mathfrak{q}$ in $O_\mathfrak{q}$. The group element $u \in U_0(\mathfrak{N}\mathfrak{p}^\alpha)/U$ in $\mathbb{Z}[U_0(\mathfrak{N}\mathfrak{p}^\alpha)/U]$ corresponds to the double coset UuU (cf. [H95] Section 2).

The double coset ring $R(U, \Delta_0(\mathfrak{N}\mathfrak{p}^\alpha))$ naturally acts on the space of modular forms on U whose definition we now recall. Recall that T_0 is the diagonal torus of $GL(2)_{/O}$; so, $T_0 = \mathbb{G}^2_{m/O}$. Since $T_0(O/\mathfrak{N}')$ is canonically a quotient of $U_0(\mathfrak{N}')$ for an ideal $\mathfrak{N}'$, a character $\epsilon : T_0(O/\mathfrak{N}') \to \mathbb{C}^\times$ can be considered as a character of $U_0(\mathfrak{N}')$. Writing $\epsilon\left(\left(\begin{smallmatrix} a & 0 \\ 0 & d \end{smallmatrix}\right)\right) = \epsilon_1(a)\epsilon_2(d)$, if $\epsilon^- = \epsilon_1^{-1}\epsilon_2$

factors through $O/\mathfrak{N}$ for $\mathfrak{N}|\mathfrak{N}'$, then we can extend the character ϵ of $U_0(\mathfrak{N}')$ to $U_0(\mathfrak{N})$ by putting $\epsilon(u) = \epsilon_1(\det(u))\epsilon^-(d)$ for $u = \left(\begin{smallmatrix} a & b \\ c & d \end{smallmatrix}\right) \in U_0(\mathfrak{N})$. In this sense, we hereafter assume that ϵ is defined modulo $\mathfrak{N}$ and regard ϵ as a character of $U_0(\mathfrak{N})$. We choose a Hecke character $\epsilon_+ : F_{\mathbb{A}}^\times/F^\times \to \mathbb{C}^\times$ with infinity type $(1-[\kappa])I$ (for an integer $[\kappa]$) such that $\epsilon_+(z) = \epsilon_1(z)\epsilon_2(z)$ for $z \in \widehat{O}^\times$. We also write ϵ_+^t for the restriction of ϵ_+ to the maximal torsion subgroup $\Delta_F(\mathfrak{N})$ of $Cl_F^+(\mathfrak{N}p^\infty)$ (the strict ray class group modulo $\mathfrak{N}p^\infty$: $\varprojlim_n Cl_F^+(\mathfrak{N}p^n)$).

Writing I for the set of all embeddings of F into $\overline{\mathbb{Q}}$ and T^2 for $\mathrm{Res}_{O/\mathbb{Z}}T_0$ (the diagonal torus of G), the group of geometric characters $X^*(T^2)$ is isomorphic to $\mathbb{Z}[I]^2$ so that $(m,n) \in \mathbb{Z}[I]^2$ send $\mathrm{diag}[x,y] \in T^2$ to $x^m y^n = \prod_{\sigma\in I}(\sigma(x)^{m_\sigma}\sigma(y)^{n_\sigma})$. Taking $\kappa = (\kappa_1,\kappa_2) \in \mathbb{Z}[I]^2$, we assume $[\kappa]I = \kappa_1+\kappa_2$, and we associate with κ a factor of automorphy:

$$J_\kappa(g,\tau) = \det(g_\infty)^{\kappa_2-I} j(g_\infty,\tau)^{\kappa_1-\kappa_2+I} \quad \text{for } g \in G(\mathbb{A}) \text{ and } \tau \in \mathfrak{Z}. \tag{2.15}$$

We define $S_\kappa(U,\epsilon;\mathbb{C})$ by the space of functions $f : G(\mathbb{A}) \to \mathbb{C}$ satisfying the following three conditions (e.g. [H96] Section 2.2 and [PAF] Section 4.3.1):

(S1) $f(\alpha x u \delta) = \epsilon(u)\epsilon_+^t(z)f(x)J_\kappa(u,\mathbf{i})^{-1}$ for all $\alpha \in G(\mathbb{Q})$ and all $u \in U \cdot C_{\mathbf{i}}$ and $z \in \Delta_F(\mathfrak{N})$ ($\Delta_F(\mathfrak{N})$ is the maximal torsion subgroup of $Cl_F^+(\mathfrak{N}p^\infty)$);

(S2) Choose $u \in G(\mathbb{R})$ with $u(\mathbf{i}) = \tau$ for $\tau \in \mathfrak{Z}$, and put $f_x(\tau) = f(xu)J_\kappa(u,\mathbf{i})$ for each $x \in G(\mathbb{A}^{(\infty)})$ (which only depends on τ). Then f_x is a holomorphic function on $\mathfrak{Z}$ for all x;

(S3) $f_x(\tau)$ for each x is rapidly decreasing as $\eta_\sigma \to \infty$ ($\tau = \xi + \mathbf{i}\eta$) for all $\sigma \in I$ uniformly.

If we replace the word "rapidly decreasing" in (S3) by "slowly increasing", we get the definition of the space $G_\kappa(U,\epsilon;\mathbb{C})$. It is easy to check (e.g. [MFG] 3.1.5) that the function f_x in (S2) satisfies the classical automorphy condition:

$$f(\gamma(\tau)) = \epsilon(x^{-1}\gamma x)f(\tau)J_\kappa(\gamma,\tau) \quad \text{for all } \gamma \in \Gamma_x(U), \tag{2.16}$$

where $\Gamma_x(U) = xUx^{-1}G(\mathbb{R})^+ \cap G(\mathbb{Q})$. Also by (S3), f_x is rapidly decreasing towards all cusps of Γ_x (e.g. [MFG] (3.22)); so, it is a cusp form. Imposing that f have the central character ϵ_+ in place of the action of $\Delta_F(\mathfrak{N})$ in (S1), we define the subspace $S_\kappa(\mathfrak{N},\epsilon_+;\mathbb{C})$ of $S_\kappa(U_0(\mathfrak{N}),\epsilon;\mathbb{C})$. *The symbols* $\kappa = (\kappa_1,\kappa_2)$ *and* $(\varepsilon_1,\varepsilon_2)$ *here correspond to* (κ_2,κ_1) *and* $(\varepsilon_2,\varepsilon_1)$ *in [PAF] Section 4.2.6 (page 171) because of a different notational convention in [PAF].*

If we restrict f as above to $SL_2(F_{\mathbb{A}})$, the determinant factor $\det(g)^{\kappa_2}$ in the factor of automorphy disappears, and the automorphy factor becomes only dependent on $k = \kappa_1 - \kappa_2 + I \in \mathbb{Z}[I]$; so, the classical modular form in G_k has single digit weight $k \in \mathbb{Z}[I]$. Via (2.3), we have an embedding of $S_\kappa(U_0(\mathfrak{N}'), \epsilon; \mathbb{C})$ into $G_k(\Gamma_0(\mathfrak{N}'), \epsilon; \mathbb{C}) = \bigoplus_{[\mathfrak{c}] \in Cl_F^+} G_k(\mathfrak{c}, \Gamma_0(\mathfrak{N}'), \epsilon; \mathbb{C})$ ($\mathfrak{c}$ running over a complete representative set for the strict ideal class group Cl_F^+) bringing f into $(f_\mathfrak{c})_{[\mathfrak{c}]}$ for $f_\mathfrak{c} = f_x$ (as in (S3)) with $x = \left(\begin{smallmatrix} cd^{-1} & 0 \\ 0 & 1 \end{smallmatrix}\right)$ (for $d \in F_\mathbb{A}^\times$ with $d\widehat{O} = \widehat{\mathfrak{d}}$). The cusp form $f_\mathfrak{c}$ is determined by the restriction of f to $x \cdot SL_2(F_\mathbb{A})$. If we vary the weight κ keeping $k = \kappa_1 - \kappa_2 + I$, the image of S_κ in $G_k(\Gamma_0(\mathfrak{N}'), \epsilon; \mathbb{C})$ transforms accordingly. By this identification, the Hecke operator $T(\mathfrak{q})$ for non-principal $\mathfrak{q}$ makes sense as an operator acting on a single space $G_\kappa(U, \epsilon; \mathbb{C})$, and its action depends on the choice of κ. In other words, we have the double digit weight $\kappa = (\kappa_1, \kappa_2)$ for adelic modular forms in order to specify the central action of $G(\mathbb{A})$. For a given $f \in S_\kappa(U, \epsilon; \mathbb{C})$ and a Hecke character $\lambda : F_\mathbb{A}^\times / F^\times \to \mathbb{C}^\times$, the tensor product $(f \otimes \lambda)(x) = f(x)\lambda(\det(x))$ gives rise to a different modular form in $S_{\kappa_\lambda}(U, \epsilon_\lambda; \mathbb{C})$ for weight κ_λ and character ϵ_λ dependent on λ, although the two modular forms have the same restriction to $SL_2(F_\mathbb{A})$.

We identify I with $\sum_\sigma \sigma$ in $\mathbb{Z}[I]$. It is known that $G_\kappa = 0$ unless $\kappa_1 + \kappa_2 = [\kappa_1 + \kappa_2]I$ for $[\kappa_1 + \kappa_2] \in \mathbb{Z}$, because $I - (\kappa_1 + \kappa_2)$ is the infinity type of the central character of automorphic representations generated by G_κ. We write simply $[\kappa]$ for $[\kappa_1 + \kappa_2] \in \mathbb{Z}$ assuming $G_\kappa \neq 0$. The $SL(2)$–weight of the central character of an irreducible automorphic representation π generated by $f \in G_\kappa(U, \epsilon; \mathbb{C})$ is given by k (which specifies the infinity type of π_∞ as a discrete series representation of $SL_2(F_\mathbb{R})$). There is a geometric meaning of the weight κ: the Hodge weight of the motive attached to π (cf. [BR]) is given by $\{(\kappa_{1,\sigma}, \kappa_{2,\sigma}), (\kappa_{2,\sigma}, \kappa_{1,\sigma})\}_\sigma$, and thus, the requirement $\kappa_1 - \kappa_2 \geq I$ is the regularity assumption for the motive (and is equivalent to the classical weight $k \geq 2I$ condition).

Choose a prime element $\varpi_\mathfrak{q}$ of $O_\mathfrak{q}$ for each prime $\mathfrak{q}$ of F. We extend $\epsilon^- : \widehat{O}^\times \to \mathbb{C}^\times$ to $F_{\mathbb{A}(\infty)}^\times \to \mathbb{C}^\times$ just by putting $\epsilon^-(\varpi_\mathfrak{q}^m) = 1$ for $m \in \mathbb{Z}$. This is possible because $F_\mathfrak{q}^\times = O_\mathfrak{q}^\times \times \varpi_\mathfrak{q}^\mathbb{Z}$ for $\varpi_\mathfrak{q}^\mathbb{Z} = \{\varpi_\mathfrak{q}^m | m \in \mathbb{Z}\}$. Similarly, we extend ϵ_1 to $F_{\mathbb{A}(\infty)}^\times$. Then we define $\epsilon(u) = \epsilon_1(\det(u))\epsilon^-(a_\mathfrak{N})$ for $u = \left(\begin{smallmatrix} a & b \\ c & d \end{smallmatrix}\right) \in \Delta_0(\mathfrak{N})$. Let $\mathcal{U}$ be the unipotent algebraic subgroup of $GL(2)_{/O}$ defined by $\mathcal{U}(A) = \left\{\left(\begin{smallmatrix} 1 & a \\ 0 & 1 \end{smallmatrix}\right) \middle| a \in A\right\}$. For each $UyU \in R(U, \Delta_0(\mathfrak{N}\mathfrak{p}^\alpha))$, we decompose $UyU = \bigsqcup_{t \in D_0, u \in \mathcal{U}(\widehat{O})} utU$ for finitely many u and t (see [IAT] Chapter 3 or [MFG] 3.1.6) and define

$$f|[UyU](x) = \sum_{t,u} \epsilon(t)^{-1} f(xut). \tag{2.17}$$

We check that this operator preserves the spaces of automorphic forms: $G_\kappa(\mathfrak{N}, \epsilon; \mathbb{C})$ and $S_\kappa(\mathfrak{N}, \epsilon; \mathbb{C})$. This action for y with $y_\mathfrak{N} = 1$ is independent of the choice of the extension of ϵ to $T_0(F_{\mathbb{A}})$. When $y_{\mathfrak{N}} \neq 1$, we may assume that $y_{\mathfrak{N}} \in D_0 \subset T_0(F_{\mathbb{A}})$, and in this case, t can be chosen so that $t_{\mathfrak{N}} = y_{\mathfrak{N}}$ (so $t_{\mathfrak{N}}$ is independent of single right cosets in the double coset). If we extend ϵ to $T_0(F_{\mathbb{A}}^{(\infty)})$ by choosing another prime element $\varpi'_{\mathfrak{q}}$ and write the extension as ϵ', then we have

$$\epsilon(t_{\mathfrak{N}})[UyU] = \epsilon'(t_{\mathfrak{N}})[UyU]',$$

where the operator on the right-hand-side is defined with respect to ϵ'. Thus the sole difference is the root of unity $\epsilon(t_{\mathfrak{N}})/\epsilon'(t_{\mathfrak{N}}) \in \mathrm{Im}(\epsilon|_{T_0(O/\mathfrak{N})})$. Since it depends on the choice of $\varpi_{\mathfrak{q}}$, we make the choice once and for all, and write $T(\mathfrak{q})$ for $\left[U\left(\begin{smallmatrix}\varpi_{\mathfrak{q}} & 0\\ 0 & 1\end{smallmatrix}\right)U\right]$ (if $\mathfrak{q}|\mathfrak{N}$). By linearity, these action of double cosets extends to the ring action of the double coset ring $R(U, \Delta_0(\mathfrak{N}\mathfrak{p}^\alpha))$.

To introduce rationality of modular forms, we recall Fourier expansion of adelic modular forms (cf. [H96] Sections 2.3-4). Recall the embedding $i_\infty : \overline{\mathbb{Q}} \hookrightarrow \mathbb{C}$, and identify $\overline{\mathbb{Q}}$ with the image of i_∞. Recall also the differental idele $d \in F_{\mathbb{A}}^\times$ with $d^{(\mathfrak{d})} = 1$ and $d\widehat{O} = \mathfrak{d}\widehat{O}$. Each member f of $S_\kappa(U, \epsilon; \mathbb{C})$ has its Fourier expansion:

$$f\left(\begin{smallmatrix} y & x\\ 0 & 1\end{smallmatrix}\right) = |y|_{\mathbb{A}} \sum_{0 \ll \xi \in F} a(\xi y d, f)(\xi y_\infty)^{-\kappa_2} \mathbf{e}_F(i\xi y_\infty)\mathbf{e}_F(\xi x), \tag{2.18}$$

where $\mathbf{e}_F : F_{\mathbb{A}}/F \to \mathbb{C}^\times$ is the additive character which has $\mathbf{e}_F(x_\infty) = \exp(2\pi i \sum_{\sigma \in I} x_\sigma)$ for $x_\infty = (x_\sigma)_\sigma \in \mathbb{R}^I = F \otimes_{\mathbb{Q}} \mathbb{R}$. Here $y \mapsto a(y, f)$ is a function defined on $y \in F_{\mathbb{A}}^\times$ only depending on its finite part $y^{(\infty)}$. The function $a(y, f)$ is supported by the set $(\widehat{O} \times F_\infty) \cap F_{\mathbb{A}}^\times$ of *integral* ideles.

Let $F[\kappa]$ be the field fixed by $\{\sigma \in \mathrm{Gal}(\overline{\mathbb{Q}}/F) | \kappa\sigma = \kappa\}$, over which the character $\kappa \in X^*(T^2)$ is rational. Write $O[\kappa]$ for the integer ring of $F[\kappa]$. We also define $O[\kappa, \epsilon]$ for the integer ring of the field $F[\kappa, \epsilon]$ generated by the values of ϵ over $F[\kappa]$. For any $F[\kappa, \epsilon]$–algebra A inside $\mathbb{C}$, we define

$$S_\kappa(U, \epsilon; A) = \left\{ f \in S_\kappa(U, \epsilon; \mathbb{C}) \middle| a(y, f) \in A \text{ as long as } y \text{ is integral} \right\}. \tag{2.19}$$

As we have seen, we can interpret $S_\kappa(U, \epsilon; A)$ as the space of A–rational global sections of a line bundle of a variety defined over A; so, we have, by the flat base-change theorem (e.g. [GME] Lemma 1.10.2),

$$S_\kappa(\mathfrak{N}, \epsilon; A) \otimes_A \mathbb{C} = S_\kappa(\mathfrak{N}, \epsilon; \mathbb{C}). \tag{2.20}$$

The Hecke operators preserve A–rational modular forms (e.g., [PAF] 4.2.9). We define the Hecke algebra $h_\kappa(U, \epsilon; A) \subset \mathrm{End}_A(S_\kappa(U, \epsilon; A))$ by the A–subalgebra generated by the Hecke operators of $R(U, \Delta_0(\mathfrak{N}\mathfrak{p}^\alpha))$.

For any $\overline{\mathbb{Q}}_p$–algebras A, we define

$$S_\kappa(U,\epsilon;A) = S_\kappa(U,\epsilon;\overline{\mathbb{Q}}) \otimes_{\overline{\mathbb{Q}},i_p} A. \tag{2.21}$$

By linearity, $y \mapsto a(y,f)$ extends to a function on $F_{\mathbb{A}}^\times \times S_\kappa(U,\epsilon;A)$ with values in A. We define the q–expansion coefficients (at p) of $f \in S_\kappa(U,\epsilon;A)$ by

$$\mathbf{a}_p(y,f) = y_p^{-\kappa_2} a(y,f) \text{ and } \mathbf{a}_{0,p}(y,f) = \mathcal{N}(yd^{-1})^{[\kappa_2]} a_0(y,f), \tag{2.22}$$

where $\mathcal{N} : F_{\mathbb{A}}^\times/F^\times \to \overline{\mathbb{Q}}_p^\times$ is the character given by $\mathcal{N}(y) = y_p^{-I}|y^{(\infty)}|_{\mathbb{A}}^{-1}$. Here we note that $a_0(y,f) = 0$ if $\kappa_2 \notin \mathbb{Z}I$. Thus, if $a_0(y,f) \neq 0$, $[\kappa_2] \in \mathbb{Z}$ is well defined. The formal q–expansion of an A–rational f has values in the space of functions on $F_{\mathbb{A}^{(\infty)}}^\times$ with values in the formal monoid algebra $A[[q^\xi]]_{\xi \in F_+}$ of the multiplicative semi-group F_+ made up of totally positive elements, which is given by

$$f(y) = \mathcal{N}(y)^{-1} \left\{ \mathbf{a}_{0,p}(yd,f) + \sum_{\xi \gg 0} \mathbf{a}_p(\xi yd,f) q^\xi \right\}. \tag{2.23}$$

We now define for any p–adically complete $O[\kappa,\epsilon]$–algebra A in $\widehat{\mathbb{Q}}_p$

$$S_\kappa(U,\epsilon;A) = \left\{ f \in S_\kappa(U,\epsilon;\widehat{\mathbb{Q}}_p) \middle| \mathbf{a}_p(y,f) \in A \text{ for integral } y \right\}. \tag{2.24}$$

As we have already seen, these spaces have geometric meaning as the space of A–integral global sections of a line bundle defined over A of the Hilbert modular variety of level U (see [PAF] Section 4.2.6), and the q–expansion above for a fixed $y = y^{(\infty)}$ gives rise to the geometric q–expansion at the infinity cusp of the classical modular form f_x for $x = \left(\begin{smallmatrix} y & 0 \\ 0 & 1 \end{smallmatrix}\right)$ (see [H91] (1.5) and [PAF] (4.63)).

We have chosen a complete representative set $\{c_i\}_{i=1,\dots,h}$ in finite ideles for the strict idele class group $F^\times \backslash F_{\mathbb{A}}^\times / \widehat{O}^\times F_{\infty+}^\times$, where h is the strict class number of F. Let $\mathfrak{c}_i = c_i O$. Write $t_i = \left(\begin{smallmatrix} c_i d^{-1} & 0 \\ 0 & 1 \end{smallmatrix}\right)$ and consider $f_i = f_{t_i}$ as defined in (S2). The collection $(f_i)_{i=1,\dots,h}$ determines f, because of the approximation theorem. Then $f(c_i d^{-1})$ gives the q–expansion of f_i at the Tate abelian variety with $\mathfrak{c}_i$–polarization $\mathrm{Tate}_{\mathfrak{c}_i^{-1},O}(q)$ ($\mathfrak{c}_i = c_i O$). By (q–exp), the q–expansion $f(y)$ determines f uniquely.

We write $T(y)$ for the Hecke operator acting on $S_\kappa(U,\epsilon;A)$ corresponding to the double coset $U \left(\begin{smallmatrix} y & 0 \\ 0 & 1 \end{smallmatrix}\right) U$ for an integral idele y. We renormalize $T(y)$ to have a p–integral operator $\mathbb{T}(y)$: $\mathbb{T}(y) = y_p^{-\kappa_2} T(y)$. Since this only affects $T(y)$ with $y_p \neq 1$, $\mathbb{T}(\mathfrak{q}) = T(\varpi_{\mathfrak{q}}) = T(\mathfrak{q})$ if $\mathfrak{q} \nmid p$. However $\mathbb{T}(\mathfrak{p}) \neq T(\mathfrak{p})$

for primes $\mathfrak{p}|p$. The renormalization is optimal to have the stability of the A–integral spaces under Hecke operators. We define $\langle\mathfrak{q}\rangle = N(\mathfrak{q})T(\mathfrak{q},\mathfrak{q})$ for $\mathfrak{q} \nmid \mathfrak{N}\mathfrak{p}^\alpha$, which is equal to the central action of a prime element $\varpi_\mathfrak{q}$ of $O_\mathfrak{q}$ times $N(\mathfrak{q}) = |\varpi_\mathfrak{q}|_\mathbb{A}^{-1}$. We have the following formula of the action of $T(\mathfrak{q})$ and $T(\mathfrak{q},\mathfrak{q})$ (e.g., [PAF] Section 4.2.10):

$$\mathbf{a}_p(y, f|\mathbb{T}(\mathfrak{q})) = \begin{cases} \mathbf{a}_p(y\varpi_\mathfrak{q}, f) + \mathbf{a}_p(y\varpi_\mathfrak{q}^{-1}, f|\langle\mathfrak{q}\rangle) & \text{if } \mathfrak{q} \text{ is outside } \mathfrak{n} \\ \mathbf{a}_p(y\varpi_\mathfrak{q}, f) & \text{otherwise,} \end{cases} \tag{2.25}$$

where the level $\mathfrak{n}$ of U is the ideal maximal under the condition: $U_1^1(\mathfrak{n}) \subset U \subset U_0(\mathfrak{N})$. Thus $\mathbb{T}(\varpi_\mathfrak{q}) = U(\mathfrak{q})$ (up to p–adic units) when $\mathfrak{q}$ is a factor of the level of U (even when $\mathfrak{q}|p$; see [PAF] (4.65–66)). Writing the level of U as $\mathfrak{N}\mathfrak{p}^\alpha$, we assume

$$\text{either } p|\mathfrak{N}\mathfrak{p}^\alpha \text{ or } [\kappa] \geq 0, \tag{2.26}$$

since $\mathbb{T}(\mathfrak{q})$ and $\langle\mathfrak{q}\rangle$ preserve the space $S_\kappa(U, \epsilon; A)$ under this condition (see [PAF] Theorem 4.28). We then define the Hecke algebra $h_\kappa(U, \epsilon; A)$ (resp. $h_\kappa(\mathfrak{N}, \epsilon_+; A)$) with coefficients in A by the A–subalgebra of the A–linear endomorphism algebra $\mathrm{End}_A(S_\kappa(U, \epsilon; A))$ (resp. $\mathrm{End}_A(S_\kappa(\mathfrak{N}, \epsilon_+; A))$) generated by the action of the finite group $U_0(\mathfrak{N}\mathfrak{p}^\alpha)/U$, $\mathbb{T}(\mathfrak{q})$ and $\langle\mathfrak{q}\rangle$ for all $\mathfrak{q}$.

We have canonical projections:

$$R(U_1^1(\mathfrak{N}\mathfrak{p}^\alpha), \Delta_0(\mathfrak{N}\mathfrak{p}^\alpha)) \twoheadrightarrow R(U, \Delta_0(\mathfrak{N}\mathfrak{p}^\alpha)) \twoheadrightarrow R(U_0(\mathfrak{N}p^\beta), \Delta_0(\mathfrak{N}p^\beta))$$

for all $\alpha \geq \beta$ ($\Leftrightarrow \alpha(\mathfrak{p}) \geq \beta(\mathfrak{p})$ for all $\mathfrak{p}|p$) taking canonical generators to the corresponding ones, which are compatible with inclusions

$$S_\kappa(U_0(\mathfrak{N}p^\beta), \epsilon; A) \hookrightarrow S_\kappa(U, \epsilon; A) \hookrightarrow S_\kappa(U_1^1(\mathfrak{N}\mathfrak{p}^\alpha), \epsilon; A).$$

We get a projective system of Hecke algebras $\{h_\kappa(U, \epsilon; A)\}_U$ (U running through open subgroups of $U_0(\mathfrak{N}p)$ containing $U_1^1(\mathfrak{N}p^\infty)$), whose projective limit (when $\kappa_1 - \kappa_2 \geq I$) gives rise to the universal Hecke algebra $\mathbf{h}(\mathfrak{N}, \epsilon; A)$ for a complete p–adic algebra A. This algebra is known to be independent of κ (as long as $\kappa_1 - \kappa_2 \geq I$) and has canonical generators $\mathbb{T}(y)$ over $A[[\mathbf{G}]]$ (for $\mathbf{G} = (O_p \times (O/\mathfrak{N}^{(p)}))^\times \times Cl_F^+(\mathfrak{N}p^\infty)$), where $\mathfrak{N}^{(p)}$ is the prime-to-p part of $\mathfrak{N}$. Here note that the operator $\langle\mathfrak{q}\rangle$ is included in the action of $\mathbf{G}$, because $\mathfrak{q} \in Cl_F^+(\mathfrak{N}p^\infty)$. We write $h_\kappa^{n.ord}(U, \epsilon; A)$, $h_\kappa^{n.ord}(\mathfrak{N}p^\alpha, \epsilon_+; A)$ and $\mathbf{h}^{n.ord} = \mathbf{h}^{n.ord}(\mathfrak{N}, \epsilon; A)$ for the image of the (nearly) ordinary projector $e = \lim_n \mathbb{T}(p)^{n!}$. The algebra $\mathbf{h}^{n.ord}$ is by definition the universal nearly ordinary Hecke algebra over $A[[\mathbf{G}]]$ of level $\mathfrak{N}$ with "Neben character" ϵ. We also note here that this algebra $\mathbf{h}^{n.ord}(\mathfrak{N}, \epsilon; A)$ is exactly the one $\mathbf{h}(\psi^+, \psi')$

employed in [HT1] page 240 (when specialized to the CM component there) if A is a complete p–adic valuation ring.

Let $\mathbf{\Lambda}_A = A[[\mathbf{\Gamma}]]$ for the maximal torsion-free quotient $\mathbf{\Gamma}$ of $\mathbf{G}$. We fix a splitting $\mathbf{G} = \mathbf{\Gamma} \times \mathbf{G}_{tor}$ for a finite group $\mathbf{G}_{tor}$. If A is a complete p–adic valuation ring, then $\mathbf{h}^{n.ord}(\mathfrak{N}, \epsilon; A)$ is a torsion-free $\mathbf{\Lambda}_A$–algebra of finite rank and is $\mathbf{\Lambda}_A$–free under some mild conditions on $\mathfrak{N}$ and ϵ ([PAF] 4.2.12). Take a point $P \in \mathrm{Spf}(\mathbf{\Lambda})(A) = \mathrm{Hom}_{cont}(\mathbf{\Gamma}, A^\times)$. Regarding P as a character of $\mathbf{G}$, we call P *arithmetic* if it is given locally by an algebraic character $\kappa(P) \in X^*(T^2)$ with $\kappa_1(P) - \kappa_2(P) \geq I$. Thus if P is arithmetic, $\epsilon_P = P\kappa(P)^{-1}$ is a character of $T^2(O/\mathfrak{p}^\alpha\mathfrak{N})$ for some multi-exponent $\alpha \geq 0$. Similarly, the restriction of P to $Cl_F^+(\mathfrak{N}p^\infty)$ is a p–adic Hecke character ϵ_{P+} induced by an arithmetic Hecke character of infinity type $(1 - [\kappa(P)])I$. As long as P is arithmetic, we have a canonical specialization morphism:

$$\mathbf{h}^{n.ord}(\mathfrak{N}, \epsilon; A) \otimes_{\mathbf{\Lambda}_A, P} A \twoheadrightarrow h^{n.ord}_{\kappa(P)}(\mathfrak{N}\mathfrak{p}^\alpha, \epsilon_{P+}; A),$$

which is an isogeny (surjective and of finite kernel) and is an isomorphism if $\mathbf{h}^{n.ord}$ is $\mathbf{\Lambda}_A$–free. The specialization morphism takes the generators $\mathbb{T}(y)$ to $\mathbb{T}(y)$.

3 Eisenstein series

We shall study the q–expansion, Hecke eigenvalues and special values at CM points of an Eisenstein series defined on $\mathfrak{M}(\mathfrak{c}; \mathfrak{N})$.

3.1 Arithmetic Hecke characters

Recall the CM type Σ ordinary at p and the prime ideal $\mathfrak{l}$ of O introduced in the introduction. We sometimes regard Σ as a character of $T_M = \mathrm{Res}_{M/\mathbb{Q}}\mathbb{G}_m$ sending $x \in M^\times$ to $x^\Sigma = \prod_{\sigma\in\Sigma}\sigma(x)$. More generally, each integral linear combination $\kappa = \sum_{\sigma\in\Sigma\sqcup\Sigma c}\kappa_\sigma\sigma$ is regarded as a character of T_M by $x \mapsto \prod_\sigma \sigma(x)^{\kappa_\sigma}$. We fix an arithmetic Hecke character λ of infinity type $k\Sigma + \kappa(1-c)$ for $\kappa = \sum_{\sigma\in\Sigma}\kappa_\sigma\sigma \in \mathbb{Z}[\Sigma]$ and an integer k. This implies, regarding λ as an idele character of $T_M(\mathbb{A})$, $\lambda(x_\infty) = x_\infty^{k\Sigma+\kappa(1-c)}$ for $x_\infty \in T_M(\mathbb{R})$. We assume the following three conditions:

(crt) *$k > 0$ and $\kappa \geq 0$, where we write $\kappa \geq 0$ if $\kappa_\sigma \geq 0$ for all σ.*

(opl) *The conductor $\mathfrak{C}$ of λ is prime to p and $(\ell) = \mathfrak{l} \cap \mathbb{Z}$.*

(spt) *The ideal $\mathfrak{C}$ is a product of primes split over F.*

3.2 Hilbert modular Eisenstein series

We shall define an Eisenstein series whose special values at CM points interpolate the values $L(0, \lambda\chi)$ for anticyclotomic characters χ of finite order.

We split the conductor $\mathfrak{C}$ in the following way: $\mathfrak{C} = \mathfrak{F}\mathfrak{F}_c$ with $\mathfrak{F} + \mathfrak{F}_c = R$ and $\mathfrak{F} \subset \mathfrak{F}_c^c$. This is possible by (spt). We then define $\mathfrak{f} = \mathfrak{F} \cap O$ and $\mathfrak{f}' = \mathfrak{F}_c \cap O$. Then $\mathfrak{f} \subset \mathfrak{f}'$. Here $X = O/\mathfrak{f} \cong R/\mathfrak{F}$ and $Y = O/\mathfrak{f}' \cong R/\mathfrak{F}_c$. Let $\phi : X \times Y \to \mathbb{C}$ be a function such that $\phi(\varepsilon^{-1}x, \varepsilon y) = N(\varepsilon)^k \phi(x, y)$ for all $\varepsilon \in O^\times$ with the integer k as above. We put $X^* = \mathfrak{f}^*/O^*$; so, X^* is naturally the Pontryagin dual module of X under the pairing $(x^*, x) = \mathbf{e}_F(x^*x) = \epsilon(\mathrm{Tr}(x^*x))$, where $\epsilon(x) = \exp(2\pi i x)$ for $x \in \mathbb{C}$. We define the partial Fourier transform $P\phi : X^* \times Y \to \mathbb{C}$ of ϕ by

$$P\phi(x, y) = N(\mathfrak{f})^{-1} \sum_{a \in X} \phi(a, y)\mathbf{e}_F(ax), \tag{3.1}$$

where $\mathbf{e}_F$ is the restriction of the standard additive character of the adele ring $F_{\mathbb{A}}$ to the local component $F_\mathfrak{f}$ at $\mathfrak{f}$.

A function ϕ as above can be interpreted as a function of $(\mathcal{L}, \Lambda, i, i')$ in 2.5. Here $i : X^* \hookrightarrow \mathfrak{f}^{-1}\mathcal{L}/\mathcal{L}$ is the level $\mathfrak{f}$–structure. We define an $O_\mathfrak{f}$–submodule $PV(\mathcal{L}) \subset \mathcal{L} \otimes_O F_\mathfrak{f}$ specified by the following conditions:

$$PV(\mathcal{L}) \supset \mathcal{L} \otimes_O O_\mathfrak{f},\ PV(\mathcal{L})/\mathcal{L}_\mathfrak{f} = \mathrm{Im}(i)\ (\mathcal{L}_\mathfrak{f} = \mathcal{L} \otimes_O O_\mathfrak{f}). \tag{PV}$$

By definition, we may regard

$$i^{-1} : PV(\mathcal{L}) \twoheadrightarrow PV(\mathcal{L})/\left(\mathcal{L} \otimes_O O_\mathfrak{f}\right) \cong \mathfrak{f}^*/O^*.$$

By Pontryagin duality under $\mathrm{Tr} \circ \lambda$, the dual map of i gives rise to $i' : PV(\mathcal{L}) \twoheadrightarrow O/\mathfrak{f}$. Taking a lift $\widetilde{i} : (\mathfrak{f}^2)^*/O^* \hookrightarrow PV(\mathcal{L})/\mathfrak{f}\mathcal{L}_\mathfrak{f}$ with $\widetilde{i} \bmod \mathcal{L}_\mathfrak{f} = i$, we have an exact sequence:

$$0 \to (\mathfrak{f}^2)^*/O^* \xrightarrow{\widetilde{i}} PV(\mathcal{L})/\mathfrak{f}\mathcal{L}_\mathfrak{f} \xrightarrow{i'} O/\mathfrak{f} \to 0.$$

This sequence is kept under $\alpha \in \mathrm{Aut}(\mathcal{L})$ with unipotent reduction modulo $\mathfrak{f}^2$, and hence, the pair (i, i') gives a level $\Gamma_1^1(\mathfrak{f}^2)$–structure: Once we have chosen a generator f of $\mathfrak{f}$ in $O_\mathfrak{f}$, by the commutativity of the following diagram:

$$\begin{array}{ccccc} (\mathfrak{f}^2)^*/O^* & \xrightarrow{\widetilde{i}} & PV(\mathcal{L})/\mathfrak{f}\mathcal{L}_\mathfrak{f} & \xrightarrow{i'} & O/\mathfrak{f} \\ \downarrow \wr & & f \downarrow \cap & & f \downarrow \cap \\ (\mathfrak{f}^2)^*/O^* & \longrightarrow & X[\mathfrak{f}^2] & \longrightarrow & O/\mathfrak{f}^2, \end{array} \tag{3.2}$$

giving (i, i') is equivalent to having the bottom sequence of maps in the above diagram. This explains why the pair (i, i') gives rise to a level $\Gamma_1^1(\mathfrak{f}^2)$–structure;

strictly speaking, the exact level group is given by:

$$\Gamma^1_{1,0}(\mathfrak{f}^2) = \left\{ \left(\begin{smallmatrix} a & b \\ c & d \end{smallmatrix}\right) \in SL_2(O_\mathfrak{f}) \middle| a \equiv d \equiv 1 \mod \mathfrak{f},\ c \equiv 0 \mod \mathfrak{f}^2 \right\}. \quad (3.3)$$

We regard $P\phi$ as a function of $\mathcal{L} \otimes_O F$ supported on $(\mathfrak{f}^{-2}\mathcal{L}) \cap PV(\mathcal{L})$ by

$$P\phi(w) = \begin{cases} P\phi(i^{-1}(w), i'(w)) & \text{if } (w \mod \mathcal{L}) \in \mathrm{Im}(i), \\ 0 & \text{otherwise.} \end{cases} \quad (3.4)$$

For each $w = (w_\sigma) \in F \otimes_\mathbb{Q} \mathbb{C} = \mathbb{C}^I$, the norm map $N(w) = \prod_{\sigma \in I} w_\sigma$ is well defined.

For any positive integer $k > 0$, we can now define the Eisenstein series E_k. Writing $\underline{\mathcal{L}} = (\mathcal{L}, \lambda, i)$ for simplicity, we define the value $E_k(\underline{\mathcal{L}}; \phi, \mathfrak{c})$ by

$$E_k(\underline{\mathcal{L}}; \phi, \mathfrak{c}) = \frac{\{(-1)^k \Gamma(k+s)\}^{[F:\mathbb{Q}]}}{\sqrt{|D_F|}} \sum_{w \in \mathfrak{f}^{-1}\mathcal{L}/O^\times}{}' \frac{P\phi(w)}{N(w)^k |N(w)|^{2s}} \Bigg|_{s=0}. \quad (3.5)$$

Here "$\sum'$" indicates that we are excluding $w = 0$ from the summation. As shown by Hecke, this type of series is convergent when the real part of s is sufficiently large and can be continued to a meromorphic function well defined at $s = 0$ (as long as either $k \geq 2$ or $\phi(a, 0) = 0$ for all a). The weight of the Eisenstein series is the parallel weight $kI = \sum_\sigma k\sigma$. If either $k \geq 2$ or $\phi(a, 0) = 0$ for all a, the function $E_k(\mathfrak{c}, \phi)$ gives an element in $G_{kI}(\mathfrak{c}, \mathfrak{f}^2; \mathbb{C})$, whose q–expansion at the cusp $(\mathfrak{a}, \mathfrak{b})$ computed in [HT1] Section 2 is given by

$$\begin{aligned} N(\mathfrak{a})^{-1} E_k(\phi, \mathfrak{c})_{\mathfrak{a},\mathfrak{b}}(q) &= 2^{-[F:\mathbb{Q}]} L(1-k; \phi, \mathfrak{a}) \\ &\quad + \sum_{0 \ll \xi \in \mathfrak{ab}} \sum_{\substack{(a,b) \in (\mathfrak{a} \times \mathfrak{b})/O^\times \\ ab = \xi}} \phi(a, b) \frac{N(a)}{|N(a)|} N(a)^{k-1} q^\xi, \end{aligned} \quad (3.6)$$

where $L(s; \phi, \mathfrak{a})$ is the partial L–function given by the Dirichlet series:

$$\sum_{\xi \in (\mathfrak{a} - \{0\})/O^\times} \phi(\xi, 0) \left(\frac{N(\xi)}{|N(\xi)|} \right)^k |N(\xi)|^{-s}.$$

If $\phi(x, y) = \phi_X(x)\phi_Y(y)$ for two functions $\phi_X : X \to \mathbb{C}$ and $\phi_Y : Y \to \mathbb{C}$ with ϕ_Y factoring through $O/\mathfrak{f}'$, then we can check easily that $E_k(\phi) \in G_{kI}(\mathfrak{c}, \mathfrak{f}\mathfrak{f}'; \mathbb{C})$.

3.3 Hecke eigenvalues

We take a Hecke character λ as in 3.1. Then the restriction $\lambda_{\mathfrak{C}}^{-1} : R_{\mathfrak{F}}^{\times} \times R_{\mathfrak{F}_c}^{\times} \to \mathcal{W}^{\times}$ induces a locally constant function $\psi : (O/\mathfrak{f}) \times (O/\mathfrak{f}') \to \mathcal{W}$ supported on $(O/\mathfrak{f})^{\times} \times (O/\mathfrak{f}')^{\times}$, because $\lambda_{\mathfrak{C}}$ factor through $(R/\mathfrak{C})^{\times}$ which is canonically isomorphic to $(O/\mathfrak{f})^{\times} \times (O/\mathfrak{f}')^{\times}$. Since λ is trivial on $M^{\times}$, ψ satisfies

$$\psi(\varepsilon x, \varepsilon y) = \varepsilon^{k\Sigma+\kappa(1-c)}\psi(x,y) = N(\varepsilon)^k \psi(x,y)$$

for any unit $\varepsilon \in O^{\times}$.

We regard the local uniformizer $\varpi_{\mathfrak{q}} \in O_{\mathfrak{q}}$ as an idele. For each ideal $\mathfrak{A}$ of F, decomposing $\mathfrak{A} = \prod_{\mathfrak{q}} \mathfrak{q}^{e(\mathfrak{q})}$ for primes $\mathfrak{q}$, we define $\varpi^{e(\mathfrak{A})} = \prod_{\mathfrak{q}} \varpi_{\mathfrak{q}}^{e(\mathfrak{q})} \in F_{\mathbb{A}}^{\times}$. We then define a partial Fourier transform $\psi^{\circ} : X \times Y \to \mathcal{W}$ by

$$\psi^{\circ}(a,b) = \sum_{u \in O/\mathfrak{f}} \psi(u,b)\mathbf{e}_F(-ua\varpi^{-e(\mathfrak{f})}). \tag{3.7}$$

By the Fourier inversion formula, we have

$$P\psi^{\circ}(x,y) = \psi(\varpi^{e(\mathfrak{f})}x, y). \tag{3.8}$$

From this and the definition of $E_k(\underline{\mathcal{L}}) = E_k(\underline{\mathcal{L}}; \psi^{\circ}, \mathfrak{c})$, we find

$$E_k(X, \Lambda, i \circ x, i' \circ y, a\omega) = N(a)^{-k}\lambda_{\mathfrak{F}}(x)\lambda_{\mathfrak{F}_c}^{-1}(y)E_k(X, \Lambda, i, i', \omega) \tag{3.9}$$

for $x \in (O/\mathfrak{f})^{\times} = (R/\mathfrak{F})^{\times}$ and $y \in (O/\mathfrak{f}')^{\times} = (R/\mathfrak{F}_c)^{\times}$. Because of this, $E_k(\psi^{\circ}, \mathfrak{c})$ actually belongs to $G_{kI}(\mathfrak{c}, \Gamma_0(\mathfrak{f}\mathfrak{f}'), \epsilon_{\lambda}; \mathbb{C})$ for $\epsilon_{\lambda,1} = \lambda_{\mathfrak{F}_c}$ and $\epsilon_{\lambda,2} = \lambda_{\mathfrak{F}}$ identifying $O_{\mathfrak{F}} = R_{\mathfrak{F}}$ and $O_{\mathfrak{F}_c} = R_{\mathfrak{F}_c}$. Recall $G_{kI}(\Gamma_0(\mathfrak{N}), \epsilon; \mathbb{C}) = \bigoplus_{\mathfrak{c} \in Cl_F^+} G_{kI}(\mathfrak{c}, \Gamma_0(\mathfrak{N}), \epsilon; \mathbb{C})$. Via this decomposition we extend each Cl_F^+–tuple $(f_{\mathfrak{c}})_{\mathfrak{c}}$ in $G_{kI}(\Gamma_0(\mathfrak{f}\mathfrak{f}'), \epsilon_{\lambda}; \mathbb{C})$ to an automorphic form $f \in G_{kI}(\mathfrak{f}\mathfrak{f}', \epsilon_{\lambda+}; \mathbb{C})$ as follows:

(i) $f(zx) = \lambda(z)|z|_{\mathbb{A}} f(x)$ for z in the center $F_{\mathbb{A}}^{\times} \subset G(\mathbb{A})$ (so $\epsilon_{\lambda+}(z) = \lambda(z)|z|_{\mathbb{A}}$);

(ii) $f(xu) = \epsilon_{\lambda}(u)f(x)$ for $u \in U_0(\mathfrak{f}\mathfrak{f}')$;

(iii) $f_x = f_{\mathfrak{c}}$ if $x = \left(\begin{smallmatrix} c & 0 \\ 0 & 1 \end{smallmatrix}\right)$ for an idele c with $cO = \mathfrak{c}$ and $c^{(\mathfrak{c})} = 1$.

We now compute the effect of the operator $\langle \mathfrak{q} \rangle$ (defined above (2.25)) on E_k for a fractional ideal $\mathfrak{q}$ prime to the level $\mathfrak{f}$. A geometric interpretation of the operator $\langle \mathfrak{q} \rangle$ is discussed, for example, in [H04b] (5.3) (or [PAF] 4.1.9), and has the following effect on an AVRM X: $X \mapsto X \otimes_O \mathfrak{q}$. The level structure i is intact under this process. The $\mathfrak{c}$–polarization Λ induces a $\mathfrak{c}\mathfrak{q}^{-2}$–polarization on $X \otimes_O \mathfrak{q}$. On the lattice side, $\langle \mathfrak{q} \rangle$ brings $\mathcal{L}$ to $\mathfrak{q}\mathcal{L}$.

To simplify our notation, we write

$$t(w;s) = \frac{\lambda_{\mathfrak{F}}^{-1}(\varpi^{e(\mathfrak{f})} i^{-1}(w))\lambda_{\mathfrak{F}_c}^{-1}(i'(w))}{N(w)^k |N(w)|^{2s}}$$

for each term of the Eisenstein series and $c(s)$ for the Gamma factor in front of the summation, where $D = N(\mathfrak{d})$ is the discriminant of F. First we compute the effect of the operator $\langle \mathfrak{q} \rangle$ when $\mathfrak{q} = (\xi)$ for $\xi \in F$ naively as follows:

$$E_k((\mathfrak{q}\mathcal{L}, \Lambda, i); \psi^\circ, \mathfrak{c}\mathfrak{q}^{-2}) = c(s) \sum_{w \in \mathfrak{f}^{-1}\mathfrak{q}\mathcal{L}/O^\times}{}' t(w;s)|_{s=0}$$
$$\overset{\mathfrak{q}=(\xi)}{=} c(s) \sum_{w \in \mathfrak{f}^{-1}\mathcal{L}/O^\times}{}' t(\xi w;s)|_{s=0} = \lambda_{\mathfrak{F}}(\xi)^{-1}\lambda_{\mathfrak{F}_c}^{-1}(\xi)N(\xi)^{-k}E_k(\underline{\mathcal{L}};\psi^\circ,\mathfrak{c}). \tag{3.10}$$

Here we agree to put $\lambda_{\mathfrak{F}}^{-1}(x) = 0$ if $xR_{\mathfrak{F}} \neq R_{\mathfrak{F}}$ and $\lambda_{\mathfrak{F}_c}^{-1}(y) = 0$ if $yR_{\mathfrak{F}_c} \neq R_{\mathfrak{F}_c}$.

The result of the above naive calculation of the eigenvalue of $\langle \mathfrak{q} \rangle$ shows that our way of extending the Eisenstein series $(E_k(\psi^\circ;\mathfrak{c}))_{\mathfrak{c}}$ to an adelic automorphic form $G(\mathbb{A})$ is correct (and canonical): This claim follows from

$$\lambda_{\mathfrak{F}}(\xi)^{-1}\lambda_{\mathfrak{F}_c}^{-1}(\xi)N(\xi)^{-k} = \lambda(\mathfrak{q}) = \lambda(\varpi_{\mathfrak{q}})|\varpi_{\mathfrak{q}}|_{\mathbb{A}}N(\mathfrak{q}),$$

because the operator $\langle \mathfrak{q} \rangle$ on $G_{kI}(U,\epsilon;\mathbb{C})$ is defined (above (2.25)) to be the central action of $\varpi_{\mathfrak{q}} \in F_{\mathbb{A}}^\times$ (that is, multiplication by $\lambda(\varpi_{\mathfrak{q}})|\varpi_{\mathfrak{q}}|_{\mathbb{A}}$) times $N(\mathfrak{q})$. We obtain

$$\begin{aligned}&\lambda_{\mathfrak{F}_c}(\xi^2)E_k|\langle \mathfrak{q}^{-1}\rangle(\underline{\mathcal{L}};\psi^\circ,\mathfrak{c})\\ &\quad = E_k((\mathfrak{q}^{-1}\mathcal{L},\Lambda,i);\psi^\circ,\mathfrak{c}\mathfrak{q}^2) = \lambda_{\mathfrak{F}}(\xi)\lambda_{\mathfrak{F}_c}(\xi)N(\xi)^k E_k(\underline{\mathcal{L}};\psi^\circ,\mathfrak{c}).\end{aligned} \tag{3.11}$$

The factor $\lambda_{\mathfrak{F}_c}(\xi^2)$ in the left-hand-side comes from the fact that i' with respect to the $\mathfrak{c}$–polarization $\xi^2\Lambda$ of $\mathfrak{q}^{-1}\mathcal{L}$ is the multiple by ξ^2 of i' with respect to $\mathfrak{c}\mathfrak{q}^2$–polarization Λ of $\mathfrak{q}^{-1}\mathcal{L}$.

We now compute the effect of the Hecke operator $T(1,\mathfrak{q}) = T(\mathfrak{q})$ for a prime $\mathfrak{q} \nmid \mathfrak{f}$. Here we write $\mathcal{L}'$ for an O-lattice with $\mathcal{L}'/\mathcal{L} \cong O/\mathfrak{q}$. Then $\mathcal{L}' \wedge \mathcal{L}' = (\mathfrak{q}\mathfrak{c})^*$; so, Λ induces a $\mathfrak{q}\mathfrak{c}$–polarization on $\mathcal{L}'$, and similarly it induces $\mathfrak{q}^2\mathfrak{c}$-polarization on $\mathfrak{q}^{-1}\mathcal{L}$. By (2.8), $E_k|T(\mathfrak{q})$ is the sum of the terms $t(\ell;s)$ with multiplicity extended over $\mathfrak{q}^{-1}\mathfrak{f}^{-1}\mathcal{L}$. The multiplicity for each $\ell \in \mathfrak{f}^{-1}\mathcal{L}$ is $N(\mathfrak{q})+1$ and only once for $\ell \in \mathfrak{q}^{-1}\mathfrak{f}^{-1}\mathcal{L} - \mathfrak{f}^{-1}\mathcal{L}$ (thus, $N(\mathfrak{q})$ times for

$\ell \in \mathfrak{f}^{-1}\mathcal{L}$ and once for $\ell \in \mathfrak{q}^{-1}\mathfrak{f}^{-1}\mathcal{L}$). This shows

$$\begin{aligned} &c(0)^{-1}N(\mathfrak{q})E_k|T(\mathfrak{q})(\underline{\mathcal{L}};\psi^\circ,\mathfrak{q}\mathfrak{c}) \\ &= \sum_{\mathcal{L}'}\left\{\sum'_{w\in\mathfrak{f}^{-1}\mathcal{L}'/O^\times} t(w;s) + \sum'_{w\in\mathfrak{f}^{-1}\mathfrak{q}^{-1}\mathcal{L}'/O^\times} t(w;s)\right\}\Bigg|_{s=0} \\ &= c(0)^{-1}\left\{N(\mathfrak{q})E_k(\underline{\mathcal{L}};\psi^\circ,\mathfrak{c}) + E_k|\langle\mathfrak{q}\rangle^{-1}(\underline{\mathcal{L}};\psi^\circ,\mathfrak{c}\mathfrak{q}^2)\right\}. \end{aligned}$$

In short, we have

$$E_k(\psi^\circ,\mathfrak{q}\mathfrak{c})|T(\mathfrak{q}) = E_k(\psi^\circ,\mathfrak{c}) + N(\mathfrak{q})^{-1}E_k(\psi^\circ,\mathfrak{c}\mathfrak{q}^2)|\langle\mathfrak{q}\rangle^{-1}. \qquad (3.12)$$

Suppose that $\mathfrak{q}$ is principal generated by a totally positive $\xi \in F$. Substituting $\xi^{-1}\Lambda$ for Λ, i' will be transformed into $\xi^{-1}i'$, and we have

$$\lambda_{\mathfrak{F}_c}(\xi)E_k(\psi^\circ,\mathfrak{c})|T(\mathfrak{q}) = E_k(\psi^\circ,\mathfrak{q}\mathfrak{c})|T(\mathfrak{q})$$

We combine this with (3.11) assuming $\mathfrak{q} = (\xi)$ with $0 \ll \xi \in F$:

$$E_k(\psi^\circ,\mathfrak{c})|T(\mathfrak{q})(\underline{\mathcal{L}}) = (\lambda_{\mathfrak{F}_c}^{-1}(\xi) + \lambda_{\mathfrak{F}}(\xi)N(\mathfrak{q})^{k-1})E_k(\psi^\circ,\mathfrak{c}), \qquad (3.13)$$

which also follows from (3.6) noting that $\psi^\circ(a,b) = G(\lambda_{\mathfrak{F}}^{-1})\lambda_{\mathfrak{F}}(a)\lambda_{\mathfrak{F}_c}^{-1}(b)$ for the Gauss sum $G(\lambda_{\mathfrak{F}}^{-1})$.

We now look into the operator $[\mathfrak{q}]$ for a prime $\mathfrak{q}$ outside the level $\mathfrak{f}$. This operator brings a level $\Gamma_0(\mathfrak{q})$–test object (X,C,i) with level $\mathfrak{f}$ structure i outside $\mathfrak{q}$ to $(X/C,i)$, where the level $\mathfrak{f}$–structure i is intact under the quotient map: $X \to X/C$. On the lattice side, taking the lattice $\mathcal{L}_C$ with $\mathcal{L}_C/\mathcal{L} = C$, it is defined as follows:

$$f|[\mathfrak{q}](\mathcal{L},C,\Lambda,i) = N(\mathfrak{q})^{-1}f(\mathcal{L}_C,\Lambda,i). \qquad (3.14)$$

The above operator is useful to relate $U(\mathfrak{q})$ and $T(\mathfrak{q})$. By definition,

$$f|U(\mathfrak{q})(\mathcal{L},\Lambda,C,i) = N(\mathfrak{q})^{-1}\sum_{\mathcal{L}',\mathcal{L}'\neq\mathcal{L}_C} f(\mathcal{L}',\Lambda,C',i)$$

for $C' = \mathcal{L}_C + \mathcal{L}'/\mathcal{L}' = \mathfrak{q}^{-1}\mathcal{L}/\mathcal{L}'$. Thus we have

$$U(\mathfrak{q}) = T(\mathfrak{q}) - [\mathfrak{q}]. \qquad (3.15)$$

A similar computation yields:

$$[\mathfrak{q}] \circ U(\mathfrak{q}) = N(\mathfrak{q})^{-1}\langle\mathfrak{q}\rangle^{-1}. \qquad (3.16)$$

Lemma 3.1 *Let $\mathfrak{q}$ be a prime outside $\mathfrak{f}$. Suppose that $\mathfrak{q}^h = (\xi)$ for a totally positive $\xi \in F$. Let $\mathbb{E}'_k(\psi,\mathfrak{c}) = E_k(\psi^\circ,\mathfrak{c}) - E_k(\psi^\circ,\mathfrak{c}\mathfrak{q})|[\mathfrak{q}]$ and $\mathbb{E}_k(\psi,\mathfrak{c}) = E_k(\psi^\circ,\mathfrak{c}) - N(\mathfrak{q})E_k(\psi^\circ,\mathfrak{c}\mathfrak{q}^{-1})|\langle\mathfrak{q}\rangle|[\mathfrak{q}]$. Then we have*

(1) $\mathbb{E}'_k(\psi,\mathfrak{c})|U(\mathfrak{q}) = E_k(\psi^\circ,\mathfrak{q}^{-1}\mathfrak{c}) - E_k(\psi^\circ,\mathfrak{c})|[\mathfrak{q}]$,
(2) $\mathbb{E}'_k(\psi,\mathfrak{c})|U(\mathfrak{q}^h) = \lambda^{-1}_{\mathfrak{F}_c}(\xi)\mathbb{E}'_k(\psi,\mathfrak{c})$,
(3) $\mathbb{E}_k(\psi,\mathfrak{c})|U(\mathfrak{q}) = (E_k(\psi^\circ,\mathfrak{q}\mathfrak{c}) - N(\mathfrak{q})E_k(\psi^\circ,\mathfrak{c})|\langle\mathfrak{q}\rangle|[\mathfrak{q}])|(N(\mathfrak{q})^{-1}\langle\mathfrak{q}\rangle^{-1})$
(4) $\mathbb{E}_k(\psi,\mathfrak{c})|U(\mathfrak{q}^h) = \lambda_{\mathfrak{F}}(\xi)N(\mathfrak{q})^{h(k-1)}\mathbb{E}_k(\psi,\mathfrak{c})$.

Proof We prove (1) and (3), because (2) and (4) follow by iteration of these formulas combined with the fact: $\lambda_{\mathfrak{F}_c}(\xi)E_k(\psi^\circ,\mathfrak{c}) = E_k(\psi^\circ,\xi\mathfrak{c})$ for a totally positive $\xi \in F$. Since (3) can be proven similarly, we describe computation to get (1), writing $E_k(\mathfrak{c}) = E_k(\psi^\circ,\mathfrak{c})$:

$$\begin{aligned}
\mathbb{E}'_k(\psi;\mathfrak{c})|U(\mathfrak{q}) &= E_k(\mathfrak{c})|U(\mathfrak{q}) - E_k(\mathfrak{c}\mathfrak{q})|[\mathfrak{q}]|U(\mathfrak{q})\\
&\overset{(3.16)}{=} E_k(\mathfrak{c})|U(\mathfrak{q}) - N(\mathfrak{q})^{-1}E_k(\mathfrak{c}\mathfrak{q})|\langle\mathfrak{q}\rangle^{-1}\\
&\overset{(3.15)}{=} E_k(\mathfrak{c})|T(\mathfrak{q}) - E_k(\mathfrak{c}\mathfrak{q})|[\mathfrak{q}] - N(\mathfrak{q})^{-1}E_k(\mathfrak{c}\mathfrak{q})|\langle\mathfrak{q}\rangle^{-1}\\
&\overset{(3.12)}{=} E_k(\mathfrak{c}\mathfrak{q}^{-1}) + N(\mathfrak{q})^{-1}E_k(\mathfrak{c}\mathfrak{q})|\langle\mathfrak{q}\rangle^{-1} - E_k(\mathfrak{c})|[\mathfrak{q}] - N(\mathfrak{q})^{-1}E_k(\mathfrak{c}\mathfrak{q})|\langle\mathfrak{q}\rangle^{-1}\\
&= E_k(\mathfrak{c}\mathfrak{q}^{-1}) - E_k(\mathfrak{c})|[\mathfrak{q}].
\end{aligned}$$

□

Remark 3.2 As follows from the formulas in [H96] 2.4 (T1) and [H91] Section 7.G, the Hecke operator $T(\mathfrak{q})$ and $U(\mathfrak{q})$ commutes with the Katz differential operator as long as $\mathfrak{q} \nmid p$. Thus for $\mathbb{E}(\lambda,\mathfrak{c}) = d^\kappa\mathbb{E}_k(\psi,\mathfrak{c})$ and $\mathbb{E}'(\lambda,\mathfrak{c}) = d^\kappa\mathbb{E}'_k(\psi,\mathfrak{c})$, we have under the notation of Lemma 3.1

$$\begin{aligned}
\mathbb{E}'(\lambda,\mathfrak{c})|U(\mathfrak{q}^h) &= \lambda^{-1}_{\mathfrak{F}_c}(\xi)\mathbb{E}'(\lambda,\mathfrak{c}),\\
\mathbb{E}(\lambda,\mathfrak{c})|U(\mathfrak{q}^h) &= \lambda_{\mathfrak{F}}(\xi)N(\mathfrak{q})^{h(k-1)}\mathbb{E}(\lambda,\mathfrak{c}).
\end{aligned} \tag{3.17}$$

3.4 Values at CM points

We take a proper R_{n+1}–ideal $\mathfrak{a}$ for $n > 0$, and regard it as a lattice in $\mathbb{C}^\Sigma$ by $a \mapsto (a^\sigma)_{\sigma\in\Sigma}$. Then $\Lambda(\mathfrak{a})$ induces a polarization of $\mathfrak{a} \subset \mathbb{C}^\Sigma$. We suppose that $\mathfrak{a}$ is prime to $\mathfrak{C}$ (the conductor of λ). For a p–adic modular form f of the form $d^\kappa g$ for classical $g \in G_{kI}(\mathfrak{c},\Gamma_{1,0}(\mathfrak{f}^2);\mathcal{W})$, we have by (K) in 2.6

$$\frac{f(x(\mathfrak{a}),\omega_p)}{\Omega_p^{k\Sigma+2\kappa}} = f(x(\mathfrak{a}),\omega(\mathfrak{a})) = \frac{f(x(\mathfrak{a}),\omega_\infty)}{\Omega_\infty^{k\Sigma+2\kappa}}.$$

Here $x(\mathfrak{a})$ is the test object: $x(\mathfrak{a}) = (X(\mathfrak{a}),\Lambda(\mathfrak{a}),i(\mathfrak{a}),i'(\mathfrak{a}))_{/\mathcal{W}}$.

We write $c_0 = (-1)^{k[F:\mathbb{Q}]}\frac{\pi^\kappa\Gamma_\Sigma(k\Sigma+\kappa)}{\mathrm{Im}(\delta)^\kappa\sqrt{D}\Omega_\infty^{k\Sigma+2\kappa}}$. Here $\Gamma_\Sigma(s) = \prod_{\sigma\in\Sigma}\Gamma(s_\sigma)$, $\Omega_\infty^s = \prod_\sigma \Omega_\sigma^{s_\sigma}$, $\mathrm{Im}(\delta)^s = \prod_\sigma \mathrm{Im}(\delta^s)^{s_\sigma}$, and so on, for $s = \sum_\sigma s_\sigma\sigma$. By definition (see [H04c] 4.2), we find, for $e = [R^\times : O^\times]$,

$$\begin{aligned}
&(c_0 e)^{-1}\delta_{kI}^{\kappa} E_k(\mathfrak{c})(x(\mathfrak{a}), \omega(\mathfrak{a})) \\
&= \lambda_{\mathfrak{C}}^{-1}(\varpi^{e(\mathfrak{F})}) \sum_{w \in \mathfrak{F}^{-1}\mathfrak{a}/R^\times}{}' \frac{\lambda_{\mathfrak{C}}^{-1}(w)\lambda(w^{(\infty)})}{N_{M/\mathbb{Q}}(w)^s}\Big|_{s=0} \\
&= \lambda_{\mathfrak{C}}^{-1}(\varpi^{e(\mathfrak{F})})\lambda(\mathfrak{a}) N_{M/\mathbb{Q}}(\mathfrak{F}\mathfrak{a}^{-1})^s \sum_{w\mathfrak{F}\mathfrak{a}^{-1} \subset R_{n+1}}{}' \frac{\lambda(w\mathfrak{F}\mathfrak{a}^{-1})}{N_{M/\mathbb{Q}}(w\mathfrak{F}\mathfrak{a}^{-1})^s}\Big|_{s=0} \\
&= \lambda_{\mathfrak{C}}^{-1}(\varpi^{e(\mathfrak{F})})\lambda(\mathfrak{a}) L^{n+1}_{[\mathfrak{F}\mathfrak{a}^{-1}]}(0, \lambda),
\end{aligned} \tag{3.18}$$

where for an ideal class $[\mathfrak{A}] \in Cl_{n+1}$ represented by a proper R_{n+1}–ideal $\mathfrak{A}$,

$$L^{n+1}_{[\mathfrak{A}]}(s, \lambda) = \sum_{\mathfrak{b} \in [\mathfrak{A}]} \lambda(\mathfrak{b}) N_{M/\mathbb{Q}}(\mathfrak{b})^{-s}$$

is the partial L–function of the class $[\mathfrak{A}]$ for $\mathfrak{b}$ running over all R_{n+1}–proper integral ideals prime to $\mathfrak{C}$ in the class $[\mathfrak{A}]$. In the second line of (3.18), we regard λ as an idele character and in the other lines as an ideal character. For an idele a with $a\widehat{R} = \mathfrak{a}\widehat{R}$ and $a_{\mathfrak{C}} = 1$, we have $\lambda(a^{(\infty)}) = \lambda(\mathfrak{a})$.

We put $\mathbb{E}(\lambda, \mathfrak{c}) = d^{\kappa}\mathbb{E}_k(\psi, \mathfrak{c})$ and $\mathbb{E}'(\lambda, \mathfrak{c}) = d^{\kappa}\mathbb{E}'_k(\psi, \mathfrak{c})$ as in Remark 3.2. We want to evaluate $\mathbb{E}(\lambda, \mathfrak{c})$ and $\mathbb{E}'(\lambda, \mathfrak{c})$ at $x = (x(\mathfrak{a}), \omega(\mathfrak{a}))$. Here $\mathfrak{c}$ is the polarization ideal of $\Lambda(\mathfrak{a})$; so, if confusion is unlikely, we often omit the reference to $\mathfrak{c}$ (which is determined by $\mathfrak{a}$). Thus we write, for example, $\mathbb{E}(\lambda)$ and $\mathbb{E}'(\lambda)$ for $\mathbb{E}(\lambda, \mathfrak{c})$ and $\mathbb{E}'(\lambda, \mathfrak{c})$. Then by definition and (K) in 2.6, we have for $x = (x(\mathfrak{a}), \omega(\mathfrak{a}))$

$$\begin{aligned}
\mathbb{E}'(\lambda)(x) &= \delta_{kI}^{\kappa} E_k(\psi^\circ, \mathfrak{c})(x) - N(\mathfrak{q})^{-1}\delta_{kI}^{\kappa} E_k(\psi^\circ, \mathfrak{c}\mathfrak{q})(x(\mathfrak{a}R_n), \omega(\mathfrak{a}R_n)) \\
\mathbb{E}(\lambda)(x) &= \delta_{kI}^{\kappa} E_k(\psi^\circ, \mathfrak{c})(x) - \delta_{kI}^{\kappa} E_k(\psi^\circ, \mathfrak{c}\mathfrak{q}^{-1})(x(\mathfrak{q}\mathfrak{a}R_n), \omega(\mathfrak{a}R_n))
\end{aligned} \tag{3.19}$$

because $C(\mathfrak{a}) = \mathfrak{a}R_n/\mathfrak{a}$ and hence $[\mathfrak{q}](x(\mathfrak{a})) = x(\mathfrak{a}R_n)$.

To simplify notation, write $\phi([\mathfrak{a}]) = \lambda(\mathfrak{a})^{-1}\phi(x(\mathfrak{a}), \omega(\mathfrak{a}))$. By (3.9), for $\phi = \mathbb{E}(\lambda)$ and $\mathbb{E}'(\lambda)$, the value $\phi([\mathfrak{a}])$ only depends on the ideal class $[\mathfrak{a}]$ but not the individual $\mathfrak{a}$. The formula (3.19) combined with (3.18) shows, for a proper R_{n+1}–ideal $\mathfrak{a}$,

$$\begin{aligned}
e^{-1}\lambda_{\mathfrak{C}}(\varpi^{e(\mathfrak{F})})\mathbb{E}'(\lambda)([\mathfrak{a}]) &= c_0\left(L^{n+1}_{[\mathfrak{F}\mathfrak{a}^{-1}]}(0, \lambda) - N(\mathfrak{q})^{-1} L^{n}_{[\mathfrak{F}\mathfrak{a}^{-1}R_n]}(0, \lambda)\right) \\
e^{-1}\lambda_{\mathfrak{C}}(\varpi^{e(\mathfrak{F})})\mathbb{E}(\lambda)([\mathfrak{a}]) &= c_0\left(L^{n+1}_{[\mathfrak{F}\mathfrak{a}^{-1}]}(0, \lambda) - \lambda(\mathfrak{q}) L^{n}_{[\mathfrak{F}\mathfrak{q}^{-1}\mathfrak{a}^{-1}R_n]}(0, \lambda)\right)
\end{aligned} \tag{3.20}$$

where $e = [R^\times : O^\times]$. Now we define

$$L^n(s,\lambda) = \sum_{\mathfrak{a}} \lambda(\mathfrak{a}) N_{M/\mathbb{Q}}(\mathfrak{a})^{-s}, \tag{3.21}$$

where $\mathfrak{a}$ runs over all proper ideals in R_n prime to $\mathfrak{C}$ and $N_{M/\mathbb{Q}}(\mathfrak{a}) = [R_n : \mathfrak{a}]$. For each primitive character $\chi : Cl_f \to \overline{\mathbb{Q}}^\times$, we pick $n+1 = mh$ so that $(m-1)h \le f \le n+1$, where $\mathfrak{q}^h = (\xi)$ for a totally positive $\xi \in F$. Then we have

$$\begin{aligned} e^{-1}\lambda_{\mathfrak{C}}(\varpi^{e(\mathfrak{F})}) &\sum_{[\mathfrak{a}]\in Cl_{n+1}} \chi(\mathfrak{a})\mathbb{E}'(\lambda)([\mathfrak{a}]) \\ &= c_0\chi(\mathfrak{F})\left(L^{n+1}(0,\lambda\chi^{-1}) - L^n(0,\lambda\chi^{-1})\right) \\ e^{-1}\lambda_{\mathfrak{C}}(\varpi^{e(\mathfrak{F})}) &\sum_{[\mathfrak{a}]\in Cl_{n+1}} \chi(\mathfrak{a})\mathbb{E}(\lambda)([\mathfrak{a}]) \\ &= c_0\chi(\mathfrak{F})\left(L^{n+1}(0,\lambda\chi^{-1}) - \lambda\chi^{-1}(\mathfrak{q})N(\mathfrak{q})L^n(0,\lambda\chi^{-1})\right). \end{aligned} \tag{3.22}$$

As computed in [H04c] 4.1 and [LAP] V.3.2, if $k \ge f$ then the Euler $\mathfrak{q}$–factor of $L^k(s,\chi^{-1}\lambda)$ is given by

$$\begin{aligned} &\sum_{j=0}^{k-f} (\chi^{-1}\lambda(\mathfrak{q}))^j N(\mathfrak{q})^{j-2sj} \quad \text{if } f > 0, \\ &\sum_{j=0}^{k-1} (\chi^{-1}\lambda(\mathfrak{q}))^j N(\mathfrak{q})^{j-2sj} \\ &\qquad + \left(N(\mathfrak{q}) - \left(\frac{M/F}{\mathfrak{q}}\right)\right)(\chi^{-1}\lambda(\mathfrak{q}))^k N(\mathfrak{q})^{k-1-2ks} L^0_{\mathfrak{q}}(s,\chi^{-1}\lambda) \quad \text{if } f = 0, \end{aligned} \tag{3.23}$$

where $\left(\frac{M/F}{\mathfrak{q}}\right)$ is 1, -1 or 0 according as $\mathfrak{q}$ splits, remains prime or ramifies in M/F, and $L^0_{\mathfrak{q}}(s,\chi^{-1}\lambda)$ is the $\mathfrak{q}$–Euler factor of the primitive L–function $L(s,\chi^{-1}\lambda)$. We define a possibly imprimitive L–function

$$L^{(\mathfrak{q})}(s,\chi^{-1}\lambda) = L_{\mathfrak{q}}(s,\chi^{-1}\lambda)L^0(s,\chi^{-1}\lambda)$$

removing the $\mathfrak{q}$–Euler factor.

Combining all these formulas, we find

$$e^{-1}\lambda\chi^{-1}(\varpi^{e(\mathfrak{F})}) \sum_{[\mathfrak{a}]\in Cl_{n+1}} \chi(\mathfrak{a})\mathbb{E}(\lambda)([\mathfrak{a}]) = c_0 L^{(\mathfrak{q})}(0,\chi^{-1}\lambda), \tag{3.24}$$

$$e^{-1}\lambda\chi^{-1}(\varpi^{e(\mathfrak{F})}) \sum_{[\mathfrak{a}]\in Cl_{n+1}} \chi(\mathfrak{a})\mathbb{E}'(\lambda)([\mathfrak{a}])$$
$$= \begin{cases} c_0 L^{(\mathfrak{q})}(0,\chi^{-1}\lambda) & \text{if } f>0, \\ c_0 \left(\frac{M/F}{\mathfrak{q}}\right) L_{\mathfrak{q}}(1,\chi^{-1}\lambda)L(0,\chi^{-1}\lambda) & \text{if } f=0 \text{ and } \left(\frac{M/F}{\mathfrak{q}}\right)\neq 0, \\ -c_0\chi^{-1}\lambda(\mathfrak{Q})L_{\mathfrak{q}}(1,\chi^{-1}\lambda)L(0,\chi^{-1}\lambda) & \text{if } \mathfrak{q}=\mathfrak{Q}^2 \text{ in } R \text{ and } f=0. \end{cases} \tag{3.25}$$

All these values are algebraic in $\overline{\mathbb{Q}}$ and integral over $\mathcal{W}$.

4 Non-vanishing modulo p of L–values

We construct an $\mathbb{F}$–valued measure ($\mathbb{F} = \overline{\mathbb{F}}_p$ as in 2.2) over the anti-cyclotomic class group $Cl_\infty = \varprojlim_n Cl_n$ modulo $\mathfrak{l}^\infty$ whose integral against a character χ is the Hecke L–value $L(0,\chi^{-1}\lambda)$ (up to a period). The idea is to translate the Hecke relation of the Eisenstein series into a distribution relation on the profinite group Cl_∞. At the end, we relate the non-triviality of the measure to the q–expansion of the Eisenstein series by the density of $\{x(\mathfrak{a})\}_\mathfrak{a}$ (see [H04c]).

4.1 Construction of a modular measure

We choose a complete representative set $\{\mathfrak{c}\}_{[\mathfrak{c}]\in Cl_F^+}$ of the strict ideal class group Cl_F^+ made up of ideals $\mathfrak{c}$ prime to $p\mathfrak{f}\mathfrak{l}$. For each proper R_n–ideal $\mathfrak{a}$, the polarization ideal $\mathfrak{c}(\mathfrak{a})$ of $x(\mathfrak{a})$ is equivalent to one of the representatives $\mathfrak{c}$ (so $[\mathfrak{c}] = [\mathfrak{c}(\mathfrak{a})]$). Writing $\mathfrak{c}_0$ for $\mathfrak{c}(R)$, we have $\mathfrak{c}(\mathfrak{a}) = \mathfrak{c}_0\mathfrak{l}^{-n}(\mathfrak{a}\mathfrak{a}^c)^{-1}$. Take a modular form g in $G_{kI}(\Gamma_0(\mathfrak{f}\mathfrak{f}'\mathfrak{l}),\epsilon_\lambda;W)$. Thus $g = (g_{[\mathfrak{c}]})$ is an h–tuple of modular forms for $h = |Cl_F^+|$. Put $f = (f_{[\mathfrak{c}]})_\mathfrak{c}$ for $f_{[\mathfrak{c}]} = d^\kappa g_{[\mathfrak{c}]}$ for the differential operator $d^\kappa = \prod_\sigma d_\sigma^{\kappa_\sigma}$ in 2.6. We write $f(x(\mathfrak{a}))$ for the value of $f_{[\mathfrak{c}(\mathfrak{a})]}(x(\mathfrak{a}))$. Similarly, we write $f(X,\Lambda,i,\omega)$ for $f_{[\mathfrak{c}]}(X,\Lambda,i,\omega)$ for the ideal class $\mathfrak{c}$ determined by $\overline{\Lambda}$. The Hecke operator $U(\mathfrak{l})$ takes the space $V(\mathfrak{c},\Gamma_0(\mathfrak{f}\mathfrak{f}'\mathfrak{l}),\epsilon_\lambda;W)$ into $V(\mathfrak{c}\mathfrak{l}^{-1},\Gamma_0(\mathfrak{f}\mathfrak{f}'\mathfrak{l}),\epsilon_\lambda;W)$. Choosing $\mathfrak{c}_\mathfrak{l}$ in the representative set equivalent to the ideal $\mathfrak{c}\mathfrak{l}^{-1}$, we have a canonical isomorphism $V(\mathfrak{c}\mathfrak{l},\Gamma_0(\mathfrak{f}\mathfrak{f}'\mathfrak{l}),\epsilon_\lambda;W) \cong V(\mathfrak{c}_\mathfrak{l},\Gamma_0(\mathfrak{f}\mathfrak{f}'\mathfrak{l}),\epsilon_\lambda;W)$ sending f to f' given by

$$f'(X,\xi\Lambda,i,i',\omega) = f(X,\Lambda,i,i',\omega)$$

for totally positive $\xi \in F$ with $\xi\mathfrak{c}_\mathfrak{l} = \mathfrak{l}^{-1}\mathfrak{c}$. This map is independent of the choice of ξ. Since the image of $\mathfrak{M}(\mathfrak{c};\mathfrak{N})$ in $Sh^{(p)}$ depends only on $\mathfrak{N}$ and the strict ideal class of $\mathfrak{c}$ as explained in 2.8, the Hecke operator $U(\mathfrak{l})$ is induced from the algebraic correspondence on the Shimura variety associated to the double coset $U\left(\begin{smallmatrix}\varpi_\mathfrak{l} & 0\\ 0 & 1\end{smallmatrix}\right)U$. So we regard $U(\mathfrak{l})$ as an operator acting on h–tuple of p–adic modular forms in $V(\Gamma_0(\mathfrak{f}\mathfrak{f}'\mathfrak{l}), \epsilon_\lambda; W) = \bigoplus_{\mathfrak{c}} V(\mathfrak{c}, \Gamma_0(\mathfrak{f}\mathfrak{f}'\mathfrak{l}), \epsilon_\lambda; W)$ inducing permutation $\mathfrak{c} \mapsto \mathfrak{c}_\mathfrak{l}$ on the polarization ideals. Suppose that $g|U(\mathfrak{l}) = ag$ with $a \in W^\times$; so, $f|U(\mathfrak{l}) = af$ (see Remark 3.2). The Eisenstein series $(\mathbb{E}(\lambda, \mathfrak{c}))_\mathfrak{c}$ satisfies this condition by Lemma 3.1. The operator $U(\mathfrak{l}^h)$ ($h = |Cl_F^+|$) takes $V(\mathfrak{c}, \Gamma_0(\mathfrak{f}\mathfrak{f}'\mathfrak{l}), \epsilon_\lambda; W)$ into itself. Thus $f_\mathfrak{c}|U(\mathfrak{l}^h) = a^h f_\mathfrak{c}$.

Choosing a base $w = (w_1, w_2)$ of $\widehat{R} = R \otimes_\mathbb{Z} \widehat{\mathbb{Z}}$, identify $T(X(R)_{/\overline{\mathbb{Q}}}) = \widehat{R}$ with $\widehat{O}^2$ by $\widehat{O} \ni (a, b) \mapsto aw_1 + bw_2 \in T(X(R))$. This gives a level structure $\eta^{(p)}(R) : F^2 \otimes_\mathbb{Q} \mathbb{A}^{(p\infty)} \cong V^{(p)}(X(R))$ defined over $\mathcal{W}$. Choose the base w satisfying the following two conditions:

(B1) $w_{2,\mathfrak{l}} = 1$ and $R_\mathfrak{l} = O_\mathfrak{l}[w_{1,\mathfrak{l}}]$;
(B2) By using the splitting: $R_\mathfrak{f} = R_\mathfrak{F} \times R_{\mathfrak{F}^c}$, $w_{1,\mathfrak{f}} = (1, 0)$ and $w_{2,\mathfrak{f}} = (0, 1)$.

Let $\mathfrak{a}$ be a proper R_n–ideal (for $R_n = O + \mathfrak{l}^n R$) prime to $\mathfrak{f}$. Recall the generator $\varpi = \varpi_\mathfrak{l}$ of $\mathfrak{l}O_\mathfrak{l}$. Regarding $\varpi \in F_\mathbb{A}^\times$, $w_n = (\varpi^n w_1, w_2)$ is a base of $\widehat{R}_n$ and gives a level structure $\eta^{(p)}(R_n) : F^2 \otimes_\mathbb{Q} \mathbb{A}^{(p\infty)} \cong V^{(p)}(X(R_n))$. We choose a complete representative set $A = \{a_1, \ldots, a_H\} \subset M_\mathbb{A}^\times$ so that $M_\mathbb{A}^\times = \bigsqcup_{j=1}^H M^\times a_j \widehat{R}_n^\times M_\infty^\times$. Then $\mathfrak{a}\widehat{R}_n = \alpha a_j \widehat{R}_n$ for $\alpha \in M^\times$ for some index j. We then define $\eta^{(p)}(\mathfrak{a}) = \alpha a_j \eta^{(p)}(R_n)$. The small ambiguity of the choice of α does not cause any trouble.

Write $x_0(\mathfrak{a}) = (X(\mathfrak{a}), \Lambda(\mathfrak{a}), i(\mathfrak{a}), i'(\mathfrak{a}), C(\mathfrak{a}), \omega(\mathfrak{a}))$. This is a test object of level $\Gamma_{1,0}^1(\mathfrak{f}^2) \cap \Gamma_0(\mathfrak{l})$ (see (3.3) for $\Gamma_{1,0}^1(\mathfrak{f}^2)$). We pick a subgroup $C \subset X(R_n)$ such that $C \cong O/\mathfrak{l}^m$ ($m > 0$) but $C \cap C(R_n) = \{0\}$. Then we define $x_0(R_n)/C$ by

$$\left(\frac{X(R_n)}{C}, \pi_*\Lambda(R_n), \pi \circ i(R_n), \pi^{-1} \circ i'(R_n), \frac{C + C(R_n)[\mathfrak{l}]}{C}, (\pi^*)^{-1}\omega(R_n)\right)$$

for the projection map $\pi : X(R_n) \twoheadrightarrow X(R_n)/C$. We can write

$$x_0(R_n)/C = x_0(\mathfrak{a}) \in \mathcal{M}(\mathfrak{c}\mathfrak{l}^{-n-m}, \mathfrak{f}^2, \Gamma_0(\mathfrak{l}))(\mathcal{W})$$

for a proper R_{n+m}–ideal $\mathfrak{a} \supset R_n$ with $(\mathfrak{a}\mathfrak{a}^c) = \mathfrak{l}^{-2m}$, and for $u \in O_\mathfrak{l}^\times$ we have

$$x_0(\mathfrak{a}) = x_0(R_n)/C = \begin{pmatrix} 1 & \frac{u}{\varpi_\mathfrak{l}^m} \\ 0 & 1 \end{pmatrix} (x_0(R_{m+n})). \tag{4.1}$$

See Section 2.8 in the text for the action of $g = \begin{pmatrix} 1 & \frac{u}{\varpi_\mathfrak{l}^m} \\ 0 & 1 \end{pmatrix}$ on the point $x_0(R_{m+n})$, and see [H04c] Section 3.1 for details of the computation leading to (4.1).

Let $T_M = \mathrm{Res}_{M/\mathbb{Q}}\mathbb{G}_m$. For each proper R_n–ideal $\mathfrak{a}$, we have an embedding $\rho_\mathfrak{a} : T_M(\mathbb{A}^{(p\infty)}) \to G(\mathbb{A}^{(p\infty)})$ given by $\alpha\eta^{(p)}(\mathfrak{a}) = \eta^{(p)}(\mathfrak{a}) \circ \rho_\mathfrak{a}(\alpha)$. Since

$\det(\rho_{\mathfrak{a}}(\alpha)) = \alpha\alpha^c \gg 0$, $\alpha \in T_M(\mathbb{Z}_{(p)})$ acts on $Sh^{(p)}$ through $\rho_{\mathfrak{a}}(\alpha) \in G(\mathbb{A})$. We have

$$\rho_{\mathfrak{a}}(\alpha)(x(\mathfrak{a})) = (X(\mathfrak{a}), (\alpha\alpha^c)\Lambda(\mathfrak{a}), \eta^{(p)}(\mathfrak{a})\rho_{\mathfrak{a}}(\alpha)) \\ = (X(\alpha\mathfrak{a}), \Lambda(\alpha\mathfrak{a}), \eta^{(p)}(\alpha\mathfrak{a}))$$

for the prime-to-p isogeny $\alpha \in \mathrm{End}_O(X(\mathfrak{a})) = R_{(p)}$. Thus $T_M(\mathbb{Z}_{(p)})$ acts on $Sh^{(p)}$ fixing the point $x(\mathfrak{a})$. We find $\rho(\alpha)^*\omega(\mathfrak{a}) = \alpha\omega(\mathfrak{a})$, and by (B2), we have

$$g(x(\alpha\mathfrak{a}), \alpha\omega(\mathfrak{a})) = g(\rho(\alpha)(x(\mathfrak{a}), \omega(\mathfrak{a}))) = \alpha^{-k\Sigma}\lambda_{\mathfrak{F}}(\alpha)\lambda_{\mathfrak{F}_c}(\alpha)g(x(\mathfrak{a}), \omega(\mathfrak{a})).$$

From this, we conclude

$$f(x(\alpha\mathfrak{a}), \alpha\omega(\mathfrak{a})) = f(\rho(\alpha)(x(\mathfrak{a}), \omega(\mathfrak{a}))) \\ = \alpha^{-k\Sigma-\kappa(1-c)}\lambda_{\mathfrak{F}}(\alpha)\lambda_{\mathfrak{F}_c}(\alpha)f(x(\mathfrak{a}), \omega(\mathfrak{a})),$$

because the effect of the differential operator d is identical with that of δ at the CM point $x(\mathfrak{a})$ by (K). By our choice of the Hecke character λ, we find

$$\lambda(\alpha\mathfrak{a}) = \alpha^{-k\Sigma-\kappa(1-c)}\lambda_{\mathfrak{F}}(\alpha)\lambda_{\mathfrak{F}_c}(\alpha)\lambda(\mathfrak{a}).$$

If $\mathfrak{a}$ and α is prime to $\mathfrak{C}p$, then the value $\alpha^{-k\Sigma-\kappa(1-c)}\lambda_{\mathfrak{C}}(\alpha)$ is determined independently of the choice of α for a given ideal $\alpha\mathfrak{a}$, and the value $\lambda(\mathfrak{a})^{-1}f(x(\mathfrak{a}), \omega(\mathfrak{a}))$ is independent of the representative set $A = \{a_j\}$ for Cl_n. Defining

$$f([\mathfrak{a}]) = \lambda(\mathfrak{a})^{-1}f(x(\mathfrak{a}), \omega(\mathfrak{a})) \text{ for a proper } R_n\text{–ideal } \mathfrak{a} \text{ prime to } \mathfrak{C}p, \quad (4.2)$$

we find that $f([\mathfrak{a}])$ only depends on the proper ideal class $[\mathfrak{a}] \in Cl_n$.

We write $x(\mathfrak{a}_u) = \begin{pmatrix} 1 & \frac{u}{\varpi_{\mathfrak{l}}} \\ 0 & 1 \end{pmatrix}(x(\mathfrak{a}))$, where $\mathfrak{l}^h = (\varpi)$ for an element $\varpi \in F$. Then $\mathfrak{a}_u$ depends only on $u \bmod \mathfrak{l}^h$, and $\{\mathfrak{a}_u\}_{u \bmod \mathfrak{l}^h}$ gives a complete representative set for proper R_{n+h}–ideal classes which project down to the ideal class $[\mathfrak{a}] \in Cl_n$. Since $\mathfrak{a}_u R_n = \varpi^{-1}\mathfrak{a}$, we find $\lambda(\mathfrak{a}_u) = \lambda(\mathfrak{l})^{-h}\lambda(\mathfrak{a})$. Then we have

$$a^h f([\mathfrak{a}]) = \lambda(\mathfrak{a})^{-1}f|U(\mathfrak{l}^h)(x(\mathfrak{a})) = \frac{1}{\lambda(\mathfrak{l})^h N(\mathfrak{l})^h}\sum_{u \bmod \mathfrak{l}^h} f([\mathfrak{a}_u]),$$

and we may define a measure φ_f on Cl_∞ with values in $\mathbb{F}$ by

$$\int_{Cl_\infty} \phi d\varphi_f = b^{-m}\sum_{\mathfrak{a}\in Cl_{mh}} \phi(\mathfrak{a}^{-1})f([\mathfrak{a}]) \text{ (for } b = a^h\lambda(\mathfrak{l})^h N(\mathfrak{l})^h). \quad (4.3)$$

4.2 Non-triviality of the modular measure

The non-triviality of the measure φ_f can be proven in exactly the same manner as in [H04c] Theorems 3.2 and 3.3. To recall the result in [H04c], we need to describe some functorial action on p–adic modular forms, commuting with $U(\mathfrak{l}^h)$. Let $\mathfrak{q}$ be a prime ideal of F. For a test object $(X, \overline{\Lambda}, \eta)$ of level $\Gamma_0(\mathfrak{N}\mathfrak{q})$, η induces a subgroup $C \cong O/\mathfrak{q}$ in X. Then we can construct canonically $[\mathfrak{q}](X, \overline{\Lambda}, \eta) = (X', \overline{\Lambda}, \eta')$ with $X' = X/C$ (see [H04b] Subsection 5.3). If $\mathfrak{q}$ splits into $\mathfrak{Q}\overline{\mathfrak{Q}}$ in M/F, choosing $\eta_\mathfrak{q}$ induced by $X(\mathfrak{a})[\mathfrak{q}^\infty] \cong M_\mathfrak{Q}/R_\mathfrak{Q} \times M_{\overline{\mathfrak{Q}}}/R_{\overline{\mathfrak{Q}}} \cong F_\mathfrak{q}/O_\mathfrak{q} \times F_\mathfrak{q}/O_\mathfrak{q}$, we always have a canonical level $\mathfrak{q}$–structure on $X(\mathfrak{a})$ induced by the choice of the factor $\mathfrak{Q}$. Then $[\mathfrak{q}](X(\mathfrak{a})) = X(\mathfrak{a}\mathfrak{Q}_n^{-1})$ for $\mathfrak{Q}_n = \mathfrak{Q} \cap R_n$ for a proper R_n–ideal $\mathfrak{a}$. When $\mathfrak{q}$ ramifies in M/F as $\mathfrak{q} = \mathfrak{Q}^2$, $X(\mathfrak{a})$ has a subgroup $C = X(\mathfrak{a})[\mathfrak{Q}_n]$ isomorphic to $O/\mathfrak{q}$; so, we can still define $[\mathfrak{q}](X(\mathfrak{a})) = X(\mathfrak{a}\mathfrak{Q}_n^{-1})$. The effect of $[\mathfrak{q}]$ on the q–expansion at the infinity cusp $(O, \mathfrak{c}^{-1})$ is computed in [H04b] (5.12) and is given by the q–expansion of f at the cusp $(\mathfrak{q}, \mathfrak{c}^{-1})$. The operator $[\mathfrak{q}]$ corresponds to the action of $g = \begin{pmatrix} 1 & 0 \\ 0 & \varpi_\mathfrak{q}^{-1} \end{pmatrix} \in GL_2(F_\mathfrak{q})$. Although the action of $[\mathfrak{q}]$ changes the polarization ideal by $\mathfrak{c} \mapsto \mathfrak{c}\mathfrak{q}$, as in the case of Hecke operator, we regard it as a linear map well defined on $V(\Gamma_0(\mathfrak{f}\mathfrak{f}'\mathfrak{l}), \epsilon_\lambda; W)$ into $V(\Gamma_0(\mathfrak{f}\mathfrak{f}'\mathfrak{l}\mathfrak{q}), \epsilon_\lambda; W)$ (inducing the permutation $\mathfrak{c} \mapsto \mathfrak{c}_\mathfrak{q}$)

For ideals $\mathfrak{A}$ in F, we can think of the association $X \mapsto X \otimes_O \mathfrak{A}$ for each AVRM X. There are a natural polarization and a level structure on $X \otimes \mathfrak{A}$ induced by those of X. Writing $(X, \Lambda, \eta) \otimes \mathfrak{A}$ for the triple made out of (X, Λ, η) after tensoring $\mathfrak{A}$, we define $f|\langle\mathfrak{A}\rangle(X, \Lambda, \eta) = f((X, \Lambda, \eta) \otimes \mathfrak{A})$. For $X(\mathfrak{a})$, we have $\langle\mathfrak{A}\rangle(X(\mathfrak{a})) = X(\mathfrak{A}\mathfrak{a})$. The effect of the operator $\langle\mathfrak{A}\rangle$ on the Fourier expansion at $(O, \mathfrak{c}^{-1})$ is given by that at $(\mathfrak{A}^{-1}, \mathfrak{A}\mathfrak{c})$ (see [H04b] (5.11) or [PAF] (4.53)). The operator $\langle\mathfrak{A}\rangle$ induces an automorphism of $V(\Gamma_0(\mathfrak{f}\mathfrak{f}'\mathfrak{l}), \epsilon_\lambda; W)$. By q–expansion principle, $f \mapsto f|[\mathfrak{q}]$ and $f \mapsto f|\langle\mathfrak{A}\rangle$ are injective on the space of (p–adic) modular forms, since the effect on the q-expansion at one cusp of the operation is described by the q-expansion of the same form at another cusp.

We fix a decomposition $Cl_\infty = \Gamma_f \times \Delta$ for a finite group Δ and a torsion-free subgroup Γ_f. Since each fractional R–ideal $\mathfrak{A}$ prime to $\mathfrak{l}$ defines a class $[\mathfrak{A}]$ in Cl_∞, we can embed the ideal group of fractional ideals prime to $\mathfrak{l}$ into Cl_∞. We write Cl_∞^{alg} for its image. Then $\Delta^{alg} = \Delta \cap Cl_\infty^{alg}$ is represented by prime ideals of M non-split over F. We choose a complete representative set for Δ^{alg} as $\{\mathfrak{s}\mathfrak{R}^{-1} | \mathfrak{s} \in \mathcal{S}, \mathfrak{r} \in \mathcal{R}\}$, where $\mathcal{S}$ contains O and ideals $\mathfrak{s}$ of F outside $p\ell\mathfrak{C}$, $\mathcal{R}$ is made of square-free product of primes in F ramifying in M/F, and $\mathfrak{R}$ is a unique ideal in M with $\mathfrak{R}^2 = \mathfrak{r}$. The set $\mathcal{S}$ is a complete representative set for the image Cl_F^0 of Cl_F in Cl_0 and $\{\mathfrak{R} | \mathfrak{r} \in \mathcal{R}\}$ is a complete representative set for 2–torsion elements in the quotient Cl_0/Cl_F^0. We fix a character $\nu : \Delta \to \mathbb{F}^\times$, and define

$$f_\nu = \sum_{\mathfrak{r}\in\mathcal{R}} \lambda\nu^{-1}(\mathfrak{R}) \left(\sum_{\mathfrak{s}\in\mathcal{S}} \nu\lambda^{-1}(\mathfrak{s}) f|\langle\mathfrak{s}\rangle\right) |[\mathfrak{r}]. \tag{4.4}$$

Choose a complete representative set $\mathcal{Q}$ for $Cl_\infty/\Gamma_f\Delta^{alg}$ made of primes of M split over F outside $p\mathfrak{l}\mathfrak{C}$. We choose $\eta_n^{(p)}$ out of the base (w_1, w_2) of $\widehat{R}_n$ so that at $\mathfrak{q} = \mathfrak{Q} \cap F$, $w_1 = (1,0) \in R_{\mathfrak{Q}} \times R_{\mathfrak{Q}^c} = R_{\mathfrak{q}}$ and $w_2 = (0,1) \in R_{\mathfrak{Q}} \times R_{\mathfrak{Q}^c} = R_{\mathfrak{q}}$. Since all operators $\langle\mathfrak{s}\rangle$, $[\mathfrak{q}]$ and $[\mathfrak{r}]$ involved in this definition commutes with $U(\mathfrak{l})$, $f_\nu|[\mathfrak{q}]$ is still an eigenform of $U(\mathfrak{l})$ with the same eigenvalue as f. Thus in particular, we have a measure φ_{f_ν}. We define another measure φ_f^ν on Γ_f by

$$\int_{\Gamma_f} \phi d\varphi_f^\nu = \sum_{\mathfrak{Q}\in\mathcal{Q}} \lambda\nu^{-1}(\mathfrak{Q}) \int_{\Gamma_f} \phi|\mathfrak{Q} d\varphi_{f_\nu|[\mathfrak{q}]},$$

where $\phi|\mathfrak{Q}(y) = \phi(y[\mathfrak{Q}]_f^{-1})$ for the projection $[\mathfrak{Q}]_f$ in Γ_f of the class $[\mathfrak{Q}] \in Cl_\infty$.

Lemma 4.1 *If $\chi : Cl_\infty \to \mathbb{F}^\times$ is a character inducing ν on Δ, we have*

$$\int_{\Gamma_f} \chi d\varphi_f^\nu = \int_{Cl_\infty} \chi d\varphi_f.$$

Proof Write $\Gamma_{f,n}$ for the image of Γ_f in Cl_n. For a proper R_n–ideal $\mathfrak{a}$, by the above definition of these operators,

$$f|\langle\mathfrak{s}\rangle|[\mathfrak{r}]|[\mathfrak{q}]([\mathfrak{a}]) = \lambda(\mathfrak{a})^{-1} f(x(\mathfrak{Q}^{-1}\mathfrak{R}^{-1}\mathfrak{a}), \omega(\mathfrak{Q}^{-1}\mathfrak{R}^{-1}\mathfrak{a})).$$

For sufficiently large n, χ factors through Cl_n. Since $\chi = \nu$ on Δ, we have

$$\int_{\Gamma_f} \chi d\varphi_f^\nu = \sum_{\mathfrak{Q}\in\mathcal{Q}} \sum_{\mathfrak{s}\in\mathcal{S}} \sum_{\mathfrak{r}\in\mathcal{R}} \sum_{\mathfrak{a}\in\Gamma_{f,n}} \lambda\chi^{-1}(\mathfrak{Q}\mathfrak{R}\mathfrak{s}^{-1}\mathfrak{a}) f|\langle\mathfrak{s}\rangle|[\mathfrak{r}]|[\mathfrak{q}]([\mathfrak{a}])$$

$$= \sum_{\mathfrak{a},\mathfrak{Q},\mathfrak{s},\mathfrak{r}} \chi(\mathfrak{Q}\mathfrak{R}\mathfrak{s}^{-1}\mathfrak{a}) f([\mathfrak{Q}^{-1}\mathfrak{R}^{-1}\mathfrak{s}\mathfrak{a}]) = \int_{Cl_\infty} \chi d\varphi_f,$$

because $Cl_\infty = \bigsqcup_{\mathfrak{Q},\mathfrak{s},\mathfrak{R}} [\mathfrak{Q}^{-1}\mathfrak{R}^{-1}\mathfrak{s}]\Gamma_f$. □

We identify $\mathrm{Hom}(\Gamma_f, \mathbb{F}^\times) \cong \mathrm{Hom}(\Gamma_f, \mu_{\ell^\infty})$ with $\mathrm{Hom}(\Gamma_f, \widehat{\mathbb{G}}_{m/\mathbb{Z}_\ell}) \cong \widehat{\mathbb{G}}_m^d$ for the formal multiplicative group $\widehat{\mathbb{G}}_m$ over $\mathbb{Z}_\ell$. Choosing a basis $\beta = \{\gamma_1, \ldots, \gamma_d\}$ of Γ_f over $\mathbb{Z}_\ell$ (so, $\mathbb{Z}^\beta = \sum_j \mathbb{Z}\gamma_j \subset \Gamma_f$) is to choose a multiplicative group $\mathbb{G}_m^\beta = \mathrm{Hom}(\mathbb{Z}^\beta, \mathbb{G}_m)$ over $\mathbb{Z}_\ell$ whose formal completion along the identity of $\mathbb{G}_m^\beta(\overline{\mathbb{F}}_\ell)$ giving rise to $\mathrm{Hom}(\Gamma_f, \widehat{\mathbb{G}}_{m/\mathbb{Z}_\ell})$. Thus we may regard

$\mathrm{Hom}(\Gamma_f, \mu_{\ell^\infty})$ as a subset of $\mathrm{Hom}(\mathbb{Z}^\beta, \mathbb{G}_m) \cong \mathbb{G}_m^\beta$. We call a subset $\mathcal{X}$ of characters of Γ_f *Zariski-dense* if it is Zariski-dense as a subset of the algebraic group $\mathbb{G}^\beta_{m/\overline{\mathbb{Q}}_\ell}$ (for any choice of β). Then we quote the following result ([H04c] Theorems 3.2 and 3.3):

Theorem 4.2 *Suppose that p is unramified in $M/\mathbb{Q}$ and Σ is ordinary for p. Let $f \neq 0$ be an eigenform defined over $\mathbb{F}$ of $U(\mathfrak{l})$ of level $(\Gamma_0(\mathfrak{f}\mathfrak{f}'\mathfrak{l}), \epsilon_\lambda)$ with non-zero eigenvalue. Fix a character $\nu : \Delta \to \mathbb{F}^\times$, and define f_ν as in* (4.4). *If f satisfies the following two conditions:*

(H1) There exists a strict ideal class $\mathfrak{c} \in Cl_F$ with the following two properties:

- (a) *the polarization ideal $\mathfrak{c}(\mathfrak{Q}^{-1}\mathfrak{R}^{-1}\mathfrak{s})$ is in $\mathfrak{c}$ for some $(\mathfrak{Q}, \mathfrak{R}, \mathfrak{s}) \in \mathcal{Q} \times \mathcal{S} \times \mathcal{R}$;*
- (b) *for any given integer $r > 0$, the $N(\mathfrak{l})^r$ modular forms $f_{\psi,\mathfrak{c}}| \left(\begin{smallmatrix} 1 & u \\ 0 & 1 \end{smallmatrix}\right)$ for $u \in \mathfrak{l}^{-r}/O$ are linearly independent over $\mathbb{F}$,*

(H2) λ and f are rational over a finite field,

then the set of characters $\chi : \Gamma_f \to \mathbb{F}^\times$ with non-vanishing $\int_{Cl_\infty} \nu\chi d\varphi_f \neq 0$ is Zariski dense. If $\mathrm{rank}_{\mathbb{Z}_\ell}\Gamma_f = 1$, under the same assumptions, the non-vanishing holds except for finitely many characters of Γ_f. Here $\nu\chi$ is the character of $Cl_\infty = \Gamma_f \times \Delta$ given by $\nu\chi(\gamma, \delta) = \nu(\delta)\chi(\gamma)$ for $\gamma \in \Gamma_f$ and $\delta \in \Delta$.

4.3 $\mathfrak{l}$–Adic Eisenstein measure modulo p

We apply Theorem 4.2 to the Eisenstein series $\mathbb{E}(\lambda)$ in (3.17) for the Hecke character λ fixed in 3.1. Choosing a generator π of $\mathfrak{m}_W$, the exact sequence $\underline{\omega}^{kI}_{/W} \xrightarrow{\varpi} \underline{\omega}^{kI}_{/W} \twoheadrightarrow \underline{\omega}^{kI}_{/\mathbb{F}}$ induces a reduction map: $H^0(\mathfrak{M}, \underline{\omega}^{kI}_{/W}) \to H^0(\mathfrak{M}, \underline{\omega}^{kI}_{/\mathbb{F}})$. We write $E_k(\psi^\circ, \mathfrak{c}) \bmod \Lambda\text{-}\mathrm{mod}\mathfrak{m}_W$ for the image of the Eisenstein series $E_k(\psi^\circ, \mathfrak{c})$. Then we put

$$f = (d^\kappa(E_k(\psi^\circ, \mathfrak{c})) \bmod \mathfrak{m}_W)_\mathfrak{c} \in V(\Gamma_0(\mathfrak{f}\mathfrak{f}'\mathfrak{l}), \epsilon_\lambda; \mathbb{F}).$$

By definition, the q–expansion of $f_{[\mathfrak{c}]}$ is the reduction modulo $\mathfrak{m}_W$ of the q–expansion of $\mathbb{E}(\lambda, \mathfrak{c})$ of characteristic 0.

We fix a character $\nu : \Delta \to \mathbb{F}^\times$ as in the previous section and write $\varphi = \varphi_f$ and $\varphi^\nu = \varphi_f^\nu$. By (3.24) combined with Lemma 4.1, we have, for a character $\chi : CL_\infty \to \mathbb{F}^\times$ with $\chi|_\Delta = \nu$,

$$\int_{\Gamma_f} \chi d\varphi^\nu = \int_{Cl_\infty} \chi d\varphi = C\chi(\mathfrak{F}) \frac{\pi^\kappa \Gamma_\Sigma(k\Sigma + \kappa) L^{(\mathfrak{l})}(0, \chi^{-1}\lambda)}{\Omega_\infty^{k\Sigma+2\kappa}} \bmod \mathfrak{m}_W, \tag{4.5}$$

where C is a non-zero constant given by the class modulo $\mathfrak{m}_{\mathcal{W}}$ of

$$\frac{(-1)^{k[F:\mathbb{Q}]}(R^\times : O^\times)\lambda^{-1}(\varpi^{e(\mathfrak{F})})}{\mathrm{Im}(\delta)^\kappa \sqrt{D}}.$$

The non-vanishing of C follows from the unramifiedness of p in $M/\mathbb{Q}$ and that $\mathfrak{F}$ is prime to p.

Theorem 4.3 *Let p be an odd prime unramified in $M/\mathbb{Q}$. Let λ be a Hecke character of M of conductor $\mathfrak{C}$ and of infinity type $k\Sigma + \kappa(1 - c)$ with $0 < k \in \mathbb{Z}$ and $0 \le \kappa \in \mathbb{Z}[\Sigma]$ for a CM type Σ that is ordinary with respect to p. Suppose* (spt) *and* (opl) *in* 3.1. *Fix a character $\nu : \Delta \to \overline{\mathbb{Q}}^\times$. Then $\frac{\pi^\kappa \Gamma_\Sigma(k\Sigma+\kappa)L^{(\mathfrak{l})}(0,\nu^{-1}\chi^{-1}\lambda)}{\Omega_\infty^{k\Sigma+2\kappa}} \in \mathcal{W}$ for all characters $\chi : Cl_\infty \to \mu_{\ell^\infty}(\overline{\mathbb{Q}})$ factoring through Γ_f. Moreover, for Zariski densely populated character χ in* $\mathrm{Hom}(\Gamma_f, \mu_{\ell^\infty})$, *we have*

$$\frac{\pi^\kappa \Gamma_\Sigma(k\Sigma+\kappa)L^{(\mathfrak{l})}(0,\nu^{-1}\chi^{-1}\lambda)}{\Omega_\infty^{k\Sigma+2\kappa}} \not\equiv 0 \bmod \mathfrak{m}_{\mathcal{W}},$$

unless the following three conditions are satisfied by ν and λ simultaneously:

- *(M1) M/F is unramified everywhere;*
- *(M2) The strict ideal class (in F) of the polarization ideal $\mathfrak{c}_0$ of $X(R)$ is not a norm class of an ideal class of M ($\Leftrightarrow \left(\frac{M/F}{\mathfrak{c}_0}\right) = -1$);*
- *(M3) The ideal character $\mathfrak{a} \mapsto (\lambda\nu^{-1}N(\mathfrak{a}) \bmod \mathfrak{m}_{\mathcal{W}}) \in \mathbb{F}^\times$ of F is equal to the character $\left(\frac{M/F}{\cdot}\right)$ of M/F.*

If $\mathfrak{l}$ is a split prime of degree 1 over $\mathbb{Q}$, under the same assumptions, the non-vanishing holds except for finitely many characters of Γ_f. If (M1-3) *are satisfied, the L–value as above vanishes modulo $\mathfrak{m}$ for all anticyclotomic characters χ.*

See [H04b] 5.4 for an example of $(M, \mathfrak{c}_0, \Sigma)$ satisfying (M1-3).

Proof By Theorem 4.2, we need to verify the condition (H1-2) for $\mathbb{E}(\lambda)$. The rationality (H2) follows from the rationality of $E_k(\psi^\circ, \mathfrak{c})$ and the differential operator d described in 2.6. For a given q–expansion $h(q) = \sum_\xi a(\xi, h)q^\xi \in \mathbb{F}[[\mathfrak{c}^{-1}_{\ge 0}]]$ at the infinity cusp $(O, \mathfrak{c}^{-1})$, we know that, for $u \in O_\mathfrak{l} \subset F_\mathbb{A}$,

$$a(\xi, h|\alpha_u) = \mathbf{e}_F(u\xi)a(\xi, h) \text{ for } \alpha_u = \left(\begin{smallmatrix}1 & u\\ 0 & 1\end{smallmatrix}\right).$$

The condition (H1) for h concerns the linear independence of $h|\alpha_u$ for $u \in \mathfrak{l}^{-r}O_\mathfrak{l}/O_\mathfrak{l}$. For any function $\phi : \mathfrak{c}^{-1}/\mathfrak{l}^r\mathfrak{c}^{-1} = O/\mathfrak{l}^r \to \mathbb{F}$, we write

$h|\phi = \sum_\xi \phi(\xi)a(\xi,h)q^\xi$. By definition, we have

$$h|R_\phi = \sum_{u\in O/\mathfrak{l}^r} \phi(u)h|\alpha_u = h|\phi^*$$

for the Fourier transform $\phi^*(v) = \sum_u \phi(u)\mathbf{e}_F(uv)$. For the characteristic function χ_v of $v \in \mathfrak{c}^{-1}/\mathfrak{l}^r\mathfrak{c}$, we compute its Fourier transform

$$\chi_v^*(u) = \sum_{a\in O/\mathfrak{l}^r} \mathbf{e}_F(au)\chi_v(a) = \mathbf{e}_F(vu).$$

Since the Fourier transform of the finite group $O/\mathfrak{l}^r$ is an automorphism (by the inversion formula), the linear independence of $\{h|\alpha_u = h|\chi_u^*\}_u$ is equivalent to the linear independence of $\{h|\chi_u\}_u$.

We recall that f_ν is a tuple $(f_{\nu,[\mathfrak{c}]})_\mathfrak{c} \in V(\Gamma_{01}(\mathfrak{f}^2), \Gamma_0(\mathfrak{l}); W)$. Thus we need to prove: there exists $\mathfrak{c}$ such that for a given congruence class $u \in \mathfrak{c}^{-1}/\mathfrak{l}^r\mathfrak{c}^{-1}$

$$a(\xi, f_{\nu,[\mathfrak{c}]}) \not\equiv 0 \bmod \mathfrak{m}_W \quad \text{for at least one } \xi \in u. \tag{4.6}$$

Since $a(\xi, d^\kappa h) = \xi^\kappa a(\xi,h)$ ((2.6)), (4.6) is achieved if

$$a(\xi, f'_{\nu,[\mathfrak{c}]}) \not\equiv 0 \bmod \mathfrak{m}_W \quad \text{for at least one } \xi \in u \text{ prime to } p \tag{4.7}$$

holds for

$$f' = (E_k(\psi^\circ, \mathfrak{c}) - N(\mathfrak{l})E_k(\psi^\circ, \mathfrak{c}\mathfrak{l}^{-1})|\langle\mathfrak{l}\rangle|[\mathfrak{l}])_\mathfrak{c},$$

because $\mathfrak{l} \nmid p$. Up to a non-zero constant, $\psi^\circ(a,b)$ in (3.7) is equal to $\phi(a,b) = \lambda_\mathfrak{F}(a)\lambda_{\mathfrak{F}_c}^{-1}(b)$ for $(a,b) \in (O/\mathfrak{f})^\times$. Thus we are going to prove, for a well chosen $\mathfrak{c}$,

$$a(\xi, f''_{\nu,[\mathfrak{c}]}) \not\equiv 0 \bmod \mathfrak{m}_W \quad \text{for at least one } \xi \in u \text{ prime to } p, \tag{4.8}$$

where $f''_{[\mathfrak{c}]} = E_k(\phi,\mathfrak{c}) - N(\mathfrak{l})E_k(\phi, \mathfrak{c}\mathfrak{l}^{-1})|\langle\mathfrak{l}\rangle|[\mathfrak{l}]$. Recall (4.4):

$$f''_\nu = \sum_{\mathfrak{r}\in\mathcal{R}} \lambda\nu^{-1}(\mathfrak{R}) \left(\sum_{\mathfrak{s}\in\mathcal{S}} \nu\lambda^{-1}(\mathfrak{s}) f''|\langle\mathfrak{s}\rangle \right) |[\mathfrak{r}]. \tag{4.9}$$

As computed in [H04b] (5.11) and (5.12), we have

$$N(\mathfrak{s}^{-1}\mathfrak{r})^{-1}E_k(\phi,\mathfrak{c})|\langle\mathfrak{s}\rangle|[\mathfrak{r}]_{O,\mathfrak{c}^{-1}}(q) = 2^{-[F:\mathbb{Q}]}L(1-k;\phi,\mathfrak{s}^{-1}\mathfrak{r})$$
$$+ \sum_{0\ll\xi\in\mathfrak{c}^{-1}\mathfrak{r}} q^\xi \sum_{\substack{(a,b)\in(\mathfrak{s}^{-1}\mathfrak{r}\times\mathfrak{c}^{-1}\mathfrak{s})/O^\times \\ ab=\xi}} \phi(a,b)\frac{N(a)}{|N(a)|}N(a)^{k-1}. \tag{4.10}$$

Thus we have, writing $t(a,b) = \phi(a,b)\frac{N(a)^k}{|N(a)|}$, $N(\mathfrak{a})^{-1}a(\xi, f''_{\mathfrak{a},\mathfrak{b}})$ is given by

$$\sum_{\substack{(a,b)\in(\mathfrak{a}\times\mathfrak{b})/O^\times \\ ab=\xi}} t(a,b) - \sum_{\substack{(a,b)\in(\mathfrak{a}\mathfrak{r}\times\mathfrak{l}\mathfrak{b})/O^\times \\ ab=\xi}} t(a,b)$$

$$= \sum_{\substack{(a,b)\in(\mathfrak{a}\times(\mathfrak{b}-\mathfrak{l}\mathfrak{b}))/O^\times \\ ab=\xi}} t(a,b). \quad (4.11)$$

We have the freedom of moving around the polarization ideal class $[\mathfrak{c}]$ in the coset $N_{M/F}(Cl_M)[\mathfrak{c}_0]$ for the polarization ideal class $[\mathfrak{c}_0]$ of $x(R)$. We first look into a single class $[\mathfrak{c}]$. We choose $\mathfrak{c}^{-1}$ to be a prime $\mathfrak{q}$ prime to $p\mathfrak{f}\mathfrak{l}$ (this is possible by changing $\mathfrak{c}^{-1}$ in its strict ideal class and choosing $\delta \in M$ suitably). We take a class $0 \ll \xi \in u$ for $u \in O/\mathfrak{l}^r$ so that $(\xi) = \mathfrak{q}\mathfrak{n}\mathfrak{l}^e$ for an integral ideal $\mathfrak{n} \nmid p\mathfrak{l}\mathfrak{C}$ prime to the relative discriminant $D(M/F)$ and $0 \le e \le r$. Since we have a freedom of choosing ξ modulo $\mathfrak{l}^r$, the ideal $\mathfrak{n}$ moves around freely in a given ray class modulo $\mathfrak{l}^{r-e}$.

We pick a pair $(a,b) \in F^2$ with $ab = \xi$ with $a \in \mathfrak{s}^{-1}$ and $b \in \mathfrak{q}\mathfrak{s}$. Then $(a) = \mathfrak{s}^{-1}\mathfrak{l}^\alpha\mathfrak{x}$ for an integral ideal $\mathfrak{x}$ prime to $\mathfrak{l}$ and $(b) = \mathfrak{s}\mathfrak{q}\mathfrak{l}^{e-\alpha}\mathfrak{x}'$ for an integral ideal $\mathfrak{x}'$ prime to $\mathfrak{l}$. Since $(ab) = \mathfrak{q}\mathfrak{n}\mathfrak{l}^e$, we find that $\mathfrak{x}\mathfrak{x}' = \mathfrak{n}$. By (4.11), b has to be prime to $\mathfrak{l}$; so, we find $\alpha = e$. Since $\mathfrak{x}\mathfrak{x}' = \mathfrak{n}$ and hence $\mathfrak{r} = O$ because $\mathfrak{n}$ is prime to $D(M/F)$. Thus for each factor $\mathfrak{x}$ of $\mathfrak{n}$, we could have two possible pairs $(a_\mathfrak{x}, b_\mathfrak{x})$ with $a_\mathfrak{x} b_\mathfrak{x} = \xi$ such that

$$((a_\mathfrak{x}) = \mathfrak{s}_\mathfrak{x}^{-1}\mathfrak{l}^e\mathfrak{x},\ (b_\mathfrak{x}) = (\xi a_\mathfrak{x}^{-1}) = \mathfrak{s}_\mathfrak{x}\mathfrak{q}\mathfrak{n}\mathfrak{x}^{-1})$$

for $\mathfrak{s}_\mathfrak{x} \in S$ representing the ideal class of the ideal $\mathfrak{l}^e\mathfrak{x}$. We put $\psi = \nu^{-1}\lambda$. We then write down the q–expansion coefficient of q^ξ at the cusp $(O, \mathfrak{q})$ (see [H04c] (4.30)):

$$G(\psi_\mathfrak{f})^{-1}a(\xi, f''_\nu) = \psi_{\mathfrak{F}_c}^{-1}(\xi)\psi(\mathfrak{n}\mathfrak{l}^e)^{-1}N(\mathfrak{n}\mathfrak{l}^e)^{-1}\prod_{\mathfrak{y}|\mathfrak{n}} \frac{1-(\psi(\mathfrak{y})N(\mathfrak{y}))^{e(\mathfrak{y})+1}}{1-\psi(\mathfrak{y})N(\mathfrak{y})}, \quad (4.12)$$

where $\mathfrak{n} = \prod_{\mathfrak{y}|\mathfrak{n}} \mathfrak{y}^{e(\mathfrak{y})}$ is the prime factorization of $\mathfrak{n}$.

We define, for the valuation v of W (normalized so that $v(p) = 1$)

$$\mu_C(\psi) = \mathrm{Inf}_\mathfrak{n} v\left(\prod_{\mathfrak{y}|\mathfrak{n}} \frac{1-(\psi(\mathfrak{y})N(\mathfrak{y}))^{e(\mathfrak{y})+1}}{1-\psi(\mathfrak{y})N(\mathfrak{y})}\right), \quad (4.13)$$

where $\mathfrak{n}$ runs over a ray class C modulo $\mathfrak{l}^{r-e}$ made of all integral ideals prime to $D\mathfrak{l}$ of the form $\mathfrak{q}^{-1}\xi\mathfrak{l}^{-e}$, $0 \ll \xi \in u$. Thus if $\mu_C(\psi) = 0$, we get the desired non-vanishing. Since $\mu_C(\psi)$ only depends on the class C, we may assume (and will assume) that $e = 0$ without losing generality; thus ξ is prime to $\mathfrak{l}$, and C is the class of $u[\mathfrak{q}^{-1}]$.

Suppose that $\mathfrak{n}$ is a prime $\mathfrak{y}$. Then by (4.12), we have

$$G(\psi_{\mathfrak{f}})^{-1}a(\xi, f''_\nu) = \psi_{\mathfrak{F}_c}^{-1}(\xi)(1 + (\psi(\mathfrak{y})N(\mathfrak{y}))^{-1}).$$

If $\psi(\mathfrak{y})N(\mathfrak{y}) \equiv -1 \bmod \mathfrak{m}_W$ for all prime ideals $\mathfrak{y}$ in the ray class C modulo $\mathfrak{l}^r$, the character $\mathfrak{a} \mapsto (\psi(\mathfrak{a})N(\mathfrak{a}) \bmod \mathfrak{m}_W)$ is of conductor $\mathfrak{l}^r$. We write $\overline{\psi}$ for the character: $\mathfrak{a} \mapsto (\psi(\mathfrak{a})N(\mathfrak{a}) \bmod \mathfrak{m}_W)$ of the ideal group of F with values in $\mathbb{F}^\times$. This character therefore has conductor $\widetilde{\mathfrak{C}}|\mathfrak{l}^r$. Since ν is anticyclotomic, its restriction to $F_{\mathbb{A}}^\times$ has conductor 1. Since λ has conductor $\mathfrak{C}$ prime to $\mathfrak{l}$, the conductor of $\overline{\psi}$ is a factor of the conductor of $\lambda \bmod \mathfrak{m}_W$, which is a factor of $p\mathfrak{C}$. Thus $\widetilde{\mathfrak{C}}|p\mathfrak{C}$. Since $\mathfrak{l} \nmid p\mathfrak{C}$, we find that $\widetilde{\mathfrak{C}} = 1$.

We are going to show that if $\mu_C(\psi) > 0$, M/F is unramified and $\overline{\psi} \equiv \left(\frac{M/F}{}\right) \bmod \mathfrak{m}_W$. We now choose two prime ideals $\mathfrak{y}$ and $\mathfrak{y}'$ so that $\mathfrak{q}\mathfrak{y}\mathfrak{y}' = (\xi)$ with $\xi \in u$. Then by (4.12), we have

$$G(\psi_{\mathfrak{f}})^{-1}a(\xi, f''_\nu) = \psi_{\mathfrak{F}_c}^{-1}(\xi)\left(1 + \frac{1}{\psi(\mathfrak{y})N(\mathfrak{y})}\right)\left(1 + \frac{1}{\psi(\mathfrak{y}')N(\mathfrak{y}')}\right). \tag{4.14}$$

Since $\overline{\psi}(\mathfrak{y}\mathfrak{y}') = \overline{\psi}(u[\mathfrak{q}^{-1}]) = \overline{\psi}(C) = -1$, we find that if $a(\xi, f''_\nu) \equiv 0 \bmod \mathfrak{m}_W$,

$$-1 = \overline{\psi}(\mathfrak{y}/\mathfrak{y}') = \overline{\psi}(\mathfrak{l}^{-1})\overline{\psi}(\mathfrak{y}^2) = -\overline{\psi}(\mathfrak{y}^2).$$

Since we can choose $\mathfrak{y}$ arbitrary, we find that $\overline{\psi}$ is quadratic. Thus $\mu_C(\psi) > 0$ if and only if $\overline{\psi}(\mathfrak{c}) = -1$, which is independent of the choice of u. Since we only need to show the existence of $\mathfrak{c}$ with $\overline{\psi}(\mathfrak{c}) = 1$, we can vary the strict ideal class $[\mathfrak{c}]$ in $[\mathfrak{c}_0]N_{M/F}(Cl_M)$. By class field theory, assuming that $\overline{\psi}$ has conductor 1, we have

$$\begin{aligned}&\overline{\psi}(\mathfrak{c}) = -1 \text{ for all } [\mathfrak{c}] \in [\mathfrak{c}_0]N_{M/F}(Cl_M)\\ &\quad\text{if and only if } \overline{\psi}(\mathfrak{c}_0) = -1 \text{ and } \overline{\psi}(\mathfrak{a}) = \left(\frac{M/F}{\mathfrak{a}}\right) \text{ for all } \mathfrak{a} \in Cl_F.\end{aligned} \tag{4.15}$$

If M/F is unramified, by definition, $2\delta\mathfrak{c}^* = 2\delta\mathfrak{d}^{-1}\mathfrak{c}^{-1} = R$. Taking squares, we find that $(\mathfrak{d}\mathfrak{c})^2 = 4\delta^2 \ll 0$. Thus $1 = \overline{\psi}(\mathfrak{d}^{-2}\mathfrak{c}^{-2}) = (-1)^{[F:\mathbb{Q}]}$, and this never happens when $[F : \mathbb{Q}]$ is odd. Thus (4.15) is equivalent to the three conditions (M1-3). The conditions (M1) and (M3) combined is equivalent to $\psi^* \equiv \psi \bmod \mathfrak{m}_W$, where the dual character ψ^* is defined by $\psi^*(x) = \psi(x^{-c})N(x)^{-1}$. Then the vanishing of $L(0, \chi^{-1}\nu^{-1}\lambda) \equiv 0$ for all anti-cyclotomic $\chi\nu$ follows from the functional equation of the p-adic Katz measure interpolating the p–adic Hecke L–values. This finishes the proof. □

5 Anticyclotomic Iwasawa series

We fix a conductor $\mathfrak{C}$ satisfying (spt) and (opl) in 3.1. We consider $Z = Z(\mathfrak{C}) = \varprojlim_n Cl_M(\mathfrak{C}p^n)$ for the ray class group $Cl_M(\mathfrak{x})$ of M modulo $\mathfrak{x}$. We split $Z(\mathfrak{C}) = \Delta_{\mathfrak{C}} \times \Gamma_{\mathfrak{C}}$ for a finite group $\Delta = \Delta_{\mathfrak{C}}$ and a torsion-free subgroup $\Gamma_{\mathfrak{C}}$. Since the projection: $Z(\mathfrak{C}) \twoheadrightarrow Z(1)$ induces an isomorphism $\Gamma_{\mathfrak{C}} = Z(\mathfrak{C})/\Delta_{\mathfrak{C}} \cong Z(1)/\Delta_1 = \Gamma_1$, we identify $\Gamma_{\mathfrak{C}}$ with Γ_1 and write it as Γ, which has a natural action of $\mathrm{Gal}(M/F)$. We define $\Gamma^+ = H^0(\mathrm{Gal}(M/F), \Gamma)$ and $\Gamma_- = \Gamma/\Gamma^+$. Write $\pi_- : Z \to \Gamma_-$ and $\pi_\Delta : Z \to \Delta$ for the two projections. Take a character $\varphi : \Delta \to \overline{\mathbb{Q}}^\times$, and regard it as a character of Z through the projection: $Z \twoheadrightarrow \Delta$. The Katz measure $\mu_{\mathfrak{C}}$ on $Z(\mathfrak{C})$ associated to the p–adic CM type Σ_p as in [HT1] Theorem II induces the anticyclotomic φ–branch μ_φ^- by

$$\int_{\Gamma_-} \phi d\mu_\varphi^- = \int_{Z(\mathfrak{C})} \phi(\pi_-(z))\varphi(\pi_\Delta(z))d\mu_{\mathfrak{C}}(z).$$

We write $L_p^-(\varphi)$ for this measure $d\mu_\varphi^-$ regarding it as an element of the algebra $\Lambda^- = W[[\Gamma_-]]$ made up of measures with values in W.

We look into the arithmetic of the unique $\mathbb{Z}_p^{[F:\mathbb{Q}]}$–extension M_∞^- of M on which we have $c\sigma c^{-1} = \sigma^{-1}$ for all $\sigma \in \mathrm{Gal}(M_\infty^-/M)$ for complex conjugation c. The extension M_∞^-/M is called the anticyclotomic tower over M. Writing $M(\mathfrak{C}p^\infty)$ for the ray class field over M modulo $\mathfrak{C}p^\infty$, we identify $Z(\mathfrak{C})$ with $\mathrm{Gal}(M(\mathfrak{C}p^\infty)/M)$ via the Artin reciprocity law. Then one has $\mathrm{Gal}(M(\mathfrak{C}p^\infty)/M_\infty^-) = \Gamma^+ \times \Delta_{\mathfrak{C}}$ and $\mathrm{Gal}(M_\infty^-/M) = \Gamma_-$. We then define M_Δ by the fixed field of $\Gamma_{\mathfrak{C}}$ in $M(\mathfrak{C}p^\infty)$; so, $\mathrm{Gal}(M_\Delta/M) = \Delta$. Since φ is a character of Δ, φ factors through $\mathrm{Gal}(M_\infty^- M_\Delta/M)$. Let $L_\infty/M_\infty^- M_\Delta$ be the maximal p–abelian extension unramified outside Σ_p. Each $\gamma \in \mathrm{Gal}(L_\infty/M)$ acts on the normal subgroup $X = \mathrm{Gal}(L_\infty/M_\infty^- M_\Delta)$ continuously by conjugation, and by the commutativity of X, this action factors through $\mathrm{Gal}(M_\Delta M_\infty^-/M)$. Then we look into the Γ_-–module: $X[\varphi] = X \otimes_{\Delta_{\mathfrak{C}},\varphi} W$.

As is well known, $X[\varphi]$ is a Λ^-–module of finite type, and in many cases, it is torsion by a result of Fujiwara (cf. [Fu], [H00] Corollary 5.4 and [HMI] Section 5.3) generalizing the fundamental work of Wiles [Wi] and Taylor-Wiles [TW]. If one assumes the Σ–Leopoldt conjecture for abelian extensions of M, we know that $X[\varphi]$ is a torsion module over Λ^- unconditionally (see [HT2] Theorem 1.2.2). If $X[\varphi]$ is a torsion Λ^-–module, we can think of the characteristic element $\mathcal{F}^-(\varphi) \in \Lambda^-$ of the module $X[\varphi]$. If $X[\varphi]$ is not of torsion over Λ^-, we simply put $\mathcal{F}^-(\varphi) = 0$. A character φ of Δ is called *anticyclotomic* if $\varphi(c\sigma c^{-1}) = \varphi^{-1}$.

We are going to prove in this section the following theorem:

Theorem 5.1 *Let ψ be an anticyclotomic character of Δ. Suppose* (spt) *and* (opl) *in* 3.1 *for the conductor $\mathfrak{C}(\psi)$ of ψ. If p is odd and unramified in $F/\mathbb{Q}$, then the anticyclotomic p–adic Hecke L–function $L_p^-(\psi)$ is a factor of $\mathcal{F}^-(\psi)$ in Λ^-.*

Regarding φ as a Galois character, we define $\varphi^-(\sigma) = \varphi(c\sigma c^{-1}\sigma^{-1})$ for $\sigma \in \mathrm{Gal}(\overline{M}/M)$. Then φ^- is anticyclotomic. By enlarging $\mathfrak{C}$ if necessary, we can find a character φ such that $\psi = \varphi^-$ for any given anticyclotomic ψ (e.g. [GME] page 339 or [HMI] Lemma 5.31). Thus we may always assume that $\psi = \varphi^-$.

It is proven in [HT1] and [HT2] that $L_p(\varphi^-)$ is a factor of $\mathcal{F}^-(\varphi^-)$ in $\Lambda^- \otimes_{\mathbb{Z}} \mathbb{Q}$. Thus the improvement concerns the p–factor of $L_p^-(\varphi^-)$, which has been shown to be trivial in [H04b]. The main point of this paper is to give another proof of this fact reducing it to Theorem 4.3. The new proof actually gives a stronger result: Corollary 5.6, which is used in our paper [H04d] to prove the identity $L_p^-(\psi) = \mathcal{F}^-(\psi)$ under suitable assumptions on ψ. The proof is a refinement of the argument in [HT1] and [HT2]. We first deduce a refinement of the result in [HT1] Section 7 using a unique Hecke eigenform (in a given automorphic representation) of minimal level. The minimal level is possibly a proper factor of the conductor of the automorphic representation. Then we proceed in the same manner as in [HT1] and [HT2].

Here we describe how to reduce Theorem 5.1 to Corollary 5.6. Since the result is known for $F = \mathbb{Q}$ by the works of Rubin and Tilouine, we may assume that $F \neq \mathbb{Q}$. Put $\Lambda = W[[\Gamma]]$. By definition, for the universal Galois character $\widetilde{\psi} : \mathrm{Gal}(M(\mathfrak{C}p^\infty)/M) \to \Lambda^\times$ sending $\delta \in \Delta_{\mathfrak{C}}$ to $\psi(\delta)$ and $\gamma \in \Gamma$ to the group element $\gamma \in \Gamma \subset \Lambda$, the Pontryagin dual of the adjoint Selmer group $\mathrm{Sel}(Ad(\mathrm{Ind}_M^F \widetilde{\psi}))$ defined in [MFG] 5.2 is isomorphic to the direct sum of $X[\psi] \otimes_{\Lambda^-} \Lambda$ and $\frac{Cl_M \otimes_{\mathbb{Z}} \Lambda}{Cl_F \otimes_{\mathbb{Z}} \Lambda}$. Thus the characteristic power series of the Selmer group is given by $(h(M)/h(F))\mathcal{F}^-(\psi)$.

To relate this power series $(h(M)/h(F))\mathcal{F}^-(\psi)$ to congruence among automorphic forms, we identify $O_{\mathfrak{f}} \cong R_{\mathfrak{F}} \cong R_{\mathfrak{F}_c}$. Recall the maximal diagonal torus $T_0 \subset GL(2)_{/O}$. Thus ψ restricted to $(R_{\mathfrak{F}} \times R_{\mathfrak{F}_c})^\times$ gives rise to the character ψ of $T_0^2(O_{\mathfrak{f}})$. We then extend ψ to a character ψ_F of $T_0^2(O_{\mathfrak{f}} \times O_{D(M/F)})$ by $\psi_F(x_{\mathfrak{f}}, y_{\mathfrak{f}}, x', y') = \psi(x_{\mathfrak{f}}, y_{\mathfrak{f}}) \left(\frac{M/F}{y'}\right)$. Then we define the level ideal $\mathfrak{N}$ by $(\mathfrak{C}(\psi^-) \cap F)D(M/F)$ and consider the Hecke algebra $\mathbf{h}^{n.ord} = \mathbf{h}^{n.ord}(\mathfrak{N}, \psi_F; W)$. It is easy to see that there is a unique $W[[\mathbf{\Gamma}]]$–algebra homomorphism $\lambda : \mathbf{h}^{n.ord} \to \Lambda$ such that the associated Galois representation ρ_λ ([H96] 2.8) is $\mathrm{Ind}_M^F \widetilde{\psi}$. Here $\mathbf{\Gamma}$ is the maximal torsion-free quotient of $\mathbf{G}$ introduced in 2.10. Note that the restriction of ρ_λ to the decomposition group $D_{\mathfrak{q}}$ at a prime $\mathfrak{q}|\mathfrak{N}$ is the diagonal representation $\begin{pmatrix} \psi_{F,1} & 0 \\ 0 & \psi_{F,2} \end{pmatrix}$ with values in $GL_2(W)$, which we write $\rho_{\mathfrak{q}}$. We write $H(\psi)$ for the congruence power series $H(\lambda)$ of λ (see [H96] Section 2.9, where $H(\lambda)$ is written as $\eta(\lambda)$). Writing $\mathbb{T}$ for the local ring of $\mathbf{h}^{n.ord}$ through which λ factors, the divisibility: $H(\psi)|(h(M)/h(F))\mathcal{F}^-(\psi)$ follows from the surjectivity onto $\mathbb{T}$ of the natural morphism from the universal nearly ordinary deformation ring $R^{n.ord}$ of $\mathrm{Ind}_F^M \psi \bmod \mathfrak{m}_W$ (without deforming $\rho_{\mathfrak{q}}$ for each $\mathfrak{q}|\mathfrak{N}$ and the restriction of the determinant character to $\Delta(\mathfrak{N})$). See [HT2] Section 6.2 for details of this implication. The surjectivity is obvious from our construction of $\mathbf{h}^{n.ord}(\mathfrak{N}, \psi_F; W)$ because it is generated by $\mathrm{Tr}(\rho_\lambda(Frob_{\mathfrak{q}}))$ for primes $\mathfrak{q}$ outside $p\mathfrak{N}$ and by the diagonal entries of ρ_λ restricted to $D_{\mathfrak{q}}$ for $\mathfrak{q}|p\mathfrak{N}$. Thus we need to prove $(h(M)/h(F))L_p^-(\psi)|H(\psi)$, which is the statement of Corollary 5.6. This corollary will be proven in the rest of this section. As a final remark, if we write $\mathbb{T}^\chi$ for the quotient of $\mathbb{T}$ which parametrizes all p–adic modular Galois representations congruent to $\mathrm{Ind}_M^F \psi$ with a given determinant character χ, we have $\mathbb{T} \cong \mathbb{T}^\chi \widehat{\otimes}_W W[[\mathbf{\Gamma}^+]] = \mathbb{T}^\chi[[\mathbf{\Gamma}^+]]$ for the maximal torsion-free quotient $\mathbf{\Gamma}^+$ of $Cl_F^+(\mathfrak{N}p^\infty)$ (cf. [MFG] Theorem 5.44). This implies $H(\psi) \in W[[\Gamma_-]]$.

5.1 Adjoint square L–values as Petersson metric

We now set $G := \mathrm{Res}_{O/\mathbb{Z}} GL(2)$. Let π be a cuspidal automorphic representation of $G(\mathbb{A})$ which is everywhere principal at finite places and holomorphic discrete series at all archimedean places. Since π is associated to holomorphic automorphic forms on $G(\mathbb{A})$, π is rational over the Hecke field generated by eigenvalues of the primitive Hecke eigenform in π. We have $\pi = \pi^{(\infty)} \otimes \pi_\infty$ for representations $\pi^{(\infty)}$ of $G(\mathbb{A}^{(\infty)})$ and π_∞ of $G(\mathbb{R})$. We further decompose

$$\pi^{(\infty)} = \otimes_{\mathfrak{q}} \pi(\epsilon_{1,\mathfrak{q}}, \epsilon_{2,\mathfrak{q}})$$

for the principal series representation $\pi(\epsilon_{1,\mathfrak{q}}, \epsilon_{2,\mathfrak{q}})$ of $GL_2(F_{\mathfrak{q}})$ with two characters $\epsilon_{1,\mathfrak{q}}, \epsilon_{2,\mathfrak{q}} : F_{\mathfrak{q}}^\times \to \overline{\mathbb{Q}}^\times$. By the rationality of π, these characters have values in $\overline{\mathbb{Q}}$. The central character of $\pi^{(\infty)}$ is given by $\epsilon_+ = \prod_{\mathfrak{q}}(\epsilon_{1,\mathfrak{q}}\epsilon_{2,\mathfrak{q}})$, which is a Hecke character of F. However $\epsilon_1 = \prod_{\mathfrak{q}} \epsilon_{1,\mathfrak{q}}$ and $\epsilon_2 = \prod_{\mathfrak{q}} \epsilon_{2,\mathfrak{q}}$ are just characters of $F_{\mathbb{A}^{(\infty)}}^\times$ and may not be Hecke characters.

In the space of automorphic forms in π, there is a unique normalized Hecke eigenform $f = f_\pi$ of minimal level satisfying the following conditions (see [H89] Corollary 2.2):

(L1) The level $\mathfrak{N}$ is the conductor of $\epsilon^- = \epsilon_2\epsilon_1^{-1}$.

(L2) Note that $\epsilon_\pi : \left(\begin{smallmatrix} a & b \\ c & d \end{smallmatrix}\right) \mapsto \epsilon_1(ad - bc)\epsilon^-(d)$ is a character of $U_0(\mathfrak{N})$ whose restriction to $U_0(C(\pi))$ for the conductor $C(\pi)$ of π induces the "Neben" character $\left(\begin{smallmatrix} a & b \\ c & d \end{smallmatrix}\right) \mapsto \epsilon_1(a)\epsilon_2(d)$. Then $f : G(\mathbb{Q})\backslash G(\mathbb{A}) \to \mathbb{C}$ satisfies

$$f(xu) = \epsilon_\pi(u)f(x).$$

(L3) The cusp form f corresponds to holomorphic cusp forms of weight $\kappa = (\kappa_1, \kappa_2) \in \mathbb{Z}[I]^2$.

In short, f_π is a cusp form in $S_\kappa(\mathfrak{N}, \epsilon_+; \mathbb{C})$. It is easy to see that $\Pi = \pi \otimes \epsilon_2^{-1}$ has conductor $\mathfrak{N}$ and that $v \otimes \epsilon_2$ is a constant multiple of f for the new vector v of Π (note here that Π may not be automorphic, but Π is an admissible irreducible representation of $G(\mathbb{A})$; so, the theory of new vectors still applies). Since the conductor $C(\pi)$ of π is given by the product of the conductors of ϵ_1 and ϵ_2, the minimal level $\mathfrak{N}$ is a factor of the conductor $C(\pi)$ and is often a proper divisor of $C(\pi)$.

By (L2), the Fourier coefficients $a(y, f)$ satisfy $a(uy, f) = \epsilon_1(u)a(y, f)$ for $u \in \widehat{O}^\times$ ($\widehat{O} = O \otimes_\mathbb{Z} \widehat{\mathbb{Z}}$). In particular, the function: $y \mapsto a(y, f)\overline{a(y, f)}$ only depends on the fractional ideal yO. Thus writing $a(\mathfrak{a}, f)\overline{a(\mathfrak{a}, f)}$ for the ideal $\mathfrak{a} = yO$, we defined in [H91] the self Rankin product by

$$D(s - [\kappa] - 1, f, f) = \sum_{\mathfrak{a} \subset O} a(\mathfrak{a}, f)\overline{a(\mathfrak{a}, f)}N(\mathfrak{a})^{-s},$$

where $N(\mathfrak{a}) = [O : \mathfrak{a}] = |O/\mathfrak{a}|$. We have a shift: $s \mapsto s - [\kappa] - 1$, because in order to normalize the L–function, we used in [H91] (4.6) the unitarization $\pi^u = \pi \otimes |\cdot|_\mathbb{A}^{([\kappa]-1)/2}$ in place of π to define the Rankin product. The weight κ^u of the unitarization satisfies $[\kappa^u] = 1$ and $\kappa^u \equiv \kappa \bmod \mathbb{Q}I$. Note that (cf. [H91] (4.2b))

$$f_\pi^u(x) := f_{\pi^u}(x) = D^{-([\kappa]+1)/2} f_\pi(x)|\det(x)|_\mathbb{A}^{([\kappa]-1)/2}. \tag{5.1}$$

We are going to define Petersson metric on the space of cusp forms satisfying (L1-3). For that, we write

$$X_0 = X_0(\mathfrak{N}) = G(\mathbb{Q})_+\backslash G(\mathbb{A})_+/U_0(\mathfrak{N})F_\mathbb{A}^\times SO_2(F_\mathbb{R}).$$

We define the inner product (f, g) by

$$(f, g)_\mathfrak{N} = \int_{X_0(\mathfrak{N})} \overline{f(x)}g(x)|\det(x)|_\mathbb{A}^{[\kappa]-1}dx \tag{5.2}$$

with respect to the invariant measure dx on X_0 as in [H91] page 342. In exactly the same manner as in [H91] (4.9), we obtain

$$D^s(4\pi)^{-I(s+1)-(\kappa_1-\kappa_2)}\Gamma_F((s+1)I+(\kappa_1-\kappa_2))\zeta_F^{(\mathfrak{N})}(2s+2)D(s,f,f)$$
$$= N(\mathfrak{N})^{-1}D^{-[\kappa]-2}(f, f\mathbb{E}_{0,0}(x,\mathbf{1},\mathbf{1};s+1))_{\mathfrak{N}},$$

where D is the discriminant $N(\mathfrak{d})$ of F, $\zeta_F^{(\mathfrak{N})}(s)=\zeta_F(s)\prod_{\mathfrak{q}|\mathfrak{N}}(1-N(\mathfrak{q})^{-s})$ for the Dedekind zeta function $\zeta_F(s)$ of F and $\mathbb{E}_{k,w}(x,\mathbf{1},\mathbf{1};s)$ ($k=\kappa_1-\kappa_2+I$ and $w=I-\kappa_2$) is the Eisenstein series of level $\mathfrak{N}$ defined above (4.8e) of [H91] for the identity characters $(\mathbf{1},\mathbf{1})$ in place of $(\chi^{-1}\psi^{-1},\theta)$ there.

By the residue formula at $s=1$ of $\zeta_F^{(\mathfrak{N})}(2s)\mathbb{E}_{0,0}(x,\mathbf{1},\mathbf{1};s)$ (e.g. (RES2) in [H99] page 173), we find

$$(4\pi)^{-I-(\kappa_1-\kappa_2)}\Gamma_F(I+(\kappa_1-\kappa_2))\mathrm{Res}_{s=0}\zeta_F^{(\mathfrak{N})}(2s+2)D(s,f,f)$$
$$= D^{-[\kappa]-2}N(\mathfrak{N})^{-1}\prod_{\mathfrak{q}|\mathfrak{N}}(1-N(\mathfrak{q})^{-1})\frac{2^{[F:\mathbb{Q}]-1}\pi^{[F:\mathbb{Q}]}R_\infty h(F)}{w\sqrt{D}}(f,f)_{\mathfrak{N}},$$
$$(5.3)$$

where $w=2$ is the number of roots of unity in F, $h(F)$ is the class number of F and R_∞ is the regulator of F.

Since f corresponds to $v\otimes\epsilon_2$ for the new vector $v\in\Pi$ of the principal series representation $\Pi^{(\infty)}$ of minimal level in its twist class $\{\Pi\otimes\eta\}$ (η running over all finite order characters of $F^\times_{\mathbb{A}^{(\infty)}}$), by making product $\overline{f}\cdot f$, the effect of tensoring ϵ_2 disappears. Thus we may compute the Euler factor of $D(s,f,f)$ as if f were a new vector of the minimal level representation (which has the "Neben" character with conductor exactly equal to that of Π). Then for each prime factor $\mathfrak{q}|\mathfrak{N}$, the Euler $\mathfrak{q}$–factor of $\zeta_F^{(\mathfrak{N})}(2s+2)D(s,f,f)$ is given by

$$\sum_{\nu=0}^{\infty}a(\mathfrak{q}^\nu,f)\overline{a(\mathfrak{q}^\nu,f)}N(\mathfrak{q})^{-\nu s}=\left(1-N(\mathfrak{q})^{[\kappa]-s}\right)^{-1},$$

because $a(\mathfrak{q},f)\overline{a(\mathfrak{q},f)}=N(\mathfrak{q})^{[\kappa]}$ by [H88] Lemma 12.2. Thus the zeta function $\zeta_F^{(\mathfrak{N})}(2s+2)D(s,f,f)$ has the single Euler factor $(1-N(\mathfrak{q})^{-s-1})^{-1}$ at $\mathfrak{q}|\mathfrak{N}$, and the zeta function $\zeta_F(s+1)L(s+1,Ad(f))$ has its square $(1-N(\mathfrak{q})^{-s-1})^{-2}$ at $\mathfrak{q}|\mathfrak{N}$, because $L(s+1,Ad(f))$ contributes one more factor $(1-N(\mathfrak{q})^{-s-1})^{-1}$. The Euler factors outside $\mathfrak{N}$ are the same by the standard computation. Therefore, the left-hand-side of (5.3) is given by

$$\zeta_F^{(\mathfrak{N})}(2s+2)D(s,f,f)$$
$$=\left(\prod_{\mathfrak{q}|\mathfrak{N}}(1-N(\mathfrak{q})^{-s-1})\right)\zeta_F(s+1)L(s+1,Ad(f)) \quad (5.4)$$

By comparing the residue at $s = 0$ of (5.4) with (5.3) (in view of (5.1)), we get

$$(f_\pi^u, f_\pi^u)_{\mathfrak{N}} = D^{-[\kappa]-1}(f_\pi, f_\pi)_{\mathfrak{N}} =$$
$$D\Gamma_F((\kappa_1-\kappa_2)+I)N(\mathfrak{N})2^{-2((\kappa_1-\kappa_2)+I)+1}\pi^{-((\kappa_1-\kappa_2)+2I)}L(1, Ad(f)) \tag{5.5}$$

for the primitive adjoint square L–function $L(s, Ad(f))$ (e.g. [H99] 2.3). Here we have written $x^s = \prod_\sigma x^{s_\sigma}$ for $s = \sum_\sigma s_\sigma \sigma \in \mathbb{C}[I]$, and $\Gamma_F(s) = \prod_\sigma \Gamma(s_\sigma)$ for the Γ–function $\Gamma(s) = \int_0^\infty e^{-t}t^{s-1}dt$. This formula is consistent with the one given in [HT1] Theorem 7.1 (but is much simpler).

5.2 Primitive p–Adic Rankin product

Let $\mathfrak{N}$ and $\mathfrak{J}$ be integral ideals of F prime to p. We shall use the notation introduced in 2.10. Thus, for a p–adically complete valuation ring $W \subset \widehat{\mathbb{Q}}_p$, $\mathbf{h}^{n.ord}(\mathfrak{N}, \psi; W)$ and $\mathbf{h}^{n.ord}(\mathfrak{J}, \chi; W)$ are the universal nearly ordinary Hecke algebra with level $(\mathfrak{N}, \psi)$ and $(\mathfrak{J}, \chi)$ respectively. The character $\psi = (\psi_1, \psi_2, \psi_+^t)$ is made of the characters of ψ_j of $T_0(O_p \times (O/\mathfrak{N}'^{(p)}))$ (for an ideal $\mathfrak{N}' \subset \mathfrak{N}$) of finite order and for the restriction ψ_+^t to $\Delta_F(\mathfrak{N})$ (the torsion part of $Cl_F^+(\mathfrak{N}'p^\infty)$) of a Hecke character ψ_+ extending $\psi_1\psi_2$. Similarly we regard χ as a character of $\mathbf{G}(\mathfrak{J}')$ for an ideal $\mathfrak{J}' \subset \mathfrak{J}$); so, $\psi^- = \psi_1^{-1}\psi_2$ and χ^- are well defined (finite order) character of $T_0(O_p \times (O/\mathfrak{N}))$ and $T_0(O_p \times (O/\mathfrak{J}))$ respectively. In particular we have $\mathfrak{C}^{(p)}(\psi^-)|\mathfrak{N}$ and $\mathfrak{C}^{(p)}(\chi^-)|\mathfrak{J}$, where $\mathfrak{C}^{(p)}(\psi^-)$ is the prime-to–p part of the conductor $\mathfrak{C}(\psi^-)$ of ψ^-. We assume that

$$\mathfrak{C}^{(p)}(\psi^-) = \mathfrak{N}, \text{ and } \mathfrak{C}^{(p)}(\chi^-) = \mathfrak{J}. \tag{5.6}$$

For the moment, we also assume for simplicity that

$$\psi_\mathfrak{q}^- \neq \chi_\mathfrak{q}^- \text{ on } O_\mathfrak{q}^\times \text{ for } \mathfrak{q}|\mathfrak{J}\mathfrak{N} \text{ and } \psi_1 = \chi_1 \text{ on } \widehat{O}^\times. \tag{5.7}$$

Let $\lambda : \mathbf{h}^{n.ord}(\mathfrak{N}, \psi; W) \to \Lambda$ and $\varphi : \mathbf{h}^{n.ord}(\mathfrak{J}, \chi; W) \to \Lambda'$ be $\mathbf{\Lambda}$–algebra homomorphisms for integral domains Λ and Λ' finite torsion-free over $\mathbf{\Lambda}$. For each arithmetic point $P \in \mathrm{Spf}(\Lambda)(\overline{\mathbb{Q}}_p)$, we let $f_P \in S_{\kappa(P)}(U_0(\mathfrak{N}p^\alpha), \psi_P; \overline{\mathbb{Q}}_p)$ denote the normalized Hecke eigenform of minimal level belonging to λ. In other words, for $\lambda_P = P \circ \lambda : \mathbf{h}^{n.ord} \to \overline{\mathbb{Q}}_p$, we have $a(y, f_P) = \lambda_P(T(y))$ for all integral ideles y with $y_p = 1$. In the automorphic representation generated by f_P, we can find a unique automorphic form f_P^{ord} with $a(y, f_P^{ord}) = \lambda(T(y))$ for all y, which we call the (nearly) *ordinary projection* of f_P. Similarly, using φ, we define $g_Q \in S_{\kappa(Q)}(U_0(\mathfrak{J}p^\beta), \chi_Q; \overline{\mathbb{Q}}_p)$

for each arithmetic point $Q \in \mathrm{Spf}(\Lambda')(\overline{\mathbb{Q}}_p)$. Recall that we have two characters $(\psi_{P,1}, \psi_{P,2})$ of $T_0(\widehat{O})$ associated to ψ_P. Recall $\psi_P = (\psi_{P,1}, \psi_{P,2}, \psi_{P+}) : T_0(\widehat{O})^2 \times (F_{\mathbb{A}}^\times/F^\times) \to \mathbb{C}^\times$. The central character ψ_{P+} of f_P coincides with $\psi_{P,1}\psi_{P,2}$ on $\widehat{O}^\times$ and has infinity type $(1-[\kappa(P)])I$. We suppose

$$\text{The character } \psi_{P,1}\chi_{Q,1}^{-1} \text{ is induced by a global finite order character } \theta. \tag{5.8}$$

This condition combined with (5.6) implies that θ is unramified outside p. As seen in [H91] 7.F, we can find an automorphic form $g_Q|\theta^{-1}$ on $G(\mathbb{A})$ whose Fourier coefficients are given by $a(y, g_Q|\theta^{-1}) = a(y, g_Q)\theta^{-1}(yO)$, where $\theta(\mathfrak{a}) = 0$ if $\mathfrak{a}$ is not prime to $\mathfrak{C}(\theta)$. The above condition implies, as explained in the previous subsection,

$$y \mapsto a(y, f_P)\overline{a(y, g_Q|\theta^{-1})\theta(y)}$$

factors through the ideal group of F. Note that

$$a(y, f_P)\overline{a(y, g_Q|\theta^{-1})\theta(y)} = a(y, f_P)\overline{a(y, g_Q)}$$

as long as y_p is a unit. We thus write $a(\mathfrak{a}, f_P)\overline{a(\mathfrak{a}, g_Q|\theta^{-1})\theta(\mathfrak{a})}$ for the above product when $yO = \mathfrak{a}$ and define

$$D(s - \frac{[\kappa(P)] + [\kappa(Q)]}{2} - 1, f_P, g_Q|\theta^{-1}, \theta^{-1}) = \sum_{\mathfrak{a}} a(\mathfrak{a}, f_P)\overline{a(\mathfrak{a}, g_Q|\theta^{-1})\theta(\mathfrak{a})}N(\mathfrak{a})^{-s}.$$

Hereafter we write $\kappa = \kappa(P)$ and $\kappa' = \kappa(Q)$ if confusion is unlikely.

Note that for $g'_Q(x) = g_Q|\theta^{-1}(x)\theta(\det(x))$,

$$D(s, f_P, g'_Q) := D(s, f_P, g'_Q, \mathbf{1}) = D(s, f_P, g_Q|\theta^{-1}, \theta^{-1}).$$

Though the introduction of the character θ further complicates our notation, we can do away with it just replacing g_Q by g'_Q, since the local component $\pi(\chi'_{Q,1,\mathfrak{q}}, \chi'_{Q,2,\mathfrak{q}})$ of the automorphic representation generated by g'_Q satisfies $\chi'_{1,Q} = \psi_{1,P}$, and hence without losing much generality, we may assume a slightly stronger condition:

$$\psi_{P,1} = \chi_{Q,1} \text{ on } \widehat{O}^\times \tag{5.9}$$

in our computation.

For each holomorphic Hecke eigenform f, we write $M(f)$ for the rank 2 motive attached to f (see [BR]), $\widetilde{M}(f)$ for its dual, ρ_f for the $\mathfrak{p}$–adic Galois representation of $M(f)$ and $\widetilde{\rho}_f$ for the contragredient of ρ_f. Here $\mathfrak{p}$ is the p–adic place of the Hecke field of f induced by $i_p : \overline{\mathbb{Q}} \hookrightarrow \overline{\mathbb{Q}}_p$. Thus

$L(s, M(f))$ coincides with the standard L–function of the automorphic representation generated by f, and the Hodge weight of $M(f_P)$ is given by $\{(\kappa_{1,\sigma}, \kappa_{2,\sigma}), (\kappa_{1,\sigma}, \kappa_{2,\sigma})\}_\sigma$ for each embedding $\sigma : F \hookrightarrow \mathbb{C}$. We have $\det(\rho_{f_P}(Frob_{\mathfrak{q}})) = \psi_P^u(\mathfrak{q})N(\mathfrak{q})^{[\kappa]}$ ($\mathfrak{p} \neq \mathfrak{q}$; see [MFG] 5.6.1).

Lemma 5.2 *Suppose* (5.6) *and* (5.8). *For primes* $\mathfrak{q} \nmid p$, *the Euler* $\mathfrak{q}$*–factor of*

$$L^{(\mathfrak{N}\mathfrak{J})}(2s - [\kappa] - [\kappa'], \psi_P^u \chi_Q^{-u}) D(s - \frac{[\kappa] + [\kappa']}{2} - 1, f_P, g'_Q)$$

is equal to the Euler $\mathfrak{q}$*–factor of* $L_{\mathfrak{q}}(s, M(f_P) \otimes \widetilde{M}(g_Q))$ *given by*

$$\det\left(1 - (\rho_{f_P} \otimes \widetilde{\rho}_{g_Q})(Frob_{\mathfrak{q}})\big|_{V^I} N(\mathfrak{q})^{-s})\right)^{-1},$$

where V *is the space of the* $\mathfrak{p}$*–adic Galois representation of the tensor product:* $\rho_{f_P} \otimes \widetilde{\rho}_{g_Q}$ *and* $V^I = H^0(I, V)$ *for the inertia group* $I \subset \mathrm{Gal}(\overline{\mathbb{Q}}/F)$ *at* $\mathfrak{q}$.

Proof As already explained, we may assume (5.9) instead of (5.8). By abusing the notation, we write $\pi(\psi_{P,1,\mathfrak{q}}, \psi_{P,2,\mathfrak{q}})$ (resp. $\pi(\chi_{Q,1,\mathfrak{q}}, \chi_{Q,2,\mathfrak{q}})$) for the $\mathfrak{q}$–factor of the representation generated by f_P (resp. g_Q). By the work of Carayol, R. Taylor and Blasius-Rogawski combined with a recent work of Blasius [B], the restriction of ρ_{f_P} to the Decomposition group at $\mathfrak{q}$ is isomorphic to $\mathrm{diag}[\psi_{P,1,\mathfrak{q}}, \psi_{P,2,\mathfrak{q}}]$ (regarding $\psi_{i,P,\mathfrak{q}}$ as Galois characters by local class field theory). The same fact is true for g_Q. If $\mathfrak{q}|\mathfrak{J}\mathfrak{N}$, then V^I is one dimensional on which $Frob_{\mathfrak{q}}$ acts by $\psi_{P,1,\mathfrak{q}}(\varpi_{\mathfrak{q}})\overline{\chi}_{Q,1,\mathfrak{q}}(\varpi_{\mathfrak{q}}) = a(\mathfrak{q}, f_P)\overline{a(\mathfrak{q}, g_Q)}$ because $\psi_{i,P,\mathfrak{q}}\overline{\chi}_{j,Q,\mathfrak{q}}$ is ramified unless $i = j = 1$ ($\Leftrightarrow \psi_1 = \chi_1$ on $\widehat{O}^\times$ and $\psi_{\mathfrak{q}}^- \neq \psi_{\mathfrak{q}}^-$). If $\mathfrak{q} \nmid \mathfrak{J}\mathfrak{N}$, both $\pi(\psi_{P,1,\mathfrak{q}}, \psi_{P,2,\mathfrak{q}}) \otimes \psi_{P,1,\mathfrak{q}}^{-1} = \pi(1, \psi_{P,\mathfrak{q}}^-)$ and $\pi(\chi_{Q,1,\mathfrak{q}}, \chi_{Q,2,\mathfrak{q}}) \otimes \chi_{Q,1,\mathfrak{q}}^{-1} = \pi(1, \chi_{Q,\mathfrak{q}}^-)$ are unramified principal series. By $\psi_{P,1,\mathfrak{q}} = \chi_{Q,1,\mathfrak{q}}$:(5.8), we have an identity:

$$\rho_{f_P} \otimes \rho_{g_Q^c} \cong (\rho_{f_P} \otimes \psi_{P,1,\mathfrak{q}}^{-1}) \otimes (\rho_{g_Q^c} \otimes \chi_{Q,1,\mathfrak{q}})$$

on the inertia group, which is unramified. Therefore V is unramified at $\mathfrak{q}$. At the same time, the L–function has full Euler factor at $\mathfrak{q} \nmid \mathfrak{J}\mathfrak{N}$. □

We would like to compute $f_P|\tau(x) := \psi_P^u(\det(x))^{-1} f_P(x\tau)$ for $\tau(N) = \left(\begin{smallmatrix} 0 & -1 \\ N & 0 \end{smallmatrix}\right) \in G(\mathbb{A}^{(\infty)})$ for an idele $N = N(P)$ with $N^{(p\mathfrak{N})} = 1$ and $NO = \mathfrak{C}(\psi_P^-)$ (whose prime-to-p factor is $\mathfrak{N}$). We continue to abuse notation and write $\pi(\psi_{P,1,\mathfrak{q}}, \psi_{P,2,\mathfrak{q}}) \otimes \psi_{P,1,\mathfrak{q}}^{-1}$ as $\pi(1, \psi_{P,\mathfrak{q}}^-)$ (thus $\psi_{P,\mathfrak{q}}^-$ is the character of $F_{\mathfrak{q}}^\times$ inducing the original $\psi_{P,\mathfrak{q}}^-$ on $O_{\mathfrak{q}}^\times$). We write $(\psi_{P,\mathfrak{q}}^-)^u = \psi_{P,\mathfrak{q}}^-/|\psi_{P,\mathfrak{q}}^-|$ (which is a unitary character). In the Whittaker model $V(1, \psi_{P,\mathfrak{q}}^-)$ of $\pi(1, \psi_{P,\mathfrak{q}}^-)$ (realized in the space of functions on $GL_2(F_{\mathfrak{q}})$), we have a unique function

$\phi_{\mathfrak{q}}$ on $GL_2(F_{\mathfrak{q}})$ whose Mellin transform gives rise to the local L–function of $\pi(1,\psi_{\mathfrak{q}}^-)$. In particular, we have (cf. [H91] (4.10b))

$$\phi_{\mathfrak{q}}|\tau_{\mathfrak{q}}(x) := (\psi_P^-)^u(\det(x))^{-1}\phi_{\mathfrak{q}}(x\tau_{\mathfrak{q}}) = W(\phi_{\mathfrak{q}})|\psi_P^-(N_{\mathfrak{q}})|^{1/2}\overline{\phi}_{\mathfrak{q}}(x),$$

where $\overline{\phi}_{\mathfrak{q}}$ is the complex conjugate of $\phi_{\mathfrak{q}}$ belonging to the space representation $V(1,\overline{\psi_{P,\mathfrak{q}}^-})$ and $W(\phi_{\mathfrak{q}})$ is the epsilon factor of the representation $\pi(1,\psi_{P,\mathfrak{q}}^-)$ as in [H91] (4.10c) (so, $|W(\phi_{\mathfrak{q}})| = 1$). Then $\phi_{\mathfrak{q}} \otimes \psi_{P,1,\mathfrak{q}}(x) := \psi_{P,1,\mathfrak{q}}(\det(x))\phi_{\mathfrak{q}}(x)$ is in $V(\psi_{P,1,\mathfrak{q}},\psi_{P,2,\mathfrak{q}})$ and gives rise to the $\mathfrak{q}$–component of the global Whittaker model of the representation π generated by f_P. The above formula then implies

$$(\phi_{\mathfrak{q}} \otimes \psi_{P,1,\mathfrak{q}})|\tau_{\mathfrak{q}} := \psi_P^u(\det(x))^{-1}(\phi_{\mathfrak{q}} \otimes \psi_{P,1,\mathfrak{q}})(x\tau_{\mathfrak{q}})$$
$$= \psi_{P,1,\mathfrak{q}}^u(N_{\mathfrak{q}})|N_{\mathfrak{q}}|_{\mathfrak{q}}^{(1-[\kappa])/2}W(\phi_{\mathfrak{q}})(\overline{\phi}_{\mathfrak{q}} \otimes \overline{\psi}_{P,1,\mathfrak{q}})(x).$$

Define the root number $W_{\mathfrak{q}}(f_P) = W(\phi_{\mathfrak{q}})$ and $W(f_P) = \prod_{\mathfrak{q}} W_{\mathfrak{q}}(f_P)$. Here note that $W_{\mathfrak{q}}(f_P) = 1$ if the prime $\mathfrak{q}$ is outside $\mathfrak{C}(\psi_{P,1})\mathfrak{C}(\psi_{P,2})D$. We conclude from the above computation the following formula:

$$f_P|\tau(x) := \psi_P^u(\det(x))^{-1}f_P(x\tau) = W(f_P)\psi_{P,1}^u(N)|N|_{\mathbb{A}}^{(1-[\kappa])/2}f_P^c(x), \tag{5.10}$$

where f_P^c is determined by $a(y, f_P^c) = \overline{a(y, f_P)}$ for all $y \in F_{\mathbb{A}}^\times$. This shows

$$W(f_P)W(f_P^c) = \psi_{P,\infty}^u(-1) = \psi_{P,\infty}^+(-1). \tag{5.11}$$

Using the formula (5.10) instead of [H91] (4.10b), we can prove in exactly the same manner as in [H91] Theorem 5.2 the following result:

Theorem 5.3 *Suppose* (5.6) *and* (5.7). *There exists a unique element $\mathcal{D}$ in the field of fractions of $\Lambda\widehat{\otimes}_W\Lambda'$ satisfying the following interpolation property: Let $(P,Q) \in \mathrm{Spf}(\Lambda) \times \mathrm{Spf}(\Lambda')$ be an arithmetic point such that*

(W) $\kappa_1(P) - \kappa_1(Q) > 0 \geq \kappa_2(P) - \kappa_2(Q)$ *and* $\psi_{P,1} = \chi_{Q,1}$ *on* $\widehat{O}^\times$.

Then $\mathcal{D}$ is finite at (P,Q) and we have

$$\mathcal{D}(P,Q) = W(P,Q)C(P,Q)S(P)^{-1}E(P,Q)\frac{L^{(p)}(1, M(f_P) \otimes \widetilde{M}(g_Q))}{(f_P, f_P)},$$

where, writing $k(P) = \kappa_1(P) - \kappa_2(P) + I$,

$$W(P,Q) =$$
$$\frac{(-1)^{k(Q)}}{(-1)^{k(P)}}\frac{N(\mathfrak{J})^{([\kappa(Q)]+1)/2}}{N(\mathfrak{N})^{([\kappa(P)]-1)/2}} \cdot \prod_{\mathfrak{p}|p} \frac{\chi_Q^u(d_{\mathfrak{p}})G(\chi_{Q,1,\mathfrak{p}}^{-1}\psi_{P,1,\mathfrak{p}})G(\chi_{Q,2,\mathfrak{p}}^{-1}\psi_{P,1,\mathfrak{p}})}{\psi_P^u(d_{\mathfrak{p}})G((\psi_{P,\mathfrak{p}}^-)^{-1})}$$

$$C(P,Q) = 2^{([\kappa(P)]-[\kappa(Q)])I-2k(P)}\pi^{2\kappa_2(P)-([\kappa(Q)]+1)I}$$
$$\times \Gamma_F(\kappa_1(Q)-\kappa_2(P)+I)\Gamma_F(\kappa_2(Q)-\kappa_2(P)+I),$$

$$S(P) =$$
$$\prod_{\mathfrak{p}\nmid\mathfrak{C}_p(\psi_P^-)} (\psi_P^-(\varpi_\mathfrak{p})-1)\,(1-\psi_P^-(\varpi_\mathfrak{p})|\varpi_\mathfrak{p}|_\mathfrak{p}) \prod_{\mathfrak{p}|\mathfrak{C}_p(\psi_P^-)} (\psi_P^-(\varpi_\mathfrak{p})|\varpi_\mathfrak{p}|_\mathfrak{p})^{\delta(\mathfrak{p})},$$

$$E(P,Q) =$$
$$\prod_{\mathfrak{p}\nmid\mathfrak{C}_p(\chi_Q^-)} \frac{(1-\chi_{Q,1}\psi_{P,1}^{-1}(\varpi_\mathfrak{p}))(1-\chi_{Q,2}\psi_{P,1}^{-1}(\varpi_\mathfrak{p}))}{(1-\chi_{Q,1}^{-1}\psi_{P,1}(\varpi_\mathfrak{p})|\varpi_\mathfrak{p}|_\mathfrak{p})(1-\chi_{Q,2}^{-1}\psi_{P,1}(\varpi_\mathfrak{p})|\varpi_\mathfrak{p}|_\mathfrak{p})}$$
$$\times \prod_{\mathfrak{p}|\mathfrak{C}_p(\chi_Q^-)} \frac{\chi_{Q,2}\psi_{P,1}^{-1}(\varpi_\mathfrak{p}^{\gamma(\mathfrak{p})})(1-\chi_{Q,1}\psi_{P,1}^{-1}(\varpi_\mathfrak{p}))}{(1-\chi_{Q,1}^{-1}\psi_{P,1}(\varpi_\mathfrak{p})|\varpi_\mathfrak{p}|_\mathfrak{p})}.$$

Here $\mathfrak{C}_p(\psi_P^-) := \mathfrak{C}(\psi_P^-) + (p) = \prod_{\mathfrak{p}|p}\mathfrak{p}^{\delta(\mathfrak{p})}$ *and* $\mathfrak{C}_p(\chi_P^-) := \mathfrak{C}(\chi_Q^-) + (p) = \prod_{\mathfrak{p}|p}\mathfrak{p}^{\gamma(\mathfrak{p})}$. *Moreover for the congruence power series* $H(\lambda)$ *of* λ, $H(\lambda)\mathcal{D} \in \Lambda\widehat{\otimes}_W\Lambda'$.

The expression of p–Euler factors and root numbers is simpler than the one given in [H91] Theorem 5.1, because automorphic representations of f_P and g_Q are everywhere principal at finite places (by (5.6)). The shape of the constant $W(P,Q)$ appears to be slightly different from [H91] Theorem 5.2. Firstly the present factor $(-1)^{k(P)+k(Q)}$ is written as $(\chi_{Q+}\psi_{P+})_\infty(-1)$ in [H91]. Secondly, in [H91], it is assumed that $\chi_{Q,1}^{-1}$ and $\psi_{P,1}^{-1}$ are both induced by a global character ψ'_P and ψ'_P unramified outside p. Thus the factor $(\chi'_{Q,\infty}\psi'_{P,\infty})(-1)$ appears there. This factor is equal to $(\chi_{Q,1,p}\psi_{P,1,p})(-1) = \theta_p(-1)$, which is trivial because of the condition (W). We do not need to assume the individual extensibility of $\chi_{Q,1}$ and $\psi_{P,1}$. This extensibility is assumed in order to have a global Hecke eigenform $f_P^\circ = f_P^u \otimes \psi'_P$. However this assumption is redundant, because all computation we have done in [H91] can be done locally using the local Whittaker model. Also $C(P,Q)$ in the above theorem is slightly different from the one in [H91] Theorem 5.2, because $(f_P, f_P) = D^{[\kappa(P)]+1}(f_P^\circ, f_P^\circ)$ for f_P° appearing in the formula of [H91] Theorem 5.2.

Proof We start with a slightly more general situation. We shall use the symbol introduced in [H91]. Suppose $\mathfrak{C}(\psi^-)|\mathfrak{N}$ and $\mathfrak{C}(\chi^-)|\mathfrak{J}$, and take normalized Hecke eigenforms $f \in S_\kappa(\mathfrak{N},\psi_+;\mathbb{C})$ and $g \in S_{\kappa'}(\mathfrak{J},\chi_+;\mathbb{C})$. Suppose

$\psi_1 = \chi_1$. We define $f^c \in S_\kappa(\mathfrak{N}, \overline{\psi}_+; \mathbb{C})$ by $a(y, f^c) = \overline{a(y, f)}$. Then $f^c(w) = \overline{f(\varepsilon^{-1} w \varepsilon)}$ for $\varepsilon = \left(\begin{smallmatrix} -1 & 0 \\ 0 & 1 \end{smallmatrix}\right)$. We put $\Phi(w) = (\overline{f^c} g^c)(w)$. Then we see $\Phi(wu) = \psi(u)\chi^{-1}(u)\Phi(w)$ for $u \in U = U_0(\mathfrak{N}') \cap U_0(\mathfrak{J}')$. Since $\psi_1 = \chi_1$, we find that $\psi(u)\chi(u)^{-1} = \psi^-(\chi^-)^{-1}(d) = \psi^u(\chi^u)^{-1}(d)$ if $u = \left(\begin{smallmatrix} a & b \\ c & d \end{smallmatrix}\right)$. We write simply ω for the central character of $\overline{f}^c g^c$, which is the Hecke character $\psi_+^u(\chi_+^u)^{-1}|\cdot|_{\mathbb{A}}^{-[\kappa']-[\kappa]}$. Then we have $\Phi(zw) = \omega(z)\Phi(w)$, and $\Phi(wu_\infty) = \overline{J_\kappa(u_\infty, \mathbf{i})}^{-1} J_{\kappa'}(u_\infty, \mathbf{i})^{-1}\Phi(w)$. We then define $\omega^*(w) = \omega(d_{\mathfrak{N}'\mathfrak{J}'})$ for $w = \left(\begin{smallmatrix} a & b \\ c & d \end{smallmatrix}\right) \in B(\mathbb{A})U \cdot G(\mathbb{R})^+$. Here B is the algebraic subgroup of G made of matrices of the form $\left(\begin{smallmatrix} y & x \\ 0 & 1 \end{smallmatrix}\right)$. We extend $\omega^* : G(\mathbb{A}) \to \mathbb{C}$ outside $B(\mathbb{A})U \cdot G(\mathbb{R})^+$ just by 0. Similarly we define $\eta : G(\mathbb{A}) \to \mathbb{C}$ by

$$\eta(w) = \begin{cases} |y|_{\mathbb{A}} & \text{if } g = \left(\begin{smallmatrix} y & x \\ 0 & 1 \end{smallmatrix}\right) zu \text{ with } z \in F_{\mathbb{A}}^\times \text{ and } u \in U \cdot SO_2(F_{\mathbb{R}}) \\ 0 & \text{otherwise.} \end{cases}$$

For each $\mathbb{Q}$–subalgebra $A \subset \mathbb{A}$, we write $B(A)_+ = B(A) \cap G(\mathbb{A}^{(\infty)}) \times G(\mathbb{R})^+$. Note that $\Phi(w)\overline{\omega}^*(w)\eta(w)^{s-1}$ for $s \in \mathbb{C}$ is left invariant under $B(\mathbb{Q})_+$. Then we compute

$$\mathcal{Z}(s, f, g) = \int_{B(\mathbb{Q})_+ \backslash B(\mathbb{A})_+} \Phi(w)\overline{\omega}^*(w)\eta(w)^{s-1} d\varphi_B(w)$$

for the measure $\varphi_B\left(\begin{smallmatrix} y & x \\ 0 & 1 \end{smallmatrix}\right) = |y|_{\mathbb{A}}^{-1} dx \otimes d^\times y$ defined in [H91] page 340. We have

$$\begin{aligned} &\mathcal{Z}(s, f, g) \\ &= \int_{F_{\mathbb{A}+}^\times} \int_{F_{\mathbb{A}}/F} \Phi\left(\begin{smallmatrix} y & x \\ 0 & 1 \end{smallmatrix}\right) dx |y|_{\mathbb{A}}^{s-1} d^\times y \\ &= D^{\frac{1}{2}} \int_{F_{\mathbb{A}}^\times} a(dy, f)\overline{a(dy, g)} \mathbf{e}_F(2\sqrt{-1} y_\infty) y_\infty^{-(\kappa_2 + \kappa_2')} |y|_{\mathbb{A}}^s d^\times y \\ &\overset{dy \mapsto y}{=} D^{s+\frac{1}{2}} \int_{F_{\mathbb{A}}^\times} a(y, f)\overline{a(y, g)} \mathbf{e}_F(2\sqrt{-1} y_\infty) y_\infty^{-(\kappa_2 + \kappa_2')} |y|_{\mathbb{A}}^s d^\times y \\ &= D^{s+\frac{1}{2}} (4\pi)^{-sI + \kappa_2 + \kappa_2'} \Gamma_F(sI - \kappa_2 - \kappa_2') D(s - \frac{[\kappa] + [\kappa']}{2} - 1, f, g). \end{aligned}$$

Define $C_{\infty+} \subset G(\mathbb{R})^+$ by the stabilizer in $G(\mathbb{R})^+$ of $\mathbf{i} \in \mathfrak{Z}$. We now choose an invariant measure φ_U on $X(U) = G(\mathbb{Q})_+ \backslash G(\mathbb{A})_+ / UC_{\infty+}$ so that

$$\int_{X(U)} \sum_{\gamma \in O^\times B(\mathbb{Q})_+ \backslash G(\mathbb{Q})_+} \phi(\gamma w) d\varphi_U(w) = \int_{B(\mathbb{Q})_+ \backslash B(\mathbb{A})_+} \phi(b) d\varphi_B(b)$$

whenever ϕ is supported on $B(\mathbb{A}^{(\infty)})u \cdot G(\mathbb{R})^+$ and the two integrals are absolutely convergent. There exists a unique invariant measure φ_U as seen in [H91] page 342 (where the measure is written as μ_U). On $B(\mathbb{A})_+$, $\Phi(w) = \overline{f}^c g^c(w)\overline{J_\kappa(w,\mathbf{i})}J_{\kappa'}(w,\mathbf{i})$ and the right-hand-side is left $C_{\infty+}$ invariant (cf. (S2) in 2.10). Then by the definition of φ_U, we have

$$\int_{B(\mathbb{Q})_+\backslash B(\mathbb{A})_+} \Phi\overline{\omega}^*\eta^{s-1}d\varphi_B = \int_{X(U)} \overline{f}^c(w)g^c(w)E(w,s-1)d\varphi_U(w), \tag{5.12}$$

where

$$E(w,s) = \sum_{\gamma\in O^\times B(\mathbb{Q})_+\backslash G(\mathbb{Q})_+} \overline{\omega}^*(\gamma w)\eta(\gamma w)^s\overline{J_\kappa(\gamma w,\mathbf{i})}J_{\kappa'}(\gamma w,\mathbf{i}).$$

Note that $E(zw,s) = (\psi_+{}^{-u}\chi_+^u)(z)E(w,s)$ for $z \in \widehat{O}^\times$. By definition, $E(\alpha x) = E(x)$ for $\alpha \in G(\mathbb{Q})_+$; in particular, it is invariant under $\alpha \in F^\times$. For $z \in F_\mathbb{R}^\times$, one has $E(zw,s) = N(z)^{[\kappa]+[\kappa']-2}E(w,s)$. It follows that $|\det(w)|_\mathbb{A}^{1-([\kappa]+[\kappa])/2}E(w,s)$ has eigenvalue $\psi_+^{-u}\chi_+^u(z)$ under the central action of $z \in F^\times\widehat{O}^\times F_\mathbb{R}^\times$. The averaged Eisenstein series:

$$\begin{aligned}\mathcal{E}(w,s) &= \sum_{a\in Cl_F} \psi_+^u\chi_+^{-u}(a)|\det(aw)|_\mathbb{A}^{1-([\kappa]+[\kappa'])/2}E(aw,s)\\ &= |\det(w)|_\mathbb{A}^{1-([\kappa]+[\kappa'])/2}\sum_{a\in Cl_F}\psi_+\overline{\chi}_+(a)E(aw,s)\end{aligned}$$

satisfies $\mathcal{E}(zw,s) = \psi_+^{-u}\chi_+^u(z)\mathcal{E}(w,s)$, where a runs over complete representative set for $F_\mathbb{A}^\times/F^\times\widehat{O}^\times F_\mathbb{R}^\times$ and ψ_+ is the central character of f and $\overline{\chi}_+$ is the central character of g^c. Defining the PGL_2 modular variety $\overline{X}(U) = X(U)/F_\mathbb{A}^\times$, by averaging (5.12), we find

$$\begin{aligned}&D^{s+\frac{1}{2}}(4\pi)^{-sI+\kappa_2+\kappa_2'}\Gamma_F(sI-\kappa_2-\kappa_2')D(s-\frac{[\kappa]+[\kappa']}{2}-1,f,g)\\ &= \int_{\overline{X}(U)}\overline{f}^c(w)g^c(w)\mathcal{E}(w,s-1)|\det(w)|_\mathbb{A}^{([\kappa]+[\kappa'])/2-1}d\varphi_U(w).\end{aligned} \tag{5.13}$$

Writing $U = U_0(\mathfrak{L})$ and writing $r = \kappa_2' - \kappa_2$, we define an Eisenstein series $\mathbb{E}_{k-k',r}(\overline{\omega}^u,\mathbf{1};s)$ by

$$N(\mathfrak{L})^{-1}\sqrt{D}|\det(w)|_\mathbb{A}^{\frac{[\kappa']-[\kappa]}{2}}L^{(\mathfrak{L})}(2s,\omega^u)\mathcal{E}(w,s+\frac{[\kappa]+[\kappa']}{2}-1),$$

where $k = \kappa_1 - \kappa_2$ and $k' = \kappa'_1 - \kappa'_2$. The ideal $\mathfrak{L}$ is given by $\mathfrak{N} \cap \mathfrak{J}$. Then, changing variable $s - \frac{[\kappa]+[\kappa']}{2} - 1 \mapsto s$, we can rewrite (5.13) as

$$D^{s+\frac{1}{2}}(4\pi)^{-sI-\frac{k+k'}{2}}\Gamma_F(sI + \frac{k+k'}{2})L^{(\mathfrak{L})}(2s+2, \psi^u\chi^{-u})D(s,f,g)$$
$$= N(\mathfrak{L})^{-1}D^{-(3+[\kappa]+[\kappa'])/2}(f^c, g^c\mathbb{E}_{k-k',r}(\overline{\omega}^u, \mathbf{1}; s+1))_{\mathfrak{L}}$$
$$= N(\mathfrak{L})^{-s-\frac{[\kappa]+[\kappa']}{2}}D^{-(3+[\kappa]+[\kappa'])/2}(f^c|\tau, (g^c|\tau)\mathbf{G}_{k-k',r}(\omega^u, \mathbf{1}; s+1))_{\mathfrak{L}}, \tag{5.14}$$

where $\mathbf{G}_{k-k',r}(\omega^u, \mathbf{1}; s) = N(\mathfrak{L})^{s-1+\frac{[\kappa']-[\kappa]}{2}}\mathbb{E}_{k-k',r}(\overline{\omega}^u, \mathbf{1}; s)|\tau$ for τ of level $\mathfrak{L}$, and

$$(\phi, \varphi)_{\mathfrak{L}} = \int_{\overline{X}(U)} \overline{\phi}(w)\varphi(w)|\det(w)|_{\mathbb{A}}^{[\kappa]-1}d\varphi_U(w)$$

is the normalized Petersson inner product on $S_\kappa(U, \psi; \mathbb{C})$. This formula is equivalent to the formula in [H91] (4.9) (although we have more general forms f and g with character ψ and χ not considered in [H91]). In [H91] (4.9), k' is written as κ and r is written as $w - \omega$.

Let E be the Eisenstein measure of level $\mathfrak{L} = \mathfrak{N} \cap \mathfrak{J}$ defined in [H91] Section 8, where $\mathfrak{L}$ is written as L. We take an idele L with $LO = \mathfrak{L}$ and $L^{(\mathfrak{L})} = 1$. Similarly we take ideles J and N replacing in the above formula $\mathfrak{L}$ by $\mathfrak{J}$ and $\mathfrak{N}$, respective;y, and L by the corresponding J and N, respectively.

The algebra homomorphism $\varphi : \mathbf{h}^{n.ord}(\mathfrak{J}, \chi; W) \to \Lambda'$ induces, by the W–duality, $\varphi^* : \Lambda'^* \hookrightarrow \mathbf{S}^{n.ord}(\mathfrak{J}, \chi; W)$, where $\mathbf{S}^{n.ord}(\mathfrak{J}, \chi; W)$ is a subspace of p–adic modular forms of level $(\mathfrak{J}, \chi)$ (see [H96] 2.6). We then consider the convolution as in [H91] Section 9 (page 382):

$$\mathcal{D} = \lambda * \varphi = \frac{E *_\lambda ([L/J] \circ \varphi^*)}{H(\lambda) \otimes 1} \text{ for } E *_\lambda ([L/J] \circ \varphi^*) \in \Lambda\widehat{\otimes}\Lambda',$$

where $[L/J]$ is the operator defined in [H91] Section 7.B and all the ingredient of the above formula is as in [H91] page 383. An important point here is that we use the congruence power series $H(\lambda) \in \Lambda$ (so $H(\lambda) \otimes 1 \in \Lambda\widehat{\otimes}\Lambda'$) defined with respect to $\mathbf{h}^{n.ord}(\mathfrak{N}, \psi; W)$ instead of $\mathbf{h}(\psi^u, \psi_1)$ considered in [H91] page 379 (so, $H(\lambda)$ is actually a factor of H in [H91] page 379, which is an improvement).

We write the minimal level of f_P^{ord} as $\mathfrak{N}\mathfrak{p}^\alpha$ for $\mathfrak{p}^\alpha = \prod_{\mathfrak{p}|p} \mathfrak{p}^{\alpha(\mathfrak{p})}$. Then we define $\varpi^\alpha = \prod_{\mathfrak{p}|p} \varpi_{\mathfrak{p}}^{\alpha(\mathfrak{p})}$. The integer $\alpha(\mathfrak{p})$ is given by the exponent of $\mathfrak{p}$ in $\mathfrak{C}(\psi_P^-)$ or 1 whichever larger. We now compute $\mathcal{D}(P, Q)$. We shall give the argument only when $j = [\kappa(P)] - [\kappa(Q)] \geq 1$, since the other case can be treated in the same manner as in [H91] Case II (page 387).

Put $\mathbf{G} = \mathbf{G}_{jI,0}\left(\chi_Q^{-u}\psi_P^u, \mathbf{1}; 1-\frac{j}{2}\right)$. We write $(\cdot,\cdot)_{\mathfrak{N}\mathfrak{p}^\alpha} = (\cdot,\cdot)_\alpha$ and put $\mathfrak{m} = L\mathfrak{p}^\alpha$. As before, $m = L\varpi^\alpha$ satisfies $mO = \mathfrak{m}$ and $m^{(\mathfrak{m})} = 1$. Put $r(P,Q) = \kappa_2(Q) - \kappa_2(P)$, which is non-negative by the weight condition (W) in the theorem. Then in exactly the same manner as in [H91] Section 10 (page 386), we find, for $c = (2\sqrt{-1})^{j[F:\mathbb{Q}]}\pi^{[F:\mathbb{Q}]}$,

$$(-1)^{k(Q)}N(\mathfrak{J}/\mathfrak{L})^{-1}N(\mathfrak{L}\mathfrak{p}^\alpha)^{[\kappa(Q)]-1}c((f_P^{ord})^c|\tau(N\varpi^\alpha), f_P^{ord})_\alpha \mathcal{D}(P,Q)$$
$$= ((f_P^{ord})^c|\tau(m), (g_Q^{ord}|\tau(J\varpi^\alpha))|\tau(m))\cdot(\delta_{jI}^{r(P,Q)}\mathbf{G}))_{\mathfrak{m}}. \quad (5.15)$$

By [H91] Corollary 6.3, we have, for $r = r(P,Q)$,

$$\delta_{jI}^r\mathbf{G} = \Gamma_F(r+I)(-4\pi)^{-r}\mathbf{G}_{jI+2r,r}\left(\chi_Q^{-u}\psi_P^u, \mathbf{1}; 1-\frac{j}{2}\right).$$

Then by (5.14), we get

$$c(-1)^{k(Q)}N(\mathfrak{J}\mathfrak{p}^\alpha)^{-1}C(P,Q)^{-1}(f_P^{ord}|\tau(N\varpi^\alpha), f_P^{ord})_\alpha\mathcal{D}(P,Q)$$
$$= L^{(\mathfrak{m})}(2-[\kappa(P)]+[\kappa(Q)], \omega_{P,Q})$$
$$\times D\left(\frac{[\kappa(Q)]-[\kappa(P)]}{2}, f_P^{ord}, (g_Q^{ord}|\tau(J\varpi^\alpha))^c\right), \quad (5.16)$$

where $\omega_{P,Q} = \chi_{Q+}^{-1}\psi_{P+}$ for the central characters χ_{Q+} of g_Q and ψ_{P+} of f_P.

Now we compute the Petersson inner product $(f_P^{ord}|\tau(N\varpi^\alpha), f_P^{ord})_\alpha$ in terms of (f_P, f_P). Note that for $f, g \in S_\kappa(U_0(N), \epsilon; \mathbb{C})$

$$(f^u)|\tau(N) = |N|_{\mathbb{A}}^{([\kappa]-1)/2}(f|\tau(N))^u \text{ and } (f,g)_{\mathfrak{N}} = D^{[\kappa]+1}(f^u, g^u). \quad (5.17)$$

The computation we have done in [H91] page 357 in the proof of Lemma 5.3 (vi) is valid without any change for each $\mathfrak{p}|p$, since at p–adic places, f_P in [H91] has the Neben type we introduced in this paper also for places outside p. The difference is that we compute the inner product in terms of (f_P, f_P) not (f_P°, f_P°) as in [H91] Lemma 5.3 (vi), where f_P° is the primitive form associated to $f_P^u \otimes \psi_{P,1}^u$ assuming that $\psi_{P,1}^u$ lifts to a global finite order character (the character $\psi_{P,1}^{-u}$ is written as ψ' in the proof of Lemma 5.3 (vi) of [H91]). Note here $f_P^\circ = f_P \otimes \psi_{P,1}^{-1}$ by definition and hence $(f_P^\circ, f_P^\circ) = (f_P^u, f_P^u)$, because tensoring a unitary character to a function does not alter the hermitian inner product. Thus we find

$$\frac{(f_P^{ord,u}|\tau(N\varpi^\alpha), f_P^{ord,u})}{(f_P^\circ, f_P^\circ)} = |N\varpi^\alpha|_{\mathbb{A}}^{([\kappa(P)]-1)/2}\frac{(f_P^{ord}|\tau(N\varpi^\alpha), f_P^{ord})}{(f_P, f_P)}. \quad (5.18)$$

A key point of the proof of Lemma 5.3 (vi) is the formula writing down $f_P^{ord,u} \otimes \psi_{P,1}^{-u}$ in terms of f_P°. Even without assuming the liftability of $\psi_{P,1}^u$

to a global character, the same formula is valid for f_P^{ord} and f_P before tensoring $\psi_{P,1}^{-1}$ (by computation using local Whittaker model). We thus have $f_P^{ord} = f_P|R$ for a product $R = \prod_{\mathfrak{p}|p} R_{\mathfrak{p}}$ of local operators $R_{\mathfrak{p}}$ given as follows: If the prime $\mathfrak{p}$ is a factor of $\mathfrak{C}(\psi_P^-)$, then $R_{\mathfrak{p}}$ is the identity operator. If $\mathfrak{p}$ is prime to $\mathfrak{C}(\psi_P^-)$ ($\Leftrightarrow \pi(1, \psi_{P,\mathfrak{p}}^-)$ is spherical), then $f|R_{\mathfrak{p}} = f - \psi_{P,2}(\varpi_{\mathfrak{p}})f\|[\varpi_{\mathfrak{p}}]$, where $f\|[\varpi_{\mathfrak{p}}](x) = |\varpi_{\mathfrak{p}}|_{\mathfrak{p}} f|g$ with $f|g(x) = f(xg)$ for $g = \left(\begin{smallmatrix} \varpi_{\mathfrak{p}}^{-1} & 0 \\ 0 & 1 \end{smallmatrix}\right)$. Writing U for the level group of f_P and $U' = U \cap U_0(\mathfrak{p})$, we note $f|T(\mathfrak{p}) = \mathrm{Tr}_{U/U'}(f|g^{-1})$. This shows

$$\begin{aligned}(f_P\|[\varpi_{\mathfrak{p}}], f_P)_{U'} &= |\varpi_{\mathfrak{p}}|_{\mathfrak{p}}^{[\kappa(P)]}(f_P, f_P|g^{-1})_{U'} \\ &= |\varpi_{\mathfrak{p}}|_{\mathfrak{p}}^{[\kappa(P)]}(f_P, \mathrm{Tr}_{U/U'}(f_P|g^{-1}))_U = (a+b)(f_P, f_P)_U,\end{aligned}$$

where $a = |\varpi_{\mathfrak{p}}|_{\mathfrak{p}}^{[\kappa(P)]}\psi_{P,1}(\varpi_{\mathfrak{p}})$, $b = |\varpi_{\mathfrak{p}}|_{\mathfrak{p}}^{[\kappa(P)]}\psi_{P,2}(\varpi_{\mathfrak{p}})$, and $(\cdot,\cdot)_U$ is the Petersson metric on $\overline{X}(U)$. Similarly we have,

$$\begin{aligned}&(f_P, f_P\|[\varpi_{\mathfrak{p}}])_{U'} = \overline{(a+b)}(f_P, f_P)_U \\ &\qquad \text{and } (f_P\|[\varpi_{\mathfrak{p}}], f_P\|[\varpi_{\mathfrak{p}}])_{U'} = |\varpi_{\mathfrak{p}}|_{\mathfrak{p}}^{[\kappa(P)]}(|\varpi_{\mathfrak{p}}|_{\mathfrak{p}} + 1)(f_P, f_P)_U.\end{aligned}$$

By (5.10) and by (5.18), we conclude from [H91] Lemma 5.3 (vi)

$$\begin{aligned}&\frac{((f_P^{ord})^c|\tau(N\varpi^\alpha), f_P^{ord})_\alpha}{(f_P, f_P)} \\ &\quad = |N|_{\mathbb{A}}^{(1-[\kappa(P)])/2}(-1)^{k(P)}\psi_P^u(d_p)W'(f_P)S(P) \\ &\qquad\qquad \times \psi_{P,2}(\varpi^\alpha) \prod_{\mathfrak{p}|((p)+\mathfrak{C}(\psi_P^-))} G(\psi_{P,2,\mathfrak{p}}^{-1}\psi_{P,1,\mathfrak{p}}) \qquad (5.19)\end{aligned}$$

for $\mathfrak{p}$ running over the prime factors of p.

We now compute the extra Euler factors: $E(P,Q)$ and $W(P,Q)$. Again the computation is the same as in [H91] Lemma 5.3 (iii)-(v), because the level structure and the Neben character at p–adic places are the same as in [H91] for f_P and g_Q and these factors only depend on p–adic places. Then we get the Euler p–factor $E(P,Q)$ and $W(P,Q)$ as in the theorem from [H91] lemma 5.3. □

Remark 5.4 We assumed the condition (5.7) to make the proof of the theorem simpler. We now remove this condition. Let $\mathcal{E}$ be the set of all prime factors $\mathfrak{q}$ of $\mathfrak{I}\mathfrak{N}$ such that $\chi_{\mathfrak{q}}^- = \psi_{\mathfrak{q}}^-$ on $O_{\mathfrak{q}}^\times$. Thus we assume that $\mathcal{E} \neq \varnothing$. Then in the proof of Lemma 5.2, the inertia group at $\mathfrak{q} \in \mathcal{E}$ fixes a two-dimensional subspace of $\rho_{f_P} \otimes \rho_{g_Q^c}$, one corresponding to $\psi_{1,\mathfrak{q}} \otimes \chi_{1,Q}^{-1}$ and the other coming from $\psi_{2,\mathfrak{q}} \otimes \chi_{2,Q}^{-1}$. The Euler factor corresponding to the latter does not appear in the Rankin product process; so, we get an imprimitive L–function, whose missing Euler factors are

$$E'(P,Q)^{-1} = \prod_{\mathfrak{q}\in\mathcal{E}} (1 - \psi_{2,\mathfrak{q}}\chi_{2,Q}^{-1}(\varpi_{\mathfrak{q}}))^{-1}.$$

Thus the final result is identical to Theorem 5.3 if we multiply $E(P,Q)$ by $E'(P,Q)$ in the statement of the theorem. In our application, λ and φ will be (Λ–adic) automorphic inductions of Λ–adic characters $\widetilde{\lambda}, \widetilde{\varphi} : \mathrm{Gal}(\overline{M}/M) \to \Lambda^\times$ for an ordinary CM field M/F. If the prime-to–p conductors of $\widetilde{\lambda}^-$ and $\widetilde{\varphi}^-$ are made of primes split in M/F, $\mathcal{E}$ is the set of primes ramifying in M/F. Then $E'(P,Q)$ is the specialization of $E' = \prod_{\mathfrak{q}\in\mathcal{E}}(1 - \widetilde{\lambda}\otimes\widetilde{\varphi}^{-1}(Frob_{\mathfrak{q}}))$ at (P,Q), and $E' \in \Lambda\widehat{\otimes}\Lambda$ is not divisible by the prime element of W (that is, the μ–invariant of E' vanishes). Actually, we can choose $\widetilde{\lambda}$ and $\widetilde{\varphi}$ so that $\widetilde{\lambda}\widetilde{\varphi}^{-1}(Frob_{\mathfrak{q}}) \not\equiv 1 \mod \mathfrak{m}_\Lambda$, and under this choice, we may assume that $E' \in (\Lambda\widehat{\otimes}\Lambda)^\times$.

5.3 Comparison of p–adic L–functions

For each character $\psi : \Delta_{\mathfrak{C}} \to W^\times$, we have the extension $\widetilde{\psi} : \mathrm{Gal}(\overline{\mathbb{Q}}/M) \to \Lambda$ sending $(\gamma, \delta) \in \Gamma \times \Delta_{\mathfrak{C}}$ to $\psi(\delta)\gamma$ for the group element $\gamma \in \Gamma$ inside the group algebra Λ. Regarding $\widetilde{\psi}$ as a character of $\mathrm{Gal}(\overline{\mathbb{Q}}/M)$, the induced representation $\mathrm{Ind}_M^F\widetilde{\psi}$ is modular nearly ordinary at p, and hence, by the universality of the nearly p–ordinary Hecke algebra $h_U^{n.ord}(W)$ defined in [H96] 2.5, we have a unique algebra homomorphism $\lambda : h_U^{n.ord} \to \Lambda$ such that $\mathrm{Ind}_M^F\widetilde{\psi} \cong \lambda \circ \rho^{Hecke}$ for the universal nearly ordinary modular Galois representation ρ^{Hecke} with coefficients in h, where $U = U_1^1(\mathfrak{N})$ with $\mathfrak{N} = D_{M/F}\mathfrak{c}(\psi)$ for the relative discriminant $D_{M/F}$ of M/F and $\mathfrak{c} = \mathfrak{C}(\psi) \cap F$ for the conductor $\mathfrak{C}(\psi)$ of ψ. Thus for each arithmetic point $P \in \mathrm{Spf}(\Lambda)(\overline{\mathbb{Q}}_p)$ (in the sense of [H96] 2.7), we have a classical Hecke eigenform $\theta(\widetilde{\psi}_P)$ of weight $\kappa(P)$.

We suppose that the conductor $\mathfrak{C}(\psi^-)$ of ψ^- consists of primes of M split over F. Then the automorphic representation $\pi(\widetilde{\psi}_P)$ of weight $\kappa(P)$ is everywhere principal at finite places. By [H96]7.1, the Hecke character $\widetilde{\psi}_P$ has infinity type

$$\infty(\widetilde{\psi}_P) = -\sum_{\sigma\in\Sigma}(\kappa_1(P)_{\sigma|_F}\sigma + \kappa_2(P)_{\sigma|_F}\sigma c).$$

In the automorphic representation $\pi(\widetilde{\psi}_P)$ generated by right translation of $\theta(\widetilde{\psi}_P)$, we have a unique normalized Hecke eigenform $f(\widetilde{\psi}_P)$ of minimal level.

The prime-to-p level of the cusp form $f(\widetilde{\psi}_P)$ is equal to

$$\mathfrak{N}(\psi) = N_{M/F}(\mathfrak{C}(\psi^-))D_{M/F},$$

and it satisfies (L1-3) in 5.1 for $\epsilon = \epsilon_P$ given by

$$\epsilon_{1,\mathfrak{q}} = \begin{cases} \widetilde{\psi}_P|_{\mathrm{Gal}(\overline{M}_{\mathfrak{L}}/M_{\mathfrak{L}})} & \text{if } \mathfrak{q} = \mathfrak{L}\overline{\mathfrak{L}}, \\ \text{an extension of } \widetilde{\psi}_P|_{\mathrm{Gal}(\overline{M}_{\mathfrak{L}}/M_{\mathfrak{L}})} \text{ to } \mathrm{Gal}(\overline{M}_{\mathfrak{L}}/F_{\mathfrak{q}}) & \text{otherwise,} \end{cases}$$

$$\epsilon_{2,\mathfrak{q}} = \begin{cases} \widetilde{\psi}_P|_{\mathrm{Gal}(\overline{M}_{\overline{\mathfrak{L}}}/M_{\overline{\mathfrak{L}}})} & \text{if } \mathfrak{q} = \mathfrak{L}\overline{\mathfrak{L}}, \\ \text{another extension of } \widetilde{\psi}_P|_{\mathrm{Gal}(\overline{M}_{\mathfrak{L}}/M_{\mathfrak{L}})} \text{ to } \mathrm{Gal}(\overline{M}_{\mathfrak{L}}/F_{\mathfrak{q}}) & \text{otherwise,} \end{cases} \tag{5.20}$$

where $\mathfrak{L}$ and $\overline{\mathfrak{L}}$ are distinct primes in M.

Write ϵ_+^t for the restriction of $\epsilon_+ = \epsilon_1\epsilon_2$ to $\Delta_F(\mathfrak{N}')$, which is independent of P (because it factors through the torsion part of $Cl_F^+(\mathfrak{N}'p^\infty)$). Since $\{f(\widetilde{\psi}_P)\}_P$ is again a p–adic analytic family of cusp forms, they are induced by a new algebra homomorphism $\lambda_\psi : \mathbf{h} = \mathbf{h}^{n.ord}(\mathfrak{N}, \epsilon; W) \to \Lambda$. Since λ_ψ is of minimal level, the congruence module $C_0(\lambda_\psi; \Lambda)$ is a well defined Λ–module of the form $\Lambda/H(\psi)\Lambda$ (see [H96] 2.9). Actually we can choose $H(\psi)$ in $\Lambda^- = W[[\Gamma_-]]$ (see [GME] Theorem 5.44). The element $H(\psi)$ is called the *congruence power series* of λ_ψ (identifying Λ^- with a power series ring over W of $[F : \mathbb{Q}]$ variables)

By Theorem 5.3 and Remark 5.4, we have the (imprimitive) p–adic Rankin product $\mathcal{D} = \lambda_\psi * \lambda_\varphi$ with missing Euler factor $E' \in (\Lambda\widehat{\otimes}\Lambda)^\times$ as in Remark 5.4 for two characters $\psi : \Delta_{\mathfrak{C}} \to W^\times$ and $\varphi : \Delta_{\mathfrak{C}'} \to W^\times$. Writing $\mathcal{R} = \mathcal{D} \cdot H(\psi) \in \Lambda\widehat{\otimes}_W\Lambda$, we have $\mathcal{D} = \frac{\mathcal{R}}{H(\psi)}$.

We define two p–adic L–functions $\mathcal{L}_p(\psi^{-1}\varphi)$ and $\mathcal{L}_p(\psi^{-1}\varphi_c)$ by

$$\mathcal{L}_p(\psi^{-1}\varphi)(P, Q) = E'(P, Q)L_p(\widetilde{\psi}_P^{-1}\widetilde{\varphi}_Q)$$

and

$$\mathcal{L}_p(\psi^{-1}\varphi_c)(P, Q) = L_p(\widetilde{\psi}_P^{-1}\widetilde{\varphi}_{Q,c})$$

for the Katz p–adic L–function L_p, where $\chi_c(\sigma) = \chi(c\sigma c^{-1})$.

We follow the argument in [H91], [HT1] and [H96] to show the following identity of p–adic L–functions, which is a more precise version of [HT1] Theorem 8.1 without the redundant factor written as $\Delta(M/F; \mathfrak{C})$ there:

Theorem 5.5 *Let ψ and φ be two characters of $\Delta_{\mathfrak{C}}$ with values in $W^\times$. Suppose the following three conditions:*

(i) *$\mathfrak{C}(\psi^-)\mathfrak{C}(\varphi^-)$ is prime to any inert or ramified prime of M;*

(ii) *At each inert or ramified prime factor $\mathfrak{q}$ of $\mathfrak{C}(\psi)\mathfrak{C}(\varphi)$, $\psi_{\mathfrak{q}} = \varphi_{\mathfrak{q}}$ on $R_{\mathfrak{q}}^\times$;*

(iii) *For each split prime* $\mathfrak{q}|\mathfrak{C}(\psi)\mathfrak{C}(\varphi)$, *we have a choice of one prime factor* $\mathfrak{Q}|\mathfrak{q}$ *so that* $\psi_{\mathfrak{Q}} = \varphi_{\mathfrak{Q}}$ *on* $R_{\mathfrak{Q}}^{\times}$.

Then we have, for a power series $\mathcal{L} \in \Lambda\widehat{\otimes}\Lambda$,

$$\frac{\mathcal{L}}{H(\psi)} = \frac{\mathcal{L}_p(\psi^{-1}\varphi)\mathcal{L}_p(\psi^{-1}\varphi_c)}{(h(M)/h(F))L_p^-(\psi^-)}.$$

The power series $\mathcal{L}$ *is equal to* $(\lambda_\psi * \lambda_\varphi) \cdot H(\psi)$ *up to units in* $\Lambda\widehat{\otimes}\Lambda$.

The improvement over [HT1] Theorem 8.1 is that our identity is exact without missing Euler factors, but we need to have the additional assumptions (i)–(iii) to ensure the matching condition (5.6) for the automorphic induction of $\widetilde{\psi}_P$ and $\widetilde{\varphi}_Q$.

The proof of the theorem is identical to the one given in [HT1] Section 10, since the factors $C(P,Q)$, $W(P,Q)$, $S(P)$ and $E(P,Q)$ appearing in Theorem 5.3 are identical to those appearing in [H91] Theorem 5.2 except for the power of the discriminant D, which is compensated by the difference of (f_P°, f_P°) appearing in [H91] Theorem 5.2 from (f_P, f_P) in Theorem 5.3. Since we do not lose any Euler factors in (5.5) (thanks to our minimal level structure and the assumption (5.6) assuring principality everywhere), we are able to remove the missing Euler factor denoted $\Delta(1)$ in [HT1] (0.6b).

If M/F is ramified at some finite place, by Theorem 4.3 and Remark 5.4, we can always choose a pair $(\mathfrak{l}, \varphi)$ of a prime ideal $\mathfrak{l}$ and a character φ of $\mathfrak{l}$–power conductor so that

(i) $\mathfrak{l}$ is a split prime of F of degree 1;
(ii) $\mathcal{L}_p(\psi^{-1}\varphi)\mathcal{L}_p(\psi^{-1}\varphi_c)$ is a unit in $\Lambda\widehat{\otimes}\Lambda$.

Even if M/F is unramified, we shall show that such choice is possible: Writing Ξ for the set of primes $\mathfrak{q}$ as specified in the condition (3) in the theorem. Then we choose a prime $\mathfrak{L}$ split in $M/\mathbb{Q}$ so that $\mathfrak{L}\overline{\mathfrak{L}}$ is outside Ξ. We then take a finite order character $\varphi' : M_{\mathbb{A}}^{\times}/M^{\times} \to W^{\times}$ of conductor $\mathfrak{L}^m$ so that $\varphi'_{\overline{\mathfrak{L}}}$ is trivial but the reduction modulo $\mathfrak{m}_W$ of $\psi^{-1}\varphi\varphi'\mathcal{N}$ and $\psi^{-1}\varphi_c\varphi'_c\mathcal{N}$ is not equal to $\left(\frac{M/F}{\cdot}\right)$. Then again by Theorem 4.3, we find (infinitely many) φ' with unit power series $\mathcal{L}_p(\psi^{-1}\varphi\varphi')\mathcal{L}_p(\psi^{-1}(\varphi\varphi')_c)$. This implies the following corollary:

Corollary 5.6 *Suppose that* $\mathfrak{C}(\psi^-)$ *is prime to any inert or ramified prime of* M. *Then in* Λ^-, *we have* $(h(M)/h(F))L_p^-(\psi^-)\big|H(\psi)$.

As a byproduct of the proof of Theorem 5.5, we can express the p–adic L–value $L_p(\widetilde{\psi}_{\overline{P}})$ (up to units in W and the period in $\mathbb{C}^{\times}$) by the Petersson metric of the normalized Hecke eigenform $f_P = f(\widetilde{\psi}_P)$ of minimal level (in the automorphic induction π_P of $\widetilde{\psi}_P$). We shall describe this fact.

We are going to express the value $L(1, Ad(f_P))$ in terms of the L–values of the Hecke character $\widetilde{\psi}_P$. The representation π_P is everywhere principal outside archimedean places if and only if ψ^- has split conductor. Then

$$L(s, Ad(f_P)) = L(1, \left(\frac{M/F}{\cdot}\right))L(0, (\widetilde{\psi}_P^-)^*).$$

The infinity type of $\widetilde{\psi}_P^-$ is $(\kappa_1(P) - \kappa_2(P)) + (\kappa_2(P) - \kappa_1(P))c = (k(P) - I)(1-c)$, where we identify $\mathbb{Z}[\Sigma]$ with $\mathbb{Z}[I]$ sending σ to $\sigma|_F$. Thus the infinity type of $(\widetilde{\psi}_P^-)^*$ is given by $k(P) - k(P)c + 2\Sigma c = 2\Sigma + (k(P) - 2I)(1-c)$. Thus we have

$$\begin{aligned} L_p((\widetilde{\psi}_P^-)^*) &= N(\mathfrak{c})\widetilde{\psi}_P^-(\mathfrak{c})W'(\widetilde{\psi}_P^-)^{-1}L_p(\widetilde{\psi}_P^-) \\ &= CW_p(\widetilde{\psi}_P^-)E(P)\frac{\pi^{k(P)-I}\Gamma_\Sigma(k(P)-I)L(0,\psi_P^-)}{\Omega_\infty^{2(k(P)-I)}} \\ &= C'W_p((\widetilde{\psi}^-)_P^*)E(P)\frac{\pi^{k(P)-2I}\Gamma_\Sigma(k(P))L(0,(\psi^-)_P^*)}{\Omega_\infty^{2(k(P)-I)}}, \end{aligned} \tag{5.21}$$

where C and C' are constants in $W^\times$ and

$$E(P) = \prod_{\mathfrak{P}\in\Sigma_p} \left((1 - \widetilde{\psi}_P^-(\mathfrak{P}^c))(1 - (\widetilde{\psi}_P^-)^*(\mathfrak{P}^c))\right).$$

Since we have, for the conductor $\mathfrak{P}^{e(\mathfrak{P})}$ of $\psi_{P,\mathfrak{P}}^-$,

$$(\widetilde{\psi}_P^-)^*(\varpi_{\mathfrak{P}}^{-e(\mathfrak{P})}) = N(\mathfrak{P})^{e(\mathfrak{P})}\widetilde{\psi}_P^-(\varpi_{\mathfrak{P}}^{-e(\mathfrak{P})}),$$

by definition, $W((\widetilde{\psi}_P^-)^*)$ is the product over $\mathfrak{P} \in \Sigma$ of $G(2\delta_{\mathfrak{P}}, \psi_{P,\mathfrak{P}}^-)$, and hence we have

$$W_p((\widetilde{\psi}_P^-)^*) = N(\mathfrak{P}^{e(\Sigma)})W_p(\widetilde{\psi}_P^-),$$

where $\mathfrak{P}^{e(\Sigma)}$ is the Σ_p–part of the conductor of ψ_P^-. Thus we have, for $h(M/F) = h(M)/h(F)$,

$$\begin{aligned} L_p((\widetilde{\psi}_P^-)^*) &= C'W_p(\widetilde{\psi}_P^-)N(\mathfrak{P}^{e(\Sigma)})E(P)\frac{\pi^{k(P)-2I}\Gamma_\Sigma(k(P))L(0,(\widetilde{\psi}^-)_P^*)}{\Omega_\infty^{2(k(P)-I)}} \\ &= C''W_p(\widetilde{\psi}_P^-)E(P)\frac{\pi^{2(\kappa_1(P)-\kappa_2(P))}(f(\widetilde{\psi}_P), f(\widetilde{\psi}_P))}{h(M/F)\Omega_\infty^{2(\kappa_1(P)-\kappa_2(P))}}, \end{aligned} \tag{5.22}$$

where C'' and C' are constants in $W^\times$.

We suppose that the conductor $\mathfrak{C}(\widetilde{\psi}_P)$ is prime to p. Then $W_p(\widetilde{\psi}_P^-) = 1$ if p is unramified in $F/\mathbb{Q}$ (and even if p ramifies in $F/\mathbb{Q}$, it is a unit in W). Let $h = h(M)$, and choose a global generator ϖ of $\mathfrak{P}^h$. Thus $\varpi^\Sigma \equiv 0 \mod \mathfrak{m}_W$ and $\varpi^{\Sigma c} \not\equiv 0 \mod \mathfrak{m}_W$. By the arithmeticity of the point P, we have $k = k(P) \geq 2I$. Then we have, up to p–adic unit,

$$\widetilde{\psi}_P^-(\mathfrak{P}^c)^h = \widetilde{\psi}_P^-(\varpi) = \varpi^{(k-I)-(k-I)c}$$
$$(\widetilde{\psi}_P^-)^*(\mathfrak{P}^c)^h = (\widetilde{\psi}_P^-)^*(\varpi) = \varpi^{(k-2I)-kc}.$$

Thus $\widetilde{\psi}_P^-(\mathfrak{P}^c) \in \mathfrak{m}_W$ if $k \geq 2I$ and $(\widetilde{\psi}_P^-)^*(\mathfrak{P}^c) \in \mathfrak{m}_W$ if $k \geq 3I$. For each $\sigma \in \Sigma$, we write $\mathfrak{P}_\sigma$ for the place in Σ_p induced by $i_p \circ \sigma$. Thus we obtain

Proposition 5.7 *Suppose that either $k(P)_\sigma = \kappa_1(P)_\sigma - \kappa_2(P)_\sigma + 1 \geq 3$ or $(\widetilde{\psi}_P^-)^*(\mathfrak{P}_\sigma^c) \not\equiv 1 \mod \mathfrak{m}_W$ for all $\sigma \in I$ and that $\widetilde{\psi}_P$ has split conductor. Write $h(M/F)$ for $h(M)/h(F)$. Then up to units in W, we have*

$$h(M/F)L_p((\widetilde{\psi}_P^-)^*) = h(M/F)L_p(\widetilde{\psi}_P^-)$$
$$= \frac{\pi^{2(k(P)-I)} W_p(\widetilde{\psi}_P^-)(f(\widetilde{\psi}_P), f(\widetilde{\psi}_P))_{\mathfrak{N}}}{\Omega_\infty^{2(k(P)-I)}},$$

where $f(\widetilde{\psi}_P)$ is the normalized Hecke eigenform of minimal level $\mathfrak{N}$ (necessarily prime to p) of the automorphic induction of $\widetilde{\psi}_P$.

5.4 A case of the anticyclotomic main conjecture

Here we describe briefly an example of a case where the divisibility: $L_p^-(\psi)|\mathcal{F}^-(\psi)$ implies the equality $L_p^-(\psi) = \mathcal{F}^-(\psi)$ (up to units), relying on the proof by Rubin of the one variable main conjecture over an imaginary quadratic field in [R] and [R1]. Thus we need to suppose

(ab) $F/\mathbb{Q}$ is abelian, $p \nmid [F : \mathbb{Q}]$, and $M = F[\sqrt{D}]$ with $0 > D \in \mathbb{Z}$,

in order to reduce our case to the imaginary quadratic case treated by Rubin. Write $E = \mathbb{Q}[\sqrt{D}]$ and suppose that D is the discriminant of $E/\mathbb{Q}$. We have $p \nmid D$ since p is supposed to be unramified in $M/\mathbb{Q}$. We suppose also

(sp) The prime ideal (p) splits into $\mathfrak{p}\overline{\mathfrak{p}}$ in $E/\mathbb{Q}$.

Then we take $\Sigma = \{\sigma : M \hookrightarrow \overline{\mathbb{Q}} | \sigma(\sqrt{D}) = \sqrt{D}\}$. By (sp), Σ is an ordinary CM type.

We fix a conductor ideal $\mathfrak{c}$ (prime to p) of E satisfying (opl) and (spt) for $E/\mathbb{Q}$ (in place of M/F). We then put $\mathfrak{C} = \mathfrak{c}R$. We consider $Z_E = Z_E(\mathfrak{c}) = \varprojlim_n Cl_E(\mathfrak{c}p^n)$. We split $Z_E(\mathfrak{c}) = \Delta_\mathfrak{c} \times \Gamma_\mathfrak{c}$ for a finite group $\Delta = \Delta_\mathfrak{c}$ and a torsion-free subgroup $\Gamma_{\mathfrak{c}}$. As before, we identify $\Gamma_{\mathfrak{c}}$ and Γ_1 for E and write it as Γ_E. We then write $\Gamma_{E,-} = \Gamma_E/\Gamma_E^+$ for $\Gamma_E^+ = H^0(\mathrm{Gal}(E/\mathbb{Q}), \Gamma_E)$. We consider the anticyclotomic $\mathbb{Z}_p$–extension E_∞^- of E on which we have $c\sigma c^{-1} = \sigma^{-1}$ for all $\sigma \in \mathrm{Gal}(E_\infty^-/E)$ for complex conjugation c. Writing $E(\mathfrak{c}p^\infty)$ (inside $M(\mathfrak{C}p^\infty)$) for the ray class field over E modulo $\mathfrak{c}p^\infty$, we identify $Z_E(\mathfrak{c})$ with $\mathrm{Gal}(E(\mathfrak{c}p^\infty)/E)$ via the Artin reciprocity law. Then $\mathrm{Gal}(E(\mathfrak{c}p^\infty)F/E_\infty^-) = \Gamma_E^+ \times \Delta_{\mathfrak{c}}$ and $\mathrm{Gal}(E_\infty^-/E) = \Gamma_{E,-}$. We then define E_Δ by the fixed field of $\Gamma_{\mathfrak{c}}$ in the composite $F \cdot E(\mathfrak{c}p^\infty)$; so, $\mathrm{Gal}(E_\Delta/E) = \Delta_{\mathfrak{c}}$ and $E_\Delta \supset F$. We have $M_\Delta \supset E_\Delta$. Thus we have the restriction maps $\mathrm{Res}_Z : Z(\mathfrak{C}) = \mathrm{Gal}(M(\mathfrak{C}p^\infty)/M) \to \mathrm{Gal}(E(\mathfrak{c}p^\infty)/M) = Z_E(\mathfrak{c})$, $\mathrm{Res}_\Gamma : \Gamma_- = \mathrm{Gal}(M_\infty^-/M) \to \mathrm{Gal}(E_\infty^-/E) = \Gamma_{E,-}$ and $\mathrm{Res}_\Delta : \Delta_{\mathfrak{C}} = \mathrm{Gal}(M_\Delta/M) \to \mathrm{Gal}(E_\Delta/E) = \Delta_{\mathfrak{c}}$. We suppose

(res) There exists an anticyclotomic character ψ_E of $\Delta_{\mathfrak{c}}$ such that $\psi = \psi_E \circ \mathrm{Res}_\Delta$.

Let $L_\infty^E/E_\infty^- E_\Delta$ be the maximal p–abelian extension unramified outside $\mathfrak{p}$. Each $\gamma \in \mathrm{Gal}(L_\infty^E/E)$ acts on the normal subgroup $X_E = \mathrm{Gal}(L_\infty^E/E_\infty^- E_\Delta)$ continuously by conjugation, and by the commutativity of X_E, this action factors through $\mathrm{Gal}(E_\Delta E_\infty^-/E)$. Then we look into the Λ_E^-–module: $X_E[\psi_E\chi] = X_E \otimes_{\Delta_{\mathfrak{c}},\psi\chi} W$ for a character χ of $\mathrm{Gal}(F/\mathbb{Q})$, where $\Lambda_E^- = W[[\Gamma_{E,-}]]$. The projection Res_Γ induces a W-algebra homomorphism $\Lambda^- \to \Lambda_E^-$ whose kernel we write as $\mathfrak{a}$.

Theorem 5.8 *Let the notation be as above. Suppose that ψ has order prime to p in addition to* (ab), (sp) *and* (res). *Then $L_p^-(\psi) = \mathcal{F}^-(\psi)$ up to units in Λ^-.*

Proof We shall use functoriality of the Fitting ideal $F_A(H)$ of an A-module H with finite presentation over a commutative ring A with identity (see [MW] Appendix for the definition and the functoriality Fitting ideals listed below):

(i) If $I \subset A$ is an ideal, we have $F_{A/I}(H/IH) = F_A(H) \otimes_A A/I$;

(ii) If A is a noetherian normal integral domain and H is a torsion A-module of finite type, the characteristic ideal $\mathrm{char}_A(H)$ is the reflexive closure of $F_A(H)$. In particular, we have $\mathrm{char}_A(H) \supset F_A(H)$.

By definition, we have

$$H_0(\mathrm{Ker}(\mathrm{Res}_\Gamma), X[\psi]) = X[\psi]/\mathfrak{a}X[\psi] \cong \bigoplus_\chi X_E[\psi_E\chi]$$

for χ running all characters of $\mathrm{Gal}(F/\mathbb{Q})$. By (i) above, we have

$$\prod_\chi F_{\Lambda_E^-}(X_E[\psi\chi]) = F_{\Lambda^-}(X[\psi]) \otimes_{\Lambda^-} \Lambda_E^-.$$

Since $\text{char}_{\Lambda^-}(X[\psi]) = \mathcal{F}^-(\psi)$, by Theorem 5.1, $\text{char}_{\Lambda^-}(X[\psi]) \subset L_p^-(\psi)$. Thus by (ii), we obtain

$$\prod_{\chi} \text{char}_{\Lambda_E^-}(X_E[\psi\chi]) \subset L_p^-(\psi)\Lambda_E^-,$$

where $L_p^-(\psi)\Lambda_E^-$ is the ideal of Λ_E^- generated by the image of $L_p^-(\psi) \in \Lambda^-$ in Λ_E^-. Write R_E for the integer ring of E, and let $X(R_E)_{/\mathcal{W}}$ be the elliptic curve with complex multiplication whose complex points give the torus $\mathbb{C}/R_E$. Since $X(R) = X(R_E) \otimes_{R_E} R$ for our choice of CM type Σ, the complex and p-adic periods of $X(R)$ and $X(R_E)$ are identical. Thus by the factorization of Hecke L-functions, we have

$$L_p^-(\psi)\Lambda_E^- = \prod_{\chi} L_p^-(\psi_E\chi)\Lambda_E^-.$$

Then by Rubin [R] Theorem 4.1 (i) applied to the $\mathbb{Z}_p$-extension E_∞^-/E, we find that

$$\text{char}_{\Lambda_E^-}(X_E[\psi_E\chi]) = L_p^-(\psi_E\chi)\Lambda_E^-.$$

Thus $(\mathcal{F}^-(\psi)/L_p^-(\psi))\Lambda_E^- = \Lambda_E^-$, and hence $\mathcal{F}^-(\psi)\Lambda^- = L_p^-(\psi)\Lambda^-$. □

Bibliography

Books

[AAF] G. Shimura, *Arithmeticity in the Theory of Automorphic Forms*, Mathematical Surveys and Monographs **82**, AMS, 2000

[ACM] G. Shimura, *Abelian Varieties with Complex Multiplication and Modular Functions*, Princeton University Press, 1998

[CRT] H. Matsumura, *Commutative Ring Theory*, Cambridge studies in advanced mathematics **8**, Cambridge Univ. Press, 1986

[GME] H. Hida, *Geometric Modular Forms and Elliptic Curves*, 2000, World Scientific Publishing Co., Singapore (a list of errata downloadable at `www.math.ucla.edu/~hida`)

[HMI] H. Hida, *Hilbert Modular Forms and Iwasawa Theory*, Oxford University Press, 2006 (a list of errata downloadable at `www.math.ucla.edu/~hida`)

[IAT] G. Shimura, *Introduction to the Arithmetic Theory of Automorphic Functions*, Princeton University Press and Iwanami Shoten, 1971, Princeton-Tokyo

[LAP] H. Yoshida, *Lectures on Absolute CM period*, AMS mathematical surveys and monographs **106**, 2003, AMS

[LFE] H. Hida, *Elementary Theory of L–functions and Eisenstein Series*, LMSST **26**, Cambridge University Press, Cambridge, 1993

[MFG] H. Hida, *Modular Forms and Galois Cohomology*, Cambridge studies in advanced mathematics **69**, Cambridge University Press, Cambridge, 2000 (a list of errata downloadable at www.math.ucla.edu/~hida)

[PAF] H. Hida, *p–Adic Automorphic Forms on Shimura Varieties*, Springer Monographs in Mathematics. Springer, New York, 2004 (a list of errata downloadable at www.math.ucla.edu/~hida)

Articles

[B] D. Blasius, Hilbert modular forms and Ramanujan conjecture, Aspects of Mathematics **E37** (2006), 35–56

[BR] D. Blasius and J. D. Rogawski, Motives for Hilbert modular forms, Inventiones Math. **114** (1993), 55–87

[C] C.-L. Chai, Arithmetic minimal compactification of the Hilbert-Blumenthal moduli spaces, Ann. of Math. **131** (1990), 541–554

[De] P. Deligne, Travaux de Shimura, Sem. Bourbaki, Exp. 389, Lecture notes in Math. **244** (1971), 123–165

[DeR] P. Deligne and K. A. Ribet, Values of abelian L–functions at negative integers over totally real fields, Inventiones Math. **59** (1980), 227–286

[Di] M. Dimitrov, Compactifications arithmétiques des variétés de Hilbert et formes modulaires de Hilbert pour $\Gamma_1(\mathfrak{c}, \mathfrak{n})$, in *Geometric aspects of Dwork theory*, Vol. I, II, 527–554, Walter de Gruyter GmbH & Co. KG, Berlin, 2004

[DiT] M. Dimitrov and J. Tilouine, Variété de Hilbert et arithmétique des formes modulaires de Hilbert pour $\Gamma_1(\mathfrak{c}, N)$, in *Geometric Aspects of Dwork's Theory*, Vol. I, II, 555–614, Walter de Gruyter GmbH & Co. KG, Berlin, 2004 Walter de Gruyter, Berlin, 2004.

[Fu] K. Fujiwara, Deformation rings and Hecke algebras in totally real case, preprint, 1999 (ArXiv. math. NT/0602606)

[H88] H. Hida, On p-adic Hecke algebras for GL_2 over totally real fields, Ann. of Math. **128** (1988), 295–384

[H89] H. Hida, Nearly ordinary Hecke algebras and Galois representations of several variables, Proc. JAMI Inaugural Conference, Supplement to Amer. J. Math. (1989), 115–134

[H91] H. Hida, On p–adic L–functions of $GL(2) \times GL(2)$ over totally real fields, Ann. Inst. Fourier **41** (1991), 311–391

[H95] H. Hida, Control theorems of p–nearly ordinary cohomology groups for $SL(n)$, Bull. Soc. math. France, **123** (1995), 425–475

[H96] H. Hida, On the search of genuine p–adic modular L–functions for $GL(n)$, Mémoire SMF **67**, 1996

[H99] H. Hida, Non-critical values of adjoint L-functions for $SL(2)$, Proc. Symp. Pure Math., 1998

[H00] H. Hida, Adjoint Selmer group as Iwasawa modules, Israel J. Math. **120** (2000), 361–427

[H02] H. Hida, Control theorems of coherent sheaves on Shimura varieties of PEL–type, Journal of the Inst. of Math. Jussieu, 2002 **1**, 1–76

[H04a] H. Hida, p–Adic automorphic forms on reductive groups, Astérisque **298** (2005), 147–254 (preprint downloadable at www.math.ucla.edu/~hida)

[H04b] H. Hida, The Iwasawa μ–invariant of p–adic Hecke L–functions, preprint, 2004 (preprint downloadable at www.math.ucla.edu/~hida)

[H04c] H. Hida, Non-vanishing modulo p of Hecke L–values, in: "*Geometric Aspects of Dwork's Theory, II*" (edited by Alan Adolphson, Francesco Baldassarri, Pierre Berthelot, Nicholas Katz, and Francois Loeser), Walter de Gruyter, 2004, pp. 735–784 (preprint downloadable at www.math.ucla.edu/~hida)

[H04d] H. Hida, Anticyclotomic main conjectures, Documenta Math. Extra volume Coates (2006), 465–532

[HT1] H. Hida and J. Tilouine, Anticyclotomic Katz p–adic L–functions and congruence modules, Ann. Sci. Ec. Norm. Sup. 4-th series **26** (1993), 189–259

[HT2] H. Hida and J. Tilouine, On the anticyclotomic main conjecture for CM fields, Inventiones Math. **117** (1994), 89–147

[K] N. M. Katz, p-adic L-functions for CM fields, Inventiones Math. **49** (1978), 199–297

[Ko] R. Kottwitz, Points on Shimura varieties over finite fields, J. Amer. Math. Soc. **5** (1992), 373–444

[MW] B. Mazur and A. Wiles, Class fields of abelian extensions of $\mathbb{Q}$, Inventiones Math. **76** (1984), 179–330

[R] K. Rubin, The "main conjectures" of Iwasawa theory for imaginary quadratic fields, Inventiones Math. **103** (1991), 25–68

[R1] K. Rubin, More"main conjectures" for imaginary quadratic fields. Elliptic curves and related topics, 23–28, CRM Proc. Lecture Notes, **4**, Amer. Math. Soc., Providence, RI, 1994.

[ST] J.-P. Serre and J. Tate, Good reduction of abelian varieties, Ann. of Math. **88** (1968), 452–517

[Sh] G. Shimura, On canonical models of arithmetic quotients of bounded symmetric domains, Ann. of Math. **91** (1970), 144-222; II, **92** (1970), 528-549

[Sh1] G. Shimura, On some arithmetic properties of modular forms of one and several variables, Ann. of Math. **102** (1975), 491–515

[Si] W. Sinnott, On a theorem of L. Washington, Astérisque **147-148** (1987), 209–224

[TW] R. Taylor and A. Wiles, Ring theoretic properties of certain Hecke modules, Ann. of Math. **141** (1995), 553–572

[Wa] L. Washington, The non-p–part of the class number in a cyclotomic $\mathbb{Z}_p$–extension, Inventiones Math. **49** (1978), 87–97

[Wi] A. Wiles, Modular elliptic curves and Fermat's last theorem, Ann. of Math. **141** (1995), 443–551

Serre's modularity conjecture: a survey of the level one case

Chandrashekhar Khare

Department of Mathematics
155 South 1400 East, Room 233
Salt Lake City, UT 84112-0090
U.S.A.
shekhar@math.utah.edu

Abstract

We give a survey of recent work on Serre's conjecture. In particular we report on joint work with Wintenberger [32] which proved the conjecture for finitely many weights and levels and set up a strategy to prove Serre's conjecture, and the author's subsequent proof in [33] of the level 1 case of the conjecture using "weight cycles". The article gives an informal survey of the ideas of the proof.

1 Introduction

We recall the statement of Serre's conjecture, state the main result of [33], and give a synopsis of earlier work on the conjecture.

1.1 Statement of Serre's conjecture and the result

Let $\bar{\rho}$: $\mathrm{Gal}(\bar{\mathbb{Q}}/\mathbb{Q}) \to \mathrm{GL}_2(\mathbb{F})$ be a continuous, absolutely irreducible, two-dimensional, odd ($\det\overline{\rho}(c) = -1$ for c a complex conjugation), mod p representation, with $\mathbb{F}$ a finite field of characteristic p. We say that such a representation is of *Serre-type*, or S-type, for short.

We denote by $N(\overline{\rho})$ the (prime to p) Artin conductor of $\overline{\rho}$, and $k(\overline{\rho})$ the weight of $\overline{\rho}$ as defined in [55]. The invariant $N(\overline{\rho})$ is made out of $(\overline{\rho}|_{I_\ell})_{\ell \neq p}$, and is divisible exactly by the primes ramified in $\overline{\rho}$ that are $\neq p$, while $k(\overline{\rho})$ is such that $2 \leq k(\overline{\rho}) \leq p^2 - 1$ if $p \neq 2$ ($2 \leq k(\overline{\rho}) \leq 4$ if $p = 2$), and is made from information of $\overline{\rho}|_{I_p}$. It is an important feature of the weight $k(\overline{\rho})$, for $p > 2$, that if $\overline{\chi_p}$ is the mod p cyclotomic character, then for some $i \in \mathbb{Z}$, $2 \leq k(\overline{\rho} \otimes \overline{\chi_p}^i) \leq p + 1$.

Serre has conjectured in [55] that such a $\bar{\rho}$ *arises* (with respect to some fixed embedding $\iota : \overline{\mathbb{Q}} \hookrightarrow \overline{\mathbb{Q}_p}$) from a newform f of weight $k(\overline{\rho})$ and level

$N(\overline{\rho})$. We fix embeddings $\iota : \overline{\mathbb{Q}} \hookrightarrow \overline{\mathbb{Q}_p}$ for all primes p hereafter, and when we say (a place above) p, we will mean the place induced by this embedding. By *arises* from we mean that the reduction of an integral model of the p-adic representation ρ_f associated to f, which is valued in $GL_2(\mathcal{O})$ for $\mathcal{O}$ the ring of integers of some finite extension of $\mathbb{Q}_p$, modulo the maximal ideal of $\mathcal{O}$ is isomorphic to $\overline{\rho}$:

$$\begin{array}{ccc} G_{\mathbb{Q}} & \xrightarrow{\rho_f} & \mathrm{GL}_2(\mathcal{O}) \\ \| & & \downarrow \\ G_{\mathbb{Q}} & \xrightarrow{\overline{\rho}} & \mathrm{GL}_2(\mathbb{F}). \end{array}$$

We have contented ourselves here with a schematic statement of the conjecture, referring to Serre's original article [55] for a beautiful, more extensive, account of his conjecture and its many consequences.

It is convenient to split the conjecture into 2 parts:

1. *Qualitative form:* In this form, one only asks that $\overline{\rho}$ arise from a newform of unspecified level and weight.

2. *Refined form:* In this form, one asks that $\overline{\rho}$ arise from a newform $f \in S_{k(\overline{\rho})}(\Gamma_1(N(\overline{\rho})))$.

A large body of difficult, important work of a large number of people, Ribet, Mazur, Carayol, Gross, Coleman-Voloch, Edixhoven, Diamond et al., see [45] and [23], proves that (for $p > 2$) the qualitative form implies the refined form. We focus only on the qualitative form of the conjecture.

In fact the conjecture, especially in its qualitative form, is much older and dates from the early 1970's. The *qualitative form* of the level 1 (i.e., $N(\overline{\rho}) = 1$) conjecture/question was officially formulated in an article that Serre wrote for the Journées Arithmétiques de Bordeaux in 1975: see article 104 of [54]. Perhaps the restriction to level 1 is merely for simplicity. The fact that it is only in a qualitative form, in the weight aspect, is for a more substantial reason. The definition of the weight $k(\overline{\rho})$ in [55] is very delicate and probably came later, motivated by resuts of Deligne and Fontaine that describe $\overline{\rho}|_{I_p}$ when $\overline{\rho}$ arises from $S_k(\Gamma_1(N))$ with $2 \leq k \leq p+1$ and $(N,p) = 1$, and Tate's analysis of Θ-cycles.

The *level 1 conjecture* was proved in [33].

Theorem 1.1 *A $\overline{\rho}$ of S-type with $N(\overline{\rho}) = 1$ arises from $S_{k(\overline{\rho})}(SL_2(\mathbb{Z}))$.*

This built on the ideas introduced in [32]. In this paper we survey some of the ideas which are used in the proof of Theorem 1.1.

1.2 Lifting theorems

One of the main techniques of [32] and [33] is *lifting theorems*.

These may be motivated as follows. The conjecture in its qualitative form predicts that given an S-type $\overline{\rho}$ (which we assume is such that $2 \leq k(\overline{\rho}) \leq p+1$), there is a lifting ρ

$$\begin{array}{ccc} G_{\mathbb{Q}} & \xrightarrow{\rho} & \mathrm{GL}_2(\mathcal{O}) \\ \Vert & & \downarrow \\ G_{\mathbb{Q}} & \xrightarrow{\overline{\rho}} & \mathrm{GL}_2(\mathbb{F}) \end{array}$$

which is finitely ramified and potentially semistable at p. This was proved by Ramakrishna, under some mild conditions, in [43] (see also [63] for some improvements): he also proved that the lift ρ can be chosen so that it is crystalline at p of weight $k(\overline{\rho})$ (i.e., of Hodge-Tate weights $(k(\overline{\rho})-1,0)$), but his method cannot control precisely the set of primes at which ρ can be ramified. The conjecture in its refined form predicts that given a $\overline{\rho}$ there is a lifting ρ thas has the same conductor outside p as ρ, and is crystalline at p of weight $k(\overline{\rho})$. Proving the existence of such a *minimal lifting* in many cases is one of the key steps of [32]. This uses crucially a result of Böckle in [2], and a result of Taylor that we state below as Theorem 1.2.

The two methods for producing liftings are quite different. This is reflected in the kinds of lifts produced. Ramakrishna's lifts will in general be ramified at more primes than $\overline{\rho}$ while having the best possible field of definition, i.e., the fraction field of $W(\mathbb{F})$ the Witt vectors of $\mathbb{F}$. The lifts produced in [32] are minimally ramified, but there is little control of the field of definition. This lack of control is not of relevance for the methods below.

1.3 Taylor's potential version of Serre's conjecture and some earlier proven cases of the conjecture

Taylor had devised (see [63], and [61], [65]) a strategy to use the Galois-theoretic lifting theorem of Ramakrishna, together with *modularity lifting results* pioneered by Wiles, to prove Serre's conjecture modulo some open question in Diophantine geometry (see Question 5.5 of [64] and also [62]). The strategy of Taylor in part relied on an observation of the author (see [35]) about using a Galois-theoretic lifting result to prove descent results for mod p Hilbert modular forms.

In [61] and [65], Taylor proved, using modularity lifting results (over totally real fields), and a result in Diophantine geometry (over "large" fields) of Moret-Bailly, Rumley and Pop, the following *potential version* of Serre's conjecture:

Theorem 1.2 *Assume $\overline{\rho}$ is of S-type in odd residue characteristic, and* $\mathrm{im}(\overline{\rho})$ *is not solvable. Then there is a totally real field F that is Galois over $\mathbb{Q}$ and unramified at p, and even split above p if $\overline{\rho}|_{D_p}$ is irreducible, such that $\overline{\rho}|_{G_F}$ arises from a cuspidal automorphic representation π of $GL_2(\mathbb{A}_F)$ that is unramified at all finite places, and is discrete series of weight $k(\overline{\rho})$ at the infinite places.*

Taylor's question in essence was if F in the theorem could be chosen so that $F/\mathbb{Q}$ was a solvable extension.

In some cases Taylor in [63], and following Taylor's method, Manoharmayum and Ellenberg, see [40] and [27], were able to control F and thus prove some (non-solvable) cases of Serre's conjecture when the image was contained in $\mathrm{GL}_2(\mathbb{F}_5)$, $\mathrm{GL}_2(\mathbb{F}_7)$ or $\mathrm{GL}_2(\mathbb{F}_9)$. The case when the image is contained in $\mathrm{GL}_2(\mathbb{F}_4)$ had been addressed by Shepherd-Baron and Taylor, by another method, in [56].

In [61] and [65], Taylor used his potential modularity result Theorem 1.2 to prove some cases of the Fontaine-Mazur conjecture, see [28]. For this control of F of Theorem 1.2, beyond the local properties say at infinity and p, was not needed. For instance, the conjectures in [28] predict that *geometric* p-adic representations $\rho : G_{\mathbb{Q}} \to \mathrm{GL}_2(\mathcal{O})$ (i.e., finitely ramified and potentially semistable at p) can be propagated into compatible systems. Taylor's results allow this to be verified in many cases, and this is a crucial input into a result of Dieulefait recalled below, see [24], and also into the method of [32] and [33] (see Section 2 below).

1.4 Measures of the complexity of $\overline{\rho}$

The difficulty in proving the modularity of a $\bar{\rho}$ can be either measured in terms of the ramification properties of $\overline{\rho}$, for example $N(\overline{\rho}), k(\overline{\rho})$, or in terms of $\mathrm{im}(\overline{\rho})$. The results proven by Tate and Serre (in [60] and page 710 of [54]) early on were for the cases of residue characteristic $p = 2, 3$ and $N(\overline{\rho}) = 1$, i.e., cases which were more tractable in terms of the first measure. The results of Langlands, and Tunnell, implied Serre's conjecture for $p > 2$ when $\mathrm{im}(\overline{\rho})$ was solvable, i.e., cases which were more tractable in terms of the second measure. The results in [63], [40] and [27] also proved cases which were more tractable as per the second measure. In [32] the "complexity" of $\overline{\rho}$ is again, like in the results of Tate and Serre, measured in terms of the ramification properties of $\overline{\rho}$.

1.5 The function field case

Lifting theorems were used by Böckle and the author, [3], when proving in many cases an (n-dimensional) analog of Serre's conjecture for function fields. This has also been proven by D. Gaitsgory in [31] using different methods. The idea of [3] was to lift the given residual representation to a characteristic 0 representation using the techniques of Ramakrishna, which sufficed because of Lafforgue's result, [39]. For 2-dimensional representations this was established by A. J. de Jong in [19], and his use of base change to prove properties of deformation rings influenced the lifting theorems proved in [4] and [32].

1.6 Modularity lifting theorems

The principal technical tool of the work of Taylor described above, and the papers [32] and [33] are the methods developed by Wiles, and Taylor and Wiles, see [67] and [66], to prove a *relative version* of the Fontaine-Mazur conjecture, i.e., *modularity lifting theorems*. Given a $\rho : G_{\mathbb{Q}} \to \mathrm{GL}_2(\mathcal{O})$ that is finitely ramified, and potentially semistable at p (i.e., *geometric*) with Hodge-Tate weights $(*, 0)$ with the integer $* > 0$, then a particular case of Fontaine-Mazur conjecture predicts that ρ arises from a modular form. Wiles, and Taylor, developed techniques in [67], [66] to show a relative version of this wherein one assumes that residually the representation arises from a newform, and then lifts this modularity property to ρ. Most subsequent work on Serre's conjecture has used this technique, that assumes mod p modularity results to prove modularity of certain p-adic lifts, to prove mod p modularity in certain instances. This almost gives an appearance of circularity to the method. What saves it from this charge is that p should be treated as a variable in the statement. The kind of lifting statement invoked is that, under appropriate conditions, if one residual representation arising from a compatible system of λ-adic representations of $G_{\mathbb{Q}}$ is modular, then so is the corresponding ℓ-adic representation. Hence so is the compatible system, and hence so are *all* the residual representations that arise from the system.

1.7 Linked compatible systems and congruences between Galois representations

The use of Theorem 1.2 in both [32] and [33] is indirect: it is used only to draw some Galois theoretic consequences like existence of lifts of various kinds. Then the various lifts of $\overline{\rho}$ can be inserted into compatible systems using Taylor's work. These are purely Galois theoretic consequences which pure algebraic number theory seems unable to prove. The following definition is crucial for us:

Definition 1.3 *We say that two E-rational, (weakly) compatible systems of representations of $G_{\mathbb{Q}}$ are* **linked** *if for some finite place λ of a number field E the semisimplifications of the corresponding residual mod λ representations arising from the two systems are isomorphic up to a twist by a (one-dimensional) character of $G_{\mathbb{Q}}$.*

(The twist will in practise always be by a power of the mod λ cyclotomic character.) The magic the definition works below is partly because being linked is not a transitive property.

The role that two linked compatible systems played in Wiles proof of Fermat's Last Theorem (the 3 - 5 switch) is well-known. The idea of using a sequence of linked compatible systems, by using the liftings of Ramakrishna and the result of Taylor that put these into compatible systems, had always been a tempting one, and had surely been considered by many people. But it was not clear perhaps how to profit by proliferating compatible systems.

In [33], in the course of proving the level one case, it is shown how Serre's conjecture in a given residue characteristic p is influenced by the conjecture in other residue characteristics. This influence is mediated through linked compatible systems and distribution properties of primes of the type proved by Chebyshev in the 1850's (we will in fact use better estimates proved in [48], although this is most likely not essential) which ensure their proximity. An important starting point for the proof of Theorem 1.1 is provided by the results proven by Fontaine [29], and subsequently by Brumer, Kramer and Schoof [11], [52] that classify abelian varieties over $\mathbb{Q}$ with some prescribed reduction properties.

The lifting results of [32] and [33] develop to a certain extent the theory of congruences between Galois representations to parallel the corresponding results known for modular forms in the work of Ribet, see [45] and [46], and subsequent work in [12], [21] and [34]. The idea of recreating in the theory of deformations of Galois representations, the theory of congruences between modular forms, already occurs in work of Boston, see [5].

We draw the attention of the reader to Ribet's notes available at `http://math.berkeley.edu/ribet/cms.pdf` which give a nice survey of some of these ideas.

1.8 Notation

For F a field, $\mathbb{Q} \subset F \subset \overline{\mathbb{Q}}$, we write G_F for the Galois group of $\overline{\mathbb{Q}}/F$. For λ a prime/place of F, we mean by D_λ (resp., I_λ) a decomposition (resp., inertia) subgroup of G_F at λ. For each place p of $\mathbb{Q}$, we fix embeddings ι_p of $\overline{\mathbb{Q}}$ in its completions $\overline{\mathbb{Q}_p}$. Denote by χ_p the p-adic cyclotomic character, and ω_p the Teichmüller lift of the mod p cyclotomic character $\overline{\chi_p}$ (the latter being the reduction mod p of χ_p). By abuse of notation we also denote by ω_p the ℓ-adic character $\iota_\ell \iota_p^{-1}(\omega_p)$ for any prime ℓ: this should not cause confusion as from the context it will be clear where the character is valued. For a number field F we denote the restriction of a character of $G_{\mathbb{Q}}$ to G_F by the same symbol. Mod p and p-adic Galois representations arising from newforms are said to be *modular*, another standard bit of terminology.

2 Compatible systems and a result of Dieulefait

We state a result of Dieulefait in [24] (this was also observed later, independently, by Wintenberger in [68]), which was a precursor to the work in [32].

Theorem 2.1 *Let $\rho : G_{\mathbb{Q}} \to \mathrm{GL}_2(\mathcal{O})$ be a representation that arises from a p-divisible group over $\mathbb{Z}$, i.e., ρ is unramified outside p and Barsotti-Tate at p. Then ρ is reducible.*

The main ingredient of the proof, both in [24] and [68], is the use of compatible systems.

Let E and F be number fields. We recall that a E-rational, 2-dimensional, strictly compatible system of representations (ρ_λ) of G_F consists of the data:

(i) for each finite place λ of E, $\rho_\lambda : G_F \to GL_2(E_\lambda)$ is a continuous, semisimple representation of G_F,

(ii) for all finite places q of F, if λ ia a place of E of residue characteristic different from q, the Frobenius semisimplification of the Weil-Deligne parameter of $\rho_\lambda|_{D_q}$ is independent of λ, and for almost all primes q this parameter is unramified.

The results of Taylor in [61] and [65] imply that ρ is part of an E-rational, strictly compatible system (ρ_λ) of 2-dimensional representations of $G_{\mathbb{Q}}$, with

E a number field, so that for each place λ above an odd prime ℓ the representation ρ_λ is unramified outside ℓ. Dieulefait checks that for each λ above an odd prime ℓ the representation ρ_λ is Barsotti-Tate at ℓ.

We recall the arguments of Taylor (see proof of Theorem 6.6 of [65] and 5.3.3 of [62]) and Dieulefait.

Strictly compatible systems: We may assume that $p \geq 5$ as otherwise the result is proved by Fontaine in [29].

We may assume that $\mathrm{im}(\overline{\rho})$ is not solvable, as otherwise we are done by Theorem 6.2 below, in conjunction with the consequence of the Langlands-Tunnell theorem spelled out in Theorem 4 of [42], for instance.

By Theorem 1.2, and the modularity lifting theorems of Skinner-Wiles in [59], there is a totally real field F, Galois over $\mathbb{Q}$ and that is unramified at p, such that $\rho|_{G_F}$ arises from a holomorphic, cuspidal automorphic representation π of $\mathrm{GL}_2(\mathbb{A}_F)$ with respect to the embedding ι_p, and with $\mathbb{A}_F$ the adeles of F. The strictly compatible system corresponding to π is such that each member is irreducible. The irreducibility is a standard consequence of the fact that $\rho|_{G_F}$ is irreducible (as $\mathrm{im}(\overline{\rho})$ is not solvable: see Lemma 2.6 of [32] for instance) and Hodge-Tate.

Let $G := \mathrm{Gal}(F/\mathbb{Q})$. Using Brauer's Theorem we get subextensions F_i of F such that $G_i = \mathrm{Gal}(F/F_i)$ is solvable, characters χ_i of G_i (that we may also regard as characters of G_{F_i}) with values in $\overline{\mathbb{Q}}$ (that we embed in $\overline{\mathbb{Q}_p}$ using ι_p), and integers n_i such that $1_G = \sum_{G_i} n_i \mathrm{Ind}_{G_i}^{G} \chi_i$. Using results on base change in [38] and [1] we also get holomorphic cuspidal automorphic representations π_i of $\mathrm{GL}_2(\mathbb{A}_{F_i})$ such that if ρ_{π_i,ι_p} is the representation of G_{F_i} corresponding to π_i w.r.t. ι_p, then $\rho_{\pi_i,\iota_p} = \rho|_{G_{F_i}}$. Thus $\rho = \sum_{G_i} n_i \mathrm{Ind}_{G_{F_i}}^{G_\mathbb{Q}} \chi_i \otimes \rho_{\pi_i,\iota_p}$. Now for any prime ℓ and any embedding $\iota : \overline{\mathbb{Q}} \to \overline{\mathbb{Q}_\ell}$, we define the virtual representation $\rho_\iota = \sum_{G_i} n_i \mathrm{Ind}_{G_{F_i}}^{G_\mathbb{Q}} \chi_i \otimes \rho_{\pi_i,\iota}$ of $\mathrm{Gal}(\bar{\mathbb{Q}}/\mathbb{Q})$ with the χ_i's now regarded as ℓ-adic characters via the embedding ι. We check that ρ_ι is a true representation by computing its inner product in the Grothendieck group of $\overline{\mathbb{Q}_\ell}$-valued (continuous, linear) representations of $G_\mathbb{Q}$. We claim that this is independent of ι. This is because, as the $\mathrm{Ind}_{G_{F_i}}^{G_\mathbb{Q}} \chi_i \otimes \rho_{\pi_i,\iota}$ are semisimple, the value of the inner product is the dimension of $\mathrm{End}(\rho_\iota)$ as a $\overline{\mathbb{Q}_\ell}$-vector space. Using Mackey's formula, and the fact that the compatible system of representations of G_F arising from π is irreducible, we see that this dimension is independent of ι. As for $\iota = \iota_p$ this dimension is 1 we see that $(\pm 1)\rho_\iota$ is an irreducible representation, and then using the trace of the identity we see that ρ_ι is a true representation of dimension 2. The representations ρ_ι together constitute the strictly compatible system we seek (see proof of Theorem 6.6 of [65]). Note that ρ_λ for λ above $\ell \neq p$ is not ramified at p. This is easily deduced using that F as above is unramified at p. (The details of the proof were explained to us by Wintenberger.)

Dieulefait's refinement: Now we recall the argument in [24] to check that for each λ above an odd prime ℓ the representation ρ_λ is crystalline at ℓ. Consider a decomposition group D in G of an odd prime ℓ. The corresponding fixed field F' has a prime λ' above ℓ that is split. By [1], $\rho|_{F'}$ arises from a cuspial automorphic representation π' of $GL_2(\mathbb{A}_{F'})$, and then as $\rho|_{F'}$ is unramified at λ' we see that π' is unramified at λ' (because of results of Carayol and Taylor in the "$\ell \neq p$ case"). This gives that $\rho_\lambda|_{G_{F'}}$ is Barsotti-Tate at λ' by a result of Breuil in [7] (in the "$\ell = p$ case") which suffices as the prime λ' is split.

After this, to prove Theorem 2.1 we proceed as in [68]. Looking at ρ_λ for λ above 7, Fontaine's work in [29] implies that ρ_λ is reducible and hence that ρ is reducible. (In [24] modularity lifting results together with results of Tate and Serre at $p = 3$ are used instead.)

3 The minimal lifting result of [32] and a sketch of its proof

We refer to [20], see also Section 2 of [32], for the definition of minimal lifting ρ of $\overline{\rho}$. Away from p, this is just a little more than requiring an equality of conductors $N(\rho) = N(\overline{\rho})$. At p the condition is that ρ be crystalline of weight $k(\overline{\rho})$ (which in the case $k(\overline{\rho}) = p+1$ is taken to be an ordinarity condition), or ρ could also be semistable of weight 2 in the $k(\overline{\rho}) = p+1$ case. (The definition of minimal lifts makes sense even for representations valued in (abstract) local $W(\mathbb{F})$-algebras with residue field $\mathbb{F}$.)

In [32] a new lifting technique to prove the existence of *minimal lifts* was introduced which used as an essential ingredient Taylor's result that proved a potential version of Serre's conjecture (see Theorem 1.2).

Theorem 3.1 *Let p be a prime* > 2. *Let* $\overline{\rho} : G_{\mathbb{Q}} \to \mathrm{GL}_2(\mathbb{F})$ *be a S-type representation with non-solvable image. We suppose that* $2 \leq k(\bar{\rho}) \leq p+1$ *and* $k(\bar{\rho}) \neq p$.

(i) If $k(\overline{\rho}) \neq p+1$, *then* $\bar{\rho}$ *has a lift* ρ *which is minimally ramified at every* ℓ.

(ii) In the case $k(\overline{\rho}) = p+1$, *there is a lifting* ρ *that is minimal at all primes* $\ell \neq p$ *and which is crystalline of weight* $p+1$ *at p. There is also a lifting* ρ' *that is minimal at all primes* $\ell \neq p$ *and is semi-stable of weight* 2 *at p.*

We give the *grandes lignes* of the proof referring to [32] for more details. The idea of the proof is rather simple. The first remark is that it will suffice to prove the flatness over $\mathbb{Z}_p$ of the corresponding minimal deformation ring $R_{\mathbb{Q}}$

which parametrises all minimal liftings. Using obstruction theory, and using Poitou-Tate exact sequences as in the work of [67], Böckle in [2] has proven that the latter has a presentation $W[[X_1, \cdots, X_r]]/(f_1, \cdots, f_s)$ where $s \leq r$. Thus it will suffice to prove that $R_{\mathbb{Q}}/(p)$ is a finite ring. An easy argument in [32] (see Lemma 2.4 of loc. cit.) shows that this will be implied by showing that an analogous ring $R_F/(p)$, the universal mod p minimal deformation ring for $\overline{\rho}|_{G_F}$ for some totally real number field F ($\overline{\rho}|_{G_F}$ remains irreducible: see Lemma 2.6 of [32]) unramified at p, is finite.

(We recall the argument: Assuming $R_F/(p)$ is finite we see that the image of the universal mod p Galois deformation $\bar{\rho}_{\text{univ}} : G_{\mathbb{Q}} \to \mathrm{GL}_2(R_{\mathbb{Q}}/(p))$ is finite. As the representation $\bar{\rho}_{\text{univ}}$ is absolutely irreducible, a theorem of Carayol implies that the Noetherian ring $R_{\mathbb{Q}}/(p)$ is generated by the traces of $\bar{\rho}_{\text{univ}}$. As $\bar{\rho}_{\text{univ}}$ has finite image, for each prime ideal $\wp$ of $R_{\mathbb{Q}}/(p)$, the images of these traces in the quotient by $\wp$ are sums of roots of unity, and there is a finite number of them. Thus we see that each of these quotients is a finite extension of $\mathbb{F}$. It follows that the noetherian ring $R_{\mathbb{Q}}/(p)$ is of dimension 0, and so is finite.)

This will be known by the type of modularity lifting results proved in [59] and [30] if we know that $\overline{\rho}|_{G_F}$ is modular. The existence of a F with this property is precisely the content of Taylor's potential modularity result Theorem 1.2.

Remarks: It is important to note the crucial use of deformation rings (introduced by Mazur: see [41] for an account) in the above "existence" proof of minimal lifts.

4 The proof in [32] of Serre's conjecture for low levels and low weights

We present one of the main results of [32].

Theorem 4.1 *Serre's conjecture is true for residue characteristic $p \geq 3$ for $\overline{\rho}$ such that $N(\overline{\rho}) = 1$ and $k(\overline{\rho}) = 2, 4, 6, 8, 12, 14$, or when $k(\overline{\rho}) = 2$ and $N(\overline{\rho}) = 1, 2, 3, 5, 7, 11, 13$, and* $\det(\overline{\rho}) = \overline{\chi_p}$.

We sketch a proof in some cases. The case of $N(\overline{\rho}) = 1, k(\overline{\rho}) = 2$ follows immediately from the result of Dieulefait in [24] (see Theorem 2.1) and the minimal lifting result in Theorem 3.1.

$k(\bar{\rho}) = 6, N(\overline{\rho}) = 1$ (see Figure 1): Suppose we have an irreducible $\bar{\rho}$ with $N(\overline{\rho}) = 1, k(\bar{\rho}) = 6$ as in the theorem (and we may assume $p > 3$ by Serre's result for $p = 3$). We use Theorem 3.1 to get a compatible system

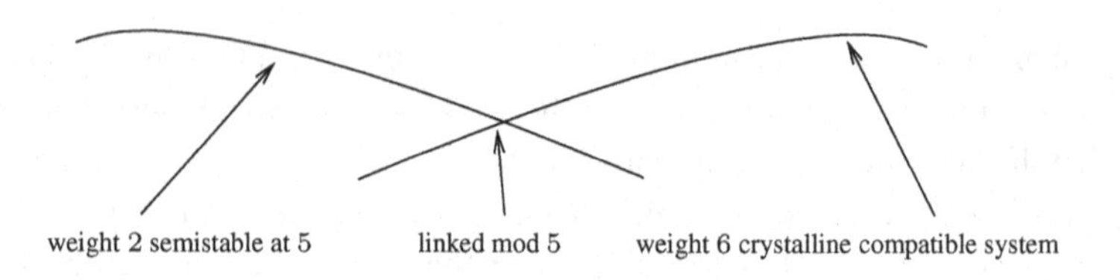

Fig. 1. Two linked compatible systems: the case of weight 6

(ρ_λ) of weight 6, i.e., Hodge-Tate of weights $(5,0)$, and with good reduction everywhere.

Consider a prime λ above 5 and assume the representation arising from the compatible system $\bar{\rho}_5$ has solvable image. The image has then to be reducible by known cases of Serre's conjecture. Then by results in [6], and as the residual representation is globally and hence of course locally reducible at 5, we see that ρ_λ is ordinary at 5, and hence by Skinner-Wiles [58] (see Theorem 6.2 (1) below) corresponds to a cusp form of level 1 and weight 6 of which there are none.

Otherwise, using Theorem 3.1, we get a minimal lift ρ' of $\bar{\rho}_5$ that is unramified outside 5, and semistable (and not crystalline) of weight 2 at 5. Such a ρ' by results of [61] arises from an abelian variety A over $\mathbb{Q}$ that is semistable and with good reduction outside 5. But by results of Brumer-Kramer [11] (see also Schoof [52]) such an A does not exist.

4.1 Even representations

Theorem 4.1 proves *inter alia* that for $p \leq 7$ there are no *odd*, irreducible representations $\bar{\rho} : G_{\mathbb{Q}} \to GL_2(\overline{\mathbb{F}_p})$ unramified outside p, as predicted by Serre's conjectures. This, together with the results of Moon-Taguchi in [42] which handle the case of even representations following the method of [60], proves that for $p \leq 7$ there are no irreducible representations $\bar{\rho} : G_{\mathbb{Q}} \to GL_2(\overline{\mathbb{F}_p})$ unramified outside p.

4.2 Explanation of Figures 1 and 4

The conceit of Figures 1 and 4 that depict linked compatibles systems is that each curve represents a compatible system and they intersect "at" a residual representation which arises (up to twist) from the 2 compatible systems.

5 A variation on Wiles' proof of Fermat's last theorem

Recall that in Section 4 of Serre's paper [55] it is shown how the Frey construction of a semistable elliptic curve E_{a^p,b^p,c^p} over $\mathbb{Q}$, associated to a Fermat triple (a, b, c), i.e., $a^p+b^p+c^p = 0$, a, b, c coprime and $abc \neq 0$, and where we may assume a is -1 mod 4, b even, and p a prime > 3, leads to a S-type representation $\overline{\rho}$ (the irreducibility is a consequence of a theorem of Mazur) with $k(\overline{\rho}) = N(\overline{\rho}) = 2$. Wiles proved Fermat's Last Theorem in [67] by showing that E is modular and hence $\overline{\rho}$ is modular and thus by Ribet's level-lowering results, $\overline{\rho}$ arises from $S_2(\Gamma_0(2))$ which gave a contradiction as the latter space is empty.(Note that for the $\overline{\rho}$ considered here, $\mathrm{im}(\overline{\rho})$ is never solvable: see Proposition 21 of article 94 of [54] and also [47].) It is remarked in [32] that one may proceed differently. We expand on that remark.

A possible way to prove Fermat's Last Theorem was to show that $\overline{\rho}$ arises from a semistable abelian variety A over $\mathbb{Q}$ with good reduction outside 2 (note that E has conductor the radical of abc). This would give a contradiction as by the results of [11] such an A does not exist.

The work in [32] now enables one to give such a proof. The minimal lifting result of [32] produces a lift of $\overline{\rho}$ that is Barsotti-Tate at p and has semistable reduction at 2, and is unramified everywhere else. Taylor's results towards the Fontaine-Mazur conjecture prove that such a $\overline{\rho}$ arises from A which gives a contradiction.

This proof while different in appearance from that of Wiles, uses all the techniques he developed in his original proof. The "simplifications" in this slightly different approach are:

(i) We do not need to use the results of Langlands-Tunnell, that prove Serre's conjecture for a $\overline{\rho}$ with solvable image, that Wiles had needed;

(ii) This altered proof does not make use of the most difficult of the level-lowering results which is due to Ribet, but instead uses the level lowering up to base change results of Skinner-Wiles [57].

6 Lifting results

The results stated in this section are of a technical nature, but are crucial to the proof of Theorem 1.1. We refer to the original papers for the proofs.

6.1 Three Galois-theoretic lifting results

Consider a S-type representation $\overline{\rho}$ with $2 \leq k(\overline{\rho}) \leq p+1$, $k(\overline{\rho}) \neq p$, $p > 2$ and assume that the image of $\overline{\rho}$ is not solvable.

When we say that for some number field E, a E-rational compatible system of 2-dimensional representations of $G_{\mathbb{Q}}$ lifts $\overline{\rho}$ we mean that for the place λ of E fixed by ι_p, the residual representation arising from ρ_λ is isomorphic to $\overline{\rho}$.

The following theorem is contained in [32] and [33]. The essential idea for constructing the lifts is contained in the sketch of the proof of the minimal liftings given in Section 3, and that for constructing compatible systems of representations of $G_{\mathbb{Q}}$ in Section 2. If the Weil-Deligne parameter at a prime q is (τ, N), with τ a 2-dimensional complex representation of the Weil group and N a nilpotent matrix in $M_2(\mathbb{C})$, by the inertial Weil-Deligne parameter we mean $(\tau|_{I_q}, N)$. In the theorem below we spell out more concretely what the property of strict compatibility means for $\rho_\lambda|_{D_q}$, when the corresponding parameter is ramified, in the instances where we use this property.

Theorem 6.1 *1. Assume $N(\overline{\rho}) = 1$. Then $\overline{\rho}$ lifts to a E-rational strictly compatible system (ρ_λ) such that for all odd primes ℓ, and λ a prime above ℓ, ρ_λ is crystalline at ℓ of weight $k(\overline{\rho})$ and unramified outside ℓ.*

2. Assume $N(\overline{\rho}) = 1$, and that $\overline{\rho}$ is ordinary at p, and thus $\bar{\rho}|_{I_p}$ is of the form

$$\begin{pmatrix} \overline{\chi_p}^{k(\overline{\rho})-1} & * \\ 0 & 1 \end{pmatrix}.$$

(i) Then $\overline{\rho}$ lifts to an E-rational strictly compatible system (ρ_λ) such that for all λ the inertial Weil-Deligne parameter of ρ_λ at p is $(\omega_p^{k(\overline{\rho})-2} \oplus 1, 0)$ if $k(\overline{\rho}) \neq p+1$ and otherwise is of the form (id, N) with N a non-zero nilpotent matrix $\in GL_2(\mathbb{C})$.

(ii) For all primes $\ell \neq p$, $\ell > 2$, and λ above ℓ, ρ_λ is unramified outside $\{\ell, p\}$ and is crystalline of weight 2 at ℓ, and $\rho_\lambda|_{I_p}$ is of the form

$$\begin{pmatrix} \omega_p^{k-2} & * \\ 0 & 1 \end{pmatrix},$$

and unramified if $k(\overline{\rho}) = 2$.

(iii) For the fixed place λ above p, ρ_λ is unramified outside p, and if $k(\overline{\rho}) \neq p+1$, then $\rho_\lambda|_{I_p}$ is of the form

$$\begin{pmatrix} \omega_p^{k-2}\chi_p & * \\ 0 & 1 \end{pmatrix},$$

and Barsotti-Tate over $\mathbb{Q}_p(\mu_p)$ (and Barsotti-Tate over $\mathbb{Q}_p$ if $k(\overline{\rho}) = 2$). If $k(\overline{\rho}) = p+1$ then $\rho_\lambda|_{I_p}$ is of the form

$$\begin{pmatrix} \chi_p & * \\ 0 & 1 \end{pmatrix}.$$

3. Assume $k(\overline{\rho}) = 2$ and $N(\overline{\rho}) = q$ an odd prime. Then $\overline{\rho}|_{I_q}$ is of the form

$$\begin{pmatrix} \overline{\chi} & * \\ 0 & 1 \end{pmatrix},$$

with $\overline{\chi}$ a character of I_q that factors through its quotient $(\mathbb{Z}/q\mathbb{Z})^*$. Let χ' be any non-trivial $\overline{\mathbb{Z}}_p^*$-valued character of I_q that factors though $(\mathbb{Z}/q\mathbb{Z})^*$ and reduces to $\overline{\chi}$. Thus we may write χ' as ω_q^i for some $0 \leq i \leq q-2$.

(i) Then there is an E-rational strictly compatible system (ρ_λ) that lifts $\overline{\rho}$ such that for λ a prime above p fixed by ι_p, ρ_λ is unramified outside $\{p, q\}$, is Barsotti-Tate at p, and $\rho_\lambda|_{I_q}$ is of the form

$$\begin{pmatrix} \chi' & * \\ 0 & 1 \end{pmatrix}.$$

(ii) For the place λ of E fixed above q, ρ_λ is unramified outside q, and either semistable of weight 2 at q or Barsotti-Tate over $\mathbb{Q}(\mu_q)$.

(iii) For the place λ of E fixed above q, the residual representation $\bar{\rho}_\lambda|_{I_q}$ is of one of the following 3 forms

(a)

$$\begin{pmatrix} \overline{\chi_q}^{i+1} & * \\ 0 & 1 \end{pmatrix},$$

(b)

$$\begin{pmatrix} \overline{\chi_q} & * \\ 0 & \overline{\chi_q}^{i} \end{pmatrix},$$

or

(c)

$$\begin{pmatrix} \psi_q^{i+1} & 0 \\ 0 & \psi_q'^{\,i+1} \end{pmatrix},$$

where ψ_q, ψ'_q are the fundamental character of level 2 of I_q.

Thus in particular the Serre weight of some twist of $\bar{\rho}_\lambda$ by a power of $\overline{\chi}_q$ is either $i+2$ or $q+1-i$.

The last part of the theorem uses the results of Breuil, Mezard and Savitt (see [10] and [50]).

The theorem can be motivated by results on modular forms. The first, which is contained in Theorem 3.1 (1), may be motivated by recalling that if $\overline{\rho}$ is modular, then it arises from $S_k(\overline{\rho})(\Gamma_1(N))$ (see [45]: this is the level and weight optimisation result).

The second and third parts may be motivated by recalling that a mod p representation that arises from $S_2(\Gamma_0(p), \overline{\chi_p}^{i})$, arises also from either

$S_{i+2}(SL_2(\mathbb{Z}))$ or $S_{p+1-i}(SL_2(\mathbb{Z}))(\overline{\chi}_p{}^{-i})$ where the spaces are suitable spaces of mod p modular forms.

The third may be motivated by Carayol's result that if $\overline{\rho}$ arises from $S_2(\Gamma_0(q), \epsilon)$, then it also arises from $S_2(\Gamma_0(q), \epsilon')$ for any ϵ' congruent to ϵ modulo the place above p we have fixed.

6.2 Three modularity lifting results

Consider a 2-dimensional mod $p > 2$ representation $\overline{\rho}$ of $G_{\mathbf{Q}}$ which is odd and with $2 \leq k(\overline{\rho}) \leq p+1$ with $p > 2$. We do not assume that $\overline{\rho}$ is irreducible, but we do assume that $\overline{\rho}$ is modular, which in the reducible case simply means odd. The following theorem is the work of many people, Wiles, Taylor, Breuil, Conrad, Diamond, Flach, Fujiwara, Guo, Kisin, Savitt, Skinner et al. (see [67], [66], [30], [16],[17], [58], [59], [49], [37]) and is absolutely vital to us.

Theorem 6.2

1. Let ρ be a lift of $\overline{\rho}$ to a p-adic representation that is unramified outside p and crystalline of weight k, with $2 \leq k \leq p+1$, at p. Then ρ is modular.

2. Let ρ be a lift of $\overline{\rho}$ to a p-adic representation that is unramified outside p and either semistable of weight 2, or Barsotti-Tate over $\mathbb{Q}_p(\mu_p)$. Then ρ is modular.

3. Let ρ be a lift of $\overline{\rho}$ to a p-adic representation that is unramified outside a finite set of primes and is Barsotti-Tate at p. Then ρ is modular.

The theorem crucially uses developments in the *modularity lifting* technology that build on [67] and [66]. In particular the results of Skinner-Wiles in the *residually degenerate* cases are crucial to us, see [58] and [59]. The simplifications of the original method of [67] and [66], due independently to Diamond and Fujiwara in one aspect, see [22] and [30], and due to Kisin in another aspect, see Proposition 3.3.1 of [37], are again vital.

The fact that we do not have the usual condition that $\overline{\rho}$ is non-degenerate, i.e., $\overline{\rho}|_{\mathbb{Q}(\mu_p)}$ is irreducible, is because when $\overline{\rho}|_{\mathbb{Q}(\mu_p)}$ is reducible in the cases envisaged in the theorem, the lift ρ is up to twist by a power of ω_p ordinary at p, and in this case as our representations are also distinguished at p, results of Skinner-Wiles in [58] and [59] apply.

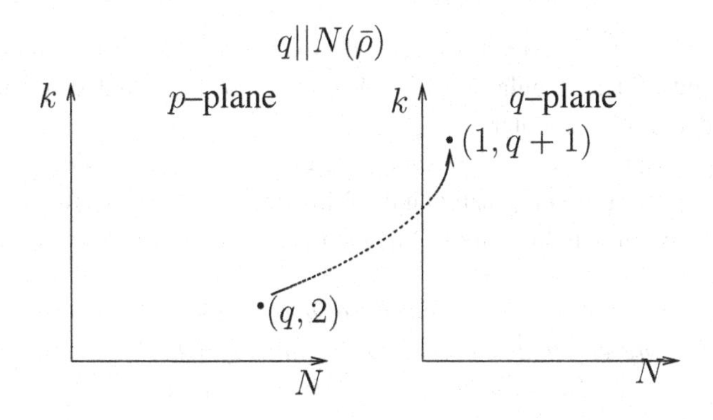

Fig. 2. Killing ramification at the expense of weight increase

7 The reductions in [32] of Serre's conjectures to modularity lifting results

In [32] it was observed that the lifting technique introduced there, together with the use of compatible systems, would yield Serre's conjecture in two steps *if* modularity lifting results were proved in sufficient generality:

1. **Reduction to level 1:** Here the key observation is that a Serre-type mod p representation with weights between 2 and $p+1$ always lifts to a geometric p-adic representation of $G_{\mathbb{Q}}$ which is minimally ramified away from p and crystalline at p of weight between 2 and $p+1$. This together with the fact that such a lift can be made part of a compatible system explains this procedure. We will explain the idea pictorially (see Figure 2, which depicts the generic case: sometimes the point plotted in the q-plane could have co-ordinates $(1,2)$) to show how the conjecture for a S-type semistable $\overline{\rho}$ when the conductor of $\overline{\rho}$ is a prime q and the weight is 2 follows from the level 1 case if one knew an appropriate lifting theorem (and one does in this case!). This idea when combined with Theorem 1.1 results in Corollary 10.1 below.

2. **The level 1 case using induction on the residue characteristic:** Given a S-type $\overline{\rho}$ of level 1 in residue characteristic p_n, the nth prime, we use Theorem 6.1 (1) to lift it to a compatible system (ρ_λ), and extract from this $\rho_{p_{n-1}}$, and consider the residual representation $\bar{\rho}_{p_{n-1}}$. By induction this would be modular (the starting point of the induction would use the results of Tate and Serre), and if one knew an appropriate lifting theorem one would be done. The lifting theorem needed would be to show that a p-adic representation ρ unramified outside p is modular, if residually it is modular (which includes the reducible, odd case), and it is crystalline at p of weight $\leq 2p$: here we are using

the Bertrand postulate, which was proved by Chebyshev, to restrict attention to these weights. Such modularity lifting theorems especially in degenerate cases seem hard as explained later.

While this approach has been superseded by the different inductive step in [33], it contained implicitly within it the following result (see Section 5 of [33]) that was used in the induction of [33] (see Figure 3):

Theorem 7.1 *(i) Given an odd prime p, if all 2-dimensional, mod p, odd, irreducible representations $\overline{\rho}$ of a given weight $k(\overline{\rho}) = k \leq p+1$, and unramified outside p, are known to be modular, then for any prime $q \geq k-1$, all 2-dimensional, mod q, odd, irreducible representations $\overline{\rho}'$ of weight $k(\overline{\rho}') = k$, and unramified outside q, are modular.*

(ii) If the level 1 case of Serre's conjecture is known for a prime $p > 2$, then for any prime q the level 1 case of Serre's conjectures is known for all 2-dimensional, mod q, odd, irreducible representations $\overline{\rho}$ of weight $k(\overline{\rho}) \leq p+1$.

Thus in particular it suffices to prove Serre's conjecture for infinitely many primes (and the argument in the level 1 case partly works because there are infinitely many non-Fermat primes!).

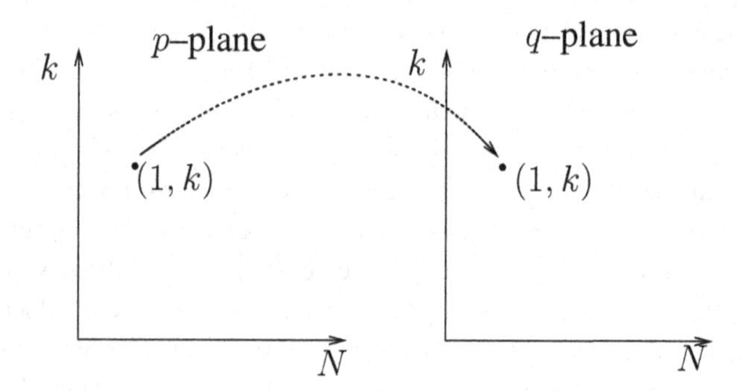

Fig. 3. Compatible systems

7.1 Explanation of Figures 2, 3 and 5

Figures 2,3 and 5 are supposed to express the following idea. Serre's conjecture for each prime p is imagined as happening on a p-plane with the axes recording the level and weight invariants (the weight axis is not semi-infinite, and stops at the ordinate p^2-1). For instance in Figure 3, plotting a point means that one knows Serre's conjecture in residue characteristic p for all S-type representations with invariants the co-ordinates of the point (*a posteriori* there are only finitely many isomorphism classes of such representations). Drawing an arrow between a p-plane and a q-plane describes a "move" made using compatible systems (we owe this terminology to Mazur).

8 The proof of the level 1 case: a sketch

We use as an input in the proof of the full level 1 case, the fact that by Theorem 4.1 we know the level 1 case in weights 2, 4 and 6 in all residue characteristics.

8.1 The weight 8 case: a trailer for the general case

Consider the weight 8 case (see Figure 4). By Theorem 7.1 it's enough to prove that an irreducible 2-dimensional, mod 7 representation $\bar{\rho}$ of level 1 and weight 8 is modular. (If the image is solvable we are done.) Such a representation is ordinary at 7 and in fact $\overline{\rho}|_{I_7}$ is of the form

$$\begin{pmatrix} \overline{\chi}_7 & * \\ 0 & 1 \end{pmatrix},$$

and is très ramifiée.

By Theorem 6.1 (2) lift $\overline{\rho}$ to a strictly compatible system (ρ_λ) such that the 7-adic representation ρ_7 (corresponding to our fixed place of $\overline{\mathbb{Q}}$ above 7) lifts $\overline{\rho}$ and is weight 2, semistable at 7 that is unramified outside 7. Thus $\rho_7|_{I_7}$ is of the form

$$\begin{pmatrix} \chi_7 & * \\ 0 & 1 \end{pmatrix}.$$

We now extract the 3-adic representation ρ_3 determined by ι_3 from (ρ_λ), and consider a corresponding residual mod 3 representation $\overline{\rho}'$.

Note that $k(\overline{\rho}') = 2$, and $\overline{\rho}'$ is unramified outside 3 and 7. If $\overline{\rho}'$ has solvable image we are done, as in that case we have a representation to which we can apply modularity lifting results of Theorem 6.2 (3) to conclude that the compatible system ρ_λ is modular and hence so is $\overline{\rho}$. Similarly if $\overline{\rho}'$ is unramified at 7 we are again done as we know by page 710 of [54] that the residual mod 3 representation is then reducible. Note that $\overline{\rho}'|_{I_7}$ is of the form

$$\begin{pmatrix} 1 & * \\ 0 & 1 \end{pmatrix}.$$

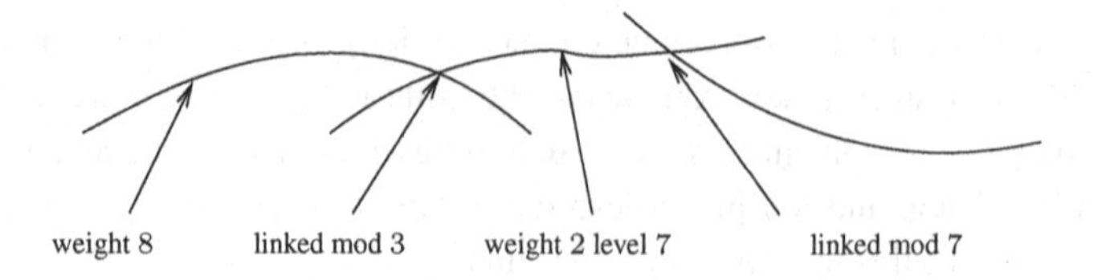

Fig. 4. Three linked compatible systems: the weight 8 case

Using Theorem 6.1 (3) get a 3-adic lift ρ' of $\overline{\rho}'$ with nebentype ω_7^2 at 7 (ω_7^4 would also work). Thus $\rho'|_{I_7}$ is of the form

$$\begin{pmatrix} \omega_7^2 & * \\ 0 & 1 \end{pmatrix}.$$

Note that the behavior of the lifts ρ_3 and ρ' of $\overline{\rho}'$ when restricted to I_7 are quite different: $\rho_3(I_3)$ has infinite image, while $\rho'(I_3)$ has image of order 3.

Using Theorem 6.1 (3) get a compatible system (ρ'_λ) with ρ' the member of this compatible system at the place corresponding to ι_3, and consider a residual representation $\overline{\rho}'_7$ arising from this system at a place λ above 7 fixed by ι_7.

If $\overline{\rho}'_7$ has solvable image, and hence known to be modular, we are done by applying Theorem 6.2 (2).

Now assume that $\overline{\rho}'_7$ has non-solvable image. We get a residual representation whose Serre weight (up to twisting) is either 4 or $7+3-4=6$ by Theorem 6.1 (3). But we know the modularity of level 1, S-type representations of weights 4 and 6. Now we again use Theorem 6.2 (2) and conclude that (ρ'_λ) is modular.

Hence so is the residual representation $\overline{\rho}'$, and hence by another application of modularity lifting theorem we conclude that the first compatible system (ρ_λ) is modular (as the compatible systems (ρ_λ) and (ρ'_λ) are linked at the place above 3 fixed by ι_3), and hence so is $\overline{\rho}$ (which in this case means that it does not exist!).

8.2 Estimates on distribution of primes

The proof of the general case is very similar except that one uses some estimates on prime numbers of the type proven by Chebyshev. In fact, we will use the better estimates proved in [48]. (The estimates quoted in [33] were incorrect: we thank Dietrich Burde for bringing to light this error.)

If $\pi(x)$ is the prime counting function, then in Theorem 1 of [48] various estimates of the form

$$A(\frac{x}{\log(x)}) < \pi(x) < B(\frac{x}{\log(x)})$$

are proven with varying values of A, B, for $x > x_0$ where x_0 depends on A, B. From such a bound we easily deduce that if we fix a real number $a > C$, with $C = \frac{B}{A}$, and denote by p_n the nth prime then $p_{n+1} \leq ap_n$ for $p_{n+1} > \max(ax_0, a^{\frac{a}{a-C}})$, .

In the arguments below we need to check, that for each non-Fermat prime $P > 7$, there is a non-Fermat prime $p < P$ (for example p the largest non-Fermat prime $< P$) and an odd prime power divisor $\ell^r || (P-1)$ so that

$$\frac{P}{p} \leq \frac{2m+1}{m+1} - (\frac{m}{m+1})(\frac{1}{p}) \tag{8.1}$$

where we have set $\ell^r = 2m+1$ with $m \geq 1$. For non-Fermat primes $7 < P \leq 31$ we see this by inspection. For $P > 31$ we in fact prove the stronger estimate $\frac{P}{p} \leq \frac{3}{2} - (\frac{1}{30}) = 1.4\overline{6}$.

In *loc. cit.* it is proven that for $x > 17$,

$$\frac{x}{\log(x)} < \pi(x) < 1.25506(\frac{x}{\log(x)}).$$

From this we deduce the estimate that $p_{n+1} \leq (1.4\overline{6})p_n$ when $p_{n+1} > 31$. Consider the only Fermat primes 257 and 65,537 (the latter being the largest known Fermat prime!) between 17 and 200,000. For $P = 263$, the next prime after 257, $p = 251$ satisfies the required estimate, and for $P = 65,539$, the next prime after 65,537, $p = 65,521$ satisfies the required estimate.

From all this we deduce the estimate $\frac{P}{p} \leq \frac{3}{2} - (\frac{1}{30})$ for $31 < P < 200,000$.

In *loc. cit.* it is also proven that

$$\pi(x) < \frac{x}{\log(x)}(1 + \frac{3}{2\log(x)}),$$

for $x > 1$, which yields the inequalities

$$\frac{x}{\log(x)} < \pi(x) < 1.130289(\frac{x}{\log(x)}),$$

for $x > 100,000$. Using $a = 1.2$ in the estimates at the beginning of the section, and noting that $(1.2)^2 < 1.4\overline{6}$, and using that there are no successive Fermat primes after the pair 3, 5, we deduce that the estimate $\frac{P}{p} \leq \frac{3}{2} - (\frac{1}{30})$ is true for all $P > 100,000$, and thus the truth of (8.1) for all $P > 7$. (As will be clear to the discerning reader, this is not necessarily the most efficient way to check this!)

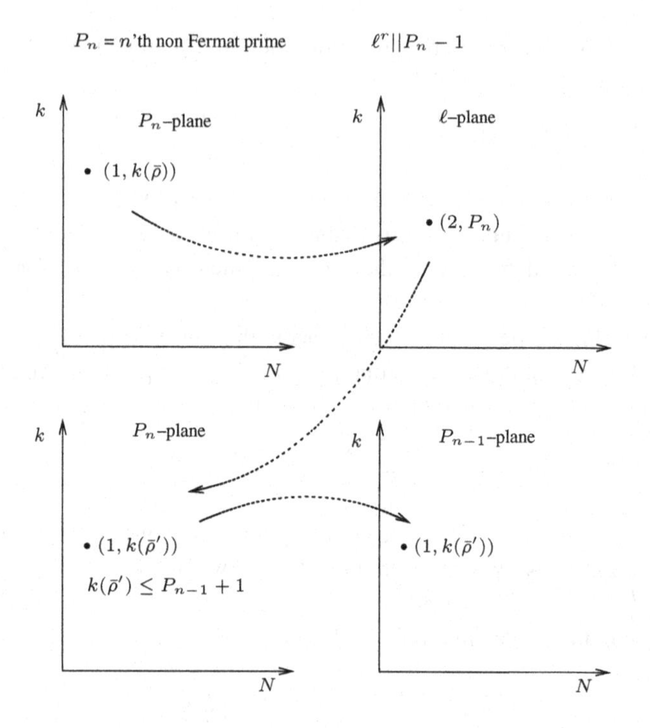

Fig. 5. Weight reduction

8.3 The general case

Now we give the general argument (see Figure 5). Rather than present the argument as an induction as we did in [33], we unravel the induction there, and lay out the skein of *linked compatible systems* that makes the argument work.

Consider a S-type representation $\overline{\rho}$ of residue characteristic $P > 7$ and such that $2 < k(\overline{\rho}) \leq P + 1$. We may assume that P is a non-Fermat prime by Theorem 7.1. By the estimates of Section 8.2, there is an odd prime power divisor $\ell^r := 2m + 1$ of $P - 1$ and a non-Fermat prime $p < P$ which satisfies the bound of (8.1). We may assume that the image is non-solvable. Besides the basic idea presented in the weight 8 case, and the use of the Chebyshev-type estimates, there is one extra argument that uses the simple fact that knowing modularity for a representation $\overline{\rho}$ is equivalent to knowing it for a twist of the representation by a power of $\overline{\chi}_P$. This allows one to accomplish the first reduction of weight rather easily in the case when $\overline{\rho}$ at P is irreducible, as in this case we see that a twist of $\overline{\rho}$ has weight $P + 3 - k$ (when $k > 2$). Now we claim that at least one of k or $P + 3 - k$ is $\leq p + 1$: otherwise we get that $P > 2p - 1$. Using (8.1) this will give a contradiction if

$$2p - 1 \geq \frac{2m+1}{m+1}p - (\frac{m}{m+1}),$$

i.e., if $\frac{p}{m+1} \geq \frac{1}{m+1}$, which is always true. Thus in the case when $\overline{\rho}$ is irreducible at P we may reduce to the case when $k(\overline{\rho}) \leq p+1$.

Otherwise we use Theorem 6.1 (2) to produce a weight 2 compatible lifting (ρ_λ), consider $\overline{\rho}_\ell$ (and stop if this either has solvable image or is unramified at P) and then choose a suitable lifting (ρ'_λ) using Theorem 6.1 (3) with the suitability dictated by the estimate (8.1). Namely, we choose the ℓ-adic lifting ρ' of $\overline{\rho}_\ell$ so that it is unramified outside ℓ, P, is Barsotti-Tate at ℓ and at P, $\rho'|_{I_P}$ is of the form:

$$\begin{pmatrix} \omega_P^i & * \\ 0 & 1 \end{pmatrix},$$

with $i \in (\frac{m}{2m+1}(P-1), (\frac{m+1}{2m+1})(P-1)]$. Then we stop if the residual mod P representation arising from (ρ'_λ), say $\overline{\rho}'$, has solvable image. Otherwise we know by our choice of i (and Theorem 6.2 (3)) that $\overline{\rho}'$ is such that, after twisting by a power of $\overline{\chi}_P$, $k(\overline{\rho}') \leq p+1$. (This is where the Chebyshev-type etimate is used: see [33] for more details. Note that it is essential to choose i to be in the "middle" of its possible range of values because of the dichotomy for the weight of $\overline{\rho}'$ in Theorem 6.1(3).)

After rechristening $\overline{\rho}'$ again as $\overline{\rho}$, and conflating now the locally at P reducible and irreducible cases, use Theorem 6.1 (1) to lift $\overline{\rho}$ to a compatible system (ρ_λ), and consider $\overline{\rho}_p$.

We repeat this procedure and in the end conclude modularity either: (i) because we got lucky early and we hit a residual representation that was considered in the procedure which happened to have solvable image, or (ii) at the end of constructing (very roughly $\frac{3p}{\log p}$) compatible systems, we know the modularity of the last system by the proven level 1 case of Serre's conjecture mod 7.

The fact that we are able to pass the modularity property through the linkages is because of the modularity lifting theorems in Theorem 6.2: it requires some thought to convince oneself about this, but we leave this as an exercise to the interested reader.

Remark: It seems possible to avoid using the fact that Serre's conjecture is known when the mod p representation has solvable image ($p > 2$), in all but the dihedral case (that case being known to Hecke).

9 Some observations about why the proof works

We make some heuristic remarks about the proof.

1. In the first remark we try to explain why the approach in [33] was able to sidestep the technical difficulties that the initially proposed inductive approach to the level 1 conjecture in [32], sketched in Section 7 of this paper, encountered.

The main innovation of [33] was to prove the level 1 case by a new method of carrying out the inductive step that used known modularity lifting results. Besides the fact that this method works, there is a "philosophical" point to be made here which might explain why the modularity lifting results that were needed by the method were indeed available.

The method of [33] introduces the mildest possible *singularities* in the process to prove Serre's conjecture. Here we define singularities to be points in the proof (*linkages*) when one encounters a residually degenerate representation. The singularity is mild when the corresponding modularity lifting theorem needed is the one proved in the work of Skinner and Wiles in [57] and [58], i.e., when the lift is (up to twist) ordinary at the residue characteristic (and which is the simplest kind of behavior at p from the view-point of p-adic Hodge theory).

When the lift is not ordinary in these globally residually degenerate cases, no modularity lifting theorems seem to be known. This seems technically like a very hard problem, although the author is no expert.

One would imagine that encountering a singularity makes some *dévissage* possible resulting in "reducing" the conjecture to a known case. While this is in a sense true, the corresponding modularity lifting theorem required to implement this intution is more involved and contained in [58], [59].

The method presented here uses modularity lifting theorems detailed in Theorem 6.2 when the geometric liftings that need to be proved modular are at p either crystalline of weight $\leq p+1$, semistable of weight 2 or Barsotti-Tate over $\mathbb{Q}_p(\mu_p)$. If the weight of a crystalline representation $\rho : G_{\mathbb{Q}_p} \to GL_2(\mathcal{O})$ is bigger than $p+1$ (the original method proposed in [32] to prove the level 1 conjecture involved liftings that were of these kind), or it is semistable of weight > 2, or Barsotti-Tate only over a wildly ramified extension or even over a tamely ramified extension of ramification index not dividing $p-1$, it is not true that if $\overline{\rho}$ is reducible, then so is ρ. Thus the method just about avoids having to deal with anything beyond mild singularities.

2. As there are no irreducible, level 1, S-type representations of weights up to 6, and this is used crucially in the proof of the level 1 conjecture, it

is important to note how crucial the modularity lifting results of Skinner and Wiles of [58] in the reducible case are to us.

3. The difference between the approach envisaged in [32] for the level 1 case, and that of [33], may also be brought out by pointing out that the former sought to prove the level 1 conjecture in residue characteristic p by using a sequence of linked compatible systems of cardinality roughly $\frac{p}{\log(p)}$, while in the latter roughly $\frac{3p}{\log(p)}$ linked compatible systems are used. The interpolation of more compatible systems in the proof ameliorates the technical difficulties that were encountered in [32].

10 A higher level result

The following result follows from the level 1 result together with the method of killing ramification:

Corollary 10.1 *If $\bar{\rho}$ is an irreducible, odd, 2-dimensional, mod p representation of $G_{\mathbb{Q}}$ with $k(\overline{\rho}) = 2$, $N(\overline{\rho}) = q$, with q prime, and $p > 2$, then it arises from $S_2(\Gamma_1(q))$.*

Characterisation of simple factors of $J_0(p)$: Theorem 1.1 also implies that a simple abelian variety A over $\mathbb{Q}$ is a factor of $J_0(p)$ for a prime p, up to isogeny, if and only if it is a simple GL_2-type semistable abelian variety over $\mathbb{Q}$ with good reduction outside p (one direction is classical needless to say).

11 Finite flat group schemes of type $(p, \cdots, p)$ over $\mathbb{Z}$

The weight 2, level 1 case of Serre's conjecture proven in [32] can be rephrased as follows and where we use the terminology of [44]:

Theorem 11.1 *For a finite field $\mathbb{F}$ of characteristic p, there are no irreducible finite, flat $\mathbb{F}$-vector space group schemes of rank 2 over $\mathbb{Z}$.*

The following problem belongs to the folklore of the subject:

Problem 11.2 *Prove that for an odd prime p all finite, flat group schemes of type $(p, \cdots, p)$ over $\mathbb{Z}$ are direct sums of μ_p's and $\mathbb{Z}/p\mathbb{Z}$'s.*

This has been proved by Fontaine in [29] for $p = 3, 5, 7, 11, 13, 17$. Perhaps one could be less ambitious and pose instead:

Problem 11.3 *For a finite field $\mathbb{F}$ of characteristic p, prove that there are no irreducible finite flat, $\mathbb{F}$-vector space group schemes over $\mathbb{Z}$ of rank $n > 1$ which are isomorphic to their Cartier duals.*

12 Perspectives: the Fontaine-Mazur conjecture

Serre's conjecture, when combined with modularity lifting results, allows one to prove more cases of the Fontaine-Mazur conjecture (in its restricted version for 2-dimensional representations of $G_{\mathbb{Q}}$) beyond the cases when the image of the representation is pro-solvable.

The work in [33], when combined with modularity lifting theorems, implies that if $\rho : G_{\mathbb{Q}} \to GL_2(\mathcal{O})$ is an irreducible p-adic representation unramified outside $p > 2$ and at p crystalline of Hodge-Tate weights $(k-1, 0)$ with k even and either $2 \leq k \leq p+1$ or ρ is ordinary at p, then ρ arises from $S_k(SL_2(\mathbb{Z}))$. This also has bearing on some of the finiteness conjectures in [28].

Even after Serre's conjecture is proved completely to prove the (restricted version of) Fontaine-Mazur conjecture one will need new modularity lifting results for lifts whose behaviour at p is complicated. Work of Berger-Breuil, see [8], and forthcoming work of Kisin, addresses some of these cases.

While we have commented on how Serre's conjecture implies modularity of certain p-adic representations, it is perhaps worth noting that in [36] it has been shown that Serre's conjecture also implies Artin's conjecture for 2-dimensional odd, irreducible, complex representations of $G_{\mathbb{Q}}$. The observation of [36] uses the modification in [26] of the definition of $k(\overline{\rho})$ which allows it to take the value 1, and the results of Coleman-Voloch about the weight using this modified definition. The observation of [36] again relies on using compatible systems, but only the "constant" systems that arise from Artin representations. As almost all of this case of Artin's conjecture is known (see [18]), the result of [36] just offers a different perspective.

The Fontaine-Mazur conjectures are related to (restricted cases of) the Langlands conjectures that seek to establish intimate relationships between motives, automorphic forms and Galois representations. Thus results and methods that prove residual automorphy of Galois representations, and modularity lifting methods, will help make progress in the Langlands program in "arithmetic cases" (see [15] and [62]) in the direction of showing that certain motives are automorphic, or that certain Galois representations are motivic/automorphic. (The association of Galois representations to automorphic forms of arithmetic type is a separate problem, and in a sense precedes the others.)

The case for 2-dimensional odd representations of $G_{\mathbb{Q}}$, which one may hope now will be understood in a not too distant future, is just a first step in this direction.

For instance it will be of interest to see if the methods presented here can be generalised to prove the analog of Serre's conjectures for totally real fields (see [9]). At first one might restrict attention to a few fields over which abelian varieties with good reduction everywhere have been classified. Even for these, while the ideas of [32] and [33] presented in this expository article will probably be useful, there are interesting technical complications to be overcome. We make a list of (not necessarily totally real) number fields over which abelian varieties with good reduction eveywhere have been classified. Fontaine proved that there are none such over $\mathbb{Q}$, $\mathbb{Q}(\sqrt{5})$, $\mathbb{Q}(i)$ and $\mathbb{Q}(\sqrt{-3})$. Schoof proved in [51] that there are none such over the cyclotomic fields $\mathbb{Q}(\zeta_f)$ for $f = 5, 7, 8, 9, 12$.

Schoof has also informed us that in unpublished work he has proven a similar result over $\mathbb{Q}(\sqrt{f})$ if discriminant f is equal to $5, 8, 12, 13, 17, 21$ (the "first" six real quadratic fields). In all cases the corresponding space of quadratic nebentypus weight 2 cuspforms for $\Gamma_1(f)$ is zero, and for other conductors f this space is non-zero: thus the result is best possible.

For $f = 24$ (and assuming the GRH for $f = 28, 29, 33$), Schoof in [53] has shown that any abelian variety over $\mathbb{Q}(\sqrt{f})$ with good reduction everywhere is isogenous to a power of the elliptic curve associated by Shimura to the space of (quadratic) nebentypus weight 2 cuspforms for $\Gamma_1(f)$ (the spaces in these cases are 2-dimensional).

The "weight cycle" idea of [33] and killing ramification idea of [32] give a method to prove Serre's conjecture once one knows (a proof of) Fontaine's result that there are no abelian varieties over $\mathbb{Z}$. Thus the real quadratic fields over which one has a "starting point" to launch the methods of [32] and [33] are $\mathbb{Q}(\sqrt{3})$, $\mathbb{Q}(\sqrt{5})$, $\mathbb{Q}(\sqrt{13})$, $\mathbb{Q}(\sqrt{17})$, $\mathbb{Q}(\sqrt{21})$ over which there are no abelian varieties with good reduction everywhere, and $\mathbb{Q}(\sqrt{6})$ (and $\mathbb{Q}(\sqrt{7})$, $\mathbb{Q}(\sqrt{29})$, $\mathbb{Q}(\sqrt{33})$ under the GRH) over which all such simple abelian varieties are classified up to isogeny (and known to arise from Hilbert modular forms that are in this case base changed from $\mathbb{Q}$). For the case of a general real quadratic field, it might be better to use some instances of functoriality and work with the group $GSp_{4/\mathbb{Q}}$.

Acknowledgements: I thank Najmuddin Fakhruddin for his help with the writing of this article. I would like to thank Dietrich Burde, Najmuddin Fakhruddin, Mark Kisin, Ravi Ramakrishna, Jean-Pierre Serre, René Schoof and Jean-Pierre Wintenberger for helpful correspondence during the writing of this expository paper. I would like to especially thank Barry Mazur for his generous, detailed comments. The presentation of the results of [33] given here follows the lines of the talks the author gave at the conference "Représentations galoisiennes" in Strasbourg in July 2005. I would like to thank the organisers, Jean-Francois Boutot, Jacques Tilouine and Jean-Pierre Wintenberger, for their invitation to speak at the conference.

Bibliography

[1] James Arthur and Laurent Clozel. *Simple algebras, base change, and the advanced theory of the trace formula*, volume 120 of *Annals of Mathematics Studies*. Princeton University Press, Princeton, NJ, 1989.

[2] Gebhard Böckle. A local-to-global principle for deformations of Galois representations. J. Reine Angew. Math., 509 (1999) 199–236.

[3] Gebhard Böckle and Chandrashekhar Khare. Mod ℓ representations of arithmetic fundamental groups I. preprint, to appear in Duke Math J.

[4] Gebhard Böckle and Chandrashekhar Khare. Mod ℓ representations of arithmetic fundamental groups II. preprint.

[5] Nigel Boston. Deformations of Galois representations: increasing the ramification Duke Math. J. 66 (1992), no. 3, 357–367.

[6] Laurent Berger, Hanfeng Li, and Hui June Zhu. Construction of some families of 2-dimensional crystalline representations. Math. Ann., 329(2):365–377, 2004.

[7] Christophe Breuil. Une remarque sur les représentations locales p-adiques et les congruences entre formes modulaires de Hilbert. Bull. Soc. Math. France, 127(3):459–472, 1999.

[8] Laurent Berger and Christophe Breuil. Sur la réduction des représentations cristallines de dimension 2 en poids moyens. preprint.

[9] Kevin Buzzard, Fred Diamond and Frazer Jarvis. On Serre's conjecture for mod ℓ Galois representations over totally real number fields. preprint.

[10] Christophe Breuil and Ariane Mézard. Multiplicités modulaires et représentations de $\mathrm{GL}_2(\mathbf{Z}_p)$ et de $\mathrm{Gal}(\overline{\mathbf{Q}}_p/\mathbf{Q}_p)$ en $l = p$. Duke Math. J., 115(2):205–310, 2002. With an appendix by Guy Henniart.

[11] Armand Brumer and Kenneth Kramer. Non-existence of certain semistable abelian varieties. Manuscripta Math., 106(3):291–304, 2001.

[12] Henri Carayol. Sur les représentations galoisiennes modulo ℓ attachées aux formes modulaires. Duke Math. J. 59 (1989), no. 3, 785–801.

[13] P. L. Chebyshev. Sur la fonction qui détermine la totalité des nombres premiers inférieurs à une limite donnée. J. de Math. 17 (1852), 341-365.

[14] P. L. Chebyshev. Mémoire sur les nombres premiers. J. de Math. 17 (1852), 366-390.

[15] Laurent Clozel. Motifs et formes automorphes: applications du principe de fonctorialité. Automorphic forms, Shimura varieties, and L-functions, Vol. I (Ann Arbor, MI, 1988), 77–159, Perspect. Math., 10, Academic Press, Boston, MA, 1990.

[16] Brian Conrad, Fred Diamond, and Richard Taylor. Modularity of certain potentially Barsotti-Tate Galois representations. J. Amer. Math. Soc., 12(2):521–567, 1999.

[17] Christophe Breuil, Brian Conrad, Fred Diamond, and Richard Taylor. On the modularity of elliptic curves over $\mathbb{Q}$: wild 3-adic exercises J. Amer. Math. Soc., 14(4):843–939, 2001.

[18] K. Buzzard, M. Dickinson, N. Shepherd-Barron and R. Taylor. On icosahedral Artin representations. Duke Math. J. 109 (2001), no. 2, 283–318.

[19] A. J. de Jong. A conjecture on arithmetic fundamental groups. Israel J. Math., 121:61–84, 2001.

[20] Fred Diamond. An extension of Wiles' results. In *Modular forms and Fermat's last theorem (Boston, MA, 1995)*, pages 475–489. Springer, New York, 1997.

[21] Fred Diamond and Richard Taylor. Lifting modular mod ℓ representations. Duke Math. J. 74 (1994), no. 2, 253–269.

[22] Fred Diamond. The Taylor-Wiles construction and multiplicity one. Invent. Math. 128 (1997), no. 2, 379–391.

[23] Fred Diamond. The refined conjecture of Serre. Elliptic curves, modular forms, & Fermat's last theorem (Hong Kong, 1993), 172–186, Ser. Number Theory, I, Internat. Press, Cambridge, MA, 1995.

[24] Luis Dieulefait. Existence of families of Galois representations and new cases of the Fontaine-Mazur conjecture. J. Reine Angew. Math. 577 (2004), 147–151.

[25] Fred Diamond, Matthias Flach, Li Guo. The Tamagawa number conjecture of adjoint motives of modular forms. Ann. Sci. École Norm. Sup. (4) 37 (2004), no. 5, 663–727.

[26] Bas Edixhoven. The weight in Serre's conjectures on modular forms. Invent. Math., 109 (1992), no. 3, 563–594.

[27] Jordan Ellenberg. Serre's conjecture over $\mathbb{F}_9$ Annals of Math 161 (2005), 1111-1142.

[28] Jean-Marc Fontaine and Barry Mazur. Geometric Galois representations. Elliptic curves, modular forms, & Fermat's last theorem (Hong Kong, 1993), 41–78, Ser. Number Theory, I, Internat. Press, Cambridge, MA, 1995.

[29] Jean-Marc Fontaine. Il n'y a pas de variété abélienne sur $\mathbf{Z}$. Invent. Math., 81(3):515–538, 1985.

[30] Kazuhiro Fujiwara. Deformation rings and Hecke algebras for totally real fields. *preprint*, 1999.

[31] Dennis Gaitsgory. On de Jong's conjecture. preprint.

[32] Chandrashekhar Khare and Jean-Pierre Wintenberger. On Serre's reciprocity conjecture for 2-dimensional mod p representations of $\mathrm{Gal}(\bar{\mathbb{Q}}/\mathbb{Q})$. preprint available at `http://xxx.lanl.gov/abs/math.NT/0412076`

[33] Chandrashekhar Khare. Serre's modularity conjecture: the level one case. Duke Math. J. 134 (2006), 557–589.

[34] Chandrashekhar Khare. A local analysis of congruences in the (p, p) case: Part II. Invent. Math., 143 (2001), no. 1, 129–155.

[35] Chandrashekhar Khare. Mod p descent for Hilbert modular forms Math. Res. Lett. 7 (2000), no. 4, 455–462.

[36] Chandrashekhar Khare. Remarks on mod p forms of weight one. Internat. Math. Res. Notices 1997, no. 3, 127–133. (Corrigendum: IMRN 1999, no. 18, pg. 1029.)

[37] Mark Kisin. Modularity of potentially Barsotti-Tate representations and moduli of finite flat group schemes. *preprint* (2004).

[38] Robert Langlands. Base change for GL_2. Annals of Math. Series, Princeton University Press, 1980.

[39] Laurent Lafforgue. Chtoucas de Drinfeld et correspondance de Langlands. Invent. Math. 147 (2002), no. 1, 1–241.

[40] Jayanta Manoharmayum. On the modularity of certain $\mathrm{GL}_2(\mathbb{F}_7)$ Galois representations. Math. Res. Lett. 8 (2001), no. 5-6, 703–712.

[41] Barry Mazur. An introduction to the deformation theory of Galois representations. In *Modular forms and Fermat's last theorem (Boston, MA, 1995)*, pages 243–311. Springer, New York, 1997.

[42] Hyunsuk Moon and Yuichiro Taguchi. Refinement of Tate's discriminant bound and non-existence theorems for mod p Galois representations. Doc. Math. 2003, Extra Vol.(Kazuya Kato's fiftieth birthday), 641–654 (electronic).

[43] Ravi Ramakrishna. Deforming Galois representations and the conjectures of Serre and Fontaine-Mazur. Ann. of Math. (2), 156(1):115–154, 2002.

[44] Michel Raynaud. Schémas en groupes de type $(p, \cdots, p)$. Bull. Soc. Math. France, 102, 459–472, 1974.

[45] Kenneth A. Ribet. Report on mod l representations of $\mathrm{Gal}(\overline{\mathbf{Q}}/\mathbf{Q})$. In *Motives (Seattle, WA, 1991)*, volume 55 of *Proc. Sympos. Pure Math.*, pages 639–676.

[46] Kenneth A. Ribet. Congruence relations between modular forms. Proceedings of the International Congress of Mathematicians, Vol. 1, 2 (Warsaw, 1983), 503–514, PWN, Warsaw, 1984.

[47] Kenneth A. Ribet. Images of semistable Galois representations. Pacific J. Math. 1997, Special Issue, 277–297.

[48] J. Barkley Rosser and Lowell Schoenfeld. Approximate formulas for some functions of prime numbers. Illinois J. Math. 6 (1962), 64–94.

[49] David Savitt. Modularity of some potentially Barsotti-Tate Galois representations. Compos. Math. 140 (2004), 31–63.

[50] David Savitt. On a Conjecture of Conrad, Diamond, and Taylor. Duke Math. J. 128 (2005), no. 1, 141–197.

[51] René Schoof. Abelian varieties over cyclotomic fields with good reduction everywhere. Math. Ann. 325 (2003), no. 3, 413–448.

[52] René Schoof. Abelian varieties over $\mathbf{Q}$ with bad reduction in one prime only. Compos. Math. 141 (2005), no. 4, 847–868.

[53] René Schoof. Abelian varieties over $\mathbf{Q}(\sqrt{6})$ with good reduction everywhere. Class field theory—its centenary and prospect (Tokyo, 1998), 287–306, Adv. Stud. Pure Math., 30, Math. Soc. Japan, Tokyo, 2001.

[54] Jean-Pierre Serre. *Œuvres. Vol. III*, 1972–1984. Springer-Verlag, Berlin, 1986.

[55] Jean-Pierre Serre. Sur les représentations modulaires de degré 2 de $\mathrm{Gal}(\overline{\mathbf{Q}}/\mathbf{Q})$. Duke Math. J., 54(1):179–230, 1987.

[56] N. Shepherd-Barron and R. Taylor. mod 2 and mod 5 icosahedral representations. J. Amer. Math. Soc. 10 (1997), no. 2, 283–298.

[57] C. M. Skinner and A. J. Wiles. Base change and a problem of Serre. Duke Math. J., 107(1):15–25, 2001

[58] C. M. Skinner and A. J. Wiles. Residually reducible representations and modular forms. Inst. Hautes Études Sci. Publ. Math., (89):5–126 (2000), 1999.

[59] C. M. Skinner and Andrew J. Wiles. Nearly ordinary deformations of irreducible residual representations. Ann. Fac. Sci. Toulouse Math. (6), 10(1):185–215, 2001.

[60] John Tate. The non-existence of certain Galois extensions of **Q** unramified outside 2, *Arithmetic geometry (Tempe, AZ, 1993)*, Contemp. Math., 174, 153–156, Amer. Math. Soc., 1994.

[61] Richard Taylor. Remarks on a conjecture of Fontaine and Mazur. J. Inst. Math. Jussieu, 1(1):125–143, 2002.

[62] Richard Taylor. Galois representations. Proceedings of the International Congress of Mathematicians, Vol. I (Beijing, 2002), 449–474, Higher Ed. Press, Beijing, 2002.

[63] Richard Taylor. On icosahedral Artin representations. II. Amer. J. Math., 125(3):549–566, 2003.

[64] Richard Taylor. Galois representations. Ann. Fac. Sci. Toulouse Math. (6) 13 (2004), no. 1, 73–119.

[65] Richard Taylor. On the meromorphic continuation of degree two L-functions. Documenta Math. Extra Volume: John H. Coates' Sixtieth Birthday (2006) 729–779.

[66] Richard Taylor and Andrew Wiles. Ring-theoretic properties of certain Hecke algebras. Ann. of Math. (2), 141(3):553–572, 1995.

[67] Andrew Wiles. Modular elliptic curves and Fermat's last theorem. Ann. of Math. (2), 141(3):443–551, 1995.

[68] Jean-Pierre Wintenberger. Sur les représentations p-adiques géométriques de conducteur 1 et de dimension 2 de $\mathrm{Gal}(\bar{\mathbb{Q}}/\mathbb{Q})$. preprint available at `http://xxx.lanl.gov/abs/math.NT/0406576`

Two p-adic L-functions and rational points on elliptic curves with supersingular reduction

Masato Kurihara
Department of Mathematics
Keio University
3-14-1 Hiyoshi, Kohoku-ku
Yokohama, 223-8522, Japan
kurihara@math.keio.ac.jp

Robert Pollack
Department of Mathematics and Statistics
Boston University
Boston, MA 02215, USA
rpollack@math.bu.edu

Introduction

Let E be an elliptic curve over $\mathbf{Q}$. We assume that E has good supersingular reduction at a prime p, and for simplicity, assume p is odd and $a_p = p + 1 - \#E(\mathbf{F}_p)$ is zero. Then, as the second author showed, the p-adic L-function $\mathcal{L}_{p,\alpha}(E)$ of E corresponding to $\alpha = \pm\sqrt{-p}$ (by Amice-Vélu and Vishik) can be written as

$$\mathcal{L}_{p,\alpha}(E) = f \log_p^+ + g \log_p^- \alpha$$

by using two Iwasawa functions f and $g \in \mathbf{Z}_p[[\mathrm{Gal}(\mathbf{Q}_\infty/\mathbf{Q})]]$ ([20] Theorem 5.1). Here $\log_p^\pm$ is the $\pm$-log function and $\mathbf{Q}_\infty/\mathbf{Q}$ is the cyclotomic $\mathbf{Z}_p$-extension (precisely, see §1.3).

In Iwasawa theory for elliptic curves, the case when p is a supersingular prime is usually regarded to be more complicated than the ordinary case, but the fact that we have two nice Iwasawa functions f and g gives us some advantage in several cases. The aim of this paper is to give such examples.

0.1. Our first application is related to the weak Birch and Swinnerton-Dyer conjecture. Let $L(E, s)$ be the L-function of E. The so called weak Birch and Swinnerton-Dyer conjecture is the statement

Conjecture (Weak BSD)

$$L(E, 1) = 0 \Longleftrightarrow \mathrm{rank} E(\mathbf{Q}) > 0. \tag{0.1}$$

We know by Kolyvagin that the right hand side implies the left hand side, but the converse is still a very difficult conjecture. For a prime number p, let $\mathrm{Sel}(E/\mathbf{Q})_{p^\infty}$ be the Selmer group of E over $\mathbf{Q}$ with respect to the p-power torsion points $E[p^\infty]$. Hence $\mathrm{Sel}(E/\mathbf{Q})_{p^\infty}$ sits in an exact sequence

$$0 \longrightarrow E(\mathbf{Q}) \otimes \mathbf{Q}_p/\mathbf{Z}_p \longrightarrow \mathrm{Sel}(E/\mathbf{Q})_{p^\infty} \longrightarrow Ш(E/\mathbf{Q})[p^\infty] \longrightarrow 0$$

where $Ш(E/\mathbf{Q})[p^\infty]$ is the p-primary component of the Tate-Shafarevich group of E over $\mathbf{Q}$. In this paper, we are interested in the following conjecture.

Conjecture 0.1

$$L(E,1) = 0 \Longleftrightarrow \#\mathrm{Sel}(E/\mathbf{Q})_{p^\infty} = \infty$$

This is equivalent to the weak Birch and Swinnerton-Dyer conjecture if we assume $\#Ш(E/\mathbf{Q})[p^\infty] < \infty$. (Of course, the problem is the implication from the left hand side to the right hand side.)

We note that Conjecture 0.1 is obtained as a corollary of the main conjecture in Iwasawa theory for E over the cyclotomic $\mathbf{Z}_p$-extension $\mathbf{Q}_\infty/\mathbf{Q}$. We also remark that if the sign of the functional equation is -1, Conjecture 0.1 was proved by Skinner-Urban [24] and Nekovář [16] in the case when p is ordinary, and by Byoung-du Kim [10] in the case when p is supersingular.

In this paper, we will give a simple condition which can be checked numerically and which implies Conjecture 0.1.

Suppose that p is an odd supersingular prime with $a_p = 0$. We identify $\mathbf{Z}_p[[\mathrm{Gal}(\mathbf{Q}_\infty/\mathbf{Q})]]$ with $\mathbf{Z}_p[[T]]$ by the usual correspondence between a generator γ of $\mathrm{Gal}(\mathbf{Q}_\infty/\mathbf{Q})$ and $1+T$. When we regard the above two Iwasawa functions f, g as elements of $\mathbf{Z}_p[[T]]$, we denote them by $f(T)$, $g(T)$. The interpolation property of $f(T)$ and $g(T)$ tells us that $f(0) = (p-1)L(E,1)/\Omega_E$ and $g(0) = 2L(E,1)/\Omega_E$ where Ω_E is the Néron period. Hence, if $L(E,1) \neq 0$, we have

$$\left.\frac{f(T)}{g(T)}\right|_{T=0} = \frac{p-1}{2}.$$

We conjecture that the converse is also true, namely

Conjecture 0.2

$$L(E,1) = 0 \Longleftrightarrow \left.\frac{f(T)}{g(T)}\right|_{T=0} \neq \frac{p-1}{2}$$

(Again, the problem is the implication from the left hand side to the right hand side.) Our first theorem says that Conjecture 0.2 implies Conjecture 0.1, namely

Theorem 0.3 *Assume that* $(f/g)(0)$ $\left(= \frac{f(T)}{g(T)}\Big|_{T=0}\right)$ *does not equal* $(p-1)/2$. *Then,* $\mathrm{Sel}(E/\mathbf{Q})_{p^\infty}$ *is infinite.*

This result in a different terminology was essentially obtained by Perrin-Riou (cf. [19] Proposition 4.10, see also §1.6 in this paper), but we will prove this theorem in §1.4 by a different and simple method, using a recent formulation of Iwasawa theory of an elliptic curve with supersingular reduction. In §1, we also review the recent formulation of such supersingular Iwasawa theory.

Conversely, assuming condition $(*)_0$ which will be introduced in §1.4 and which should always be true, we will show in §1.4 that the weak BSD conjecture (0.1) implies Conjecture 0.2, namely

Theorem 0.4 *Assume condition* $(*)_0$ *in* §1.4, *and* $\mathrm{rank} E(\mathbf{Q}) > 0$. *Then,* $(f/g)(0) \neq (p-1)/2$ *holds.*

Combining Theorems 0.3 and 0.4, we get

Corollary 0.5 *Assume* $(*)_0$ *in* §1.4 *and* $\#\text{Ш}(E/\mathbf{Q})[p^\infty] < \infty$. *Then, we have*

$$\mathrm{rank} E(\mathbf{Q}) > 0 \Longleftrightarrow \frac{f(T)}{g(T)}\Big|_{T=0} \neq \frac{p-1}{2}.$$

Note that the left hand side is algebraic information and the right hand side is p-adic analytic information. The weak BSD conjecture (0.1) is usually regarded to be a typical relation between algebraic and analytic information. The above Corollary 0.5 also gives such a relation, but in a different form. (Concerning the meaning of $(f/g)(0) \neq \frac{p-1}{2}$, see also §1.6.)

We note that $f(T)$ and $g(T)$ can be computed numerically and the condition $(f/g)(0) \neq (p-1)/2$ can be checked numerically. For example, for $N = 17$ and 32, we considered the quadratic twist $E = X_0(N)_d$ of the elliptic curve $X_0(N)$ by the Dirichlet character χ_d of conductor $d > 0$. For $p = 3$ (which is a supersingular prime in both cases with $a_p = 0$), we checked the condition $(f/g)(0) \neq (p-1)/2$ for all E_d such that $L(E_d, 1) = 0$ with $0 < d < 500$ and d prime to $3N$. We did this by computing $(f/g)(0) - (p-1)/2$ mod 3^n; the biggest n we needed was 7 for $E = X_0(32)_d$ with $d = 485$. For $N = 17$ and $d = 76, 104, 145, 157, 185, \ldots$ (resp. $N = 32$ and $d = 41, 65, 137, 145, 161, \ldots$) $\mathrm{ord}_T(f(T)) = \mathrm{ord}_T(g(T)) = 2$, so the corank of $\mathrm{Sel}(E/\mathbf{Q})_{p^\infty}$ should be 2. Our computation together with Theorem 0.3 implies that this corank is ≥ 1.

We give here one simple condition which implies $(f/g)(0) \neq (p-1)/2$. Put $r = \min\{\mathrm{ord}_T f(T), \mathrm{ord}_T g(T)\}$, and set $f^*(T) = T^{-r} f(T)$ and

$g^*(T) = T^{-r}g(T)$. If $\mathrm{ord}_p(f^*(0)) \neq \mathrm{ord}_p(g^*(0))$, then $(f^*/g^*)(0)$ is not in $\mathbf{Z}_p^\times$ and hence cannot be $(p-1)/2$. Therefore, by Theorem 0.3 we have

Corollary 0.6 *If* $\mathrm{ord}_p(f^*(0)) \neq \mathrm{ord}_p(g^*(0))$, *then* $\mathrm{Sel}(E/\mathbf{Q})_{p^\infty}$ *is infinite. In particular, if one of* $f^*(T)$ *or* $g^*(T)$ *has* λ*-invariant zero and the other has non-zero* λ*-invariant, then* $\mathrm{Sel}(E/\mathbf{Q})_{p^\infty}$ *is infinite.*

For example, we again consider $E = X_0(17)_d$ with $d > 0$ and $p = 3$. Then, the condition on the λ-invariants in Corollary 0.6 is satisfied by many examples, namely for $d = 29, 37, 40, 41, 44, 56, 65, \ldots$ with $r = 1$, and for $d = 145, 157, 185, 293, 409, \ldots$ with $r = 2$.

Corollary 0.6 implies a small result on the Main Conjecture (Proposition 1.5, see §1.5). The case $r = 1$ will be treated in detail in §2.

0.2. In 0.1, we explained that the computation of the value $(f/g)(0) - (p-1)/2$, or $f^{(r)}(0) - \frac{p-1}{2}g^{(r)}(0)$, yields information on the Selmer group $\mathrm{Sel}(E/\mathbf{Q})_{p^\infty}$ where $r = \min\{\mathrm{ord}_T f(T), \mathrm{ord}_T g(T)\}$. It is natural to ask what this value means. In the case $r = 1$, we can interpret this value very explicitly by using the p-adic Birch and Swinnerton-Dyer conjecture (cf. Bernardi and Perrin-Riou [1] and Colmez [4]).

In this case, for a generator P of $E(\mathbf{Q})/E(\mathbf{Q})_{\mathrm{tor}}$, the p-adic Birch and Swinnerton-Dyer conjecture predicts that $\log_{\hat{E}}(P)$ is related to the quantity $f'(0) - \frac{p-1}{2}g'(0)$ where $\log_{\hat{E}}$ is the logarithm of the formal group $\hat{E}$ (see §2.6 (2.5)). Using this formula for $\log_{\hat{E}}(P)$, we can find P numerically. More precisely, we compute

$$\exp_{\hat{E}}\left(\sqrt{-\frac{(f'(0) - \frac{p-1}{2}g'(0))\, 2p\log(\kappa(\gamma))[\varphi(\omega_E), \omega_E]}{\mathrm{Tam}(E)} \cdot \frac{\#E(\mathbf{Q})_{\mathrm{tor}}}{p+1}}\right) \tag{0.2}$$

which is a point on $E(\mathbf{Q}_p)$, and which would produce a point on $E(\mathbf{Q})$ with a slight modification (see §2.7). Namely, we can construct a rational point of infinite order p-adically in the case $r = 1$ as Rubin did in his paper [22] §3 for a CM elliptic curve.

Note here that we have an advantage in the supersingular case in that the two p-adic L-functions together encode $\log_{\hat{E}}(P)$ (and not just the p-adic height of a point).

We did the above computation for quadratic twists of the curve $X_0(17)$ with $p = 3$. We *found* a rational point on $X_0(17)_d$ *by this method* for *all* d such that $0 < d < 250$ except for $d = 197$, $\gcd(d, 3 \cdot 17) = 1$, and the rank of $X_0(17)_d$ is 1. For example, for the curve

$$X_0(17)_{193} : y^2 + xy + y = x^3 - x^2 - 25609x - 99966422$$

we found the rational point

$$\left(\frac{915394662845247271}{25061097283236}, \frac{-878088421712236204458830141}{125458509476191439016}\right)$$

by this method, namely 3-adically. To get this rational point, we had to compute the value (0.2) modulo 3^{80} (to 80 3-adic digits) to recognize that the modified point constructed from the value (0.2) is a point on $E(\mathbf{Q})$. For the curve $X_0(32)$ and $p = 3$ we did a similar computation and found rational points on $X_0(32)_d$ for all d with $0 < d < 150$, $\gcd(d, 6)$, and the rank of $X_0(32)_d$ is 1. In §4, there are tables listing points for both of these curves.

To compute $f'(0)$ and $g'(0)$ to high accuracy, the usual definition of the p-adic L-function (namely, the computation of the Riemann sums to approximate $f'(0)$ and $g'(0)$) is not at all suitable. We use the theory of overconvergent modular symbols as in [21] and [6]. We will explain in detail in §2 this theory and the method to compute a rational point in practice.

0.3. Next, we study a certain important subgroup of the Selmer group over $\mathbf{Q}_\infty$. This is related to studying common divisors of $f(T)$ and $g(T)$. For any algebraic extension F of $\mathbf{Q}$, we define the fine Selmer group $\mathrm{Sel}_0(E/F)$ by

$$\mathrm{Sel}_0(E/F) = \mathrm{Ker}(\mathrm{Sel}(E/F)_{p^\infty} \longrightarrow \prod_{v|p} H^1(F_v, E[p^\infty]))$$

where $\mathrm{Sel}(E/F)_{p^\infty}$ is the Selmer group of E over F with respect to $E[p^\infty]$, and v ranges over primes of F lying over p. (The name "fine Selmer group" is due to J. Coates.) Our interest in this subsection is in $\mathrm{Sel}_0(E/\mathbf{Q}_\infty)$.

Let $\mathbf{Q}_n$ be the intermediate field of $\mathbf{Q}_\infty/\mathbf{Q}$ with degree p^n. We put $e_0 = \mathrm{rank}E(\mathbf{Q})$ and $\Phi_0(T) = T$. For $n \geq 1$, we define

$$e_n = \frac{\mathrm{rank}E(\mathbf{Q}_n) - \mathrm{rank}E(\mathbf{Q}_{n-1})}{p^{n-1}(p-1)}$$

which is a non-negative integer, $\omega_n(T) = (1+T)^{p^n} - 1$, and $\Phi_n(T) = \omega_n(T)/\omega_{n-1}(T)$.

The Pontryagin dual $\mathrm{Sel}_0(E/\mathbf{Q}_\infty)^\vee$ of the fine Selmer group over $\mathbf{Q}_\infty$ is a finitely generated torsion $\mathbf{Z}_p[[\mathrm{Gal}(\mathbf{Q}_\infty/\mathbf{Q})]]$-module by Kato [9]. Concerning the characteristic ideal, Greenberg raised the following problem (conjecture) (see §3.1)

Problem 0.7

$$\mathrm{char}(\mathrm{Sel}_0(E/\mathbf{Q}_\infty)^\vee) = \Big(\prod_{\substack{e_n \geq 1 \\ n \geq 0}} \Phi_n^{e_n - 1} \Big).$$

We remark that this "conjecture" has the same flavor as his famous conjecture on the vanishing of the λ-invariants for class groups of totally real fields.

Let $\mathrm{Sel}^\pm(E/\mathbf{Q}_\infty)$ be Kobayashi's $\pm$-Selmer groups ([12], or see 1.5). By definition we have $\mathrm{Sel}_0(E/\mathbf{Q}_\infty) \subset \mathrm{Sel}^\pm(E/\mathbf{Q}_\infty)$. In this subsection, we assume the μ-invariant of $\mathrm{Sel}^\pm(E/\mathbf{Q}_\infty)$ vanishes. Using Kato's result [9] we know $g(T) \in \mathrm{char}(\mathrm{Sel}^+(E/\mathbf{Q}_\infty)^\vee)$, and $f(T) \in \mathrm{char}(\mathrm{Sel}^-(E/\mathbf{Q}_\infty)^\vee)$ (cf. Kobayashi [12] Theorem 1.3, see also 1.5). Hence, any generator of $\mathrm{char}(\mathrm{Sel}_0(E/\mathbf{Q}_\infty)^\vee)$ divides both $f(T)$ and $g(T)$. Thus, in the supersingular case, we can check this conjecture (Problem 0.7) numerically in many cases, by computing $f(T)$ and $g(T)$.

For example, suppose that $\mathrm{rank} E(\mathbf{Q}) = e_0$ and $\min\{\lambda(f(T)), \lambda(g(T))\} = e_0$. Then, we can show that the above "conjecture" is true and, moreover, $\mathrm{char}(\mathrm{Sel}_0(E/\mathbf{Q}_\infty)^\vee) = (T^{e_0'})$ where $e_0' = \max\{0, e_0 - 1\}$ (see Proposition 3.1). For $E = X_0(17)_d$ and $p = 3$, the condition of Proposition 3.1 is satisfied for all d such that $0 < d < 250$ except for $d = 104, 193, 233$. For these exceptional values, we also checked Problem 0.7 holds (see §3.3).

In §3.2 we raise a question on the greatest common divisor of $f(T)$ and $g(T)$ (Problem 3.2), and study a relation with the above Greenberg "conjecture" (see Propositions 3.3 and 3.4).

We would like to heartily thank R. Greenberg for fruitful discussions on all subjects in this paper, and for his hospitality when both of us were invited to the University of Washington in May 2004. Furthermore, we learned to use Wingberg's result from him when we studied the problem in Proposition 3.4 (1). We would also like to express our hearty thanks to G. Stevens for his helpful suggestion when we studied the example (the case $d = 193$) in §3.3.

1 Iwasawa theory of an elliptic curve with supersingular reduction

1.1. $\pm$-Coleman homomorphisms. Kobayashi defined in [12] §8 $\pm$-Coleman homomorphisms. We will give here a slightly different construction of these homomorphisms using the results of the first author in [13].

Suppose that E has good supersingular reduction at an odd prime p with $a_p = p + 1 - \#E(\mathbf{F}_p) = 0$. We denote by $T = T_p(E)$ the Tate module, and set $V = T \otimes \mathbf{Q}_p$. For $n \geq 0$, let $\mathbf{Q}_{p,n}$ denote the intermediate field of the cyclotomic $\mathbf{Z}_p$-extension $\mathbf{Q}_{p,\infty}/\mathbf{Q}_p$ of the p-adic field $\mathbf{Q}_p$ such that $[\mathbf{Q}_{p,n} : \mathbf{Q}_p] = p^n$. We set $\Lambda = \mathbf{Z}_p[[\mathrm{Gal}(\mathbf{Q}_\infty/\mathbf{Q})]] = \mathbf{Z}_p[[\mathrm{Gal}(\mathbf{Q}_{p,\infty}/\mathbf{Q}_p)]]$, and identify Λ with $\mathbf{Z}_p[[T]]$ by identifying γ with $1+T$. We put

$$\mathbf{H}^1_{\mathrm{loc}} = \varprojlim H^1(\mathbf{Q}_{p,n}, T)$$

where the limit is taken with respect to the corestriction maps. We will define two Λ-homomorphisms

$$\mathrm{Col}^{\pm} : \mathbf{H}^1_{\mathrm{loc}} \longrightarrow \Lambda.$$

Let $D = D_{dR}(V)$ be the Dieudonné module which is a two dimensional $\mathbf{Q}_p$-vector space. Let ω_E be the Néron differential which we regard as an element of D. Since D is isomorphic to the crystalline cohomology space $H^1_{cris}(E \bmod p/\mathbf{Q}_p)$, the Frobenius operator φ acts on D and satisfies $\varphi^{-2} - a_p\varphi^{-1} + p = \varphi^{-2} + p = 0$.

We take a generator (ζ_{p^n}) of $\mathbf{Z}_p(1)$; namely, ζ_{p^n} is a primitive p^n-th root of unity, and $\zeta_{p^{n+1}}^p = \zeta_{p^n}$ for any $n \geq 1$. For $n \geq 1$ and $x \in D$, put

$$\gamma_n(x) = \sum_{i=0}^{n-1} \varphi^{i-n}(x) \otimes \zeta_{p^{n-i}} + (1-\varphi)^{-1}(x) \in D \otimes \mathbf{Q}_p(\mu_{p^n}).$$

Putting $\mathcal{G}_{n+1} = \mathrm{Gal}(\mathbf{Q}_p(\mu_{p^{n+1}})/\mathbf{Q}_p)$, we define

$$P_n : H^1(\mathbf{Q}_p(\mu_{p^{n+1}}), T) \longrightarrow \mathbf{Q}_p[\mathcal{G}_{n+1}]$$

by setting

$$P_n(z) := \frac{1}{[\varphi(\omega_E), \omega_E]} \sum_{\sigma \in \mathcal{G}_{n+1}} \mathrm{Tr}_{\mathbf{Q}_p(\mu_{p^{n+1}})/\mathbf{Q}_p}([\gamma_{n+1}(\varphi^{n+2}(\omega_E))^\sigma, \exp^*(z)])\sigma.$$

Here

$$\exp^* : H^1(\mathbf{Q}_p(\mu_{p^{n+1}}), T) \longrightarrow D \otimes \mathbf{Q}_p(\mu_{p^{n+1}})$$

is the dual exponential map of Bloch and Kato (which is the dual of the exponential map: $D \otimes \mathbf{Q}_p(\mu_{p^{n+1}}) \longrightarrow H^1(\mathbf{Q}_p(\mu_{p^{n+1}}), V)$), and $[x, y] \in D \otimes \mathbf{Q}_p(\mu_{p^{n+1}})$ is the cup product of the de Rham cohomology for x, $y \in D \otimes \mathbf{Q}_p(\mu_{p^{n+1}})$. In this, $[\varphi(\omega_E), \omega_E] \in \mathbf{Z}_p$ plays the role of a p-adic period. By Proposition 3.6 in [13], we have

$$P_n(z) \in \mathbf{Z}_p[\mathcal{G}_{n+1}]$$

(note that we slightly changed the notation γ_n, P_n from [13]).

Put $G_n = \mathrm{Gal}(\mathbf{Q}_{p,n}/\mathbf{Q}_p)$, and let $i : H^1(\mathbf{Q}_{p,n}, T) \longrightarrow H^1(\mathbf{Q}_p(\mu_{p^{n+1}}), T)$ and $\pi : \mathbf{Z}_p[\mathcal{G}_{n+1}] \longrightarrow \mathbf{Z}_p[G_n]$ be the natural maps. We define

$$\mathcal{P}_n : H^1(\mathbf{Q}_{p,n}, T) \longrightarrow \mathbf{Z}_p[G_n]$$

by

$$\mathcal{P}_n(z) = \frac{1}{p-1}\pi \circ P_n(i(z)).$$

These elements satisfy a distribution property; namely, we have

$$\pi_{n,n-1}\mathcal{P}_n(z) = -\nu_{n-2,n-1}\mathcal{P}_{n-2}(N_{n,n-2}(z))$$

where $\pi_{n,n-1} : \mathbf{Z}_p[G_n] \longrightarrow \mathbf{Z}_p[G_{n-1}]$ is the natural projection, $\nu_{n-2,n-1} : \mathbf{Z}_p[G_{n-2}] \longrightarrow \mathbf{Z}_p[G_{n-1}]$ is the norm map such that $\sigma \mapsto \Sigma\tau$ (for $\sigma \in G_{n-2}$, τ ranges over elements in G_{n-1} such that $\pi_{n-1,n-2}(\tau) = \sigma$), and $N_{n,n-2} : H^1(\mathbf{Q}_{p,n}, T) \longrightarrow H^1(\mathbf{Q}_{p,n-2}, T)$ is the corestriction map. This relation can be proved by showing $\psi(\pi_{n,n-1}\mathcal{P}_n(z)) = \psi(-\nu_{n-2,n-1}\mathcal{P}_{n-2}(N_{n,n-2}(z)))$ for any character ψ of G_{n-1} (cf. the proof of Lemma 7.2 in [13]).

Our identification of γ with $1 + T$, gives an identification of $\mathbf{Z}_p[G_n]$ with $\mathbf{Z}_p[T]/((1+T)^{p^n} - 1)$. Set $\omega_m = (1+T)^{p^m} - 1$, and $\Phi_m = \omega_m/\omega_{m-1}$ which is the p^m-th cyclotomic polynomial evaluated at $1 + T$. The above distribution relation implies that Φ_{n-1} divides $\mathcal{P}_n(z)$. By induction on n, we can show that $\Phi_{n-1}\Phi_{n-3} \cdot \ldots \cdot \Phi_1$ divides $\mathcal{P}_n(z)$ if n is even, and $\Phi_{n-1}\Phi_{n-3} \cdot \ldots \cdot \Phi_2$ divides $\mathcal{P}_n(z)$ if n is odd. Put

$$\omega_n^+ = \prod_{2 \le m \le n, 2|m} \Phi_m, \qquad \omega_n^- = \prod_{1 \le m \le n, 2 \nmid m} \Phi_m.$$

Suppose that z is an element in $\mathbf{H}^1_{\mathrm{loc}}$, and $z_n \in H^1(\mathbf{Q}_{p,n}, T)$ is its image. Suppose at first n is odd, and write $\mathcal{P}_n(z_n) = \omega_n^+ h_n(T)$ with $h_n(T) \in \mathbf{Z}_p[T]/(\omega_n)$. Then, $h_n(T)$ is uniquely determined in $\mathbf{Z}_p[T]/(T\omega_n^-)$ because $\omega_n^+\omega_n^- T = \omega_n$; so we regard $h_n(T)$ as an element in this ring. By the above distribution property, we know that $((-1)^{(n+1)/2}h_n(T))_{n:\mathrm{odd}\ge 1}$ is a projective system with respect to the natural maps $\mathbf{Z}_p[T]/(T\omega_{n+2}^-) \longrightarrow \mathbf{Z}_p[T]/(T\omega_n^-)$. Hence, it defines an element $h(T) \in \varprojlim \mathbf{Z}_p[T]/(T\omega_n^-) = \mathbf{Z}_p[[T]] = \Lambda$. We define $\mathrm{Col}^+(z) = h(T)$.

Next, suppose n is even. We write $\mathcal{P}_n(z_n) = \omega_n^- k_n(T)$ with $k_n(T) \in \mathbf{Z}_p[T]/(T\omega_n^+)$. By the same method as above, the distribution property implies that $((-1)^{(n+2)/2}k_n(T))_{n:\mathrm{even}\ge 1}$ is a projective system, so it defines $k(T) \in \varprojlim \mathbf{Z}_p[T]/(T\omega_n^+) = \mathbf{Z}_p[[T]] = \Lambda$. We define $\mathrm{Col}^-(z) = k(T)$. Thus, we have obtained two power series from $z \in \mathbf{H}^1_{\text{loc}}$. We define Col : $\mathbf{H}^1_{\text{loc}} \longrightarrow \Lambda \oplus \Lambda$ by $\text{Col}(z) = (\text{Col}^+(z), \text{Col}^-(z)) = (h(T), k(T))$.

The next lemma will be useful in what follows.

Lemma 1.1 *Suppose* $z \in \mathbf{H}^1_{\text{loc}}$, $\text{Col}^+(z) = h(T)$, *and* $\text{Col}^-(z) = k(T)$. *Then,*

$$h(0) = \frac{p(p-1)}{p+1}\frac{\exp^*(z_0)}{\omega_E} \quad \textit{and} \quad k(0) = \frac{2p}{p+1}\frac{\exp^*(z_0)}{\omega_E}$$

where z_0 *is the image of* z *in* $H^1(\mathbf{Q}_p, T)$, *and* $\exp^*(z_0)/\omega_E$ *is the element* $a \in \mathbf{Q}_p$ *such that* $\exp^*(z_0) = a\omega_E$.

We note that $\exp^*(z_0)/\omega_E$ is known to be in $p^{-1}\mathbf{Z}_p$ (cf. [23] Proposition 5.2), hence the right hand side of the above formula is in $\mathbf{Z}_p$.

Proof. This follows from the construction of $\text{Col}^{\pm}(z)$ and Lemma 3.5 in [13] (cf. the proof of Lemma 7.2 in [13], pg. 220).

1.2. An exact sequence. We have defined

$$\text{Col} = \text{Col}^+ \oplus \text{Col}^- : \mathbf{H}^1_{\text{loc}} \longrightarrow \Lambda \oplus \Lambda.$$

This homomorphism induces

Proposition 1.2 *We have an exact sequence*

$$0 \longrightarrow \mathbf{H}^1_{\text{loc}} \xrightarrow{\text{Col}} \Lambda \oplus \Lambda \xrightarrow{\rho} \mathbf{Z}_p \longrightarrow 0$$

where ρ *is the map defined by* $\rho(h(T), k(T)) = h(0) - \frac{p-1}{2}k(0)$.

Proof. First of all, we note that $\mathbf{H}^1_{\text{loc}}$ is a free Λ-module of rank 2. In fact, since $H^0(\mathbf{Q}_{p,\infty}, E[p]) = 0$, it follows that

$$H^1(\mathbf{Q}_p, E[p^\infty]) \xrightarrow{\simeq} H^1(\mathbf{Q}_{p,\infty}, E[p^\infty])^{\text{Gal}(\mathbf{Q}_{p,\infty}/\mathbf{Q}_p)}$$

is bijective. Taking the dual, we get an isomorphism $(\mathbf{H}^1_{\text{loc}})_{\text{Gal}(\mathbf{Q}_{p,\infty}/\mathbf{Q}_p)} \simeq H^1(\mathbf{Q}_p, T)$. Since $H^0(\mathbf{Q}_p, E[p]) = H^2(\mathbf{Q}_p, E[p]) = 0$, $H^1(\mathbf{Q}_p, T)$ is a free $\mathbf{Z}_p$-module of rank 2, and so $\mathbf{H}^1_{\text{loc}}$ is a free Λ-module of rank 2.

By Lemma 1.1, we know $\rho \circ \text{Col} = 0$. Hence, to prove Proposition 1.2, it suffices to show that the cokernel of Col is isomorphic to $\mathbf{Z}_p$.

Kobayashi defined two subgroups $E^{\pm}(\mathbf{Q}_{p,n})$ of $E(\mathbf{Q}_{p,n}) \otimes \mathbf{Z}_p$ ([12] Definition 8.16). We will explain these subgroups in a slightly different way (this idea is due to R. Greenberg). Since $E(\mathbf{Q}_{p,n}) \otimes \mathbf{Q}_p$ is the regular representation of G_n, it decomposes into $\bigoplus_{i=0}^{n} V_i$ where the V_i's are irreducible representations such that $\dim_{\mathbf{Q}_p} V_0 = 1$, and $\dim_{\mathbf{Q}_p} V_i = p^{i-1}(p-1)$ for $i > 0$. Then $E^+(\mathbf{Q}_{p,n})$ (resp. $E^-(\mathbf{Q}_{p,n})$) is defined to be the subgroup consisting of all points $P \in E(\mathbf{Q}_{p,n}) \otimes \mathbf{Z}_p$ such that the image of P in V_i is zero for every odd i (resp. for every positive even i). We define $E^\pm(\mathbf{Q}_{p,\infty})$ as the direct limit of $E^\pm(\mathbf{Q}_{p,n})$. By definition, the sequence

$$0 \longrightarrow E(\mathbf{Q}_p) \otimes \mathbf{Q}_p/\mathbf{Z}_p \longrightarrow E^+(\mathbf{Q}_{p,\infty}) \otimes \mathbf{Q}_p/\mathbf{Z}_p \oplus E^-(\mathbf{Q}_{p,\infty}) \otimes \mathbf{Q}_p/\mathbf{Z}_p$$
$$\longrightarrow E(\mathbf{Q}_{p,\infty}) \otimes \mathbf{Q}_p/\mathbf{Z}_p \longrightarrow 0$$

is exact. Since p is supersingular, $E(\mathbf{Q}_{p,\infty}) \otimes \mathbf{Q}_p/\mathbf{Z}_p = H^1(\mathbf{Q}_{p,\infty}, E[p^\infty])$. We also know that $E^\pm(\mathbf{Q}_{p,\infty}) \otimes \mathbf{Q}_p/\mathbf{Z}_p$ is the exact annihilator of the kernel of $\mathrm{Col}^\pm$ with respect to the cup product ([12] Proposition 8.18). Hence, taking the dual of the above exact sequence, we get $\mathrm{Coker}(\mathrm{Col}) \simeq (E(\mathbf{Q}_p) \otimes \mathbf{Q}_p/\mathbf{Z}_p)^\vee \simeq \mathbf{Z}_p$, which completes the proof.

1.3. p-adic L-functions and Kato's zeta elements. We now consider global cohomology groups. For $n \geq 0$, let $\mathbf{Q}_n$ denote the intermediate field of $\mathbf{Q}_\infty/\mathbf{Q}$ with degree p^n. We define

$$\mathbf{H}^1_{\mathrm{glob}} = \varprojlim H^1(\mathbf{Q}_n, T) = \varprojlim H^1_{et}(O_{\mathbf{Q}_n}[1/S], T)$$

where the limit is taken with respect to the corestriction maps, and S is the product of the primes of bad reduction and p. The image of $z \in \mathbf{H}^1_{\mathrm{glob}}$ in $\mathbf{H}^1_{\mathrm{loc}}$ we continue to denote by z. In our situation, $\mathbf{H}^1_{\mathrm{glob}}$ was proved to be a free Λ-module of rank 1 (Kato [9] Theorem 12.4). Kato constructed an element $z_K = ((z_K)_n)_{n\geq 0} \in \mathbf{H}^1_{\mathrm{glob}}$ with the following properties [9]. For a faithful character ψ of $G_n = \mathrm{Gal}(\mathbf{Q}_n/\mathbf{Q})$ with $n > 0$,

$$\sum_{\sigma \in G_n} \psi(\sigma) \exp^*(\sigma(z_K)_n) = \omega_E \frac{L(E, \psi, 1)}{\Omega_E},$$

and

$$\exp^*((z_K)_0) = \omega_E(1 - a_p p^{-1} + p^{-1}) \frac{L(E,1)}{\Omega_E}$$

where $\exp^* : H^1(\mathbf{Q}_{p,n}, T) \longrightarrow D \otimes \mathbf{Q}_{p,n}$ is the dual exponential map.

Suppose that $\theta_{\mathbf{Q}_n} \in \mathbf{Z}_p[G_n]$ is the modular element of Mazur and Tate [15], which satisfies the distribution property $\pi_{n,n-1}\theta_{\mathbf{Q}_n} = -\nu_{n-2,n-1}\theta_{\mathbf{Q}_{n-2}}$, and the property that for a faithful character ψ of $G_n = \mathrm{Gal}(\mathbf{Q}_n/\mathbf{Q})$ with $n > 0$,

$$\psi(\theta_{\mathbf{Q}_n}) = \tau(\psi) \frac{L(E, \psi^{-1}, 1)}{\Omega_E}$$

where $\tau(\psi)$ is the Gauss sum, and $\psi : \mathbf{Z}_p[G_n] \longrightarrow \mathbf{Z}_p[\mathrm{Image}\,\psi]$ is the ring homomorphism induced by ψ.

Let α and β be two roots of $t^2+p=0$. We have two p-adic L-functions $\mathcal{L}_{p,\alpha}$ and $\mathcal{L}_{p,\beta}$ by Amice-Vélu and Vishik, which are in $\mathcal{H}_\infty \otimes \mathbf{Q}_p(\sqrt{-p})$ where

$$\mathcal{H}_\infty = \left\{ \sum_{n=0}^{\infty} a_n T^n \in \mathbf{Q}_p[[T]];\ \lim_{n\to\infty} |a_n|_p n^{-h} = 0 \ \text{ for some } h \in \mathbf{Z}_{>0} \right\}.$$

As the second author proved in [20], there are two Iwasawa functions $f(T)$ and $g(T)$ in $\mathbf{Z}_p[[T]]$ such that

$$\mathcal{L}_{p,\alpha}(T) = f(T)\log^+(T) + \alpha g(T)\log^-(T) \tag{1.1}$$

and

$$\mathcal{L}_{p,\beta}(T) = f(T)\log^+(T) + \beta g(T)\log^-(T) \tag{1.2}$$

where $\log^+(T) = p^{-1}\Pi_{n>0}\frac{1}{p}\Phi_{2n}(T)$ and $\log^-(T) = p^{-1}\Pi_{n>0}\frac{1}{p}\Phi_{2n-1}(T)$.

Let $\mathcal{P}_n$ be as in §1.1, and $z_K = ((z_K)_n)$ be the zeta element of Kato. By Lemma 7.2 in [13] by the first author, we have

$$\mathcal{P}_n((z_K)_n) = \theta_{\mathbf{Q}_n} \tag{1.3}$$

(we note that we need no assumption (for example on $L(E,1)$) to get (1.3)). Since we know that $\mathcal{L}_{p,\alpha}$ is also obtained as the limit of

$$\alpha^{-n-1}(\theta_{\mathbf{Q}_n} - \alpha^{-1}\nu_{n-1,n}\theta_{\mathbf{Q}_{n-1}}),$$

using (1.3), we obtain

$$\mathcal{L}_{p,\alpha}(T) = \mathrm{Col}^+(z_K)\log^+(T) + \alpha\mathrm{Col}^-(z_K)\log^-(T).$$

Comparing this formula with (1.1), we have proved

Theorem 1.3 (Kobayashi [12] Theorem 6.3) *Let z_K be Kato's zeta element, and f, g be the Iwasawa functions as in* (1.1). *Then, we have* $\mathrm{Col}^+(z_K) = f(T)$ *and* $\mathrm{Col}^-(z_K) = g(T)$.

1.4. Proofs of Theorems 0.3 and 0.4. We begin by proving Theorem 0.3. For a non-zero element z in $\mathbf{H}^1_{\mathrm{glob}}$, we write $\mathrm{Col}(z) = (h_z(T), k_z(T))$. Since $\mathbf{H}^1_{\mathrm{glob}}$ is free of rank 1 over Λ ([9] Theorem 12.4), $h_z(T)/k_z(T)$ does not depend on the choice of $z \in \mathbf{H}^1_{\mathrm{glob}}$. So choosing $z = z_K$, we have by Theorem 1.3, that $h_z(T)/k_z(T) = f(T)/g(T)$ for all non-zero $z \in \mathbf{H}^1_{\mathrm{glob}}$.

Let ξ be a generator of $\mathbf{H}^1_{\mathrm{glob}}$. Suppose that $h_\xi(0) \neq 0$. By Proposition 1.2, $\rho(\mathrm{Col}(\xi)) = 0$, so we get $k_\xi(0) \neq 0$ and $h_\xi(0)/k_\xi(0) = (f/g)(0) = (p-1)/2$. This contradicts our assumption. Hence, $h_\xi(0) = 0$. This implies that the image $\xi_{\mathbf{Q}_p}$ of ξ in $H^1(\mathbf{Q}_p, T)$ satisfies $\exp^*(\xi_{\mathbf{Q}_p}) = 0$ by Lemma 1.1, so $\xi_{\mathbf{Q}_p}$ is in $E(\mathbf{Q}_p) \otimes \mathbf{Z}_p$. Hence, the image $\xi_{\mathbf{Q}}$ of ξ in $H^1(\mathbf{Q}, T)$ is in $\mathrm{Sel}(E/\mathbf{Q}, T)$ which is the Selmer group of $E/\mathbf{Q}$ with respect to T. Since $\mathbf{H}^1_{\mathrm{glob}} = H^1(\mathbf{Q}, \Lambda \otimes T)$, from an exact sequence $0 \to \Lambda \otimes T \to \Lambda \otimes T \to T \to 0$, the natural map $(\mathbf{H}^1_{\mathrm{glob}})_{\mathrm{Gal}(\mathbf{Q}_\infty/\mathbf{Q})} \longrightarrow H^1(\mathbf{Q}, T)$ is injective ($(\mathbf{H}^1_{\mathrm{glob}})_{\mathrm{Gal}(\mathbf{Q}_\infty/\mathbf{Q})}$ is the $\mathrm{Gal}(\mathbf{Q}_\infty/\mathbf{Q})$-coinvariants of $\mathbf{H}^1_{\mathrm{glob}}$). Thus, $\xi_{\mathbf{Q}} \in \mathrm{Sel}(E/\mathbf{Q}, T)$ is of infinite order. Therefore, $\#\mathrm{Sel}(E/\mathbf{Q}, T) = \infty$, which implies $\#\mathrm{Sel}(E/\mathbf{Q})_{p^\infty} = \infty$ and establishes Theorem 0.3.

Next, we introduce condition $(*)_0$. We consider the composite $\mathbf{H}^1_{\mathrm{glob}} \longrightarrow H^1(\mathbf{Q}, T) \longrightarrow H^1(\mathbf{Q}_p, T)$ of natural maps, and the property

$(*)_0$ $\qquad \mathbf{H}^1_{\mathrm{glob}} \longrightarrow H^1(\mathbf{Q}_p, T)$ is not the zero map.

This property $(*)_0$ should always be true. In fact, it is a consequence of the p-adic Birch and Swinnerton-Dyer conjecture. More precisely, it follows from a conjecture that a certain p-adic height pairing is non-degenerate (see Perrin-Riou [17] pg. 979 Conjecture 3.3.7 B and Remarque iii)).

We now prove Theorem 0.4. As we saw in the proof of Proposition 1.2, $\mathbf{H}^1_{\mathrm{loc}}$ is a free Λ-module of rank 2. The p-adic rational points $E(\mathbf{Q}_p) \otimes \mathbf{Z}_p$ is a direct summand of $H^1(\mathbf{Q}_p, T)$; hence, we can take a basis e_1, e_2 of $\mathbf{H}^1_{\mathrm{loc}}$ such that the image e_1^0 of e_1 in $H^1(\mathbf{Q}_p, T)$ is not in $E(\mathbf{Q}_p) \otimes \mathbf{Z}_p$, and the image e_2^0 of e_2 in $H^1(\mathbf{Q}_p, T)$ generates $E(\mathbf{Q}_p) \otimes \mathbf{Z}_p$. Since e_1^0 is not in $E(\mathbf{Q}_p) \otimes \mathbf{Z}_p$, $\exp^*(e_1^0) \neq 0$, and, by Lemma 1.1, $\mathrm{Col}^+(e_1)(0) \neq 0$, and $\mathrm{Col}^-(e_1)(0) \neq 0$. Since e_2^0 is in $E(\mathbf{Q}_p) \otimes \mathbf{Z}_p$, by Lemma 1.1, $\mathrm{Col}^+(e_2)(0) = \mathrm{Col}^-(e_2)(0) = 0$. We also have that $\mathrm{Col}^+(e_2)'(0) \neq 0$ and $\mathrm{Col}^-(e_2)'(0) \neq 0$. This follows from the fact that the determinant of the Λ-homomorphism $\mathrm{Col} : \mathbf{H}^1_{\mathrm{loc}} \longrightarrow \Lambda \oplus \Lambda$ is T modulo units by Proposition 1.2. We also note that $\mathrm{Col}^+(e_2)'(0) - \frac{p-1}{2}\mathrm{Col}^-(e_2)'(0) \neq 0$. Indeed, since $(\mathrm{Col}^+(e_1)/\mathrm{Col}^-(e_1))(0) = (p-1)/2$, if we had $(\mathrm{Col}^+(e_2)/\mathrm{Col}^-(e_2))(0) = (p-1)/2$, we would have

$$\mathrm{Image}(\mathrm{Col}) \cap T(\Lambda \oplus \Lambda) \subset ((p-1)/2, 1)T\Lambda + T^2(\Lambda \oplus \Lambda),$$

which contradicts Proposition 1.2.

Now we assume $(*)_0$ and $\mathrm{rank}E(\mathbf{Q}) > 0$. Let ξ, $h_\xi(T)$, $k_\xi(T)$, ... be as in the proof of Theorem 0.3. We write $\xi = a(T)e_1 + b(T)e_2$ with $a(T), b(T) \in \Lambda$. Since $\mathrm{rank}E(\mathbf{Q}) > 0$, $E(\mathbf{Q}) \otimes \mathbf{Q}_p \longrightarrow E(\mathbf{Q}_p) \otimes \mathbf{Q}_p$ is surjective. So the image of $H^1(\mathbf{Q}, T) \longrightarrow H^1(\mathbf{Q}_p, T)$ is in $E(\mathbf{Q}_p) \otimes \mathbf{Z}_p$ by Lemma 1.4 below which we will prove later. Therefore, $a(0) = 0$. Hence, $(*)_0$ implies that $b(0) \neq 0$. Thus, we get

$$h'_\xi(0) - \frac{p-1}{2} k'_\xi(0) = b(0) \left(\mathrm{Col}^+(e_2)'(0) - \frac{p-1}{2} \mathrm{Col}^-(e_2)'(0) \right) \neq 0.$$

Hence,

$$\left. \frac{f(T)}{g(T)} \right|_{T=0} = \frac{h'_\xi(0)}{k'_\xi(0)} \neq \frac{p-1}{2},$$

which completes the proof.

Our last task in this subsection is to prove the following well-known property.

Lemma 1.4 . *Let $V = T \otimes \mathbf{Q}$. The image of $H^1(\mathbf{Q}, V) \longrightarrow H^1(\mathbf{Q}_p, V)$ is a one dimensional $\mathbf{Q}_p$-vector space.*

Proof of Lemma 1.4. We first note that $H^1(\mathbf{Q}, V) = H^1(\mathbf{Z}[1/S], V)$ where S is the product of the primes of bad reduction and p. Since V is self-dual, by the Tate-Poitou duality we have an exact sequence

$$H^1(\mathbf{Q}, V) \xrightarrow{i} H^1(\mathbf{Q}_p, V) \xrightarrow{i^\vee} H^1(\mathbf{Q}, V)^\vee$$

where $i^\vee : H^1(\mathbf{Q}_p, V) = H^1(\mathbf{Q}_p, V)^\vee \longrightarrow H^1(\mathbf{Q}, V)^\vee$ is obtained as the dual of $i : H^1(\mathbf{Q}, V) \longrightarrow H^1(\mathbf{Q}_p, V)$. This shows that $\dim(\mathrm{Image}(i)) = \dim(\mathrm{Image}(i^\vee)) = \dim(\mathrm{Coker}(i))$. Thus, we obtain $\dim(\mathrm{Image}(i)) = 1$ from $\dim(H^1(\mathbf{Q}_p, V)) = 2$.

1.5. Main Conjecture. We review the main conjectures in our case. We define $\mathrm{Sel}^\pm(E/\mathbf{Q}_\infty)$ to be the Selmer group which is defined by replacing the local condition at p with $E^\pm(\mathbf{Q}_{p,\infty}) \otimes \mathbf{Q}_p/\mathbf{Z}_p$ (see the proof of Proposition 1.2). Then, the main conjecture is formulated as

$$\mathrm{char}(\mathrm{Sel}^+(E/\mathbf{Q}_\infty)^\vee) = (g(T))$$

and

$$\mathrm{char}(\mathrm{Sel}^-(E/\mathbf{Q}_\infty)^\vee) = (f(T))$$

where $(*)^\vee$ is the Pontryagin dual (Kobayashi [12]). These two conjectures are equivalent to each other ([12] Theorem 7.4) and, furthermore, each of them is equivalent to

$$\mathrm{char}(\mathrm{Sel}_0(E/\mathbf{Q}_\infty)^\vee) = \mathrm{char}(\mathbf{H}^1_{\mathrm{glob}}/ < z_K >)$$

where $\mathrm{Sel}_0(E/\mathbf{Q}_\infty)$ is defined as in 0.3. The inclusion $\supset$ is proved by using Kato's result [9] up to μ-invariants (Kobayashi [12] Theorem 1.3).

We give here a corollary of Corollary 0.6 in §0.1.

Proposition 1.5 *Assume that* $L(E,1)=0$, $\mu(f(T))=\mu(g(T))=0$, *and*

$$1=\min\{\lambda(f(T)),\lambda(g(T))\}<\max\{\lambda(f(T)),\lambda(g(T))\}.$$

Then, the Iwasawa main conjecture for E *is true; namely,* $\mathrm{char}(\mathrm{Sel}^+(E/\mathbf{Q}_\infty)^\vee)=(g(T))$ *and* $\mathrm{char}(\mathrm{Sel}^-(E/\mathbf{Q}_\infty)^\vee)=(f(T))$.

Proof. Corollary 0.6 implies that $\mathrm{Sel}(E/\mathbf{Q})_{p^\infty}$ is infinite. Hence, by the control theorem for $\mathrm{Sel}^\pm(E/\mathbf{Q}_\infty)$ ([12] Theorem 9.3), T divides the characteristic power series of $\mathrm{Sel}^\pm(E/\mathbf{Q}_\infty)^\vee$, and thus, T divides $f(T)$ and $g(T)$. Then, by our assumption, one of $(f(T))$ or $(g(T))$ equals (T). Hence, the main conjecture for one of $\mathrm{Sel}^+(E/\mathbf{Q}_\infty)$ or $\mathrm{Sel}^-(E/\mathbf{Q}_\infty)$ holds. This implies that both statements are true ([12] Theorem 7.4).

As an example, we consider the quadratic twist $E=X_0(17)_d$ as in §0.1. Then, the condition in Proposition 1.5 is satisfied by $X_0(17)_d$ for $d=29,37,40,41,44,56,65,\ldots$. For example, when $d=37$, we have

$$\mathrm{char}(\mathrm{Sel}^+(E/\mathbf{Q}_\infty)^\vee)=(T)\text{ and }\mathrm{char}(\mathrm{Sel}^-(E/\mathbf{Q}_\infty)^\vee)=((1+T)^3-1).$$

Hence, if we assume the finiteness of $\text{Ш}(E/\mathbf{Q}(\cos 2\pi/9))[3^\infty]$, we know that $\mathrm{rank}E(\mathbf{Q})=1$ and $\mathrm{rank}E(\mathbf{Q}(\cos 2\pi/9))=3$. Note that we get this conclusion just from analytic information (the computation of modular symbols).

1.6. Remark. We consider in this subsection a more general case. We assume that E has good reduction at an odd prime p. If p is ordinary, we assume that E does not have complex multiplication. Let α, β be the two roots of $x^2-a_px+p=0$ in an algebraic closure of $\mathbf{Q}_p$ where $a_p=p+1-\#E(\mathbf{F}_p)$. If p is ordinary, we take α to be a unit in $\mathbf{Z}_p$ as usual. If p is a supersingular prime, we have two p-adic L-functions $\mathcal{L}_{p,\alpha}$ and $\mathcal{L}_{p,\beta}$ by Amice-Vélu and Vishik.

If p is ordinary, $\mathcal{L}_{p,\alpha}$ is the p-adic L-function of Mazur and Swinnerton-Dyer. The other function $\mathcal{L}_{p,\beta}$ is defined in the following way. Since E does not have complex multiplication, if p is ordinary, we can write $\omega_E=e_\alpha+e_\beta$ in $D\otimes\mathbf{Q}_p(\alpha)$ where e_α (resp. e_β) is an eigenvector of the Frobenius operator φ corresponding to the eigenvalue α^{-1} (resp. β^{-1}). Perrin-Riou constructed a map which interpolates the dual exponential map (cf. [18] Theorem 3.2.3,

[9] Theorem 16.4) $\mathbf{H}^1_{\mathrm{loc}} \otimes_\Lambda \mathcal{H}_\infty \longrightarrow D \otimes_{\mathbf{Q}_p} \mathcal{H}_\infty$. The image of Kato's element z_K can be written as $\mathcal{L}_{p,\alpha} e_\alpha + \mathcal{L}_{p,\beta} e_\beta$ if p is supersingular, and the e_α component of the image of z_K is $\mathcal{L}_{p,\alpha}$ in the ordinary case. We simply define $\mathcal{L}_{p,\beta}$ by $z_K \mapsto \mathcal{L}_{p,\alpha} e_\alpha + \mathcal{L}_{p,\beta} e_\beta$ in the ordinary case.

Set

$$r = \min\{\mathrm{ord}_{T=0}\, \mathcal{L}_{p,\alpha}, \mathrm{ord}_{T=0}\, \mathcal{L}_{p,\beta}\}.$$

We conjecture that if $r > 0$,

$$\frac{\mathcal{L}^{(r)}_{p,\alpha}(0)}{(1-\frac{1}{\alpha})^2} \neq \frac{\mathcal{L}^{(r)}_{p,\beta}(0)}{(1-\frac{1}{\beta})^2} \tag{1.4}$$

which is equivalent to Conjecture 0.2 in the case $a_p = 0$. Note that if $r = 0$, we have

$$\frac{\mathcal{L}_{p,\alpha}(0)}{(1-\frac{1}{\alpha})^2} = \frac{\mathcal{L}_{p,\beta}(0)}{(1-\frac{1}{\beta})^2} = \frac{L(E,1)}{\Omega_E}.$$

So the above conjecture (1.4) asserts that $\mathcal{L}^{(r)}_{p,\alpha}(0)/(1-\alpha^{-1})^2 = \mathcal{L}^{(r)}_{p,\beta}(0)/(1-\beta^{-1})^2$ if and only if $r = 0$.

We remark that the p-adic Birch and Swinnerton-Dyer conjecture would imply $r = \mathrm{ord}_{T=0}\, \mathcal{L}_{p,\alpha} = \mathrm{ord}_{T=0}\, \mathcal{L}_{p,\beta}$, but if $\mathrm{ord}_{T=0}\, \mathcal{L}_{p,\alpha} \neq \mathrm{ord}_{T=0}\, \mathcal{L}_{p,\beta}$, then Conjecture (1.4) would follow automatically. The p-adic Birch and Swinnerton-Dyer conjecture predicts

$$\left(\frac{d}{ds}\right)^r \mathcal{L}_{p,\alpha}(\kappa(\gamma)^{s-1}-1)\big|_{s=1} = \left(1-\frac{1}{\alpha}\right)^2 \frac{\#\text{Ш}(E/\mathbf{Q})\,\mathrm{Tam}(E/\mathbf{Q})}{(\#E(\mathbf{Q})_{\mathrm{tor}})^2} R_{p,\alpha}$$

(cf. Colmez [4]) where $\kappa : \mathrm{Gal}(\mathbf{Q}_\infty/\mathbf{Q}) \longrightarrow \mathbf{Z}_p^\times$ is the cyclotomic character, and $R_{p,\alpha}$ (resp. $R_{p,\beta}$) is the p-adic α-regulator (β-regulator) of E. Hence, if we admit this conjecture, Conjecture (1.4) means that $R_{p,\alpha} \neq R_{p,\beta}$.

We now establish that $\mathcal{L}^{(r)}_{p,\alpha}(0)/(1-\alpha^{-1})^2 \neq \mathcal{L}^{(r)}_{p,\beta}(0)/(1-\beta^{-1})^2$ implies that $\#\mathrm{Sel}(E/\mathbf{Q})_{p^\infty} = \infty$. Namely, Conjecture (1.4) implies Conjecture 0.1. We know $(\mathbf{H}^1_{\mathrm{glob}})_{\mathbf{Q}} = \mathbf{H}^1_{\mathrm{glob}} \otimes \mathbf{Q}$ is a free $\Lambda_{\mathbf{Q}} = \Lambda \otimes \mathbf{Q}$-module of rank 1 (Kato [9] Theorem 12.4). Suppose that T^s divides Kato's element z_K and T^{s+1} does not divide z_K in $(\mathbf{H}^1_{\mathrm{glob}})_{\mathbf{Q}}$. We denote by ξ_i the image of z_K/T^i in $H^1(\mathbf{Q}, V)$ for $i = 1, \ldots, s$. Clearly, $\xi_1 = \ldots = \xi_{s-1} = 0$, and $\xi_s \neq 0$ because the natural map $(\mathbf{H}^1_{\mathrm{glob}})_{\mathrm{Gal}(\mathbf{Q}_\infty/\mathbf{Q})} \otimes \mathbf{Q} \longrightarrow H^1(\mathbf{Q}, V)$ is injective ($(\mathbf{H}^1_{\mathrm{glob}})_{\mathrm{Gal}(\mathbf{Q}_\infty/\mathbf{Q})}$ is the $\mathrm{Gal}(\mathbf{Q}_\infty/\mathbf{Q})$-coinvariants of $\mathbf{H}^1_{\mathrm{glob}}$). We assume $\mathcal{L}^{(r)}_{p,\alpha}(0)/(1-\alpha^{-1})^2 \neq \mathcal{L}^{(r)}_{p,\beta}(0)/(1-\beta^{-1})^2$, which implies that $L(E,1) = 0$. We will show that the image of ξ_s in $H^1(\mathbf{Q}_p, V)$ is in $H^1_f(\mathbf{Q}_p, V) = E(\mathbf{Q}_p) \otimes \mathbf{Q}_p$. Put $D^0 = \mathbf{Q}_p \omega_E$

and

$$L(i) = (1-\alpha^{-1})^{-2}\mathcal{L}_{p,\alpha}^{(i)}(0)e_\alpha + (1-\beta^{-1})^{-2}\mathcal{L}_{p,\beta}^{(i)}(0)e_\beta \in D.$$

We know that the image of ξ_i by $\exp^*$ is $L(i)$ times a non-zero element, and if ξ_i is in $H^1_f(\mathbf{Q}_p, V)$, the image of ξ_i by $\log : H^1_f(\mathbf{Q}_p, V) \longrightarrow D/D^0$ is $L(i+1)$ times a non-zero element modulo D^0 by Perrin-Riou's formulas [17] Propositions 2.1.4 and 2.2.2. (Note that Conjecture Réc(V) in [17] was proved by Colmez [3].)

Suppose that $r \le s$. By our assumption, $L(r)$ is not in D^0, hence $\log(\xi_{r-1})$ is not in D^0. Thus, ξ_{r-1} is a non-zero element in $H^1(\mathbf{Q}_p, V)$. But this is a contradiction because $\xi_{r-1} = 0$ in $H^1(\mathbf{Q}, V)$ by the definition of s. Thus, we have $r > s$. This implies that $L(s) = 0$, and hence, $\exp^*(\xi_s) = 0$. So ξ_s is in $H^1_f(\mathbf{Q}_p, V)$. Therefore, ξ_s is in the Selmer group $\mathrm{Sel}(E/\mathbf{Q}, V)$ with respect to V, and $\#\mathrm{Sel}(E/\mathbf{Q})_{p^\infty} = \infty$. This can be also obtained from Proposition 4.10 in Perrin-Riou [19].

Next, we will show that $\mathrm{rank}E(\mathbf{Q}) > 0$ and $(*)_0$ imply (1.4). These conditions imply that ξ_s is in $H^1_f(\mathbf{Q}_p, V)$ (by Lemma 1.4), and also is non-zero. Therefore, $L(s+1)$ is not in D^0. This shows that $r = s+1$ and (1.4) holds.

2 Constructing a rational point in $E(\mathbf{Q})$

We saw in the previous section that the value $(f/g)(0) - (p-1)/2$ is important to understand the Selmer group. In this section, we consider the case $r = \mathrm{ord}_{T=0} f(T) = \mathrm{ord}_{T=0} g(T) = 1$. We will see that the computation of the value $f'(0) - \frac{p-1}{2}g'(0)$ helps to produce a rational point in $E(\mathbf{Q})$ numerically. To do this, we have to compute the value $f'(0) - \frac{p-1}{2}g'(0)$ to high accuracy, which we do using the theory of overconvergent modular symbols.

2.1. Overconvergent modular symbols. Let Δ_0 denote the space of degree zero divisors on $\mathbf{P}^1(\mathbf{Q})$ which naturally are a left $\mathrm{GL}_2(\mathbf{Q})$-module under linear fractional transformations. Let $\Sigma_0(p)$ be the semigroup of matrices $\left(\begin{smallmatrix} a & b \\ c & d \end{smallmatrix}\right) \in M_2(\mathbf{Z}_p)$ such that p divides c, $\gcd(a, p) = 1$ and $ad - bc \neq 0$. If V is some right $\mathbf{Z}_p[\Sigma_0(p)]$-module, then the space $\mathrm{Hom}(\Delta_0, V)$ is naturally a right $\Sigma_0(p)$-module by

$$(\varphi|\gamma)(D) = \varphi(\gamma D)|\gamma.$$

For a congruence subgroup $\Gamma \subset \Gamma_0(p) \subset \mathrm{SL}_2(\mathbf{Z})$, we set

$$\mathrm{Symb}_\Gamma(V) = \{\varphi \in \mathrm{Hom}(\Delta_0, V) \ : \ \varphi|\gamma = \varphi\},$$

the subspace of Γ-invariant maps which we refer to as the space of *V-valued modular symbols of level Γ*.

We note that this space is naturally a Hecke module. For instance, U_p is defined by $\sum_{a=1}^{p-1}\left(\begin{smallmatrix}1 & a\\ 0 & p\end{smallmatrix}\right)$. Also, the action of the matrix $\left(\begin{smallmatrix}-1 & 0\\ 0 & 1\end{smallmatrix}\right)$ decomposes $\mathrm{Symb}_\Gamma(V)$ into plus/minus subspaces $\mathrm{Symb}_\Gamma(V)^\pm$.

If we take $V = \mathbf{Q}_p$, $\mathrm{Symb}_\Gamma(\mathbf{Q}_p)$ is the classical space of modular symbols of level Γ over $\mathbf{Q}_p$. By Eichler-Shimura theory, each eigenform f over $\mathbf{Q}_p$ of level Γ gives rise to an eigensymbol $\varphi_f^\pm$ in $\mathrm{Symb}_\Gamma(\mathbf{Q}_p)^\pm$ with the same Hecke-eigenvalues as f.

Let $\mathcal{D}(\mathbf{Z}_p)$ denote the space of (locally analytic) distributions on $\mathbf{Z}_p$. Then $\mathcal{D}(\mathbf{Z}_p)$ inherits a right $\Sigma_0(p)$-action defined by:

$$(\mu|\gamma)(f(x)) = \mu\left(f\left(\frac{b+dx}{a+cx}\right)\right).$$

Then $\mathrm{Symb}_\Gamma(\mathcal{D}(\mathbf{Z}_p))$ is Steven's space of *overconvergent modular symbols.* This space admits a Hecke-equivariant map to the space of classical modular symbols

$$\rho : \mathrm{Symb}_\Gamma(\mathcal{D}(\mathbf{Z}_p)) \longrightarrow \mathrm{Symb}_\Gamma(\mathbf{Q}_p)$$

by taking total measure. That is, $\rho(\Phi)(D) = \Phi(D)(\mathbf{1}_{\mathbf{Z}_p})$. We refer to this map as the *specialization* map.

Theorem 2.1 (Stevens) *The operator U_p is completely continuous on $\mathrm{Symb}_\Gamma(\mathcal{D}(\mathbf{Z}_p))$. Moreover, the Hecke-equivariant map*

$$\rho : \mathrm{Symb}_\Gamma(\mathcal{D}(\mathbf{Z}_p))^{(<1)} \xrightarrow{\sim} \mathrm{Symb}_\Gamma(\mathbf{Q}_p)^{(<1)}$$

is an isomorphism. Here the superscript (< 1) refers to the subspace where U_p acts with slope less than 1.

See [26] Theorem 7.1 for a proof of this theorem.

2.2. Connection to p-adic L-functions. Now consider an elliptic curve $E/\mathbf{Q}$ of level N with good supersingular reduction at p. By the Modularity theorem, there is some modular form $f = f_E$ on $\Gamma_0(N)$ corresponding to E. If α is a root of $x^2 - a_p x + p = 0$, let $f_\alpha(\tau) = f(\tau) + \alpha f(p\tau)$ denote the p-stabilization of f to level $\Gamma_0(Np)$. By Eichler-Shimura theory, there exists a Hecke-eigensymbol $\varphi_{f,\alpha} = \varphi_{f,\alpha}^+ \in \mathrm{Symb}_\Gamma(\mathbf{Q}_p)^+ \otimes \mathbf{Q}_p(\alpha)$ with the same Hecke-eigenvalues as f_α. Explicitly, we have

$$\varphi_{f,\alpha}^+(\{r\} - \{s\}) = \pi i\left(\int_s^r f_\alpha + \int_{-s}^{-r} f_\alpha\right)\Omega_E^{-1}$$

where Ω_E is the Néron period of $E/\mathbf{Q}$.

By Stevens' comparison theorem (Theorem 2.1), there exists a unique overconvergent Hecke-eigensymbol $\Phi_\alpha = \Phi_\alpha^+ \in \mathrm{Symb}_\Gamma(\mathcal{D}(\mathbf{Z}_p))^+ \otimes \mathbf{Q}_p(\alpha)$ such that $\rho(\Phi_\alpha) = \varphi_{f,\alpha}$. (Note that since $\varphi_{f,\alpha}\big|U_p = \alpha \cdot \varphi_{f,\alpha}$, the symbol $\varphi_{f,\alpha}$ has slope 1/2 and the theorem applies.)

The overconvergent symbol Φ_α is intimately connected to the p-adic L-function of E. Indeed,

$$\Phi_\alpha(\{0\} - \{\infty\}) = L_{p,\alpha}(E), \tag{2.1}$$

the p-adic L-function of E viewed as a (locally analytic) distribution on $\mathbf{Z}_p^\times$. To verify this, one uses the fact that Φ_α lifts $\varphi_{f,\alpha}$, that $\Phi_\alpha\big|U_p = \alpha \cdot \Phi_\alpha$ and that, by definition,

$$L_{p,\alpha}(E)(\mathbf{1}_{a+p^n\mathbf{Z}_p}) = \tfrac{1}{\alpha^n} \cdot \varphi_{f,\alpha}\left(\left\{\tfrac{a}{p^n}\right\} - \{\infty\}\right).$$

See [26] Theorem 8.3.

2.3. Computing p-adic L-functions. As in [21] and [6], one can use the theory of overconvergent modular symbols to very efficiently compute the p- adic L-function of an elliptic curve. Indeed, by (2.1), it suffices to compute the corresponding overconvergent modular symbol Φ_α.

To do this, we first lift $\varphi_{f,\alpha}$ to any overconvergent symbol Φ (not necessarily a Hecke-eigensymbol). Then, using Theorem 2.1, one can verify that the sequence $\{\alpha^{-n}\Phi\big|U_p^n\}$ converges to Φ_α. Thus, as long as we can efficiently compute U_p on spaces of overconvergent modular symbols, we can form approximations to the symbol Φ_α.

To actually perform such a computation, one must work modulo various powers of p and thus we must make a careful look at the denominators that are present. If $\mathcal{O}$ denotes the ring of integers of $\mathbf{Q}_p(\alpha)$, then $\varphi_{f,\alpha} \in \frac{1}{\alpha}\mathrm{Symb}_\Gamma(\mathbf{Z}_p)$. (The factor of $\frac{1}{\alpha}$ comes about from the p-stabilization of f from level N to level Np.) Let $\mathcal{D}^0(\mathbf{Z}_p)$ denote the set of distributions whose moments are all integral; that is,

$$\mathcal{D}^0(\mathbf{Z}_p) = \{\mu \in \mathcal{D}(\mathbf{Z}_p) : \mu(x^j) \in \mathbf{Z}_p\}.$$

It is then possible to lift $\varphi_{f,\alpha}$ to a symbol Φ in $\frac{1}{\alpha^2}\mathrm{Symb}_\Gamma(\mathcal{D}^0(\mathbf{Z}_p)) \otimes \mathcal{O}$.

As mentioned above, $\{\alpha^{-n}\Phi\big|U_p^n\}$ converges to Φ_α. However, the space $\mathrm{Symb}_\Gamma(\mathcal{D}^0(\mathbf{Z}_p)) \otimes \mathcal{O}$ is not preserved by the operator $\frac{1}{\alpha}U_p$. However, the subspace

$$X := \{\Phi \in \mathrm{Symb}_\Gamma(\mathcal{D}^0(\mathbf{Z}_p)) \otimes \mathcal{O} : \rho(\Phi)\big|U_p = \alpha\rho(\Phi), \rho(\Phi) \in \alpha\mathrm{Symb}_\Gamma(\mathcal{O})\}$$

is preserved by $\frac{1}{\alpha}U_p$. (Note that the two conditions defining this set are clearly preserved by this operator. The key point here is that overconvergent symbols with *integral* moments satisfying these two conditions will still have integral moments after applying $\frac{1}{\alpha}U_p$.)

Thus, to form our desired overconvergent symbol using only symbols with integral moments, we begin with $\alpha^2\varphi_{f,\alpha} \in \alpha\mathrm{Symb}_\Gamma(\mathcal{O})$ and lift this symbol to an overconvergent symbol Φ' in X. Then $\frac{1}{\alpha^n}\Phi'\big|U_p^n$ is in X for all n and converges to $\alpha^2\Phi_\alpha$.

To perform this computation on a computer, we need a method of approximating an overconvergent modular symbol with a finite amount of data. Furthermore, we must ensure that our approximations are stable under the action of $\Sigma_0(p)$ so that Hecke operators can be computed. A method of approximating distributions using "finite approximation modules" is given in [21] and [6] Section 2.4 which we will now describe.

Consider the set

$$\mathcal{F}(M) = \mathcal{O}/p^M \times \mathcal{O}/p^{M-1} \times \cdots \times \mathcal{O}/p.$$

We then have a map

$$\mathcal{D}^0(\mathbf{Z}_p) \otimes \mathcal{O} \longrightarrow \mathcal{F}(M) \text{ given by } \mu \mapsto \{\mu(x^j) \bmod p^{M-j}\}_{j=0}^{M-1}.$$

Moreover, one can check that the kernel of this map is stable under the action of $\Sigma_0(p)$. This allows us to give $\mathcal{F}(M)$ the structure of a $\Sigma_0(p)$-module.

As $\mathcal{F}(M)$ is a finite set, the space $\mathrm{Symb}_\Gamma(\mathcal{F}(M))$ can be represented on a computer. Indeed, there is a finite set of divisors such that any modular symbol of level Γ is uniquely determined by its values on these divisors. Thus, any element of $\mathrm{Symb}_\Gamma(\mathcal{F}(M))$ can be represented by a finite list of elements in $\mathcal{F}(M)$.

Moreover, since

$$\mathrm{Symb}_\Gamma(\mathcal{D}^0(\mathbf{Z}_p)) \otimes \mathcal{O} \cong \varprojlim \mathrm{Symb}_\Gamma(\mathcal{F}(M)),$$

these spaces provide a natural setting to perform computations with overconvergent modular symbols.

To compute the p-adic L-function of E, we fix some large integer M and consider the image of $\alpha^2\varphi_{f,\alpha}$ in $\mathrm{Symb}_\Gamma(\mathcal{O}/p^M)$. One can explicitly lift this symbol to a symbol $\overline{\Phi}'$ in $\mathrm{Symb}_\Gamma(\mathcal{F}(M))$. (See [21] for explicit formulas related to such liftings.)

In the ordinary case, one then proceeds by simply computing the sequence $\{\alpha^{-n}\overline{\Phi}'\big|U_p^n\}$ which will eventually stabilize to the image of $\alpha^2\Phi_\alpha$ in $\mathrm{Symb}_\Gamma(\mathcal{F}(M))$.

However, in the supersingular case, α is not a unit and thus division by α causes a loss of accuracy. The symbol $\alpha^{-2n}\overline{\Phi}'\big|U_p^{2n}$ is naturally a symbol in $\mathcal{F}(M-n)$ and is congruent to $\alpha^2\Phi_\alpha$ modulo p^n. By choosing M and n appropriately, one can then produce a U_p-eigensymbol to any given desired accuracy.

2.4. Twists. Let χ_d denote the quadratic character of conductor d and let E_d be the quadratic twist of E by χ_d. Let $\alpha^* = \chi_d(p)\alpha$ and let Φ_α be the overconvergent eigensymbol whose special value at $\{0\}-\{\infty\}$ equals $L_{p,\alpha}(E)$. If $\gcd(d, Np) = 1$, then

$$\Phi_{\alpha^*,d} := \sum_{a=1}^{|d|} \chi_d(a)\cdot\Phi_\alpha\bigg|\begin{pmatrix}1 & \frac{a}{d}\\ 0 & 1\end{pmatrix}$$

is a U_p-eigensymbol of level $\Gamma_0(d^2Np)$ with eigenvalue α^*. Moreover, if $d > 0$, then

$$\Phi_{\alpha^*,d}(\{0\}-\{\infty\}) = L_{p,\alpha^*}(E_d).$$

In particular, once we have constructed an eigensymbol that computes $L_{p,\alpha}(E)$, we can use this symbol to compute the p-adic L-function of all real quadratic twists of E. (To compute imaginary quadratic twists one needs to instead use the overconvergent symbol that corresponds to $\varphi_{f,\alpha}^-$.)

2.5. Computing derivatives of p-adic L-functions. Once the image of Φ_α has been computed in $\mathrm{Symb}_\Gamma(\mathcal{F}(M))$ for some M, by evaluating at $D = \{0\}-\{\infty\}$ we can recover the first M moments of the p-adic L-function modulo certain powers of p. As in [6] pg. 17, we can also recover

$$\int_{a+p\mathbf{Z}_p} (x-\{a\})^j \, dL_{p,\alpha}(E) \mod p^M$$

for $0 \le j \le M-1$ where $\{a\}$ denotes the Teichmüller lift of a.

Using these values one can compute the derivative of this p-adic L-function. Indeed, if we let $L_{p,\alpha}(E,s)$ be the function of $L_{p,\alpha}(E)$ in the s-variable, then

$$\begin{aligned} L'_{p,\alpha}(E,s)\Big|_{s=1} &= \int_{\mathbf{Z}_p^\times} \log_p(x)\, dL_{p,\alpha}(E) \\ &= \sum_{a=1}^{p-1}\int_{a+p\mathbf{Z}_p} \log_p(x/\{a\})\, dL_{p,\alpha}(E) \end{aligned}$$

$$= \sum_{a=1}^{p-1} \int_{a+p\mathbf{Z}_p} \sum_{j=1}^{\infty} \frac{(-1)^{j+1}}{j} (x/\{a\} - 1)^j \, dL_{p,\alpha}(E)$$

$$= \sum_{a=1}^{p-1} \sum_{j=1}^{\infty} \frac{(-1)^{j+1}}{j} \{a\}^{-j} \int_{a+p\mathbf{Z}_p} (x - \{a\})^j \, dL_{p,\alpha}(E).$$

Since $\int_{a+p\mathbf{Z}_p} (x - \{a\})^j \, dL_{p,\alpha}(E)$ is divisible by p^j, by calculating the above expression for j between 0 and $M-1$, one obtains an approximation to $L'_{p,\alpha}(E,1)$ which is correct modulo p^{M-r} where $p^r \leq M < p^{r+1}$.

Let $\mathcal{L}_{p,\alpha}(T)$ be the p-adic L-function in the T-variable as in §1. Then $\mathcal{L}_{p,\alpha}(\kappa(\gamma)^{s-1} - 1) = L_{p,\alpha}(E,s)$, and $\mathcal{L}_{p,\alpha}(T) = f(T)\log^+(T) + g(T)\log^-(T)\alpha$. Hence,

$$\begin{aligned} L'_{p,\alpha}(E,s)\Big|_{s=1} &= \mathcal{L}'_{p,\alpha}(T)\Big|_{T=0} \cdot \log(\kappa(\gamma)) \\ &= \left(f'(0)\log^+(0) + g'(0)\log^-(0)\alpha\right) \cdot \log(\kappa(\gamma)) \\ &= \frac{1}{p} \cdot (f'(0) + g'(0)\alpha) \cdot \log(\kappa(\gamma)). \end{aligned} \tag{2.2}$$

So from the computation of the derivative of $L_{p,\alpha}(E,s)$, we can easily compute the values of $f'(0)$ and $g'(0)$.

2.6. p-adic Birch and Swinnerton-Dyer conjecture. We now give a precise statement of the p-adic Birch and Swinnerton-Dyer conjecture as in Bernardi and Perrin-Riou [1]. As before, $E/\mathbf{Q}$ is an elliptic curve for which p is a good supersingular prime with $a_p = 0$. If r is the order of $L_{p,\alpha}(E/\mathbf{Q},s)$ at $s=1$, then this conjecture asserts that $r = \mathrm{rank} E(\mathbf{Q})$ and

$$\frac{1}{r!} L^{(r)}_{p,\alpha}(E/\mathbf{Q},s)\big|_{s=1} = \left(1 - \frac{1}{\alpha}\right)^2 C_p(E/\mathbf{Q}) \cdot R_{p,\alpha}(E/\mathbf{Q}) \tag{2.3}$$

where $R_{p,\alpha}(E/\mathbf{Q})$ is the p-adic α-regulator of $E/\mathbf{Q}$, and

$$C_p(E/\mathbf{Q}) = \frac{\#\text{Ш}(E/\mathbf{Q}) \cdot \mathrm{Tam}(E/\mathbf{Q})}{(\#E(\mathbf{Q})_{\mathrm{tor}})^2}$$

(cf. also [4], [19]). In the case $r=1$, we have

$$R_{p,\alpha}(E/\mathbf{Q}) = -\frac{1}{[\varphi(\omega_E), \omega_E]} \cdot \left(h_{\varphi(\omega_E)}(P) + \frac{h_{\omega_E}(P)}{p}\alpha\right) \tag{2.4}$$

where P is some generator of $E(\mathbf{Q})/E(\mathbf{Q})_{\mathrm{tor}}$, and ω_E, φ, $[\varphi(\omega_E), \omega_E]$ are as in §1. Here, for each $\nu \in D = D_{dR}(V_p(E))$, there is a p-adic height function $h_\nu : E(\mathbf{Q}) \longrightarrow \mathbf{Q}_p$ attached to ν. If $\nu = \omega_E$ and if P is in the kernel of reduction modulo p, then $h_{\omega_E}(P) = -\log_{\hat{E}}(P)^2$ where $\log_{\hat{E}} : \hat{E}(\mathbf{Q}_p) \longrightarrow \mathbf{Q}_p$ is the logarithm of the formal group $\hat{E}/\mathbf{Q}_p$ attached to E.

Assuming the p-adic Birch and Swinnerton-Dyer conjecture (with $r = 1$), equations (2.2), (2.3) and (2.4) yield

$$\left(1 - \frac{1}{\alpha}\right)^{-2} (f'(0) + g'(0)\alpha) \log(\kappa(\gamma))$$
$$= \frac{p(p-1) - 2p\alpha}{(p+1)^2} (f'(0) + g'(0)\alpha) \log(\kappa(\gamma))$$
$$= -\frac{p}{[\varphi(\omega_E), \omega_E]} \cdot \left(h_{\varphi(\omega_E)}(P) + \frac{h_{\omega_E}(P)}{p}\alpha\right) C_p(E/\mathbf{Q}).$$

Equating α-coefficients of both sides gives

$$\frac{2p\log(\kappa(\gamma))}{(p+1)^2} \cdot \left(f'(0) - \frac{p-1}{2}g'(0)\right) = \frac{h_{\omega_E}(P) \cdot C_p(E/\mathbf{Q})}{[\varphi(\omega_E), \omega_E]}.$$

Since $\hat{E}(\mathbf{Q}_p) = E^1(\mathbf{Q}_p)$ is the kernel of the reduction map $E(\mathbf{Q}_p) \longrightarrow E(\mathbf{F}_p)$, $(E(\mathbf{Q}_p) : \hat{E}(\mathbf{Q}_p)) = \#E(\mathbf{F}_p) = p + 1$, and $(p+1)P$ is in $\hat{E}(\mathbf{Q}_p)$. The above formula can be written as

$$\log_{\hat{E}}((p+1)P)^2 = -\frac{f'(0) - \frac{p-1}{2}g'(0)}{C_p(E/\mathbf{Q})} 2p\log(\kappa(\gamma))[\varphi(\omega_E), \omega_E]. \quad (2.5)$$

Note that the fact that we have two p-adic L-functions which are computable, gives us the advantage that $\log_{\hat{E}}((p+1)P)$ is expressed by computable terms.

We apply this formula to a quadratic twist of E, and have a twisted version of this equation. Let $\mathcal{L}_{p,\alpha^*}(E_d, T)$ be the function in the T-variable corresponding to $L_{p,\alpha^*}(E_d)$ defined in 2.4. We assume that the rank of $E_d(\mathbf{Q})$ equals 1 and P_d is a generator of the Mordell-Weil group modulo torsion. Also, we define $f_d(T)$ and $g_d(T)$ by

$$\mathcal{L}_{p,\alpha^*}(E_d, T) = f_d(T)\log^+(T) + g_d(T)\log^-(T)\alpha^*.$$

Then, we have a twisted equation

$$\log_{\hat{E}_d}((p+1)P_d)^2 = -\frac{\eta_d^2\left(f_d'(0) - \frac{p-1}{2}g_d'(0)\right)}{d \cdot C_p(E_d/\mathbf{Q})} 2p\log(\kappa(\gamma))[\varphi(\omega_E), \omega_E]. \quad (2.6)$$

Here, η_d is defined by $\omega_{E_d} = \frac{\eta_d}{\sqrt{d}}\omega_E$. (Note that η_d is written down explicitly in [1] pg. 230. In most cases, $\eta_d = 1$.)

2.7. Computing rational points. Using equation (2.5) we can attempt to use the p-adic L-function to p-adically compute global points on $E(\mathbf{Q})$. Suppose that we do not know $Ш(E/\mathbf{Q})$. We first compute

$$z_p(E) = \exp_{\hat{E}}\left(\sqrt{-\frac{\left(f'(0) - \frac{p-1}{2}g'(0)\right) 2p\log(\kappa(\gamma))[\varphi(\omega_E), \omega_E]}{\mathrm{Tam}(E)} \cdot \frac{\#E(\mathbf{Q})_{\mathrm{tor}}}{p+1}}\right) \tag{2.7}$$

where $\exp_{\hat{E}}$ is the inverse of $\log_{\hat{E}}$ (the formal exponential of $\hat{E}$). The quantity in the parentheses should be in $p\mathbf{Z}_p$ because of the p-adic Birch and Swinnerton -Dyer conjecture (cf. (2.5)), so the right hand side should converge. If $\hat{x}(t), \hat{y}(t)$ represent the formal x and y coordinate functions of $\hat{E}/\mathbf{Q}_p$, then

$$\tilde{P}' := (\hat{x}(z_p(E)), \hat{y}(z_p(E)))$$

is a point in $\hat{E}(\mathbf{Q}_p)$. The p-adic Birch and Swinnerton-Dyer conjecture (2.5) predicts that there is a global point $P' \in E(\mathbf{Q})$ such that

$$(p+1)P' = (p+1)\tilde{P}' \text{ in } E(\mathbf{Q}_p).$$

(The point P' should be $\sqrt{\#Ш(E/\mathbf{Q})}P$ in the terminology of (2.5)). Thus, there is a point $Q \in E(\mathbf{Q}_p)$ of order dividing $p+1$ such that $P' = \tilde{P}' + Q$. So we can proceed in the following way. We first compute p-adically all Q of order dividing $p+1$ in $E(\mathbf{Q}_p)$. Then, for each such Q, we check to see if $\tilde{P}' + Q$ appears to be a global point.

2.8. Computing in practice. We explain our method in the case of computing points on a quadratic twist of E. Considering (2.6), we define

$$z_p(E,d) := \exp_{\hat{E}}\left(\sqrt{-\frac{\left(f_d'(0) - \frac{p-1}{2}g_d'(0)\right) \cdot 2p\log(\kappa(\gamma))[\varphi(\omega_E), \omega_E]}{d\,\mathrm{Tam}(E_d)} \cdot \tau_d}\right) \tag{2.8}$$

where $\tau_d := \frac{\eta_d \#E_d(\mathbf{Q})_{\mathrm{tor}}}{p+1}$, and also

$$\tilde{P}_d' := (\hat{x}(z_p(E,d)), \hat{y}(z_p(E,d))).$$

To carry out this computation, we need to first get a good p-adic approximation of $z_p(E,d)$ and thus a p-adic approximation of $\tilde{P}_d'$. Then, we compute p-adic approximations of the $(p+1)$-torsion in $E_d(\mathbf{Q}_p)$. To find a global point, we translate $\tilde{P}_d'$ around by these torsion points with the hope of finding some rational point that is very close p-adically to one of these translates.

Fortunately, if we find a candidate global point, we can simply go back to the equation of E_d to see if the point actually sits on our curve.

The key terms that need to be computed in order to determine $\tilde{P}'_d$ are $f'_d(0)$, $g'_d(0)$, $\exp_{\hat{E}}(t)$, $\hat{x}(t)$, $\hat{y}(t)$, $\mathrm{Tam}(E_d/\mathbf{Q})$, $\#E_d(\mathbf{Q})_{\mathrm{tor}}$, and $[\varphi(\omega_E), \omega_E]$. The most difficult of these terms to compute are $f'_d(0)$ and $g'_d(0)$. We cannot use Riemann sums to approximate $f'_d(0)$ and $g'_d(0)$, since in practice we will need a fairly high level of p-adic accuracy in order to recognize global points. (To get n digits of p-adic accuracy, one needs to sum together approximately p^n modular symbols which becomes implausible for large n.) Instead, we use the theory of overconvergent modular symbols explained above to compute $f'_d(0)$ and $g'_d(0)$ to high accuracy.

For the remaining terms, computing invariants of the formal group $\hat{E}/\mathbf{Q}_p$ is standard. (We used the package [25].) The arithmetic invariants ($\mathrm{Tam}(E_d/\mathbf{Q})$ and $\#E_d(\mathbf{Q})_{\mathrm{tor}}$) of the elliptic curve E_d are easy to compute. (We used the intrinsic functions of MAGMA [14].)

Lastly, an algorithm to compute $[\varphi(\omega_E), \omega_E]$ is outlined in [1] pg. 232. However, we sidestepped this issue in the following way. For one particular twist E_d with $\text{Ш}(E_d/\mathbf{Q}) = 0$, we found a generator P_d of $E_d(\mathbf{Q})/E_d(\mathbf{Q})_{\mathrm{tor}}$ directly using `mwrank` [5]. Then, using (2.5), we can determine what the value of $[\varphi(\omega_E), \omega_E]$ should be to high accuracy since every other expression in (2.5) is computable to high accuracy. (We actually repeated this computation for several different twists to make sure that the predicted value of $[\varphi(\omega_E), \omega_E]$ was always the same.)

Lastly, to recognize the coordinates of $\tilde{P}'_d$ as rational numbers, we used the method of rational reconstruction as explained in [11] and, in practice, we used the recognition function in [7].

2.9. The computations. We performed the above described computations for the curves $X_0(17)$ and $X_0(32)$ and the prime $p = 3$. (These computations were done on William Stein's meccah cluster.) For $X_0(17)$ (resp. $X_0(32)$) we computed the associated overconvergent symbol Φ_α modulo 3^{200} (resp. 3^{100}). Because of the presence of the square root in (2.8), we then were only able to compute $\tilde{P}_d$ to 100 (resp. 50) 3-adic digits for $X_0(17)$ (resp. $X_0(32)$).

For $X_0(17)$ (resp. $X_0(32)$), we could find a global point on all quadratic twists $0 < d < 250$ except for $d = 197$ (resp. $0 < d < 150$) with $\gcd(d, 3N) = 1$. (For $X_0(17)_{197}$ one could find a point via several different methods, but our method would require a more accurate computation of overconvergent modular symbols.) We made a table of these points in §4.

Another interesting example whose d is not in the above mentioned range is $E_d = X_0(17)_d$ with $d = 328$. We found a global point

$$P' = (28069/25, 3626247/125)$$

on the curve

$$E_{328} : y^2 = x^3 - 73964x - 490717520$$

by the method which we described above. We also computed

$$\text{“}\frac{1}{2}\text{”} z_p(E,d) :=$$

$$\exp_{\hat{E}} \left(\frac{1}{2} \sqrt{ - \frac{(f_d'(0) - \frac{p-1}{2} g_d'(0)) \cdot 2p \log(\kappa(\gamma)) [\varphi(\omega_E), \omega_E]}{d \, \mathrm{Tam}(E_d)} } \cdot \tau_d \right).$$

From this local point, we produced a global point $P = (1398, -46240)$ on the curve E_{328}. This reflects well the fact that $\#\text{Ш}(E/\mathbf{Q}) = 4$ in this case. The point $P = (1398, -46240)$ should be a generator of $E(\mathbf{Q})$ modulo torsion.

3 The structure of fine Selmer groups and the gcd of $f(T)$ and $g(T)$

In this section, we will study the problem mentioned in §0.3, and the greatest common divisor of $f(T)$ and $g(T)$.

3.1. Greenberg's "conjecture" (problem). Suppose that $\mathrm{Sel}_0(E/\mathbf{Q}_\infty)$, e_n, $f(T)$, $g(T)$, ... are as in §0.3. As we explained in §0.3, we are interested in the following problem (conjecture) by Greenberg.

Problem 0.7

$$\mathrm{char}(\mathrm{Sel}_0(E/\mathbf{Q}_\infty)^\vee) = (\prod_{\substack{e_n \geq 1 \\ n \geq 0}} \Phi_n^{e_n - 1}).$$

First of all, since $E(\mathbf{Q}_{p,n}) \otimes \mathbf{Q}_p$ is a one dimensional regular representation of $G_n = \mathrm{Gal}(\mathbf{Q}_n/\mathbf{Q})$, by the definition of e_n, $\mathrm{Ker}(E(\mathbf{Q}_n) \otimes \mathbf{Q}_p \longrightarrow E(\mathbf{Q}_{p,n}) \otimes \mathbf{Q}_p)$ contains $(\mathbf{Q}_p[T]/\Phi_n(T))^{e_n - 1}$ if $e_n \geq 1$. Hence, $\mathrm{Sel}_0(E/\mathbf{Q}_n)$ contains $((\mathbf{Z}_p[T]/\Phi_n(T)) \otimes \mathbf{Q}_p/\mathbf{Z}_p)^{e_n - 1}$, and we always have

$$\mathrm{char}(\mathrm{Sel}_0(E/\mathbf{Q}_\infty)^\vee) \subset (\prod_{\substack{e_n \geq 1 \\ n \geq 0}} \Phi_n^{e_n - 1}).$$

Coates and Sujatha conjectured in [2] that $\mu(\mathrm{Sel}_0(E/\mathbf{Q}_\infty)^\vee) = 0$. In fact, they showed that $\mu^{class}_{\mathbf{Q}(E[p])} = 0$ implies $\mu(\mathrm{Sel}_0(E/\mathbf{Q}_\infty)^\vee) = 0$ ([2] Theorem 3.4) where $\mu^{class}_{\mathbf{Q}(E[p])}$ is the classical Iwasawa μ-invariant for the class group of cyclotomic $\mathbf{Z}_p$-extension of $\mathbf{Q}(E[p])$ which is the field obtained by adjoining all p-torsion points of E. The proof of this fact can be described simply and slightly differently from [2], so we give here the proof. Put $F = \mathbf{Q}(E[p])$, and denote by F_∞/F the cyclotomic $\mathbf{Z}_p$-extension. By Iwasawa [8] Theorem 2, $\mu^{class}_F = 0$ implies $H^2(O_{F_\infty}[1/S], \mathbf{Z}/p\mathbf{Z}) = H^2_{et}(\mathrm{Spec}\, O_{F_\infty}[1/S], \mathbf{Z}/p\mathbf{Z}) = 0$ (S is the product of the primes of bad reduction and p). Hence, assuming $\mu^{class}_F = 0$, we have $H^2(O_{F_\infty}[1/S], E[p]) = 0$. Since the p-cohomological dimension of $\mathrm{Spec}\, O_{F_\infty}[1/S]$ is 2, the corestriction map $H^2(O_{F_\infty}[1/S], E[p]) \longrightarrow H^2(O_{\mathbf{Q}_\infty}[1/S], E[p])$ is surjective, so we get $H^2(O_{\mathbf{Q}_\infty}[1/S], E[p]) = 0$, which implies $\mu(\mathrm{Sel}_0(E/\mathbf{Q}_\infty)^\vee) = 0$.

In the following, we assume p is supersingular and $a_p = 0$. As we explained in §0.3, a generator of $\mathrm{char}(\mathrm{Sel}_0(E/\mathbf{Q}_\infty)^\vee)$ divides both $f(T)$ and $g(T)$ (at least up to μ-invariants). Using this fact, we will prove what we mentioned in §0.3.

Proposition 3.1 *Assume that* $\mathrm{rank} E(\mathbf{Q}) = e_0$, $\min\{\lambda(f(T)), \lambda(g(T))\} = e_0$, *and* $\mu(\mathrm{Sel}_0(E/\mathbf{Q}_\infty)^\vee) = 0$. *Then,* $\mathrm{char}(\mathrm{Sel}_0(E/\mathbf{Q}_\infty)^\vee) = (T^{e_0'})$ *where* $e_0' = \max\{0, e_0 - 1\}$, *and* Problem 0.7 *holds.*

Proof. Suppose, for example, $\lambda(f(T)) = e_0$. Since one divisibility of the main conjecture was proved (cf. 1.5), this assumption implies $(f(T)) = (T^{e_0})$ as ideals of Λ. If $e_0 = 0$, then $f(T)$ is a unit. Thus, $\mathrm{Sel}_0(E/\mathbf{Q}_\infty)$ is finite and we get the conclusion. Hence, we may assume $e_0 > 0$. Then, $\mathrm{Sel}^-(E/\mathbf{Q}_\infty)^\vee \sim (\Lambda/(T))^{e_0}$ (pseudo-isomorphic), and by the control theorem for $\mathrm{Sel}^-(E/\mathbf{Q}_\infty)$ ([12] Theorem 9.3), $\mathrm{Sel}(E/\mathbf{Q})_{p^\infty}$ is of corank e_0. Since $E(\mathbf{Q}) \otimes \mathbf{Q}_p \longrightarrow E(\mathbf{Q}_p) \otimes \mathbf{Q}_p$ is surjective (because $e_0 > 0$), $\mathrm{Sel}_0(E/\mathbf{Q})$ is of corank $e_0 - 1$. By the control theorem for $\mathrm{Sel}_0(E/\mathbf{Q}_\infty)$ (cf. [13] Remark 4.4), we get $\mathrm{Sel}_0(E/\mathbf{Q}_\infty)^\vee \sim (\Lambda/(T))^{e_0-1}$, which implies the conclusion.

3.2. The gcd of $f(T)$ and $g(T)$.

We use the convention that we always express the greatest common divisor of elements in Λ of the form $p^\mu h(T)$ where $h(T)$ is a distinguished polynomial. Concerning the greatest common divisor of $f(T)$ and $g(T)$, we propose

Problem 3.2

$$\gcd(f(T), g(T)) = T^{e_0} \prod_{\substack{e_n \geq 1 \\ n \geq 1}} \Phi_n^{e_n - 1}.$$

Proposition 3.3 *Assume* $\mu(\mathrm{Sel}_0(E/\mathbf{Q}_\infty)^\vee) = 0$. *Then,* Problem 3.2 *implies* Problem 0.7.

We note that if we assume Problem 3.2, we have $\min\{\mu(f(T)), \mu(g(T))\} = 0$, hence if the Galois representation on the p-torsion points of E is surjective, $\mu(\mathrm{Sel}_0(E/\mathbf{Q}_\infty)^\vee) = 0$ holds by [9] Theorem 13.4.

Proof of Proposition 3.3. As we explained in §0.3, $\mathrm{Sel}_0(E/\mathbf{Q}_\infty) \subset \mathrm{Sel}^\pm(E/\mathbf{Q}_\infty)$ which implies $\gcd(f(T), g(T)) \in \mathrm{char}(\mathrm{Sel}_0(E/\mathbf{Q}_\infty)^\vee)$. So if Problem 3.2 is true, we have

$$(T^{e_0} \prod_{\substack{e_n \geq 1 \\ n \geq 1}} \Phi_n^{e_n - 1}) \subset \mathrm{char}(\mathrm{Sel}_0(E/\mathbf{Q}_\infty)^\vee) \subset (\prod_{\substack{e_n \geq 1 \\ n \geq 0}} \Phi_n^{e_n - 1}).$$

The rest of the proof is the same as that of Proposition 3.1. We may assume $e_0 > 0$. By the control theorem for $\mathrm{Sel}^\pm(E/\mathbf{Q}_\infty)$ ([12] Theorem 9.3) and our assumption that Problem 3.2 is true, $\mathrm{Sel}(E/\mathbf{Q})_{p^\infty}$ is of corank e_0. Hence, $\mathrm{Sel}_0(E/\mathbf{Q})$ is of corank $e_0 - 1$. Hence, if $\mathcal{F}$ is a generator of $\mathrm{char}(\mathrm{Sel}_0(E/\mathbf{Q}_\infty)^\vee)$, by the control theorem for $\mathrm{Sel}_0(E/\mathbf{Q}_\infty)$ (cf. [13] Remark 4.4), we have $\mathrm{ord}_T(\mathcal{F}) = e_0 - 1$. This implies that Problem 0.7 is true.

Next, we will assume Problem 0.7 and deduce Problem 3.2 under certain assumptions. Let $\mathbf{H}^1_{\mathrm{glob}}$ and $\mathbf{H}^1_{\mathrm{loc}}$ be as in §1. We consider the natural map $\mathbf{H}^1_{\mathrm{glob}} \longrightarrow H^1(\mathbf{Q}_n, T) \longrightarrow H^1(\mathbf{Q}_{p,n}, T) \longrightarrow H^1(\mathbf{Q}_{p,n}, T)/(\Phi_n)$, and assume

$$(*)_n \qquad \mathbf{H}^1_{\mathrm{glob}} \longrightarrow H^1(\mathbf{Q}_{p,n}, T)/(\Phi_n) \quad \text{is not the zero map}$$

for all $n \geq 0$. In particular, condition $(*)_0$ coincides with the condition we considered in §1.4.

Proposition 3.4 *We assume* $(*)_n$ *for all* $n \geq 0$ *and* Problem 0.7. *Then,*
(1) *The cokernel of the natural map* $\mathbf{H}^1_{\mathrm{glob}} \longrightarrow \mathbf{H}^1_{\mathrm{loc}}$ *is pseudo-isomorphic to* Λ.
(2) *If we also assume the Main Conjecture for* E *(cf. 1.6) and that the* p*-primary component* $\text{Ш}(E/\mathbf{Q})[p^\infty]$ *of the Tate-Shafarevich group of* $E/\mathbf{Q}$ *is finite, then* Problem 3.2 *holds.*

Proof. (1) First of all, $\mathbf{H}^1_{\text{glob}}$ is isomorphic to Λ ([9] Theorem 12.4), and as we saw in §1.4, $\mathbf{H}^1_{\text{loc}}$ is isomorphic to $\Lambda \oplus \Lambda$. If we denote by (a, b) the image of a generator of $\mathbf{H}^1_{\text{glob}} \simeq \Lambda$ in $\mathbf{H}^1_{\text{loc}} \simeq \Lambda \oplus \Lambda$, Φ_n does not divide the gcd of a and b. Indeed, if Φ_n divided the gcd of a and b, the map $\mathbf{H}^1_{\text{glob}} \longrightarrow H^1(\mathbf{Q}_{p,n}, T)/(\Phi_n)$ would be the zero map, which contradicts $(*)_n$.

On the other hand, we have an exact sequence

$$0 \longrightarrow \mathbf{H}^1_{\text{loc}}/\mathbf{H}^1_{\text{glob}} \longrightarrow \text{Sel}(E/\mathbf{Q}_\infty)^\vee_{p^\infty} \longrightarrow \text{Sel}_0(E/\mathbf{Q}_\infty)^\vee \longrightarrow 0;$$

hence, taking the Λ-torsion parts, we get an exact sequence

$$0 \longrightarrow (\mathbf{H}^1_{\text{loc}}/\mathbf{H}^1_{\text{glob}})_{\Lambda-\text{tor}} \longrightarrow (\text{Sel}(E/\mathbf{Q}_\infty)^\vee_{p^\infty})_{\Lambda-\text{tor}} \longrightarrow (\text{Sel}_0(E/\mathbf{Q}_\infty)^\vee)_{\Lambda-\text{tor}}.$$

We write $\text{char}((\mathbf{H}^1_{\text{loc}}/\mathbf{H}^1_{\text{glob}})_{\Lambda-\text{tor}}) = (\epsilon(T))$. By Wingberg [27] Corollary 2.5, we have

$$\text{char}((\text{Sel}(E/\mathbf{Q}_\infty)^\vee_{p^\infty})_{\Lambda-\text{tor}}) = \text{char}((\text{Sel}_0(E/\mathbf{Q}_\infty)^\vee)_{\Lambda-\text{tor}}).$$

Hence, by Problem 0.7, any irreducible factor of $\epsilon(T)$ is of the form Φ_n. But Φ_n does not divide the gcd of a and b, so does not divide $\epsilon(T)$. Therefore, $\text{char}((\mathbf{H}^1_{\text{loc}}/\mathbf{H}^1_{\text{glob}})_{\Lambda-\text{tor}}) = (1)$, and $(\mathbf{H}^1_{\text{loc}}/\mathbf{H}^1_{\text{glob}})_{\Lambda-\text{tor}}$ is pseudo-null. This implies $\mathbf{H}^1_{\text{loc}}/\mathbf{H}^1_{\text{glob}} \sim \Lambda$.

(2) Put $h(T) = \Pi_{n \geq 0, e_n \geq 1} \Phi_n^{e_n - 1}$. By the Main Conjecture and Problem 0.7, we have

$$\text{char}(\mathbf{H}^1_{\text{glob}}/ < z_K >) = \text{char}(\text{Sel}_0(E/\mathbf{Q}_\infty)^\vee) = (h(T)).$$

Hence, we can write $z_K = h(T)\xi$ where ξ is a generator of $\mathbf{H}^1_{\text{glob}}$. By Theorem 1.3, we have $f(T) = \text{Col}^+(\xi)h(T)$ and $g(T) = \text{Col}^-(\xi)h(T)$. We take $a = a(T)$, $b = b(T)$, e_1, e_2 as in the proof of Theorem 0.4 in §1.4, namely, $\xi = ae_1 + be_2$ with $a, b \in \Lambda$. By the proof of Proposition 3.4 (1), a is prime to b. Hence, by Proposition 1.2, the greatest common divisor of $\text{Col}^+(\xi)$ and $\text{Col}^-(\xi)$ is T or 1.

Suppose at first that $e_0 = 0$. Since $Ш(E/\mathbf{Q})[p^\infty]$ is finite, $\text{Sel}(E/\mathbf{Q})_{p^\infty}$ is finite. Hence, by the control theorem for $\text{Sel}^\pm(E/\mathbf{Q}_\infty)$, $\text{char}(\text{Sel}^\pm(E/\mathbf{Q}_\infty)^\vee) \not\subset (T)$. Hence, by the Main Conjecture, $f(0) \neq 0$ and $g(0) \neq 0$. Thus, $\text{Col}^\pm(\xi)(0) \neq 0$. Therefore, the gcd of $\text{Col}^+(\xi)$ and $\text{Col}^-(\xi)$ is 1, and the gcd of $f(T)$ and $g(T)$ is $h(T)$.

Next, suppose $e_0 > 0$. Then, $E(\mathbf{Q}) \otimes \mathbf{Q}_p \longrightarrow E(\mathbf{Q}_p) \otimes \mathbf{Q}_p$ is surjective, so the image of $H^1(\mathbf{Q}, T) \longrightarrow H^1(\mathbf{Q}_p, T)$ is in $E(\mathbf{Q}_p) \otimes \mathbf{Z}_p$ by Lemma 1.4. In particular, the image of ξ is in $E(\mathbf{Q}_p) \otimes \mathbf{Z}_p$. Recall that $\xi = a(T)e_1 + b(T)e_2$.

Since $\xi \in E(\mathbf{Q}_p) \otimes \mathbf{Z}_p$, we get $a(0) = 0$. It follows from $T \mid \mathrm{Col}^{\pm}(e_2)$ (cf. 1.4) that T divides $\mathrm{Col}^{\pm}(\xi)$. Thus, the gcd of $\mathrm{Col}^{+}(\xi)$ and $\mathrm{Col}^{-}(\xi)$ is T, and the gcd of $f(T)$ and $g(T)$ is $Th(T)$, which completes the proof.

3.3. Examples. Let $E = X_0(17)$ and consider the quadratic twist E_d. The condition of Proposition 3.1 is satisfied for all d such that $0 < d < 250$ except for $d = 104, 193, 233$.

For $d = 104$, we have $r = 2$ and $\lambda(f_d(T)) = \lambda(g_d(T)) = 4$. We know that T^2 divides both $f_d(T)$ and $g_d(T)$, and a computer computation shows that $f_d(T)$ has two addition zeroes of slope 1, and $g_d(T)$ has two addition zeroes of slope $1/2$. Hence, the greatest common divisor of $f_d(T)$ and $g_d(T)$ is T^2 as predicted by Problem 3.2. The computer computations also show that $\mu(f_d(T)) = \mu(g_d(T)) = 0$. Since the Galois representation on the 3-torsion of E_d is surjective, [9] Theorem 13.4 yields that $\mu(\mathrm{Sel}_0(E/\mathbf{Q}_\infty)^\vee) = 0$. Therefore, Proposition 3.3 applies and we have $\mathrm{char}(\mathrm{Sel}_0(E/\mathbf{Q}_\infty)^\vee) = (T)$.

For $d = 233$, we have $r = 1$ and $\min\{\lambda(f_d(T)), \lambda(g_d(T))\} = 3$. In this case, a similar computation can be done and, simply by computing slopes of the zeroes, we can conclude that the gcd of $f_d(T)$ and $g_d(T)$ is T. Again, the relevant μ-invariants are zero and thus Proposition 3.3 yields that $\mathrm{Sel}_0(E/\mathbf{Q}_\infty)$ is finite.

Finally, we consider the case $d = 193$. We have that $\lambda(f_d(T)) = \lambda(g_d(T)) = 7$. However, in this case, both power series have 6 roots of valuation $1/6$ and a simple zero at 0. Thus, simply by looking at slopes of roots, we cannot conclude that their greatest common divisor is T.

To analyze this situation more carefully, we set

$$A(T) = \frac{f_d(T)}{f_d(0)T} \cdot \log^{+}(T) \text{ and } B(T) = \frac{g_d(T)}{g_d(0)T} \cdot \frac{\log^{-}(T)}{\Phi_1(T)}$$

which are both convergent power series on the open unit disc. We divide here by $\Phi_1(T)$ so that every root of both $A(T)$ and $B(T)$ has valuation at most $1/6$.

Let π be some 6-th root of p in $\overline{\mathbf{Q}}_p$ and set

$$A'(T) = A(\pi T) \text{ and } B'(T) = B(\pi T).$$

Then both A' and B' are convergent power series on the closed unit disc and are thus in the Tate algebra

$$\mathcal{O}\langle T\rangle = \left\{ f(T) = \sum_{n=0}^{\infty} a_n T^n \ : \ |a_n|_p \to 0 \right\}$$

where $\mathcal{O} = \mathbf{Z}_p[\pi]$.

We consider the image of A' and B' in

$$\mathcal{O}\langle T\rangle/\pi^m \cong (\mathcal{O}/\pi^m\mathcal{O})[T]$$

for various m with the hopes of noticing that the image of these power series do not share a common root in this small polynomial ring.

For $m = 1$, by a computer computation, we have

$$A'(T) \equiv 1 + 2T^6 \quad\text{and}\quad B'(T) \equiv 1 + 2T^6,$$

from which we deduce nothing.

For $m = 2$, we have

$$A'(T) \equiv 1 + 2T^6 \quad\text{and}\quad B'(T) \equiv 1 + 2\pi T + 2T^6 + \pi T^7.$$

From this, we again deduce nothing as

$$A'(T) \cdot (1 + 2\pi T) \equiv B'(T).$$

For $m = 3$ though, we have

$$A'(T) \equiv 1+\pi^2T^2+2T^6 \quad\text{and}\quad B'(T) \equiv 1+2\pi T+\pi^2T^2+2T^6+\pi T^7+2\pi^2T^8.$$

Now, one computes that

$$B'(T) \equiv A'(T) \cdot (1 + 2\pi T + \pi^2T^2) + 2\pi^2T^2.$$

If $\gcd(f_d, g_d) \neq T$, then there exists some common root α in $\mathcal{O}_{\overline{\mathbf{Q}}_p}$ with valuation $1/6$. If we write $\alpha = \pi u$ with u a p-adic unit, then $A'(u) = B'(u) = 0$. But the above identity then forces that $2\pi^2u^2$ is divisible by π^3, which is impossible! Thus, $\gcd(f_d, g_d) = T$ and $\mathrm{Sel}_0(E/\mathbf{Q}_\infty)$ is finite.

4 Tables

In this section, we present two tables listing the rational points we constructed 3-adically on quadratic twists of $X_0(17)$ and $X_0(32)$. For each curve $X_0(N)$, we considered d such that $d > 0$, $\gcd(d, 3N) = 1$, and $\mathrm{rank} X_0(N)_d = 1$. In each case, we used the curve's globally minimal Weierstrass equation. For the curve $X_0(17)_d$, we included this equation in the table (by listing the coefficients of $y^2 + a_1xy + a_3y = x^3 + a_2x^2 + a_4x + a_6$). For the curve $X_0(32)_d$, the globally minimal Weierstrass equation is simply $y^2 = x^3 + 4d^2x$.

These computations were done for the curve $X_0(17)$ (resp. $X_0(32)$) using an overconvergent modular symbol that was accurate mod 3^{200} (resp. mod 3^{100}).

Table of global points P_d found on $X_0(17)_d$

d	P_d	$[a_1, a_2, a_3, a_4, a_6]$
5	$(14, -32)$	$[1, -1, 0, -17, -1734]$
28	$(\frac{9895}{81}, \frac{-878410}{729})$	$[0, 0, 0, -539, -305270]$
29	$(\frac{139064}{1225}, \frac{-46339707}{42875})$	[1,-1,0,-578,-339015]
37	$(\frac{150455134}{974169}, \frac{1539419885296}{961504803})$	[1,-1,0,-941,-704158]
40	$(190, -2400)$	$[0, 0, 0, -1100, -890000]$
41	$(\frac{1067}{4}, \frac{-34691}{8})$	$[1, -1, 1, -1156, -958144]$
44	$(\frac{49351}{441}, \frac{-2413090}{9261})$	$[0, 0, 0, -1331, -1184590]$
56	$(\frac{19244}{121}, \frac{-1480800}{1331})$	$[0, 0, 0, -2156, -2442160]$
61	$(\frac{3952}{9}, \frac{235937}{27})$	$[1, -1, 0, -2558, -3155815]$
65	$(959, 29095)$	$[1, -1, 1, -2905, -3818278]$
73	$(\frac{111991}{36}, \frac{37126789}{216})$	$[1, -1, 1, -3664, -5408852]$
88	$(\frac{2140023710080}{8942160969}, \frac{1453764842693104220}{845597567711547})$	$[0, 0, 0, -5324, -9476720]$
92	$(\frac{3242226594695}{62457409}, \frac{-5838006309203270250}{493600903327})$	$[0, 0, 0, -5819, -10828630]$
97	$(\frac{123907127}{7396}, \frac{1373877614721}{636056})$	$[1, -1, 1, -6469, -12690242]$
109	$(\frac{117267845674060}{1957797009}, \frac{-1272481700834989645855}{86626644257223})$	$[1, -1, 0, -8168, -18006955]$
113	$(\frac{23663936531}{729316}, \frac{3630080811299531}{622835864})$	$[1, -1, 1, -8779, -20063092]$
124	$(\frac{2195359}{1089}, \frac{-3243299390}{35937})$	$[0, 0, 0, -10571, -26513990]$
133	$(\frac{673327635832141}{308624691600}, \frac{-17605988238521415099109}{171453361171464000})$	$[1, -1, 0, -12161, -32713318]$
173	$(\frac{9693514595778788}{18312896333449}, \frac{-653587463274626644218393}{78367421114819312293})$	$[1, -1, 0, -20576, -71997483]$
181	$(\frac{56621266}{50625}, \frac{402821517014}{11390625})$	$[1, -1, 0, -22523, -82454830]$
184	$(\frac{33961175037484}{73808392329}, \frac{-5571288550874575360}{20052042602765733})$	$[0, 0, 0, -23276, -86629040]$
193	$(\frac{915394662845247271}{25061097283236}, \frac{-878088421712236204458830141}{125458509476191439016})$	$[1, -1, 1, -25609, -99966422]$
197	not found	$[1, -1, 0, -26681, -106311798]$
209	$(\frac{472269}{400}, \frac{303288257}{8000})$	$[1, -1, 1, -30031, -126947224]$
232	$(\frac{1106304238}{1117249}, \frac{-32569057816800}{1180932193})$	$[0, 0, 0, -37004, -173649680]$
233	$(\frac{847800975918361973}{716449376631184}, \frac{-738030759904517944421609807}{19176893823953703981248})$	$[1, -1, 1, -37324, -175895512]$
241	$(\frac{1313533}{1296}, \frac{-1347623317}{46656})$	$[1, -1, 1, -39931, -194643044]$
248	$(\frac{3841589315420}{407192041}, \frac{-7526764618576173000}{8216728195339})$	$[0, 0, 0, -42284, -212111920]$

Table of global points P_d found on $X_0(32)_d$

d	P_d
5	$(5, 25)$
13	$(\frac{13}{9}, \frac{845}{27})$
29	$(\frac{1421}{25}, \frac{76531}{125})$
37	$(\frac{37}{441}, \frac{198505}{9261})$
53	$(\frac{750533}{20449}, \frac{1987241095}{2924207})$
61	$(\frac{102541}{1521}, \frac{67889645}{59319})$
77	$(\frac{275625}{719104}, \frac{58139738475}{609800192})$
85	$(765, 21675)$
101	$(\frac{42672500}{9409}, \frac{279031013300}{912673})$
109	$(\frac{5341}{9}, \frac{415835}{27})$
133	$(\frac{314109807025}{1937936484}, \frac{338319926884539145}{85311839898648})$
149	$(\frac{43061}{49}, \frac{9435425}{343})$

Bibliography

[1] Bernardi, D. et Perrin-Riou B., Variante p-adique de la conjecture de Birch et Swinnerton-Dyer (le cas supersingulier), C. R. Acad. Sci. Paris, 317 Sér. I (1993), 227-232.

[2] Coates, J. and Sujatha, R., Fine Selmer groups of elliptic curves over p-adic Lie extensions, Math. Annalen 331 (2005) 809-839.

[3] Colmez, P., Théorie d'Iwasawa des representations de de Rham d'un corps local, Ann. of Math. 148 (1998), 485-571.

[4] Colmez, P., La conjecture de Birch et Swinnerton-Dyer p-adique, Séminaire Bourbaki 919, Astérisque 294 (2004), 251-319.

[5] Cremona, J., `mwrank`, `http://www.maths.nott.ac.uk/personal/jec/ftp/progs/mwrank.info`.

[6] Darmon, H. and Pollack, R., The efficient calculation of Stark-Heegner points via overconvergent modular symbols, Israel Journal of Mathematics, **153** (2006), 319-354.

[7] Darmon, H. and Pollack, R., Stark-Heegner point computational package, `http://www.math.mcgill.ca/darmon/programs/shp/shp.html`.

[8] Iwasawa, K., Riemann-Hurwitz formula and p-adic Galois representations for number fields, Tôhoku Math. J. 33 (1981), 263-288.

[9] Kato, K., p-adic Hodge theory and values of zeta functions of modular forms, in *Cohomologies p-adiques et applications arithmétiques III*, Astérisque 295 (2004), 117-290.

[10] Kim, B.-D., The parity conjecture for elliptic curves at supersingular reduction primes, Compos. Math. 143 (2007), 47-72.

[11] Knuth, D. E., *The art of computer programming*, Vol 2., 3rd ed, Addison-Wesley, Reading, Mass. 1997.

[12] Kobayashi, S., Iwasawa theory for elliptic curves at supersingular primes, Invent. math. 152 (2003), 1-36.

[13] Kurihara, M., On the Tate-Shafarevich groups over cyclotomic fields of an elliptic curve with supersingular reduction I, Invent. math. 149 (2002), 195-224.

[14] The MAGMA computational algebra system, `http://magma.maths.usyd.edu.au`.

[15] Mazur, B. and Tate, J., Refined conjectures of the "Birch and Swinnerton-Dyer type", Duke Math. J. 54 No 2 (1987), 711-750.

[16] Nekovář, J., On the parity of ranks of Selmer groups II, C. R. Acad. Sci. Paris Sér. I 332 (2001), 99-104.

[17] Perrin-Riou, B., Fonctions L p-adiques d'une courbe elliptique et points rationnels, Ann. Inst. Fourier 43, 4 (1993), 945-995.

[18] Perrin-Riou, B., Théorie d'Iwasawa des representations p-adiques sur un corps local, Invent. math. 115 (1994), 81-149.

[19] Perrin-Riou, B., Arithmétique des courbes elliptiques à reduction supersingulière en p, Experimental Mathematics 12 (2003), 155-186.

[20] Pollack, R., On the p-adic L-functions of a modular form at a supersingular prime, Duke Math. J. 118 (2003), 523-558.

[21] Pollack, R. and Stevens, G., Explicit computations with overconvergent modular symbols, preprint.

[22] Rubin, K., p-adic variants of the Birch and Swinnerton-Dyer conjecture with complex multiplication, Contemporary Math. 165 (1994), 71-80.

[23] Rubin, K., Euler systems and modular elliptic curves, in *Galois representations in Arithmetic Algebraic Geometry*, London Math. Soc., Lecture Note Series 254 (1998), 351-367.

[24] Skinner, C. and Urban E., Sur les déformations p-adiques des formes de Saito-Kurokawa, C. R. Acad. Sci. Paris, Sér. I 335 (2002), 581-586.

[25] Stein, W., Computing p-adic heights, `http://www.williamstein.org/talks/harvard-talk-2004-12-08/`.

[26] Stevens, G., Rigid analytic modular symbols, preprint `http://math.bu.edu/people/ghs/research.d`

[27] Wingberg, K., Duality theorems for abelian varieties over $\mathbf{Z}_p$-extensions, in *Algebraic Number Theory - in honor of K. Iwasawa*, Advanced Studies in Pure Mathematics 17 (1989), 471-492.

From the Birch and Swinnerton-Dyer Conjecture to non-commutative Iwasawa theory via the Equivariant Tamagawa Number Conjecture - a survey

Otmar Venjakob
Universität Heidelberg
Mathematisches Institut
Im Neuenheimer Feld 288
69120 Heidelberg, Germany.
otmar@mathi.uni-heidelberg.de

Introduction

This paper aims to give a survey on Fukaya and Kato's article [23] which establishes the relation between the Equivariant Tamagawa Number Conjecture (ETNC) of Burns and Flach [9] and the noncommutative Iwasawa Main Conjecture (MC) (with p-adic L-function) as formulated by Coates, Fukaya, Kato, Sujatha and the author [14]. Moreover, we compare their approach with that of Huber and Kings [24] who formulate an Iwasawa Main Conjecture (without p-adic L-functions). We do not discuss these conjectures in full generality here, in fact we are mainly interested in the case of an abelian variety defined over $\mathbb{Q}$. Nevertheless we formulate the conjectures for general motives over $\mathbb{Q}$ as far as possible. We follow closely the approach of Fukaya and Kato but our notation is sometimes inspired by [9, 24]. In particular, this article does not contain any new result, but hopefully serves as introduction to the original articles. See [47] for a more down to earth introduction to the GL_2 Main Conjecture for an elliptic curve without complex multiplication. There we had pointed out that the Iwasawa main conjecture for an elliptic curve is morally the same as the (refined) Birch and Swinnerton Dyer (BSD) Conjecture for a whole tower of number fields. The work of Fukaya and Kato makes this statement precise as we are going to explain in these notes. For the convenience of the reader we have given some of the proofs here which had been left as an exercise in [23] whenever we had the feeling that the presentation of the material becomes more transparent thereby.
Since the whole paper bears an expository style we omit a lengthy introduction and just state briefly the content of the different sections:

In section 1 we recall the fundamental formalism of (non-commutative) determinants which were introduced first by Burns and Flach to formulate equivariant versions of the TNC. In section 2 we briefly discuss the setting of (realisations of) motives as they are used to formulate the conjectures concerning their L-functions, which are defined in section 3. There, also the absolute version of the TNC is discussed, which predicts the order of vanishing of the L-function at $s = 0$, the rationality and finally the precise value of the leading coefficient at $s = 0$ up to the period and regulator. In subsection 3.1 we sketch how one retrieves the BSD conjecture in its classical formulation if one applies the TNC to the motive $h^1(A)(1)$ of an abelian variety $A/\mathbb{Q}$. Though well known to the experts this is not very explicit in the literature. In section 4 we consider a p-adic Lie extension of $\mathbb{Q}$ with Galois group G. In this context the TNC is extended to an equivariant version using the absolute version for all twists of the motive by certain representations of G. The compatibility of the ETNC with respect to Artin-Verdier/Poitou-Tate duality and the functional equation of the L-function is studied in section 5. A refinement leads to the formulation of the local ϵ-conjecture in subsection 5.1. In order to involve p-adic L-functions one has to introduce Selmer groups or better complexes. The necessary modifications of the L-function and the Galois cohomology - in a way that respects the functional equation - are described in section 6. From this the MC in the form of [14] is derived in subsection 6.2 after a short interlude concerning the new "localized K_1". In the Appendix we collect basic facts about Galois cohomology on the level of complexes.

Acknowledgements: I am very grateful to Takako Fukaya and Kazuya Kato for providing me with updated copies of their work and for answering many questions. I also would like to thank John Coates for his interest and Ramdorai Sujatha for her valuable comments. The referee is heartily acknowledged for a very careful reading and many useful suggestions. Finally I am indebted to David Burns, Pedro Luis del Angel, Matthias Flach, Annette Huber, Adrian Iovita, Bruno Kahn and Guido Kings for helpful discussions. This survey was written during a stay at Centro de Investigacion en Matematicas (CIMAT), Mexico, as a Heisenberg fellow of the Deutsche Forschungsgemeinschaft (DFG) and I want to thank these institutions for their hospitality and financial support.

1 Noncommutative determinants

The (absolute) TNC compares integral structures of Galois cohomology with values of complex L-functions. For this purpose the determinant is the adequate tool as is illustrated by the following basic

Example 1.1 Let T be a $\mathbb{Z}_p$-lattice in a finite dimensional $\mathbb{Q}_p$-vector space V and $f : T \to T$ a $\mathbb{Z}_p$-linear map which induces an automorphism of V. Then the cokernel of f is finite with cardinality $|\det(f)|_p^{-1}$ where $| - |_p$ denotes the p-adic valuation normalized as usual: $|p|_p = 1/p$. This is an immediate consequence of the elementary divisor theorem.

Since the equivariant TNC involves the action of a possibly non-commutative ring R one needs a determinant formalism over an arbitrary (associative) ring R (with unit). This can be achieved by either using virtual objects a la Deligne as Burns and Flach [9, §2] do or by Fukaya and Kato's adhoc construction [23, 1.2], both approaches lead to an equivalent description.

Let $\mathrm{P}(R)$ denote the category of finitely generated projective R-modules and $(\mathrm{P}(R), is)$ its subcategory of isomorphisms, i.e. with the same objects, but whose morphisms are precisely the isomorphisms. Then there exists a category $\mathcal{C}_R$ and a functor

$$\mathbf{d}_R : (\mathrm{P}(R), is) \to \mathcal{C}_R$$

which satisfies the following properties:

a) $\mathcal{C}_R$ has an associative and commutative product structure $(M, N) \mapsto M \cdot N$ or written just MN with unit object $\mathbf{1}_R = \mathbf{d}_R(0)$ and inverses. All objects are of the form $\mathbf{d}_R(P)\mathbf{d}_R(Q)^{-1}$ for some $P, Q \in \mathrm{P}(R)$.
b) all morphisms of $\mathcal{C}_R$ are isomorphisms, $\mathbf{d}_R(P)$ and $\mathbf{d}_R(Q)$ are isomorphic if and only if their classes in $K_0(R)$ coincide. There is an identification of groups $\mathrm{Aut}_{\mathcal{C}_R}(\mathbf{1}_R) = K_1(R)$ and $\mathrm{Mor}_{\mathcal{C}_R}(M, N)$ is either empty or a $K_1(R)$-torsor where $\alpha : \mathbf{1}_R \to \mathbf{1}_R \in K_1(R)$ acts on $\phi : M \to N$ as $\alpha\phi : M = \mathbf{1}_R \cdot M \overset{\alpha \cdot \phi}{\to} \mathbf{1}_R \cdot N = N$.
c) $\mathbf{d}_R$ preserves the "product" structures: $\mathbf{d}_R(P \oplus Q) = \mathbf{d}_R(P) \cdot \mathbf{d}_R(Q)$.

This functor can be naturally extended to complexes. Let $\mathrm{C}^p(R)$ be the category of bounded (cohomological) complexes in $\mathrm{P}(R)$ and $(\mathrm{C}^p(R), quasi)$ its subcategory of quasi-isomorphisms. For $C \in \mathrm{C}^p(R)$ we set $C^+ = \bigoplus_{i\ even} C^i$ and $C^- = \bigoplus_{i\ odd} C^i$ and define $\mathbf{d}_R(C) := \mathbf{d}_R(C^+)\mathbf{d}_R(C^-)^{-1}$ and thus we obtain a functor

$$\mathbf{d}_R : (\mathrm{C}^p(R), quasi) \to \mathcal{C}_R$$

with the following properties for all objects $C, C', C'' \in \mathrm{C}^p(R)$:

d) If $0 \to C' \to C \to C'' \to 0$ is a short exact sequence of complexes, then there is a canonical isomorphism

$$\mathbf{d}_R(C) \cong \mathbf{d}_R(C')\mathbf{d}_R(C'')$$

which we take to be an identification;

e) If C is acyclic, then the quasi-isomorphism $0 \to C$ induces a canonical isomorphism in $\mathcal{C}_R$ of the form

$$\mathbf{1}_R = \mathbf{d}_R(0) \to \mathbf{d}_R(C);$$

f) For any integer r there is a canonical morphism $\mathbf{d}_R(C[r]) \cong \mathbf{d}_R(C)^{(-1)^r}$ in $\mathcal{C}_R$ which we take to be an identification, here $C[r]$ denotes the r-fold shift of C;

g) the functor $\mathbf{d}_R$ factorizes through the image of $\mathrm{C}^p(R)$ in the category $\mathrm{D}^p(R)$ of perfect complexes (as full triangulated subcategory of the derived category $\mathrm{D}^b(R)$ of the homotopy category of bounded complexes of R-modules), and extends to $(\mathrm{D}^p(R), is)$ (uniquely up to unique isomorphisms) [1];

h) If $C \in \mathrm{D}^p(R)$ has the property that all cohomology groups $\mathrm{H}^i(C)$ belong again to $\mathrm{D}^p(R)$, then there is a canonical isomorphism

$$\mathbf{d}_R(C) \cong \prod_i \mathbf{d}_R(\mathrm{H}^i(C))^{(-1)^i}.$$

Moreover, if R' is another ring, Y a finitely generated projective R'-module endowed with a structure as right R-module such that the actions of R and R' on Y commute, then the functor $Y \otimes_R - : \mathrm{P}(R) \to P(R')$ extends to give a diagram

$$\begin{array}{ccc} (\mathrm{D}^p(R), is) & \xrightarrow{\mathbf{d}_R} & \mathcal{C}_R \\ {\scriptstyle Y\otimes^{\mathbb{L}}_R -}\downarrow & & \downarrow{\scriptstyle Y\otimes_R -} \\ (\mathrm{D}^p(R'), is) & \xrightarrow{\mathbf{d}_{R'}} & \mathcal{C}_{R'} \end{array}$$

which commutes (up to canonical isomorphism). In particular, if $R \to R'$ is a ring homomorphism and $C \in \mathrm{D}^p(R)$, we just write $\mathbf{d}_R(C)_{R'}$ for $R' \otimes_R \mathbf{d}_R(C)$.

Now let R° be the opposite ring of R. Then the functor $\mathrm{Hom}_R(-, R)$ induces an anti-equivalence between $\mathcal{C}_R$ and $\mathcal{C}_{R^\circ}$ with quasi-inverse induced by $\mathrm{Hom}_{R^\circ}(-, R^\circ)$; both functors will be denoted by $-^*$. This extends to give a diagram

1 But property d) does not in general extend to arbitrary distinguished triangles, thus from a technical point of view all constructions involving complexes will have to be made carefully avoiding this problem. We will neglect this problem but see [9] for details.

$$\begin{array}{ccc} (\mathrm{D}^p(R), is) & \xrightarrow{\mathbf{d}_R} & \mathcal{C}_R \\ \Big\downarrow{\scriptstyle \mathrm{RHom}_R(-,R)} & & \Big\downarrow{\scriptstyle -^*} \\ (\mathrm{D}^p(R^\circ), is) & \xrightarrow{\mathbf{d}_{R^\circ}} & \mathcal{C}_{R^\circ} \end{array}$$

which commutes (up to unique isomorphism); similarly we have such a commutative diagram for $\mathrm{RHom}_{R^\circ}(-, R^\circ)$.

For the handling of the determinant functor in practice the following considerations are quite important:

Remark 1.2 (i) For objects $A, B \in \mathcal{C}_R$ we often identify a morphism $f : A \to B$ with the induced morphism

$$\mathbf{1}_R = A \cdot A^{-1} \xrightarrow{f \cdot \mathrm{id}_{A^{-1}}} B \cdot A^{-1}.$$

Then for morphisms $f : A \to B$ and $g : B \to C$ in $\mathcal{C}_R$, the composition $g \circ f : A \to C$ is identified with the product $g \cdot f : \mathbf{1}_R \to C \cdot A^{-1}$ of $g : \mathbf{1}_R \to C \cdot B^{-1}$ and $f : \mathbf{1}_R \to B \cdot A^{-1}$. Also, by this identification a map $f : A \to A$ corresponds uniquely to an element in $K_1(R) = \mathrm{Aut}_{\mathcal{C}_R}(\mathbf{1}_R)$. Furthermore, for a map $f : A \to B$ in $\mathcal{C}_R$, we write $\overline{f} : B \to A$ for its inverse with respect to composition and $f^{-1} =: \overline{\mathrm{id}_{B^{-1}} \cdot \phi \cdot \mathrm{id}_{A^{-1}}} : A^{-1} \to B^{-1}$ for its inverse with respect to the multiplication in $\mathcal{C}_R$, i.e. $f \cdot f^{-1} = \mathrm{id}_{\mathbf{1}_R}$. Obviously, for a map $f : A \to A$ both inverses $\overline{f}$ and f^{-1} coincide if all maps are considered as elements of $K_1(R)$ as above.

Convention: If $f : \mathbf{1}_R \to A$ is a morphism and B an object in $\mathcal{C}_R$, then we write $B \xrightarrow{\cdot f} B \cdot A$ for the morphism $\mathrm{id}_B \cdot f$. In particular, any morphism $B \xrightarrow{f} A$ can be written as $B \xrightarrow{\cdot (\mathrm{id}_{B^{-1}} \cdot f)} A$.

(ii) The determinant of the complex $C = [P_0 \xrightarrow{\phi} P_1]$ (in degree 0 and 1) with $P_0 = P_1 = P$ is by definition $\mathbf{d}_R(C) \overset{def}{=\!=} \mathbf{1}_R$ and is defined even if ϕ is not an isomorphism (in contrast to $\mathbf{d}_R(\phi)$). But if ϕ happens to be an isomorphism, i.e. if C is acyclic, then by e) there is also a canonical map $\mathbf{1}_R \xrightarrow{acyc} \mathbf{d}_R(C)$, which is in fact nothing else than

$$\mathbf{1}_R = \mathbf{d}_R(P_1)\mathbf{d}_R(P_1)^{-1} \xrightarrow{\mathbf{d}(\phi)^{-1} \cdot \mathrm{id}_{\mathbf{d}(P_1)^{-1}}} \mathbf{d}_R(P_0)\mathbf{d}_R(P_1)^{-1} = \mathbf{d}_R(C)$$

(and which depends in contrast to the first identification on ϕ). Hence, the composite $\mathbf{1}_R \xrightarrow{acyc} \mathbf{d}_R(C) \overset{def}{=\!=} \mathbf{1}_R$ corresponds to $\mathbf{d}_R(\phi)^{-1} \in K_1(R)$ according to the first remark. In order to distinguish the above identifications between

$\mathbf{1}_R$ and $\mathbf{d}_R(C)$ we also say that C is *trivialized by the identity* when we refer to $\mathbf{d}_R(C) \overset{def}{=\!=} \mathbf{1}_R$ (or its inverse with respect to composition). For $\phi = \mathrm{id}_P$ both identifications agree obviously.

We end this section by considering the example where $R = K$ is a field and V a finite dimensional vector space over K. Then, according to [23, 1.2.4], $\mathbf{d}_K(V)$ can be identified with the highest exterior product $\bigwedge^{top} V$ of V and for an automorphism $\phi : V \to V$ the determinant $\mathbf{d}_K(\phi) \in K^\times = K_1(K)$ can be identified with the usual determinant $\det_K(\phi)$. In particular, we identify $\mathbf{1}_K = K$ with canonical basis 1. Then a map $\mathbf{1}_K \xrightarrow{\psi} \mathbf{1}_K$ corresponds uniquely to the value $\psi(1) \in K^\times$.

Remark 1.3 Note that every *finite* $\mathbb{Z}_p$-module A possesses a free resolution C as in Remark 1.2 (ii), i.e. $\mathbf{d}_{\mathbb{Z}_p}(A) \cong \mathbf{d}_{\mathbb{Z}_p}(C)^{-1} = \mathbf{1}_{\mathbb{Z}_p}$. Taking into account the above and Example 1.1 we see that modulo $\mathbb{Z}_p^\times$ the composite

$$\mathbf{1}_{\mathbb{Q}_p} \xrightarrow{acyc} \mathbf{d}_{\mathbb{Z}_p}(C)_{\mathbb{Q}_p} \overset{def}{=\!=} \mathbf{1}_{\mathbb{Q}_p}$$

corresponds to the cardinality $|A| \in \mathbb{Q}_p^\times$.

2 K-Motives over $\mathbb{Q}$

In this survey we will be mainly interested in the Tamagawa Number Conjecture and Iwasawa theory for the motive $M = h^1(E)(1)$ of an elliptic curve E or the slightly more general $M = h^1(A)(1)$ of an abelian variety A defined over $\mathbb{Q}$. But as it will be important to consider certain twists of M we also recall basic facts on the Tate motive $\mathbb{Q}(1)$ and Artin motives. We shall simply view (pure) motives in the naive sense, as being defined by a collection of realizations satisfying certain axioms, together with their motivic cohomology groups. The archetypical motive is $h^i(X)$ for a smooth projective variety X over $\mathbb{Q}$ with its obvious étale cohomology $\mathrm{H}^i_{ét}(X \times_{\mathbb{Q}} \overline{\mathbb{Q}}, \mathbb{Q}_l)$, singular cohomology $\mathrm{H}^i(X(\mathbb{C}), \mathbb{Q})$ and de Rham cohomology $\mathrm{H}^i_{dR}(X/\mathbb{Q})$, their additional structures and comparison isomorphisms. More generally, let K be a finite extension of $\mathbb{Q}$. A K-motive M over $\mathbb{Q}$, i.e. a motive over $\mathbb{Q}$ with an action of K, will be given by the following data, which for $M = h^n(X)_K$ arise by tensoring the above cohomology groups by K over $\mathbb{Q}$:

2.1 The l-adic realisation M_l of M (for every prime number l)

For a place λ of K lying above l we denote by K_λ the completion of K with respect to λ. Then M_λ is a continuous finite dimensional K_λ-linear representation of the absolute Galois group $G_\mathbb{Q}$ of $\mathbb{Q}$. We put $K_l := K \otimes_\mathbb{Q} \mathbb{Q}_l = \prod_{\lambda|l} K_\lambda$ and we denote by M_l the free K_l-module $\prod_{\lambda|l} M_\lambda$.

2.2 The Betti realisation M_B of M

Attached to M is a finite dimensional K-vector space M_B which carries an action of complex conjugation ι and a $\mathbb{Q}$-Hodge structure (of *weight* $w(M)$)

$$M_B \otimes_\mathbb{Q} \mathbb{C} \cong \bigoplus \mathcal{H}^{i,j}$$

(over $\mathbf{R}$) with $\iota\mathcal{H}^{i,j} = \mathcal{H}^{j,i}$ and $\mathcal{H}^{i,j} = 0$ if $i + j \neq w(M)$, where $\mathcal{H}^{i,j}$ are free $K_\mathbb{C} := K \otimes_\mathbb{Q} \mathbb{C} \cong \mathbb{C}^{\Sigma_K}$-modules and where Σ_K denotes the set of all embeddings $K \to \mathbb{C}$. E.g. the motive $M = h^n(X)$ has weight $w(M) = n$.

2.3 The de Rham realisation M_{dR} of M

M_{dR} is a finite dimensional K-vector space with a decreasing exhaustive filtration M^k_{dR}, $k \in \mathbb{Z}$. The quotient $t_M = M_{dR}/M^0_{dR}$ is called the *tangent space* of M.

2.4 Comparison between M_B and M_l

For each prime number l there is an isomorphism of K_l-modules

$$K_l \otimes_K M_B \xrightarrow[\cong]{g_l} M_l \tag{2.1}$$

which respects the action of complex conjugation; in particular it induces canonical isomorphisms

$$g^+_\lambda : K_\lambda \otimes_K M^+_B \cong M^+_\lambda \text{ and } g^+_l : K_l \otimes_K M^+_B \cong M^+_l. \tag{2.2}$$

Here and in what follows, for any commutative ring R and $R[G(\mathbb{C}/\mathbb{R})]$-module X we denote by X^+ and X^- the R-submodule of X on which ι acts by $+1$ and -1, respectively.

2.5 Comparison between M_B and M_{dR}

There is a $G(\mathbb{C}/\mathbb{R})$-invariant isomorphism of $K_{\mathbb{C}}$-modules

$$\mathbb{C} \otimes_{\mathbb{Q}} M_B \xrightarrow[\cong]{g_\infty} \mathbb{C} \otimes_{\mathbb{Q}} M_{dR} \tag{2.3}$$

(on the left hand side ι acts diagonally while on the right hand side only on $\mathbb{C}$) such that for all $k \in \mathbb{Z}$

$$g_\infty\Big(\bigoplus_{i \geq k} \mathcal{H}^{i,j}(M)\Big) = \mathbb{C} \otimes_{\mathbb{Q}} M^k_{dR}.$$

This induces an isomorphism

$$(\mathbb{C} \otimes_{\mathbb{Q}} M_B)^+ \cong \mathbb{R} \otimes_{\mathbb{Q}} M_{dR} \tag{2.4}$$

and the period map

$$\mathbb{R} \otimes_{\mathbb{Q}} M_B^+ \xrightarrow{\alpha_M} \mathbb{R} \otimes_{\mathbb{Q}} t_M. \tag{2.5}$$

We say that M is *critical* if this happens to be an isomorphism[2].

2.6 Comparison between M_p and M_{dR}

Let B_{dR} be the filtered field of de Rham periods with respect to $\overline{\mathbb{Q}_p}/\mathbb{Q}_p$, which is endowed with a continuous action of the absolute Galois group $G_{\mathbb{Q}_p}$ of $\mathbb{Q}_p$, and set as usual $D_{dR}(V) = (B_{dR} \otimes_{\mathbb{Q}_p} V)^{G_{\mathbb{Q}_p}}$ for a finite-dimensional $\mathbb{Q}_p$-vector space V endowed with a continuous action of $G_{\mathbb{Q}_p}$. The (decreasing) filtration B^i_{dR} of B_{dR} induces a filtration $D^i_{dR}(V) = (B^i_{dR} \otimes_{\mathbb{Q}_p} V)^{G_{\mathbb{Q}_p}}$ of $D_{dR}(V)$. Then there is an $G_{\mathbb{Q}_p}$-invariant isomorphism of filtered $K_p \otimes_{\mathbb{Q}_p} B_{dR}$-modules

$$B_{dR} \otimes_{\mathbb{Q}_p} M_p \xrightarrow[\cong]{g_{dR}} B_{dR} \otimes_{\mathbb{Q}} M_{dR} \tag{2.6}$$

which induces an isomorphism of filtered K_p-modules by taking $G_{\mathbb{Q}}$-invariants

$$D_{dR}(M_p) \xrightarrow[\cong]{g_{dR}} K_p \otimes_K M_{dR}, \tag{2.7}$$

an isomorphism of K_p-modules

$$t(M_p) := D_{dR}(M_p)/D^0_{dR}(M_p) \xrightarrow[\cong]{g^t_{dR}} K_p \otimes_K t_M \tag{2.8}$$

2 By [13, Lem. 3] M is critical if and only if one of the following equivalent conditions holds: a) both infinite Euler factors $L_\infty(M, s)$ and $L_\infty(M^*(1), -s)$ (see section 5) are holomorphic at $s = 0$, b) if $j < k$ and $\mathcal{H}^{j,k} \neq \{0\}$ then $j < 0$ and $k \geq 0$, and, in addition, if $\mathcal{H}^{k,k} \neq \{0\}$, then ι acts on this space as $+1$ if $k < 0$ and by -1 if $k \geq 0$. See also [15, Lem. 2.3] for another criterion.

and, for each place λ of K over p, an isomorphism of K_λ-vector spaces

$$t(M_\lambda) := D_{dR}(M_\lambda)/D^0_{dR}(M_\lambda) \xrightarrow[\cong]{g^t_{dR}} K_\lambda \otimes_K t_M. \tag{2.9}$$

The tensor product $M \otimes_K N$ of two K-motives is given by the data which arises from the tensor products of all realizations and their additional structures. Similarly the dual M^* of the K-motive M is given by the duals of the corresponding realizations. In particular, we denote by $M(n)$, $n \in \mathbb{Z}$, the twist of M by the $|n|$-fold tensor product $\mathbb{Q}(n) = \mathbb{Q}(1)^{\otimes n}$ of the Tate motive if $n \geq 0$ and of its dual $\mathbb{Q}(-1) = \mathbb{Q}(1)^*$ if $n < 0$. By $\det(M)$ we denote the K-motive whose realisations arise as the highest exterior products of the realisations of the K-motive M.

For the motive $M = h^i(X)(j)$ where the dimension of X is d, Poincaré duality gives a perfect pairing

$$h^i(X)(j) \times h^{2d-i}(X)(d-j) \to h^{2d}(X)(d) \cong \mathbb{Q}$$

which identifies M^* with $h^{2d-i}(X)(d-j)$. Here $\mathbb{Q} = h^0(\mathrm{Spec}(\mathbb{Q}))(0)$ denotes the trivial $\mathbb{Q}$-motive.

Example 2.1 A) The Tate motive $\mathbb{Q}(1) = h^2(\mathbb{P}^1)^*$ should be thought of as $h_1(\mathbb{G}_m)$ even though the multiplicative group $\mathbb{G}_m$ is not proper. Its l-adic realisation is the usual Tate module $\mathbb{Q}_l(1) = \mathbb{Q}_l$ on which $G_\mathbb{Q}$ acts via the cyclotomic character $\chi_l : G_\mathbb{Q} \to \mathbb{Z}_l^\times$. The action of complex conjugation on $\mathbb{Q}(1)_B = \mathbb{Q}$ is by -1, its Hodge structure is of weight $w(M) = -2$ and given by $\mathcal{H}^{-1,-1}$. The filtration $\mathbb{Q}(1)^k_{dR}$ of $\mathbb{Q}(1)_{dR} = \mathbb{Q}$ is either $\mathbb{Q}$ or 0, according as $k \leq -1$ or $k > -1$, in particular we have $t_{\mathbb{Q}(1)} = \mathbb{Q}$. Finally, g_∞ sends $1 \otimes 1$ to $2\pi i \otimes 1$ while g_{dR} sends $1 \otimes 1$ to $t \otimes 1$, where $t =$ "$2\pi i$" is the p-adic period analogous to $2\pi i$.

B) For the $\mathbb{Q}$-motive $M = h^1(A)(1)$ of an abelian variety A over $\mathbb{Q}$ we have $M_l = H^1_{ét}(A_{\bar{\mathbb{Q}}}, \mathbb{Q}_l(1)) = \mathrm{Hom}_{\mathbb{Q}_l}(V_l A, \mathbb{Q}_l(1)) \cong V_l(A^\vee)$ via the Weil pairing. More generally, the Poincare bundle on $A \times A^\vee$ induces isomorphisms $M^*(1) \cong h^1(A^\vee)(1)$ and $M \cong h^1(A^\vee)^*$, while by fixing a (very) ample symmetric line bundle on A, whose existence is granted by [33, cor. 7.2], it is sometimes convenient to identify M with $h_1(A) := h^1(A)^*$ using the hard Lefschetz theorem ([43, 1.15, Thm. 5.2 (iii)], see also [30]) (but in general better to work with the dual abelian variety $A^\vee$). Then M_l can be identified with $V_l(A)$, while $M_B = H^1(A(\mathbb{C}), \mathbb{Q})(1)$ can be identified with $H_1(A(\mathbb{C}), \mathbb{Q})$, the Hodge-decomposition (pure of weight -1) is given by $\mathcal{H}^{0,-1} = H^0(A(\mathbb{C}), \Omega^1_A) (\cong \mathrm{Hom}_\mathbb{C}(\mathrm{H}^1(A(\mathbb{C}), \Omega^0_A), \mathbb{C}))$ and $\mathcal{H}^{-1,0} = \mathrm{H}^1(A(\mathbb{C}), \Omega^0_A) (\cong \mathrm{Hom}_\mathbb{C}(\Omega^1(A), \mathbb{C}))$. Furthermore, we have $M^{-1}_{dR} = M_{dR}$,

M^0_{dR} is the image of $\Omega^1_{A/\mathbb{Q}}(A)(\cong \mathrm{H}^1(A,\Omega^0_{A/\mathbb{Q}})^*)$ and $M^1_{dR}=0$. In particular, $t_M = \mathrm{H}^1(A,\Omega^0_{A/\mathbb{Q}}) = \mathrm{Lie}(A^\vee)$ (e.g. [31, Thm. 5.11]) the Lie-algebra of $A^\vee$, can be identified with $t_{h_1(A)} = \mathrm{Hom}_{\mathbb{Q}}(\Omega^1_{A/\mathbb{Q}}(A),\mathbb{Q}) = \mathrm{Lie}(A)$. The map α_M for the motive $M = h_1(A)$, which is in fact an isomorphism, is induced by sending a 1-cycle $\gamma \in \mathrm{H}_1(A(\mathbb{C}),\mathbb{Q})^+$ to $\int_\gamma \in \mathrm{Hom}_{\mathbb{Q}}(\Omega^1_{A/\mathbb{Q}}(A),\mathbb{R}) = \mathrm{Lie}(A)_{\mathbb{R}}$ which sends a 1-form ω to $\int_\gamma \omega \in \mathbb{R}$.

C) Artin motives $[\rho]$ (with coefficients in a finite extension K of $\mathbb{Q}$) are direct summands of the K-motive $h^0(\mathrm{Spec}(F)) \otimes_{\mathbb{Q}} K$ but can also be identified with the category of finite-dimensional K-vector spaces V with an action by $G_{\mathbb{Q}}$, i.e. representations $\rho : G_{\mathbb{Q}} \to \mathrm{Aut}_K(V)$ with finite image. We write $[\rho]$ for the corresponding motive and have $[\rho]_l = V \otimes_K K_l$ with $G_{\mathbb{Q}}$ acting just on V, $[\rho]_B = V$ with Hodge-Structure pure of Type $(0,0)$ and $[\rho]_{dR} = (V \otimes_{\mathbb{Q}} \bar{\mathbb{Q}})^{G_{\mathbb{Q}}}$, where $G_{\mathbb{Q}}$ acts diagonally. Since $[\rho]^k_{dR}$ is either $[\rho]_{dR}$ or 0 according as $k \le 0$ of $k > 0$, we have $t_{[\rho]} = 0$. The inverse of g_∞ is induced by the natural inclusion $(V \otimes_{\mathbb{Q}} \bar{\mathbb{Q}})^{G_{\mathbb{Q}}} \subseteq V \otimes_{\mathbb{Q}} \bar{\mathbb{Q}}$. E.g. if ψ denotes a Dirichlet character of conductor f considered via $(\mathbb{Z}/f\mathbb{Z})^* \cong G(\mathbb{Q}(\zeta_f)/\mathbb{Q})$ as a character $G \to K^\times$ where $K = \mathbb{Q}(\zeta_{\varphi(f)})$ and φ denotes the Euler φ-function, then we obtain a basis of $[\psi]_{dR}$ over K given by the Gauss sum

$$\sum_{1\le n<f,(n,f)=1} \psi(n) \otimes e^{-2\pi i n/f} \quad \in (K(\psi) \otimes_{\mathbb{Q}} \bar{\mathbb{Q}})^{G_{\mathbb{Q}}},$$

where $K(\psi)$ denotes the 1-dimensional K-vector space on which $G_{\mathbb{Q}}$ acts via ψ.

Of course, $h^0(\mathrm{Spec}(F)) \otimes_{\mathbb{Q}} K$ corresponds to the regular representation of $G(F/\mathbb{Q})$ on $K[G(F/\mathbb{Q})]$ considered as representation of $G_{\mathbb{Q}}$.

Other examples arise by taking symmetric products or tensor products of the above examples. In particular, we will be concerned with the motives

D) $[\rho] \otimes h^1(A)(1)$, where ρ runs through all Artin representations.

E) Finally, the motive $M(f)$ of a modular form f is a prominent example, see [18, §7] and [42].

2.7 Motivic cohomology

The motivic cohomology K-vector spaces $\mathrm{H}^0_f(M) := \mathrm{H}^0(M)$ and $\mathrm{H}^1_f(M)$ may be defined by algebraic K-theory or motivic cohomology a la Voevodsky. They are conjectured to be finite dimensional. Instead of a general definition we just describe them in our standard examples.

Example 2.2 A) For the Tate motive we have $\mathrm{H}^0_f(\mathbb{Q}(1)) = \mathrm{H}^1_f(\mathbb{Q}(1)) = 0$ and for its Kummer dual $\mathrm{H}^0_f(\mathbb{Q}) = \mathbb{Q}$ while $\mathrm{H}^1_f(\mathbb{Q}) = 0$.

B) If $M = h^1(A)(1)$ for an abelian variety A over $\mathbb{Q}$ one has $\mathrm{H}^0_f(M) = 0$ and $\mathrm{H}^1_f(M) = A^\vee(\mathbb{Q}) \otimes_{\mathbb{Z}} \mathbb{Q}$.

C) For $M = h^0(\mathrm{Spec}(F))$ we have $\mathrm{H}^0_f(M) = \mathbb{Q}$ and $\mathrm{H}^1_f(M) = 0$ while for $M^*(1) = h^0(\mathrm{Spec}(F))(1)$ one has $\mathrm{H}^0_f(M^*(1)) = 0$ and $\mathrm{H}^1_f(M^*(1)) = \mathcal{O}_F^\times \otimes_{\mathbb{Z}} \mathbb{Q}$. More generally, for a K-Artin motive $[\rho]$ one has $\mathrm{H}^0_f([\rho]) = K^n$, where n is the multiplicity with which K occurs in $[\rho]$.

Unfortunately the functor H^i_f does not behave well with tensor products, i.e. in general one cannot derive $\mathrm{H}^*_f([\rho] \otimes h^1(A)(1))$ from $\mathrm{H}^*_f([\rho])$ and $\mathrm{H}^*_f(h^1(A)(1))$ (e.g. in the form of a Künneth formula).

3 The Tamagawa Number Conjecture - absolute version

In [5] Bloch and Kato formulated a vast generalization of the analytic class number formula and the BSD-conjecture. While the conjecture of Deligne and Beilinson links the order of vanishing of the L-function attached to a motive M to its motivic cohomology and claims rationality of special L-values or more general leading coefficients (up to periods and regulators) the Tamagawa number conjecture of Bloch and Kato predicts the precise L-value in terms of Galois cohomology (assuming the conjecture of Deligne-Beilinson).

Later, Fontaine and Perrin-Riou [22] found an equivalent formulation using (commutative) determinants instead of (Tamagawa) measures[3]. In this section we follow closely their approach.

Let us first recall the definition of the complex L-function attached to a K-motive M. We fix a place λ of K lying over l and an embedding $K \to \mathbb{C}$. For every prime p take a prime $l \neq p$ and set

$$P_p(M_\lambda, X) = \det_{K_\lambda}(1 - \varphi_p X | (M_\lambda)^{I_p}) \in K_\lambda[X],$$

where φ_p denotes the geometric Frobenius automorphism of p in $G_{\mathbb{Q}_p}/I_p$ and I_p is the inertia subgroup of p in $G_{\mathbb{Q}_p} \subseteq G_{\mathbb{Q}}$. It is conjectured that $P_p(X)$ belongs to $K[X]$ and is independent of the choices of l and λ. For example this is known by the work of Deligne proving the Weil conjectures for $M = h^i(X)$ for places p where X has good reduction; by the compatibility of the system of l-adic realisations for abelian varieties [21, Rem. 2.4.6(ii)] and Artin motives it is also clear for our examples A)-D). Then we have the L-function of M as Euler product

$$L_K(M, s) = \prod_p P_p(M_\lambda, p^{-s})^{-1},$$

defined and analytic for $\Re(s)$ large enough.

3 The name comes from an analogy with the theory of algebraic groups, see [5].

Example 3.1 A) The L-function $L_{\mathbb{Q}}(\mathbb{Q}(1), s-1)$ of the Tate motive is just the Riemann zeta function $\zeta(s)$. In general, one has $L_K(M(n), s) = L_K(M, s+n)$ for any K-motive M and any integer n.
B) If $M = h^1(A)(1)$ for an abelian variety A over $\mathbb{Q}$, then $L_{\mathbb{Q}}(M, s-1)$ is the classical Hasse-Weil L-function of $A^\vee$, which coincides with that for A because A and $A^\vee$ are isogenous.
C) $L_K([\rho], s)$ coincides with the usual Artin L-function of ρ, in particular we retrieve the Dedekind zeta-function $\zeta_F(s)$ as $L_{\mathbb{Q}}(h^0(\mathrm{Spec}(F)), s)$.
D) The L-functions $L_K([\rho] \otimes h^1(A)(1), s)$ will play a crucial role for the interpolation property of the p-adic L-function.

Also, the meromorphic continuation to the whole plane $\mathbb{C}$ is part of the conjectural framework. The Taylor expansion

$$L_K(M, s) = L_K^*(M) s^{r(M)} + \dots$$

defines the leading coefficient $L_K^*(M) \in \mathbb{C}^\times$ and the order of vanishing $r(M) \in \mathbb{Z}$ of $L_K(M, s)$ at $s = 0$. The aim of the conjectures to be formulated now is to express $L_K^*(M)$ and $r(M)$ in terms of motivic and Galois cohomology.

Conjecture 3.2 (Order of Vanishing; Deligne-Beilinson)

$$r(M) = \dim_K \mathrm{H}^1_f(M^*(1)) - \dim_K \mathrm{H}^0_f(M^*(1))$$

According to the remark in [19] the duals of $\mathrm{H}^i_f(M^*(1))$ should be considered as "motivic cohomology with compact support $\mathrm{H}^{2-i}_c(M)$" and thus $r(M)$ is just their Euler characteristic. This explains why the Kummer duals $M^*(1)$ are involved here.

The link between the complex world, where the values $L_K^*(M)$ live, and the p-adic world, where the Galois cohomology lives, is formed by the *fundamental line* in $\mathcal{C}_K$ following the formulation of Fontaine and Perrin-Riou [22]:

$$\begin{aligned}\Delta_K(M) := \mathbf{d}_K(\mathrm{H}^0_f(M))^{-1}\mathbf{d}_K(\mathrm{H}^1_f(M))\mathbf{d}_K(\mathrm{H}^0_f(M^*(1))^*) \\ \mathbf{d}_K(\mathrm{H}^1_f(M^*(1))^*)^{-1}\mathbf{d}_K(M_B^+)\mathbf{d}_K(t_M)^{-1}.\end{aligned}$$

The relation of $\Delta_K(M)$ with the Betti and de Rham realisation of M is given by the following

Conjecture 3.3 (Fontaine/Perrin-Riou) *There exist an exact sequence of $K_{\mathbb{R}} := \mathbb{R} \otimes_{\mathbb{Q}} K$-modules*

$$0 \longrightarrow \mathrm{H}^0_f(M)_{\mathbb{R}} \xrightarrow{c} \ker(\alpha_M) \xrightarrow{r_B^*} (\mathrm{H}^1_f(M^*(1))_{\mathbb{R}})^*$$

$$\xrightarrow{h} \mathrm{H}^1_f(M)_{\mathbb{R}} \xrightarrow{r_B} \mathrm{Coker}(\alpha_M) \xrightarrow{c^*} (\mathrm{H}^0_f(M^*(1))_{\mathbb{R}})^* \longrightarrow 0$$

where by $-_{\mathbb{R}}$ we denote the base change from $\mathbb{Q}$ to $\mathbb{R}$ (respectively from K to $K_{\mathbb{R}}$), c is the cycle class map into singular cohomology, r_B is the Beilinson regulator map and (if both $\mathrm{H}^1_f(M)$ and $\mathrm{H}^1_f(M^*(1))$ are nonzero so that M is of weight -1, then) h is a height pairing.

Example 3.4 A),C) For the motive $M = h^0(\mathrm{Spec}(F))$ the above exact sequence is just the $\mathbb{R}$-dual of the following

$$0 \longrightarrow \mathcal{O}_F^{\times} \otimes_{\mathbb{Z}} \mathbb{R} \xrightarrow{r} \mathbb{R}^{r_1} \times \mathbb{R}^{r_2} \xrightarrow{\Sigma} \mathbb{R} \longrightarrow 0\,,$$

where r is the Dirichlet(=Borel) regulator map (see [6] for a comparison of the Beilinson and Borel regulator maps) and r_1 and r_2 denote the number of real and complex places of F, respectively.
B) The Néron height pairing (see [4] and the references therein)

$$<,>: A^{\vee}(\mathbb{Q}) \otimes_{\mathbb{Z}} \mathbb{R} \times A(\mathbb{Q}) \otimes_{\mathbb{Z}} \mathbb{R} \to \mathbb{R}$$

induces an isomorphism $A^{\vee}(\mathbb{Q}) \otimes_{\mathbb{Z}} \mathbb{R} \to \mathrm{Hom}_{\mathbb{Z}}(A(\mathbb{Q}), \mathbb{R})$, the inverse of which gives the exact sequence for the motive $M = h^1(A)(1)$.

We assume this conjecture. Using property e) and change of rings of the functor $\mathbf{d}$, it induces a canonical isomorphism (period-regulator map)

$$\vartheta_\infty : K_{\mathbb{R}} \otimes_K \Delta_K(M) \cong \mathbf{1}_{K_{\mathbb{R}}}. \tag{3.1}$$

Conjecture 3.5 (Rationality; Deligne-Beilinson) *There is a unique isomorphism*

$$\zeta_K(M) : \mathbf{1}_K \to \Delta_K(M)$$

such that for every embedding $K \to \mathbb{C}$ we have

$$L_K^*(M) : \mathbf{1}_{\mathbb{C}} \xrightarrow{\zeta_K(M)_{\mathbb{C}}} \Delta_K(M)_{\mathbb{C}} \xrightarrow{(\vartheta_\infty)_{\mathbb{C}}} \mathbf{1}_{\mathbb{C}}$$

In other words, the preimage $\overline{\vartheta_\infty}(L_K^*(M))$ of $L_K^*(M)$ with respect to ϑ_∞ generates the K-vector space $\Delta_K(M)$ if the determinant functor is identified with the highest exterior product. Thus up to a period and a regulator (the determinant of ϑ_∞ with respect to a K-rational basis) the value $L^*_K(M)$ belongs to K.

The rationality enables us to relate $L^*_K(M)$ to the p-adic world which we will describe now.

Let S be a finite set of places of $\mathbb{Q}$ containing p, ∞ and the places of bad reduction of M, i.e. $U := \mathrm{Spec}(\mathbb{Z}[\frac{1}{S}])$ is an open dense subset of $\mathrm{Spec}(\mathbb{Z})$. Then we have complexes $\mathrm{R}\Gamma_c(U, M_p)$, $\mathrm{R}\Gamma_f(\mathbb{Q}, M_p)$ and $\mathrm{R}\Gamma_f(\mathbb{Q}_v, M_p)$ calculating the (global) cohomology $\mathrm{H}^i_c(U, M_p)$ with compact support, the finite part of global and local cohomology, $\mathrm{H}^i_f(\mathbb{Q}, M_p)$ and $\mathrm{H}^i_f(\mathbb{Q}_v, M_p)$, respectively, see appendix, section 7. These complexes fit into a distinguished triangle (see (7.5))

$$\mathrm{R}\Gamma_c(U, M_p) \longrightarrow \mathrm{R}\Gamma_f(\mathbb{Q}, M_p) \longrightarrow \bigoplus_{v\in S} \mathrm{R}\Gamma_f(\mathbb{Q}_v, M_p) \longrightarrow \cdot \tag{3.2}$$

On the other hand motivic cohomology specializes to the finite parts of global Galois cohomology:

Conjecture 3.6 *There are natural isomorphisms* $\mathrm{H}^0_f(M)_{\mathbb{Q}_l} \cong \mathrm{H}^0_f(\mathbb{Q}, M_l)$ *(cycle class maps) and* $\mathrm{H}^1_f(M)_{\mathbb{Q}_l} \cong \mathrm{H}^1_f(\mathbb{Q}, M_l)$ *(Chern class maps).*

Hence, as there is a duality $\mathrm{H}^i_f(\mathbb{Q}, M_l) \cong \mathrm{H}^{3-i}_f(\mathbb{Q}, M_l^*(1))^*$ for all i, this conjecture determines all cohomology groups $\mathrm{H}^i_f(\mathbb{Q}, M_l)$.

Using properties d), g) and change of rings of the determinant functor, Conjecture 3.6 for $l = p$, the canonical isomorphisms (see appendix (7.9))

$$\eta_p(M_p): \ \mathbf{1}_{K_p} \ \to \mathbf{d}_{K_p}(\mathrm{R}\Gamma_f(\mathbb{Q}_p, M_p)) \cdot \mathbf{d}_{K_p}(t(M_p)), \tag{3.3}$$

$$\eta_l(M_p): \ \mathbf{1}_{K_p} \ \to \mathbf{d}_{K_p}(\mathrm{R}\Gamma_f(\mathbb{Q}_l, M_p)),\ l \neq p, \tag{3.4}$$

and the comparison isomorphisms (2.2) and (2.8) as well as (3.2), we obtain a canonical isomorphism in $\mathcal{C}_{K_p}$ (p-adic period-regulator map)

$$\vartheta_p(M) : \Delta_K(M)_{K_p} \cong \mathbf{d}_{K_p}(\mathrm{R}\Gamma_c(U, M_p))^{-1}, \tag{3.5}$$

which induces for any place λ above p an isomorphism in $\mathcal{C}_{K_\lambda}$

$$\vartheta_\lambda(M) : \Delta_K(M)_{K_\lambda} \cong \mathbf{d}_{K_\lambda}(\mathrm{R}\Gamma_c(U, M_\lambda))^{-1}. \tag{3.6}$$

Now let T_λ be a Galois stable $\mathcal{O}_\lambda$-lattice of M_λ and $\mathrm{R}\Gamma_c(U, T_\lambda)$ its Galois cohomology with compact support, see section 7. Here, $\mathcal{O}_\lambda$ denotes the valuation ring of K_λ. Note that by Artin-Verdier/Poitou-Tate duality (see (7.6)) the "cohomology" $\mathrm{R}\Gamma_c(U, T_\lambda)$ with compact support can also be replaced by the complex $\mathrm{R}\Gamma(U, T_\lambda^*(1))^* \oplus (T_\lambda^*(1))^+$ where $\mathrm{R}\Gamma(U, T_\lambda^*(1))$ calculates as usual the global Galois cohomology with restricted ramification.

The following Conjecture, for every prime p, gives a precise description of the special L-value $L^*_K(M) \in \mathbb{C}^\times$ up to $\mathcal{O}_K^\times$, i.e. up to sign if $K = \mathbb{Q}$, where O_K denotes the ring of integers in K:

Conjecture 3.7 (Integrality; Bloch/Kato, Fontaine/Perrin-Riou) *Assume Conjecture 3.5. Then for every place λ above p there exist a (unique) isomorphism*

$$\zeta_{\mathcal{O}_\lambda}(T_\lambda) : \mathbf{1}_{\mathcal{O}_\lambda} \to \mathbf{d}_{\mathcal{O}_\lambda}(\mathrm{R}\Gamma_c(U, T_\lambda))^{-1}$$

which induces via $K_\lambda \otimes_{\mathcal{O}_\lambda} -$ the following map

$$\zeta_{\mathcal{O}_\lambda}(T_\lambda)_{K_\lambda} : \mathbf{1}_{K_\lambda} \xrightarrow{\zeta_K(M)_{K_\lambda}} \Delta_K(M)_{K_\lambda} \xrightarrow{\vartheta_\lambda(M)} \mathbf{d}_{K_\lambda}(\mathrm{R}\Gamma_c(U, M_\lambda))^{-1}. \tag{3.7}$$

If we identify again the determinant functor with the highest exterior product, this conjecture can be rephrased as follows: $\vartheta_\lambda \overline{\vartheta_\infty}(L^*_K(M))$ generates the $\mathcal{O}_\lambda$-lattice $\mathbf{d}_{\mathcal{O}_\lambda}(\mathrm{R}\Gamma_c(U, T_\lambda))^{-1}$ of $\mathbf{d}_{K_\lambda}(\mathrm{R}\Gamma_c(U, M_\lambda))^{-1}$. In other words, this generator is determined up to a unit in $\mathcal{O}_\lambda$.

It can be shown that this conjecture is independent of the choice of S and T_λ.

Example 3.8 *(Analytic class number formula)* For the motive $M = h^0(\mathrm{Spec}(F))$ we have that $r(M) = r_1 + r_2 - 1$ if $F \otimes_{\mathbb{Q}} \mathbb{R} \cong \mathbb{R}^{r_1} \times \mathbb{C}^{r_2}$ and that $L^*_{\mathbb{Q}}(M) = \frac{-|Cl(\mathcal{O}_F)|R}{|\mu(F)|}$ for the unit regulator R. Thus Conjectures 3.2, 3.5 and 3.7 are all theorems in this case!

For other known cases of these conjectures we refer the reader to the excellent survey article [19], where in particular the results of Burns-Greither [10] and Huber-Kings [25] are discussed.

Another example will be discussed in the following section.

3.1 Equivalence to classical formulation of BSD

I am very grateful to Matthias Flach for some advice concerning this section, in which we assume $p \neq 2$ for simplicity. In order to see that the above conjectures for the motive $M = h^1(A)(1)$ of an abelian variety A are equivalent to the classical formulation involving all the arithmetic invariants of A one has to consider also integral structures for the finite parts of global and local Galois cohomology. For T_p we take the Tate-module $T_p(A^\vee)$ of $A^\vee$. In particular one can define perfect complexes of $\mathbb{Z}_p$-modules $\mathrm{R}\Gamma_f(\mathbb{Q}, T_p)$ and $\mathrm{R}\Gamma_f(\mathbb{Q}_v, T_p)$

such that the analogue of (3.2) holds, see [8, §1.5]. We just state some results concerning their cohomology groups H^i_f in the following

Proposition 3.9 ([8, (1.35)-(1.37)]) *(a)(global) If the Tate-Shafarevich group* $Ш(A/\mathbb{Q})$ *is finite, then one has*

$$\mathrm{H}^0_f(\mathbb{Q},T_p)=0 \qquad \mathrm{H}^3_f(\mathbb{Q},T_p)\cong \mathrm{Hom}_{\mathbb{Z}}(A(\mathbb{Q})_{\mathrm{tors}},\mathbb{Q}_p/\mathbb{Z}_p),$$
$$\mathrm{H}^1_f(\mathbb{Q},T_p)\cong A^\vee(\mathbb{Q})\otimes_{\mathbb{Z}}\mathbb{Z}_p \qquad \mathrm{H}^i_f(\mathbb{Q},T_p)=0\, for\ \ i\neq 0,1,2,3$$

and an exact sequence of $\mathbb{Z}_p$*-modules*

$$0\longrightarrow Ш(A/\mathbb{Q})(p)\longrightarrow \mathrm{H}^2_f(\mathbb{Q},T_p)\longrightarrow \mathrm{Hom}_{\mathbb{Z}}(A(\mathbb{Q}),\mathbb{Z}_p)\longrightarrow 0.$$

(b)(local) For all primes l one has

$$\mathrm{H}^0_f(\mathbb{Q}_l,T_p)=0,\mathrm{H}^1_f(\mathbb{Q}_l,T_p)\cong A^\vee(\mathbb{Q}_l)^{\wedge p},\mathrm{H}^i_f(\mathbb{Q}_l,T_p)=0\, for\, i\neq 0,1,$$

where $A^\vee(\mathbb{Q}_l)^{\wedge p}$ denotes the p-adic completion of $A^\vee(\mathbb{Q}_l)$.

Note that one has $\mathrm{H}^i_f(-,T_p)\otimes_{\mathbb{Z}_p}\mathbb{Q}_p\cong \mathrm{H}^i_f(-,M_p)$ for both the local and global versions.

Recall from Example 2.2 B) that we have

$$\Delta_{\mathbb{Q}}(M)=\mathbf{d}_{\mathbb{Q}}(A^\vee(\mathbb{Q})\otimes_{\mathbb{Z}}\mathbb{Q}))\mathbf{d}_{\mathbb{Q}}(\mathrm{Hom}_{\mathbb{Z}}(A(\mathbb{Q}),\mathbb{Q}))^{-1}$$
$$\mathbf{d}_{\mathbb{Q}}(H_1(A^\vee(\mathbb{C}),\mathbb{Q})^+)\mathbf{d}_{\mathbb{Q}}(\mathrm{Lie}(A^\vee))^{-1}$$

In order to define the period and regulator we have to choose bases: we first fix $P_1^\vee,\ldots,P_r^\vee\in A^\vee(\mathbb{Q})$ (respectively $P_1,\ldots,P_r\in A(\mathbb{Q})$), where $r=\mathrm{rk}_{\mathbb{Z}}(A^\vee(\mathbb{Q}))=\mathrm{rk}_{\mathbb{Z}}(A(\mathbb{Q}))$, such that setting $T_{A^\vee}:=\bigoplus\mathbb{Z}P_i^\vee$ (respectively $T_A^d:=\bigoplus\mathbb{Z}P_i^d\subseteq\mathrm{Hom}_{\mathbb{Z}}(A(\mathbb{Q}),\mathbb{Z})$, where P_i^d denotes the obvious dual "basis") we obtain

$$A^\vee(\mathbb{Q})\cong A^\vee(\mathbb{Q})_{tor}\oplus T_{A^\vee} \qquad \mathrm{Hom}_{\mathbb{Z}}(A(\mathbb{Q}),\mathbb{Z})=T_A^d. \tag{3.8}$$

Similarly we fix a $\mathbb{Z}$-basis $\gamma^+=(\gamma_1^+,\ldots,\gamma_{d_+}^+)$ of $T_B^+:=H_1(A^\vee(\mathbb{C}),\mathbb{Z})^+$ and a $\mathbb{Z}$-basis $\delta=(\delta_1,\ldots,\delta_{d_+})$ of the $\mathbb{Z}$-lattice $\mathrm{Lie}_{\mathbb{Z}}(A^\vee):=\mathrm{Lie}(\mathcal{B})=\mathrm{Hom}_{\mathbb{Z}}(\Omega^1_{\mathcal{B}/\mathbb{Z}}(\mathcal{B}),\mathbb{Z})$ of $\mathrm{Lie}(A^\vee)$, respectively. Here $\mathcal{B}/\mathbb{Z}$ denotes the (smooth, but not proper) Néron model of $A^\vee$ over $\mathbb{Z}$. Thus we obtain an integral structure of $\Delta_{\mathbb{Q}}(M)$:

$$\Delta_{\mathbb{Z}}(M):=\mathbf{d}_{\mathbb{Z}}(T_{A^\vee})\mathbf{d}_{\mathbb{Z}}(T_A^d)^{-1}\mathbf{d}_{\mathbb{Z}}(T_B^+)\mathbf{d}_{\mathbb{Z}}(\mathrm{Lie}_{\mathbb{Z}}(A^\vee))^{-1} \tag{3.9}$$

together with a canonical isomorphism

$$\mathbf{1}_{\mathbb{Z}} \xrightarrow{can_{\mathbb{Z}}} \Delta_{\mathbb{Z}}(M) \tag{3.10}$$

induced by the above choices of bases.[4]

Define the period $\Omega_\infty^+(A)$ and the regulator R_A of A to be the determinant of the maps α_M and h with respect to the bases chosen above, respectively. Then Conjecture 3.5 tells us that

$$\zeta_{\mathbb{Q}}(M) = \frac{L_{\mathbb{Q}}^*(M)}{\Omega_\infty^+(A) \cdot R_A} \cdot can_{\mathbb{Q}}, \tag{3.11}$$

where $can_{\mathbb{Q}} : \mathbf{1}_{\mathbb{Q}} \to \Delta_{\mathbb{Q}}(M)$ is induced from $can_{\mathbb{Z}}$ by base change. Indeed, we have by definition of the period and regulator

$$\Omega_\infty^+(A) R_A = (\vartheta_\infty)_{\mathbb{C}} \circ (can_{\mathbb{Q}})_{\mathbb{C}}$$

and by Conjecture 3.5

$$L_{\mathbb{Q}}^*(M) = (\vartheta_\infty)_{\mathbb{C}} \circ (\zeta_{\mathbb{Q}}(M))_{\mathbb{C}}$$

in $\mathrm{Aut}_{\mathcal{C}_{\mathbb{C}}}(\mathbf{1}_{\mathbb{C}}) = \mathbb{C}^\times$ and thus $\zeta_{\mathbb{Q}}(M)$ differs from $can_{\mathbb{Q}}$ by $\frac{L_{\mathbb{Q}}^*(M)}{\Omega_\infty^+(A) \cdot R_A}$.

On the other hand, using among others property h) of the determinant functor, Proposition 3.9 and the identification $T_B^+ \otimes_{\mathbb{Z}} \mathbb{Z}_p \cong T_p^+$ (induced from (2.2)) one easily verifies that there is an isomorphism

$$\begin{aligned}\Delta_{\mathbb{Z}_p}(M) := \Delta_{\mathbb{Z}}(M)_{\mathbb{Z}_p} \cong \\ &\mathbf{d}_{\mathbb{Z}_p}(\mathrm{R}\Gamma_f(\mathbb{Q}, T_p))^{-1} \mathbf{d}_{\mathbb{Z}_p}(T_p^+) \mathbf{d}_{\mathbb{Z}_p}(\mathrm{Lie}_{\mathbb{Z}_p}(A^\vee))^{-1} \\ &\cdot \mathbf{d}_{\mathbb{Z}_p}(\text{Ш}(A/\mathbb{Q})(p)) \mathbf{d}_{\mathbb{Z}_p}(A(\mathbb{Q})(p))^{-1} \mathbf{d}_{\mathbb{Z}_p}(A^\vee(\mathbb{Q})(p))^{-1}\end{aligned}$$

where $\mathrm{Lie}_{\mathbb{Z}_p}(A^\vee) := \mathrm{Lie}_{\mathbb{Z}}(A^\vee) \otimes_{\mathbb{Z}} \mathbb{Z}_p$ is a $\mathbb{Z}_p$-sublattice of $t(M_p) \cong \mathrm{H}^1(A, \mathcal{O}_A) \otimes_{\mathbb{Q}} \mathbb{Q}_p$.

In order to compare this with the integral structure $\mathrm{R}\Gamma_c(U, T_p)$ of $\mathrm{R}\Gamma_c(U, M_p)$ we have to introduce the *local Tamagawa numbers* $c_l(M_p)$ [22, I §4].

We first assume $l \neq p$ and we write $\phi_l \in G(\bar{\mathbb{F}}_l/\mathbb{F}_l)$ for the geometric Frobenius automorphism at l. Then there is an exact sequence (cf. [22, I §4.2])

$$0 \longrightarrow T_p^{I_l} \xrightarrow{1-\phi_l} T_p^{I_l} \longrightarrow \mathrm{H}_f^1(\mathbb{Q}_l, T_p) \longrightarrow \mathrm{H}^1(I_l, T_p)_{\mathrm{tors}}^{G_{\mathbb{Q}_l}} \longrightarrow 0$$

4 The choice of the basis $P_i^\vee$ of $T_{A^\vee}$ induces a map $\mathbb{Z}^r \to T_{A^\vee}$ and, taking determinants, the isomorphism $can_{P^\vee} : \mathbf{d}_{\mathbb{Z}}(\mathbb{Z}^r) \to \mathbf{d}_{\mathbb{Z}}(T_{A^\vee})$. Similarly, we obtain canonical isomorphisms can_{P^d}, can_{γ^+} and can_δ for T_A^d, T_B^+ and $\mathrm{Lie}_{\mathbb{Z}}(A^\vee)$, respectively. Set $can_{\mathbb{Z}} := can_{P^\vee} \cdot can_{P^d}^{-1} \cdot can_{\gamma^+} \cdot can_\delta^{-1}$.

which induces an isomorphism

$$\psi_l : \mathbf{1}_{\mathbb{Z}_p} \to \mathbf{d}_{\mathbb{Z}_p}([\, T_p^{I_l} \overset{1-\phi_l}{\longrightarrow} T_p^{I_l} \,]) \cong$$
$$\mathbf{d}_{\mathbb{Z}_p}(\mathrm{R}\Gamma_f(\mathbb{Q}_l, T_p))\mathbf{d}_{\mathbb{Z}_p}(\mathrm{H}^1(I_l, T_p)_{\mathrm{tors}}^{G_{\mathbb{Q}_l}})^{-1} \cong \mathbf{d}_{\mathbb{Z}_p}(\mathrm{R}\Gamma_f(\mathbb{Q}_l, T_p)).$$

Here the first map arises as trivialization by the identity, the second comes from the above exact sequence (interpreted as short exact sequence of complexes) while the last comes again from trivializing by the identity according to Remark 1.3.

We define $c_l(M_p) := |\mathrm{H}^1(I_l, T_p)_{\mathrm{tors}}^{G_{\mathbb{Q}_l}}|$ and remark that $(\psi_l)_{\mathbb{Q}_p}$ differs from $\eta_l(M_p)$ precisely by the map $\mathbf{1}_{\mathbb{Q}_p} \overset{acyc}{\longrightarrow} \mathbf{d}_{\mathbb{Z}_p}(\mathrm{H}^1(I_l, T_p)_{\mathrm{tors}}^{G_{\mathbb{Q}_l}})_{\mathbb{Q}_p} \overset{def}{=\!=} \mathbf{1}_{\mathbb{Q}_p}$, which appealing to Remark 1.3 we also denote by $c_l(M_p)$. In other words we have

$$(\psi_l)_{\mathbb{Q}_p} = c_l(M_p) \cdot \eta_l(M_p). \tag{3.12}$$

Note also, that one has $|A^\vee(\mathbb{Q}_l) \otimes_{\mathbb{Z}} \mathbb{Z}_p| = |P_l(M_p, 1)|_p^{-1} \cdot c_l(M_p)$ and that $c_l(M_p) = 1$ whenever A has good reduction at l.

It can be shown [44, Exp. IX,(11.3.8)] that $c_l(M_p)$ is the order of the p-primary part of the group of $\mathbb{F}_l$-rational components $(\mathcal{E}/\mathcal{E}^0)(\mathbb{F}_l) \cong \mathcal{E}(\mathbb{F}_l)/\mathcal{E}^0(\mathbb{F}_l) \cong A(\mathbb{Q}_l)/A_0(\mathbb{Q}_l)$ of the special fibre $\mathcal{E} := \mathcal{A}_{\mathbb{F}_l}$ of the smooth (but not necessarily proper) Néron model $\mathcal{A}$ of A over $\mathbb{Z}$.[5] Here we write $\mathcal{G}^0$ for the identity component of an algebraic group $\mathcal{G}$ while $A_0(\mathbb{Q}_l)$ denotes the image of $\mathcal{A}^0(\mathbb{Z}_l)$ under the canonical isomorphism $\mathcal{A}(\mathbb{Z}_l) \cong A(\mathbb{Q}_l)$. For an elliptic curve E over $\mathbb{Q}$ this group can be identified with the set of points $\mathcal{W}^0(\mathbb{Z}_l)$ of $\mathcal{W}(\mathbb{Z}_l) \cong E(\mathbb{Q}_l)$ with non-singular reduction, where $\mathcal{W} \subseteq \mathbb{P}^2_{\mathbb{Z}}$ is the closed subscheme defined by a minimal Weierstrass equation over $\mathbb{Z}$ for $E/\mathbb{Q}$ and $\mathcal{W}^0/\mathbb{Z}$ is the smooth part of $\mathcal{W}/\mathbb{Z}$ (cf. [46, IV cor. 9.1,9.2]).

Now let $l = p$. Similarly one defines maps (both depending on the choice of δ)

$$\psi_p : \mathbf{1}_{\mathbb{Z}_p} \to \mathbf{d}_{\mathbb{Z}_p}(\mathrm{R}\Gamma_f(\mathbb{Q}_p, T_p))\mathbf{d}_{\mathbb{Z}_p}(\mathrm{Lie}_{\mathbb{Z}_p}(A^\vee)),$$

$$c_p(M_p) : \mathbf{1}_{\mathbb{Q}_p} \to \mathbf{1}_{\mathbb{Q}_p},$$

5 The first isomorphism is a consequence of the theorem of Lang [32] that the map $x \mapsto \phi(x)x^{-1}$ on the $\overline{k}$-rational points of a connected algebraic group over a finite field k (with Frobenius ϕ) is surjective.

such that

$$(\psi_p)_{\mathbb{Q}_p} = c_p(M_p) \cdot \eta_p(M_p)$$

holds.[6]

Using the integral version of (3.2), the maps ψ_l (analogously as η_l for ϑ_p (3.5)) induce a canonical map

$$\kappa_p : \Delta_{\mathbb{Z}_p}(M) \to \mathbf{d}_{\mathbb{Z}_p}(\mathrm{R}\Gamma_c(\mathbb{Q}, T_p))^{-1}$$

where all of the terms $\mathbf{d}_{\mathbb{Z}_p}(\text{Ш}(A/\mathbb{Q})(p))$, $\mathbf{d}_{\mathbb{Z}_p}(A(\mathbb{Q})(p))^{-1}$ and $\mathbf{d}_{\mathbb{Z}_p}(A^\vee(\mathbb{Q})(p))^{-1}$ are trivialized by the identity. Hence, using again Remark 1.3, we have

$$(\kappa_p)_{\mathbb{Q}_p} = \frac{|\text{Ш}(A/\mathbb{Q})|}{|A(\mathbb{Q})_{\mathrm{tors}}||A^\vee(\mathbb{Q})_{\mathrm{tors}}|} \prod c_l(M_p) \cdot \vartheta_p \tag{3.13}$$

modulo $\mathbb{Z}_p^\times$. Since $\zeta_{\mathbb{Z}_p}(T_p)$ equals $\kappa_p \circ can_{\mathbb{Z}_p}$ up to an element in $\mathbb{Z}_p^\times$, it follows immediately from (3.13),(3.11) and (3.7) that

$$\frac{L^*_{\mathbb{Q}}(M)}{\Omega^+_\infty(A) \cdot R_A} \sim \frac{|\text{Ш}(A/\mathbb{Q})|}{|A(\mathbb{Q})_{\mathrm{tors}}||A^\vee(\mathbb{Q})_{\mathrm{tors}}|} \prod c_l(M_p) \mod \mathbb{Z}_p^\times. \tag{3.14}$$

6 Assume that A has dimension d and let $\widehat{B}$ be the formal group of $A^\vee$ over $\mathbb{Z}_p$, i.e. the formal completion of $\mathcal{B}$ along the zero-section in the fibre over p. Note that $\mathrm{Lie}_{\mathbb{Z}_p}(A^\vee)$ can be identified with the tangent space $t_{\widehat{B}}(\mathbb{Z}_p)$ of $\widehat{B}$ with values in $\mathbb{Z}_p$ (a good reference for formal groups is [17]). Furthermore we write $\widehat{\mathbb{G}}_a$ for the formal additive group over $\mathbb{Z}_p$, $\mathcal{B}^\circ$ and $\widetilde{B}^\circ$ for the connected component of the identity of $\mathcal{B}$ and its fibre $\widetilde{B}$ over p, respectively, and Φ for the group of connected components of $\widetilde{B}$. Again by Lang's theorem we have $\Phi(\mathbb{F}_p) = \widetilde{B}(\mathbb{F}_p)/\widetilde{B}^\circ(\mathbb{F}_p)$. Moreover there are exact sequences

$$0 \longrightarrow \mathcal{B}^\circ(\mathbb{Z}_p) \longrightarrow \mathcal{B}(\mathbb{Z}_p) \longrightarrow \Phi(\mathbb{F}_p) \longrightarrow 0$$

and

$$0 \longrightarrow \widehat{B}(\mathbb{Z}_p) \longrightarrow \mathcal{B}^\circ(\mathbb{Z}_p) \longrightarrow \widetilde{B}^\circ(\mathbb{F}_p) \longrightarrow 0.$$

Now the logarithm map

$$\mathrm{Lie}_{\mathbb{Z}_p}(A^\vee) \supseteq (p\mathbb{Z}_p)^d = \widehat{\mathbb{G}}_a(\mathbb{Z}_p)^d \xleftarrow[\log]{\cong} \widehat{B}(\mathbb{Z}_p) \subseteq A^\vee(\mathbb{Q}_p)^{\wedge p} = \mathrm{H}^1_f(\mathbb{Q}_p, T_p)$$

induces the map ψ_p by trivializing all finite subquotients of the above line by the identity. Note that the first subquotient on the left has order p^d. Using [5, ex. 3.11], which says that the Bloch-Kato exponential map coincides, up to the identification induced by the Kummer map, with the usual exponential map of the corresponding formal group, it is easy to see that $c_p(M_p) := \eta_p^{-1} \cdot (\psi_p)_{\mathbb{Q}_p} = \bar{\eta_p} \circ (\psi_p)_{\mathbb{Q}_p}$ equals modulo $\mathbb{Z}_p^\times$

$$c_p(M_p) \quad = \quad p^{-d}|P(M_p,1)|_p^{-1}\#\widetilde{B}^\circ(\mathbb{F}_p)(p)\#\Phi(\mathbb{F}_p)(p) = \#\Phi(\mathbb{F}_p)(p),$$

where we used the relation $|P(M_p,1)|_p = |P(M_l,1)|_p = p^{-d}\#\widetilde{B}^\circ(\mathbb{F}_p)(p)$. For elliptic curves this is well known [45, appendix §16], the general case is an exercise using the description of the reduction of abelian varieties in [44, Exp. IX].

For all primes p this implies the classical statement of the BSD-Conjecture up to sign (and a power of 2 due to our restriction $p \neq 2$).

4 The TNC - equivariant version

The first equivariant version of the TNC with commutative coefficients (other than number fields) was given by Kato [28, 27] observing that classical Iwasawa theory is, roughly speaking, nothing else than the ETNC for a "big" coefficient ring. Inspired by Kato's work Burns and Flach formulated an ETNC where the coefficients of the motive are allowed to be (possibly non-commutative) finite-dimensional $\mathbb{Q}$-algebras, using for the first time the general determinant functor described in section 1 and relative algebraic K-groups. Their systematic approach recovers all previous versions of the TNC and more over all central conjectures of Galois module theory. Huber and Kings [24] were the first to realize that the formulation of the ETNC by relative K-groups is equivalent to the perhaps more suggestive use of "generators," i.e. maps of the form $\mathbf{1}_R \to \mathbf{d}_R(?)$ in the category $\mathcal{C}_R$ for various rings R instead, see also Flach's survey [19, §6]. They used this approach to give - for motives of the form $M^*(1-k)$ with k big enough, i.e. with very negative weight - the first version of a ETNC over general p-adic Lie extensions, which they call Iwasawa Main Conjecture (while in this survey we reserve this name for versions involving p-adic L-functions). While Burns and Flach use "equivariant" motives and L-functions in their general formalism, Fukaya and Kato realized that, at least for the connection with Iwasawa theory which we have in mind, it is sufficient to use non-commutative coefficients only for the Galois cohomology, but to stick to number fields as coefficients for the involved motives. In this survey we closely follow their approach.

To be more precise, consider for any motive M the motive $h^0(\mathrm{Spec}(F)) \otimes M$ (both defined over $\mathbb{Q}$) for some finite Galois extension F of $\mathbb{Q}$ with Galois group $G = G(F/\mathbb{Q})$. This motive has a natural action by the group algebra $\mathbb{Z}[G]$ and thus will be of particular interest for Iwasawa theory where a whole tower of finite extensions F_n of $\mathbb{Q}$ is considered simultaneously. Since there is an isomorphism of K-motives (for K sufficiently big)

$$h^0(\mathrm{Spec}(F))_K \otimes M \cong \bigoplus_{\rho \in \widehat{G}} [\rho^*]^{n_\rho} \otimes M$$

where ρ runs through all absolutely irreducible representations of G and n_ρ denotes the multiplicity with which it occurs in the regular representation of G on $K[G]$, it suffices - on the complex side - to consider the collection of K-motives $[\rho^*] \otimes M$ and their L-functions or more precisely the corresponding leading terms and vanishing orders. Indeed, the $\mathbb{C}$-algebra $\mathbb{C}[G]$ can be identified with $\prod_{\rho\in\widehat{G}} M_{n_\rho}(\mathbb{C})$ and thus its first K-group identifies with $\prod_{\rho\in\widehat{G}} \mathbb{C}^\times \cong \mathrm{center}(\mathbb{C}[G])^\times$. In contrast, on the p-adic side, even more when integrality is concerned, such a decomposition for $\mathbb{Z}_p[G]$ is impossible in general.

This motivated Fukaya and Kato to choose the following form of the ETNC. In fact, in order to keep the presentation concise, we will only describe a small extract of their complex and much more general treatment.

Let F be a p-adic Lie extension of $\mathbb{Q}$ with Galois group $G = G(F/\mathbb{Q})$. By $\Lambda = \Lambda(G)$ we denote its Iwasawa algebra. For a $\mathbb{Q}$-motive M over $\mathbb{Q}$ we fix a $G_{\mathbb{Q}}$-stable $\mathbb{Z}_p$ -lattice $T_p = T_p(M)$ of M_p and define a left Λ-module

$$\mathbb{T} := \Lambda \otimes_{\mathbb{Z}_p} T_p$$

on which Λ acts via multiplication on the left factor from the left while $G_{\mathbb{Q}}$ acts diagonally via $g(x \otimes y) = x\bar{g}^{-1} \otimes g(y)$, where $\bar{g}$ denotes the image of $g \in G_{\mathbb{Q}}$ in G. This is a "big Galois representation" in the sense of Nekovar [34]. Choose S as in the previous section and such that $\mathbb{T}$ is unramified outside S and denote, for any number field F', by $G_S(F')$ the Galois group of the maximal outside S unramified extension of F'. Then by Shapiro's Lemma the cohomology of $\mathrm{R}\Gamma(U,\mathbb{T})$ with $U = Spec(\mathbb{Z}[1/S])$ for example is nothing else than the perhaps more familiar $\Lambda(G)$-module $\mathrm{H}^i_{Iw}(F,T_p) := \varprojlim_{F'} \mathrm{H}^i(G_S(F'),T_p)$ where the limit is taken with respect to corestriction and F' runs over all finite subextensions of $F/\mathbb{Q}$.

Let K be a finite extension of $\mathbb{Q}$, λ a finite place of K above p, $\mathcal{O}_\lambda$ the ring of integers of the completion K_λ of K at λ and assume that $\rho : G \to \mathrm{GL}_n(\mathcal{O}_\lambda)$ is a continuous representation of G which, for some suitable choice of a basis, is the λ-adic realisation N_λ of some K-motive N. We also write ρ for the induced ring homomorphism $\Lambda \to M_n(\mathcal{O}_\lambda)$ and we consider $\mathcal{O}_\lambda^n$ as a right Λ-module via the action by the transpose ρ^t on the left, viewing $\mathcal{O}_\lambda^n$ as set of column vectors (contained in K_λ^n). Note that, setting $M(\rho^*) := N^* \otimes M$, we obtain an isomorphism of Galois representations

$$\mathcal{O}_\lambda^n \otimes_\Lambda \mathbb{T} \cong T_\lambda(M(\rho^*)),$$

where $T_\lambda(M(\rho^*))$ is the $\mathcal{O}_\lambda$-lattice $\rho^* \otimes T_p := \mathcal{O}_\lambda^n \otimes_{\mathbb{Z}_p} T_p$ of $M(\rho^*)_\lambda$ on which $G_{\mathbb{Q}}$ acts diagonally, via the contragredient(=dual) representation ρ^* of ρ on the left factor.

Now the equivariant version of Conjecture 3.7 reads as follows

Conjecture 4.1 (Equivariant Integrality; Fukaya/Kato) *There exists a (unique)*[7] *isomorphism*

$$\zeta_\Lambda(M) := \zeta_\Lambda(\mathbb{T}) : \mathbf{1}_\Lambda \to \mathbf{d}_\Lambda(\mathrm{R}\Gamma_c(U, \mathbb{T}))^{-1}$$

with the following property:

For all K, λ and ρ as above the (generalized) base change $\mathcal{O}_\lambda^n \otimes_\Lambda -$ sends $\zeta_\Lambda(M)$ to $\zeta_{\mathcal{O}_\lambda}(T_\lambda(M(\rho^*)))$.

Note that this conjecture assumes Conjecture 3.7 for all K-motives $M(\rho^*)$ with varying K. Furthermore, it is independent of the choice of S and of the lattices $T_p(M)$ and $T_\lambda(M(\rho^*))$.

One obtains a slight modification - to which we will refer as the *Artin-version* - of the above conjecture by restricting the representations ρ in question to the class of all Artin representations of G, (i.e. having finite image). If $F/\mathbb{Q}$ is finite, both versions coincide. Moreover, it is easy to see[8] that in this situation the conjecture is equivalent to (the p-part of) Burns and Flach's equivariant integrality conjecture [9, Conj. 6] for the $\mathbb{Q}$-algebra $\mathbb{Q}[G]$ with $\mathbb{Z}$-order $\mathbb{Z}[G]$. Also, $\mathbb{T} = \mathbb{Z}_p[G] \otimes T_p(M)$ identifies with the induced representation $\mathrm{Ind}_{G_\mathbb{Q}}^{G_F} T_p(M)$.

Assume now that $F = \bigcup_n F_n$ is the union of finite extensions F_n of $\mathbb{Q}$ with Galois groups G_n. Putting $\zeta_{\mathbb{Q}_p[G_n]}(M) = \mathbb{Q}_p[G_n] \otimes_\Lambda \zeta_\Lambda(M)$ one recovers the "generator" $\delta_p(G_n, M, k)$[9] (for k big enough) in [24] as $\zeta_{\mathbb{Q}_p[G_n]}(M^*(1-k))$. Hence, up to shifting and Kummer duality, the Artin-version of Conjecture 4.1

7 In fact, Fukaya and Kato assign such an isomorphism to each pair $(R, \mathbb{T})$ where R belongs to a certain class of rings containing the Iwasawa algebras for arbitrary p-adic Lie extensions of $\mathbb{Q}$ as well as the valuation rings of finite extensions of $\mathbb{Q}_p$ and where $\mathbb{T}$ is a projective R-module endowed with a continuous $G_\mathbb{Q}$-action. Then $\zeta_?(?)$ is supposed to behave well under arbitrary change of rings for such pairs. Moreover they require that the assignment $\mathbb{T} \mapsto \zeta_R(\mathbb{T})$ is multiplicative for short exact sequences. Only this full set of conditions leads to the uniqueness [23, §2.3.5], while e.g. for finite a group G the map $K_1(\mathbb{Z}_p[G]) \to K_1(\mathbb{Q}_p[G])$ need not be injective and thus $\zeta_{\mathbb{Z}_p[G]}(?)$ might not be unique if considered alone.

8 If we assume that K is big enough such that $K[G]$ decomposes completely into matrix algebras with coefficients in K, then the equivariant integrality statement (inducing (absolute) integrality for $M(\rho^*)$ for all Artin representations of G) amounts to an integrality statement for the generator $\zeta_{K_\lambda[G]}(M) := K_\lambda[G] \otimes_{\mathbb{Z}_p[G]} \zeta_{\mathbb{Z}_p[G]}(M)$ and thus Burns and Flach's version for the $\mathbb{Q}$-algebra $K[G]$ with order $\mathcal{O}_K[G]$. Using the functorialities of their construction [9, Thm. 4.1], it is immediate that taking norms leads to the conjecture for the pair $(\mathbb{Q}[G], \mathbb{Z}[G])$.

9 To be precise, this is only morally true, since Huber and Kings take for the definition of their generators the leading coefficients of the modified L-function without the Euler factors in S. It is not clear to what extent this is compatible with our formulation above.

for F is (morally) equivalent to [24, Conj. 3.2.1]. Hence, using [24, Lem. 6.0.2] we obtain the following

Proposition 4.2 (Huber/Kings) *Assume Conjectures 3.2, 3.5 and 3.7 for all $M(\rho^*)$ where ρ varies over all absolutely irreducible Artin representations of G. Then the existence of $\zeta_{\Lambda(G)}(M)$ satisfying the Artin-version of Conjecture 4.1 is equivalent to the existence of $\zeta_{\Lambda(G_n)}(M)$ for all n.*

In general, as remarked in footnote 7, the isomorphism $\zeta_{\mathbb{Z}_p[G_n]}(M)$ might not be unique (if it is considered alone). But it is realistic to hope uniqueness for infinite G (cf. [29]) and then the previous zeta isomorphism would be unique by the requirement that $\zeta_{\mathbb{Z}_p[G_n]}(M)$ is equal to $\mathbb{Z}_p[G_n] \otimes_{\Lambda(G)} \zeta_{\Lambda(G)}(M)$. Indeed, this is true at least if G is big enough, see [23, Prop. 2.3.7]. Moreover, as Huber and Kings [24, §3.3] pointed out, by twist invariance (over trivializing extensions $F/\mathbb{Q}$ for a given motive) arbitrary zeta-isomorphisms $\zeta_\Lambda(M)$ are reduced to those of the form $\zeta_{\mathbb{Z}_p[G(F/\mathbb{Q})]}(\mathbb{Q})$ for the trivial motive $\mathbb{Q}$ and where F runs through all finite extensions of the field $\mathbb{Q}$.

Question: Does the Artin-version imply the full version of Conjecture 4.1 ?

5 The functional equation and ϵ-isomorphisms

The L-function of a $\mathbb{Q}$-motive satisfies conjecturally a functional equation, which we want to state in the following way

$$L_{\mathbb{Q}}(M,s) = \epsilon(M,s)\frac{L_\infty(M^*(1),-s)}{L_\infty(M,s)}L_{\mathbb{Q}}(M^*(1),-s)$$

where the factor L_∞ at infinity is built up by certain Γ-factors and certain powers of 2 and π depending on the Hodge structure of M_B. The ϵ-factor decomposes into local factors

$$\epsilon(M,s) = \prod_{v \in S} \epsilon_v(M,s),$$

whose definition for finite places v is recalled in footnotes 11 and 14; $\epsilon_\infty(M,s)$ is a constant equal to a power of i.[10] We assume this conjecture. Then, taking leading coefficients induces

$$L^*_{\mathbb{Q}}(M) = (-1)^\eta \epsilon(M)\frac{L^*_\infty(M^*(1))}{L^*_\infty(M)}L^*_{\mathbb{Q}}(M^*(1))$$

where $\epsilon(M) = \prod \epsilon_v(M)$ with $\epsilon_v(M) = \epsilon(M,0)$ and η denotes the order of vanishing at $s = 0$ of the completed L-function $L_\infty(M^*(1),s)L_{\mathbb{Q}}(M^*(1),s)$.

10 We fix once and for all the complex period $2\pi i$, i.e. a square root of -1, and, for every l, the l-adic period $t =$ "$2\pi i$", i.e. a generator of $\mathbb{Z}_l(1)$.

Example 5.1 For the motive $M = h^1(A)(1)$ of an abelian variety one has $L_\infty(M,s) = L_\infty(M^*(1),s) = 2(2\pi)^{-(s+1)}\Gamma(s+1)$, $L^*_\infty(M) = L^*_\infty(M^*(1)) = \pi^{-1}$, $\epsilon_\infty(M) = -1$ and $\eta = 0$.

It is in no way obvious that the ETNC is compatible with the functional equation and Artin-Verdier/Poitou-Tate duality. The following discussion is a combination and reformulation of [9, section 5] and [37, Appendix C]. In order to formulate the precise condition under which the compatibility holds we first return to the absolute case and define "difference" terms

$$L^*_{dif}(M) := L^*_{\mathbb{Q}}(M)L^*_{\mathbb{Q}}(M^*(1))^{-1} = (-1)^\eta\epsilon(M)\frac{L^*_\infty(M^*(1))}{L^*_\infty(M)}$$

and

$$\Delta_{dif}(M) := \mathbf{d}_{\mathbb{Q}}(M_B)\mathbf{d}_{\mathbb{Q}}(M_{dR})^{-1}.$$

We obtain an isomorphism

$$\vartheta^{PD} : \Delta_{\mathbb{Q}}(M) \cdot \Delta_{\mathbb{Q}}(M^*(1))^* \cong \Delta_{dif}(M)$$

which arises from the mutual cancellation of the terms arising from motivic cohomology, the following isomorphism

$$M_B^+ \oplus (M_B^*(1)^+)^* \cong M_B^+ \oplus M_B(-1)^+ \cong M_B,$$

where the last map is $(x,y) \mapsto x + 2\pi i y$, and from the Poincare duality exact sequence

$$0 \longrightarrow (t_{M^*(1)})^* \longrightarrow M_{dR} \longrightarrow t_M \longrightarrow 0. \qquad (5.1)$$

On the other side define an isomorphism

$$\vartheta_\infty^{dif} : \Delta_{dif}(M)_{\mathbb{R}} \cong \mathbf{1}_{\mathbb{R}}$$

applying the determinant to (2.4) and to the following isomorphism

$$(\mathbb{C} \otimes_{\mathbb{Q}} M_B)^+ = (\mathbb{R} \otimes_{\mathbb{Q}} M_B^+) \oplus (\mathbb{R}(2\pi i)^{-1} \otimes_{\mathbb{Q}} M_B^-) \cong (M_B)_{\mathbb{R}} \qquad (5.2)$$

where the last map is induced by $\mathbb{R}(2\pi i)^{-1} \to \mathbb{R}, x \mapsto 2\pi i x$.

Due to the autoduality of the exact sequence of Conjecture 3.3 (see [9, Lem. 12]) we have a commutative diagram

$$\begin{array}{ccc} \Delta_{\mathbb{Q}}(M)_{\mathbb{R}} \cdot \Delta_{\mathbb{Q}}(M^*(1))^*_{\mathbb{R}} & \xrightarrow{\vartheta^{PD}_{\mathbb{R}}} & \Delta_{dif}(M)_{\mathbb{R}} \\ {\scriptstyle \vartheta_\infty(M)\cdot\overline{\vartheta_\infty(M^*(1))^*}} \downarrow & & \downarrow {\scriptstyle \vartheta_\infty^{dif}} \\ \mathbf{1}_{\mathbb{R}} & \xrightarrow{\mathrm{id}_{\mathbf{1}_{\mathbb{R}}}} & \mathbf{1}_{\mathbb{R}} \end{array}$$

Thus we obtain the following

Proposition 5.2 (Rationality) *Assume that Conjecture 3.5 is valid for the $\mathbb{Q}$-motive M. Then it is also valid for its Kummer dual $M^*(1)$ if and only if there exists a (unique) isomorphism*

$$\zeta^{dif}(M) : \mathbf{1}_{\mathbb{Q}} \to \Delta_{dif}(M)$$

such that we have

$$L^*_{dif}(M) : \mathbf{1}_{\mathbb{C}} \xrightarrow{\zeta^{dif}(M)_{\mathbb{C}}} \Delta_{dif}(M)_{\mathbb{C}} \xrightarrow{(\vartheta^{dif}_{\infty})_{\mathbb{C}}} \mathbf{1}_{\mathbb{C}}.$$

Putting $t_H(M) := \sum_{r\in\mathbb{Z}} rh(r)$ with $h(r) := \dim_K gr^r(M_{dR})$ $(= \dim_{K_\lambda} gr^r(D_{dR}(M_\lambda)))$ for a K-motive M and noting that $t_H(M) = t_H(\det(M))$, we have in fact the following

Theorem 5.3 (Deligne [18, Thm. 5.6], Burns-Flach [9, Thm. 5.2]) *If the motive* $\det(M)$ *is of the form* $\mathbb{Q}(-t_H(M))$ *twisted by a Dirichlet character, then* $\zeta^{dif}(M)$ *exists.*

Deligne [18] conjectured that the condition of the theorem is satisfied for all motives. It is known to hold in all examples A)-E).

See (5.6) below for the rationality statement which is hidden in the formulation of this theorem. Now we have to check the compatibility with respect to the p-adic realizations. To this aim we define the isomorphism

$$\vartheta^{dif}_p : \Delta_{dif}(M)_{\mathbb{Q}_p} \cong \mathbf{d}_{\mathbb{Q}_p}(M_p) \cdot \prod_{S\setminus S_\infty} \mathbf{d}_{\mathbb{Q}_p}(\mathrm{R}\Gamma(\mathbb{Q}_l, M_p))$$

as follows: Apply the determinant to (2.1) and multiply the resulting isomorphism by

$$\mathrm{id}_{\mathbf{d}_{\mathbb{Q}}(M_{dR})^{-1}_{\mathbb{Q}_p}} \cdot \prod_{l\in S\setminus S_\infty} \Theta_l(M_p) : \mathbf{d}_{\mathbb{Q}}(M_{dR})^{-1}_{\mathbb{Q}_p} \to \prod_{S\setminus S_\infty} \mathbf{d}_{\mathbb{Q}_p}(\mathrm{R}\Gamma(\mathbb{Q}_l, M_p))$$

where $\Theta_l(M_p) = \eta_l(M) \cdot \eta_l(M^*(1))$ is defined in the appendix, section 7.

On the other hand Artin-Verdier/Poitou-Tate Duality induces the following isomorphism

$$\begin{aligned}
\mathbf{d}_{\mathbb{Q}_p}(\mathrm{R}\Gamma_c(U, M_p))^{-1} &\cong \mathbf{d}_{\mathbb{Q}_p}(\mathrm{R}\Gamma(U, M_p))^{-1}\mathbf{d}_{\mathbb{Q}_p}(\bigoplus_{v\in S} \mathrm{R}\Gamma(\mathbb{Q}_v, M_p))\\
&\cong \mathbf{d}_{\mathbb{Q}_p}(\mathrm{R}\Gamma_c(U, M_p^*(1))^*)\mathbf{d}_{\mathbb{Q}_p}((M_p^*(1)^+)^*) \prod_{v\in S} \mathbf{d}_{\mathbb{Q}_p}(\mathrm{R}\Gamma(\mathbb{Q}_v, M_p))\\
&\cong \mathbf{d}_{\mathbb{Q}_p}(\mathrm{R}\Gamma_c(U, M_p^*(1))^*)\mathbf{d}_{\mathbb{Q}_p}(M_p(-1)^+)\mathbf{d}_{\mathbb{Q}_p}(M_p^+) \prod_{l\in S\setminus S_\infty} \mathbf{d}_{\mathbb{Q}_p}(\mathrm{R}\Gamma(\mathbb{Q}_l, M_p)).
\end{aligned}$$

Using the identification

$$M_p^+ \oplus M_p(-1)^+ = M_p^+ \oplus M_p^-(-1) \cong M_p, \tag{5.3}$$

where the last map is induced by multiplication with the p-adic period $t =$ "$2\pi i$" $: M_p^-(-1) \to M_p^-$, we obtain

$$\vartheta_p^{AV} : \mathbf{d}_{\mathbb{Q}_p}(\mathrm{R}\Gamma_c(U, M_p))^{-1} \cdot \mathbf{d}_{\mathbb{Q}_p}(\mathrm{R}\Gamma_c(U, M_p^*(1))^*)^{-1} \cong \\ \mathbf{d}_{\mathbb{Q}_p}(M_p) \cdot \prod_{S \setminus S_\infty} \mathbf{d}_{\mathbb{Q}_p}(\mathrm{R}\Gamma(\mathbb{Q}_l, M_p)).$$

Again one has to check the commutativity of the following diagram (cf. [9, Lem. 12])

$$\begin{array}{ccc} \Delta_{\mathbb{Q}}(M)_{\mathbb{Q}_p} \cdot \Delta_{\mathbb{Q}}(M^*(1))^*_{\mathbb{Q}_p} & \xrightarrow{(\vartheta^{PD})_{\mathbb{Q}_p}} & \Delta_{dif}(M)_{\mathbb{Q}_p} \\ \Big\downarrow{\scriptstyle \vartheta_p(M) \cdot \overline{\vartheta_p(M^*(1))^*}} & & \Big\downarrow{\scriptstyle \vartheta_p^{dif}} \\ \mathbf{d}_{\mathbb{Q}_p}(\mathrm{R}\Gamma_c(U,M_p))^{-1} \mathbf{d}_{\mathbb{Q}_p}(\mathrm{R}\Gamma_c(U,M_p^*(1))^*)^{-1} & \xrightarrow{\vartheta_p^{AV}} & \mathbf{d}_{\mathbb{Q}_p}(M_p) \prod_{S \setminus S_\infty} \mathbf{d}_{\mathbb{Q}_p}(\mathrm{R}\Gamma(\mathbb{Q}_l, M_p)) \end{array}$$

Note that analogous maps exist and analogous properties hold also if we replace M_p by a Galois stable $\mathbb{Z}_p$-lattice T_p or even by the free Λ-module $\mathbb{T}$. Thus we obtain the following

Proposition 5.4 (Integrality) *Assume that Conjecture 3.7 is valid for the $\mathbb{Q}$-motive M. Then it is also valid for its Kummer dual $M^*(1)$ if and only if there exists a (unique) isomorphism*

$$\zeta_{\mathbb{Z}_p}^{dif}(T_p) : \mathbf{1}_{\mathbb{Z}_p} \to \mathbf{d}_{\mathbb{Z}_p}(T_p) \cdot \prod_{S \setminus S_\infty} \mathbf{d}_{\mathbb{Z}_p}(\mathrm{R}\Gamma(\mathbb{Q}_l, T_p)).$$

which induces via $\mathbb{Q}_p \otimes_{\mathbb{Z}_p} -$ the following map

$$\mathbf{1}_{\mathbb{Q}_p} \xrightarrow{\zeta^{dif}(M)_{\mathbb{Q}_p}} \Delta_{dif}(M)_{\mathbb{Q}_p} \xrightarrow{\vartheta_p^{dif}} \mathbf{d}_{\mathbb{Q}_p}(M_p) \cdot \prod_{S \setminus S_\infty} \mathbf{d}_{\mathbb{Q}_p}(\mathrm{R}\Gamma(\mathbb{Q}_l, M_p)).$$

If this holds we have, using the above identifications, the functional equation

$$\zeta_{\mathbb{Z}_p}(T_p) = (\overline{\zeta_{\mathbb{Z}_p}(T_p^*(1))^*})^{-1} \cdot \zeta_{\mathbb{Z}_p}^{dif}(T_p)$$

Note that $(\overline{\zeta_{\mathbb{Z}_p}(T_p^*(1))^*})^{-1}$ is the same as $\zeta_{\mathbb{Z}_p}(T_p^*(1))^* \cdot \mathrm{id}_{\mathbf{d}_{\mathbb{Z}_p}(\mathrm{R}\Gamma_c(U,T_p^*(1))^*)}$ according to Remark 1.2(i). Needless to say, all of the above has an analogous version for K-motives whose formulation we leave to the reader. Then it is clear how the equivariant version of this proposition looks like:

Proposition 5.5 (Equivariant Integrality) *Assume that Conjecture 4.1 is valid for the $\mathbb{Q}$-motive M. Then it is also valid for its Kummer dual $M^*(1)$ if and only if there exists a (unique) isomorphism*

$$\zeta_{\Lambda}^{dif}(M) : \mathbf{1}_{\Lambda} \to \mathbf{d}_{\Lambda}(\mathbb{T}) \cdot \prod_{S \setminus S_{\infty}} \mathbf{d}_{\Lambda}(\mathrm{R}\Gamma(\mathbb{Q}_l, \mathbb{T}))$$

with the following property $(*)$:

For all K, λ and ρ as before Conjecture 4.1 the (generalized) base change $\mathcal{O}_{\lambda}^{n} \otimes_{\Lambda} -$ sends $\zeta_{\Lambda}^{dif}(M)$ to $\zeta_{\mathcal{O}_{\lambda}}^{dif}(T_{\lambda}(M(\rho^*)))$, the analogue of $\zeta_{\mathbb{Z}_p}^{dif}(T_p)$ for the K-motive $M(\rho^*)$.

If this holds we have the functional equation

$$\zeta_{\Lambda}(M) = (\overline{\zeta_{\Lambda}(M^*(1))^*})^{-1} \cdot \zeta_{\Lambda}^{dif}(M).$$

Thus we formulate the

Conjecture 5.6 (Local Equivariant Tamagawa Number Conjecture) *The isomorphism $\zeta_{\Lambda}^{dif}(M)$ in the previous Proposition exists (uniquely).*

5.1 ϵ-isomorphisms

One obtains a refinement of the above functional equation if one looks more closely at which part of the Galois cohomology (and comparison isomorphisms) the factors occurring in $L_{dif}^*(M)$ belong precisely. We first recall from [37] the equality

$$\frac{L_{\infty}^*(M^*(1))}{L_{\infty}^*(M)} = \pm 2^{d_-(M)-d_+(M)}(2\pi)^{-(d_-(M)+t_H(M))} \prod_{j\in\mathbb{Z}} \Gamma^*(-j)^{-h_j(M)} \tag{5.4}$$

where $\Gamma^*(-j)$ is defined to be $\Gamma(j) = (j-1)!$ if $j > 0$ and $\lim_{s\to j}(s-j)\Gamma(s) = (-1)^j((-j)!)^{-1}$ otherwise.

The factor $(2\pi)^{-(d_-(M)+t_H(M))}$ arises as follows. Assume for simplicity that $\det(M) = \mathbb{Q}(-t_H(M))$. Then fixing a $\mathbb{Q}$-basis $\gamma = (\gamma^+, \gamma^-)$ of M_B and $\omega = (\delta_M, \delta_{M^*(1)})$ of M_{dR} which induce the canonical basis (cf. Example A)) of $\det(M)_B$ and $\det(M)_{dR}$, respectively, gives rise to a map

$$\mathbf{1}_{\mathbb{Q}} \xrightarrow{can_{\gamma,\omega}} \mathbf{d}_{\mathbb{Q}}(M_B)\mathbf{d}_{\mathbb{Q}}(M_{dR})^{-1}. \tag{5.5}$$

Base change and the comparison isomorphism (2.3) induce

$$\mathbf{1}_{\mathbb{C}} \xrightarrow{(can_{\gamma,\omega})_{\mathbb{C}}} \mathbf{d}_{\mathbb{Q}}(M_B)_{\mathbb{C}}\mathbf{d}_{\mathbb{Q}}(M_{dR})_{\mathbb{C}}^{-1} \xrightarrow{\mathbf{d}(g_{\infty})} \mathbf{1}_{\mathbb{C}}$$

whose value Ω_∞^{dif} in $\mathbb{C}^\times$ is nothing else than the inverse of the determinant over $\mathbb{C}$ of the comparison isomorphism

$$\mathbb{C} \otimes_{\mathbb{Q}} \det(M)_B \to \mathbb{C} \otimes_{\mathbb{Q}} \det(M)_{dR}$$

and thus $\Omega_\infty^{dif} = (2\pi i)^{-t_H(M)}$. But note that due to the definition of (5.2) the above map differs from $(\vartheta_\infty^{dif})_{\mathbb{C}} \circ (can_{\gamma,\omega})_{\mathbb{C}}$ by the factor $(2\pi i)^{-d_-(M)}$. Thus we obtain an explanation of the factor $(2\pi i)^{-(t_H(M)+d_-(M))}$. Moreover, Theorem 5.3 and Proposition 5.2 tell us that

$$\frac{L^*_{dif}(M)}{(2\pi i)^{-d_-(M)}\Omega_\infty^{dif}} = \pm 2^{d_-(M)-d_+(M)} \frac{\epsilon_\infty(M)}{i^{-(t_H(M)+d_-(M))}} \cdot \prod_{l\in S\smallsetminus S_\infty} \epsilon_l(M) \prod_{j\in\mathbb{Z}} \Gamma^*(-j)^{-h_j(M)} \tag{5.6}$$

is rational and that $\zeta_{dif}(M)$ is the map $can_{\gamma,\omega}$ multiplied by this rational number.

The factor $2^{d_-(M)-d_+(M)}$ arises as quotient of the Tamagawa factors of M and $M^*(1)$ at infinity. One can either cover it by defining Θ_∞ (see below) or changing the last map of the identification in (5.3) as follows: on the summand M_p^+ multiply with 2 and on M_p^- by $\frac{1}{2}$ (as Fukaya and Kato do).

The map (5.5) induces $\mathbf{1}_{\mathbb{Q}_p} \to \Delta_{dif}(M)_{\mathbb{Q}_p} \cong \mathbf{d}_{\mathbb{Q}_p}(M_p)\mathbf{d}_{\mathbb{Q}_p}(D_{dR}(M_p))^{-1}$ and furthermore

$$\mathbf{1}_{B_{dR}} \xrightarrow{(can_{\gamma,\omega})_{B_{dR}}} \Delta_{dif}(M)_{B_{dR}} \xrightarrow{\mathbf{d}_{\mathbb{Q}_p}(g_{dR})_{B_{dR}}} \mathbf{1}_{B_{dR}}$$

whose value Ω_p^{dif} in $B_{dR}^\times$ is nothing else than the inverse of the determinant over B_{dR} of the comparison isomorphism

$$B_{dR} \otimes_{\mathbb{Q}_p} D_{dR}(\det(M)_p) \to B_{dR} \otimes_{\mathbb{Q}_p} \det(M)_p$$

and thus $\Omega_p^{dif} = (2\pi i)^{-t_H(M)}$, where we consider the p-adic period $t =$ "$2\pi i$" as an element of B_{dR}. Note that $(B_{dR})^{I_p} = \widehat{\mathbb{Q}_p^{nr}}$, the completion of the maximal unramified extension $\mathbb{Q}_p^{nr}$ of $\mathbb{Q}_p$. We need the following

Lemma 5.7 ([23, Prop. 3.3.5],[37, C.2.8]) *The map $\epsilon_p(M) \cdot \Omega_p^{dif} \cdot (can)_{B_{dR}}$ comes from a map*

$$\epsilon_{dR}(M_p) : \mathbf{1}_{\widehat{\mathbb{Q}_p^{nr}}} \to \mathbf{d}_{\widehat{\mathbb{Q}_p^{nr}}}(M_p)\mathbf{d}_{\widehat{\mathbb{Q}_p^{nr}}}(D_{dR}(M_p))^{-1}.$$

Moreover, let L be any finite extension of $\mathbb{Q}_p$. Then a similar statement holds for any finite dimensional L-vector space V with continuous $G_{\mathbb{Q}_p}$-action *instead of* $\underline{M_p}$[11]. *We write* $\epsilon_{dR}(V)$ *for the corresponding map, which is defined over* $\widetilde{L} := \widehat{\mathbb{Q}_p^{nr}} \otimes_{\mathbb{Q}_p} L$ *(see [23, Prop. 3.3.5] for details).*

For any V as in the lemma we define an isomorphism

$$\epsilon_{p,L}(V) : \mathbf{1}_{\widetilde{L}} \to \left(\mathbf{d}_L(\mathrm{R}\Gamma(\mathbb{Q}_p, V))\mathbf{d}_L(V)\right)_{\widetilde{L}}$$

as product of $\Gamma_L(V) := \prod_{\mathbb{Z}} \Gamma^*(j)^{-h(-j)}$, $\Theta_p(V)$ (see (7.10) in the appendix) and $\epsilon_{dR}(V)$, where $h(j) = \dim_L gr^j D_{dR}(V)$.

Now let T be a Galois stable $\mathcal{O} := \mathcal{O}_L$-lattice of V and set $\widetilde{\mathcal{O}} := W(\overline{\mathbb{F}_p}) \otimes_{\mathbb{Z}_p} \mathcal{O}$, where $W(\overline{\mathbb{F}_p})$ denotes the Wittring of $\overline{\mathbb{F}_p}$. The following conjecture is a local integrality statement

Conjecture 5.8 (Absolute ϵ-isomorphism) *There exists a (unique) isomorphism*

$$\epsilon_{p,\mathcal{O}}(T) : \mathbf{1}_{\widetilde{\mathcal{O}}} \to \left(\mathbf{d}_{\mathcal{O}}(\mathrm{R}\Gamma(\mathbb{Q}_p, T))\mathbf{d}_{\mathcal{O}}(T)\right)_{\widetilde{\mathcal{O}}}$$

which induces $\epsilon_{p,L}(V)$ *by base change* $L \otimes_{\mathcal{O}} -$.

This conjecture, which is equivalent to conjecture $C_{EP}(V)$ in [22, III 4.5.4], or more precisely its equivariant version below is closely related to the conjecture $\delta_{\mathbb{Z}_p}(V)$ [37] via the explicit reciprocity law $Réc(V)$, which was conjectured by Perrin-Riou and proven independently by Benois [1], Colmez [16], and Kurihara/Kato/Tsuji [26]. In particular, the above conjecture is known for ordinary crystalline p-adic representations [37, 1.28,C.2.10] and for certain semi-stable representations, see [3].

To formulate an equivariant version, define

$$\widetilde{\Lambda} := \widehat{\mathbb{Z}_p^{nr}}[\![G]\!] = \varprojlim_n \left(W(\overline{\mathbb{F}_p}) \otimes_{\mathbb{Z}_p} \mathbb{Z}_p[G/G_n]\right),$$

where $\widehat{\mathbb{Z}_p^{nr}} = W(\overline{\mathbb{F}_p})$ denotes the ring of integers of $\widehat{\mathbb{Q}_p^{nr}}$. We assume $L = \mathbb{Q}_p$ and set as before $\mathbb{T} := \Lambda \otimes_{\mathbb{Z}_p} T$ (but later T might differ from our global T_p). We write $T(\rho^*)$ for the $\mathcal{O}$-lattice $\rho^* \otimes T$ of $\rho^* \otimes V$, which we assume de Rham.

11 $\epsilon_p(V) = \epsilon(D_{pst}(V))$ where $D_{pst}(V)$ is endowed with the linearized action of the Weil-group and thereby considered as a representation of the Weil-Deligne group, see [20]. Furthermore, we suppress the dependence of the choice of a Haar measure and of $t =$ "$2\pi i$" in the notation. The choice of $t = (t_n) \in \mathbb{Z}_p(1)$ determines a homomorphism $\psi_p : \mathbb{Q}_p \to \overline{\mathbb{Q}_p}^\times$ with $\ker(\psi_p) = \mathbb{Z}_p$ sending $\frac{1}{p^n}$ to $t_n \in \mu_{p^n}$.

Conjecture 5.9 (Equivariant ϵ-isomorphism) *There exists a (unique)*[12] *isomorphism*

$$\epsilon_{p,\Lambda}(\mathbb{T}) : \mathbf{1}_{\widetilde{\Lambda}} \to \left(\mathbf{d}_\Lambda(R\Gamma(\mathbb{Q}_p, \mathbb{T}))\mathbf{d}_\Lambda(\mathbb{T})\right)_{\widetilde{\Lambda}}$$

such that for all $\rho : G \to \mathrm{GL}_n(\mathcal{O}) \subseteq \mathrm{GL}_n(L)$, L a finite extension of $\mathbb{Q}_p$ with valuation ring $\mathcal{O}$, we have

$$\mathcal{O}^n \otimes_\Lambda \epsilon_{p,\Lambda}(\mathbb{T}) = \epsilon_{p,\mathcal{O}}(T(\rho^*)).$$

If $T = T_p \subseteq M_p$ is fixed we also write $\epsilon_{p,\Lambda}(M)$ for $\epsilon_{p,\Lambda}(\mathbb{T})$. This equivariant version has recently been proved by Benois and Berger [2] in the cyclotomic situation for crystalline representations.

Similarly we proceed in the case $l \neq p$, formulating the analogues of Conjectures 5.8 and 5.9 just in one

Conjecture 5.10 *There exists a (unique)* [13] *isomorphism*

$$\epsilon_{l,\Lambda}(\mathbb{T}) : \mathbf{1}_{\widetilde{\Lambda}} \to \mathbf{d}_\Lambda(R\Gamma(\mathbb{Q}_l, \mathbb{T}))_{\widetilde{\Lambda}}$$

such that for all $\rho : G \to \mathrm{GL}_n(L)$, L a finite extension of $\mathbb{Q}_p$, we have

$$\mathcal{O}^n \otimes_\Lambda \epsilon_{l,\Lambda}(\mathbb{T}) = \epsilon_{l,\mathcal{O}}(T(\rho^*)).$$

Here $\epsilon_{l,\mathcal{O}}(T(\rho^*))$ is the analogue of the above with respect to $\mathcal{O}$ instead of Λ and required to induce

$$L \otimes_\mathcal{O} \epsilon_{l,\mathcal{O}}(T(\rho^*))) = \Theta_l(V) \cdot \epsilon_l(V(\rho^*))$$

and its existence is part of the conjecture[14]*.*

For commutative Λ this conjecture was proved by S. Yasuda [48] and it seems that he can extend his methods to cover the non-commutative case, too.

12 Again, Fukaya and Kato assign such an isomorphism to each triple $(R, \mathbb{T}, t)$, where R is as before, $\mathbb{T}$ is a projective R-module endowed with a continuous $G_{\mathbb{Q}_p}$-action and t is a generator of $\mathbb{Z}_p(1)$. Then $\epsilon_{p,?}(?)$ is supposed to behave well under arbitrary change of rings for such pairs. Moreover they require that the assignment $\mathbb{T} \mapsto \epsilon_{p,R}(\mathbb{T})$ is multiplicative for short exact sequences, that it satisfies a duality relation when replacing $\mathbb{T}$ by $\mathbb{T}^*(1)$, that the group $G_{\mathbb{Q}_p}^{ab}$ acts on a predetermined way (modifying t) compatible in a certain sense with the Frobenius ring homomorphism on $\widetilde{\Lambda}$ induced from the absolute Frobenius of $\overline{\mathbb{F}_p}$. Only this full set of conditions may lead to the uniqueness in general.

13 A similar comment to that in footnote 12 applies here.

14 $\epsilon_l(V) = \epsilon(V)$ where V is considered as representation of the Weil-Deligne group of $\mathbb{Q}_l$ and where we suppress the dependence of the choice of a Haar measure and of $t = 2\pi i$ in the notation. The choice of $t = (t_n) \in \mathbb{Z}_l(1)$ determines a homomorphism $\psi_l : \mathbb{Q}_l \to \bar{\mathbb{Q}_l}^\times$ with $Ker(\psi_l) = \mathbb{Z}_l$ sending $\frac{1}{l^n}$ to $t_n \in \mu_{l^n}$. The formulation of this conjecture is equivalent to [23, conj. 3.5.2] where the constants ϵ_0 are used instead of ϵ and where θ_l does not occur. More precisely, our $\epsilon_{l,\Lambda}(\mathbb{T})$ equals $\epsilon_{0,\Lambda}(\mathbb{Q}_l, T, \xi) \cdot s_l(T)$ in [23, 3.5.2,3.5.4].

If $T = T_p \subseteq M_p$ we also write $\epsilon_{l,\Lambda}(M)$ for $\epsilon_{l,\Lambda}(\mathbb{T})$.

Finally we set $\epsilon_{\infty,\Lambda}(M) = \pm 2^{d_-(M)-d_+(M)} \frac{\epsilon_\infty(M)}{i^{-(t_H(M)+d_-(M))}}$ where the sign is that which makes (5.4) correct. The following result is now immediate.

Theorem 5.11 (cf. [23, Conj. 3.5.5]) *Assume Conjectures 5.8, 5.9 and 5.10. Then Conjecture 5.6 holds, $\zeta_\Lambda^{dif}(M) = \prod_{v\in S} \epsilon_{v,\Lambda}(M)$ and we have the functional equation*

$$\zeta_\Lambda(M) = (\overline{\zeta_\Lambda(M^*(1))^*})^{-1} \cdot \prod_{v\in S} \epsilon_{v,\Lambda}(M).$$

6 p-adic L-functions and the Iwasawa main conjecture

A p-adic L-function attached to a $\mathbb{Q}$-motive M should be considered as a map on a certain class of representations of G which interpolates the L-values of the twists $M(\rho^*)$ at $s = 0$. The experience from those cases where such p-adic L-functions are known to exist, shows that one has to modify the complex L-values by certain factors before one can hope to obtain a p-adic interpolation (cf. [12, 13] or [37]). The reason for this becomes clearer if one considers the Galois cohomology involved together with the functional equation; in fact, that was the main motivation of the previous section.

In order to evaluate e.g. the ζ-isomorphism or a modification of it at a representation ρ over a finite extension L of $\mathbb{Q}_p$, one needs that the complex $\rho \otimes_\Lambda C$, where C is (a modification of) $\mathrm{R}\Gamma_c(U, \mathbb{T})$, becomes acyclic: then the induced map $\mathbf{1}_L \to \mathbf{d}_L(\rho \otimes^{\mathbb{L}}_\Lambda C) \to \mathbf{1}_L$ can be considered as value in $K_1(L) = L^\times$ at ρ.

In general, $\mathrm{R}\Gamma_c(U, \mathbb{T})$ does not behave well enough and will have to be replaced by some Selmer complex, which we well achieve in two steps. This modification corresponds to a shifting of certain Euler- and ϵ-factors from one side of the functional equation to the other such that both sides are 'balanced'.

Though the following part of the theory holds in much greater generality (e.g. in the *ordinary* good reduction case, but not in the *supersingular* good reduction case) we just discuss the case of abelian varieties in order to keep the situation as concise as possible. Thus let A be an abelian variety over $\mathbb{Q}$ with good ordinary reduction at a fixed prime $p \neq 2$ and set $M = h^1(A)(1)$ as before. Let F_∞ be an infinite p-adic Lie extension of $\mathbb{Q}$ with Galois group G. For simplicity we assume also that G has no element of order p, hence its Iwasawa algebra $\Lambda = \Lambda(G)$ is a regular ring.

Due to our assumption on the reduction type of A, we have the following fact: There is a unique $\mathbb{Q}_p$-subspace $\hat{V}$ of $V = M_p$ which is stable under the action of $G_{\mathbb{Q}_p}$ and such that

$$D_{dR}(\hat{V}) \cong D_{dR}(V)/D^0_{dR}(V). \tag{6.1}$$

More precisely, $\hat{V} = V_p(\widehat{A^\vee})$ where $\widehat{A^\vee}$ denotes the formal group of the dual abelian variety $A^\vee$, i.e. the formal completion of the Néron model $\mathcal{A}/\mathbb{Z}_p$ of $A^\vee$ along the zero section of the special fibre $\widetilde{\mathcal{A}}$. Then (6.1) arises from the *unit root splitting* $D^0_{dR}(V_p(\widetilde{\mathcal{A}})) \cong D^0_{dR}(V) \subseteq D_{dR}(V)$ (see [35, 1.31]) which is induced by applying $D^0_{dR}(-)$ to the exact sequence of $G_{\mathbb{Q}_p}$-modules

$$0 \longrightarrow V_p(\widehat{A^\vee}) \longrightarrow V_p(A^\vee) \longrightarrow V_p(\widetilde{\mathcal{A}}) \longrightarrow 0.$$

Let T be the $G_{\mathbb{Q}}$-stable $\mathbb{Z}_p$-lattice $T_p(A^\vee)$ of V and set

$$\hat{T} := T \cap \hat{V},$$

a $G_{\mathbb{Q}_p}$-stable $\mathbb{Z}_p$-lattice of $\hat{V}$. As before let $\mathbb{T}$ denote the big Galois representation $\Lambda \otimes_{\mathbb{Z}_p} T$ and put $\hat{\mathbb{T}} := \Lambda \otimes_{\mathbb{Z}_p} \hat{T}$ similarly. Then $\hat{\mathbb{T}}$ is a $G_{\mathbb{Q}_p}$-stable sub-Λ-module of $\mathbb{T}$. In fact, it is a direct summand of $\mathbb{T}$ and we have an isomorphism of Λ-modules[15]

$$\beta : \mathbf{d}_\Lambda(\mathbb{T}^+) \xrightarrow{\cong} \mathbf{d}_\Lambda(\hat{\mathbb{T}}). \tag{6.5}$$

In this good ordinary case one can now first replace $R\Gamma_c(U, \mathbb{T})$ by the Selmer complex $SC_U := SC_U(\hat{\mathbb{T}}, \mathbb{T})$ (see (7.11)) which fits into the following

15 which arises as follows: Choose a basis $\gamma^+ = (\gamma_1^+, \ldots, \gamma_r^+)$ of $\mathrm{H}_1(A^\vee(\mathbb{C}), \mathbb{Q})^+$ and $\gamma^- = (\gamma_1^-, \ldots, \gamma_r^-)$ of $\mathrm{H}_1(A^\vee(\mathbb{C}), \mathbb{Q})^-$, which gives rise to a $\mathbb{Z}_p$-basis of $T^+ \cong (\mathrm{H}_1(A^\vee(\mathbb{C}), \mathbb{Z}) \otimes_{\mathbb{Z}} \mathbb{Z}_p)^+$ and $T^- \cong (\mathrm{H}_1(A^\vee(\mathbb{C}), \mathbb{Z}) \otimes_{\mathbb{Z}} \mathbb{Z}_p)^-$ respectively, where $r = d_+(M) = d_-(M)$. Then we obtain isomorphisms

$$\Lambda^r \cong \mathbb{T}^+ \quad \text{and} \quad \phi : \mathbf{d}_\Lambda(\Lambda^r) \to \mathbf{d}_\Lambda(T^+) \tag{6.2}$$

using the Λ-basis $\frac{1+\iota}{2} \otimes \gamma_j^+ + \frac{1-\iota}{2} \otimes \gamma_j^-$, $1 \le j \le r$, of $\mathbb{T}^+$. On the other hand one can choose a $\mathbb{Q}$-basis $\delta = (\delta_1, \ldots, \delta_r)$ of $\mathrm{Lie}(A^\vee)$ (e.g. a $\mathbb{Z}$-basis of $\mathrm{Lie}_{\mathbb{Z}}(A^\vee)$ as in section 3.1) such that the isomorphism

$$\mathbf{d}_{\mathbb{Q}_p}(\mathbb{Q}_p^r)_{\widehat{\mathbb{Q}_p^{nr}}} \cong \mathbf{d}_{\mathbb{Q}_p}(\mathbb{Q}_p \otimes_{\mathbb{Q}} t_M)_{\widehat{\mathbb{Q}_p^{nr}}} \cong \mathbf{d}_{\mathbb{Q}_p}(D_{dR}(\hat{V}))_{\widehat{\mathbb{Q}_p^{nr}}} \cong \mathbf{d}_{\mathbb{Q}_p}(\hat{V})_{\widehat{\mathbb{Q}_p^{nr}}}, \tag{6.3}$$

which is induced by δ, (2.8), (6.1) and $\epsilon_{dR}(\hat{V})$ (according to Lemma 5.7, but note that $\epsilon_p(\hat{V}) = 1$ due to the good reduction), comes from an isomorphism

$$\mathbf{d}_{\mathbb{Z}_p}(\mathbb{Z}_p^r)_{\widehat{\mathbb{Z}_p^{nr}}} \cong \mathbf{d}_{\mathbb{Z}_p}(\hat{T})_{\widehat{\mathbb{Z}_p^{nr}}},$$

where $\widehat{\mathbb{Z}_p^{nr}} := W(\overline{\mathbb{F}_p})$. Then base change $\widetilde{\Lambda} \otimes_{\widehat{\mathbb{Z}_p^{nr}}} -$ induces an isomorphism

$$\psi : \mathbf{d}_\Lambda(\Lambda^r)_{\widetilde{\Lambda}} \cong \mathbf{d}_\Lambda(\hat{\mathbb{T}})_{\widetilde{\Lambda}}. \tag{6.4}$$

Now $\beta = \beta_{\gamma,\delta}$ is $\psi \circ \overline{\phi}$.

distinguished triangle

$$\mathrm{R}\Gamma_c(U,\mathbb{T}) \longrightarrow SC_U \longrightarrow \mathbb{T}^+ \oplus \mathrm{R}\Gamma(\mathbb{Q}_p,\hat{\mathbb{T}}) \longrightarrow \tag{6.6}$$

and thus induces an isomorphism

$$\mathbf{d}_\Lambda(\mathrm{R}\Gamma_c(U,\mathbb{T}))^{-1} \cong \mathbf{d}_\Lambda(SC_U)^{-1}\mathbf{d}_\Lambda(\mathbb{T}^+)\mathbf{d}_\Lambda(\mathrm{R}\Gamma(\mathbb{Q}_p,\hat{\mathbb{T}})).$$

The p-adic L-function $\mathcal{L}_U = \mathcal{L}_{U,\beta}(M, F_\infty/\mathbb{Q})$ arises from the zeta-isomorphism $\zeta_\Lambda(M)$ by a suitable cancellation of the two last terms, compatible with the functional equation. This is achieved by putting

$$\begin{aligned}&\mathcal{L}_U := \mathcal{L}_{U,\beta} := \\ &\quad (\beta \cdot \mathrm{id}_{\mathbf{d}_\Lambda(SC_U)^{-1}\mathbf{d}_\Lambda(\hat{\mathbb{T}})^{-1}}) \circ (\epsilon_{p,\Lambda}(\hat{\mathbb{T}})^{-1} \cdot \zeta_\Lambda(M)) : \mathbf{1}_{\tilde{\Lambda}} \to \mathbf{d}_\Lambda(SC_U)^{-1}_{\tilde{\Lambda}}.\end{aligned} \tag{6.7}$$

In order to arrive at a p-adic L-function which is independent of U one has to replace SC_U by another Selmer complex $SC := SC(\hat{\mathbb{T}},\mathbb{T})$ (see (7.12)), which fits into a distinguished triangle

$$SC_U \longrightarrow SC \longrightarrow \bigoplus_{l\in S\setminus\{p,\infty\}} \mathrm{R}\Gamma_f(\mathbb{Q}_l,\hat{\mathbb{T}}) \longrightarrow, \tag{6.8}$$

where for $l \neq p$

$$\mathrm{R}\Gamma_f(\mathbb{Q}_l,\hat{\mathbb{T}}) \cong [\hat{\mathbb{T}}^{I_l} \xrightarrow{1-\varphi_l} \hat{\mathbb{T}}^{I_l}] \tag{6.9}$$

in the derived category. Replacing $\hat{\mathbb{T}}^{I_l}$ by a projective resolution if necessary and using the identity isomorphism of it we obtain isomorphisms

$$\zeta_l(M) = \zeta_l(M, F_\infty/\mathbb{Q}) : \mathbf{1}_\Lambda \to \mathbf{d}_\Lambda(\mathrm{R}\Gamma_f(\mathbb{Q}_l,\hat{\mathbb{T}}))^{-1} \tag{6.10}$$

and we define the p-adic L-function as

$$\mathcal{L} = \mathcal{L}_\beta(M) = \mathcal{L}_{U,\beta} \cdot \prod_{l\in S\setminus\{p,\infty\}} \zeta_l(M) : \mathbf{1}_{\tilde{\Lambda}} \to \mathbf{d}_\Lambda(SC)^{-1}_{\tilde{\Lambda}}. \tag{6.11}$$

Let Υ be the set of all primes $l \neq p$ such that the ramification index of l in $F_\infty/\mathbb{Q}$ is infinite. Note that Υ is empty if G has a commutative open subgroup.

Lemma 6.1 *[23, Prop. 4.2.14(3)]* $\hat{\mathbb{T}}^{I_l} = 0$ *and thus* $\zeta_l(M) = 1$ *in* $K_1(\Lambda)$ *for all* l *in* Υ.

Let us derive the *interpolation property* of $\mathcal{L}_U$ and $\mathcal{L}$. Whenever $L^n \otimes^{\mathbb{L}}_\Lambda SC_U$ is acyclic for a continuous representation $\rho : G \to \mathrm{GL}_n(\mathcal{O}_L)$,

L a finite extension of $\mathbb{Q}_p$, we obtain an element $\mathcal{L}_U(\rho) \in \widehat{L^{nr}}^\times$ from the isomorphism

$$\mathbf{1}_{\tilde{L}} \xrightarrow{L^n \otimes_\Lambda \mathcal{L}_U} \mathbf{d}_{\tilde{L}}(L^n \otimes^{\mathbb{L}}_\Lambda SC_U)^{-1} \xrightarrow{acyclic} \mathbf{1}_{\tilde{L}} \tag{6.12}$$

which via $K_1(\tilde{L}) \to K_1(\widehat{L^{nr}})$ can be considered as an element of $\widehat{L^{nr}}^\times$.

Let K be a finite extension of $\mathbb{Q}$, $\rho : G \to \mathrm{GL}_n(\mathcal{O}_K)$ an Artin representation, $[\rho^*]$ the Artin motive corresponding to ρ^*. Fix a place λ of K above p, put $L := K_\lambda$ and consider the L-linear representation of $G_\mathbb{Q}$ or its restriction to $G_{\mathbb{Q}_l}$

$$W := M(\rho^*)_\lambda = [\rho^*]_\lambda \otimes_{\mathbb{Q}_p} M_p$$

and the $G_{\mathbb{Q}_p}$-representation

$$\hat{W} := [\rho^*]_\lambda \otimes_{\mathbb{Q}_p} \hat{V}.$$

For a $G_{\mathbb{Q}_p}$-representation V define $P_{L,l}(V,u) := \det_L(1-\varphi_l u|V^{I_l}) \in L[u]$ if $l \neq p$ and $P_{L,p}(V,u) := \det_L(1-\varphi_p u|D_{cris}(V)) \in L[u]$ otherwise.

Some conditions for acyclicity are summarized in the next

Proposition 6.2 ([23, 4.2.21, 4.1.6-8]) *Assume the following conditions:*

(i) $H^j_f(\mathbb{Q}, W) = H^j_f(\mathbb{Q}, W^*(1)) = 0$ *for* $j = 0, 1$,

(ii) $P_{L,l}(W, 1) \neq 0$ *for any* $l \in \Upsilon$ *(respectively for any* $l \in S \setminus \{p, \infty\}$*).*

(iii) $\{P_{L,p}(W,u)P_{L,p}(\hat{W},u)^{-1}\}_{u=1} \neq 0$ *and* $P_{L,p}(\hat{W}^*(1),1) \neq 0$.

Then the following complexes are acyclic: $L^n \otimes^{\mathbb{L}}_{\Lambda,\rho} SC$ *(respectively* $L^n \otimes^{\mathbb{L}}_{\Lambda,\rho} SC_U$*),* $\mathrm{R}\Gamma_f(\mathbb{Q}_l, W) = L^n \otimes^{\mathbb{L}}_{\Lambda,\rho} \mathrm{R}\Gamma_f(\mathbb{Q}_l, \mathbb{T})$, *for any* $l \in \Upsilon$ *(respectively for any* $l \in S \setminus \{p\}$*). Furthermore, there are quasi-isomorphisms*

$$\mathrm{R}\Gamma(\mathbb{Q}_p, \hat{W}) \to \mathrm{R}\Gamma_f(\mathbb{Q}_p, W) \ \ and \ \ \mathrm{R}\Gamma_f(\mathbb{Q}_p, \hat{W}) \to \mathrm{R}\Gamma(\mathbb{Q}_p, \hat{W}).$$

Finally, assuming Conjectures 3.2 and 3.6, $L_K(M(\rho^*), s)$ *has neither a zero nor a pole at* $s = 0$.

Henceforth we assume the conditions (i)-(iii).

We define $\Omega_\infty(M(\rho^*)) \in \mathbb{C}^\times$ to be the determinant of the period map $\mathbb{C} \otimes_\mathbb{R} \alpha_{M(\rho^*)}$ with respect to the K-basis which arise from γ (respectively δ) and the basis given by ρ. It is easy to see that we have

$$\Omega_\infty(M(\rho^*)) = \Omega^+_\infty(M)^{d_+(\rho)}\Omega^-_\infty(M)^{d_-(\rho)}, \tag{6.13}$$

where $d_\pm(\rho) = \dim_K([\rho]^\pm_B)$ and $\Omega^\pm_\infty(M)$ is the determinant of $\mathbb{C} \otimes_\mathbb{Q} M^\pm_B \xrightarrow{\cong} \mathbb{C} \otimes_\mathbb{Q} t_M$ with respect to the basis $\gamma^\pm$ and δ. Assuming Conjecture 3.5 we have

$$\frac{L_K(M(\rho^*), 0)}{\Omega_\infty(M(\rho^*))} \in K^\times.$$

We claim that, using Proposition 7.2, the isomorphism $\mathcal{L}_U(\rho)$

$$\mathbf{1} \xrightarrow[\vartheta_\lambda\circ\zeta(M(\rho^*))]{\zeta_\Lambda(M)(\rho)=} \mathbf{d}(R\Gamma_c(U,W))^{-1} \xrightarrow[=\cdot\,\epsilon_{p,L}(\hat{W})^{-1}]{\cdot\,\epsilon_{p,\Lambda}(\hat{\mathbb{T}})^{-1}(\rho)}$$

$$\mathbf{d}(SC_U(\hat{W},W))^{-1}\mathbf{d}(\hat{W})^{-1}\mathbf{d}(W^+) \xrightarrow{\mathrm{id}\cdot\beta(\rho)} \mathbf{d}(SC_U(\hat{W},W))^{-1} \xrightarrow{acyc} \mathbf{1}$$

(we suppress for ease of notation the subscripts and remind the reader of our convention in Remark 1.2) is the product of the following automorphisms of $\mathbf{1}$:

(1) $L_K(M(\rho^*),0)\Omega_\infty(M(\rho^*))^{-1}$,

(2) $\Gamma_L(\hat{W})^{-1} = \Gamma_{\mathbb{Q}_p}(\hat{V})^{-1}$,

(3) $\Omega_p(M(\rho^*))$ which is, by definition, the composite

$$\begin{aligned} &\mathbf{d}(\hat{W}) \xrightarrow{\cdot\,\epsilon_{dR}(\hat{W})^{-1}} \mathbf{d}(D_{dR}(\hat{W})) \xrightarrow{\mathbf{d}(g^t_{dR})} \mathbf{d}(t_{M(\rho^*)}) \xrightarrow{\cdot\,can_{\gamma,\delta}} \\ &\mathbf{d}((M(\rho^*)^+_B)_L) \xrightarrow{\mathbf{d}(g^+_\lambda)} \mathbf{d}(W^+) \xrightarrow{\beta(\rho)} \mathbf{d}(\hat{W}) \end{aligned} \tag{6.14}$$

where we use $D^0_{dR}(\hat{W}) = 0$ for the second isomorphism and where we apply Remark 1.2 to obtain an automorphism of $\mathbf{1}$,[16]

(4) $\prod_{l\in S\setminus\{p,\infty\}} P_{L,l}(W,1)$: $\mathbf{1} \xrightarrow{\prod \eta_l(W)} \prod \mathbf{d}(R\Gamma_f(\mathbb{Q}_l,W)) \xrightarrow{acyc} \mathbf{1}$
where the first map comes from the trivialization by the identity and the second from the acyclicity,

(5) $\{P_{L,p}(W,u)P_{L,p}(\hat{W},u)^{-1}\}_{u=1}$:

$$\mathbf{1} \xrightarrow{\eta_p(W)\cdot\eta_p(\hat{W})^{-1}} \mathbf{d}(R\Gamma_f(\mathbb{Q}_p,W))\mathbf{d}(R\Gamma(\mathbb{Q}_p,\hat{W}))^{-1} \xrightarrow{quasi} \mathbf{1},$$

where we use that $t(W) = D_{dR}(\hat{W}) = t(\hat{W})$ and the quasi-isomorphisms mentioned in Proposition 6.2, and

(6) $P_{L,p}(\hat{W}^*(1),1) : \mathbf{1} \xrightarrow{\overline{\eta_p(\hat{W}^*(1))^*}} \mathbf{d}(R\Gamma_f(\mathbb{Q}_p,\hat{W}^*(1))) \xrightarrow{acyc} \mathbf{1}$,
where we use again that $t(\hat{W}^*(1)) = D^0_{dR}(\hat{W}) = 0$.

16 Using Remark 1.2(i) it is easy to see that this amounts to taking the product of the following isomorphisms and identifying the target with $\mathbf{1}$ afterwards

$$\mathbf{1} \xrightarrow{can_{\gamma,\delta}} \mathbf{d}((M(\rho^*)^+_B)_L)\mathbf{d}(t_{M(\rho^*)})^{-1}, \qquad \mathbf{1} \xrightarrow{\mathrm{id}_-\cdot\mathbf{d}(g^+_\lambda)} \mathbf{d}(W^+)\mathbf{d}((M(\rho^*)^+_B)_L)^{-1},$$

$$\mathbf{1} \xrightarrow{\mathrm{id}_-\cdot\mathbf{d}(g^t_{dR})} \mathbf{d}(D_{dR}(\hat{W}))^{-1}\mathbf{d}(t_{M(\rho^*)}), \qquad \mathbf{1} \xrightarrow{\epsilon_{dR}(\hat{W})^{-1}} \mathbf{d}(\hat{W})^{-1}\mathbf{d}(D_{dR}(\hat{W})),$$

$$\mathbf{1} \xrightarrow{\mathrm{id}_-\cdot\beta(\rho)} \mathbf{d}(\hat{W})\mathbf{d}(W^+)^{-1},$$

where the identity maps are those of $\mathbf{d}((M(\rho^*)^+_B)_L)^{-1}$, $\mathbf{d}(D_{dR}(\hat{W}))^{-1}$ and $\mathbf{d}(W^+)^{-1}$, respectively.

In order to describe the interpolation property for $\mathcal{L}$ we need in addition to (6.8) and Lemma 6.1 another

Lemma 6.3 ([23, Lem. 4.2.23]) *Let $l \neq p$ be a prime that does not belong to Υ. Then $L^n \otimes^{\mathbb{L}}_{\Lambda,\rho} R\Gamma_f(\mathbb{Q}_l, \hat{\mathbb{T}})$ is acyclic if and only if $P_{L,l}(W,1) \neq 0$. If this holds then we have $\zeta_l(M)(\rho) = P_{L,l}(W,1)^{-1}$.*

Thus we obtain the following

Theorem 6.4 ([23, Thm. 4.2.26]) *Under the conditions (i)-(iii) from Proposition 6.2 and assuming Conjecture 4.1 for M and Conjecture 5.9 for $\hat{\mathbb{T}}$ the value $\mathcal{L}(\rho)$ (respectively $\mathcal{L}_U(\rho)$) is equal to*

$$\frac{L_K(M(\rho^*),0)}{\Omega_\infty(M(\rho^*))} \cdot \Omega_p(M(\rho^*)) \cdot \Gamma_{\mathbb{Q}_p}(\hat{V})^{-1} \cdot \\ \cdot \{P_{L,p}(W,u) P_{L,p}(\hat{W},u)^{-1}\}_{u=1} \cdot P_{L,p}(\hat{W}^*(1),1) \cdot \prod_{l \in B} P_{L,l}(W,1),$$

where $B = \Upsilon \subseteq S \setminus \{p,\infty\}$ (respectively $B = S \setminus \{p,\infty\}$).

Remark 6.5 Note that conditions (ii) and (iii) are satisfied in the case of an abelian variety with good ordinary reduction at p. Furthermore, the quotient $\Omega_p(M(\rho^*))/\Omega_\infty(M(\rho^*))$ is independent of the choice of basis γ and δ. Also, it is easy to see[17] that for some suitable choice we have $\Omega_p(M(\rho^*)) = \epsilon_p(\hat{W})^{-1}$ which, according to standard properties of ϵ-constants (cf. [23, §3.2]) using that $\hat{V}$ is unramified as module under the Weil-group, is equal to $\epsilon_p(\rho^*)^{-k} \cdot \nu^{-f_p(\rho)}$ where $k = \dim_{\mathbb{Q}}(t_M) = \dim_{\mathbb{Q}_p}(\hat{V})$, $\nu = \det_{\mathbb{Q}_p}(\varphi_p | D_{cris}(\hat{V}))$ and where $f_p(\rho)$ is the p-adic order of the Artin-conductor of ρ. Due to the compatibility conjecture C_{WD} in [21, 2.4.3], which is known for abelian varieties (loc.cit., Rem. 2.4.6(ii)) and for Artin motives, one obtains the ϵ- and Euler-factors either from $D_{pst}(W)$ or from the corresponding l-adic realisations with $l \neq p$. Furthermore, we have $P_{L,p}(W,1) \neq 0$ and $P_{L,p}(\hat{W},1) \neq 0$ for weight reasons. Thus, noting that for abelian varieties $\Gamma(\hat{V}) = 1$, the above formula becomes

$$\frac{L_{K,\Upsilon'}(M(\rho^*),0)}{\Omega_\infty(M(\rho^*))} \cdot \epsilon_p(\rho^*)^{-k} \cdot \nu^{-f_p(\rho)} \cdot \frac{P_{L,p}(\hat{W}^*(1),1)}{P_{L,p}(\hat{W},1)}, \qquad \text{(Int)}$$

17 Note that in the definition of β and thus in $\beta(\rho)$ the epsilon factor $\epsilon_p(\hat{V})$ in $\epsilon_{dR}(\hat{V})$ equals 1 and thus $\Omega_p(M(\rho^*)) = \beta(\rho) \circ (\epsilon_p(\hat{W})^{-1} \cdot \overline{\beta(\rho)})$.

where $L_{K,\Upsilon'}$ denotes the modified L-function without the Euler-factors in $\Upsilon' := \Upsilon \cup \{p\}$.[18]

Proof We consider the case $\mathcal{L}_U$. First observe that due to the vanishing of the motivic cohomology the map

$$\zeta_K(M(\rho^*)) : \mathbf{1}_K \to \mathbf{d}(\Delta(M(\rho^*))) = \mathbf{d}(M(\rho^*)_B^+)\mathbf{d}(t_{M(\rho^*)})^{-1}$$

is just the map $can_{\gamma,\delta} : \mathbf{1} \cong \mathbf{d}(M(\rho^*)_B^+)\mathbf{d}(t_{M(\rho^*)})^{-1}$, induced by the bases arising from γ and δ, multiplied with $L_K(M(\rho^*),0)\Omega_\infty(M(\rho^*))^{-1}$. Secondly, since $\mathbf{d}(\mathrm{R}\Gamma_f(\mathbb{Q}, W)) = \mathbf{1}$, the isomorphism $\vartheta_\lambda(M(\rho^*)) : \mathbf{d}(\Delta(M(\rho^*)))_L \cong \mathbf{d}(\mathrm{R}\Gamma_c(U, W))^{-1}$ corresponds up to the identification $\mathbf{d}(M(\rho^*)_B^+)_L \cong \mathbf{d}(W^+)$ to the product of

$$\mathbf{d}(t_{M(\rho^*)})_L^{-1} \xrightarrow{\overline{\mathbf{d}(g_{dR}^t)}^{-1}} \mathbf{d}(t(W))^{-1} \xrightarrow{\cdot \eta_\lambda(W)} \mathbf{d}(\mathrm{R}\Gamma_f(\mathbb{Q}_p, W))$$

with $\underline{\prod_B P_{L,l}(W,1)}$. Thirdly, the contribution from $\epsilon_{p,L}(\hat{W})$ is equal to $\eta_p(\hat{W}) \cdot \eta_p(\hat{W}^*(1))^* \cdot \Gamma_L(\hat{W}) \cdot \epsilon_{dR}(\hat{W})$ up to the canonical local duality isomorphism. Together with $\beta(\rho)$ we thus obtain all the factors (1)-(6) above, i.e. after revealing all definitions and identifications, in particular all comparison isomorphisms, the same constituents show up in both expressions, the product of the factors (1)-(6) and $\mathcal{L}_U(\rho)$; hence by Remark 1.2 which says that all compositions of maps in $\mathcal{C}_{\tilde{L}}$ can be rephrased in terms of products, which do not depend on any particular order, we have proven the interpolation property for $\mathcal{L}_U$. To finish the proof in the case $\mathcal{L}$ use Lemmata 6.1, 6.3 and equation (7.14). □

6.1 Leading terms

In the interpolation formula of Theorem 6.4 both sides are equal to zero if the (analytic and algebraic) rank of the motive $M(\rho^*)$ is strictly positive (assuming here and henceforth the validity of the conjectures involved in that theorem). In this case one might still wonder whether one can describe the 'leading term' of the p-adic L-function $\mathcal{L}$ in terms of the leading term $L_K^*(M(\rho^*))$ of the complex L-function of $M(\rho^*)$. To this end in [11] the notion of the 'leading term at ρ' for elements of a suitable localized K_1-groups and for representations ρ has

18 In order to compare this formula with (107) in [14] we remark that, with the notation of (loc. cit.), $u = \det(\phi_l|\widehat{V}(-1)) = p\nu = p\omega^{-1}$. Then by [23, Rem. 4.2.27] one has $\epsilon(\rho^*)^{-d_\nu - f_p(\rho)} = \epsilon(\rho)^{d_u - f_p(\rho)}$ (strictly speaking one has to replace the period t by $-t$ in the second epsilon factor). But it seems that one has to interchange ρ and $\widehat{\rho}$ on the right hand side of (107).

been introduced in terms of the Bockstein homomorphisms that have already played significant roles (either implicitly or explicitly) in work of Perrin-Riou [36, 37], of Schneider [41, 40, 39, 38] and of Burns and Greither [10, 7] and have been systematically incorporated into Nekovář's theory of Selmer complexes [34]. We briefly sketch the result in the case $M = h^1(A)(1)$ of an abelian variety A over $\mathbb{Q}$ with good ordinary reduction at a fixed prime $p \neq 2$, for more general results and the details see [11].

We assume that the Galois group G of F_∞ over $\mathbb{Q}$ has a quotient isomorphic to $\mathbb{Z}_p$. For an arbitrary Artin representation $\rho : G \to \mathrm{GL}_n(\mathcal{O}_L)$, L a finite extension of $\mathbb{Q}_p$, the complex $L^n \otimes^{\mathbb{L}}_\Lambda SC_U(\hat{\mathbb{T}}, \mathbb{T})$ need not be acyclic. But nevertheless the above assumption on G grants the existence of Bockstein homomorphisms

$$\mathfrak{B}_i = \mathfrak{B}_i(SC_U) : \mathbb{H}_i(G, \rho^*, SC_U) \to \mathbb{H}_{i-1}(G, \rho^*, SC_U)$$

for the (co)homology groups $\mathbb{H}_i(G, \rho^*, SC_U) := \mathrm{H}^{-i}(L^n \otimes^{\mathbb{L}}_\Lambda SC_U(\hat{\mathbb{T}}, \mathbb{T}))$ giving rise to a complex $(\mathbb{H}_\bullet(G, \rho^*, SC_U), \mathfrak{B}_\bullet)$ and we say that $SC_U(\hat{\mathbb{T}}, \mathbb{T})$ is *semisimple at* ρ if this complex is acyclic. If this condition holds, using property h) of the determinant functor one easily constructs a trivialization

$$t(\rho^*, SC_U) : \mathbf{d}_L(L^n \otimes^{\mathbb{L}}_\Lambda SC_U(\hat{\mathbb{T}}, \mathbb{T})) \to \mathbf{1}_L.$$

Analogously to the evaluation $\mathcal{L}_U(\rho)$ of $\mathcal{L}_U$ at ρ using (6.12) we thus can define the *leading term* $\mathcal{L}^*_U(\rho)$ *of* $\mathcal{L}_U$ *at* ρ as the element in $\in \widehat{L^{nr}}^\times$ given by the isomorphism

$$\mathbf{1}_{\tilde{L}} \xrightarrow{L^n \otimes^{\mathbb{L}}_\Lambda \mathcal{L}_U} \mathbf{d}_{\tilde{L}}(L^n \otimes^{\mathbb{L}}_\Lambda SC_U(\hat{\mathbb{T}}, \mathbb{T}))^{-1} \xrightarrow{t(\rho^*, SC_U)^{-1}} \mathbf{1}_{\tilde{L}}$$

and similarly for $\mathcal{L}^*(\rho)$.

On the other hand one can show that there exists an isomorphism in $D^p(L)$ of the form

$$L^n \otimes^{\mathbb{L}}_\Lambda SC_U(\hat{\mathbb{T}}, \mathbb{T}) \cong SC_U(\hat{W}, W) \cong \mathrm{R}\Gamma_f(\mathbb{Q}, W)$$

under which the Bockstein map $\mathfrak{B}_{-1}(SC_U)$ coincides with the map

$$\mathrm{ad}(h_p(W)) : \mathrm{H}^1_f(\mathbb{Q}, W) \to \mathrm{H}^2_f(\mathbb{Q}, W) \cong \mathrm{H}^1_f(\mathbb{Q}, Z)^*$$

which is induced from Nekovář's p-adic height pairing

$$h_p(W) : \mathrm{H}^1_f(\mathbb{Q}, W) \times \mathrm{H}^1_f(\mathbb{Q}, Z) \to L$$

and global duality with $Z = W^*(1)$. Since the cohomology of $\mathrm{R}\Gamma_f(\mathbb{Q}, W)$ is concentrated in degrees 1 and 2, it turns out that $SC_U(\hat{\mathbb{T}}, \mathbb{T})$ is semisimple at ρ if and only if the pairing $h_p(W)$ is non-degenerate, which we will assume henceforth. We also assume that the archimedean height pairing for $N = M(\rho^*)$ is non-degenerate. Fixing K-bases of $H^1_f(N)$ and $H^1_f(N^*(1))^*$ we define the regulators $R_\infty(N)$ and $R_p(N)$ as the determinant of the inverse of

$$h_\infty(N) : \left(H^1_f(N^*(1))^*\right)_{\mathbb{C}} \to H^1_f(N)_{\mathbb{C}}$$

and as the determinant of $\mathrm{ad}(h_p(W))$ with respect to the induced bases, respectively; then the ratio $R_p(N)R_\infty(N)^{-1}$ is independent of the choice of the bases. Finally, we set $r := r(N) := \dim_L \mathrm{H}^1_f(\mathbb{Q}, W)$.

Theorem 6.6 ([11, Thm. 6.5]) *Let $M = h^1(A)(1)$ be the motive of an abelian variety A over $\mathbb{Q}$ with good ordinary reduction at a fixed prime $p \neq 2$. We assume that the archimedean and p-adic height pairing for the motive $M(\rho^*)$ are non-degenerate and that the morphisms $\zeta_\Lambda(M)$ and $\epsilon_{p,\Lambda}(\hat{\mathbb{T}})$ that are described in Conjecture 4.1 and Conjecture 5.9 exist.*

Then both $SC_U(\hat{\mathbb{T}}, \mathbb{T})$ and $SC(\hat{\mathbb{T}}, \mathbb{T})$ are semisimple at ρ and the leading term $\mathcal{L}^*(\rho)$ (respectively $\mathcal{L}^*_U(\rho)$) is equal to the product

$$(-1)^r \frac{L^*_{K,B}(M(\rho^*))}{\Omega_\infty(M(\rho^*)) \cdot R_\infty(M(\rho^*))} \cdot \Omega_p(M(\rho^*)) \cdot R_p(M(\rho^*)) \cdot \frac{P_{L,p}(\hat{W}^*(1), 1)}{P_{L,p}(\hat{W}, 1)},$$

where $L^*_{K,B}(M(\rho^*))$ denotes the leading term at $s = 0$ of the B-truncated complex L-function of $M(\rho^*)$ with $B := \Upsilon \cup S_p$ (respectively $B := S \setminus S_\infty$).

6.2 Interlude - Localised K_1

The following construction of a localized K_1 by Fukaya and Kato is one of the differences to the approach of Huber and Kings [24][19]. For a moment let Λ be an arbitrary ring with unit and let Σ be a full subcategory of $C^p(\Lambda)$ satisfying (i) if C is quasi-isomorphic to an object in Σ then it belongs to Σ, too, (ii) Σ contains the trivial complex, (iii) all translations of objects in Σ belong again to Σ and (iv) any extension C in $C^p(\Lambda)$ (by an exact sequence of complexes) of $C', C'' \in \Sigma$ is again in Σ. Then Fukaya and Kato construct a group $K_1(\Lambda, \Sigma)$ whose objects are all of the form $[C, a]$ with $C \in \Sigma$ and an isomorphism $a : \mathbf{1}_\Lambda \to \mathbf{d}_\Lambda(C)$ (in particular, $[C] := \sum_{i \in \mathbb{Z}} [C]^{(-1)^i} = 0$ in $K_0(\Lambda)$) satisfying certain relations, see [23, 1.3]. This group fits into an exact sequence

$$K_1(\Lambda) \longrightarrow K_1(\Lambda, \Sigma) \xrightarrow{\partial} K_0(\Sigma) \longrightarrow K_0(\Lambda), \qquad (6.15)$$

19 Instead of the localised K_1 they work with K_1 of the ring $\varprojlim_n \mathbb{Q}_p[G/G_n]$, which occurs in the context of distributions, see [16].

where $K_0(\Sigma)$ is the abelian group generated by $[[C]]$, $C \in \Sigma$ and satisfying certain relations, see (loc. cit.). Here the first map is given by sending the class of an automorphism $\Lambda^r \to \Lambda^r$ to $[[\Lambda^r \to \Lambda^r], can]$, where can denotes the trivialization of the complex $[\Lambda^r \to \Lambda^r]$ by the identity according to Remark 1.2, ∂ maps $[C, a]$ to $-[[C]]$ while the last map is given by $[[C]] \mapsto [C]$. If S is an left denominator set of Λ, $\Lambda_S := S^{-1}\Lambda$ the corresponding localization and Σ_S the full subcategory of $C^p(\Lambda)$ consisting of all complexes C such that $\Lambda_S \otimes_\Lambda C$ is acyclic, then $K_1(\Lambda, \Sigma_S)$ and $K_0(\Sigma_S)$ can be identified with $K_1(\Lambda_S)$ and $K_0(S\text{-tor}^{pd})$, respectively. Here S-tor^{pd} denotes the category of finitely generated S-torsion Λ-modules with finite projective dimension.

6.3 Iwasawa main conjecture I

Let $\mathcal{O}$ be the ring of integers of the completion at any place λ above p of the maximal abelian outside p unramified extension $F_\infty^{ab,p}$ of $\mathbb{Q}$ inside F_∞. Note that the latter extension is finite because every non-finite abelian p-adic Lie extension of $\mathbb{Q}$ contains the cyclotomic $\mathbb{Z}_p$-extension, which is ramified at p. Then by [23, Thm. 4.2.26(2)] $\epsilon_{p,\Lambda}(\hat{\mathbb{T}})$ and thus $\mathcal{L}$ are already defined over $\Lambda_\mathcal{O} := \mathcal{O} \otimes_{\mathbb{Z}_p} \Lambda(G)$ instead of $\widetilde{\Lambda}$.

Now let $\Sigma = \Sigma_{SC}$ be the smallest full subcategory of $C^p(\Lambda_\mathcal{O})$ containing SC and satisfying the conditions (i)-(iv) above. Then the evaluation of $\mathcal{L}$ factorizes over its class in $K_1(\Lambda_\mathcal{O}, \Sigma)$ which we still denote by $\mathcal{L}$. By the construction of $\mathcal{L}$ we have

Theorem 6.7 ([23, Thm. 4.2.22]) *Assume Conjectures 4.1 for (M, Λ) and 5.9 for $(\hat{\mathbb{T}}, \Lambda)$. Then the following holds:*

(i) $\partial(\mathcal{L}) = [[SC]]$

(ii) $\mathcal{L}$ satisfies the interpolation property (Int).

Question: If one knows the existence of $\mathcal{L} \in K_1(\Lambda_\mathcal{O}, \Sigma)$ with the above properties, what is missing to obtain the zeta-isomorphism?

6.4 The canonical Ore set

Now assume that the cyclotomic $\mathbb{Z}_p$-extension $\mathbb{Q}_{cyc}$ is contained in F_∞ and set $H := G(F_\infty/\mathbb{Q}_{cyc})$. In this situation there exists a canonical left and right denominator set of $\Lambda_\mathcal{O}$

$$S^* = \bigcup_{i \geq 0} p^i S$$

with

$$S = \{\lambda \in \Lambda_{\mathcal{O}} | \Lambda_{\mathcal{O}}/\Lambda_{\mathcal{O}}\lambda \text{ is a finitely generated } \Lambda_{\mathcal{O}}(H)\text{-module}\}$$

as was shown in [14].

In this case we write $\mathfrak{M}_H(G)$ for the category of S^*-torsion modules and identify $K_0(\mathfrak{M}_H(G))$ with $K_0(\Sigma_{S^*})$ recalling that $\Lambda_{\mathcal{O}}$ is regular.

We write

$$X = Sel(A/F_\infty)^\vee$$

for the Pontryagin dual of the classical Selmer group of A over F_∞, see [14].

Conjecture 6.8 ([14, Conj. 5.1]) $X \in \mathfrak{M}_H(G)$.

It is shown in [23, Prop. 4.3.7] that Conjecture 6.8 is equivalent to SC belonging to Σ_{S^*}. We assume this conjecture. Observe that then $\Sigma \subseteq \Sigma_{S^*}$, which induces a commutative diagram

$$\begin{array}{ccccccc}
K_1(\Lambda_{\mathcal{O}}) & \longrightarrow & K_1(\Lambda_{\mathcal{O}}, \Sigma) & \xrightarrow{\partial} & K_0(\Sigma) & \longrightarrow & K_0(\Lambda_{\mathcal{O}}) \\
\| & & \downarrow & & \downarrow & & \| \\
K_1(\Lambda_{\mathcal{O}}) & \longrightarrow & K_1((\Lambda_{\mathcal{O}})_{S^*}) & \xrightarrow{\partial} & K_0(\mathfrak{M}_H(G)) & \longrightarrow & K_0(\Lambda_{\mathcal{O}}).
\end{array}$$

Question: What is the (co)kernel of the middle vertical maps?

6.5 *Iwasawa main conjecture II*

In [14, §3] it is explained how to evaluate elements of $K_1((\Lambda_{\mathcal{O}})_{S^*})$ at representations. By [23, Lem. 4.3.10] this is compatible with the evaluation of elements in $K_1(\Lambda_{\mathcal{O}}, \Sigma)$ as explained above. The following version of a Main Conjecture was formulated in [14].

Conjecture 6.9 (Noncommutative Iwasawa Main Conjecture) *There exists a (unique) element $\mathcal{L}$ in $K_1((\Lambda_{\mathcal{O}})_{S^*})$ such that*

(i) $\partial\mathcal{L} = [X_{\mathcal{O}}]$ in $K_0(\mathfrak{M}_H(G))$ and

(ii) $\mathcal{L}$ satisfies the interpolation property (Int).

The connection with the previous version is given by the following

Proposition 6.10 *Let F_∞ be e.g. $\mathbb{Q}(A(p))$ or $\mathbb{Q}(\mu(p), \sqrt[p^\infty]{\alpha})$ for some $\alpha \in \mathbb{Q}^\times \setminus \mu$ (false Tate curve)*[20]. *Then*

$$[[X]] = [[SC]]$$

in $K_0(\Sigma_{S^*})$. In particular, Conjecture 6.9 is a consequence of Conjecture 4.1 for (M, Λ) and Conjecture 5.9 for $(\hat{\mathbb{T}}, \Lambda)$.

The advantage of the localisation $(\Lambda_{\mathcal{O}})_{S^*}$ relies on the fact that one has an explicit description of its first K-group since the natural map $(\Lambda_{\mathcal{O}})_{S^*}^\times \to K_1((\Lambda_{\mathcal{O}})_{S^*})$ induces quite often an isomorphism of the maximal abelian quotient of $(\Lambda_{\mathcal{O}})_{S^*}^\times$ onto $K_1((\Lambda_{\mathcal{O}})_{S^*})$, see [14, Thm. 4.4]. On the other hand, the localized $K_1(\Lambda_{\mathcal{O}}, \Sigma)$ exists without the assumption that G maps surjectively onto $\mathbb{Z}_p$, e.g. if $G = SL_n(\mathbb{Z}_p)$. Also, if G has p-torsion elements, i.e. if $\Lambda_{\mathcal{O}}(G)$ is *not* regular, one can still formulate the Main Conjecture using the complex SC instead of the classical Selmer group X (which could have infinite projective dimension).

Question: To which extent does a p-adic L-function $\mathcal{L}$ together with Conjecture 6.9 determine the ζ-isomorphism in Conjecture 4.1? In other words, does the Main conjecture imply the ETNC?

7 Appendix: Galois cohomology

The main reference for this appendix is [23, §1.6], but see also [9, 8]. For simplicity we assume $p \neq 2$ throughout this section. Let $U = \mathrm{Spec}(\mathbb{Z}[\frac{1}{S}])$ be a dense open subset of $\mathrm{Spec}(\mathbb{Z})$ where S contains $S_p := \{p\}$ and $S_\infty := \{\infty\}$ (by abuse of notation). We write G_S for the Galois group of the maximal outside S unramified extension of $\mathbb{Q}$. Let X be a topological abelian group with a continuous action of G_S. Examples we have in mind are $X = T_p, M_p, \mathbb{T}$, etc. Using continuous cochains one defines a complex $\mathrm{R}\Gamma(U, X)$[21] whose cohomology is $\mathrm{H}^n(G_S, X)$. Then $\mathrm{R}\Gamma_c(U, X)$ is defined by the exact triangle

$$\mathrm{R}\Gamma_c(U, X) \longrightarrow \mathrm{R}\Gamma(U, X) \longrightarrow \bigoplus_{v \in S} \mathrm{R}\Gamma(\mathbb{Q}_v, X) \longrightarrow \qquad (7.1)$$

where the $\mathrm{R}\Gamma(\mathbb{Q}_l, X)$ and $\mathrm{R}\Gamma(\mathbb{R}, X)$ denote the continuous cochain complexes calculating the local Galois groups $\mathrm{H}^n(\mathbb{Q}_l, X)$ and $\mathrm{H}^n(\mathbb{R}, X)$. Its cohomology is concentrated in degrees $0, 1, 2, 3$.

20 See [23, Prop. 4.3.15-17] for a more general statement.

21 For ease of notation we do not distinguish between complexes and their image in the derived category, though this is sometimes necessary in view of the correct use of the determinant functor and exact sequences of complexes.

Let L be a finite extension of $\mathbb{Q}_p$ with ring of integers $\mathcal{O}$. Now we define the local and global *"finite parts"* for a finite dimensional L-vector space V with continuous $G_{\mathbb{Q}_v}$- and $G_{\mathbb{Q}}$-action, respectively. For $\mathbb{Q}_v = \mathbb{R}$ we set

$$\mathrm{R}\Gamma_f(\mathbb{R}, V) := \mathrm{R}\Gamma(\mathbb{R}, V)$$

while for a finite place $\mathrm{R}\Gamma_f(\mathbb{Q}_l, V)$ is defined as a certain subcomplex of $\mathrm{R}\Gamma(\mathbb{Q}_l, V)$, concentrated in degree 0 and 1, whose image in the derived category is isomorphic to

$$\mathrm{R}\Gamma_f(\mathbb{Q}_l, V) \cong \begin{cases} [\, V^{I_l} \xrightarrow{1-\varphi_l} V^{I_l} \,] & \text{if } l \neq p, \\ [\, D_{cris}(V) \xrightarrow{(1-\varphi_p, 1)} D_{cris}(V) \oplus D_{dR}(V)/D^0_{dR}(V) \,] & \text{if } l = p. \end{cases} \tag{7.2}$$

Here φ_l denotes the geometric Frobenius (inverse of the arithmetic) and the induced map $D_{dR}(V)/D^0_{dR}(V) \to H^1_f(\mathbb{Q}_p, V)$ is called the *exponential map* $\exp_{BK}(V)$ of Bloch-Kato, where we write $\mathrm{H}^n_f(\mathbb{Q}_l, V)$ for the cohomology of $\mathrm{R}\Gamma_f(\mathbb{Q}_l, V)$.

Defining $\mathrm{R}\Gamma_{/f}(\mathbb{Q}_l, V)$ as a mapping cone

$$\mathrm{R}\Gamma_f(\mathbb{Q}_l, V) \longrightarrow \mathrm{R}\Gamma(\mathbb{Q}_l, V) \longrightarrow \mathrm{R}\Gamma_{/f}(\mathbb{Q}_l, V) \longrightarrow \tag{7.3}$$

we finally define $\mathrm{R}\Gamma_f(\mathbb{Q}, V)$, whose cohomology is concentrated in degrees $0, 1, 2, 3$, as mapping fibre

$$\mathrm{R}\Gamma_f(\mathbb{Q}, V) \longrightarrow \mathrm{R}\Gamma(U, V) \longrightarrow \bigoplus_{l \in S \setminus S_\infty} \mathrm{R}\Gamma_{/f}(\mathbb{Q}_l, V) \longrightarrow . \tag{7.4}$$

This is independent of the choice of U. The octahedral axiom induces an exact triangle

$$\mathrm{R}\Gamma_c(U, V) \longrightarrow \mathrm{R}\Gamma_f(\mathbb{Q}, V) \longrightarrow \bigoplus_{v \in S} \mathrm{R}\Gamma_f(\mathbb{Q}_v, V) \longrightarrow . \tag{7.5}$$

7.1 Duality

Let G, $\Lambda = \Lambda(G)$, $\mathbb{T}$ as in section 4. By abuse of notation we write $-^*$ for both (derived) functors $\mathbf{R}\mathrm{Hom}_\Lambda(-, \Lambda)$ and $\mathbf{R}\mathrm{Hom}_{\Lambda^\circ}(-, \Lambda^\circ)$. Then Artin-Verdier/Poitou-Tate duality induces the existence of the following distinguished triangle in the derived category of Λ-modules

$$\mathrm{R}\Gamma_c(U, \mathbb{T}) \longrightarrow \mathrm{R}\Gamma(U, \mathbb{T}^*(1))^*[-3] \longrightarrow \mathbb{T}^+ \longrightarrow \tag{7.6}$$

and similarly for T (a Galois stable $\mathcal{O}$-lattice of V) and V as coefficients (with Λ replaced by $\mathcal{O}$ and L, respectively).[22]

For the finite parts one obtains from Artin-Verdier/Poitou-Tate and local Tate-duality the following isomorphisms

$$\mathrm{R}\Gamma_f(\mathbb{Q}_l, V) \cong (\mathrm{R}\Gamma(\mathbb{Q}_l, V^*(1))/\mathrm{R}\Gamma_f(\mathbb{Q}_l, V^*(1)))^*[-2], \tag{7.7}$$

$$\mathrm{R}\Gamma_f(\mathbb{Q}, V) \cong \mathrm{R}\Gamma_f(\mathbb{Q}, V^*(1))^*[-3]. \tag{7.8}$$

Set $t(V) := D_{dR}(V)/D^0_{dR}(V)$ if $l = p$ and $t(V) = 0$ otherwise. Trivializing V^{I_l} and $D_{cris}(V)$, respectively, in (7.2) by the identity induces, for each l, an isomorphism

$$\eta_l(V): \quad \mathbf{1}_L \quad \to \mathbf{d}_L(\mathrm{R}\Gamma_f(\mathbb{Q}_l, V))\mathbf{d}_L(t(V)). \tag{7.9}$$

Then, setting $D_l(V) = D_{dR}(V)$ if $l = p$ and $D_l(V) = 0$ otherwise, the isomorphism

$$\Theta_l(V): \mathbf{1}_L \to \mathbf{d}_L(\mathrm{R}\Gamma(\mathbb{Q}_l, V)) \cdot \mathbf{d}_L(D_l(V)) \tag{7.10}$$

is by definition induced from $\eta_l(V)\cdot\overline{(\eta_l(V^*(1))^*)}$ followed by an isomorphism induced by local duality (7.7) and using the analogue $D^0_{dR}(V) = t(V^*(1))^*$ of (5.1) if $l = p$.[23]

7.2 Selmer complexes

For $l \neq p$ we define $\mathrm{R}\Gamma_f(\mathbb{Q}_l, \mathbb{T})$ as in (7.2) and $\mathrm{R}\Gamma_{/f}(\mathbb{Q}_l, \mathbb{T})$ as in (7.3) with V replaced by $\mathbb{T}$, see also (6.9). We do *not* define $\mathrm{R}\Gamma_f(\mathbb{Q}_p, \mathbb{T})$ since there is in general no integral version of $D_{cris}(V)$.

22 A more precise form to state the duality is the following. Let $\mathrm{R}\Gamma_{(c)}(U, \mathbb{T})$ be defined like $\mathrm{R}\Gamma_c(U, \mathbb{T})$ but using Tate cohomology $\widehat{\mathrm{R}\Gamma}(\mathbb{R}, \mathbb{T})$ instead of the usual group cohomology $\mathrm{R}\Gamma(\mathbb{R}, \mathbb{T})$. Then one has isomorphisms

$$\mathrm{R}\Gamma(U, \mathbb{T}^*(1))^* \cong \mathrm{R}\Gamma_{(c)}(U, \mathbb{T})[3] \cong \mathrm{R}\Gamma(U, \mathbb{T}^\vee(1))^\vee$$

where $-^\vee = \mathrm{Hom}_{cont}(-\mathbb{Q}_p/\mathbb{Z}_p)$ denotes the Pontryagin dual.

23 More explicitly, $\theta_p(V)$ is obtained from applying the determinant functor to the following exact sequence

$$0 \longrightarrow \mathrm{H}^0(\mathbb{Q}_p, V) \longrightarrow D_{cris}(V) \longrightarrow D_{cris}(V) \oplus t(V) \xrightarrow{\exp_{BK}(V)}$$

$$\mathrm{H}^1(\mathbb{Q}_p, V) \xrightarrow{\exp_{BK}(V^*(1))^*} D_{cris}(V^*(1))^* \oplus t(V^*(1))^* \longrightarrow D_{cris}(V^*(1))^*$$

$$\longrightarrow \mathrm{H}^2(\mathbb{Q}_p, V) \longrightarrow 0$$

which arises from joining the defining sequences of $\exp_{BK}(V)$ with the dual sequence for $\exp_{BK}(V^*(1))$ by local duality (7.7).

The Selmer complex $SC_U(\hat{\mathbb{T}},\mathbb{T})$ is by definition the mapping fibre

$$SC_U(\hat{\mathbb{T}},\mathbb{T}) \longrightarrow \mathrm{R}\Gamma(U,\mathbb{T}) \longrightarrow \mathrm{R}\Gamma(\mathbb{Q}_p,\mathbb{T}/\hat{\mathbb{T}}) \oplus \bigoplus_{l\in S\setminus(S_p\cup S_\infty)} \mathrm{R}\Gamma(\mathbb{Q}_l,\mathbb{T}) \tag{7.11}$$

while $SC(\hat{\mathbb{T}},\mathbb{T})$ is the mapping fibre

$$SC(\hat{\mathbb{T}},\mathbb{T}) \to \mathrm{R}\Gamma(U,\mathbb{T}) \to \mathrm{R}\Gamma(\mathbb{Q}_p,\mathbb{T}/\hat{\mathbb{T}}) \oplus \bigoplus_{l\in S\setminus(S_p\cup S_\infty)} \mathrm{R}\Gamma_{/f}(\mathbb{Q}_l,\mathbb{T})\,. \tag{7.12}$$

Thus by the octahedral axiom one obtains the distinguished triangles (6.6), (6.8) and, using Artin-Verdier/Poitou-Tate duality,

$$\mathrm{R}\Gamma(U,\mathbb{T}^\vee(1))^\vee \to SC(\hat{\mathbb{T}},\mathbb{T}) \to \mathrm{R}\Gamma(\mathbb{Q}_p,\hat{\mathbb{T}}) \oplus \bigoplus_{l\in S\setminus(S_p\cup S_\infty)} \mathrm{R}\Gamma_f(\mathbb{Q}_l,\mathbb{T})\,. \tag{7.13}$$

With the notation of section 6 the Selmer complexes $SC_U(\hat{W},W)$ and $SC(\hat{W},W)$ are defined analogously and satisfy analogous properties.

The following properties [23, (4.2),propositions 1.6.5, 2.1.3 and 4.2.15] are necessary conditions for the existence of the zeta-isomorphism $\zeta_\Lambda(\mathbb{T})$ in Conjecture 4.1 and the p-adic L-functions $\mathcal{L}_U$ (6.11) and $\mathcal{L}$ (6.7).

Proposition 7.1 *The complexes* $\mathrm{R}\Gamma_c(U,\mathbb{T})$ *and* $SC_U(\hat{\mathbb{T}},\mathbb{T})$ *are perfect*[24] *and in* $K_0(\Lambda)$ *we have*

$$[\mathrm{R}\Gamma_c(U,\mathbb{T})] = [SC_U(\hat{\mathbb{T}},\mathbb{T})] = 0.$$

If G does not have p-torsion, then $SC(\hat{\mathbb{T}},\mathbb{T})$ *is also perfect and we have* $[SC(\hat{\mathbb{T}},\mathbb{T})] = 0$.

7.3 Descent properties

For the evaluation at representations one needs good descent properties of the complexes involved.

Proposition 7.2 *[23, Prop. 1.6.5] With the notation as in section 6 we have canonical isomorphisms (for all primes l)*

$$L^n \otimes^{\mathbb{L}}_{\Lambda,\rho} \mathrm{R}\Gamma(U,\dot{\mathbb{T}}) \cong \mathrm{R}\Gamma(U,W), \qquad L^n \otimes^{\mathbb{L}}_{\Lambda,\rho} \mathrm{R}\Gamma_c(U,\mathbb{T}) \cong \mathrm{R}\Gamma_c(U,W),$$
$$L^n \otimes^{\mathbb{L}}_{\Lambda,\rho} \mathrm{R}\Gamma_{(c)}(U,\mathbb{T}) \cong \mathrm{R}\Gamma_{(c)}(U,W), \quad L^n \otimes^{\mathbb{L}}_{\Lambda,\rho} \mathrm{R}\Gamma(\mathbb{Q}_l,\mathbb{T}) \cong \mathrm{R}\Gamma(\mathbb{Q}_l,W),$$
$$L^n \otimes^{\mathbb{L}}_{\Lambda,\rho} SC_U(\hat{\mathbb{T}},\mathbb{T}) \cong SC_U(\hat{W},W).$$

For $l \notin \Upsilon \cup S_p$ we also have: $L^n \otimes^{\mathbb{L}}_{\Lambda,\rho} \mathrm{R}\Gamma_f(\mathbb{Q}_l,\mathbb{T}) \cong \mathrm{R}\Gamma_f(\mathbb{Q}_l,W)$.

24 $\mathrm{R}\Gamma_c(U,\mathbb{T})$ is even perfect for $p=2$, this is the reason that it is better for the formulation of the ETNC than $\mathrm{R}\Gamma(U,\mathbb{T})$.

But note that the complex $\mathrm{R}\Gamma_f(\mathbb{Q}_l, \hat{\mathbb{T}})$ for $l \in \Upsilon$, and thus also $SC(\hat{\mathbb{T}}, \mathbb{T})$, does *not* descend like this in general. Instead, according to [23, Prop. 4.2.17] one has a distinguished triangle

$$L^n \otimes^{\mathbb{L}}_{\Lambda,\rho} SC(\hat{\mathbb{T}}, \mathbb{T}) \longrightarrow SC(\hat{W}, W) \longrightarrow \bigoplus_{l\in\Upsilon} \mathrm{R}\Gamma_f(\mathbb{Q}_l, W) . \tag{7.14}$$

Bibliography

[1] D. Benois, *On Iwasawa theory of crystalline representations*, Duke Math. J. **104** (2000), no. 2, 211–267.

[2] D. Benois and L. Berger, *Théorie d'Iwasawa des Représentations Cristallines II,* preprint 2005, http://arxiv.org/abs/math/0509623.

[3] L. Berger, *Tamagawa numbers of some crystalline representations*, arXiv:math.NT/0209233.

[4] S. Bloch, *A note on height pairings, Tamagawa numbers, and the Birch and Swinnerton-Dyer conjecture*, Invent. Math. **58** (1980), no. 1, 65–76.

[5] S. Bloch and K. Kato, *L-functions and Tamagawa numbers of motives*, The Grothendieck Festschrift, Vol. I, Progr. Math., vol. 86, Birkhäuser Boston, Boston, MA, 1990, pp. 333–400.

[6] J. I. Burgos Gil, *The regulators of Beilinson and Borel*, CRM Monograph Series, vol. 15, American Mathematical Society, Providence, RI, 2002.

[7] D. Burns, *On the values of equivariant Zeta functions of curves over finite fields*, Documenta Math. **9** (2004), 357–399.

[8] D. Burns and M. Flach, *Motivic L-functions and Galois module structures*, Math. Ann. **305** (1996), no. 1, 65–102.

[9] D. Burns and M. Flach, *Tamagawa numbers for motives with (non-commutative) coefficients*, Documenta Math. **6** (2001), 501–570.

[10] D. Burns and C. Greither, *On the equivariant Tamagawa number conjecture for Tate motives*, Invent. Math. **153** (2003), no. 2, 303–359.

[11] D. Burns and O. Venjakob, *On the leading terms of zeta isomorphisms and p-adic L-functions in non-commutative Iwasawa theory*, Documenta Math. Extra Vol.: John H. Coates' Sixtieth Birthday, (2006), pp. 165-209.

[12] J. Coates, *On p-adic L-functions*, Astérisque **177-178** (1989), no. Exp. No. 701, 33–59.

[13] J. Coates, *Motivic p-adic L-functions*, L-functions and arithmetic (Durham, 1989), London Math. Soc. Lecture Note Ser., vol. 153, Cambridge Univ. Press, Cambridge, 1991, pp. 141–172.

[14] J. Coates, T. Fukaya, K. Kato, R. Sujatha, and O. Venjakob, *The* GL_2 *main conjecture for elliptic curves without complex multiplication*, Publ. Math. IHES. **101** (2005), no. 1, 163 – 208.

[15] J. Coates and B. Perrin-Riou, *On p-adic L-functions attached to motives over* **Q**, Algebraic number theory, Adv. Stud. Pure Math., vol. 17, Academic Press, Boston, MA, 1989, pp. 23–54.

[16] P. Colmez, *Théorie d'Iwasawa des représentations de de Rham d'un corps local*, Ann. of Math. (2) **148** (1998), no. 2, 485–571.

[17] B. Conrad and M. Lieblich, *Galois representations arising from p-divisible groups*, http://www.math.lsa.umich.edu/~ bdconrad/papers/pdivbook.pdf.

[18] P. Deligne, *Valeurs de fonctions L et périodes d'intégrales*, Automorphic forms, representations and L-functions (Proc. Sympos. Pure Math., Oregon State Univ., Corvallis, Ore., 1977), Part 2, Proc. Sympos. Pure Math., XXXIII, Amer. Math. Soc., Providence, R.I., 1979, pp. 313–346.

[19] M. Flach, *The equivariant Tamagawa number conjecture: a survey*, Stark's conjectures: recent work and new directions, Contemp. Math., vol. 358, Amer. Math. Soc., Providence, RI, 2004, With an appendix by C. Greither, pp. 79–125.

[20] J.-M. Fontaine, *Représentations l-adiques potentiellement semi-stables*, Astérisque (1994), no. 223, 321–347.

[21] J.-M. Fontaine, *Représentations p-adiques semi-stables*, Astérisque (1994), no. 223, 113–184.

[22] J.-M. Fontaine and B. Perrin-Riou, *Autour des conjectures de Bloch et Kato: cohomologie galoisienne et valeurs de fonctions L*, Motives (Seattle, WA, 1991), Proc. Sympos. Pure Math., vol. 55, Amer. Math. Soc., Providence, RI, 1994, pp. 599–706.

[23] T. Fukaya and K. Kato, *A formulation of conjectures on p-adic zeta functions in non-commutative Iwasawa theory*, Proceedings of the St. Petersburg Mathematical Society, Vol. XII (Providence, RI), Amer. Math. Soc. Transl. Ser. 2, vol. 219, Amer. Math. Soc., 2006, pp. 1–86.

[24] A. Huber and G. Kings, *Equivariant Bloch-Kato conjecture and non-abelian Iwasawa main conjecture*, Proceedings of the International Congress of Mathematicians, Vol. II (Beijing, 2002) (Beijing), Higher Ed. Press, 2002, pp. 149–162.

[25] A. Huber and G. Kings, *Bloch Kato Conjecture and Main Conjecture of Iwasawa Theory for Dirichlet Characters*, Duke Math. Journal **119** (2003), no. 3, 393–464.

[26] K. Kato, M. Kurihara and T. Tsuji, *Local Iwasawa theory of Perrin-Riou and syntomic complexes*, 1996.

[27] K. Kato, *Iwasawa theory and p-adic Hodge theory*, Kodai Math. J. **16** (1993), no. 1, 1–31.

[28] K. Kato, *Lectures on the approach to Iwasawa theory for Hasse-Weil L-functions via B_{dR}. I*, Arithmetic algebraic geometry (Trento, 1991), Lecture Notes in Math., vol. 1553, Springer, Berlin, 1993, pp. 50–163.

[29] K. Kato, *K_1 of some non-commutative completed group rings*, K-Theory **34** (2005), no. 2, 99–140.

[30] S. L. Kleiman, *Algebraic cycles and the Weil conjectures*, Dix esposés sur la cohomologie des schémas, North-Holland, Amsterdam, 1968, pp. 359–386.

[31] S. L. Kleiman, *The picard scheme*, arXiv:math.AG/0504020v1 (2005).

[32] S. Lang, *Algebraic groups over finite fields*, Amer. J. Math. **78** (1956), 555–563.

[33] J. S. Milne, *Abelian varieties*, Arithmetic geometry (Storrs, Conn., 1984), Springer, New York, 1986, pp. 103–150.

[34] J. Nekovář, *Selmer complexes*, to appear in Astérisque. Available at `http://www.math.jussieu.fr/~nekovar/pu/`

[35] J. Nekovář, *On p-adic height pairings*, Séminaire de Théorie des Nombres, Paris, 1990–91, Progr. Math., vol. 108, Birkhäuser Boston, Boston, MA, 1993, pp. 127–202.

[36] B. Perrin-Riou, *Théorie d'Iwasawa et hauteurs p-adiques (cas des variétés abéliennes)*, Séminaire de Théorie des Nombres, Paris, 1990–91, Progr. Math., vol. 108, Birkhäuser Boston, Boston, MA, 1993, pp. 203–220.

[37] B. Perrin-Riou, *p-adic L-functions and p-adic representations*, SMF/AMS Texts and Monographs, vol. 3, American Mathematical Society, Providence, RI, 2000.

[38] P. Schneider, *Height pairings in the Iwasawa theory of abelian varieties.*, Theorie des nombres, Semin. Delange-Pisot-Poitou, Paris 1980-81, Prog. Math. 22, 1982, pp. 309–316.

[39] P. Schneider, *p-adic height pairings. I.*, Invent. Math. **69** (1982), 401–409.

[40] P. Schneider, *Iwasawa L-functions of varieties over algebraic number fields. A first approach*, Invent. Math. **71** (1983), no. 2, 251–293.

[41] P. Schneider, *p-adic height pairings. II.*, Invent. Math. **79** (1985), 329–374.

[42] A. J. Scholl, *Motives for modular forms*, Invent. Math. **100** (1990), no. 2, 419–430.

[43] A. J. Scholl, *Classical motives*, Motives (Seattle, WA, 1991), Proc. Sympos. Pure Math., vol. 55, Amer. Math. Soc., Providence, RI, 1994, pp. 163–187.

[44] SGA7, *Groupes de monodromie en géométrie algébrique. I*, Springer-Verlag, Berlin, 1972, Séminaire de Géométrie Algébrique du Bois-Marie 1967–1969 (SGA 7 I), Dirigé par A. Grothendieck. Avec la collaboration de M. Raynaud et D. S. Rim, Lecture Notes in Mathematics, Vol. 288.

[45] J. H. Silverman, *The arithmetic of elliptic curves*, Graduate Texts in Mathematics, vol. 106, Springer-Verlag, New York, 1986.

[46] J. H. Silverman, *Advanced topics in the arithmetic of elliptic curves*, Graduate Texts in Mathematics, vol. 151, Springer-Verlag, New York, 1994.

[47] O. Venjakob, *From classical to non-commutative Iwasawa theory - an introduction to the* GL_2 *main conjecture*, 4ECM Stockholm 2004, EMS, 2005.

[48] S. Yasuda, *Local constants in torsion rings*, Ph.D. thesis, University of Tokyo, 2001.

The André-Oort conjecture - a survey

Andrei Yafaev
University College London,
Dept. of Mathematics,
25 Gordon Street,
WC1H 0AH, U.K.
yafaev@math.ucl.ac.uk[a]

[a] *The author was supported by an EPSRC grant and the European Research Training Network "Arithmetic Algebraic Geometry".*

1 Introduction

The purpose of this paper is to review the strategies and methods developed over the past few years in relation to the André-Oort conjecture. We would like bring the attention of the reader to a survey by Rutger Noot (see [20]) based on his talk at the Séminaire Bourbaki in November 2004. In this paper, we have tried to avoid overlapping too much with Noot's survey. This paper is based on the author's talk at the Durham Symposium in the summer of 2004. Laurent Clozel gave a talk on an approach to the André-Oort conjecture involving ergodic-theoretic methods. We will touch upon the contents of his lecture in the last section of this paper.

Let us recall the statement of the André-Oort conjecture.

Conjecture 1.1 (André-Oort) *Let S be a Shimura variety and let Σ be a set of special points in S. The irreducible components of the Zariski closure of Σ are special subvarieties (or subvarieties of Hodge type).*

In the next section we will review the notions of Shimura varieties, special points and special subvarieties. In this introduction we review some of the results on this conjecture obtained so far. This conjecture was stated by Yves André in 1989 (Problem 9 in [1]) for one dimensional subvarieties of Shimura varieties and in 1995 by Frans Oort in [22] for subvarieties of the moduli space $\mathcal{A}_g$ of principally polarised abelian varieties of dimension g. The statement above is the obvious generalisation of these two conjectures and is now refered to as the André-Oort conjecture. One of the motivations for the André-Oort conjecture was the Manin-Mumford conjecture about the distribution of torsion points on subvarieties of abelian varieties. The Manin-Mumford conjecture has been first proved by Raynaud in 1983 and since then, the number of new proofs of this conjecture has been increasingly growing. One may consult [26], [28], [36], [13] or [25] for some of the proofs of this conjecture. The strategies to attack the André-Oort conjecture that we will review in this survey are inspired by some of the proofs of the Manin-Mumford conjecture.

Although the conjecture remains open, a number of results on this conjecture have been obtained in the course of the past ten years or so. Let us review the most significant of them.

(i) Moonen proved the following result (see [17] and [18]). Let Σ be a set of special points in $\mathcal{A}_g$. Suppose that there exists a prime p such that for every s in Σ, the corresponding abelian variety A_s has good ordinary reduction at some place above p of which A_s is the canonical lift. Then the components of the Zariski closure of Σ are special.

This result has been generalised by Yafaev in 2003 to the case of arbitrary Shimura varieties, see [34]. We will come back to this result and its proof in the section 5.

(ii) Edixhoven (see [10]) and André (see [3]) proved the André-Oort conjecture for products of two modular curves. Edixhoven's proof is conditional on the Generalised Riemann Hypothesis (GRH) for imaginary quadratic fields while André's proof is unconditionnal (it relies on some diophantine approximation results on the j-function, see section 5.2 of [20] for a sketch of his proof). However Edixhoven's proof offers a definite strategy suitable for further generalisations. This generalisation has been subsequently carried out by Edixhoven and the author of the present article.

(iii) Edixhoven proved, assuming the GRH, the André-Oort conjecture for subvarieties of products of an arbitrary number of modular curves and for curves contained in the Hilbert modular surfaces.

In 2005, Ullmo and Yafaev proved the André-Oort conjecture for products of modular curves, also assuming the GRH using a different method. The paper [31] will be released shortly.

(iv) Edixhoven and Yafaev proved the André-Oort conjecture for curves in Shimura varieties containing an infinite set of special points lying in one Hecke orbit. See [12]. One of the motivations for doing this was an application to a problem in transcendence theory of hypergeometric functions. One can consult, for example, the section 4.2 of [20] for more on this application.

(v) In 2002 Yafaev proved the André-Oort conjecture for curves in Shimura varieties assuming the generalised Riemann hypothesis (see [35]). This was the original question of Yves André.

(vi) Clozel and Ullmo (see [5]) proved a result on equidistribution of strongly special subvarieties which has implications for the André-Oort conjecture. This result has very recently been generalised by Ullmo (see [33]), thus obtaining a nearly optimal result that can be reached using ergodic theoretic methods. This will be explained in more detail in the section 5.

The strategy, orginated by Edixhoven and used in [10], [11], [12], [35] and [34] to obtain results on the André-Oort conjecture will be reviewed in detail in this survey. This strategy relies on two principles. The first, is that a subvariety Z of a Shimura variety (satisfying certain conditions, but we are not being specific here) is special if and only if it is contained in its image by some suitable Hecke correspondence. This characterisation will be reviewed in the third section. With this characterisation in hand, one tries to find a suitable Hecke correspondence by using the Galois action on special points. Key points here are that the Galois conjugates of a special point lie in its Hecke orbit and that Hecke correspondences commute with the Galois action. The important ingredient here is a theorem on lower bounds for Galois orbits of special points. In the fourth section we will review a more general theorem, on lower bounds for non-strongly special subvarities.

The most recent developments on the conjecture exploit the dichotomy between the Galois-theoretic properties of special subvarieties and equidistribution discovered by Ullmo and Yafaev in [31]. This is the object of papers [31] and [14] in preparation.

Finally a proof of the André-Oort conjecture assuming the Generalised Riemann Hypothesis and unconditionally under the assumption that the set Σ lies in one (generalised) Hecke orbit has been announced by Bruno Klingler and the author of this survey. We will review the strategy of the announced proof in the last section.

We would like to point out that there is a generalisation of the André-Oort conjecture to mixed Shimura varieties which combines both the André-Oort conjecture and the Mordell-Lang conjecture. We will not touch upon this topic in this paper but we refer the interested reader to the papers [23] and [24] by Richard Pink available on his web-page.

Acknowledgements

The author would like to express his gratitude to the organisers of the Durham Symposium for having given him the opportunity to give a lecture and write this survey as well as for stimulating atmosphere during the conference. The author thanks the referee for pointing out many inaccuracies and suggesting improvements.

2 Preliminaries.

The purpose of this section is to briefly review the notions of Shimura varieties, special subvarieties, special points and Hecke correspondences. The detailed exposition of these topics can be found in [17] but one can also consult [11] or [20]. The exposition of the general theory of Shimura varieties can be found in [7], [8] or [16].

Let $\mathbb{S}$ denote the real torus $\mathrm{Res}_{\mathbb{C}/\mathbb{R}}\mathbb{G}_{m\mathbb{C}}$. A *Shimura datum* is a pair (G,X) where G is a connected reductive group over $\mathbb{Q}$ and X is a $G(\mathbb{R})$-orbit in $\mathrm{Hom}(\mathbb{S}, G_{\mathbb{R}})$ satisfying the conditions 2.1.1(1-3) of [8]. These conditions imply that connected components of X are hermitian symmetric domains and faithful representations of G induce variations of polarisable $\mathbb{Q}$-Hodge structures. Let K be a compact open subgroup of $G(\mathbb{A}_{\mathrm{f}})$. Let $\mathrm{Sh}_K(G,X)$ be the following double coset set

$$\mathrm{Sh}_K(G,X) := G(\mathbb{Q})\backslash(X \times G(\mathbb{A}_{\mathrm{f}})/K)$$

Let X^+ be a connected component of X and let $G(\mathbb{Q})^+$ be the stabiliser of X^+ in $G(\mathbb{Q})$. It is easy to see that $\mathrm{Sh}_K(G,X)$ is a disjoint union of quotients of X^+ by the arithmetic groups $\Gamma_g := G(\mathbb{Q})^+ \cap gKg^{-1}$ where g ranges through a set of representatives for the finite double coset set $G(\mathbb{Q})^+\backslash G(\mathbb{A}_{\mathrm{f}})/K$. A well-known theorem of Baily and Borel implies that $\mathrm{Sh}_K(G,X)$ has a structure of a quasiprojective complex algebraic variety.

The varieties $\mathrm{Sh}_K(G,X)$ form a projective system indexed by compact open subgroups $K \subset G(\mathbb{A}_{\mathrm{f}})$: the inclusions $K \subset K'$ of compact open subgroups induce finite morphisms $\mathrm{Sh}_K(G,X) \longrightarrow \mathrm{Sh}_{K'}(G,X)$. Let $\mathrm{Sh}(G,X)$ denote the projective limit of the $\mathrm{Sh}_K(G,X)$, it is a scheme naturally endowed with the action of $G(\mathbb{A}_{\mathrm{f}})$. Let Z be the centre of G. The set of complex points of $\mathrm{Sh}(G,X)$ is

$$\mathrm{Sh}(G,X)(\mathbb{C}) = \frac{G(\mathbb{Q})}{Z(\mathbb{Q})}\backslash(X \times G(\mathbb{A}_{\mathrm{f}})/Z(\mathbb{Q})^-)$$

where $Z(\mathbb{Q})^-$ denotes the closure of $Z(\mathbb{Q})$ in $Z(\mathbb{A}_{\mathrm{f}})$. Let s be a point of $\mathrm{Sh}(G,X)$ and $\mathrm{Sh}_K(G,X)$. There exists an element (x,t) of $X \times G(\mathbb{A}_{\mathrm{f}})$ such that s is the image $\overline{(x,t)}$ in the quotient. We will use this notation throughout this paper.

An element g of $G(\mathbb{A}_{\mathrm{f}})$ acts on the point $\overline{(x,t)}$ by sending this point to $\overline{(x,tg)}$. It is important to note that the action of g on $\mathrm{Sh}(G,X)$ induces an algebraic correspondence T_g on $\mathrm{Sh}_K(G,X)$ for every compact open subgroup K of $G(\mathbb{A}_{\mathrm{f}})$. These correspondences are called Hecke correspondences. One can describe T_g more explicitly. Let π_1 be the morphism $\mathrm{Sh}_{K\cap gKg^{-1}}(G,X) \longrightarrow$

$\mathrm{Sh}_K(G, X)$ induced by the inclusion $K \cap gKg^{-1} \subset K$ and let π_2 be the morphism induced by the inclusion preceeded by right multiplication by g on the second factor. Then, for every subvariety Z of $\mathrm{Sh}_K(G, X)$, by definition

$$T_g(Z) := \pi_2 \pi_1^{-1}(Z).$$

Let $S = \Gamma \backslash X^+$ be an irreducible component of $\mathrm{Sh}_K(G, X)$. Elements of $G(\mathbb{Q})^+$ induce Hecke correspondences on S in an analogous way except for left multiplication by elements of $G(\mathbb{Q})^+$ versus right multiplication by elements of $G(\mathbb{A}_f)$.

We can now define special subvarieties and special points. Let $f\colon (H, Y) \to (G, X)$ be a morphism of Shimura data. By this we mean a morphism of algebraic groups $H \longrightarrow G$ which, by composition induces a map $Y \longrightarrow X$. Such a morphism induces a morphism of algebraic varieties

$$\mathrm{Sh}(f)\colon \mathrm{Sh}(H, Y) \longrightarrow \mathrm{Sh}(G, X).$$

Definition An irreducible subvariety Z of $\mathrm{Sh}_K(G, X)$ is called special or of Hodge type if there exists a morphism of Shimura data $f\colon (H, Y) \longrightarrow (G, X)$ and an element g of $G(\mathbb{A}_f)$ such that Z is an irreducible component of the image of the following composition

$$\mathrm{Sh}(H, Y) \xrightarrow{\mathrm{Sh}(f)} \mathrm{Sh}(G, X) \xrightarrow{\cdot g} \mathrm{Sh}(G, X) \longrightarrow \mathrm{Sh}_K(G, X).$$

The terminology 'of Hodge type' comes from the fact that these subvarieties are loci of Hodge classes in the variations of Hodge structure coming from rational representations of G. Details on this can be found in [17].

By definition, a special point is a special subvariety of dimension zero. We can give an equivalent (and more useful) definition involving Mumford-Tate groups.

Definition Let x be a point of X. View x as a morphism $x\colon \mathbb{S} \longrightarrow G_{\mathbb{R}}$. By definition the Mumford-Tate group of x, denoted $\mathrm{MT}(x)$, is the smallest algebraic subgroup H of G such that $x(\mathbb{S}) \subset H_{\mathbb{R}}$.

A point $s = \overline{(x, g)}$ of $\mathrm{Sh}_K(G, X)$ is called special if $\mathrm{MT}(x)$ is a torus. Note that this is independent of g.

Given a special subvariety Z of $\mathrm{Sh}_K(G, X)$ the special points of $\mathrm{Sh}_K(G, X)$ contained in Z form a dense subset for the archimedean (and in particular Zariski) topology. This follows from two facts: the first is that a special subvariety contains a special point and the second is that for a reductive group H, $H(\mathbb{Q})$ is dense in $H(\mathbb{R})$.

We now make a couple of important remarks regarding special subvarieties. The first property is that the open compact subgroup K does not play any role. More precisely, let Z be a subvariety of $\mathrm{Sh}_K(G,X)$ and K' a compact open subgroup of K. Then Z is special if and only if one (equivalently any) component of the preimage of Z in $\mathrm{Sh}_{K'}(G,X)$ is special.

The second one is that irreducible components of intersections of special subvarieties are special. This is easily seen by using the definition of special subvarieties as 'loci of Hodge classes'. An obvious consequence of this fact is that, given a subvariety Z of $\mathrm{Sh}_K(G,X)$, there exists a smallest special subvariety of $\mathrm{Sh}_K(G,X)$ containing Z. One can describe the Shimura datum defining this special subvariety. This involves the notion of generic Mumford-Tate group. Suppose that Z is contained in the component $\Gamma\backslash X^+$ of $\mathrm{Sh}_K(G,X)$. Choose a faithful representation of G on a $\mathbb{Q}$-vector space V, this induces a variation of Hodge structures over X^+. Let $\widetilde{Z}$ be the preimage of Z in X^+. For every point x of $\widetilde{Z}$, let $\mathrm{MT}(V_x)$ be the Mumford-Tate group of the Hodge structure on V defined by x. There exists a countable union of analytically closed subsets $\Sigma \subset Z$, such that $\mathrm{MT}(V_x)$ is independent of x in $Z - \Sigma$. We denote this constant group by $\mathrm{MT}(Z)$ and call it the generic Mumford-Tate group. Points of Z lying outside the image of Σ are called Hodge generic. Let X_Z be the $\mathrm{MT}(Z)(\mathbb{R})$-orbit of a Hodge generic point x. This defines a morphism of Shimura data $\mathrm{Sh}_{K\cap \mathrm{MT}(Z)(\mathbb{A}_\mathrm{f})}(\mathrm{MT}(Z),X_Z) \longrightarrow \mathrm{Sh}_K(G,X)$ and the image of this morphism is the smallest special subvariety containing Z. The subvariety Z of $\Gamma\backslash X^+$ is called *Hodge generic* if it is not contained in a proper special subvariety. In other words, the smallest special subvariety containing Z is $\Gamma\backslash X^+$.

The last important thing that we need to mention is the existence of canonical models of Shimura varieties over certain number fields. We will not go into detail here and refer the reader to the section 6.1 of [20] and references therein for a more detailed exposition. To a Shimura datum is associated a number field $E(G,X)$ called the reflex field. The Shimura variety $\mathrm{Sh}(G,X)$ (projective limit) admits a canonical model over $E := E(G,X)$. What is meant by this is that there exists an E-scheme $\mathrm{Sh}(G,X)_E$ such that $\mathrm{Sh}(G,X)_E \times \mathbb{C} = \mathrm{Sh}(G,X)$ and the action of $G(\mathbb{A}_\mathrm{f})$ on $\mathrm{Sh}(G,X)$ is defined over E. It follows that for every compact open subgroup K of $G(\mathbb{A}_\mathrm{f})$, the variety $\mathrm{Sh}_K(G,X)$ admits a canonical model over E in such a way that Hecke correspondences commute with the Galois action. Furthermore, special points are $\overline{\mathbb{Q}}$-valued points and the Galois action on them is explicitly described via the reciprocity morphism. For details on the reciprocity morphism, we refer to the section 6.1 of [20]. Let us give a brief overview. Consider a special Shimura datum (T,x), where T is a torus and x is a morphism $x\colon \mathbb{S} \longrightarrow T_\mathbb{R}$ satisfying the axioms for Shimura datum. Suppose that T is the Mumford-Tate group of x.

One constructs a surjective morphism of $\mathbb{Q}$-tori

$$r_x\colon \mathrm{Res}_{E/\mathbb{Q}}\mathbb{G}_{mE} \longrightarrow T.$$

Consider now a special point $\overline{(x,g)}$ of $\mathrm{Sh}_K(G,X)$. Let $E_x := E(\mathrm{MT}(x),x)$. The action of $\mathrm{Gal}(\overline{E_x}/E_x)$ on $\overline{(x,g)}$ factors through the maximal abelian quotient $\mathrm{Gal}(E_x^{\mathrm{ab}}/E_x)$. Class field theory gives a surjection $(\mathbb{A}\otimes E_x)^*$ to $\mathrm{Gal}(E_x^{\mathrm{ab}}/E_x)$. Let σ be an element of $\mathrm{Gal}(E_x^{\mathrm{ab}}/E_x)$ and let $s = (s_f, s_\infty)$ be an idele in the preimage of σ in $(\mathbb{A}\otimes E_x)^*$ where $s_f \in (\mathbb{A}_{\mathrm{f}}\otimes E_x)^*$ and $s_\infty \in (\mathbb{R}\otimes E_x)^*$. Then, by definition

$$\sigma(\overline{(x,g)}) = \overline{(x, r_x(s_f)\cdot g)}$$

In particular $\sigma(\overline{(x,g)}) \in T_{g^{-1}r_x(s_f)g}(\overline{(x,g)})$ Let us note, that in the case of Shimura varieties parametrising families of abelian varieties, special points correspond to abelian varieties of CM type. The Galois action on them is then described by the main theorem of complex multiplication by Shimura-Taniyama. In particular, Galois conjugates of a CM abelian variety lie in one isogeny class. The reciprocity morphism is the one induced by "reflex type norm". A reference for the Shimura-Taniyama theorem is [15].

3 Images under Hecke correspondences.

The strategy developed by Edixhoven and the author of the present survey relies on the characterisation of special subvarieties in terms of their images by some Hecke correspondences. In this section we suppose that we are in the following situation.

Let (G,X) be a Shimura datum with G semi-simple of adjoint type. Let K be a neat open compact subgroup of $G(\mathbb{A}_{\mathrm{f}})$ which is the product of compact open subgroups K_l of $G(\mathbb{Q}_l)$ and let X^+ be a connected component of X. Let S be the image of $X^+\times\{1\}$ in $\mathrm{Sh}_K(G,X)$, then $S = \Gamma\backslash X^+$ where $\Gamma := G(\mathbb{Q})^+\cap K$ (with $G(\mathbb{Q})^+$ being the stabiliser of X^+ in $G(\mathbb{Q})$). Let Z be an irreducible Hodge generic subvariety of S containing a smooth special point.

First one proves the following irreducibility result which combines the results of section 5 of [12] and section 3 of [34].

Theorem 3.1 *There exists an integer M depending on G, X, Z and K such that the following holds.*

Let l be a prime. There is a compact open subgroup K' of K (K' is again a product of compact open subgroups K'_p of $G(\mathbb{Q}_p)$ and $K'_p = K_p$ for $p\neq l$)

and a component Z' of the preimage of Z in $S' := \Gamma'\backslash X^+$ (with $\Gamma' := K' \cap G(\mathbb{Q})^+$) with the property that for any q in $(G(\mathbb{Q}_l) \times \prod_{p\neq l} K'_p) \cap G(\mathbb{Q})^+$ (the intersection is being taken in $G(\mathbb{A}_f)$) the image $T_q Z'$ of Z' by the Hecke correspondence T_q is irreducible.

Furthermore, if l does not divide M, then $K' = K$.

Note that this theorem is far from being effective : one has no *a priori* control over M or over K'. This causes significant difficulties while trying to deal with the general case of the André-Oort conjecture. In the case of a product of two modular curves (or in the case of arbitrary products of modular curves) however, one can gain control over these quantities. Indeed, Edixhoven in [10] proves the following theorem.

Theorem 3.2 *Let Z be a Hodge generic curve contained in $\mathbb{C}^2$, viewed as the Shimura variety associated to the Shimura datum $(\mathrm{GL}_2 \times \mathrm{GL}_2, \mathbb{H}^{\pm} \times \mathbb{H}^{\pm})$. Suppose that both projections of Z are dominant. Let l be a prime larger than 13 and the degrees of the two projections of Z. Then $T_l Z$ is irreducible.*

Let us now briefly sketch the proof of Theorem 3.1. Let $\xi\colon G \longrightarrow \mathrm{GL}(V)$ be a faithful rational representation of G where V is a finite dimensional $\mathbb{Q}$-vector space. Choose a Γ-invariant lattice $V_{\mathbb{Z}}$ in V. This induces a variation of $\mathbb{Z}$-Hodge structures over S and over the smooth locus Z^{sm} of Z (this is where the assumption that K is neat is used). Let $\pi_1(Z^{\mathrm{sm}})$ be the topological fundamental group of Z^{sm} with respect to some Hodge generic point s of Z^{sm}. There is a monodromy representation $\rho\colon \pi_1(Z^{\mathrm{sm}}) \longrightarrow \mathrm{GL}(V_{\mathbb{Z}})$ associated to the local system underlying the variation of Hodge structure. The image of ρ is contained in $\xi(\Gamma)$. A theorem of Yves André (see [2, Thm. 1.4]) implies that $\rho(\pi_1(Z^{\mathrm{sm}}))$ is Zariski dense in $\xi(G)$. Let l be a prime. By a theorem of Nori ([21]), the closure of $\rho(\pi_1(Z^{\mathrm{sm}}))$ in $\xi(K_l)$ is $\xi(K'_l)$ for a compact open subgroup K'_l of $G(\mathbb{Q}_l)$.

Let K' be a compact open subgroup of $G(\mathbb{A}_f)$ obtained as follows : K' is a product $K' = \prod_p K'_p$ where K'_p are compact open subroups of $G(\mathbb{Q}_p)$ and $K'_p = K_p$ for $p \neq l$ and K'_l is as above. Let Γ' and S' be as in the statement of the theorem. There exists a component Z' of the preimage of Z such that $\rho(\pi_1(Z'^{\mathrm{sm}}))$ is equal to $\pi_1(Z^{\mathrm{sm}})$. Let, for every integer $k \geq 0$, $\Gamma'(l^k)$ be the kernel of $\Gamma' \longrightarrow \mathrm{GL}(V_{\mathbb{Z}}/l^k V_{\mathbb{Z}})$. Let $S'_k := \Gamma'(l^k)\backslash X^+$. By construction, for every $k > 0$, the preimage of Z'^{sm} in S'_k is irreducible. The conclusion of the theorem follows. As for the second statement, one uses again Nori's theorem which implies that there exists an integer M such that for every l not dividing M, the images of Γ and $\pi_1(Z^{\mathrm{sm}})$ in $\mathrm{GL}_n(V_{\mathbb{Z}}/l^k V_{\mathbb{Z}})$ are equal for all k. This finishes the proof.

The second ingredient in the characterisation is the density of Hecke orbits. Strictly speaking, one considers the orbits for the correspondence $T_q + T_{q^{-1}}$ where q is some well-chosen element of $G(\mathbb{Q})^+$.

Theorem 3.3 *Let p be a prime. Let q be an element of $(G(\mathbb{Q}_p) \times \prod_{l \neq p} K_l) \cap G(\mathbb{Q})^+$ such that for every simple factor G_i of G, the image of q in $G_i(\mathbb{Q}_p)$ is not contained in a compact subgroup.*

Then for every s in S, the $T_q + T_{q^{-1}}$-orbit of s is dense for the archimedian topology.

Proof We only sketch the proof. One reduces the proof to showing that the index of Γ in the group Γ_q generated by Γ and q is infinite. If this index is finite, then the index of the closure $\overline{\Gamma}$ of Γ in $G(\mathbb{Q}_p)$ in the group $\overline{\Gamma}_q$ generated by $\overline{\Gamma}$ and q is also finite. This implies that the group $\overline{\Gamma}_q$ is compact which contradicts our assumption. □

Regarding Hecke orbits, an equidistribution result of T_q-orbits is now available thanks to the work of Clozel, Oh and Ullmo (see [4]) and may be used in the charactersation instead of the density result above.

Theorem 3.4 (Clozel, Oh, Ullmo) *Let G be an almost simple simply connected algebraic group over $\mathbb{Q}$ such that $G(\mathbb{R})$ is not compact. Let K_∞ be a maximal compact subgroup of the neutral component $G(\mathbb{R})^+$ and let X^+ be $G(\mathbb{R})^+/K_\infty$. Let $d\mu$ be the natural normalised measure on $\Gamma\backslash X^+$ induced by the Haar measure on $G(\mathbb{R})^+$ and let q_n be a sequence of elements of $G(\mathbb{Q})^+$ such that the index $[\Gamma : q_n \Gamma q_n^{-1} \cap \Gamma]$ tends to infinity as n tends to infinity. Then the T_{q_n}-orbits are equidistributed for the measure μ in the sense that for every x in $\Gamma\backslash X^+$ and every continuous compactly supported function f on $\Gamma\backslash X^+$,*

$$\lim_{n \longrightarrow \infty} \frac{1}{[\Gamma : q_n \Gamma q_n^{-1} \cap \Gamma]} \sum_{y \in T_{q_n}(x)} f(y) = \int_{\Gamma\backslash X^+} f \, d\mu$$

In particular, for every x in $\Gamma\backslash X^+$, the set $\bigcup_{n \geq 0} T_{q_n}(x)$ is dense in $\Gamma\backslash X^+$.

From the irreducibility and density results, one derives the following characterisation of special subvarieties.

Theorem 3.5 *Let Z be a Hodge generic subvariety of $S = \Gamma\backslash X^+$. Suppose that there exists a prime l and an element m of $G(\mathbb{Q}_p)$ such that*

(i) *For every simple factor G_i of G, m is not contained in a compact subgroup of $G(\mathbb{Q}_p)$.*

(ii) *The variety Z' is contained in $T_m Z'$ where Z' is as in Theorem 3.1.*

Then Z is of Hodge type.

Note that if l does not divide M with M as in Theorem 3.1 then $K = K'$ and the second condition becomes that Z is contained in $T_m Z$.

To prove this theorem, one simply notes that the inclusion $Z' \subset T_m Z'$ implies that Z' is contained in its image by T_q and by $T_{q^{-1}}$ where q is some element of $G(\mathbb{Q})^+$ such that $T_q Z'$ and $T_{q^{-1}} Z'$ are both irreducible and the orbits of $T_q + T_{q^{-1}}$ are dense in S. A Hodge generic subvariety Z of S is special if and only if $Z = S$.

4 Lower bounds for Galois orbits of special subvarieties.

The purpose of this section is to present results on lower bounds for Galois orbits of *non-strongly* special subvarieties of Shimura varieties. In [12] lower bounds for Galois orbits of sets of special points lying in one Hecke orbits have been obtained. This result was subsequently generalised in [35] to arbitrary sets of special points and used to prove the André-Oort conjecture for curves in Shimura varieties.

The result we present in this section is one of the main results of [31] which generalises these previous results. We begin by recalling some definitions regarding special subvarieties.

As in the previous section, we let (G, X) be a Shimura datum where G is semisimple of adjoint type. Via a faithful representation, we view G as a closed subgroup of some $\mathrm{GL}_{n\mathbb{Q}}$. Let K be a compact open subgroup of $G(\mathbb{A}_f)$ contained in $\mathrm{GL}_n(\widehat{\mathbb{Z}})$. We also assume that K is a product of compact open subgroups K_p of $G(\mathbb{Q}_p)$.

Let (H, X_H) be a sub-Shimura datum of (G, X). Let $H' = \mathrm{MT}(X_H)$ be the generic Mumford-Tate group on X_H. We have $H'^{\mathrm{der}} = H^{\mathrm{der}}$. Let x be an element of X_H and let $X_{H'}$ be the $H'(\mathbb{R})$-orbit of x. Then $X_{H'} = X_H$ and (H', X_H) is a sub-Shimura datum of (H, X_H).

Definition A special subvariety $Z \subset \mathrm{Sh}_K(G, X)$ is called strongly special if the Shimura datum (H, X_H) defining Z is such that $\mathrm{MT}(X_H)$ is semisimple (in other words the connected centre of $\mathrm{MT}(X_H)$ is trivial).

Examples of strongly special subvarieties abound : the modular curves $Y_0(n)$ (and products thereof) embedded in products of modular curves, Hilbert modular varieties as subvarieties of $\mathcal{A}_g$, etc. Typically these varieties admit canonical models over 'small fields'. At the other end of the scale are the special points. Recall that the Mumford-Tate group of a special point is a torus.

It has been shown in [12] and [35] that Galois orbits of the special points tend to get very large. Examples of non-strongly special subvarieties which are not zero-dimensional include, for example subvarieties of a product of three modular curves of the form $\{x\} \times Y_0(n)$ where x is a special point of a modular curve and $Y_0(n)$ is embedded in a product of two modular curves. One can easily see that these subvarieties also have large Galois orbits (what is meant by the Galois orbit in this case is the number of Galois conjugates of geometrically irreducible components).

Let us now explain a general result on Galois orbits of non-strongly special subvarieties obtained in the forthcoming paper [31]. So let (G, X), (H, X_H) and H' be as above. We suppose that H' is not semisimple. Let T be the connected centre of H', so that H' is an almost direct product TH^{der}. Let K'_H be the compact open subgroup $H'(\mathbb{A}_{\mathrm{f}}) \cap K$ of $H'(\mathbb{A}_{\mathrm{f}})$. Our aim is to give a lower bound for the number of Galois conjugates of geometrically irreducible components of the image of $\mathrm{Sh}_{K_H}(H, X_H)$ in $\mathrm{Sh}_K(G, X)$.

Let us briefly recall how Galois acts on the geometrically irreducible components of Shimura varieties.

We refer to the sections 2.4-2.6 of [8] for details and proofs. Let $\pi_0(H', K'_H)$ be the set of geometrically irreducible components of $\mathrm{Sh}_{K'_H}(H', X_H)$. As a set, $\pi_0(H', X_H)$ is $H'(\mathbb{Q})^+ \backslash H'(\mathbb{A}_{\mathrm{f}})/K'_H$ where $H'(\mathbb{Q})^+$ is the stabiliser of a connected component of X_H in $H'(\mathbb{Q})$. Let E_H be the reflex field of (H', X_H). We refer to [8] for details on the definitions of reflex fields. We just mention here that E_H is the number field of definition of the $H(\mathbb{C})$-conjugacy class of $h_{\mathbb{C}}(z, 1)$ for h in X_H. In particular, one notes that $E_H = E_{H'}$. What follows is a generalisation of the Galois action on special points, overviewed in the second section.

The action of $\mathrm{Gal}(\overline{\mathbb{Q}}/E_H)$ on $\pi_0(H', K'_H)$ is given by the reciprocity morphism

$$r_{(H', X_H)} \colon \mathrm{Gal}(\overline{\mathbb{Q}}/E_H) \longrightarrow \pi(H')$$

where $\pi(H')$ is $H'(\mathbb{A}_{\mathrm{f}})/H'(\mathbb{Q})\rho\widetilde{H}'(\mathbb{A}_{\mathrm{f}})$. Here $\rho \colon \widetilde{H}' \longrightarrow H^{\mathrm{der}}$ is the universal covering of H^{der}. The morphism $r_{(H', X_H)}$ factors through $\mathrm{Gal}(E_H^{\mathrm{ab}}/E_H)$ which is identified with a quotient of $\pi(\mathrm{Res}_{E_H/\mathbb{Q}} \mathbb{G}_{mE_H})$ by class field theory. Let C be the torus H'/H^{der}. To (H', X_H) one associates two Shimura data $(C, \{x\})$ and $(H^{\mathrm{ad}}, X_{H^{\mathrm{ad}}})$ where x is induced by any element of X_H. The field E_H is the composite of $E(C, \{x\})$ and $E(H^{\mathrm{ad}}, X_{H^{\mathrm{ad}}})$. There are morphisms

$$\theta^{\mathrm{ab}} \colon (H', X_H) \longrightarrow (C, \{x\}) \text{ and } \theta^{\mathrm{ad}} \colon (H', X_H) \longrightarrow (H^{\mathrm{ad}}, X_{H^{\mathrm{ad}}}).$$

Note that $(C, \{x\})$ is a special Shimura datum. Let $r_{(C,\{x\})}$ be the reciprocity morphism associated to $(C, \{x\})$. The morphism θ^{ab} induces a morphism $\pi(H') \to \pi(C)$. This morphism, preceeded by $r_{(H',X_H)}$, is $r_{(C,\{x\})}$. We let F be the Galois closure of E_H. Note that the degree of F over $\mathbb{Q}$ is bounded uniformly on (H, X_H).

Note that there is an isogeny $T \longrightarrow C$ with kernel $T \cap H^{\mathrm{der}}$. One can easily see that the order of the group $T \cap H^{\mathrm{der}}$ is uniformly bounded as (H, X_H) ranges through the sub-Shimura data of (G, X). Let K_H be the compact open subgroup $H(\mathbb{A}_{\mathrm{f}}) \cap K$ of $H(\mathbb{A}_{\mathrm{f}})$. The following lemma is proved in the forthcoming paper [31], lemma 4.2. The proof is essentially elementary.

Lemma 4.1 *Let $f\colon \mathrm{Sh}_{K_H}(H, X_H) \longrightarrow \mathrm{Sh}_K(G, X)$ be the morphism induced by the inclusion (H, X_H) into (G, X).*

The morphism f is finite onto its image of uniformly (on (H, X_H)) bounded degree. Furthermore, if K is neat, then f is generically injective.

In particular, the number of geometrically irreducible components of $\mathrm{Sh}_{K_H}(H, X_H)$ is, up to a uniform (on (H, X_H)) constant, equal to the number of components of its image in $\mathrm{Sh}_K(G, X)$.

This lemma reduces the problem of giving a lower bound for the Galois orbit of a geometrically irreducible component of the image of $\mathrm{Sh}_{K_H}(H, X_H)$ in $\mathrm{Sh}_K(G, X)$ to that of giving a lower bound for the Galois orbit of a component of the Shimura variety $\mathrm{Sh}_{K_H}(H, X_H)$ itself. We can now forget about $\mathrm{Sh}_K(G, X)$ and work with $\mathrm{Sh}_{K_H}(H, X_H)$. The main theorem on the Galois orbits is the following.

Theorem 4.2 *Assume the GRH for CM fields. Let N be a positive integer. There exists a real $B > 0$ such that the following holds. Let (H, X_H) be a sub-Shimura datum of (G, X) and suppose that H is the generic Mumford-Tate group on X_H. Let T be the connected centre of H and let K_T^m be the maximal compact open subgroup of $T(\mathbb{A}_{\mathrm{f}})$ and K_T be the compact open subgroup $T(\mathbb{A}_{\mathrm{f}}) \cap K$. Let $i(T)$ be the number of primes p such that $K_{T,p}^m \neq K_{T,p}$. Let F be the composite of the reflex field (H, X_H) with the splitting field of T. The following inequality holds*

$$|\{\sigma(Z), \sigma \in \mathrm{Gal}(\overline{\mathbb{Q}}/F)\}| \gg B^{i(T)} |K_T^m / K_T| (\log(d_T))^N.$$

where d_T denotes the absolute value of the discriminant of the splitting field of T and $\gg$ stands for 'larger up to a uniform constant'.

We now give some ingredients of the proof of this theorem. The lower bound naturally splits into two parts. The first corresponds to the factor $|K_T^m/K_T|$ and the second to the term in $\log(d_T)$.

Let $\mathcal{G}$ be the image of $\mathrm{Gal}(\overline{\mathbb{Q}}/F)$ (the field F being as in the statement) in $\pi(H')/K_{H'}$ by $r_{(H',X_{H'})}$. The number of components of the Galois orbit of V is at least the size of $\mathcal{G}$. The number of components of the Galois orbit of V is at least the size of $\mathcal{G}$.

Let K_T be the compact open subgroup $T(\mathbb{A}_{\mathrm{f}}) \cap K_{H'}$ of $T(\mathbb{A}_{\mathrm{f}})$ and let K_C^m and K_T^m be the maximal compact open subgroups of $C(\mathbb{A}_{\mathrm{f}})$ and $T(\mathbb{A}_{\mathrm{f}})$ respectively. The following lemma, proved in [31], lemma 4.3 realises the splitting and explains the presence of the factor $B^{i(T)}$.

Lemma 4.3 *There exists a uniform constant B and an exact sequence*

$$0 \longrightarrow W \longrightarrow K_T^m/K_T \longrightarrow \pi(H')/K_{H'}$$

where W is a finite group of order at most $B^{i(T)}$.

Let x be a Hodge generic point of X_H (recall that this means that $\mathrm{MT}(x)$ is H'). View, as usual, x as a morphism $x\colon \mathbb{S} \longrightarrow H'_{\mathbb{R}}$. We let μ be the character $x_{\mathbb{C}}(z,1)\colon \mathbb{G}_{m\mathbb{C}} \longrightarrow H'_{\mathbb{C}}$ and let μ_C be the character μ composed with the map $H'_{\mathbb{C}} \longrightarrow C_{\mathbb{C}}$. The Lemma 4.4 of [31] proves a uniformity property of the characters defining this morphism

Lemma 4.4 *There is a basis (χ_i) of the character group $X^*(C)$ such that, with respect to the natural pairing*

$$<,>\colon X^*(L) \times X_*(L) \longrightarrow \mathbb{Z}$$

the $<\chi_i, \sigma(\mu_C)>$ are bounded uniformly on (H, X_H).

Previous lemmas allow to deal with the first part of the splitting.

Lemma 4.5 *There is a uniform integer B' such that the size of the intersection of $\mathcal{G}$ with the image of K_T^m/K_T in $\pi(H')/K_{H'}$ is at least $B'^{i(T)} \cdot |K_T^m/K_T|$*

To prove this lemma, one first notes that, using the lemma 4.3, it is enough to give a lower bound for the preimage of $r_{(C,\{x\})}((\hat{\mathbb{Z}} \otimes O_F)^*)$ in K_T^m/K_T. The group K_T^m/K_T is the direct product of the $|K_{T,p}^m/K_{T,p}|$ for the $i(T)$ primes such that $K_{T,p}^m \neq K_{T,p}$. It follows that it is enough to give a lower bound for the preimage of $r_{(C,\{x\})}((\mathbb{Z}_p \otimes O_F)^*)$ in $K_{T,p}^m/K_{T,p}$ for one prime p. Using the fact that the isogeny $T \longrightarrow C$ has uniformly bounded kernel and the lemma 4.4, one proves that the size of the preimage of $r_{(C,\{x\})}((\mathbb{Z}_p \otimes O_F)^*)$ in $K_{T,p}^m/K_{T,p}$ is, up to a uniform constant, $|K_{T,p}^m/K_{T,p}|$.

Next, one deals with the image of $r_C((\mathbb{A}_{\mathrm{f}} \otimes F)^*)$ in $C(\mathbb{Q})\backslash C(\mathbb{A}_{\mathrm{f}})/K_C^m$ where K_C^m denotes the maximal compact open subgroup of $C(\mathbb{A}_{\mathrm{f}})$.

Proposition 4.6 *Assume the GRH for CM fields. Let N be a positive integer. The size of the image of $r_{(C,\{x\})}((\mathbb{A}_{\mathrm{f}} \otimes L_C)^*)$ in $C(\mathbb{Q})\backslash C(\mathbb{A}_{\mathrm{f}})/K_C^m$ is at least a constant depending on N only times* $\log(d_T)^N$.

We will not sketch the proof of this proposition, but we refer to [35] for a detailed proof. The idea of the proof is to choose sufficiently many prime ideals P of the ring of inetegers O_F of F lying over primes splitting completely in O_F and small with respect to large power of $\log(d_T)$ (this is where the GRH is used). And then use the images of the corresponding ideles by $r_{(C,\{x\})}$ to generate a sufficiently large subgroup of $C(\mathbb{Q})\backslash C(\mathbb{A}_{\mathrm{f}})/K_C^m$.

5 Some cases of the André-Oort conjecture.

In this section we sketch the results on the André-Oort conjecture that have been obtained by using the results from the previous sections and the strategy outlined in the end of the introduction. These theorems are proved in [12], [35] and [34] and in this section we will give an overview of the proofs. The first theorem deals with the one-dimensional subvarieties.

Theorem 5.1 *Let S be a Shimura variety defined by a Shimura datum (G, X) and a compact open subgroup K. Let Z be an irreducible algebraic curve in S. Make one of the following assumptions.*

(i) *Assume the GRH for CM fields and that Z contains an infinite set of special points.*
(ii) *Assume that Z contains an infinite set of special points Σ such that with respect to some faithful representation V of G, the $\mathbb{Q}$-Hodge structures V_s with s ranging through Σ lie in one isomorphism class.*

Then Z is of Hodge type.

Remark The second condition of this statement is satisfied if the points in Σ lie in one Hecke orbit. For example, if the Shimura variety is a moduli space for some family of abelian varieties, then the condition would be satisfied for sets of abelian varieties of CM type lying in one isogeny class. Even more explicitly, one can think of ellipic curves with complex multiplication by a fixed imaginary quadratic field K and varying conductor. The proof of this theorem under the first assumption is contained in [35] and under the second in [12].

The second theorem deals with subvarieties of arbitrary dimension. The proof of this theorem is contained in the note [34].

Theorem 5.2 *Let (G, X) be a Shimura datum and let K be a compact open subgroup of $G(\mathbb{A}_{\mathrm{f}})$. Let Σ be a set of special points of X. Let V be a faithful representation of G. Via this representation, we view G as a closed subgroup of $\mathrm{GL}_{n\mathbb{Q}}$. For each x in Σ we let $\mathrm{MT}(x)$ be its Mumford-Tate group. We suppose that there is a prime p such that for any s in Σ, the torus $\mathrm{MT}(x)_{\mathbb{Q}_p}$ has a subtorus M_s isomorphic to $\mathbb{G}_{m\mathbb{Q}_p}$ such that the weights in the representation $V_{\mathbb{Q}_p}$ of M_x are bounded uniformly on x and for every nontrivial quotient T of $\mathrm{MT}(x)$, the image of M_x in $T_{\mathbb{Q}_p}$ is nontrivial. Furthermore, we assume that the Zariski closure in $\mathrm{GL}_{n\mathbb{Z}_p}$ of each M_x is isomorphic to $\mathbb{G}_{m\mathbb{Z}_p}$.*

Then for any g in $G(\mathbb{A}_{\mathrm{f}})$ the irreducible components of the Zariski closure of the image of $\Sigma \times \{g\}$ in $\mathrm{Sh}_K(G, X)$ are of Hodge type.

This theorem is a generalisation of a theorem of Moonen (see [18]) which deals with the case where the Shimura variety is $\mathcal{A}_g$. However the proof of this theorem is very different from Moonen's. In fact the proof of this theorem uses the same ideas as the proof of the previous one. For the sake of completeness, we reproduce Moonen's statement below. A survey of the ingredients of Moonen's proof can be found in the section 5.1 of [20].

Theorem 5.3 *Let $g \geq 1$ be an integer and let $\mathcal{A}_g$ be the moduli space of principally polarised abelian varieties of dimension g. Let Σ be a set of special points in $\mathcal{A}_g(\overline{\mathbb{Q}})$ such that there exists a prime p with the property that every s in Σ has good ordinary reduction at a place lying over p of which s is the canonical lift. Then all irreducible components of the Zariski closure of Σ are of Hodge type.*

In the second section of [34] it is proved that the theorem 5.2 above does indeed imply the result of Moonen. Although, there is no restriction on the dimension of the Zariski closure of Σ, the condition imposed on the set Σ is quite restrictive. As an exercise, an interested reader may want to try to construct an infinite set of elliptic curves with complex multiplication such that the condition of the statement is not satisfied for every prime p.

The method used to prove the two theorems is basically the same. One uses the characterisation of Theorem 3.5 i.e. one tries to prove that Z is contained in its image by some suitable Hecke correspondence. To find such a Hecke correspondence, one uses the Galois action on special points.

First one reduces oneself to the following case.

(i) The group G is semisimple of adjoint type.
(ii) The group K is neat and is a product of compact open subgroups K_p of $G(\mathbb{Q}_p)$.
(iii) The variety Z is Hodge generic and is contained in $S := \Gamma\backslash X^+$.

Let us briefly explain why these assumptions can be made. One replaces the ambient Shimura variety by the smallest special subvariety containing Z so that Z is Hodge generic.

Given a Shimura datum (G,X), there exists an adjoint Shimura datum $(G^{\mathrm{ad}}, X^{\mathrm{ad}})$ where X^{ad} is the $G^{\mathrm{ad}}(\mathbb{R})$-conjugacy class of the morphism h^{ad} obtained by composing any $h \in X$ with $G_{\mathbb{R}} \longrightarrow G^{\mathrm{ad}}_{\mathbb{R}}$. There is a natural morphism of Shimura data $(G,X) \longrightarrow (G^{\mathrm{ad}}, X^{\mathrm{ad}})$. By choosing a compact open subgroup K^{ad} of $G^{\mathrm{ad}}(\mathbb{A}_{\mathrm{f}})$ containing the image of K, one obtains a finite morphism of Shimura varieties

$$\mathrm{Sh}_K(G,X) \longrightarrow \mathrm{Sh}_{K^{\mathrm{ad}}}(G^{\mathrm{ad}}, X^{\mathrm{ad}}).$$

It is not hard to see that a subvariety Z of $\mathrm{Sh}_K(G,X)$ is of Hodge type if and only if its image in $\mathrm{Sh}_{K^{\mathrm{ad}}}(G^{\mathrm{ad}}, X^{\mathrm{ad}})$ is. This allows us to replace G with G^{ad}. However, one has to be a bit careful when the set Σ consists of special points such that the Hodge structures associated to these points with respect to some representation of G are isomorphic. In this case, one needs to choose a faithful representation of G^{ad} such that this property for the image of Σ in $\mathrm{Sh}_{K^{\mathrm{ad}}}(G^{\mathrm{ad}}, X^{\mathrm{ad}})$ is conserved. This is done in [12], construction 2.3. We do not reproduce it here. Lastly, as we have already noted, replacing K by another compact open subgroup changes nothing to the property of subvarieties being special. It follows that K can be chosen to be neat and to be a direct product of compact open subgroups of $G(\mathbb{Q}_p)$. By possibly replacing Z with its image by a suitable Hecke correspondence, which again changes nothing to the property of Z being special, one can assume that Z is contained in $\Gamma\backslash X^+$. It follows that every point s of Σ can be represented as $s := \overline{(x,1)}$ with x some point in X^+. We define the Mumford-Tate group of s as $\mathrm{MT}(s) := \mathrm{MT}(x)$. Note that this is defined up to conjugation by Γ.

As Z contains a Zariski dense set of special points which are $\overline{\mathbb{Q}}$-valued points, we can (and do) choose a number field F sufficiently large so that S admits a canonical model over F and Z is defined over F as an absolutely irreducible subscheme of S_F. Suppose now that Z is a curve containing an infinite set of special points. Let us recall lower bounds for Galois orbits of special points (we specialise the theorem from the previous section to the case of special points).

Theorem 5.4 *Let $\Sigma \subset S$ be an infinite set of special points. Assume either the GRH for CM fields or that the points in Σ are such that the V_s lie in one isomorphism class of $\mathbb{Q}$-Hodge structures as s ranges through Σ. For every point s in Σ, let $\mathrm{MT}(s)$ be its Mumford-Tate group and let L_s be the splitting field of $\mathrm{MT}(s)$. We let d_s be the absolute value of the discriminant of $\mathrm{MT}(s)$. We also let K_s^m be the maximal compact open subgroup of $\mathrm{MT}(s)(\mathbb{A}_{\mathrm{f}})$ and K_s be the compact open subgroup $K \cap \mathrm{MT}(s)(\mathbb{A}_{\mathrm{f}})$. These groups are products of compact open subgroups of $\mathrm{MT}(s)(\mathbb{Q}_p)$ and we denote by $i(\mathrm{MT}(s))$ the number of primes p such that $K_{s,p}^m \neq K_{s,p}$.*

Let N be an integer. There exists a constant B such that the following inequality holds

$$|\mathrm{Gal}(\overline{\mathbb{Q}}/F) \cdot s| \gg B^{i(\mathrm{MT}(s))}|K_s^m/K_s| \log(d_s)^N.$$

Recall that to every s in Σ is associated a reciprocity morphism

$$r_s \colon \mathrm{Res}_{F/\mathbb{Q}}\mathbb{G}_{mF} \longrightarrow \mathrm{MT}(s)$$

which is a surjective morphism of algebraic tori. One has the following theorem which gives the 'candidates' for the Hecke correspondences such that $Z \subset T_m Z$.

Theorem 5.5 *There exists an integer k such that the following holds. Let s be a point of Σ and p a prime splitting $\mathrm{MT}(s)$ and such that $K_{s,p}^m = K_{s,p}$. Then there exists an element m in $r_s((\mathbb{Q}_p \otimes F)^*) \subset \mathrm{MT}(s)(\mathbb{Q}_p)$ satisfying*

(i) *For every simple factor G_i of G, the image of m in $G_i(\mathbb{Q}_p)$ is not contained in a compact subgroup.*
(ii) *$[K_p : mK_p m^{-1} \cap K_p] \ll p^k$.*

Another important theorem is the following. Again, we are not giving a proof here. This theorem can be thought of as a kind of Bezout's theorem.

Theorem 5.6 ([11], Thm. 7.2) *Let Z be an irreducible algebraic curve in S and m an element of $G(\mathbb{A}_{\mathrm{f}})$. Suppose that the intersection $Z \cap T_m Z$ is proper, then*

$$|Z \cap T_m Z| \ll [K : mKm^{-1} \cap K].$$

The important observation that we are making now is that if a point s of Σ is contained in the intersection $Z \cap T_m Z$, then the Galois orbit $\mathrm{Gal}(\overline{\mathbb{Q}}/F) \cdot s$ is contained in the intersection $Z \cap T_m Z$. Hence if the size of $\mathrm{Gal}(\overline{\mathbb{Q}}/F) \cdot s$ is large compared to $[K : mKm^{-1} \cap K]$, then the intersection $Z \cap T_m Z$ can not be proper and if m is as in Theorem 3.5, then we are finished.

Suppose that Z is a curve. The three theorems stated above show that one needs to find a prime p and a point s satisfying the following inequalities.

(i) The prime p splits $\mathrm{MT}(s)$ and p is not among the $i(\mathrm{MT}(s))$ primes such that $\mathrm{MT}(s)_{\mathbb{F}_p}$ is not a torus.

(ii)
$$B^{i(\mathrm{MT}(s))}|K_s^m/K_s|\log(d_s)^N \gg p^k.$$

To find such a prime, one uses the Chebotarev density theorem. For any real $x > 0$, define

$$\pi(x) = |\{p \leq x : p \text{ splits } L_s\}|.$$

Assume the GRH. If x is larger than some absolute constant and larger than $\log(d_s)^3$, then

$$\pi(x) \gg \frac{x}{\log(x)}.$$

If the points s in Σ are such that the V_s lie in one isomorphism class of $\mathbb{Q}$-Hodge structures, then the assumption of GRH can be dropped because in this case d_s is constant and the usual Chebotarev theorem can be used.

Is is shown in [31], that $x := B^{i(\mathrm{MT}(s))}|K_s^m/K_s|\log(d_s)^N$ is unbounded as s ranges through Σ. In particular, if x is large enough, then $\log(x) < x^{1/2k}$ and consequently, $\pi(x) \gg x^{1/2k}$. It is also shown in [12], that if p is unramified in L_s and $K_{s,p}^m \neq K_{s,p}$, then

$$|K_{s,p}^m/K_{s,p}| \gg p^{n_p}$$

with $n_p \geq 1$. If $i(\mathrm{MT}(s))$ is bounded then for at least one p, the n_p's occuring in the inequalities, as s varies, are unbounded (this is proved in [31]). Using the obvious inequality, that the ith prime is larger than i, it is now easy to show that, provided that N is larger than $6k$, a prime p satisfying the conditions above exists. This finishes the proof of the first theorem.

Let us now turn to the second theorem. Suppose that Z is a subvariety (of arbitrary dimension) containing a Zariski dense set Σ of special points satisfying the condition from the second theorem (in the sense, Z is a component of the Zariski closure of the set Σ of special points satisfying the condition of the second theorem). We sketch the proof under a simplified condition, namely, that for some prime p, the groups $\mathrm{MT}(x)_{\mathbb{F}_p}$ are split tori. The proof is essentially the same. We start by replacing the group K with K' as in Theorem 3.1 and Z with Z'. We can now choose an element m of $r_x((\mathbb{Q}_p \otimes F)^*)$ for some x in Σ in such a way that for every s in a Zariski dense subset $\Sigma' \subset \Sigma$, some Galois conjugate of s is in $T_m(s)$. It follows that $Z \cap T_m Z$ contains Σ' hence $Z \subset T_m Z$ and hence Z is special.

6 The equidistribution approach.

In this section we explain another approach to the "André-Oort-type conjectures". We will start by briefly recalling the principle (mainly following Ullmo's survey [29]) and then we will review some of the results obtained so far in this direction. Ullmo has very recently written a more comprehensive set of notes [32] on the equidistribution approach to this kind of question which will appear in the proceedings of the summer school on equidistribution in number theory that took place at the Université de Montreal in July 2005. We will only very briefly summarise it.

Let S be a complex algebraic variety and E a finite subset of S. For every x in S, let δ_x be the Dirac measure at x. One defines a normalised measure Δ_E associated to E by

$$\Delta_E := \frac{1}{|E|}\sum_{x \in E} \delta_x.$$

Let μ be a probability measure on S. A sequence of finite subsets E_n is said to be equidistributed with respect to μ if for every continuous bounded function f in S, one has

$$\int_S f d\Delta_{E_n} = \frac{1}{|E_n|}\sum_{x \in E_n} f(x) \longrightarrow \int_S f d\mu.$$

A sequence x_n of points of S is said to be *generic* if for every proper subvariety Z of S, the set $\{n : x_n \in Z\}$ is a finite set. The reader may want to prove, as an exercise, that this is equivalent to saying that the sequence x_n converges to the generic point of S for the Zariski topology.

Let us now suppose that S is a Shimura variety. One says that a sequence of special points x_n is *strict* if for every *special* subvariety Z, the set $\{n : x_n \in Z\}$ is finite. It is obvious that the André-Oort conjecture is equivalent to the statement that every strict sequence of special points is generic.

In the introduction, we have mentioned an "abelian analog" of the André-Oort conjecture, the Manin-Mumford conjecture stating that the irreducible components of the Zariski closure of a set of torsion points in an abelian variety are translates of abelian subvarieties by torsion points. There is a number of proofs of this conjecture. One of the most striking is due to Ullmo [28] (subsequently generalised by Zhang [36]). They actually prove Bogomolov's conjecture which is a stronger statement than Manin-Mumford. Their proof relies on equidistribution of generic sequences of points whose Néron-Tate height tends to zero for two different measures. The proof crucially relies on Arakelov theory.

Let us now go back to the case where S is a Shimura variety. As has already been mentioned in the third section, Clozel, Oh and Ullmo proved an equidistribution result regarding Hecke orbits.

Consider a sequence x_n of special points. Let $O(x_n)$ denote the Galois orbit of x_n. It is natural to ask the question whether or not the sequence of the $O(x_n)$ is equidistributed with respect to the normalised measure on S induced by the Haar measure of $G(\mathbb{R})^+$. So far the only case where this question is known to have a positive answer is the case of a sequence of special points of a modular curve. This is due to Duke (see [9]).

Let us briefly review Duke's results. Let K be an imaginary quadratic field and let $O_{K,f} := \mathbb{Z} + fO_K$ be an order in K. We let $\Sigma_{K,f}$ be the set of isomorphism classes of pairs (E, α) where E is an elliptic curve and $\alpha\colon O_{K,f} \longrightarrow \mathrm{End}(E)$ is a ring isomorphism. The set $\Sigma_{K,f}$ is a $\mathrm{Pic}(O_{K,f})$-torsor and the action of $\mathrm{Gal}(\overline{\mathbb{Q}}/K)$ on the set $\Sigma_{K,f}$ factors through the quotient $\mathrm{Gal}(H_{K,f}/K)$ where $H_{K,f}$ is the ring class field of $O_{K,f}$. The action of $\mathrm{Gal}(H_{K,f}/K)$ is given by the class field theory isomorphism $\mathrm{Gal}(H_{K,f}/K) \cong \mathrm{Pic}(O_{K,f})$, in particular $\mathrm{Gal}(\overline{\mathbb{Q}}/K)$ acts transitively on $\Sigma_{K,f}$.

The size of the Galois orbit of the point $x_{K,f}$ of $\mathrm{SL}_2(\mathbb{Z})\backslash\mathbb{H}$ representing the isomorphism class of (E, α) is exactly $h_{K,f} := |\mathrm{Pic}(O_{K,f})|$. The Brauer-Siegel theorem implies that for every $\epsilon > 0$, one has

$$d_{K,f}^{1/2-\epsilon} \ll_\epsilon |\Sigma_{K,f}| \ll_\epsilon d_{K,f}^{1/2+\epsilon}$$

where $d_{K,f}$ denotes the absolute value of the discriminant of $O_{K,f}$. Let μ be the probability measure on $\mathrm{SL}_2(\mathbb{Z})\backslash\mathbb{H}$ induced by the Poincaré measure on $\mathbb{H}$. Duke proved the following theorem.

Theorem 6.1 *As $d_{K,f} \longrightarrow \infty$, the sets $\Sigma_{K,f}$ are equidistributed with respect to μ.*

A simpler question is whether the *toric* orbits are equidistributed. Note that in Duke's case the Galois and toric orbits are equal. A very recent result regarding toric orbits is due to Zhang. Let $s := \overline{(x, g)} \in \mathrm{Sh}_K(G, X)$ be a special point of $\mathrm{Sh}_K(G, X)$. The toric orbit of s is, by definition, the finite set $\overline{(x, \mathrm{MT}(x)(\mathbb{A}_f) \cdot g)}$. It is obvious that the Galois orbit of s is contained in its toric orbit. However, the difference between the Galois orbit and the toric orbit is not understood well enough to deduce the equidistribution of Galois orbits from that of toric orbits. Regarding the toric orbits, Zhang recently announced equidistribution of toric orbits in Hilbert modular varieties or, more generally in Shimura varieties associated to quaternion algebras over totally real fields. Very recently Clozel and Ullmo proved that in the cases where the reciprocity morphisms attached to special points have connected kernels, toric equidistribution implies the equidistribution of Galois orbits. See [6].

The Haar measure on $G(\mathbb{R})^+$ induces a measure μ on $S := \Gamma\backslash X^+$ that we normalise by $\mu(S) = 1$ (μ is a probability measure). To any special subvariety $Z \subset S$, is similarly associated a canonical probability measure μ_Z. A sequence μ_n of probability measures is said to be weakly convergent to μ if for any continuous function with compact support f on S, the sequence of real numbers $\mu_n(f)$ converges to $\mu(f)$. In general, one can formulate the following conjecture that one can call "equidistribution conjecture of Galois orbits of special subvarieties" which implies the André-Oort conjecture but actually is much stronger. The only known case of this conjecture is the case where S is a modular curve, thanks to Duke's result.

Conjecture 6.2 *Let S be a Shimura variety and fix a number field F over which S admits a canonical model.*

To a special subvariety Z of S, one associates the canonical probability measure Δ_Z defined by

$$\Delta_Z = \frac{1}{|\{\sigma(Z) : \sigma \in \mathrm{Gal}(\overline{F}/F)\}|} \sum_{\sigma \in \mathrm{Gal}(\overline{F}/F)} \mu_{\sigma(Z)}.$$

Let $Q(S)$ be the set of measures Δ_Z with Z ranging through the set of special subvarieties of S. The set $Q(S)$ is compact : every sequence of measures in $Q(S)$ admits a weakly convergent subsequence. Furthermore, if Δ_{Z_n} is a sequence in $Q(S)$ weakly convergent to μ_Z, then for all $n \gg 0$, the support of Δ_{Z_n} is contained in the support of Δ_Z.

Note that it is essential to consider measures associated to Galois orbits of special subvarieties in the statement above. If one considers sequences of measures μ_Z with Z special then the conclusion of the conjecture above is wrong even in the case of a modular curve. A sequence of measures δ_{x_n} where x_n are special points on a modular curve can converge to δ_x where x is a non-special point. However, for some classes of special subvarities, one can prove the conclusion of the conjecture above. This is precisely the object of the work of Clozel and Ullmo. Clozel and Ullmo prove the following theorem (see [5]).

Theorem 6.3 *Let Z_n be a sequence of strongly special subvarieties of S with associated measures μ_n (as explained above). There exists a strongly special subvariety Z of S and a subsequence μ_{n_k} weakly convergent to μ_Z. Furthermore, Z contains S_{n_k} for every k large enough.*

The proof of the theorem of Clozel and Ullmo relies on some theorems from ergodic theory due to Ratner, Mozes-Shah and Dani-Margulis. An obvious consequence of this theorem is the following.

Corollary 6.4 *Let Y be a proper subvariety of the Shimura variety S. There exists a finite set $\{Z_1, \dots, Z_n\}$ of strongly special subvarieties of Y such that if Z is a strongly special subvariety of Y, then $Z \subset Z_i$ for some i.*

Let us prove that the theorem 6.3 implies the corollary above. So let Y be a proper subvariety of S. We will need to see that the set of *maximal* strongly special subvarieties of Y is finite. Consider the sequence of *maximal* strongly special subvarieties Z_n contained in Y and let μ_n be the associated sequence of measures. After possibly replacing μ_n with a subsequence, we can assume that μ_n converges weakly to μ_Z for some strongly special subvariety Z of S. As the support of all of the Z_n is contained in Y, Z is also contained in Y. From the facts that the Z_n are maximal and that Z_n is contained in Z for n large enough, we deduce that the sequence Z_n is stationary. As an obvious consequence of the last corollary, we get the following.

Corollary 6.5 *Let Σ be a set of strongly special subvarieties of S. The components of the Zariski closure of Σ are special.*

The theorem 6.3 has been recently further generalised by Ullmo (see [33]). He defined a class of NF-special subvarieties (NF stands for Non-Factor). The definition is as follows.

Definition A special subvariety Z of S is called a NF-special subvariety if there does not exist a special subvariety direct product $Z_1 \times Z_2$ of S such that Z is a subvariety of the form $\{x\} \times Z_2$ of $Z_1 \times Z_2$ where x is a special point of Z_1.

It is clear that a strongly special subvariety is NF-special. As an easy exercise, the reader may prove that in the case where S is a product of modular or Shimura curves, the notions of strongly special and NF-special subvarieties are equivalent. In [33], Ullmo proved that the NF-special subvarieties are equidistributed in the above sense.

We finish this section by explaining an equidistribution result of Ullmo and the author of this survey, which is one of the main results of [31]. Before we state the theorem, we need to introduce a definition.

Definition Let (G, X) be a Shimura datum with G semisimple of adjoint type. Let T be a subtorus of G such that $T(\mathbb{R})$ is compact.

A T-sub-Shimura datum (H, X_H) of (G, X) is a sub-Shimura datum such that $H^{\rm der}$ is non-trivial and T is the connected centre of the generic Mumford-Tate group on X_H.

Let K be compact open subgroup of $G(\mathbb{A}_{\rm f})$, a T-special subvariety of $\mathrm{Sh}_K(G, X)$ is a special subvariety defined by a T-Shimura datum.

We show the following theorem, which is the equidistribution property for sequences of T-special subvarieties.

Theorem 6.6 *Let S be a connected component of $\mathrm{Sh}_K(G, X)$ with G semisimple of adjoint type. Fix T with $T(\mathbb{R})$ compact. Let Z_n be a sequence of T-special subvarieties of S and let μ_n be the associated sequence of probability measures. There exists a T-special subvariety Z of S and a subsequence Z_{n_k} such that μ_{n_k} converges weakly to μ_Z. Furthermore, Z contains Z_{n_k} for all k large enough.*

We derive this theorem from the Clozel-Ullmo theorem. The idea is that T-special subvarieties are geometrically very similar to strongly special ones. The only thing which counts in their geometry is $H^{\rm der}$, the connected centre T does not count as it is constant.

Theorem 6.7 *Let S be a connected Shimura variety and F a number field over which S admits a canonical model.*

Assume the GRH for CM fields and let N be an integer. There exists a finite set $\{T_1, \ldots, T_r\}$ of $\mathbb{Q}$-subtori of G with the following property. Let Z be a special subvariety of S defined and geometrically irreducible over a finite extension F' of F of degree at most N. Then Z_i is a T_i-special subvariety for some $i \in \{1, \ldots, r\}$.

To prove this theorem we use the lower bounds for Galois orbits. We express the fact that these lower bounds are bounded. This in particular implies that the discriminants $\log(d_T)$ are bounded and this implies that the connected centres T of H lie in finitely many $\mathrm{GL}_n(\mathbb{Q})$ conjugacy classes with respect to some faithful representation $G \hookrightarrow \mathrm{GL}_n$.

Then we derive from the fact that the quantities $B^{i(T)}|K_T^m/K_T|$ are bounded, the fact that the tori T lie in finitely many $G(\mathbb{Q}) \cap \mathrm{GL}_n(\mathbb{Z})$-classes. This uses results of Gille and Moret-Bailly on algebraic actions of arithmetic groups (see the appendix to [31]). The two theorems above imply that a sequence of special subvarieties geometrically irreducible over a field extension of fixed degree, is equidistributed. We'll explain in the next section how this fact yields a strategy for proving the André-Oort conjecture.

7 The alternative and the strategy for proving the André-Oort conjecture.

The proof of the André-Oort conjecture has been very recently announced by Bruno Klingler and the author of the present paper. The idea is a combination of the Galois-theoretic approach and of the equidistribution approach. The results of [31] explained in the end of the previous section, show that the following alternative occurs.

As usual, let S be a connected component of a special subvariety. Let Z_n be a sequence of irreducible special subvarieties of S. Let F be a number field over which S admits a canonical model. After possibly replacing Z_n by a subsequence and assuming the GRH for CM-fields, at least one of the following cases occurs.

(i) The cardinality of the sets $\{\sigma(Z_n), \sigma \in \mathrm{Gal}(\overline{\mathbb{Q}}/F)\}$ is unbounded as $n \to \infty$ (and therefore Galois-theoretic techniques can be used).

(ii) The sequence of probability measures μ_n canonically associated to Z_n weakly converges to some μ_Z, the probability measure canonically associated to a special subvariety Z of S. Moreover, for every n large enough, Z_n is contained in Z.

We now very briefly explain how this alternative gives a strategy for proving the André-Oort conjecture. Suppose now that we are given a Hodge generic subvariety X containing a Zariski dense sequence Z_n of special subvarieties. Our aim to prove that X is special.

We fix a number field F such that S admits a canonical model over F and such that X is defined over F. If the cardinality of sets $\{\sigma(Z_n), \sigma \in \mathrm{Gal}(\overline{\mathbb{Q}}/F)\}$ is bounded then the second case of the alternative occurs and X is special. If the sets $\{\sigma(Z_n), \sigma \in \mathrm{Gal}(\overline{\mathbb{Q}}/F)\}$ are unbounded, then the techniques of Edixhoven and the author apply.

Using the induction on the dimension of subvarieties of X, we produce a subsequence Z_{n_k} of Z_n such that for all k, Z_{n_k} is contained in a special subvariety Z'_{n_k} with $\dim Z'_{n_k} > \dim Z_{n_k}$. We replace Z_n with Z'_{n_k} and re-iterate the process. We continue until we get either $\dim Z_n = \dim X$ or that the sets $\{\sigma(Z_n), \sigma \in \mathrm{Gal}(\overline{\mathbb{Q}}/F)\}$ are bounded. In each case, X is special. Technically, the induction is quite difficult, we refer to [14] for details.

Bibliography

[1] Y. André, *G-functions and geometry.* Aspects of Mathematics, E13, Vieweg, 1989.

[2] Y. André, *Mumford-Tate groups of mixed Hodge structures and the theorems of fixed part.* Compositio Math., **82**, (1992), p. 1-24.

[3] Y. André, *Finitude de couples d'invariants modulaires singuliers sur une courbe algebrique plane non-modulaire.* J. Reine Angew. Math., **505**, p. 203-208, (1998).
[4] L. Clozel, H. Oh, E. Ullmo, *Hecke operators and equidistribution of Hecke points.* Invent. Math. **144**, (2001), p. 327-351.
[5] L. Clozel, E. Ullmo, *Equidistribution de sous-variétés spéciales*, Ann. of Math. (2) **161** (2005), no. 3, p. 1571–1588.
[6] L. Clozel, E. Ullmo, *Equidistribution adélique des tores et equidistribution des points CM.* Preprint 2005. Available at
`http://www.math.u-psud.fr/ ullmo/liste-prepub.html`
[7] P. Deligne. *Travaux de Shimura*, Séminaire Bourbaki, Exposé 389, Fevrier 1971, Lecture Notes in Maths. **244**, Springer-Verlag, Berlin 1971, p. 123-165.
[8] P. Deligne, *Variétés de Shimura: interprétation modulaire et techniques de construction de modèles canoniques*, dans *Automorphic Forms, Representations, and L-functions* part. **2**; Editeurs: A. Borel et W Casselman; Proc. of Symp. in Pure Math. **33**, American Mathematical Society, 1979, p. 247–290.
[9] W. Duke, *Hyperbolic distribution problems and half integral weights Maas-forms*, Invent. Math. **92**, no .1 (1988), p. 73-90
[10] B. Edixhoven, *Special points on products of modular curves* Duke Math. J. **126** (2005), no. 2, 325–348.
[11] B. Edixhoven, *On the André-Oort conjecture for Hilbert modular surfaces*, Moduli of abelian varieties (Texel Island, 1999), 133–155, Progr. Math., 195, Birkhuser, Basel, 2001.
[12] B. Edixhoven, A. Yafaev, *Subvarieties of Shimura varieties*, Ann. Math. (2) **157**, (2003), p. 621–645.
[13] M. Hindry, *Autour d'une conjecture de Serge Lang.* Invent. Math. **94** (1988) p. 575-603.
[14] B. Klingler, A. Yafaev, *The Andre-Oort Conjecture.* Preprint 2006, submitted.
[15] S. Lang, *Complex Multiplication.* Springer-Verlag, 1983.
[16] J. Milne, *Shimura varieties.* Lecture notes. Available at
`http://www.jmilne.org/math/`
[17] B. Moonen, *Linearity properties of Shimura varieties I* , Journal of Algebraic Geom. **7** (1998), p. 539–567.
[18] B. Moonen, *Linearity properties of Shimura varieties II* , Compositio Math. **114** (1998), no. 1, 3–35.
[19] S. Mozes, N. Shah, On the space of ergodic invariant measures of unipotent flows. Ergod. Th. and Dynam. Sys. **15** (1995), p. 149-159.
[20] R. Noot, *Orbites de Galois, correspondances de Hecke et la conjecture de André-Oort (d'après Edixhoven et Yafaev)* Seminaire Bourbaki 2004.
[21] M. Nori, *On subgroups of* $\mathrm{GL}_n(\mathbb{F}_p)$, Invent. Math. **88** (1987), no. 2, 257–275.
[22] F. Oort, *Canonical liftings and dense sets of CM points.* In Arithmetic Geometry, 1994, Vol. XXXVII of Sympos. Math.
[23] R. Pink, *A Combination of the Conjectures of Mordell-Lang and André-Oort.* In Geometric Methods in Algebra and Number Theory. Progress in Mathematics 253, Birkhauser (2005), 251-282.
[24] R. Pink, *A Common Generalisation of the Conjectures of André-Oort, Manin-Mumford and Mordell-Lang.* Preprint (April 2005), 13p. Available on author's web-page.

[25] R. Pink, D. Roessler, *On ψ-invariant subvarieties of abelian varieties and the Manin-Mumford conjecture.* J. Algebraic Geom. **13** (2004), n. 4, 771-798.

[26] M. Raynaud, *Sous-variétés d'une varieté abélienne et points de torsion.* Arithmetic and Geometry, Vol. I, Progr. Math. 35, Birkhauser Boston, MA, (1988).

[27] L. Szpiro, E. Ullmo, S.W. Zhang, *Equirépartition des petits points.* Invent. Math, **127** (1997), p. 337-347.

[28] E. Ullmo, *Positivité et discretion des points algébriques des courbes.* Ann. Math., **147** (1998), p. 167-179.

[29] E. Ullmo, Théorie ergodique et géometrie arithmétique. ICM 2002. Vol III, 1-3.

[30] E. Ullmo, *Equidistribution des sous-variétés spéciales II.* Preprint. 2005

[31] E. Ullmo, A. Yafaev, *Galois orbits and equidistribution of special subvarieties: towards the Andre-Oort conjecture* With an appendix by P. Gille and L. Moret-Bailly. Preprint 2006. Submitted.

[32] E. Ullmo, *Manin-Mumford, André-Oort, the equidistribution viewpoint.* preprint, 35 pages. http://www.math.u-psud.fr/ ullmo/liste-prepub.html

[33] E. Ullmo, *Equidistribution des sous-varietés speciales II.* Preprint, 2005.

[34] A. Yafaev, *On a result of Moonen on the moduli space of principally polarised abelian varieties.* Compositio Math. **141** (2005) 1103-1108.

[35] A. Yafaev, *A conjecture of Yves André's*, Duke Math. J. 132 (2006), no. 3, 393–407.

[36] S.W. Zhang, *Equidistribution of small points on abelian varieties*, Ann. Math., **147**, (1998). p.159-165.

[37] S.W. Zhang, *Equidistribution of CM-points on quaternion Shimura varieties*, Int. Math. Res. Not. (2005), no. 59, 3657–3689.

Locally analytic representation theory of p-adic reductive groups: a summary of some recent developments

Matthew Emerton

Northwestern University
Department of Mathematics
2033 Sheridan Rd.
Evanston, IL 60208-2730
USA
emerton@math.northwestern.edu

The purpose of this short note is to summarize some recent progress in the theory of locally analytic representations of reductive groups over p-adic fields. This theory has begun to find applications to number theory, for example to the arithmetic theory of automorphic forms, as well as to the "p-adic Langlands programme" (see [3, 4, 5, 10, 11, 12]). I hope that this note can serve as an introduction to the theory for those interested in pursuing such applications.

The theory of locally analytic representations relies for its foundations on notions and techniques of functional analysis. We recall some of these notions in Section 1. In Section 2 we describe some important categories of locally analytic representations (originally introduced in [20], [23] and [8]). In Section 3, we discuss the construction of locally analytic representations by applying the functor "pass to locally analytic vectors" to certain continuous Banach space representations. In Section 4 we briefly describe the process of parabolic induction in the locally analytic situation, which allows one to pass from representations of a Levi subgroup of a reductive group to representations of the reductive group itself, and in Section 5 we describe the Jacquet module construction of [9], which provides functors mapping in the opposite direction. Parabolic induction and the Jacquet module functors are "almost" adjoint to one another. (See Theorem 5.19 for a precise statement.)

Acknowledgments. I would like to thank David Ben-Zvi for his helpful remarks on an earlier draft of this note, as well as the anonymous referee, whose comments led to the clarification of some points of the text.

The author would like to acknowledge the support of the National Science Foundation (award numbers DMS-0070711 and DMS-0401545)

1. Functional analysis

We begin by recalling some notions of non-archimedean functional analysis. A more detailed exposition of the basic concepts is available in [17], which provides an excellent introduction to the subject.

Let K be a complete discretely valued field of characteristic zero. A topological K-vector space V is said to be locally convex if its topology can be defined by a basis of neighbourhoods of the origin that are $\mathcal{O}_K$-submodules of V; or equivalently, by a collection of non-archimedean semi-norms. (We will often refer to V simply as a convex space, or a convex K-space if we which to emphasize the coefficient field K.) The space V is called complete if it is complete as a topological group under addition.

If V is any locally convex K-space, then we may complete V to obtain a complete Hausdorff convex K-space $\hat{V}$, equipped with a continuous K-linear map $V \to \hat{V}$, which is universal for continuous K-linear maps from V to complete Hausdorff K-spaces. (See [17, Prop. 7.5] for a construction of $\hat{V}$. Note that in this reference $\hat{V}$ is referred to as the Hausdorff completion of V.)

If V is a convex K-space, then we let V' denote the space of K-valued continuous K-linear functionals on V, and let V'_b denote V' equipped with its strong topology (the "bounded-open" topology – see [17, Def., p. 58]; the subscript "b" stands for "bounded"). We refer to V'_b as the *strong dual* of V. There is a natural K-linear "double duality" map $V \to (V'_b)'$; we say that V is *reflexive* if this map induces a topological isomorphism $V \to (V'_b)'_b$.

If V and W are two convex K-spaces, then we always equip $V \otimes_K W$ with the projective tensor product topology. This topology is characterized by the requirement that the map $V \times W \to V \otimes_K W$ defined by $(v, w) \mapsto v \otimes w$ should be universal for continuous K-bilinear maps from $V \times W$ to convex K-spaces. (See [17, §17] for more details about the construction and properties of this topology.) We let $V \hat{\otimes}_K W$ denote the completion of $V \otimes_K W$.

A complete convex space V is called a Fréchet space if it is metrizable, or equivalently, if its topology can be defined by a countable set of semi-norms. If the topology of the complete convex space V can be defined by a single norm, then we say that V is a Banach space. Note that we don't regard a Fréchet space or a Banach space as being equipped with any particular choice of metric, or norm.

If V and W are Banach spaces, then the space $\mathcal{L}(V, W)$ of continuous linear maps from V to W again becomes a Banach space, when equipped with its strong topology. (Concretely, if we fix norms defining the topologies of V and W respectively, then we may define a norm on $\mathcal{L}(V, W)$

as follows (we denote all norms by $|| \ ||$): for any $T \in \mathcal{L}(V, W)$, set $||T|| = \sup_{v \in V \text{ s.t. } ||v||=1} ||T(v)||$.) We say that an element $T \in \mathcal{L}(V, W)$ is compact if it may be written as a limit (with respect to the strong topology) of a sequence of maps with finite dimensional range (see [17, Rem. 18.10]).

If V is a Fréchet space, then completing V with respect to each of the members of an increasing sequence of semi-norms that define its topology, we obtain a projective sequence of Banach spaces $\{V_n\}_{n \geq 1}$, and an isomorphism of topological K-vector spaces

$$V \xrightarrow{\sim} \varprojlim_n V_n,$$

where $\{V_n\}_{n \geq 1}$ is a projective system of Banach spaces over K, and the right hand side is equipped with the projective limit topology. Conversely, any such projective limit is a Fréchet space over K.

Definition 1.1. A nuclear Fréchet space over K is a K-space which admits an isomorphism of topological K-vector spaces

$$V \xrightarrow{\sim} \varprojlim_n V_n,$$

where $\{V_n\}_{n \geq 1}$ is a projective system of Banach spaces over K with compact transition maps (and the right hand side is equipped with the projective limit topology).

In fact there is a more intrinsic definition of nuclearity for any convex space [17, Def., p. 120], which is equivalent to the above definition when applied to a Fréchet space (as follows from the discussion of [17, §16] together with [20, Thm. 1.3]).

Proposition 1.2. *Let V be a nuclear Fréchet space.*

(i) V is reflexive.

(ii) Any closed subspace or Hausdorff quotient of V is again a nuclear Fréchet space.

Proof. See [17, Prop. 19.4]. □

We now introduce another very important class of locally convex spaces.

Definition 1.3. We say that a convex K-space V is of compact type if there is an isomorphism of topological K-vector spaces

$$V \xrightarrow{\sim} \varinjlim_n V_n,$$

where $\{V_n\}_{n \geq 1}$ is an inductive system of Banach spaces over K with compact and injective transition maps (and the right hand side is equipped with the locally convex inductive limit topology).

Proposition 1.4. *Let V be a space of compact type.*

(i) V is complete and Hausdorff.

(ii) V is reflexive.

(iii) Any closed subspace or Hausdorff quotient of V is again of compact type.

Proof. See [20, Thm. 1.1, Prop. 1.2]. □

Proposition 1.5. *Passing to strong duals yields an anti-equivalence of categories between the category of spaces of compact type and the category of nuclear Fréchet spaces.*

Proof. This is [20, Thm. 1.3]. A proof can also be extracted from the discussion of [17, §16]. □

We now define an important class of topological algebras over K (originally introduced in [23]).

Definition 1.6. Let A be a topological K-algebra. We say that A is a nuclear Fréchet-Stein algebra if we may find an isomorphism $A \xrightarrow{\sim} \varprojlim_{n} A_n$, where $\{A_n\}_{n\geq 1}$ is a sequence of Noetherian K-Banach algebras, for which the transition maps $A_{n+1} \to A_n$ are compact (as maps of K-Banach spaces) and flat (as maps of K-algebras), and such that each of the maps $A \to A_n$ has dense image (or equivalently, by [2, II §3.5 Thm. 1], such that each of the maps $A_{n+1} \to A_n$ has dense image).

If A is a nuclear Fréchet-Stein algebra over K, then A is certainly a nuclear Fréchet space. If A is a topological K-algebra, then any two representations of A as a projective limit as in Definition 1.6 are equivalent in an obvious sense. (See [8, Prop. 1.2.7].)

Example 1.7. Let us explain the motivating example of a nuclear Fréchet-Stein algebra. Suppose that $\mathbb{X}$ is a rigid analytic space over K that may be written as a union $\mathbb{X} = \bigcup_{n=1}^{\infty} \mathbb{X}_n$, where $\{\mathbb{X}_n\}_{n\geq 1}$ is an increasing sequence of open affinoid subdomains of $\mathbb{X}$, for which the inclusions $\mathbb{X}_n \to \mathbb{X}_{n+1}$ are admissible and relatively compact (in the sense of [1, 9.6.2]), and such that for each n the restriction map $\mathcal{C}^{\mathrm{an}}(\mathbb{X}_{n+1}, K) \to \mathcal{C}^{\mathrm{an}}(\mathbb{X}_n, K)$ has dense image. (Here $\mathcal{C}^{\mathrm{an}}(\mathbb{X}_n, K)$ denotes the Tate algebra of rigid analytic K-valued functions on $\mathbb{X}_n$.) We will say that such a rigid analytic space $\mathbb{X}$ is *strictly quasi-stein.* (If one omits the requirement that the inclusions be relatively compact, one obtains the notion of a quasi-stein rigid analytic space, as defined by Kiehl.) Since $\mathbb{X}_n \to \mathbb{X}_{n+1}$ is an admissible open immersion for each $n \geq 1$, the restriction map $\mathcal{C}^{\mathrm{an}}(\mathbb{X}_{n+1}, K) \to \mathcal{C}^{\mathrm{an}}(\mathbb{X}_n, K)$ is flat. The relative compactness assumption implies that it is furthermore compact, and hence that the space

$$\mathcal{C}^{\mathrm{an}}(\mathbb{X}, K) \xrightarrow{\sim} \varprojlim_{n} \mathcal{C}^{\mathrm{an}}(\mathbb{X}_n, K)$$

of rigid analytic functions on $\mathbb{X}$ is naturally a nuclear Fréchet-Stein algebra.

Definition 1.8. Let A be a nuclear Fréchet-Stein algebra over K, and write $A \xrightarrow{\sim} \varprojlim_{n} A_n$ as in Definition 1.6. We say that a Hausdorff topological A-module M is coadmissible if the following two conditions are satisfied:

(i) The tensor product $M_n := A_n \otimes_A M$ is a finitely generated A_n-Banach module, for each n. (We regard the tensor product $A_n \otimes_A M$ as being a quotient of $A_n \otimes_K M$, and endow it with the quotient topology induced by the projective tensor product topology on $A_n \otimes_K M$.)

(ii) The natural map $M \to \varprojlim_{n} M_n$ is an isomorphism of topological A-modules.

The preceding definition is a variation of [23, Def., p. 152], to which it is equivalent, as the results of [23, §3] show.

Theorem 1.9. *Let A be a nuclear Fréchet-Stein algebra over K.*

(i) Any coadmissible topological A-module is a nuclear Fréchet space.

(ii) Any A-linear map between coadmissible topological A-modules is automatically continuous, with closed image.

(iii) The category of coadmissible topological A-modules (with morphisms being A-linear maps, which by (ii) are automatically continuous) is closed under taking finite direct sums, passing to closed submodules, and passing to Hausdorff quotients.

Proof. This summarizes the results of [23, §3]. □

Remark 1.10. The category of all locally convex Hausdorff topological A-modules is an additive category that admits kernels, cokernels, images and coimages. More precisely, if $f : M \to N$ is a continuous A-linear morphism between such modules, then its categorical kernel is the usual kernel of f, its categorical image is the closure of its set-theoretic image (regarded as a submodule of N), its categorical coimage is its set-theoretical image (regarded as a quotient module of M), and its categorical cokernel is the quotient of N by its categorical image.

Part (ii) of Theorem 1.9 implies that if M and N in the preceding paragraph are coadmissible, then the image and coimage of f coincide. Part (iii) of the Theorem then implies that the kernel, cokernel, and image of f are again coadmissible. Thus the category of coadmissible topological A-modules is an abelian subcategory of the additive category of locally convex Hausdorff topological A-modules.

Remark 1.11. If B is a Noetherian K-Banach algebra (for example, one of the algebras A_n appearing in Definitions 1.6 and 1.8), then the results of [1, 3.7.3] show that the natural functor from the category of finitely generated B-Banach modules (with morphisms being continuous B-linear maps) to the abelian category of finitely generated B-modules, given by forgetting topologies, is an equivalence of categories. Theorem 1.9 is an analogue of this result for the nuclear Fréchet-Stein algebra A. It shows that forgetting topologies yields a fully faithful embedding of the category of coadmissible topological A-modules as an abelian subcategory of the abelian category of all A-modules. In light of this, one can suppress all mention of topologies in defining this category (as is done in the definitions of [23, p. 152]).

Definition 1.12. If A is a nuclear Fréchet-Stein algebra over K, we say that a topological A-module M is strongly coadmissible if it is a Hausdorff quotient of A^n, for some natural number n.

Since A is obviously a coadmissible module over itself, Theorem 1.9 implies that any strongly coadmissible topological A-module is a coadmissible topological A-module.

Example 1.13. Suppose that $\mathbb{X}$ is a strictly quasi-Stein rigid analytic space over K, as in Example 1.7. If $\mathcal{M}$ is any rigid analytic coherent sheaf on $\mathbb{X}$, then the space M of global sections of $\mathcal{M}$ is naturally a coadmissible $\mathcal{C}^{\mathrm{an}}(\mathbb{X}, K)$-module, and passing to global sections in fact yields an equivalence of categories between the category of coherent sheaves on $\mathbb{X}$ and the category of coadmissible $\mathcal{C}^{\mathrm{an}}(\mathbb{X}, K)$-modules. The $\mathcal{C}^{\mathrm{an}}(\mathbb{X}, K)$-module M of global sections of the coherent sheaf $\mathcal{M}$ is strongly coadmissible if and only if $\mathcal{M}$ is generated by a finite number of global sections.

Fix a complete subfield L of K. We close this section by recalling the definition of the space of locally analytic functions on a locally L-analytic manifold with values in a convex space. (More detailed discussions may be found in [14, §2.1.10], [20, p. 447], and [8, §2.1].)

Definition 1.14. If $\mathbb{X}$ is an affinoid rigid analytic space over L, and if W is a K-Banach space, then we write $\mathcal{C}^{\mathrm{an}}(\mathbb{X}, W) := \mathcal{C}^{\mathrm{an}}(\mathbb{X}, K) \hat{\otimes}_K W$. (Here, as above, we let $\mathcal{C}^{\mathrm{an}}(\mathbb{X}, K)$ denote the Tate algebra of K-valued rigid analytic functions on $\mathbb{X}$, equipped with its natural K-Banach algebra structure.)

If the set $X := \mathbb{X}(L)$ of L-valued points of $\mathbb{X}$ is Zariski dense in $\mathbb{X}$, then $\mathcal{C}^{\mathrm{an}}(\mathbb{X}, W)$ may be identified with the space of W-valued functions on X that can be described by convergent power series with coefficients in W.

Now let X be a locally L-analytic manifold. A chart of X is a compact open subset X_0 of X together with a locally analytic isomorphism between X_0 and the set of L-valued points of a closed ball. We let $\mathbb{X}_0$ denote this ball (thought of as a rigid L-analytic space), so that $X_0 \xrightarrow{\sim} \mathbb{X}_0(L)$. By an analytic partition of X we mean a partition $\{X_i\}_{i\in I}$ of X into a disjoint union of charts X_i. We assume that X is paracompact; then any covering of X by charts may be refined to an analytic partition of X. (Here we are using a result of Schneider [18, Satz 8.6], which shows that any paracompact locally L-analytic manifold is in fact strictly paracompact, in the sense of the discussion of [20, p. 446].)

If V is a Hausdorff convex space, then we say that a function $f : X \to V$ is locally analytic if for each point $x \in X$, there is a chart X_0 containing x, a Banach space W equipped with a continuous K-linear map $\phi : W \to V$, and a rigid analytic function $f_0 \in \mathcal{C}^{\mathrm{an}}(\mathbb{X}_0, W)$ such that $f = \phi \circ f_0$. (Replacing W by its quotient by the kernel of ϕ, we see that it is no loss of generality to require that ϕ be injective.) We let $\mathcal{C}^{\mathrm{la}}(X, V)$ denote the K-vector space of locally analytic V-valued functions on X, and let $\mathcal{C}^{\mathrm{la}}_c(X, V)$ denote the subspace consisting of compactly supported locally analytic functions.

It follows from the definition that there are K-isomorphisms of vector spaces

$$\mathcal{C}^{\mathrm{la}}(X, V) \xrightarrow{\sim} \varinjlim_{\{X_i, W_i, \phi_i\}_{i\in I}} \prod_{i\in I} \mathcal{C}^{\mathrm{an}}(\mathbb{X}_i, W_i)$$

and

$$\mathcal{C}^{\mathrm{la}}_c(X, V) \xrightarrow{\sim} \varinjlim_{\{X_i, W_i, \phi_i\}_{i\in I}} \bigoplus_{i\in I} \mathcal{C}^{\mathrm{an}}(\mathbb{X}_i, W_i),$$

where in both cases the inductive limit is taken over the directed set of collections of triples $\{X_i, W_i, \phi_i\}_{i\in I}$, where $\{X_i\}_{i\in I}$ is an analytic partition of X, each W_i is a K-Banach space, and $\phi_i : W_i \to V$ is a continuous injection. We regard $\mathcal{C}^{\mathrm{la}}(X, V)$ and $\mathcal{C}^{\mathrm{la}}_c(X, V)$ as Hausdorff convex spaces by equipping them with the locally convex inductive limit topologies arising from the targets of these isomorphisms. Note that the inclusion $\mathcal{C}^{\mathrm{la}}_c(X, V) \to \mathcal{C}^{\mathrm{la}}(X, V)$ is continuous, but unless X is compact (in which case it is an equality) it is typically not a topological embedding.

Given any collection $\{X_i, W_i, \phi_i\}_{i\in I}$ as above, there is a natural map

$$\bigoplus_{i\in I} \mathcal{C}^{\mathrm{an}}(\mathbb{X}_i, W_i) = \bigoplus_{i\in I} \mathcal{C}^{\mathrm{an}}(\mathbb{X}_i, K) \mathbin{\hat{\otimes}}_K W_i$$

$$\xrightarrow{\oplus\, \mathrm{id} \hat{\otimes} \phi_i} \bigoplus_{i\in I} \mathcal{C}^{\mathrm{an}}(\mathbb{X}_i, K) \mathbin{\hat{\otimes}}_K V \longrightarrow \Big(\bigoplus_{i\in I} \mathcal{C}^{\mathrm{an}}(\mathbb{X}_i, K)\Big) \mathbin{\hat{\otimes}}_K V.$$

(Note that if we were working with inductive, rather than projective, tensor product topologies, then the last map would be an isomorphism.) Passing to the inductive limit over all such collections yields a continuous map

$$\mathcal{C}^{\mathrm{la}}_c(X, V) \to \mathcal{C}^{\mathrm{la}}_c(X, K) \hat{\otimes}_K V. \tag{1.15}$$

Proposition 1.16. *If X is σ-compact (i.e. the union of a countable number of compact open subsets) and V is of compact type then the map (1.15) is a topological isomorphism and $\mathcal{C}^{\mathrm{la}}_c(X, V)$ is again of compact type.*

Proof. If X is compact (so that $\mathcal{C}^{\mathrm{la}}_c(X, V) = \mathcal{C}^{\mathrm{la}}(X, V)$) then this is [8, Prop. 2.1.28]. The proof in the general case is similar. □

2. Categories of locally analytic representations

Fix a finite extension L of $\mathbb{Q}_p$, for some prime p, as well a field K that extends L and is complete with respect to a discrete valuation extending that on L. Let G be a locally L-analytic group (an analytic group over L, in the sense of [25, p. LG 4.1]). The identity element of G then has a neighbourhood basis consisting of compact open subgroups of G [25, Cor. 2, p. LG 4.23].

If H is any compact open subgroup of G, then Proposition 1.16 shows that the space $\mathcal{C}^{\mathrm{la}}(H, K)$ of locally L-analytic K-valued functions on H is a compact type convex K-space, and hence its strong dual is a nuclear Fréchet space, which we will denote by $\mathcal{D}^{\mathrm{la}}(H, K)$. Any element $h \in H$ gives rise to a "Dirac delta function" supported at h, which is an element $\delta_h \in \mathcal{D}^{\mathrm{la}}(H, K)$. In this way we obtain an embedding $K[H] \to \mathcal{D}^{\mathrm{la}}(H, K)$ (where $K[H]$ denotes the group ring of H over K). The image of $K[H]$ is dense in $\mathcal{D}^{\mathrm{la}}(H, K)$, and the K-algebra structure on $K[H]$ extends (in a necessarily unique fashion) to a topological K-algebra structure on $\mathcal{D}^{\mathrm{la}}(H, K)$ [20, Prop. 2.3, Lem. 3.1].

Theorem 2.1. *The topological K-algebra $\mathcal{D}^{\mathrm{la}}(H, K)$ is a nuclear Fréchet-Stein algebra.*

Proof. This is the main result of [23]. A different proof is given in [8, §5.3]. □

We will now consider various convex K-spaces V equipped with actions of G by K-linear automorphisms. There are (at least) three kinds of continuity conditions on such an action that one can consider. Firstly, one may consider a situation in which G acts by continuous automorphisms of V. (Such an action is referred to as a *topological action* in [8];

note that this condition does not make any reference to the topology of G.) Secondly, one may consider the case when the action map $G \times V \to V$ is separately continuous. Thirdly, one may consider the case when the action map $G \times V \to V$ is continuous. If V is *barrelled* (see [17, Def., p. 39]; for example a Banach space, a Fréchet space, or a space of compact type) then any separately continuous action is automatically continuous, by the Banach-Steinhaus theorem.

Proposition 2.2. *If V is a compact type convex space, equipped with an action of G by continuous K-linear automorphisms, then the following are equivalent:*

(i) For some compact open subgroup H of G, the $K[H]$-module structure on V extends to a (necessarily unique) $\mathcal{D}^{\mathrm{la}}(H,K)$-module structure on V, for which the map $\mathcal{D}^{\mathrm{la}}(H,K) \times V \to V$ describing this module structure is separately continuous.

(i′) For every compact open subgroup H of G, the $K[H]$-module structure on V extends to a (necessarily unique) $\mathcal{D}^{\mathrm{la}}(H,K)$-module structure on V, for which the map $\mathcal{D}^{\mathrm{la}}(H,K) \times V \to V$ describing this module structure is separately continuous.

(ii) For some compact open subgroup H of G, the $K[H]$-module structure on V_b' arising from the contragredient H-action on V_b' extends to a (necessarily unique) topological $\mathcal{D}^{\mathrm{la}}(H,K)$-module structure on V_b'.

(ii′) For every compact open subgroup H of G, the $K[H]$-module structure on V_b' arising from the contragredient H-action on V_b' extends to a (necessarily unique) topological $\mathcal{D}^{\mathrm{la}}(H,K)$-module structure on V_b'.

(iii) There is a compact open subgroup H of G such that for any $v \in V$, the orbit map $o_v : H \to V$, defined via $h \mapsto hv$, lies in $\mathcal{C}^{\mathrm{la}}(H,V)$.

(iii′) For any $v \in V$, the orbit map $o_v : G \to V$, defined via $g \mapsto gv$, lies in $\mathcal{C}^{\mathrm{la}}(G,V)$.

Proof. The uniqueness statement in each of the first four conditions is a consequence of the fact that $K[H]$ is dense in $\mathcal{D}^{\mathrm{la}}(H,K)$, for any compact locally analytic L-analytic group. The equivalence of (i), (ii) and (iii) follows from [20, Cor. 3.3] and the accompanying discussion at the top of p. 453 of this reference. The equivalence of (iii) and (iii′) is straightforward. (See for example [8, Prop. 3.6.11].) Since (iii′) is independent of H, we see that (i′) and (ii′) are each equivalent to the other four conditions. □

Definition 2.3. If V is a compact type convex space equipped with an action of G by continuous K-linear automorphisms, then we say that V is a locally analytic representation of G if the equivalent conditions of Proposition 2.2 hold.

We let $\mathrm{Rep}_{\mathrm{la.c}}(G)$ denote the category of compact type convex spaces equipped with a locally analytic representation of G (the morphisms being continuous G-equivariant K-linear maps).

Example 2.4. If G is compact, so that $\mathcal{C}^{\mathrm{la}}(G, K)$ is a compact type convex space (by Proposition 1.16), then the left regular action of G on $\mathcal{C}^{\mathrm{la}}(G, K)$ equips this space with a locally analytic G-representation. This is perhaps most easily seen by applying the criterion of Proposition 2.2 (ii). Indeed, the strong dual of $\mathcal{C}^{\mathrm{la}}(G, K)$ is equal to $\mathcal{D}^{\mathrm{la}}(G, K)$, and under the contragredient action to the left regular representation, an element $g \in G$ acts as left multiplication by δ_g on $\mathcal{D}^{\mathrm{la}}(G, K)$. Thus the required topological $\mathcal{D}^{\mathrm{la}}(G, K)$-module structure on the strong dual of $\mathcal{C}^{\mathrm{la}}(G, K)$ is obtained by regarding the topological algebra $\mathcal{D}^{\mathrm{la}}(G, K)$ as a left module over itself in the tautological manner.

Similarly, the right regular action of G on $\mathcal{C}^{\mathrm{la}}(G, K)$ makes $\mathcal{C}^{\mathrm{la}}(G, K)$ a locally analytic G-representation. (Indeed, the topological automorphism $f(g) \mapsto f(g^{-1})$ of $\mathcal{C}^{\mathrm{la}}(G, K)$ intertwines the left and right regular representations.)

Remark 2.5. If V is an object of $\mathrm{Rep}_{\mathrm{la.c}}(G)$, then since the orbit maps o_v lie in $\mathcal{C}^{\mathrm{la}}(G, V)$ for all $v \in V$ they are in particular continuous on G. Thus the G-action on V is separately continuous, and hence (as was remarked above) continuous, by the Banach-Steinhaus theorem. Furthermore, we may differentiate the G-action on V and so make G a module over the Lie algebra $\mathfrak{g}$ of G (or equivalently, over its universal enveloping algebra $\mathrm{U}(\mathfrak{g})$). The action $\mathfrak{g} \times V \to V$ is again seen to be separately continuous (since the derivatives along the elements of $\mathfrak{g}$ of a function in $\mathcal{C}^{\mathrm{la}}(G, V)$ again lie in $\mathcal{C}^{\mathrm{la}}(G, V)$), and hence (applying the Banach-Steinhaus theorem once more) is continuous.

The $\mathrm{U}(\mathfrak{g})$-module structure on V admits an alternative description. Indeed, for any compact open subgroup H of G, there is a natural embedding $\mathrm{U}(\mathfrak{g}) \to \mathcal{D}^{\mathrm{la}}(H, K)$, given by mapping $X \in \mathrm{U}(\mathfrak{g})$ to the functional $f \mapsto (Xf)(e)$. (Here X acts on f as a differential operator,[1] and e denotes the identity of G.) Since V is an object of $\mathrm{Rep}_{\mathrm{la.c}}(V)$, it is a $\mathcal{D}^{\mathrm{la}}(H, K)$-module (by part (i) of Theorem 2.1), and so in particular is a $\mathrm{U}(\mathfrak{g})$-module. This $\mathrm{U}(\mathfrak{g})$-module structure on V coincides with the one described in the preceding paragraph.

---

[1]More precisely, the $\mathfrak{g}$ action on $\mathcal{C}^{\mathrm{la}}(H, K)$ that we have in mind is the one obtained via differentiating the right regular action of H on $\mathcal{C}^{\mathrm{la}}(H, K)$. (By applying Example 2.4 to H, we find that this H-action is locally analytic, and so may indeed be differentiated to yield a $\mathfrak{g}$-action.) It is given explicitly by the formula $(Xf)(h) = \frac{d}{dt}_{|t=0} f(h \exp(tX))$, for any $X \in \mathfrak{g}$.

Now suppose that Z is a topologically finitely generated abelian locally L-analytic group. If E is any finite extension of L, then we may consider the set $\hat{Z}(E)$ of $E^\times$-valued locally L-analytic characters on Z.

Proposition 2.6. *There is a strictly quasi-stein rigid analytic space $\hat{Z}$ over L that represents the functor $E \mapsto \hat{Z}(E)$.*

Proof. This is [8, Prop. 6.4.5]. □

Example 2.7. Suppose that $L = \mathbb{Q}_p$, and that Z is the group $\mathbb{Z}_p$. Then $\hat{Z}$ is isomorphic to the open unit disk centered at 1. (A character of $\hat{Z}$ may be identified with its value on the topological generator 1 of $\mathbb{Z}_p$.)

Example 2.8. Suppose that $L = \mathbb{Q}_p$, and that Z is the multiplicative group $\mathbb{Q}_p^\times$. There is an isomorphism

$$\mathbb{Q}_p^\times \xrightarrow{\sim} \mathbb{Z}_p^\times \times p^{\mathbb{Z}} \xrightarrow{\sim} \mu \times \Gamma \times p^{\mathbb{Z}},$$

where μ denotes the subgroup of roots of unity in $\mathbb{Q}_p^\times$, Γ denotes the subgroup of $\mathbb{Z}_p^\times$ consisting of elements congruent to 1 modulo p (respectively p^2 if $p = 2$), and $p^{\mathbb{Z}}$ denotes the cyclic group generated by $p \in \mathbb{Q}_p^\times$. The group Γ is isomorphic to $\mathbb{Z}_p$, and so there is an isomorphism

$$\hat{Z} \xrightarrow{\sim} \mathrm{Hom}(\mu, \mathbb{Q}_p^\times) \times \text{ open unit disk around 1 } \times \mathbb{G}_m.$$

Here $\mathrm{Hom}(\mu, \mathbb{Q}_p^\times)$ is the character group of the finite group μ, the open unit disk around 1 is the character group of Γ (see the preceding example), and $\mathbb{G}_m$ is the character group of $p^{\mathbb{Z}}$. (A character of the cyclic group $p^{\mathbb{Z}}$ may be identified with its value on p).

The discussion of Example 1.7 shows that the K-algebra $\mathcal{C}^{\mathrm{an}}(\hat{Z}, K)$ of rigid analytic functions on $\hat{Z}$ is a nuclear Fréchet-Stein algebra. Evaluation of characters at elements of Z induces an embedding of K-algebras $K[Z] \to \mathcal{C}^{\mathrm{an}}(\hat{Z}, K)$, with dense image (by [8, Prop. 6.4.6] and [20, Lem. 3.1]), and we have the following analogue of Proposition 2.2.

Proposition 2.9. *If V is a compact type convex space, equipped with an action of Z by continuous K-linear automorphisms, then the following are equivalent:*

(i) The $K[Z]$-module structure on V extends to a (necessarily unique) $\mathcal{C}^{\mathrm{an}}(\hat{Z}, K)$-module structure on V, for which the map $\mathcal{C}^{\mathrm{an}}(\hat{Z}, K) \times V \to V$ describing this module structure is separately continuous.

(ii) The $K[Z]$-module structure on V_b' arising from the contragredient Z-action on V_b' extends to a (necessarily unique) topological $\mathcal{C}^{\mathrm{an}}(\hat{Z}, K)$-module structure on V_b'.

Proof. This follows from [8, Prop. 6.4.7]. □

If the Z-action on V satisfies the equivalent conditions of the preceding proposition, then it is separately continuous (as follows from condition (i)), and so is in fact continuous.

If Z is a compact abelian locally L-analytic group (which is then necessarily topologically finitely generated [8, Prop. 6.4.1]), then we have the two nuclear Fréchet algebras $\mathcal{D}^{\mathrm{la}}(Z,K)$ and $\mathcal{C}^{\mathrm{an}}(\hat{Z},K)$, each containing the group ring $K[Z]$ as a dense subalgebra.

Proposition 2.10. *If Z is a compact abelian locally L-analytic group, then there is an isomorphism of topological K-algebras $\mathcal{D}^{\mathrm{la}}(Z,K) \xrightarrow{\sim} \mathcal{C}^{\mathrm{an}}(\hat{Z},K)$, uniquely determined by the condition that it reduces to the identity on $K[Z]$ (regarded as a subalgebra of the source and target in the natural manner).*

Proof. This is [8, Prop. 6.4.6]. It is proved using the p-adic Fourier theory of [22]. □

We now wish to tie together the two strands of the preceding discussion. We begin with the following strengthening of Theorem 2.1.

Theorem 2.11. *If H is a compact locally L-analytic group and Z is a topological finitely generated abelian locally L-analytic group, then the completed tensor product $\mathcal{C}^{\mathrm{an}}(\hat{Z},K)\,\hat{\otimes}_K\,\mathcal{D}^{\mathrm{la}}(H,K)$ (which by [17, p. 107] is a K-Fréchet algebra) is a nuclear Fréchet-Stein algebra.*

Proof. This follows from [8, Prop. 5.3.22], together with the remark following [8, Def. 5.3.21]. □

Suppose now that G is a locally L-analytic group, whose centre Z (an abelian locally L-analytic group) is topologically finitely generated.

Definition 2.12. We let $\mathrm{Rep}^{z}_{\mathrm{la.c}}(G)$ denote the full subcategory of $\mathrm{Rep}_{\mathrm{la.c}}(G)$ consisting of locally analytic representations V of G, the induced Z-action on which satisfies the equivalent conditions of Proposition 2.9.

It follows from Propositions 2.2 and 2.9 that if V is a compact type convex space equipped with an action of G by continuous K-linear automorphisms, then the following are equivalent:

(i) V is an object of $\mathrm{Rep}^{z}_{\mathrm{la.c}}(G)$.

(ii) For some (equivalently, every) compact open subgroup H of G, the G-action on V induces a (uniquely determined) $\mathcal{C}^{\mathrm{an}}(\hat{Z},K)\hat{\otimes}_K\mathcal{D}^{\mathrm{la}}(H,K)$-module structure on V for which the corresponding map

$$\mathcal{C}^{\mathrm{an}}(\hat{Z},K)\,\hat{\otimes}_K\,\mathcal{D}^{\mathrm{la}}(H,K)\times V\to V$$

is separately continuous.

(iii) For some (equivalently, every) compact open subgroup H of G, the contragredient G-action on V_b' induces a (uniquely determined) structure of topological $\mathcal{C}^{\mathrm{an}}(\hat{Z}, K) \hat{\otimes}_K \mathcal{D}^{\mathrm{la}}(H, K)$-module on V_b'.

We can now define some important subcategories of the category $\mathrm{Rep}^z_{\mathrm{la.c}}(G)$.

Definition 2.13. Let V be an object of $\mathrm{Rep}^z_{\mathrm{la.c}}(G)$.

(i) We say that V is an essentially admissible locally analytic representation of G if V_b' is a coadmissible $\mathcal{C}^{\mathrm{an}}(\hat{Z}, K) \hat{\otimes}_K \mathcal{D}^{\mathrm{la}}(H, K)$-module for some (equivalently, every) compact open subgroup H of G.

(ii) We say that V is an admissible locally analytic representation of G if V_b' is a coadmissible $\mathcal{D}^{\mathrm{la}}(H, K)$-module for some (equivalently, every) compact open subgroup H of G.

(iii) We say that V is a strongly admissible locally analytic representation of G if V_b' is a strongly coadmissible $\mathcal{D}^{\mathrm{la}}(H, K)$-module for some (equivalently, every) compact open subgroup H of G.

The equivalence of "some" and "every" in each of these definitions follows from the fact that if $H' \subset H$ is an inclusion of compact open subgroups of G then the algebra $\mathcal{D}^{\mathrm{la}}(H, K)$ is free of finite rank as a $\mathcal{D}^{\mathrm{la}}(H', K)$-module (since H' has finite index in H). Clearly, any strongly admissible locally analytic G-representation is admissible, and any admissible locally analytic G-representation is essentially admissible. The notion of strongly admissible (respectively admissible, respectively essentially admissible) locally analytic G-representation was first introduced in [20] (respectively [23], respectively [8]). (Let us remark that any object V of $\mathrm{Rep}_{\mathrm{la.c}}(G)$ for which V_b' satisfies condition (ii) of Definition 2.13 automatically lies in $\mathrm{Rep}^z_{\mathrm{la.c}}(G)$, by [8, Prop. 6.4.10], and so the definitions of admissible and strongly admissible locally analytic representations of G given above do coincide with those of [23] and [20].)

We let $\mathrm{Rep}_{\mathrm{es}}(G)$ denote the full subcategory of $\mathrm{Rep}^z_{\mathrm{la.c}}(G)$ consisting of essentially admissible locally analytic representations, let $\mathrm{Rep}_{\mathrm{ad}}(G)$ denote the full subcategory of $\mathrm{Rep}_{\mathrm{es}}(G)$ consisting of admissible locally analytic representations, and let $\mathrm{Rep}_{\mathrm{sa}}(G)$ denote the full subcategory of $\mathrm{Rep}_{\mathrm{ad}}(G)$ consisting of strongly admissible locally analytic representations. These various categories lie in the following sequence of full embeddings:

$$\mathrm{Rep}_{\mathrm{sa}}(G) \subset \mathrm{Rep}_{\mathrm{ad}}(G) \subset \mathrm{Rep}_{\mathrm{es}}(G) \subset \mathrm{Rep}^z_{\mathrm{la.c}}(G) \subset \mathrm{Rep}_{\mathrm{la.c}}(G).$$

Both of the categories $\mathrm{Rep}_{\mathrm{la.c}}(G)$ and $\mathrm{Rep}^z_{\mathrm{la.c}}(G)$ are closed under passing to countable direct sums (and more generally to Hausdorff countable locally convex inductive limits), closed subrepresentations, Hausdorff quotients, and completed tensor products [9, Lems. 3.1.2, 3.1.4].

Theorem 2.14. *Each of* $\mathrm{Rep}_{\mathrm{es}}(G)$ *and* $\mathrm{Rep}_{\mathrm{ad}}(G)$ *is an abelian category, closed under the passage to closed* G*-subrepresentations, and to Hausdorff quotient* G*-representations.*

Proof. This follows from Theorem 1.9. □

The subcategory $\mathrm{Rep}_{\mathrm{sa}}(G)$ of $\mathrm{Rep}_{\mathrm{ad}}(G)$ is closed under passing to finite direct sums and closed subrepresentations, but in general it is not closed under passing to Hausdorff quotients.

Remark 2.15. Let Z_0 denote the maximal compact subgroup of Z, and let H be a compact open subgroup of G. Replacing H by Z_0H if necessary, we may assume that H contains Z_0 (so that then $Z_0 = H \bigcap Z$). The K-algebra $\mathcal{C}^{\mathrm{an}}(\hat{Z}_0, K) \xrightarrow{\sim} \mathcal{D}^{\mathrm{la}}(Z_0, K)$ is a subalgebra of each of $\mathcal{C}^{\mathrm{an}}(\hat{Z}, K)$ and $\mathcal{D}^{\mathrm{la}}(H, K)$. If V is an object of $\mathrm{Rep}^{z}_{\mathrm{la.c}}(G)$, then the two actions of $\mathcal{C}^{\mathrm{an}}(\hat{Z}_0, K)$ on each of V and V'_b (obtained by regarding it as a subalgebra of $\mathcal{C}^{\mathrm{an}}(\hat{Z}, K)$ or $\mathcal{D}^{\mathrm{la}}(H, K)$ respectively) coincide (since both are obtained from the one action of Z_0 on V). Thus the $\mathcal{C}^{\mathrm{an}}(\hat{Z}, K) \hat{\otimes}_K \mathcal{D}^{\mathrm{la}}(H, K)$-action on each of V and V'_b factors through the quotient algebra $\mathcal{C}^{\mathrm{an}}(\hat{Z}, K) \hat{\otimes}_{\mathcal{C}^{\mathrm{an}}(\hat{Z}_0, K)} \mathcal{D}^{\mathrm{la}}(H, K)$. We take particular note of two consequences of this remark.

Example 2.16. If Z is compact (and so equals Z_0), and if V lies in $\mathrm{Rep}^{z}_{\mathrm{la.c}}(G)$, then the preceding remark shows that the action of $\mathcal{C}^{\mathrm{an}}(\hat{Z}, K) \hat{\otimes}_K \mathcal{D}^{\mathrm{la}}(H, K)$ on each of V and V'_b factors through $\mathcal{D}^{\mathrm{la}}(H, K)$. Thus any essentially admissible locally analytic G-representation is in fact admissible. Also, in this situation, the categories $\mathrm{Rep}_{\mathrm{la.c}}(G)$ and $\mathrm{Rep}^{z}_{\mathrm{la.c}}(G)$ are equal. Thus if the centre Z of G is compact, it can be neglected entirely throughout the preceding discussion.

Example 2.17. If G is abelian, then $G = Z$. The preceding remark shows that if V lies in $\mathrm{Rep}^{z}_{\mathrm{la.c}}(G)$, then the $\mathcal{C}^{\mathrm{an}}(\hat{Z}, K) \hat{\otimes}_K \mathcal{D}^{\mathrm{la}}(H, K)$-action on each of V and V'_b factors through $\mathcal{C}^{\mathrm{an}}(\hat{Z}, K)$. Example 1.13 then shows that passing to strong duals induces an antiequivalence of categories between the category $\mathrm{Rep}_{\mathrm{es}}(Z)$ and the category of coherent rigid analytic sheaves on $\hat{Z}$. Under this antiequivalence, the subcategory $\mathrm{Rep}_{\mathrm{ad}}(Z)$ of $\mathrm{Rep}_{\mathrm{es}}(Z)$ corresponds to the subcategory consisting of those coherent sheaves on $\hat{Z}$ whose pushforward to $\hat{Z}_0$ under the surjection $\hat{Z} \to \hat{Z}_0$ (induced by the inclusion $Z_0 \subset Z$) is again coherent. (The point is that on the level of global sections, this pushforward corresponds to regarding a $\mathcal{C}^{\mathrm{an}}(\hat{Z}, K)$-module as a $\mathcal{C}^{\mathrm{an}}(\hat{Z}_0, K)$-module, via the embedding $\mathcal{C}^{\mathrm{an}}(\hat{Z}_0, K) \to \mathcal{C}^{\mathrm{an}}(\hat{Z}, K)$.)

Example 2.18. If G is compact, then $\mathcal{C}^{\mathrm{la}}(G, K)$ is an object of $\mathrm{Rep}_{\mathrm{sa}}(G)$, and furthermore, any object of $\mathrm{Rep}_{\mathrm{sa}}(G)$ is a closed subrepresentation of $\mathcal{C}^{\mathrm{la}}(G, K)^n$, for some $n \geq 0$. (This follows directly from

Definitions 2.13 (iii) and 1.12, and the fact that passing to strong duals takes closed subrepresentations of $\mathcal{C}^{\mathrm{la}}(G,K)^n$ to Hausdorff quotient modules of $\mathcal{D}^{\mathrm{la}}(G,K)^n$.)

The following result connects the locally analytic representation theory discussed in this note with the more traditional theory of smooth representations of locally L-analytic groups.

Theorem 2.19. *If V is an admissible smooth representation of G on a K-vector space (in the usual sense), and if we equip V with its finest locally convex topology, then V becomes an element of* $\mathrm{Rep}_{\mathrm{ad}}(G)$. *Conversely, any object V of* $\mathrm{Rep}_{\mathrm{ad}}(G)$ *on which the G-action is smooth is an admissible smooth representation of G, equipped with its finest locally convex topology.*

Proof. See [8, Prop. 6.3.2] or [23, Thm. 6.5]. □

In the applications to the theory of automorphic forms, one typically assumes that G is the group of L-valued points of a connected reductive linear algebraic group $\mathbb{G}$ defined over L. (Any such group certainly has topologically finitely generated centre.) In this case, we can make the following definition.

Definition 2.20. If W is a finite dimensional algebraic representation of $\mathbb{G}$ defined over K, then we say that a representation of G on a K-vector space V is locally W-algebraic if, for each vector $v \in V$, there exists an open subgroup H of G, a natural number n, and an H-equivariant homomorphism $W^n \to V$ whose image contains the vector v.

When W is the trivial representation of V, we recover the notion of a smooth representation of G. The following result generalizes Theorem 2.19.

Theorem 2.21. *Suppose that $G = \mathbb{G}(L)$, for some connected reductive linear algebraic group over L. If V is an object of* $\mathrm{Rep}_{\mathrm{ad}}(G)$ *that is also locally W-algebraic, for some finite dimensional algebraic representation W of $\mathbb{G}$ over K, then V is isomorphic to a representation of the form $U \otimes_B W$, where B denotes the semi-simple K-algebra* $\mathrm{End}_{\mathbb{G}}(W)$, *and U is an admissible smooth representation of G defined over B, equipped with its finest locally convex topology. Conversely, any such tensor product is a locally W-algebraic representation in* $\mathrm{Rep}_{\mathrm{ad}}(G)$.

Proof. This is [8, Prop. 6.3.10]. □

Remark 2.22. Taking the tensor product of finite dimensional representations and smooth representations is something that is quite unthinkable in the classical theory of smooth representations of G (in which the field of coefficients typically is taken to be $\mathbb{C}$, or an ℓ-adic field, with

$\ell \neq p$). In the arithmetic theory of automorphic forms, the role of smooth representations of p-adic reductive groups is to carry information about representations of the absolute Galois group of L on ℓ-adic vector spaces. (This is a very vague description of the local Langlands conjecture.) The consideration of locally algebraic representations of the type considered in Theorem 2.21 opens up the possibility of finding representations of p-adic reductive groups that can carry information about the representations of the absolute Galois group of L on p-adic vector spaces; in this optic, the role of the finite dimensional factor is to remember the "p-adic Hodge numbers" of such a representation. (See the introductory discussion of [3] for a lengthier account of this possibility.)

3. Locally analytic vectors in continuous admissible representations

Let L, K and G be as in the preceding section. In this section we discuss an important method for constructing strongly admissible locally analytic representations of G, which involves applying the functor "pass to locally analytic vectors" to certain Banach space representations of G. We will begin by defining that functor, but first we must recall the notion of an *analytic open subgroup* of G.

Suppose that H is a compact open subgroup of G that admits the structure of a "chart" of G; that is, a locally analytic isomorphism with the space of L-valued points of a closed ball. We let $\mathbb{H}$ denote the corresponding rigid analytic space (isomorphic to a closed ball) that has H as its space of L-valued points. If furthermore the group structure on H extends to a rigid analytic group structure on $\mathbb{H}$, then, suppressing the choice of chart structure on H, we will refer to H as an analytic open subgroup of G. Since G is locally L-analytic, it has a basis of neighbourhoods consisting of analytic open subgroups. (See the introduction of [8, §3.5] for a more detailed discussion of the notion of analytic open subgroup.)

Suppose now that U is a Banach space over K, equipped with a continuous G-action. If H is an analytic open subgroup of H, then we let $U_{\mathbb{H}-\mathrm{an}}$ denote the subspace of U consisting of vectors u for which the orbit map $o_u : H \to U$ defined by $o_u(h) = hu$ is (the restriction to H of) a rigid analytic U-valued function on $\mathbb{H}$. Via the association of o_u to a vector $u \in U_{\mathbb{H}-\mathrm{an}}$, we may regard $U_{\mathbb{H}-\mathrm{an}}$ as a subspace of $\mathcal{C}^{\mathrm{an}}(\mathbb{H}, U)$, the Banach space of rigid analytic U-valued functions on $\mathbb{H}$.

Lemma 3.1. *For any analytic open subgroup H of G, the space $U_{\mathbb{H}-\mathrm{an}}$ is a closed subspace of $\mathcal{C}^{\mathrm{an}}(\mathbb{H}, U)$.*

Proof. A rigid analytic function ϕ in $\mathcal{C}^{\mathrm{an}}(\mathbb{H}, U)$ belongs to $U_{\mathbb{H}-\mathrm{an}}$ if and only if its restriction to H is in fact of the form o_u, for some $u \in U$

(which will then certainly lie in $U_{\mathbb{H}-\mathrm{an}}$). This is the case if and only if ϕ satisfies the equation $\phi(h) = h\phi(e)$ for all $h \in H$. (Here e denotes the identity element in H). These equations cut out a closed subspace of $\mathcal{C}^{\mathrm{an}}(\mathbb{H}, U)$, as claimed. □

We will always regard $U_{\mathbb{H}-\mathrm{an}}$ as being endowed with the Banach space topology it inherits by being considered as a closed subspace of $\mathcal{C}^{\mathrm{an}}(\mathbb{H}, U)$, as in the preceding lemma. The inclusion $U_{\mathbb{H}-\mathrm{an}} \to U$ is thus continuous, but typically is not a topological embedding.

Definition 3.2. We say that a vector u in U is locally analytic if the orbit map o_u lies in $\mathcal{C}^{\mathrm{la}}(G, U)$. (In fact, it suffices to require that o_u be locally analytic in a neighbourhood of the identity, since the G-action on U is by continuous automorphisms). We let U_{la} denote the subspace of U consisting of locally analytic vectors; the preceding parenthetical remark shows that $U_{\mathrm{la}} = \bigcup_H U_{\mathbb{H}-\mathrm{an}}$, where H runs over all analytic open subgroups of G. We topologize U_{la} by endowing it with the locally convex inductive limit topology arising from the isomorphism $U_{\mathrm{la}} \xrightarrow{\sim} \varinjlim_H U_{\mathbb{H}-\mathrm{an}}$ (the inductive limit being taken over the directed set of analytic open subgroups of G).

This definition exhibits U_{la} as the locally convex inductive limit of a sequence of Banach spaces (and thus U_{la} is a so-called LB-space). The inclusion $U_{\mathrm{la}} \to U$ is continuous, but typically is not a topological embedding.

The map $u \mapsto o_u$ defines a continuous injection

$$U_{\mathrm{la}} \to \mathcal{C}^{\mathrm{la}}(G, U). \tag{3.3}$$

Note that in [19] and [23], the topology on U_{la} is defined to be that induced by regarding it as a subspace of $\mathcal{C}^{\mathrm{la}}(G, U)$. In general, this is coarser than the inductive limit topology of Definition 3.2.

We next introduce some terminology related to lattices in convex spaces.

Definition 3.4. A separated, open lattice $\mathcal{L}$ in a convex K-space U is an open $\mathcal{O}_K$-submodule of U that is p-adically separated. We let $\mathcal{L}(U)$ denote the set of all separated open lattices in U.

Definition 3.5. If U is a convex space, then we say that two lattices $\mathcal{L}_1, \mathcal{L}_2 \in \mathcal{L}(U)$ are commensurable if $a\mathcal{L}_1 \subset \mathcal{L}_2 \subset a^{-1}\mathcal{L}_1$ for some $a \in K^\times$.

Clearly commensurability defines an equivalence relation on $\mathcal{L}(U)$.

Definition 3.6. If $\mathcal{L} \in \mathcal{L}(U)$ then we let $\{\mathcal{L}\}$ denote the commensurability class of $\mathcal{L}$ (i.e. the equivalence class of $\mathcal{L}$ under the relation of commensurability). We let $\overline{\mathcal{L}}(U)$ denote the set of commensurability classes of elements of $\mathcal{L}(U)$.

Example 3.7. If U is a Banach space over K, then $\mathcal{L}(U)$ is non-empty, and in fact the elements of $\mathcal{L}(U)$ form a neighbourhood basis of U. Furthermore, any two elements of $\mathcal{L}(U)$ are commensurable, and so $\overline{\mathcal{L}}(U)$ consists of a single element.

In general, if $\mathcal{L} \in \mathcal{L}(U)$, then $\mathcal{L}$ gives rise to a continuous norm $s_{\mathcal{L}}$ on U, its gauge, uniquely determined by the requirement that $\mathcal{L}$ is the unit ball of $s_{\mathcal{L}}$. We let $U_{\mathcal{L}}$ denote U equipped with the topology induced by $s_{\mathcal{L}}$, and let $\hat{U}_{\mathcal{L}}$ the Banach space obtained by completing $U_{\mathcal{L}}$ with respect to the norm $s_{\mathcal{L}}$. The identity map on the underlying vector space of U induces a continuous bijection $U \to U_{\mathcal{L}}$, and hence a continuous injection $U \to \hat{U}_{\mathcal{L}}$. Given a pair of elements $\mathcal{L}_1, \mathcal{L}_2 \in \mathcal{L}(U)$, the topologies on $U_{\mathcal{L}_1}$ and $U_{\mathcal{L}_2}$ coincide if and only if $\mathcal{L}_1$ and $\mathcal{L}_2$ are commensurable.

Suppose now that U is equipped with a continuous G-action. There is then an induced action of G on $\mathcal{L}(U)$, defined by $(g, \mathcal{L}) \mapsto g\mathcal{L}$ for $g \in G$ and $\mathcal{L} \in \mathcal{L}(U)$. This action evidently respects the relation of commensurability, and so descends to an action on $\overline{\mathcal{L}}(U)$. We write $\mathcal{L}(U)^G$ (respectively $\overline{\mathcal{L}}(U)^G$) to denote the subset of $\mathcal{L}(U)$ (respectively of $\overline{\mathcal{L}}(U)$) consisting of elements that are fixed under the action of G. Passing to commensurability classes induces a map $\mathcal{L}(U)^G \to \overline{\mathcal{L}}(U)^G$.

Lemma 3.8. *If $\mathcal{L}$ is an element of $\mathcal{L}(U)$, then the G-action on U induces a continuous G-action on $U_{\mathcal{L}}$ (and hence on $\hat{U}_{\mathcal{L}}$) if and only if the commensurability class $\{\mathcal{L}\}$ is G-invariant.*

Proof. It is immediate from the definitions that G acts on $U_{\mathcal{L}}$ via continuous automorphisms if and only if $\{\mathcal{L}\}$ is G-invariant. Since the G-action on U is continuous by assumption, and since the natural bijection $U \to U_{\mathcal{L}}$ is continuous, the G-action on $U_{\mathcal{L}}$ automatically satisfies conditions (i) and (iii) of [8, Lem. 3.1.1]. It thus follows from that lemma that if G acts on $U_{\mathcal{L}}$ via continuous automorphisms, then the G-action on $U_{\mathcal{L}}$ is in fact continuous. □

Lemma 3.9. *Let H be an open subgroup of G.*

(i) If H is compact, then the map $\mathcal{L}(U)^H \to \overline{\mathcal{L}}(U)^H$ is surjective.

(ii) If $\mathcal{L} \in \mathcal{L}(U)$ is such that $\{\mathcal{L}\} \in \overline{\mathcal{L}}(U)^H$, then there is an open subgroup H' of H such that $\mathcal{L} \in \mathcal{L}(U)^{H'}$.

Proof. Suppose that $\mathcal{L} \in \overline{\mathcal{L}}(U)$ is H-invariant. The H-action on U then induces a continuous H-action on $U_{\mathcal{L}}$, by Lemma 3.8. Part (i) of the present lemma is now seen to follow from [8, Lem. 6.5.3], while part (ii) follows immediately from the fact that the H-action on $U_{\mathcal{L}}$ is continuous. □

In contrast to part (i) of the preceding lemma, if G is not compact then the map $\mathcal{L}(U)^G \to \overline{\mathcal{L}}(U)^G$ is typically not surjective. For example, if U is a Banach space, then $\overline{\mathcal{L}}(U)^G = \overline{\mathcal{L}}(U)$ (since the set on the right is a singleton). On the other hand, asking that $\mathcal{L}(U)^G$ be non-empty is a rather stringent condition.

Definition 3.10. A continuous representation of G on a Banach space is said to be unitary if $\mathcal{L}(U)^G \neq \emptyset$, that is, if U contains an open, separated lattice that is invariant under the entire group G (or equivalently, if its topology can be defined by a G-invariant norm).

Suppose now that $\mathcal{L} \in \mathcal{L}(U)^H$ for some open subgroup H of G. If π denotes a uniformizer of $\mathcal{O}_K$, then $\mathcal{L}/\pi\mathcal{L}$ is a vector space over the residue field $\mathcal{O}_K/\pi\mathcal{O}_K$, equipped with a smooth representation of H.

Definition 3.11. If U is a convex space, equipped with a continuous G-action of G, then we say that $\mathcal{L} \in \mathcal{L}(U)$ is admissible if it is H-invariant, for some compact open subgroup H of G, and if the resulting smooth H-representation on $\mathcal{L}/\pi\mathcal{L}$ is admissible.

Note that if $\mathcal{L} \in \mathcal{L}(U)$ is admissible, and if $H \subset G$ is a compact open subgroup that satisfies the conditions of the preceding definition with respect to $\mathcal{L}$, then any open subgroup $H' \subset H$ also satisfies these conditions.

Lemma 3.12. *If $\mathcal{L} \in \mathcal{L}(U)$ is admissible, then every lattice in $\{\mathcal{L}\}$ is admissible.*

Proof. Let H be a compact open subgroup of G that satisfies the conditions of Definition 3.11 with respect to $\mathcal{L}$. If $\mathcal{L}'$ is an element of $\{\mathcal{L}\}$, then by Lemma 3.9 (ii) (and replacing H by an open subgroup if necessary) we may assume that $\mathcal{L}'$ is again H-invariant. Since $\mathcal{L}'$ and $\mathcal{L}$ are commensurable, we may also assume (replacing $\mathcal{L}'$ by a scalar multiple if necessary) that $\pi^n\mathcal{L} \subset \mathcal{L}' \subset \mathcal{L}$ for some $n > 0$. Thus $\mathcal{L}'/\pi\mathcal{L}'$ is an H-invariant subquotient of $\mathcal{L}/\pi^{n+1}\mathcal{L}$. The latter H-representation is a successive extension of copies of $\mathcal{L}/\pi\mathcal{L}$, and so by assumption is an admissible smooth representation of H over $\mathcal{O}_K/\pi^{n+1}\mathcal{O}_K$. Any subquotient of an admissible smooth H-representation over $\mathcal{O}_K/\pi^{n+1}\mathcal{O}_K$ is again admissible. (This uses the fact that the category of such representations is anti-equivalent – via passing to $\mathcal{O}_K/\pi^{n+1}\mathcal{O}_K$-duals – to the category of finitely generated modules over the completed group ring $(\mathcal{O}_K/\pi^{n+1}\mathcal{O}_K)[[H]]$, together with a theorem of Lazard to the effect that this completed group ring is Noetherian [16, V.2.2.4].[2]) In particular we conclude that $\mathcal{L}'/\pi\mathcal{L}'$ is admissible. □

We say that a commensurability class $\{\mathcal{L}\} \in \overline{\mathcal{L}}(U)$ is admissible if one (or equivalently every, by Lemma 3.12) member of the class is admissible in the sense of Definition 3.11.

Proposition 3.13. *If U is an object of $\mathrm{Rep}_{\mathrm{es}}(G)$, then $\mathcal{L}(U)$ contains an admissible lattice if and only if U is strongly admissible. Furthermore, if U is strongly admissible, then for any compact open subgroup H of G, we may find an admissible H-invariant lattice in $\mathcal{L}(U)$.*

Proof. See [8, Prop. 6.5.9]. □

Definition 3.14. Let U be a Banach space over K, equipped with a continuous action of G. We say that U is an admissible continuous representation of G, or an admissible Banach space representation of G, if one (or equivalently every, by Lemma 3.12) lattice in $\mathcal{L}(U)$ is admissible, in the sense of the Definition 3.11.

Theorem 3.15. *The category of admissible continuous representations of G (with morphisms being continuous G-equivariant K-linear maps) is an abelian category, closed under passing to closed G-subrepresentations and Hausdorff quotient G-representations.*

Proof. This is the main result of [21]. (See [8, Cor. 6.2.16] for the case when K is not local.) The key point is that if H is any compact open subgroup of G, then the completed group ring $\mathcal{O}_K[[H]]$ is Noetherian [16, V.2.2.4].[3] □

We let $\mathrm{Rep}_{\mathrm{b.ad}}(G)$ denote the abelian category of admissible continuous representations of G. One important aspect of the preceding result is that maps in $\mathrm{Rep}_{\mathrm{b.ad}}(G)$ are necessarily strict, with closed image.

Example 3.16. If G is compact, then the space $\mathcal{C}(G, K)$ of continuous K-valued functions on G, made into a Banach space via the sup norm, and equipped with the left regular G-action, is an admissible continuous G-representation. Furthermore any object of $\mathrm{Rep}_{\mathrm{b.ad}}(G)$ is a closed subrepresentation of $\mathcal{C}(G, K)^n$ for some $n \geq 0$. (See [8, Prop.-Def. 6.2.3].)

If G is (the group of $\mathbb{Q}_p$-points of) a p-adic reductive group over $\mathbb{Q}_p$, then the admissible G-representations that are also unitary are perhaps the most important objects in the category $\mathrm{Rep}_{\mathrm{b.ad}}(G)$. In [3, §1.3],

[2] Strictly speaking, this reference only applies to the case when $K = \mathbb{Q}_p$, so that $\mathcal{O}_K = \mathbb{Z}_p$. However, the result is easily extended to the case of general K; see for example the proof of [8, Thm. 6.2.8].

[3] See the preceding note.

Breuil explains the role that he expects these representations to play in a hoped-for "p-adic local Langlands" correspondence, in the case of the group $\mathrm{GL}_2(\mathbb{Q}_p)$. For a discussion of how some of Breuil's ideas might generalize to the case of a general reductive group, see [24, §5].

The following result provides a basic technique for producing strongly admissible locally analytic representations of G.

Proposition 3.17. *If U is an object of $\mathrm{Rep}_{\mathrm{b.ad}}(G)$, then U_{la} is a strongly admissible locally analytic representation of G.*

Proof. This follows from the discussions of Examples 2.18 and 3.16, and the following two (easily verified) facts: (i) for any compact open subgroup H of G, there is a natural isomorphism $\mathcal{C}^{\mathrm{la}}(H,K) \xrightarrow{\sim} \mathcal{C}(H,K)_{\mathrm{la}}$ [8, Prop. 3.5.11]; (ii) if U and V are Banach spaces equipped with continuous G-representations, if $U \to V$ is a G-equivariant closed embedding, and if V_{la} is of compact type, then the diagram

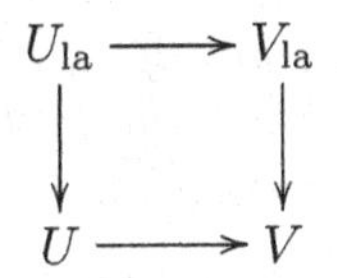

is Cartesian in the category of convex spaces; in particular, the map $U_{\mathrm{la}} \to V_{\mathrm{la}}$ is again a closed embedding [8, Prop. 3.5.10]. See [8, Prop. 6.2.4] for the details of the argument. □

A version of the preceding theorem, working with the topology obtained on U_{la} by regarding it as a closed subspace of $\mathcal{C}^{\mathrm{la}}(G,U)$, is given in [23, Thm. 7.1 (ii)]. We remark that if U is an object of $\mathrm{Rep}_{\mathrm{b.ad}}(G)$, then the map (3.3) is in fact a topological embedding (see [5, Rem. A.1.1]). Thus, for such U, the topology on U_{la} induced by regarding it as a subspace of $\mathcal{C}^{\mathrm{la}}(G,U)$ coincides with the inductive limit topology given by Definition 3.2.

Lemma 3.18. *If U is a convex space equipped with a continuous action of G, and if H is an open subgroup of G, then there exists a continuous H-equivariant injection $U \to W$ for some admissible continuous H-representation W if and only if $\overline{\mathcal{L}}(U)$ contains an H-invariant admissible commensurability class.*

Proof. Given such a map $U \to W$, the preimage of any lattice in W determines a commensurability class in $\overline{\mathcal{L}}(U)$ with the required properties. Conversely, given such a commensurability class $\{\mathcal{L}\}$, it follows from Lemma 3.8 that the H-action on U extends to a continuous H-action on $\hat{U}_{\mathcal{L}}$, and so we may take $W = \hat{U}_{\mathcal{L}}$. □

Definition 3.19. An object V of $\mathrm{Rep}_{\mathrm{ad}}(G)$ is called very strongly admissible if V admits a G-equivariant continuous K-linear injection into an object of $\mathrm{Rep}_{\mathrm{b.ad}}(G)$, or equivalently (by Lemma 3.18), if $\overline{\mathcal{L}}(V)$ contains a G-invariant admissible commensurability class.

We let $\mathrm{Rep}_{\mathrm{vsa}}(G)$ denote the full subcategory of $\mathrm{Rep}_{\mathrm{ad}}(G)$ consisting of very strongly admissible locally analytic G-representations. It is evidently closed under passing to subobjects and finite direct sums. Proposition 3.13 shows that it is a full subcategory of $\mathrm{Rep}_{\mathrm{sa}}(G)$.

It also follows from Proposition 3.13 that if G is compact, then every strongly admissible locally analytic G-representation is in fact very strongly admissible. The author knows no example of a strongly admissible, but not very strongly admissible, locally analytic G-representation (for any G).

The following theorem of Schneider and Teitelbaum is fundamental to the theory of admissible continuous representations.

Theorem 3.20. *If $L = \mathbb{Q}_p$ and if K is a finite extension of L then the map $U \mapsto U_{\mathrm{la}}$ yields an exact and faithful functor from the category $\mathrm{Rep}_{\mathrm{b.ad}}(G)$ to the category $\mathrm{Rep}_{\mathrm{vsa}}(G)$.*

Proof. See [23, Thm. 7.1]. (That the image of this functor lies in $\mathrm{Rep}_{\mathrm{vsa}}(G)$ follows from Proposition 3.17 and the definition of $\mathrm{Rep}_{\mathrm{vsa}}(G)$.) □

Given the exactness statement in the preceding result, the faithfulness statement is equivalent to the fact that U_{la} is dense as a subspace of U.

In the context of Theorem 3.20, the functor $U \mapsto U_{\mathrm{la}}$ is not full, in general, as we now explain. If U is an object of $\mathrm{Rep}_{\mathrm{b.ad}}(G)$, if $\mathcal{L}$ is an element of $\mathcal{L}(U)$, and if we write $\mathcal{L}_{\mathrm{la}} = \mathcal{L} \bigcap U_{\mathrm{la}}$, then $\{\mathcal{L}_{\mathrm{la}}\}$ is a G-invariant and admissible commensurability class in U_{la}, which is evidently well-defined independent of the choice of $\mathcal{L}$ (since all lattices in $\mathcal{L}(U)$ are commensurable).

Conversely, if V is an object of $\mathrm{Rep}_{\mathrm{vsa}}(G)$, equipped with a G-invariant and admissible commensurability class $\{\mathcal{M}\} \in \overline{\mathcal{L}}(V)$, then the completion $\hat{V}_{\mathcal{M}}$ is an object of $\mathrm{Rep}_{\mathrm{b.ad}}(G)$. In the case when $(V, \{\mathcal{M}\}) = (U_{\mathrm{la}}, \{\mathcal{L}_{\mathrm{la}}\})$ (in the notation of the previous paragraph), it follows from Theorem 3.20 (and the remark following that theorem) that $\hat{V}_{\mathcal{M}} \xrightarrow{\sim} U$.

Thus, if we let $\mathcal{C}$ denote the category whose objects consist of pairs $(V, \{\mathcal{M}\})$, where V is an object of $\mathrm{Rep}_{\mathrm{vsa}}(G)$ and $\{\mathcal{M}\} \in \overline{\mathcal{L}}(V)$ is a G-invariant admissible commensurability class (and whose morphisms are defined in the obvious way), then the preceding discussion shows that $U \mapsto (U_{\mathrm{la}}, \{\mathcal{L}_{\mathrm{la}}\})$ is a fully faithful functor $\mathrm{Rep}_{\mathrm{b.ad}}(G) \to \mathcal{C}$, to which the functor $(V, \{\mathcal{M}\}) \mapsto \hat{V}_{\mathcal{M}}$ is left adjoint, and left quasi-inverse.

On the other hand, the obvious forgetful functor $\mathcal{C} \to \mathrm{Rep}_{\mathrm{vsa}}(G)$ (forget the commensurability class of lattices), while faithful, is not full. This amounts to the fact that a given very strongly admissible locally analytic representation of G can admit more than one G-invariant commensurability class of admissible lattices. Explicit examples are provided by the results of [3] (which show that the same irreducible admissible locally algebraic representation of $\mathrm{GL}_2(\mathbb{Q}_p)$ can admit non-isomorphic admissible continuous completions, which are even unitary, in the sense of Definition 3.10).

4. Parabolic induction

This section provides a brief account of parabolic induction in the locally analytic context. We let L and K be as in the preceding sections, and we suppose that G is (the group of L-valued points of) a connected reductive linear algebraic group over L. We let P be a parabolic subgroup of G, and let M be the Levi quotient of P.

If V is an object of $\mathrm{Rep}_{\mathrm{la.c}}(M)$ (regarded as a P-representation through the projection of P onto M), then we make the following definition:

$$\mathrm{Ind}_P^G V = \{f \in \mathcal{C}^{\mathrm{la}}(G, V) \mid f(pg) = pf(g) \text{ for all } p \in P, g \in G\},$$

equipped with its right regular G-action. (We topologize $\mathrm{Ind}_P^G V$ by regarding it as a closed subspace of $\mathcal{C}^{\mathrm{la}}(G, V)$.)

Proposition 4.1. *If V lies in $\mathrm{Rep}_{\mathrm{la.c}}(M)$ (respectively $\mathrm{Rep}^z_{\mathrm{la.c}}(M)$, $\mathrm{Rep}_{\mathrm{ad}}(M)$, $\mathrm{Rep}_{\mathrm{sa}}(M)$, $\mathrm{Rep}_{\mathrm{vsa}}(M)$), then $\mathrm{Ind}_P^G V$ lies in $\mathrm{Rep}_{\mathrm{la.c}}(G)$ (respectively $\mathrm{Rep}^z_{\mathrm{la.c}}(G)$, $\mathrm{Rep}_{\mathrm{ad}}(G)$, $\mathrm{Rep}_{\mathrm{sa}}(G)$, $\mathrm{Rep}_{\mathrm{vsa}}(G)$).*

Proof. Although the proof of each of these statements is straightforward, altogether they are a little lengthy, and we omit them. □

Locally analytic parabolic induction satisfies Frobenius reciprocity.

Proposition 4.2. *If U and V are objects of $\mathrm{Rep}_{\mathrm{la.c}}(G)$ and $\mathrm{Rep}_{\mathrm{la.c}}(M)$ respectively, then the P-equivariant map $\mathrm{Ind}_P^G V \to V$ induced by evaluation at the identity of G yields a natural isomorphism $\mathcal{L}_G(U, \mathrm{Ind}_P^G V) \xrightarrow{\sim} \mathcal{L}_P(U, V)$. (Here $\mathcal{L}_G(-,-)$ and $\mathcal{L}_P(-,-)$ denote respectively the space of continuous G-equivariant K-linear maps and the space of continuous P-equivariant K-linear maps between the indicated source and target.)*

Proof. This is a particular case of [14, Thm. 4.2.6], and also follows from [8, Prop. 5.1.1 (iii)]. □

Just as in other representation theoretic contexts, parabolic induction provides a way to obtain interesting new representations from old. The following result is due to H. Frommer [15]. (The case when $G = \mathrm{GL}_2(\mathbb{Q}_p)$ was first treated in [20].)

Theorem 4.3. *Suppose that $L = \mathbb{Q}_p$ and that G is split, and let G_0 be a hyperspecial maximal compact subgroup of G. If U is a finite dimensional irreducible object of $\mathrm{Rep}_{\mathrm{la.c}}(M)$ for which $\mathrm{U}(\mathfrak{g}) \otimes_{\mathrm{U}(\mathfrak{p})} U'$ is irreducible as $\mathrm{U}(\mathfrak{g})$-module, then $\mathrm{Ind}_P^G U$ is topologically irreducible as a G_0-representation, and so in particular as a G-representation. (Here U' denotes the contragredient to U, and $\mathrm{U}(\mathfrak{p})$ is the universal enveloping algebra of the Lie algebra $\mathfrak{p}$ of P.)*

One surprising aspect of this result is that it shows (in contrast to the cases of smooth representations of compact p-adic groups, and continuous representations of compact real Lie groups) that the compact group G_0 can admit topologically irreducible infinite dimensional locally analytic representations.

5. Jacquet modules

Let L, K and G be as in the previous section, let P be a parabolic subgroup of G, and choose an opposite parabolic $\overline{P}$ to P. The intersection $M := P \bigcap \overline{P}$ is then a Levi subgroup of each of P and $\overline{P}$. Let N denote the unipotent radical of P.

If U is an object of $\mathrm{Rep}_{\mathrm{la.c}}(M)$, then let $\mathcal{C}_c^{\mathrm{sm}}(N, U)$ denote the closed subspace of $\mathcal{C}_c^{\mathrm{la}}(N, U)$ consisting of compactly supported, locally constant (= smooth) U-valued functions on N. The projection map $G \to \overline{P}\backslash G$ restricts to an open immersion of locally analytic spaces $N \to \overline{P}\backslash G$, and this immersion allows us to identify $\mathcal{C}_c^{\mathrm{la}}(N, U)$ with the subspace of $\mathrm{Ind}_{\overline{P}}^G U$ consisting of functions whose support is contained in $\overline{P}N$. In this way $\mathcal{C}_c^{\mathrm{la}}(N, U)$ becomes a closed $(\mathrm{U}(\mathfrak{g}), P)$-submodule of $\mathrm{Ind}_{\overline{P}}^G U$, and $\mathcal{C}_c^{\mathrm{sm}}(N, U)$ is identified with the closed P-submodule of $\mathcal{C}_c^{\mathrm{la}}(N, U)$ consisting of elements annihilated by $\mathfrak{n}$ (the Lie algebra of N).

Proposition 5.1. *If U is an object of $\mathrm{Rep}_{\mathrm{la.c}}(M)$ then $\mathcal{C}_c^{\mathrm{sm}}(N, U)$ is an object of $\mathrm{Rep}_{\mathrm{la.c}}(P)$.*

Proof. This follows from the identification of $\mathcal{C}_c^{\mathrm{sm}}(N, U)$ with a closed P-invariant subspace of $\mathrm{Ind}_{\overline{P}}^G U$, which Proposition 4.1 shows to be an object of $\mathrm{Rep}_{\mathrm{la.c}}(G)$. □

The formation of $\mathcal{C}_c^{\mathrm{sm}}(N, U)$ is clearly functorial in U, and so we obtain a functor $\mathcal{C}_c^{\mathrm{sm}}(N, -)$ from $\mathrm{Rep}_{\mathrm{la.c}}(M)$ to $\mathrm{Rep}_{\mathrm{la.c}}(P)$.

Proposition 5.2. *The restriction of $\mathcal{C}_c^{\mathrm{sm}}(N, -)$ to $\mathrm{Rep}_{\mathrm{la.c}}^z(M)$ (which is thus a functor from $\mathrm{Rep}_{\mathrm{la.c}}^z(M)$ to $\mathrm{Rep}_{\mathrm{la.c}}(P)$) admits a right adjoint.*

Proof. See [9, Thm. 3.5.6]. □

As usual, let δ denote the smooth character of M that describes how right multiplication by elements of M affects left-invariant Haar measure on P. Concretely, if $m \in M$, then $\delta(m)$ is equal to $[N_0 : mN_0m^{-1}]^{-1}$, for any compact open subgroup N_0 of N. If U is an object of $\mathrm{Rep}^z_{\mathrm{la.c}}(M)$, then let $U(\delta)$ denote the twist of U by δ.

Definition 5.3. We let J_P denote the functor from $\mathrm{Rep}_{\mathrm{la.c}}(P)$ to $\mathrm{Rep}^z_{\mathrm{la.c}}(M)$ obtained by twisting by δ the right adjoint to the functor $\mathcal{C}^{\mathrm{sm}}_c(N, -)$. If V is an object of $\mathrm{Rep}_{\mathrm{la.c}}(P)$, we refer to $J_P(V)$ as the Jacquet module of V.

Thus for any objects U of $\mathrm{Rep}^z_{\mathrm{la.c}}(M)$ and V of $\mathrm{Rep}_{\mathrm{la.c}}(P)$ there is a natural isomorphism

$$\mathcal{L}_P(\mathcal{C}^{\mathrm{sm}}_c(N, U), V) \xrightarrow{\sim} \mathcal{L}_M(U(\delta), J_P(V)). \tag{5.4}$$

Remark 5.5. If U is an object of $\mathrm{Rep}^z_{\mathrm{la.c}}(M)$, then the natural map $U(\delta) \to J_P(\mathcal{C}^{\mathrm{sm}}_c(N, U))$ in $\mathrm{Rep}^z_{\mathrm{la.c}}(M)$, corresponding via the adjointness isomorphism (5.4) to the identity automorphism of $\mathcal{C}^{\mathrm{sm}}_c(N, U)$, is an isomorphism [9, Lem. 3.5.2]. Thus the isomorphism (5.4) is induced by passing to Jacquet modules (i.e. applying the functor J_P).

Remark 5.6. Regarding a G-representation as a P-representation yields a forgetful functor from $\mathrm{Rep}_{\mathrm{la.c}}(G)$ to $\mathrm{Rep}_{\mathrm{la.c}}(P)$. Composing this functor with the functor J_P yields a functor from $\mathrm{Rep}_{\mathrm{la.c}}(G)$ to $\mathrm{Rep}^z_{\mathrm{la.c}}(M)$, which we again denote by J_P.

Theorem 5.7. *The functor J_P restricts to give a functor* $\mathrm{Rep}_{\mathrm{es}}(G) \to \mathrm{Rep}_{\mathrm{es}}(M)$.

Proof. See [9, Thm. 0.5]. □

This theorem provides the primary motivation for introducing the notion of essentially admissible locally analytic representations. Indeed, even if V is an object of $\mathrm{Rep}_{\mathrm{ad}}(G)$, it need not be the case that $J_P(V)$ lies in $\mathrm{Rep}_{\mathrm{ad}}(M)$; however, see Corollary 5.24 below.

Example 5.8. If G is quasi-split (that is, has a Borel subgroup defined over L), and if we take P to be a Borel subgroup of G, then M is a torus, and so $\mathrm{Rep}_{\mathrm{es}}(M)$ is antiequivalent to the category of coherent sheaves on the rigid analytic space of characters $\hat{M}$. Thus if V is an object of $\mathrm{Rep}_{\mathrm{es}}(G)$, then we may regard $J_P(V)$ as giving rise to a coherent sheaf on $\hat{M}$. This fact underlies the approach followed in [10] to the construction of the eigencurve of [7], and of more general eigenvarieties.

Example 5.9. If V is an admissible smooth representation of G, then there is a natural isomorphism between $J_P(V)$ and V_N, the space of N-coinvariants of V [9, Prop. 4.3.4]. This space of coinvariants is what is traditionally referred to as the Jacquet module of V in the theory of smooth representations.

More generally, if $V = U \otimes_B W$ is an admissible locally W-algebraic representation of G, as in Theorem 2.21, then there is a natural isomorphism $J_P(V) \xrightarrow{\sim} U_N \otimes_B W^N$ (where W^N denotes the space of N-invariants in W) [9, Prop. 4.3.6]. Since U is an admissible smooth G-representation, the space U_N is an admissible smooth M-representation [6, Thm. 3.3.1]. Thus J_P takes admissible locally W-algebraic G-representations to admissible locally W^N-algebraic M-representations.

The remainder of this section is devoted to explaining the relation between the functor J_P on $\mathrm{Rep}_{\mathrm{la.c}}(G)$ and the process of locally analytic parabolic induction. We begin with the following remark.

Remark 5.10. If V is an object of $\mathrm{Rep}_{\mathrm{la.c}}(G)$, then the universal property of tensor products yields a natural isomorphism

$$\mathcal{L}_P(\mathcal{C}_c^{\mathrm{sm}}(N, U), V) \xrightarrow{\sim} \mathcal{L}_{(\mathfrak{g},P)}(\mathrm{U}(\mathfrak{g}) \otimes_{\mathrm{U}(\mathfrak{p})} \mathcal{C}_c^{\mathrm{sm}}(N, U), V). \tag{5.11}$$

Thus for such V, the adjointness isomorphism (5.4) induces an isomorphism

$$\mathcal{L}_{(\mathfrak{g},P)}(\mathrm{U}(\mathfrak{g}) \otimes_{\mathrm{U}(\mathfrak{p})} \mathcal{C}_c^{\mathrm{sm}}(N, U), V) \xrightarrow{\sim} \mathcal{L}_M(U(\delta), J_P(V).) \tag{5.12}$$

Definition 5.13. As above, we regard $\mathcal{C}_c^{\mathrm{sm}}(N, U)$ as a closed subspace of $\mathrm{Ind}_{\overline{P}}^G(U)$. We let $I_{\overline{P}}^G(U)$ (respectively $I_{\overline{\mathfrak{p}}}^{\mathfrak{g}}(U)$) denote the closed G-subrepresentation (respectively the $\mathrm{U}(\mathfrak{g})$-submodule) of $\mathrm{Ind}_{\overline{P}}^G U$ that it generates.

Note that $I_{\overline{\mathfrak{p}}}^{\mathfrak{g}}(U)$ admits the following alternative description: taking V to be $\mathrm{Ind}_{\overline{P}}^G(U)$, the isomorphism (5.11), applied to the inclusion $\mathcal{C}_c^{\mathrm{sm}}(N, U) \subset \mathrm{Ind}_{\overline{P}}^G U$, induces a $(\mathfrak{g}, P)$-equivariant map

$$\mathrm{U}(\mathfrak{g}) \otimes_{\mathrm{U}(\mathfrak{p})} \mathcal{C}_c^{\mathrm{sm}}(N, U) \to \mathrm{Ind}_{\overline{P}}^G(U),$$

whose image coincides with $I_{\overline{\mathfrak{p}}}^{\mathfrak{g}}(U)$. In particular, there is a $(\mathfrak{g}, P)$-equivariant surjection

$$\mathrm{U}(\mathfrak{g}) \otimes_{\mathrm{U}(\mathfrak{p})} \mathcal{C}_c^{\mathrm{sm}}(N, U) \to I_{\overline{\mathfrak{p}}}^{\mathfrak{g}}(U). \tag{5.14}$$

Remark 5.15. The isomorphism of Remark 5.5 yields a closed embedding $U(\delta) \to J_P(I_{\overline{P}}^G(U))$, and hence for each object V of $\mathrm{Rep}_{\mathrm{la.c}}(G)$, passage to Jacquet modules induces a morphism

$$\mathcal{L}_G(I_{\overline{P}}^G(U), V) \to \mathcal{L}_M(U(\delta), J_P(V)), \tag{5.16}$$

which is injective, by the construction of $I_{\overline{P}}^{G}(U)$. Restricting elements in the source of this map to $I_{\overline{\mathfrak{p}}}^{\mathfrak{g}}(U)$ yields the left hand vertical arrow in the following commutative diagram

$$\begin{array}{ccc} \mathcal{L}_G(I_{\overline{P}}^{G}(U), V) & \xrightarrow{(5.16)} & \mathcal{L}_M(U(\delta), J_P(V)) \\ \downarrow & & \uparrow \wr\, (5.12) \\ \mathcal{L}_{(\mathfrak{g},P)}(I_{\overline{\mathfrak{p}}}^{\mathfrak{g}}(U), V) & \longrightarrow & \mathcal{L}_{(\mathfrak{g},P)}(\mathrm{U}(\mathfrak{g}) \otimes_{\mathrm{U}(\mathfrak{p})} \mathcal{C}_c^{\mathrm{sm}}(N, U), V) \end{array}$$

whose bottom horizontal arrow is induced by composition with (5.14).

Definition 5.17. Let U and V be objects of $\mathrm{Rep}_{\mathrm{la.c}}^{z}(M)$ and $\mathrm{Rep}_{\mathrm{la.c}}(G)$ respectively, and suppose given an element $\psi \in \mathcal{L}_M(U(\delta), J_P(V))$, corresponding via the adjointness map (5.12) to an element $\phi \in \mathcal{L}_{(\mathfrak{g},P)}(\mathrm{U}(\mathfrak{g}) \otimes_{\mathrm{U}(\mathfrak{p})} \mathcal{C}_c^{\mathrm{sm}}(N, U), V)$. We say that ψ is balanced if ϕ factors through the surjection (5.14), and we let $\mathcal{L}_M(U(\delta), J_P(V))^{\mathrm{bal}}$ denote the subspace of $\mathcal{L}_M(U(\delta), J_P(V))$ consisting of balanced maps. (The property of a morphism being balanced depends not just on $J_P(V)$ as an M-representation, but on its particular realization as the Jacquet module of the G-representation V.)

Equivalently, $\mathcal{L}_M(U(\delta), J_P(V))^{\mathrm{bal}}$ is the image of the injection

$$\mathcal{L}_{(\mathfrak{g},P)}(I_{\overline{\mathfrak{p}}}^{\mathfrak{g}}(U), V) \to \mathcal{L}_M(U(\delta), J_P(V))$$

given by composing the right hand vertical arrow and bottom horizontal arrow in the commutative diagram of Remark 5.15. A consideration of this diagram thus shows that the image of (5.16) lies in $\mathcal{L}_M(U(\delta), J_P(V))^{\mathrm{bal}}$.

Definition 5.18. Let U be an object of $\mathrm{Rep}_{\mathrm{la.c}}^{z}(M)$, and let $\mathcal{H}$ denote the space of linear M-equivariant endomorphisms of U. We say that U is allowable if for any pair of finite dimensional algebraic $\mathbb{M}$-representations W_1 and W_2, each element of $\mathcal{L}_{\mathcal{H}[M]}(U \otimes_K W_1, U \otimes_K W_2)$ is strict (i.e. has closed image). (Here each $U \otimes_K W_i$ is regarded as an $\mathcal{H}[M]$-module via the $\mathcal{H}$ action on the left hand factor along with the diagonal M-action.)

It is easily checked that if U is an object of $\mathrm{Rep}_{\mathrm{es}}(M)$ and W is a finite dimensional algebraic M-representation, then $M \otimes_K W$ is again an object of $\mathrm{Rep}_{\mathrm{es}}(M)$. Thus objects of $\mathrm{Rep}_{\mathrm{es}}(M)$ are allowable in the sense of Definition 5.18.

Theorem 5.19. *If U is an allowable object of $\mathrm{Rep}_{\mathrm{la.c}}^{z}(M)$ (in the sense of Definition 5.18) and if V is an object of $\mathrm{Rep}_{\mathrm{vsa}}(G)$ (see Definition 3.19) then the morphism*

$$\mathcal{L}_G(I_{\overline{P}}^{G}(U), V) \to \mathcal{L}_M(U(\delta), J_P(V))^{\mathrm{bal}}$$

induced by (5.16) is an isomorphism.

The proof of Theorem 5.19 will appear in [13].

Remark 5.20. An equivalent phrasing of Theorem 5.19 is that (under the hypotheses of the theorem) the left hand vertical arrow in the commutative diagram of Remark 5.15 is an isomorphism.

Remark 5.21. If U and V are admissible smooth representations of M and G respectively, then $I_{\overline{P}}^{G}(U)$ coincides with the smooth parabolic induction of U, while any M-equivariant morphism $U(\delta) \to J_P(V)$ is balanced. The isomorphism of Theorem 5.19 in this case follows from Casselman's Duality Theorem [6, §4].

Example 5.22. We consider the case when $G = \mathrm{GL}_2(\mathbb{Q}_p)$ in some detail. We take P (respectively $\overline{P}$) to be the Borel subgroup of upper triangular matrices (respectively lower triangular matrices) of G, so that M is the maximal torus consisting of diagonal matrices in G.

Let χ be a locally analytic K-valued character of $\mathbb{Q}_p^{\times}$, and let U denote the one dimensional representation of M over K on which M acts through the character $\begin{pmatrix} a & 0 \\ 0 & d \end{pmatrix} \mapsto \chi(a)$. Let $k \in K$ denote the derivative of the character χ.

Suppose first that k is a non-negative integer. Let W_k denote the irreducible representation $\mathrm{Sym}^k K^2$ of $\mathrm{GL}_2(\mathbb{Q}_p)$ over K, and let χ_k denote the highest weight of W_k with respect to P (so χ_k is the character $\begin{pmatrix} a & 0 \\ 0 & d \end{pmatrix} \mapsto a^k$ of M). If $U(\chi_k^{-1})$ denotes the twist of U by the inverse of χ_k, then $U(\chi_k^{-1})$ is a smooth representation of M.

The G-representation $I_{\overline{P}}^{G}(U)$ is a proper subrepresentation of $\mathrm{Ind}_{\overline{P}}^{G} U$; it coincides with the subspace of functions that are locally polynomial of degree $\leq k$ when restricted to $N = \mathbb{Q}_p$ under the open immersion $N \to \overline{P}\backslash G = \mathbb{P}^1(\mathbb{Q})$, and may also be characterized more intrinsically as the subspace of locally algebraic vectors in $\mathrm{Ind}_{\overline{P}}^{G} U$. It decomposes as a tensor product in the following manner:

$$I_{\overline{P}}^{G}(U) \cong (\mathrm{Ind}_{\overline{P}}^{G} U(\chi_k^{-1}))_{\mathrm{sm}} \otimes_K W_k,$$

where the subscript "sm" indicates that we are forming the smooth parabolic induction of the smooth representation $U(\chi_k^{-1})$.

If V is any object of $\mathrm{Rep}_{\mathrm{vsa}}(G)$, then we let $V_{W_k-\mathrm{lalg}}$ denote the closed subspace of W_k-locally algebraic vectors in V. (See Proposition-Definition
4.2.2 and Proposition 4.2.10 of [8].) The closed embedding $V_{W_k-\mathrm{lalg}} \to V$ induces a corresponding morphism on Jacquet modules (which is again a closed embedding; see [9, Lem. 3.4.7 (iii)]), which in turn induces an

injection $\mathcal{L}_M(U(\delta), J_P(V_{W_k-\mathrm{lalg}})) \to \mathcal{L}_M(U(\delta), J_P(V))$. It is not hard to check that $\mathcal{L}_M(U(\delta), J_P(V))^{\mathrm{bal}}$ is precisely the image of this injection.

Now the space $V_{W_k-\mathrm{lalg}}$ admits a factorization $V_{W_k-\mathrm{lalg}} \cong X \otimes_K W_k$, where X is an admissible smooth locally analytic $\mathrm{GL}_2(\mathbb{Q}_p)$-representation [8, Prop.4.2.4], and so by Example 5.9 there is an isomorphism $J_P(V_{W_k-\mathrm{lalg}}) \cong J_P(X)(\chi_k)$. Thus Theorem 5.19 reduces to the claim that the natural map

$$\mathcal{L}_G((\mathrm{Ind}_{\overline{P}}^G U(\chi_k^{-1}))_{\mathrm{sm}} \otimes_K W_k, X \otimes_K W_k) \to \mathcal{L}_M(U(\delta), J_P(X)(\chi_k))$$

induced by passing to Jacquet modules is an isomorphism. This map sits in the commutative diagram

$$\begin{array}{ccc} \mathcal{L}_G((\mathrm{Ind}_{\overline{P}}^G U(\chi_k^{-1}))_{\mathrm{sm}} \otimes_K W_k, X \otimes_K W_k) & \longrightarrow & \mathcal{L}_M(U(\delta), J_P(X)(\chi_k)) \\ \downarrow \wr & & \downarrow \wr \\ \mathcal{L}_G((\mathrm{Ind}_{\overline{P}}^G U(\chi_k^{-1}))_{\mathrm{sm}}, X) & \longrightarrow & \mathcal{L}_M(U(\chi_k^{-1})(\delta), J_P(X)), \end{array}$$

where the bottom arrow is again induced by applying J_P. Thus we are reduced to considering the case of Theorem 5.19 when U and V are both smooth. As noted in the preceding remark, this case of Theorem 5.19 follows from Casselman's Duality Theorem.

If k is not a non-negative integer, on the other hand, then $I_{\overline{P}}^G(U)$ coincides with $\mathrm{Ind}_{\overline{P}}^G U$, and every element of $\mathcal{L}_M(U(\delta), J_P(V))$ is balanced. In this case the proof of Theorem 5.19 is given in [5, Prop. 2.1.4]. (More precisely, the cited result shows that the left hand vertical arrow of the commutative diagram of Remark 5.15 is an isomorphism.)

Corollary 5.23. *Suppose that G is quasi-split, and that P is a Borel subgroup of G. If V is an absolutely topologically irreducible*[4] *very strongly admissible locally analytic representation of G for which $J_P(V) \neq 0$, then V is a quotient of $I_{\overline{P}}^G(\chi)$ for some locally L-analytic K-valued character χ of the maximal torus M of G.*

Proof. We sketch the proof; full details will appear in [13]. Since $J_P(V)$ is a non-zero object of $\mathrm{Rep}_{\mathrm{es}}(M)$, we may find a character $\psi \in \hat{M}(E)$ for some finite extension E of K for which the ψ-eigenspace of $J_P(V \otimes_K E)$ is non-zero. Taking U to be $\psi\delta^{-1}$ in Definition 5.17, we let W denote the image of the map $\mathrm{U}(\mathfrak{g}) \otimes_{\mathrm{U}(\mathfrak{p})} \mathcal{C}_c^{\mathrm{sm}}(N, U) \to V \otimes_K E$ corresponding via (5.12) to the inclusion of $U(\delta)$ into $J_P(V \otimes_K E)$. If $d\psi$ denotes the

[4]That is, $E \otimes_K V$ is topologically irreducible as a G-representation, for every finite extension E of K.

derivative of ψ (regarded as a weight of the Lie algebra $\mathfrak{m}$ of M) then $\mathcal{C}_c^{\mathrm{sm}}(N, U)$ is isomorphic to a direct sum of copies of $d\psi$ as a $\mathrm{U}(\mathfrak{p})$-module, and so W is a direct sum of copies of a quotient of the Verma module $\mathrm{U}(\mathfrak{g}) \otimes_{\mathrm{U}(\mathfrak{p})} d\psi$.

Let $W[\mathfrak{n}]$ denote the set of elements of W killed by $\mathfrak{n}$; this space decomposes as a direct sum of weights of $\mathfrak{m}$. Furthermore, for every weight α of $\mathfrak{m}$ that appears, there is a corresponding character $\tilde{\psi}$ appearing in $J_P(V \otimes_K E)$ for which $d\tilde{\psi} = \alpha$. (Compare the proof of [9, Prop. 4.4.4].) The theory of Verma modules shows that we may find a weight α of $\mathfrak{m}$ appearing in $W[\mathfrak{n}]$ such that $\alpha - \beta$ does not appear in $W[\mathfrak{n}]$ for any element β in the positive cone of the root lattice of $\mathfrak{m}$. Let $\tilde{\psi}$ be a character of M appearing in $J_P(V \otimes_K E)$ for which $\alpha = d\tilde{\psi}$, and set $\tilde{U} = \tilde{\psi}\delta^{-1}$. Our choice of α ensures that the resulting inclusion $\tilde{U}(\delta) \to J_P(V \otimes_K E)$ is balanced, and so Theorem 5.19 yields a non-zero map $I_{\overline{P}}^G(\tilde{U}) \to V \otimes_K E$. Since $V \otimes_K E$ is irreducible by assumption, this map must be surjective. Since V is defined over K, a simple argument shows that $\tilde{\psi}\delta^{-1}$ must also be defined over K. □

Corollary 5.24. *Let G and P be as in Corollary 5.23. If V is an admissible locally analytic representation of G of finite length, whose composition factors are very strongly admissible, then $J_P(V)$ is a finite dimensional M-representation.*

Proof. The functor J_P is left exact (see [9, Thm. 4.2.32]), and so it suffices to prove the result for topologically irreducible objects of $\mathrm{Rep}_{\mathrm{vsa}}(G)$. One easily reduces to the case when V is furthermore an absolutely topologically irreducible object of $\mathrm{Rep}_{\mathrm{vsa}}(G)$. If $J_P(V)$ is non-zero then Corollary 5.23 yields a surjection $I_{\overline{P}}^G(\chi) \to V$ for some $\chi \in \hat{M}(K)$. Although J_P is not right exact in general, one can show that the induced map $J_P(I_{\overline{P}}^G(\chi)) \to J_P(V)$ is surjective. Thus it suffices to prove that the source of this map is finite dimensional. This is shown by a direct calculation. The details will appear in [13]. □

References

1. S. Bosch, U. Güntzer and R. Remmert, *Non-archimedean analysis*, Springer-Verlag, 1984.
2. N. Bourbaki, *General Topology. Chapters 1-4*, Springer-Verlag, 1989.
3. C. Breuil, *Invariant $\mathcal{L}$ et série spéciale p-adique*, Ann. Scient. E.N.S. **37** (2004), 559–610.
4. C. Breuil, *Série spéciale p-adique et cohomologie étale complétée*, preprint (2003).
5. C. Breuil, M. Emerton, *Représentations p-adiques ordinaires de* $\mathrm{GL}_2(\mathbb{Q}_p)$ *et compatibilité local-global*, preprint (2005).
6. W. Casselman, *Introduction to the theory of admissible representations of p-adic reductive groups*, unpublished notes distributed by P. Sally, draft May 7, 1993.

7. R. Coleman, B. Mazur, *The eigencurve*, Galois representations in arithmetic algebraic geometry (Durham, 1996) (A. J. Scholl and R. L. Taylor, eds.), London Math. Soc. Lecture Note Ser., vol. 254, Cambridge Univ. Press, 1998, pp. 1–113.
8. M. Emerton, *Locally analytic vectors in representations of locally p-adic analytic groups*, to appear in Memoirs of the AMS.
9. M. Emerton, *Jacquet modules for locally analytic representations of p-adic reductive groups I. Construction and first properties*, Ann. Scient. E.N.S. **39** (2006), 775-839.
10. M. Emerton, *On the interpolation of systems of eigenvalues attached to automorphic Hecke eigenforms*, Invent. Math. **164** (2006), 1-84.
11. M. Emerton, *p-adic L-functions and unitary completions of representations of p-adic reductive groups*, Duke Math. J. **130** (2005), 353–392.
12. M. Emerton, *A local-global compatibility conjecture in the p-adic Langlands programme for* $\mathrm{GL}_{2/\mathbb{Q}}$, Pure and Appl. Math. Quarterly **2** (2006), no. 2 (Special issue: In honor of John H. Coates, Part 2 of 2), 1-115.
13. M. Emerton, *Jacquet modules for locally analytic representations of p-adic reductive groups II. The relation to parabolic induction*, in preparation.
14. C. T. Féaux de Lacroix, *Einige Resultate über die topologischen Darstellungen p-adischer Liegruppen auf unendlich dimensionalen Vektorräumen über einem p-adischen Körper*, Thesis, Köln 1997, Schriftenreihe Math. Inst. Univ. Münster, 3. Serie, Heft 23 (1999), 1-111.
15. H. Frommer, *The locally analytic principal series of split reductive groups*, Preprintreihe des SFB 478 – Geometrische Strukturen in der Mathematik **265** (2003).
16. M. Lazard, *Groupes analytiques* $\mathfrak{p}$*-adiques*, Publ. Math. IHES **26** (1965).
17. P. Schneider, *Nonarchimedean functional analysis*, Springer Monographs in Math., Springer-Verlag, 2002.
18. P. Schneider, *p-adische analysis*, Vorlesung in Münster, 2000; available electronically at `http://www.math.uni-muenster.de/math/u/schneider/publ/lectnotes`.
19. P. Schneider, J. Teitelbaum, *p-adic boundary values*, Astérisque **278** (2002), 51–123.
20. P. Schneider, J. Teitelbaum, *Locally analytic distributions and p-adic representation theory, with applications to* GL_2, J. Amer. Math. Soc. **15** (2001), 443–468.
21. P. Schneider, J. Teitelbaum, *Banach space representations and Iwasawa theory*, Israel J. Math. **127** (2002), 359–380.
22. P. Schneider, J. Teitelbaum, *p-adic Fourier theory*, Documenta Math. **6** (2001), 447–481.
23. P. Schneider, J. Teitelbaum, *Algebras of p-adic distributions and admissible representations*, Invent. Math **153** (2003), 145–196.
24. P. Schneider, J. Teitelbaum, *Banach-Hecke algebras and p-adic Galois representations*, Documenta Math. Extra Volume: John H. Coates' Sixtieth Birthday (2006), 631-684.
25. J.-P. Serre, *Lie algebras and Lie groups*, W. A. Benjamin, 1965.

Modularity for some geometric Galois representations

Mark Kisin
Department of Mathematics
University of Chicago
USA
kisin@math.uchicago.edu

with an appendix by Ofer Gabber

CONTENTS

INTRODUCTION

In our previous paper [Ki 1] we introduced a new technique for studying flat deformation rings of local Galois representations, and we applied this to prove the following modularity lifting theorem for two dimensional potentially Barsotti-Tate representations:

Theorem. *Let $p > 2$, S a finite set of primes of $\mathbb{Q}$, and $G_{\mathbb{Q},S}$ the Galois group of the maximal extension of $\mathbb{Q}$ unramified outside S. Let $E/\mathbb{Q}_p$ be a finite extension, $\mathcal{O}$ its ring of integers, and denote by χ the p-adic cyclotomic character of $G_{\mathbb{Q},S}$. Suppose $\rho : G_{\mathbb{Q},S} \to \mathrm{GL}_2(\mathcal{O})$ is a continuous representation such that*

(1) *ρ is potentially Barsotti-Tate at each p, and $\det \rho \cdot \chi^{-1}$ has finite order.*
(2) *The mod p representation $\bar{\rho}$ obtained by reducing ρ modulo the radical of $\mathcal{O}$ is modular.*
(3) *$\bar{\rho}|_{\mathbb{Q}(\zeta_p)}$ is absolutely irreducible.*

Then ρ is modular.

The author was partially supported by NSF grant DMS-0400666 and a Sloan Research Fellowship

Typeset by $\mathcal{A}_{\mathcal{M}}\mathcal{S}$-TEX

The purpose of this note is to give an overview of the methods of [Ki 1], and to explain how, when combined with recent work of Berger-Breuil [BB 1], they can also be used to prove a modularity lifting theorem for crystalline representations of "intermediate weight". Specifically, we show the following

Theorem. *With the notation above, let* $\rho : G_{\mathbb{Q},S} \to \mathrm{GL}_2(\mathcal{O})$ *be a continuous representation. Suppose that for some* $2 \leqslant k \leqslant 2p-1$

(1) ρ *is crystalline with Hodge-Tate weights* 0 *and* $k-1$ *at* p.
(2) $\bar{\rho}$ *arises from a modular form of weight* k *and prime to* p *level.*[1]
(3) $\bar{\rho}|_{\mathbb{Q}(\zeta_p)}$ *is absolutely irreducible, and* $\bar{\rho}|_{G_{\mathbb{Q}_p}}$ *has only scalar endomorphisms.*

Then ρ *is modular.*

In fact we prove a version of both the above theorems in the case where $\mathbb{Q}$ is replaced by a totally real field in which p is totally split. This gives a mild extension of the main result of [Ki 1].

The paper is organized as follows. In the first section, we explain the modification of the Taylor-Wiles patching argument which was used in [Ki 1]. The main new idea is that one should patch global deformation rings as algebras over local deformation rings, and not just as $\mathcal{O}$-algebras. This has the effect of reducing difficulties in deformation theory to local questions. The most important consequence is that the method has a chance to work even if the local deformation ring at p is highly singular, whereas the original method of Taylor-Wiles could succeed only if this ring was a power series ring over $\mathcal{O}$. Another upshot is that the minimal and non-minimal cases can be treated using the same argument. In particular the commutative algebra arguments of Wiles [W] comparing Fitting and congruence ideals completely disappear.

The patching argument reduces modularity lifting theorems to a question about the components of a suitable local deformation ring R. In §2 we explain the method used in [Ki 1] to resolve this question for Barsotti-Tate representations (over a possibly ramified extension of $\mathbb{Q}_p$). The basic idea is that rather than studying the deformation ring directly, one builds over it a kind of resolution $\mathscr{GR} \to \operatorname{Spec} R$, which parameterizes finite flat models of deformations. Although this resolution is not quite smooth, its singularities can be controlled, and the analysis of its connected components can eventually be reduced to a computation inside an affine Grassmannian. Using this, we give a slight extension of the main result of [Ki 1] to the case of a totally real field.

[1] Using the weight part of Serre's conjecture one can show that it suffices to assume simply that $\bar{\rho}$ is modular of any weight and level.

In §3, we study the question of components in the case of local deformation rings corresponding to crystalline deformations with weight $\leqslant 2p-2$. The basic idea is that, by Berger-Breuil, one knows the reductions of crystalline representations of medium weight. This result suggests that the generic fibre of the corresponding deformation ring should be either a disc or an annulus, and so in particular connected. To really show this one first constructs a map from Spec $R[1/p]$ to the corresponding disc (resp. annulus). For the construction one uses an analogue of the space $\mathscr{GR}$ which appears in the Barsotti-Tate case. In that situation $\mathscr{GR}$ is a kind of moduli space for finite flat models of deformations. In the higher weight case the role of the finite flat model is played by a certain integral structure in the (φ, Γ)-module corresponding to a deformation. We call this structure a *Wach lattice.* The definition is motivated by the work of Berger [Be] which asserts that the (φ, Γ)-module associated to a $\mathbb{Z}_p$-lattice in a crystalline representation admits a unique Wach lattice.

Having constructed the required map of complete local rings, one needs to show that it induces an isomorphism on generic fibres over $\mathcal{O}$. The results of Berger-Breuil guarantee that it induces a bijection on points with values in any finite extension of $\mathbb{Q}_p$. That this is enough follows from a general result about maps between complete local rings, whose proof is given in the appendix by Ofer Gabber.

Acknowledgment: It is a pleasure to thank K. Buzzard, M. Emerton, T. Gee, and F. Herzig for useful remarks on various versions of this article. I would also like to thank the referee for a very thorough reading of the paper.

§1 The Yoga of patching

(1.1) Let $p > 2$ be a prime, F a totally real field, and D be a quaternion algebra with center F which is ramified at all the infinite places of F and at a set of finite places Σ, which does not contain any primes dividing p. We fix a maximal order $\mathcal{O}_D$ of D, and for each finite place $v \notin \Sigma$, an isomorphism $(\mathcal{O}_D)_v \xrightarrow{\sim} M_2(\mathcal{O}_{F_v})$.

Let $U = \prod_v U_v \subset (D \otimes_F \mathbb{A}_F^f)^\times$ be a compact open subgroup contained in $\prod_v (\mathcal{O}_D)_v^\times$. We assume that if $v \in \Sigma$, then $U_v = (\mathcal{O}_D)_v^\times$, and that $U_v = \mathrm{GL}_2(\mathcal{O}_{F_v})$ for $v|p$.

Fix an algebraic closure $\bar{\mathbb{Q}}_p$ of $\mathbb{Q}_p$, and let $E \subset \bar{\mathbb{Q}}_p$ be a finite extension of $\mathbb{Q}_p$, with ring of integer $\mathcal{O}$, and residue field $\mathbb{F}$. We assume that E contains the images of all embeddings $F \hookrightarrow \bar{\mathbb{Q}}_p$. Write $W_k = \otimes_{F \hookrightarrow E} \mathrm{Sym}^{k-2} \mathcal{O}^2$, where $k \geq 2$ is an integer. Then W_k is naturally a $\prod_{v|p} \mathrm{GL}_2(\mathcal{O}_{F_v})$-module. We regard it as a representation of U by letting U_v act trivially for $v \nmid p$. We also fix a continuous character $\psi : (\mathbb{A}_F^f)^\times / F^\times \to \mathcal{O}^\times$ such that the action of $U \cap (\mathbb{A}_F^f)^\times$ on W_k is given by multiplication by ψ^{-1}. We think of $(\mathbb{A}_F^f)^\times$ as acting on W_k via ψ^{-1}, so that W_k becomes a $U(\mathbb{A}_F^f)^\times$-module. Note that for a given U there may be no such ψ, however such a ψ will exist if U is sufficiently small.

Let $S_{k,\psi}(U,\mathcal{O})$ denote the $\mathcal{O}$-module of continuous functions

$$f : D^\times \backslash (D \otimes_F \mathbb{A}_F^f)^\times \to W_k$$

such that for $g \in (D \otimes_F \mathbb{A}_F^f)^\times$ we have $f(gu) = u^{-1} \cdot f(g)$ for $u \in U$, and $f(gz) = \psi(z)f(g)$ for $z \in (\mathbb{A}_F^f)^\times$. If we write $(D \otimes_F \mathbb{A}_F^f)^\times = \coprod_{i\in I} D^\times t_i U(\mathbb{A}_F^f)^\times$ for some $t_i \in (D \otimes_F \mathbb{A}_F^f)^\times$ and some finite index set I, then we have

$$S_{k,\psi}(U,\mathcal{O}) \xrightarrow{\sim} \oplus_{i\in I} W_k^{(U(\mathbb{A}_F^f)^\times \cap t_i^{-1} D^\times t_i)/F^\times}.$$

For each finite prime v of F we fix a uniformiser π_v of F_v. Let S be a set of primes containing Σ, the primes dividing p, the infinite primes, and the primes v of F such that $U_v \subset D_v^\times$ is not a maximal compact subgroup. We denote by S^p the complement in S of the primes dividing p. If $v \notin S$ is a finite prime we consider the left action of $(D \otimes_F \mathbb{A}_F^f)^\times$ on W_k-valued functions on $(D \otimes_F \mathbb{A}_F^f)^\times$ given by the formula $(gf)(z) = f(zg)$. This induces an action of the double cosets $U_v \begin{pmatrix} \pi_v & 0 \\ 0 & \pi_v \end{pmatrix} U_v$ and $U_v \begin{pmatrix} \pi_v & 0 \\ 0 & 1 \end{pmatrix} U_v$ on $S_{k,\psi}(U,\mathcal{O})$. We denote these operators by S_v and T_v respectively. They do not depend on the choice of π_v. We denote by $\mathbb{T}_\psi(U)'$ the $\mathcal{O}$-algebra generated by the endomorphisms T_v and S_v for $v \notin S$. It is commutative and finite over $\mathcal{O}$.

Similarly, for $v \mid p$ there is a natural action of $M_2(\mathcal{O}_{F_v})$ on W_k-valued functions on $(D \otimes_F \mathbb{A}_F^f)^\times$, given by $(u^{\mathrm{sm}} \cdot f)(g) = u \cdot f(gu)$. This action is smooth on functions in $S_{k,\psi}(U,\mathcal{O})$, and its restriction to U_v leaves $S_{k,\psi}(U,\mathcal{O})$ invariant, and hence induces a well defined action of the double cosets $U_v \begin{pmatrix} \pi_v & 0 \\ 0 & \pi_v \end{pmatrix} U_v$ and $U_v \begin{pmatrix} \pi_v & 0 \\ 0 & 1 \end{pmatrix} U_v$ on $S_{k,\psi}(U,\mathcal{O})$. We again denote these operators by T_v and S_v, and we write $\mathbb{T}_\psi(U)$ for the $\mathbb{T}_\psi(U)'$-algebra generated by these endomorphisms. It is again commutative, and finite over $\mathcal{O}$.

(1.2) Fix an algebraic closure $\bar{F}$ of F, and let $G_{F,S}$ be the Galois group of the maximal subfield of $\bar{F}$ which is unramified over F outside of S. A maximal ideal $\mathfrak{m}' \subset \mathbb{T}_\psi(U)'$ is called *Eisenstein* if there exists a finite abelian extension F' of F such that $T_v - 2 \in \mathfrak{m}'$ for each prime v which splits completely in F'. Equivalently, if $\bar{\mathbb{F}}$ denotes an algebraic closure of $\mathbb{F}$, then $\mathfrak{m}'$ is the kernel of an $\mathcal{O}$-algebra map $\theta_{\mathfrak{m}'} : \mathbb{T}_\psi(U)' \to \bar{\mathbb{F}}$, and $\mathfrak{m}'$ is Eisenstein if there is a two-dimensional, reducible representation of $G_{F,S}$ on an $\bar{\mathbb{F}}$-vector space such for $v \notin S$, the trace of a Frobenius at v is given by $\theta_{\mathfrak{m}'}(T_v)$. We say that a maximal ideal $\mathfrak{m}$ of $\mathbb{T}_\psi(U)$ is Eisenstein if its intersection with $\mathbb{T}_\psi(U)'$ is.

Suppose that $\mathfrak{m}' \subset \mathbb{T}_\psi(U)'$ is a non-Eisenstein maximal ideal. Using the existence of Galois representations attached to Hilbert modular eigenforms [Ta 2], and the Jacquet-Langlands correspondence (see [Ta 1, 1.3]), one sees that there exists a continuous representation

$$\rho_{\mathfrak{m}'} : G_{F,S} \to \mathrm{GL}_2(\mathbb{T}_\psi(U)'_{\mathfrak{m}'})$$

such that for $v \notin S$, the characteristic polynomial of $\rho_{\mathfrak{m}'}(\mathrm{Frob}_v)$ is $X^2 - T_vX + \mathbf{N}(v)S_v$. Here Frob_v denotes an *arithmetic* Frobenius at v, and $\mathbf{N}(v)$ denotes the order of the residue field at v. After replacing $\mathcal{O}$ by a finite extension, we may assume that the residue field at $\mathfrak{m}'$ is $\mathbb{F}$, and we denote by $\bar{\rho}_{\mathfrak{m}'} : G_{F,S} \to \mathrm{GL}_2(\mathbb{F})$ the representation obtained by reducing $\rho_{\mathfrak{m}'}$ modulo $\mathfrak{m}'$.

Since $\mathfrak{m}'$ is non-Eisenstein, $\bar{\rho}_{\mathfrak{m}'}$ is an absolutely irreducible representation. In particular it admits a universal deformation $\mathcal{O}$-algebra $R_{F,S}$, and we denote by $R^\psi_{F,S}$ the quotient of $R_{F,S}$ corresponding to deformations with determinant $\psi\chi$, where χ denotes the p-adic cyclotomic character. Here we regard ψ as a character of $G_{F,S}$ via the class field theory isomorphism, normalized the take uniformizers to arithmetic Frobenii. The representation $\rho_{\mathfrak{m}'}$ induces a map $R^\psi_{F,S} \to \mathbb{T}_\psi(U)'_{\mathfrak{m}'}$. The formula for the characteristic polynomial of Frobenius at $v \notin S$ shows that this map is surjective.

We now fix, once and for all, a basis for the underlying $\mathbb{F}$-vector space $V_\mathbb{F}$ of $\bar{\rho}_{\mathfrak{m}'}$. Let $\Sigma_p = \Sigma \cup \{v\}_{v|p}$. For any $v \in \Sigma_p$ we fix a decomposition group $G_{F_v} \subset \mathrm{Gal}(\bar{F}/F)$, and we denote by $D^\square_{V_\mathbb{F},v}$ the functor which assigns to a local Artinian $\mathcal{O}$-algebra A with residue field $\mathbb{F}$ the set of isomorphism classes of pairs $(V_A, \beta_{v,A})$ where V_A is a lift of the G_{F_v}-representation $V_\mathbb{F}$ to a continuous representation on a finite free A-module, and $\beta_{v,A}$ is an A-basis for V_A lifting the chosen basis of $V_\mathbb{F}$. Then $D^\square_{V_\mathbb{F},v}$ is pro-representable by a complete local $\mathcal{O}$-algebra $R^\square_v$ called the *universal framed deformation ring* of $V_\mathbb{F}$. We denote by $R^{\square,\psi}_v$ the quotient of $R^\square_v$ corresponding to deformations with determinant $\psi\chi$. We set

$$R^{\square,\psi}_{\Sigma,p} = \widehat{\otimes}_{v\in\Sigma_p} R^{\square,\psi}_v$$

where all the completed tensor products are over $\mathcal{O}$. Similarly, we denote by $R^{\square,\psi}_{F,S}$ the complete local $\mathcal{O}$-algebra, which represents the functor obtained by assigning to A as above the set of isomorphism classes of tuples $(V_A, \{\beta_{v,A}\}_{v\in\Sigma_p})$, where V_A is a deformation of the $G_{F,S}$-representation $V_\mathbb{F}$ to A having determinant ψ, and for $v \in \Sigma_p$ $\beta_{v,A}$ is a lifting of the chosen basis to V_A. The $\mathcal{O}$-algebra $R^{\square,\psi}_{F,S}$ has an obvious structure of a formally smooth $R^{\psi}_{F,S}$-algebra.

Choose a maximal ideal $\mathfrak{m}$ of $\mathbb{T}_\psi(U)$ whose restriction to $\mathbb{T}_\psi(U)'$ is $\mathfrak{m}'$. We will slightly abuse notation, and write $\mathbb{T}_\psi(U)'_{\mathfrak{m}}$ for the image of $\mathbb{T}_\psi(U)'_{\mathfrak{m}'}$ in $\mathbb{T}_\psi(U)_{\mathfrak{m}}$. (The map $\mathbb{T}_\psi(U)'_{\mathfrak{m}'} \to \mathbb{T}_\psi(U)_{\mathfrak{m}}$ is not in general injective.) We set $\mathbb{T}^{\square}_\psi(U)'_{\mathfrak{m}} = \mathbb{T}_\psi(U)'_{\mathfrak{m}} \otimes_{R^{\psi}_{F,S}} R^{\square,\psi}_{F,S}$. The tensor product is formally smooth over $\mathbb{T}_\psi(U)'_{\mathfrak{m}}$, and hence $\mathcal{O}$-flat.

We now make the following assumptions on $\bar{\rho}_{\mathfrak{m}'}$

(1) $\bar{\rho}_{\mathfrak{m}'}$ is unramified outside primes dividing p.
(2) The restriction of $\bar{\rho}_{\mathfrak{m}'}$ to $G_{F(\zeta_p)}$ is absolutely irreducible.
(3) If $p = 5$, then $[F(\zeta_p) : F] > 2$.
(4) If $v \in S \setminus \Sigma_p$, then

$$(1 - \mathbf{N}(v))((1 + \mathbf{N}(v))^2 \det \bar{\rho}(\mathrm{Frob}_v) - (\mathbf{N}(v))(\mathrm{tr}\bar{\rho}(\mathrm{Frob}_v))^2) \in \mathbb{F}^\times.$$

The conditions (2) and (3) are needed to ensure that in the patching method of Taylor-Wiles, one can find enough auxiliary primes to kill the dual Selmer group. The order of the Selmer group which controls the number of (topological) generators of $R^{\square,\psi}_{F,S}$ as an $R^{\square,\psi}_{\Sigma,p}$-algebra is then given as a product of local terms, and (1) and (4) ensure that this order is sufficiently small that the dimension of the patched deformation ring (denoted R_∞ in (1.3) below) is equal to that of the patched Hecke algebra.

We also make the following assumption on U :

(1.2.1) For all $t \in (D \otimes_F \mathbb{A}^f_F)^\times$ the group $(U(\mathbb{A}^f_F)^\times \cap t^{-1}D^\times t)/F^\times$ has order prime to p.

This condition is always satisfied if U is sufficiently small [Ta 1, 1.1]. For example, it holds if U is a normal subgroup and $U \prod_{v\nmid\infty} \mathcal{O}^\times_{F_v} \cap D^{\det=1}$ contains no non-trivial p-power roots of unity.

Proposition (1.3). *Let $d = [F : \mathbb{Q}] + 3|\Sigma_p|$. Then there exists an integer $h \geq d$, a commutative diagram of complete local $\mathcal{O}$-algebras*

$$\begin{array}{ccccc} & & \mathcal{O}[[y_1, \ldots, y_h]] & & \\ & & \downarrow & & \\ R^{\square,\psi}_{\Sigma,p}[[x_1, \ldots, x_{h-d}]] & \longrightarrow & R_\infty & \longrightarrow & R^{\square,\psi}_{F,S} \end{array}$$

and an R_∞-module M_∞ such that

(1) *The horizontal maps are surjective and the map on the right induces an isomorphism $R_\infty/(y_1,\dots,y_h)R_\infty \xrightarrow{\sim} R^{\square,\psi}_{F,S}$.*

(2) *M_∞ is a finite flat $\mathcal{O}[\![y_1,\dots,y_h]\!]$-module, and the action of R_∞ on the quotient $M_\infty/(y_1,\dots,y_h)M_\infty$ factors through $\mathbb{T}^\square_\psi(U)'_{\mathfrak{m}}$, and makes it into a faithful $\mathbb{T}^\square_\psi(U)'_{\mathfrak{m}}$-module.*

Proof. This follows from [Ki 1, 3.2.5, 3.1.13], and the arguments of [Ki 1, 3.3.1, 3.4.11]. Let us just indicate the ingredients. The $\mathcal{O}$-algebra R_∞ and the R_∞-module M_∞ are constructed by patching deformation rings and Hecke modules at auxiliary levels as in the argument of Taylor-Wiles [TW], as modified by Diamond [Di]. The only difference is the observation that all the rings involved are $R^{\square,\psi}_{\Sigma,p}$-algebras and we patch $R^{\square,\psi}_{\Sigma,p}$-algebras rather than just $\mathcal{O}$-algebras.

More precisely, for suitably chosen collections of primes $Q_n = \{v_1,\dots, v_h\}$ which are disjoint from Σ_p and at which U is maximal, and such that $\mathbf{N}(v_i) \equiv 1\,(p^n)$ for all i, one defines a compact open subgroup U_{Q_n} of $(D\otimes_F \mathbb{A}^f_F)^\times$ by $(U_{Q_n})_v = U_v$ if $v \notin Q_n$, and

$$(U_{Q_n})_v = \{g \in \mathrm{GL}_2(\mathcal{O}_{F_v}) : g \equiv \left(\begin{smallmatrix} a & b \\ 0 & d\end{smallmatrix}\right)(\pi_v),\ ad^{-1} \mapsto 1 \in \Delta_v\}$$

for $v \in Q_n$, where π_v is a uniformizer at v, and Δ_v denotes the maximal pro-p quotient of $(\mathcal{O}_{F_v}/\pi_v)^\times$.

Set $S_{Q_n} = S \cup Q_n$, and denote by $\mathfrak{m}_{Q_n}$ the preimage of $\mathfrak{m}$ under the natural map $\mathbb{T}_\psi(U_{Q_n}) \to \mathbb{T}_\psi(U)$. Then R_∞ is defined as an inverse limit of suitable finite length $R^{\square,\psi}_{\Sigma,p}$-algebra quotients of $R^{\square,\psi}_{F,S_{Q_n}}$, and M_∞ is obtained by taking an inverse limit over certain finite length quotients of $S_{k,\psi}(U_{Q_n},\mathcal{O})_{\mathfrak{m}_{Q_n}} \otimes_{R^\psi_{F,S_{Q_n}}} R^{\square,\psi}_{F,S_{Q_n}}$.

The freeness M_∞ over $\mathcal{O}[\![y_1,\dots,y_h]\!]$ may be deduced from the fact that under the assumption (1.2.1) $S_{k,\psi}(U_{Q_n},\mathcal{O})$ is a free $\mathcal{O}[\Delta_{Q_n}]$-module, because if we write $(D\otimes_F \mathbb{A}^f_F)^\times = \coprod_{i\in I_{Q_n}} D^\times t_i U_{Q_n}(\mathbb{A}^f_F)^\times$, then Δ_{Q_n} acts freely on I_{Q_n} (see also [Ta 1, 2.3]). $\square$

Corollary (1.4). *Suppose that for $v \in \Sigma_p$ there exists a quotient $\bar{R}^{\square,\psi}_v$ of $R^{\square,\psi}_v$ such that*

(1) *The action of $R^{\square,\psi}_v$ on M_∞ factors through $\bar{R}^{\square,\psi}_v$.*

(2) *$\bar{R}^{\square,\psi}_v$ is an $\mathcal{O}$-flat domain and $\bar{R}^{\square,\psi}_v[1/p]$ is formally smooth over E and geometrically integral (that is it remains a domain after tensoring by any finite extension of E).*

(3) *If $v \in \Sigma$ then $\bar{R}^{\square,\psi}_v$ has relative dimension 3 over $\mathcal{O}$, while if $v|p$ it has relative dimension $3 + [F_v : \mathbb{Q}_p]$.*

Let $\bar{R}^{\square,\psi}_{\Sigma,p} = \widehat{\otimes}_{v\in\Sigma_p}\bar{R}^{\square,\psi}_v$, *and set* $\bar{R}^{\square,\psi}_{F,S} = R^{\square,\psi}_{F,S} \otimes_{R^{\square,\psi}_{\Sigma,p}} \bar{R}^{\square,\psi}_{\Sigma,p}$. *Then the natural map* $\bar{R}^{\square,\psi}_{F,S} \to \mathbb{T}^{\square}_{\psi}(U)'_{\mathfrak{m}}$ *is surjective with p-power torsion kernel.*

Proof. First note that (1) ensures that the map $R^{\square,\psi}_{F,S} \to \mathbb{T}^{\square}_{\psi}(U)'_{\mathfrak{m}}$ factors through $\bar{R}^{\square,\psi}_{F,S}$. We have to show that the resulting surjection $\bar{R}^{\square,\psi}_{F,S} \to \mathbb{T}^{\square}_{\psi}(U)'_{\mathfrak{m}}$ has p-power torsion kernel.

The image of $\bar{R}^{\square,\psi}_{\Sigma,p}[\![x_1,\dots,x_{h-d}]\!]$ in the endomorphisms of M_∞ is a faithful $\mathcal{O}[\![y_1,\dots,y_h]\!]$-module and hence of dimension at least $h+1$. The condition (2) ensures that $\bar{R}^{\square,\psi}_{\Sigma,p}$ is a domain.[2] Since the dimension of $\bar{R}^{\square,\psi}_{\Sigma,p}[\![x_1,\dots,x_{h-d}]\!]$ is $h+1$, M_∞ is a faithful $\bar{R}^{\square,\psi}_{\Sigma,p}[\![x_1,\dots,x_{h-d}]\!]$-module, and this ring is a finite faithful $\mathcal{O}[\![y_1,\dots,y_h]\!]$-module.

At any maximal ideal of $\bar{R}^{\square,\psi}_{\Sigma,p}[\![x_1,\dots,x_{h-d}]\!][1/p]$, M_∞ has depth $\geq h$, since it is non-zero and finite free over $\mathcal{O}[\![y_1,\dots,y_h]\!]$. It follows by the Auslander-Buchsbaum theorem that the $\bar{R}^{\square,\psi}_{\Sigma,p}[\![x_1,\dots,x_{h-d}]\!][1/p]$-module $M_\infty \otimes_{\mathbb{Z}_p} \mathbb{Q}_p$ is projective. In particular $M_\infty/(y_1,\dots,y_h)M_\infty \otimes_{\mathbb{Z}_p} \mathbb{Q}_p$ is a faithful module over

$$\bar{R}^{\square,\psi}_{\Sigma,p}[\![x_1,\dots,x_{h-d}]\!][1/p]/(y_1,\dots,y_h)\bar{R}^{\square,\psi}_{\Sigma,p}[\![x_1,\dots,x_{h-d}]\!][1/p] \xrightarrow{\sim} \bar{R}^{\square,\psi}_{F,S}[1/p].$$

Since the action of $R^{\square,\psi}_{F,S}$ on $M_\infty/(y_1,\dots,y_h)M_\infty$ factors through $\mathbb{T}^{\square}_{\psi}(U)'_{\mathfrak{m}}$, the corollary follows (cf. [Ki 1, §3.3]). □

(1.5) We now explain how to construct the quotient $\bar{R}^{\square,\psi}_v$ for $v\in\Sigma$. For $v|p$, a quotient $\bar{R}^{\square,\psi}_v$ with the required properties is known to exist only in certain situations. When $k=2$, the construction of such a quotient is one of the main points of [Ki 1], and will be explained in the next section. When p is totally split in F the existence of $\bar{R}^{\square,\psi}_v$ can sometimes be deduced from results of Berger and Breuil if $k \leqslant 2p-1$.

Proposition (1.6). *Suppose* $v\in\Sigma$, *and let* $\gamma : G_{F_v} \to \mathcal{O}^\times$ *be a continuous character. We denote by* $\bar{\gamma}$ *the composite of* γ *and the projection* $\mathcal{O}^\times \to \mathbb{F}^\times$. *Suppose that* $\bar{\rho}_{\mathfrak{m}'}|_{G_{F_v}}$ *is an extension of* $\bar{\gamma}$ *by* $\bar{\gamma}\chi$. *Then there exists a quotient* $R^{\chi\gamma,\gamma,\square}_v$ *of* $R^{\square}_v$ *with the following properties.*

(1) $R^{\chi\gamma,\gamma,\square}_v$ *is an* $\mathcal{O}$*-flat domain, and* $R^{\chi\gamma,\gamma,\square}_v[1/p]$ *is formally smooth over* E *and of dimension* 3.

[2] To see this it is enough to show that $\bar{R}^{\square,\psi}_{\Sigma,p}[1/p]$ has connected spectrum. After enlarging E we may assume that each of the rings $\bar{R}^{\square,\psi}_v[1/p]$ has an E-valued point. Then the result follows from the fact that if $Y\to X$ is a map of topological spaces, which has connected fibres and admits a section, then Y is connected if X is connected.

(2) *If E'/E is a finite extension, and $\xi : R_v^{\square} \to E'$ a map of $\mathcal{O}$-algebras, then ξ factors through $R_v^{\chi\gamma,\gamma,\square}$ if and only if the corresponding E'-representation V_ξ of G_{F_v} is an extension of γ by $\gamma\chi$.*

Proof. This is proved in [Ki 1, §2.6]. The argument already illustrates one of the ideas behind the construction $\bar{R}_v^{\square,\psi}$ for $v|p$, which is that one can study a deformation ring by studying a resolution of it. We give a sketch.

Define a functor L_v^γ on $\mathcal{O}$-algebras, as follows: If A is an $\mathcal{O}$-algebra, and $\xi : R_v^{\square} \to A$ a morphism of $\mathcal{O}$-algebras, then ξ gives rise to a representation of G_{F_v} on a finite free A-module V_A. We denote by $L_v^\gamma(A)$ the set consisting of pairs (ξ, L_A), where ξ is a morphism as above, and $L_A \subset V_A$ is a projective A-submodule of rank 1 such that V_A/L_A is projective over A.

We have an obvious morphism of functors $L_v^\gamma \to D_{V_{\mathbb{F}},v}^{\square}$ given by sending (ξ, L_A) to ξ, and this morphism is easily seen to be represented by a projective map $\Theta : \mathscr{L}_v^\gamma \to \operatorname{Spec} R_v^{\square}$. This map becomes an closed immersion after inverting p, because if p is invertible in A then L_A, if it exists, is unique. We define $R_v^{\chi\gamma,\gamma,\square}$ to be the quotient of $\operatorname{Spec} R_v^{\square}$ corresponding to the scheme theoretic image of Θ.

From the definition one sees that $R_v^{\chi\gamma,\gamma,\square}$ satisfies (2). To see that $\mathscr{L}_v^\gamma$ is formally smooth over $\mathcal{O}$ we remark that $H^2(G_{F_v}, \mathbb{Z}_p)$ has no p-torsion, so that for any finite length $\mathbb{Z}_p$-module M, equipped with a continuous action of G_{F_v}, the map

$$H^1(G_{F_v}, \mathbb{Z}_p(\chi)) \otimes_{\mathbb{Z}_p} M \to H^1(G_{F_v}, M(\chi))$$

is surjective. (Here $M(\chi)$ denotes M with G_{F_v} acting via χ.) In particular, if A is an Artin ring with residue field a finite extension of $\mathbb{F}$, the natural map

$$\operatorname{Ext}^1_{\mathcal{O}[G_{F_v}]}(\mathcal{O}(\gamma), \mathcal{O}(\gamma\chi)) \otimes_{\mathcal{O}} A \to \operatorname{Ext}^1_{A[G_{F_v}]}(A(\gamma), A(\gamma\chi))$$

is a surjection and L_v^γ is formally smooth. Hence $R_v^{\chi\gamma,\gamma,\square}$ satisfies (1). The dimension of $\mathscr{L}_v^\gamma[1/p]$ can be computed using Galois cohomology.

It remains to show that $\mathscr{L}_v^\gamma$ is connected. Since $\mathscr{L}_v^\gamma$ is formally smooth over $\mathcal{O}$ it suffices to show that the fibre of Θ over the closed point of $\operatorname{Spec} R_v^{\square}$ is connected. However this fibre is either a single point or a projective line. The second case occurs precisely, when the action of G_{F_v} on $V_{\mathbb{F}}$ is scalar, in which case we must have $\bar{\gamma} = \bar{\gamma}(1)$. □

Corollary (1.7). *For each $v \in \Sigma$, there is a unique $\gamma : G_{F_v} \to \mathcal{O}^\times$ as above such that the quotient $R_v^{\chi\gamma,\gamma,\square}$ satisfies the conditions of (1.4).*

Proof. By the local Langlands correspondence, $\bar{\rho}_{\mathfrak{m}'}$ is an extension of $\bar{\gamma}$ by $\bar{\gamma}(1)$ for some unramified character $G_{F_v} \to \mathcal{O}^\times$. We necessarily have $\bar{\gamma}^2 = \psi$. Since $p \neq 2$, $\bar{\gamma}$ lifts to a unique unramified character $\gamma : G_{F_v} \to \mathcal{O}^\times$ such that $\gamma^2 = \psi$. We claim the corollary holds with this choice of γ.

The conditions (2) and (3) of (1.4) follow from (1.6)(1), while (1.4)(1) follows from (1.6)(2) and the local Langlands correspondence, the point being that D is ramified at each $v \in \Sigma$, while M_∞ is constructed using quaternionic forms on compact open subgroups of $(D \otimes_F \mathbb{A}_F^f)^\times$ which are maximal compact at each such v. The last remark also shows that γ is the only character such that $R_v^{\chi_\gamma,\gamma,\square}$ satisfies (1.4)(1). □

§2 Barsotti-Tate representations

(2.1) In this section we explain how to construct the quotient $\bar{R}_v^{\square,\psi}$ in (1.4) in the case that $v|p$ and $k = 2$. Since this is essentially a local question, we work locally.

Let k be a finite extension of $\mathbb{F}_p$. Write $W = W(k)$ for the ring of Witt vectors of k, $K_0 = W[1/p]$ and K for a finite totally ramified extension of K_0 of degree e. Let $\mathcal{O}_K$ be the ring of integers in K, $\pi \in \mathcal{O}_K$ a fixed uniformiser, and $E(u)$ the Eisenstein polynomial of π. We fix an algebraic closure $\bar{K}$ of K, and write $\mathcal{O}_{\bar{K}}$ for its ring of integers. We set $G_K = \mathrm{Gal}(\bar{K}/K)$. Let $\sqrt[p^n]{\pi} \in \bar{K}$ be a root of π, such that $(\sqrt[p^{n+1}]{\pi})^p = \sqrt[p^n]{\pi}$ for all $n \geq 0$. Set $K_\infty = \cup_{n\geq 1} K(\sqrt[p^n]{\pi})$, and write $G_{K_\infty} = \mathrm{Gal}(\bar{K}/K_\infty)$.

Let $\mathfrak{S} = W[\![u]\!]$. We equip $\mathfrak{S}$ with a Frobenius endomorphism φ which acts as the usual Frobenius on W and takes u to u^p. Write $\mathcal{O}_\mathcal{E}$ for the p-adic completion of $\mathfrak{S}[1/u]$. The ring $\mathcal{O}_\mathcal{E}$ is a complete discrete valuation ring with uniformiser p, and residue field $k((u))$. The endomorphism φ extends to $\mathcal{O}_\mathcal{E}$ by continuity.

An *étale φ-module* over $\mathcal{O}_\mathcal{E}$ is a finite $\mathcal{O}_\mathcal{E}$-module M together with an isomorphism of $\mathcal{O}_\mathcal{E}$-modules $\varphi^*(M) \xrightarrow{\sim} M$. There is an equivalence of categories between finite length, étale $\mathcal{O}_\mathcal{E}$-modules and continuous representations of G_{K_∞} on finite length $\mathbb{Z}_p$-modules. This is essentially a consequence of [Fo, §A].

The results of *loc. cit* are more often applied to the p-cyclotomic extension of K, when they give rise to the theory of (φ, Γ)-modules. Here we use the Kummer extension K_∞. Since this field is far from being Galois over K, there is no immediate analogue of Γ, and this suggests that the Kummer extension may be less useful for studying representations of G_K. However a result of Breuil [Br 3, 3.4.3], asserts that for representations arising from finite flat $\mathcal{O}_K$-group schemes, the restriction functor from G_K to G_{K_∞} is fully faithful.

To explain how to describe finite flat models of Galois representations in terms étale φ-modules, we introduced the category (Mod FI/$\mathfrak{S}$). An object of this category consists of a finitely generated $\mathfrak{S}$-module $\mathfrak{M}$ equipped with a $\mathfrak{S}$-linear map $1 \otimes \varphi : \varphi^*(\mathfrak{M}) \to \mathfrak{M}$ such that $\mathfrak{M}$ is a finite direct sum of modules of the form $\mathfrak{S}/p^n\mathfrak{S}$ for various $n \geq 1$, and the cokernel of $1 \otimes \varphi$ is killed by $E(u)$. Then we have the following result [Ki 1, 1.1.13], a large part of which is due to Breuil [Br 4, 3.1.2, 3.1.3], [Br 3, 3.3.2].

Theorem (2.2). *We have a commutative diagram of functors*

$$\begin{array}{ccc}
(\text{Mod FI}/\mathfrak{S}) & \longrightarrow & \{\text{finite flat } \mathcal{O}_K\text{-group schemes}\} \\
\downarrow {\scriptstyle \mathfrak{M} \mapsto \mathfrak{M}[1/u]} & & \downarrow {\scriptstyle \mathcal{G} \mapsto \mathcal{G}(\bar{K})(-1)} \\
\{\text{finite length, étale } \mathcal{O}_{\mathcal{E}}\text{-modules}\} & \xrightarrow{\sim} & \{\text{finite length } G_{K_\infty}\text{-representations}\}
\end{array}$$

where the lower horizontal functor has been discussed above. The top horizontal functor is exact and fully faithful, and an equivalence on objects killed by p.

Using the theorem, and the full faithfulness of restriction from G_K to G_{K_∞} for finite flat Galois representations, we see in particular, that two modules $\mathfrak{M}$ and $\mathfrak{M}'$ give rise to finite flat group schemes with isomorphic generic fibres if and only if $\mathfrak{M}[1/u] \xrightarrow{\sim} \mathfrak{M}'[1/u]$. Our method for understanding flat deformation rings involves studying the modules in (Mod FI/$\mathfrak{S}$) which give rise to a given finite length étale $\mathcal{O}_{\mathcal{E}}$-module. The above discussion shows that this is closely related to studying finite flat models of a given Galois representations.

We remark that the category (Mod FI/$\mathfrak{S}$) in the above theorem can be slightly enlarged so that the above functor extends to an equivalence with all finite flat $\mathcal{O}_K$-group schemes [Ki 2, 2.3.5]. We denote this enlarged category by (Mod/$\mathfrak{S}$).

(2.3) Now let $\mathbb{F}$ be a finite extension of $\mathbb{F}_p$, as above, and $V_{\mathbb{F}}$ a finite dimensional $\mathbb{F}$-vector space of dimension d, equipped with a continuous action of G_K. We denote by $M_{\mathbb{F}}$ the étale φ-module attached to $V_{\mathbb{F}}(-1)|_{G_{K_\infty}}$ by the correspondence explained in (2.1). Then $M_{\mathbb{F}}$ is naturally a free $\mathbb{F} \otimes_{\mathbb{F}_p} k((u))$-module.

We will assume that $\mathrm{End}_{G_K} V_{\mathbb{F}} = \mathbb{F}$. This assumptions is made only to simplify the exposition. The general case requires the language of groupoids.

We denote by $\mathfrak{AR}_{W(\mathbb{F})}$ the category of Artinian, local $W(\mathbb{F})$-algebras with residue field $\mathbb{F}$. We denote by $\mathfrak{Aug}_{W(\mathbb{F})}$ the category of $W(\mathbb{F})$-algebras A equipped with a nilpotent ideal I such that $pA \subset I$. Note that if A is in $\mathfrak{AR}_{W(\mathbb{F})}$ then we may regard A as an object of $\mathfrak{Aug}_{W(\mathbb{F})}$

by taking I to be the maximal ideal of A. This makes $\mathfrak{AR}_{W(\mathbb{F})}$ into a full subcategory of $\mathfrak{Aug}_{W(\mathbb{F})}$.

Let A be a $\mathbb{Z}_p$-algebra. We denote by $(\mathrm{Mod\ FI}/\mathfrak{S})_A$ the category whose objects consist of a finite projective $\mathfrak{S}_A = \mathfrak{S} \otimes_{\mathbb{Z}_p} A$-module $\mathfrak{M}$ together with a map of $\mathfrak{S}_A$-modules $\varphi^*(\mathfrak{M}) \to \mathfrak{M}$ whose cokernel is killed by $E(u)$. Here we extend φ to $\mathfrak{S}_A$ by A-linearity. For (A, I) in $\mathfrak{Aug}_{W(\mathbb{F})}$, we denote by $D_{\mathfrak{S},M_{\mathbb{F}}}(A, I)$ the set of isomorphism classes of pairs $(\mathfrak{M}_A, \psi)$, where $\mathfrak{M}_A$ is in $(\mathrm{Mod\ FI}/\mathfrak{S})_A$, and $\psi : \mathcal{O}_{\mathcal{E}} \otimes_{\mathfrak{S}} \mathfrak{M}_A \otimes_A A/I \xrightarrow{\sim} M_{\mathbb{F}} \otimes_{\mathbb{F}} A/I$ is an isomorphism of $A/I \otimes_{\mathbb{F}_p} k((u))$-modules compatible with the action of φ.

For A in $\mathfrak{AR}_{W(\mathbb{F})}$ we denote by $D_{V_{\mathbb{F}}}(A)$ the set of isomorphism classes of a finite free A-module V_A equipped with a continuous action of G_K and an isomorphism of $\mathbb{F}[G_K]$-modules $V_A \otimes_A \mathbb{F} \xrightarrow{\sim} V_{\mathbb{F}}$. We denote by $D^{\mathrm{fl}}_{V_{\mathbb{F}}}(A) \subset D_{V_{\mathbb{F}}}(A)$ the subset consisting of those V_A which are the generic fibres of a finite flat group scheme over $\mathcal{O}_K$. Both $D_{V_{\mathbb{F}}}$ and $D^{\mathrm{fl}}_{V_{\mathbb{F}}}$ can be extended to functors on $\mathfrak{Aug}_{W(\mathbb{F})}$, by taking $D_{V_{\mathbb{F}}}(A, I)$ and $D^{\mathrm{fl}}_{V_{\mathbb{F}}}(A, I)$ to be the direct limit of $D_{V_{\mathbb{F}}}(A')$ and $D^{\mathrm{fl}}_{V_{\mathbb{F}}}(A')$ respectively, where $A' \subset A$ runs over Artinian local $W(\mathbb{F})$-algebras with residue field $\mathbb{F}$ whose radical is contained in I. Our assumption on $V_{\mathbb{F}}$ implies that the functors $D_{V_{\mathbb{F}}}$ and $D^{\mathrm{fl}}_{V_{\mathbb{F}}}$ are pro-representable by complete local $W(\mathbb{F})$-algebras, $R_{V_{\mathbb{F}}}$ and $R^{\mathrm{fl}}_{V_{\mathbb{F}}}$ respectively [Ma], [Ram].

Finally, for A in $\mathfrak{AR}_{W(\mathbb{F})}$ we denote by $D_{M_{\mathbb{F}}}(A)$ the set of isomorphism classes of finite free $\mathcal{O}_{\mathcal{E}} \otimes_{\mathbb{Z}_p} A$-modules M_A equipped with an isomorphism $\varphi^*(M_A) \xrightarrow{\sim} M_A$ and an $\mathbb{F}$-linear isomorphism $M_A \otimes_A \mathbb{F} \xrightarrow{\sim} M_{\mathbb{F}}$ respecting φ. As above, we may extend this to a functor on $\mathfrak{Aug}_{W(\mathbb{F})}$.

The relationship among these functors is given by the following results [Ki 1, 2.1.4, 2.1.11, 2.4.8].

Proposition (2.4). *We have a commutative diagram of functors on* $\mathfrak{Aug}_{W(\mathbb{F})}$.

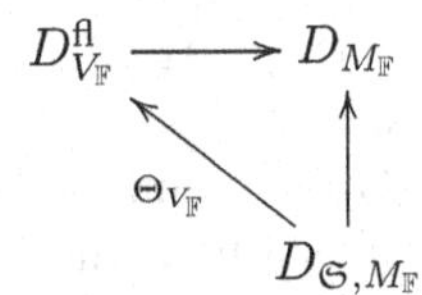

where the vertical map is given by $\mathfrak{M}_A \mapsto \mathcal{O}_{\mathcal{E}} \otimes_{\mathfrak{S}} \mathfrak{M}_A$. *The horizontal map is injective, so that* $\Theta_{V_{\mathbb{F}}}$ *is uniquely determined by requiring the commutativity of the diagram.*

Proof. This is [Ki 1, 2.1.4]. The first step in the proof is to show that the association by $\mathfrak{M}_A \mapsto \mathcal{O}_{\mathcal{E}} \otimes_{\mathfrak{S}} \mathfrak{M}_A$ gives a well defined object of $D_{M_{\mathbb{F}}}$. This is not completely obvious, since for (A, I) in $\mathfrak{Aug}_{W(\mathbb{F})}$, $D_{M_{\mathbb{F}}}(A, I)$ is defined by a limit process. One has to show that the module $\mathcal{O}_{\mathcal{E}} \otimes_{\mathfrak{S}} \mathfrak{M}_A$

can be descended to an object $M_{A'}$ in $D_{M_{\mathbb{F}}}(A')$ for some subring $A' \subset A$ with A' in $\mathfrak{AR}_{W(\mathbb{F})}$.

Using (2.2) above, $M_{A'}$ corresponds to a representation of G_{K_∞} on a finite free A'-module $V_{A'}$. One checks that $M_{A'}$ actually arises from a module in (Mod/$\mathfrak{S}$). This implies that $V_{A'}$ extends to finite flat representation of G_K. This extension is unique by Breuil's result mentioned above. □

Proposition (2.5). *The morphism $\Theta_{V_{\mathbb{F}}}$ is represented by a projective morphism*

$$\Theta_{V_{\mathbb{F}}} : \mathscr{GR}_{V_{\mathbb{F}}} \to \operatorname{Spec} R^{\mathrm{fl}}_{V_{\mathbb{F}}}.$$

After inverting p both sides of this map become formally smooth over $W(\mathbb{F})[1/p]$, and $\Theta_{V_{\mathbb{F}}}$ becomes an isomorphism.

Proof. The statement is proved in several steps. First consider any $(A, \mathfrak{m}_A)$ in $\mathfrak{AR}_{W(\mathbb{F})}$, and fix $\xi = [V_A] \in D^{\mathrm{fl}}_{V_{\mathbb{F}}}(A)$. Let M_A be the corresponding object in $D_{M_{\mathbb{F}}}(A)$. Regarding ξ as a functor on $\mathfrak{Aug}_{W(\mathbb{F})}$ (represented by the ring A equipped with its radical), we can form the product

$$D_{\mathfrak{S},M_{\mathbb{F}},\xi} = \xi \times_{D^{\mathrm{fl}}_{V_{\mathbb{F}}}} D_{\mathfrak{S},M_{\mathbb{F}}} \xrightarrow{\sim} \xi \times_{D_{M_{\mathbb{F}}}} D_{\mathfrak{S},M_{\mathbb{F}}}.$$

(The fact that this product - as defined - has reasonable properties depends crucially on the assumption that $V_{\mathbb{F}}$ has only scalar endomorphisms. Without this assumption one needs to define it as a product of groupoids.)

Fix an isomorphism of $(A \otimes_{\mathbb{Z}_p} W)((u))$-modules $M_A \xrightarrow{\sim} (A \otimes_{\mathbb{Z}_p} W)((u))^d$. Recall that the affine Grassmannian corresponding to M_A is an Ind-projective A-scheme which to an A-algebra B associates the collection of $(B \otimes W)[\![u]\!]$-lattices in $(B \otimes_{\mathbb{Z}_p} W)((u))^d$. (A $(B \otimes W)[\![u]\!]$-*lattice* is a finite projective $(B \otimes W)[\![u]\!]$-submodule of rank d, which spans the $(B \otimes_{\mathbb{Z}_p} W)((u))$-module $(B \otimes_{\mathbb{Z}_p} W)((u))^d$.) The first step in proving the proposition is to show that $D_{\mathfrak{S},M_{\mathbb{F}},\xi}$ is representable by a closed Ind-subscheme of the affine Grassmannian. This is not surprising since if $(A, \mathfrak{m}_A) \to (B, I)$ is a morphism in $\mathfrak{Aug}_{W(\mathbb{F})}$ and $\mathfrak{M}_B$ in $D_{\mathfrak{S},M_{\mathbb{F}}}(B, I)$ maps to $M_A \otimes_A B$ under the vertical functor in (2.3), then $\mathfrak{M}_B$ can be regarded as such a lattice. Next one shows that $D_{\mathfrak{S},M_{\mathbb{F}},\xi}$ is actually representable by a closed subscheme, the key point being that multiplying a φ-stable lattice by u multiplies φ by u^{p-1}. Since for any module $\mathfrak{M}_B$ in $D_{\mathfrak{S},M_{\mathbb{F}}}(B, I)$ the cokernel of $\varphi^*(\mathfrak{M}_B) \to \mathfrak{M}_B$ is killed by $E(u)$, if $\mathfrak{M}_B$ and $\mathfrak{M}'_B$ are two lattices arising in this way then one finds that $\mathfrak{M}_B \subset u^{-i}\mathfrak{M}'_B$ for some $i \leqslant \frac{ek}{p-1}$, where k is the least integer such that $p^k \cdot A = 0$.

Applying the above discussion to finite length quotients of $R^{\mathrm{fl}}_{V_{\mathbb{F}}}$ and passing to the limit gives rise to a proper map of formal schemes $\widehat{\mathscr{GR}}_{V_{\mathbb{F}}} \to \operatorname{Spf} R^{\mathrm{fl}}_{V_{\mathbb{F}}}$. That it arises from a projective map of schemes follows from formal GAGA, and the fact that the affine Grassmannian is equipped with a line bundle whose restriction to any closed subscheme of finite type is very ample.

The claims regarding smoothness can be proved by giving a functor of points style description of the generic fibres of $\mathscr{GR}_{V_{\mathbb{F}}}$ and $R^{\mathrm{fl}}_{V_{\mathbb{F}}}$. (At least for points with values in Artin rings). In the case of $R^{\mathrm{fl}}_{V_{\mathbb{F}}}$ this description is in terms of weakly admissible modules. Its availability is based on Breuil's results [Br 2] which, in particular, imply that a weakly admissible module with Hodge-Tate weights equal to $(0,1)$ arises from a p-divisible group (see also [Ki 2]).

Finally, the proof that $\Theta_{V_{\mathbb{F}}}$ becomes an isomorphism after inverting p rests on two facts: that the category $(\mathrm{Mod\ FI}/\mathfrak{S})_{\mathbb{Z}_p}$ is equivalent to the category of p-divisible groups over $\mathcal{O}_K$ (this is proved in [Ki 1, 2.2.22] using the results of [Br 2]; another proof is given in [Ki 2]) and on Tate's theorem that a p-divisible group is determined by its generic fibre. In fact, given that the morphism $\Theta_{V_{\mathbb{F}}}$ is intended to classify finite flat models of Galois representations, the fact that the generic fibre of $\Theta_{V_{\mathbb{F}}}$ is an isomorphism can be viewed as a geometric incarnation of Tate's theorem. Actually, the proof given in [Ki 1] also uses the smoothness discussed above, but this can probably be avoided. □

(2.6) We now specialize to the case $\dim V_{\mathbb{F}} = 2$. For (A, I) in $\mathfrak{Aug}_{W(\mathbb{F})}$, we define $D^{0,1}_{\mathfrak{S},M_{\mathbb{F}}}(A,I) \subset D_{\mathfrak{S},M_{\mathbb{F}}}(A,I)$ as the subset of modules $\mathfrak{M}_A$ such that the $A \otimes_{\mathbb{Z}_p} \mathcal{O}_K$-submodule $\varphi^*(\mathfrak{M}_A)/E(u)\mathfrak{M}_A \subset \mathfrak{M}_A/E(u)\mathfrak{M}_A$ is Lagrangian. When $\mathfrak{M}_A/E(u)\mathfrak{M}_A$ is free over $A \otimes_{\mathbb{Z}_p} \mathcal{O}_K$, this means that under some (and hence any) non-degenerate symplectic pairing on $\mathfrak{M}_A/E(u)\mathfrak{M}_A$, the submodule $\varphi^*(\mathfrak{M}_A)/E(u)\mathfrak{M}_A$ is equal to its own annihilator. (Here we are identifying $\varphi^*(\mathfrak{M}_A)$ with a submodule of $\mathfrak{M}_A$ via $1 \otimes \varphi$.) In general we require that this condition hold locally on $\operatorname{Spec} A \otimes_{\mathbb{Z}_p} \mathcal{O}_K$.

For A in $\mathfrak{AR}_{W(\mathbb{F})}$ we denote by $D^{0,1}_{V_{\mathbb{F}}}(A) \subset D^{\mathrm{fl}}_{V_{\mathbb{F}}}(A)$ the subset consisting of deformations V_A such that the action of inertia on $\det V_A$ is given by the cyclotomic character. As usual we can extend this subfunctor to $\mathfrak{Aug}_{W(\mathbb{F})}$.

We have the following result describing the relationship between these two subfunctors [Ki 1, 2.4.8].

Proposition (2.7). *$\Theta_{V_{\mathbb{F}}}$ induces a morphism $\Theta^{0,1}_{V_{\mathbb{F}}} : D^{0,1}_{\mathfrak{S},M_{\mathbb{F}}} \to D^{0,1}_{V_{\mathbb{F}}}$, which is represented by a projective morphism*

$$\Theta^{0,1}_{V_{\mathbb{F}}} : \mathscr{GR}^{0,1}_{V_{\mathbb{F}}} \to \operatorname{Spec} R^{0,1}_{V_{\mathbb{F}}}. \tag{2.7.1}$$

After inverting p, $\Theta^{0,1}_{V_{\mathbb{F}}}$ becomes an isomorphism of formally smooth $W(\mathbb{F})[1/p]$-schemes of pure dimension $[K:\mathbb{Q}_p]+1$. The complete local rings at closed points on $\mathcal{GR}^{0,1}_{V_{\mathbb{F}}}$ are isomorphic to those on Hilbert modular schemes. In particular, $\mathcal{GR}^{0,1}_{V_{\mathbb{F}}}$ is a local complete intersection over $W(\mathbb{F})$, and the reduction $\mathcal{GR}^{0,1}_{V_{\mathbb{F}}} \otimes_{\mathbb{Z}} \mathbb{Z}/p\mathbb{Z}$ is normal.

Proof. The statements in the first paragraph are not hard to deduce from (2.5). To see the claim on dimensions one describes the tangent space at a closed point x of $\operatorname{Spec} R^{0,1}_{V_{\mathbb{F}}}[1/p]$ in terms of crystalline self extensions of the corresponding Galois representation V_x. The dimension of the corresponding Galois cohomology group can be computed using formulas found in [Ne, 1.24].

Consider some $\mathbb{F}$-valued point $y \in \mathcal{GR}^{0,1}_{V_{\mathbb{F}}}(\mathbb{F})$. This corresponds to a lattice $\mathfrak{M}_{\mathbb{F}} \subset M_{\mathbb{F}}$. To describe the complete local ring at y one studies the problem of deforming $\mathfrak{M}_{\mathbb{F}}$ to an object $\mathfrak{M}_A$ in $(\text{Mod FI}/\mathfrak{S})_{\mathfrak{M}_A}$, for A in $\mathfrak{AR}_{W(\mathbb{F})}$, subject to the condition that $\varphi^*(\mathfrak{M}_A)/E(u)\mathfrak{M}_A$ is Lagrangian. This problem can be broken into two parts: First one specifies a Lagrangian submodule $L_A \subset \mathfrak{M}_A/E(u)\mathfrak{M}_A$, and then a surjection $\varphi^*(\mathfrak{M}_A) \to \tilde{L}_A$, where $\tilde{L}_A$ denotes the preimage of L_A in $\mathfrak{M}_A$. The problem of specifying L_A is visibly related to the moduli problem describing integral models of Hilbert modular varieties [DP]. The extra data of the map $\varphi^*(\mathfrak{M}_A) \to \tilde{L}_A$ introduces only smooth parameters because $\varphi^*(\mathfrak{M}_A)$ is a free $\mathfrak{S}_A$-module. To see this one deduces the lifting criterion for smoothness from the lifting property for maps with source a free module.

The final two statements follow from the known local structure of Hilbert modular schemes [DP]. $\square$

(2.8) If $E/W(\mathbb{F})[1/p]$ is a finite extension, and $x : R_{V_{\mathbb{F}}} \to E$ a map of $W(\mathbb{F})$-algebras, we denote by V_x the two dimensional E-representation of G_K obtained by specializing the universal representation over $R_{V_{\mathbb{F}}}$ by x. Recall that a two dimensional, Barsotti-Tate representation is called *ordinary* if its restriction to inertia is an extension of E by $E(1)$.

Theorem (2.9). *Let $E/W(\mathbb{F})[1/p]$ be a finite extension and consider E-valued points $x_1, x_2 : R^{0,1}_{V_{\mathbb{F}}} \to E$. If the images of x_1 and x_2 in* $\operatorname{Spec} R^{0,1}_{V_{\mathbb{F}}}[1/p]$ *lie in the same connected component then V_{x_1} and V_{x_2} are either both ordinary or both non-ordinary. The converse statement holds under any of the following conditions*

1. *V_{x_1} and V_{x_2} are ordinary.*
2. *The residue field of K is equal to $\mathbb{F}_p$.*
3. *The ramification index $e(K/\mathbb{Q}_p)$ is at most $p-1$.*

Proof. For a space X, denote by $H_0(X)$ its set of connected components.

Using (2.7) we have

$$H_0(\operatorname{Spec} R^{0,1}_{V_{\mathbb{F}}}[1/p]) \xrightarrow{\sim} H_0(\mathscr{G}\mathscr{R}^{0,1}_{V_{\mathbb{F}}} \otimes_{\mathbb{Z}_p} \mathbb{Q}_p) \xrightarrow{\sim} H_0(\mathscr{G}\mathscr{R}^{0,1}_{V_{\mathbb{F}}} \otimes_{\mathbb{Z}_p} \mathbb{Z}/p\mathbb{Z})$$

where the second isomorphism is easily deduced from the fact that $\mathscr{G}\mathscr{R}^{0,1}_{V_{\mathbb{F}}} \otimes_{\mathbb{Z}_p} \mathbb{Z}/p\mathbb{Z}$ is reduced, and $\mathscr{G}\mathscr{R}^{0,1}_{V_{\mathbb{F}}}$ is $\mathbb{Z}_p$-flat and proper over $\operatorname{Spec} R^{0,1}_{V_{\mathbb{F}}}$. By the theorem on formal functions, the final term is also equal to $H_0(\mathscr{G}\mathscr{R}^{0,1}_{V_{\mathbb{F}},0})$ where $\mathscr{G}\mathscr{R}^{0,1}_{V_{\mathbb{F}},0}$ denotes the fibre of $\Theta^{0,1}_{V_{\mathbb{F}}}$ over the closed point of $\operatorname{Spec} R^{0,1}_{V_{\mathbb{F}}}$.

The first statement in the theorem can be proved by considering the quotient of $R^{0,1}_{V_{\mathbb{F}}}$ corresponding to ordinary deformations, and showing that after inverting p, the corresponding inclusion map on spectra becomes a local isomorphism. One can also show the corresponding statement for the scheme $\mathscr{G}\mathscr{R}^{0,1}_{V_{\mathbb{F}}}$, by defining the notion of an ordinary module in $D_{\mathfrak{S},M_{\mathbb{F}}}(A, I)$. The corresponding subscheme of $\mathscr{G}\mathscr{R}^{0,1}_{V_{\mathbb{F}}}$ is a connected component, even without inverting p.

To show the converse one has to compute $H_0(\mathscr{G}\mathscr{R}^{0,1}_{V_{\mathbb{F}},0})$, and show that it consists of at most two components. More precisely, one has to show that two points which are both either ordinary or non-ordinary lie on the same component. In the ordinary case one finds that, if $\operatorname{End}_{G_K} V_{\mathbb{F}} = \mathbb{F}$, then the ordinary subscheme $\mathscr{G}\mathscr{R}^{\mathrm{ord}}_{V_{\mathbb{F}},0} \subset \mathscr{G}\mathscr{R}^{0,1}_{V_{\mathbb{F}},0}$ consists of at most a single point. If one works without this assumption, then $\mathscr{G}\mathscr{R}^{\mathrm{ord}}_{V_{\mathbb{F}},0}$ can also consist of two points (in certain cases where $V_{\mathbb{F}}$ is a direct sum of two distinct characters) or it may be equal to a projective line (when the action on $V_{\mathbb{F}}$ is scalar). In applications one can always reduce to the last situation by making a base change, so the case with two points does not cause a problem.

In the case that the residue field of K is $\mathbb{F}_p$, one shows that the non-ordinary subset of $\mathscr{G}\mathscr{R}^{0,1}_{V_{\mathbb{F}},0}$ is connected by constructing chains of rational curves between any two points. The connectedness result should be true without the condition on the residue field, but we are unable to prove it.

It is not hard to show that closed points on the non-ordinary part of $\mathscr{G}\mathscr{R}^{0,1}_{V_{\mathbb{F}},0}$ correspond to bi-connected (i.e connected with connected dual) finite flat models of $V_{\mathbb{F}} \otimes_{\mathbb{F}} \mathbb{F}'$, for $\mathbb{F}'$ a finite extension of $\mathbb{F}$. If the ramification degree $e(K/\mathbb{Q}_p)$ is at most $p-1$ then $V_{\mathbb{F}}$ has a unique bi-connected model by results of Raynaud [Ra, 3.3.5, 3.3.6], so in this case the non-ordinary subset consists of a single point. $\square$

(2.10) Now returning to the situation of §1, suppose that $k = 2$, and that $v|p$, and $V_{\mathbb{F}}|_{G_{F_v}}$ has only scalar endomorphisms. Taking $K = F_v$ above, the ordinary (resp. non-ordinary) points on $\operatorname{Spec} R^{0,1}_{V_{\mathbb{F}}}[1/p]$ are a union of connected components by (2.9). If $T_v \notin \mathfrak{m}$ (resp. $T_v \in \mathfrak{m}$), we denote by $\bar{R}_v$ the quotient of $R^{0,1}_{V_{\mathbb{F}}}$ corresponding to the closure of the ordinary (resp. non-ordinary) components in $\operatorname{Spec} R^{0,1}_{V_{\mathbb{F}}}[1/p]$.

Let $R^{0,1,\psi}_{V_{\mathbb{F}}}$ denote the quotient of $R^{0,1}_{V_{\mathbb{F}},\mathcal{O}} = R^{0,1}_{V_{\mathbb{F}}} \otimes_{W(\mathbb{F})} \mathcal{O}$ corresponding to deformations with determinant $\psi\chi$. Note that since $k = 2$, ψ has finite order. We have $R^{0,1}_{V_{\mathbb{F}},\mathcal{O}} \xrightarrow{\sim} R^{0,1,\psi}_{V_{\mathbb{F}}}[\![X]\!]$, the isomorphism depending on a choice of topological generator for the Galois group of the maximal unramified extension of F_v. Setting $\bar{R}^{\psi}_v = \bar{R}_v \otimes_{R^{0,1}_{V_{\mathbb{F}}}} R^{0,1,\psi}_{V_{\mathbb{F}},\mathcal{O}}$, one easily sees that $\bar{R}_v \otimes_{W(\mathbb{F})} \mathcal{O} \xrightarrow{\sim} \bar{R}^{\psi}_v[\![X]\!]$. Hence by (2.9) and (2.7), $\bar{R}^{\psi}_v[1/p]$ is formally smooth over $\mathcal{O}[1/p]$ of dimension $[F_v : \mathbb{Q}_p]$, and it is a domain either if $T_v \notin \mathfrak{m}$ or F_v has residue field $\mathbb{F}_p$. Finally we denote by $\bar{R}^{\square,\psi}_v$ the ring representing the functor which to an Artinian $\mathcal{O}$-algebra A with residue field $\mathbb{F}$, assigns the set of isomorphism classes of pairs (V_A, β_v) where V_A is a deformation of $V_{\mathbb{F}}$ isomorphic to one obtained from a morphism $\bar{R}^{\psi}_v \to A$, and β_v is an A-basis for V_A lifting some fixed basis of $V_{\mathbb{F}}$.

It is clear that $\bar{R}^{\square,\psi}_v$ satisfies the conditions (2) and (3) of (1.4). It remains to check the condition (1). The fact that the action of $R_{V_{\mathbb{F}}}$ on M_∞ factors through $R^{0,1,\psi}_{V_{\mathbb{F}}}$ follows from the construction of the Galois representations associated to Hilbert modular forms [Ca], [Ta 2], [Ta 3]; they are limits of Galois representations found in abelian varieties with good reduction at primes dividing p (because U is assumed maximal at p). That this action actually factors through $\bar{R}_v$ follows from the fact that these representations satisfy a crystalline version of the Eichler-Shimura relation. Namely, if V_f is the representation associated to a Hilbert eigenform f, then the action of T_v on f is given by the trace of Frobenius on $D_{\mathrm{cris}}(V_f)$ [Ki 1, 3.4.2]. Hence if f corresponds to a maximal ideal $\mathfrak{m}$ as in §1, then $T_v \in \mathfrak{m}$ if and only if V_f is non-ordinary at v.

Finally we remark that if $V_{\mathbb{F}}|_{G_{F_v}}$ has non-scalar endomorphisms one can still carry out a variant of the theory of this section and define a ring $\bar{R}^{\square,\psi}_v$. It is a domain except in certain cases where $T_v \notin \mathfrak{m}$ and $V_{\mathbb{F}}|_{G_{F_v}}$ is a sum of two distinct characters. In applications this does not cause any problems, since after a suitable base change we can assume that the action of G_{F_v} is trivial. In this case $\bar{R}^{\square,\psi}_v$ is a domain; the scheme $\mathscr{GR}^{\mathrm{ord}}_{V_{\mathbb{F}},0}$ of (2.9) is a projective line.

Using (1.4), results on raising and lowering the level of quaternionic forms in the style of [DT] and [SW], as well as some base change arguments, one deduces the following modularity lifting theorem. To state it, we fix embeddings of $\bar{\mathbb{Q}} = \bar{F}$ into $\bar{\mathbb{Q}}_p$ and $\mathbb{C}$. If f is a Hilbert modular eigenform, then we obtain a corresponding representation $\rho_f : G_{F,S} \to \mathrm{GL}_2(\mathcal{O})$ where $\mathcal{O}$ is the ring of integers of some sufficiently large extension E of $\mathbb{Q}_p$. In the following we will always assume that $\mathcal{O}$ is large enough. Given any such representation ρ we will denote by $\bar{\rho}$ the corresponding representation into $\mathrm{GL}_2(\mathbb{F})$, where $\mathbb{F}$ is the residue field of $\mathcal{O}$.

Theorem (2.11). *Let $\rho : G_{F,S} \to \mathrm{GL}_2(\mathcal{O})$ be a continuous representation. Suppose that*

(1) *ρ is Barsotti-Tate at each $v|p$ and $\det\rho\cdot\chi^{-1}$ has finite order.*
(2) *There exists a Hilbert modular eigenform f over F of parallel weight 2 such that $\bar{\rho}\sim\bar{\rho}_f$, and for each prime $v|p$ of F, the representation of $\mathrm{GL}_2(\mathbb{A}_F)$ generated by f is spherical at v, and the eigenvalue of T_v acting on f is a p-adic unit if and only if $\rho|_{G_{F_v}}$ is ordinary.*
(3) *If $v|p$ and $\rho|_{G_{F_v}}$ is non-ordinary then either the residue field of v is $\mathbb{F}_p$ or the ramification index $e(F_v/\mathbb{Q}_p)$ is at most $p-1$.*
(4) *$\bar{\rho}|_{F(\zeta_p)}$ is absolutely irreducible, and $[F(\zeta_p):F]>2$ if $p=5$.*

Then $\rho\sim\rho_g$ for some Hilbert modular eigenform g over F.

(2.12) By base change, the above also implies a modularity lifting theorem for potentially Barsotti-Tate representations. The condition that the eigenvalue of T_v is a unit if and only if $\rho|_{G_{F_v}}$ is ordinary is equivalent to asking that ρ is ordinary at v if and only if ρ_f is. This condition is sometimes inconvenient in applications. It is partially ameliorated by the following result.

Corollary (2.13). *Let F be a totally real field in which p is totally split, and $\rho : G_{F,S}\to\mathrm{GL}_2(\mathcal{O})$ a continuous representation. Suppose that*

(1) *ρ is potentially Barsotti-Tate at each $v|p$ and $\det\rho\cdot\chi^{-1}$ has finite order.*
(2) *There exists a Hilbert modular form over F of parallel weight 2 such that $\bar{\rho}\sim\bar{\rho}_f$.*
(3) *$\bar{\rho}|_{F(\zeta_p)}$ is absolutely irreducible.*

Then $\rho\sim\rho_g$ for some Hilbert modular eigenform g over F.

Proof. Let $\mathcal{S}$ denote the set of primes $v|p$ of F such that $\rho|_{G_{F_v}}$ is not potentially ordinary (i.e. is absolutely irreducible), and $\mathcal{S}'$ the set of $v|p$ such that $\rho|_{G_{F_v}}$ is potentially ordinary. By (2.14) below we may choose f so that at each $v\in\mathcal{S}'$, ρ_f becomes ordinary and Barsotti-Tate over an extension F_v' of F_v of degree $\leqslant p-1$. Thus after replacing F by a finite solvable extension we may assume that for each $v\in\mathcal{S}'$, ρ_f is ordinary and Barsotti-Tate. We, of course, no longer assume that p is split in F, and this hypothesis is needed only to apply (2.14).

Now an argument using the mod p reductions of irreducible representations of $\mathrm{GL}_2(\mathbb{Z}_p)$ as in [Ki 1, 3.5.7, 3.1.6] shows that there exists a Hilbert modular form f' over F of parallel weight 2, such that $\bar{\rho}_{f'}\sim\bar{\rho}_f$, $\rho_{f'}$ is ordinary and Barsotti-Tate at each $v\in\mathcal{S}'$, and such that the corresponding $\mathrm{GL}_2(\mathbb{A}_F)$-representations $\pi_{f'}$ is cuspidal at each $v\in\mathcal{S}$. After again replacing F by a solvable extension, we find that $\rho_{f'}$ is Barsotti-Tate at each $v|p$, and ordinary if and only if ρ_f is ordinary. Thus the corollary follows from (2.11), because we can choose the finite extensions of F we have made so that the conditions (2.11)(3) and (2.11)(4) are satisfied. □

Lemma (2.14). *Let F be a totally real field in which p is totally split, and $\mathcal{S}'$ a collection of places of F dividing p. Let f be a Hilbert modular cusp form over F of parallel weight 2, and suppose that $\bar\rho_f$ is absolutely irreducible, and that for $v \in \mathcal{S}'$, $\bar\rho_f|_{G_{F_v}}$ is the reduction of a potentially ordinary, potentially Barsotti-Tate representation of G_{F_v}.*

Then there exists a Hilbert modular form f' over F of parallel weight 2 and such that

(1) $\bar\rho_{f'} \sim \bar\rho_f$.

(2) $\rho_{f'}|_{G_{F_v}}$ *becomes ordinary and Barsotti-Tate over an extension of degree $\leqslant p-1$, for each $v \in \mathcal{S}'$.*

Proof. Choose f' satisfying (1). Let $\pi_{f'}$ denote the representation of $\mathrm{GL}_2(\mathbb{A}_F)$ generated by f', and let $\mathcal{S}'' \subset \mathcal{S}'$ denote the subset of places at which $\pi_{f'}$ is a twist by a character of a spherical representation. We may assume that f' has been chosen so that $\mathcal{S}''$ has largest possible order.

An argument similar to that in [Ki 1, 3.5.7, 3.1.6], using [CDT, 3.1.1], shows that one may choose f' such that for $v \in \mathcal{S}'$ the local factor at v of $\pi_{f'}$ has a corresponding Weil group representation which is tamely ramified and abelian (so that the inertia acts through a quotient of order $p-1$), and is scalar for $v \in \mathcal{S}''$. The point is that if σ is an irreducible mod p representation of $\mathrm{GL}_2(\mathbb{F}_p)$ then σ appears as a subquotient of the mod p reduction of one of the following:

• The character $\theta \circ \det$, where $\theta : \mathbb{F}_p^\times \to \mathcal{O}^\times$.

• The subspace of $\mathrm{Ind}_B^{\mathrm{GL}_2(\mathbb{F}_p)}\theta$ consisting of functions with average value 0, where B denotes the subgroup of upper triangular matrices, and θ is as above.

• The representations $\mathrm{Ind}_B^{\mathrm{GL}_2(\mathbb{F}_p)}\theta$, where $\theta : B \to E^\times$ is a character given by $\theta\begin{pmatrix} a & b \\ 0 & d\end{pmatrix} = \theta_1(a)\theta_2(d)$ with $\theta_1 \neq \theta_2$.

We claim that $\pi_{f'}$ is not special at any $v \in \mathcal{S}'$. To see this, suppose that $\pi_{f'}$ is special at some such v. Then $\bar\rho_{f'} \sim \bar\rho_f$ has the form $\begin{pmatrix} \epsilon\chi & * \\ 0 & \epsilon\end{pmatrix}$ on inertia at v, where χ is the mod p cyclotomic character. Our assumption that $\bar\rho_f|_{G_{F_v}}$ is the reduction of a potentially ordinary, potentially Barsotti-Tate representation (which necessarily becomes Barsotti-Tate over an abelian extension of $F_v = \mathbb{Q}_p$) implies that $*$ is peu ramifié, because a très ramifié cocycle cannot become peu ramifié over an abelian extension of $\mathbb{Q}_p$. Then Jarvis' generalization of Mazur's principle

[Ja, Thm. 6.1] applied to a suitable twist of f', implies that there exists a Hilbert modular form f'' of parallel weight 2 such that $\bar{\rho}_{f'} \sim \bar{\rho}_{f''}$ and at each place of $\{v\} \cup \mathcal{S}''$ $\pi_{f''}$ has a spherical twist. This contradicts the maximality of $|\mathcal{S}''|$.

It follows that at each $v \in \mathcal{S}'$, $\rho_{f'}$ becomes Barsotti-Tate over $F'_v = F_v(\zeta_p)$. We claim that $\rho_{f'}|_{G_{F'_v}}$ is ordinary. To see this we use the fact that $\rho_{f'}|_{G_{F'_v}}$ arises as the generic fibre of a p-divisible group $\mathcal{G}$. Then either $\mathcal{G}$ is ordinary, or connected with connected dual. In the latter case $\mathcal{G}[p]$ would be the unique finite flat model of $\bar{\rho}_{f'}|_{G_{F'_v}}$ [Ra, 3.3.5, 3.3.6]. However, since $\bar{\rho}_f|_{G_{F'_v}}$ is the reduction of a potentially ordinary, potentially Barsotti-Tate representation, its restriction to inertia has the form $\left(\begin{smallmatrix} \chi & * \\ 0 & 1 \end{smallmatrix}\right)$ with $*$ peu ramifié, as before. (Of course $\chi|_{G_{F'_v}}$ is actually trivial). Hence $\bar{\rho}_f|_{G_{F'_v}}$ has an ordinary finite flat model. It follows that $\mathcal{G}$ is ordinary, and so is $\rho_{f'}|_{G_{F'_v}}$. $\square$

§3 Crystalline representations in intermediate weight

(3.1) We continue to use the notation of the previous section, but we assume that $K = K_0$. We briefly recall some facts from the theory of (φ, Γ)-modules [Fo].

As in the previous section we denote by $\mathcal{O}_{\mathcal{E}}$ the p-adic completion of $W[\![u]\!][1/u]$, however we now equip this ring with a continuous Frobenius endomorphism given by $\varphi(u) = (1+u)^p - 1$, and the usual Frobenius on W. The subring $\mathfrak{S}$ is again stable by φ.

We also equip $\mathcal{O}_{\mathcal{E}}$ with a continuous action of $\Gamma = \mathrm{Gal}(\mathbb{Q}_p(\mu_{p^\infty})/\mathbb{Q}_p)$ by setting $\gamma(u) = (1+u)^{\chi(\gamma)} - 1$ where $\chi : \Gamma \xrightarrow{\sim} \mathbb{Z}_p^\times$ denotes the cyclotomic character. The actions of φ and Γ on $\mathcal{O}_{\mathcal{E}}$ are easily seen to commute, and $\mathfrak{S}$ is Γ-stable.

Recall that an *étale* (φ, Γ)*-module* is an $\mathcal{O}_{\mathcal{E}}$-module M of finite length, equipped with commuting semi-linear actions of φ and Γ, such that the $\mathcal{O}_{\mathcal{E}}$-linear map $\varphi^*(M) \stackrel{1\otimes\varphi}{\rightarrow} M$ is an isomorphism. The basic result of Fontaine asserts that the category of étale (φ, Γ)-modules is equivalent to the category of continuous representations of G_K on finite length $\mathbb{Z}_p$-modules. Given an étale (φ, Γ)-module M, we denote by $V_{\mathcal{O}_{\mathcal{E}}}(M)$ the corresponding G_K-representation. Conversely given a G_K-representation V we denote by $D_{\mathcal{O}_{\mathcal{E}}}(V)$ the corresponding étale (φ, Γ)-module.

There is an analogous result for (φ, Γ)-modules which are finite free over $\mathcal{O}_{\mathcal{E}}$. They correspond to continuous representations of G_K on finite free $\mathbb{Z}_p$-modules. We will again denote by $V_{\mathcal{O}_{\mathcal{E}}}$ and $D_{\mathcal{O}_{\mathcal{E}}}$ the functors which give rise to this equivalence of categories.

(3.2) Our task in this section is to construct and analyze certain crystalline deformation rings. The existence of these rings follows immediately from the main result of [Be], however it will be useful to give another construction (again using the results of [Be]) which provides some extra crucial information about these rings. To this end we make the following definition:

Fix a non-negative integer k. Suppose that A is a finite local $\mathbb{Z}_p$-algebra, and that V_A is a finite free A-module equipped with a continuous action of G_K. An argument as in [Ki 1, 1.2.7(4)] shows that $M_A = D_{\mathcal{O}_\mathcal{E}}(V_A)$ is a free $\mathcal{O}_{\mathcal{E},A} = \mathcal{O}_\mathcal{E} \otimes_{\mathbb{Z}_p} A$-module. Now let B be any A-algebra. We extend the action of φ and Γ to $M_B := M_A \otimes_A B$ by B-linearity. A *Wach lattice* with weights in $[0, k]$ is a finite projective $\mathfrak{S}_B$-submodule $\mathfrak{M}_B \subset M_B$ such that

(1) $\mathfrak{M}_B$ is stable under the action of φ and Γ.
(2) $\mathfrak{M}_B$ spans M_B as a $\mathcal{O}_\mathcal{E} \otimes_{\mathbb{Z}_p} B$-module.
(3) The cokernel of the induced map $\varphi^*(\mathfrak{M}_B) \to \mathfrak{M}_B$ is killed by q^k, where $q(u) = \frac{(1+u)^p-1}{u}$.
(4) The action of Γ on $\mathfrak{M}_B/u\mathfrak{M}_B$ is trivial.

The motivation for this definition is the following result of Berger [Be, III, 4.2, 4.5] (which depends on the hypothesis that $K = K_0$), where we take $A = B = \mathbb{Z}_p$.

Proposition (3.3). *If V is a finite dimensional $\mathbb{Q}_p$-vector space with a continuous action of G_K, and $T \subset V$ is a G_K-stable lattice, then V is crystalline with Hodge-Tate weights in $[-k, 0]$ if and only if the (φ, Γ)-module $D_{\mathcal{O}_\mathcal{E}}(T)$ contains a Wach lattice with weights in $[0, k]$.*

Moreover in this case, $D_{\mathcal{O}_\mathcal{E}}(T)$ contains a unique Wach lattice $\mathfrak{M}$, and if $D_{\mathrm{cris}}(V)$ denotes the weakly admissible module associated to V, then there exists a natural isomorphism $D_{\mathrm{cris}}(V) \xrightarrow{\sim} \mathfrak{M}/u\mathfrak{M} \otimes_{\mathbb{Z}_p} \mathbb{Q}_p$.

(3.4) Now let $\mathbb{F}$ and $V_\mathbb{F}$ be as in §2, A in $\mathfrak{AR}_{W(\mathbb{F})}$, and V_A a deformation of $V_\mathbb{F}$ to A. We set $M_A = D_{\mathcal{O}_\mathcal{E}}(V_A^*)$, where V_A^* denotes the A-dual of V_A. For an A-algebra B, we set $M_B = M_A \otimes_A B$ as above, and we denote by $W_{V_A,\leqslant k}(B)$ the set of Wach lattices $\mathfrak{M}_B \subset M_B$ with weights in $[0, k]$.

Proposition (3.5). *The functor $W_{V_A,\leqslant k}$ is represented by a projective A-scheme $\mathscr{W}_{V_A,\leqslant k}$. If R is any complete local Noetherian ring with residue field $\mathbb{F}$, and V_R is a deformation of $V_\mathbb{F}$ to R, then there exists a projective R scheme $\mathscr{W}_{V_R,\leqslant k}$ such that*

(1) *If A is in $\mathfrak{AR}_{W(\mathbb{F})}$ and $R \to A$ is a map inducing the identity on residue fields, then there is a canonical isomorphism*

$$\mathscr{W}_{V_R,\leqslant k} \otimes_R A \xrightarrow{\sim} \mathscr{W}_{V_A,\leqslant k}$$

where $V_A = V_R \otimes_R A$.

(2) *The map* $\Theta_R : \mathscr{W}_{V_R,\leqslant k} \to \operatorname{Spec} R$ *becomes a closed embedding after inverting* p. *If* A *is a finite* $W(\mathbb{F})[1/p]$*-algebra, then a map* $R \to A$ *factors through the scheme theoretic image of* Θ_R *if and only if* $V_A = V_R \otimes_R A$ *is a crystalline representation with Hodge-Tate weights in* $[0, k]$.

Proof. The proof of the first claim is completely analogous to that of [Ki 1, 2.1.7]. The existence of $\mathscr{W}_{V_R,\leqslant k}$ satisfying (1) follows using formal GAGA as in [Ki 1, 2.1.10].

The properties in (2) can be deduced from Berger's theorem (3.3): The fact that Θ_R is a closed embedding follows from the uniqueness of the Wach lattice, while the description of the scheme theoretic image follows (after a little work) from the criterion for the existence of a Wach lattice. (cf. [Ki 3, §1]). □

(3.6) As in §2 we fix a basis of $V_{\mathbb{F}}$. We denote by $R^{\square}_{V_{\mathbb{F}}}$ the universal framed deformation ring of $V_{\mathbb{F}}$ as in (1.2), and if $\operatorname{End}_{\mathbb{F}[G_K]} V_{\mathbb{F}} = \mathbb{F}$, we denote by $R_{V_{\mathbb{F}}}$ the universal deformation ring of $V_{\mathbb{F}}$.

Suppose now that A is a finite local $\mathbb{Q}_p$-algebra, and that V_A is a finite free A-module of rank d equipped with a continuous action of G_K. We suppose that V_A is crystalline with non-negative Hodge-Tate weights. Then $D_{\mathrm{cris}}(V_A^*)$ is a free $K_0 \otimes_{\mathbb{Q}_p} A$-module [Ki 1, 1.3.2] on which $\varphi^{[K_0:\mathbb{Q}_p]}$ acts linearly. We set

$$P_{V_A}(T) = \det{}_{K_0 \otimes_{\mathbb{Q}_p} A}(T - \varphi^{[K_0:\mathbb{Q}_p]} | D_{\mathrm{cris}}(V_A^*)),$$

and for $i = 0, 1, \dots, d-1$, we denote by $c_{V_A,i}$ the coefficient of T^i in $P_{V_A}(T)$.

Corollary (3.7). *Suppose that* $R = R^{\square}_{V_{\mathbb{F}}}$ *or* $R_{V_{\mathbb{F}}}$, *and let* $R_{\leqslant k}$ *denote the scheme theoretic image of the map* Θ_R *of (3.5). Then*

(1) $R_{\leqslant k}[1/p]$ *is formally smooth over* $W(\mathbb{F})$.

(2) *For* $i = 0, 1, \dots d-1$ *there is a unique* $c_i \in R_{\leqslant k}[1/p]$ *such that for any finite* $W(\mathbb{F})[1/p]$*-algebra* A, *and any* $W(\mathbb{F})$*-algebra map* $h : R_{\leqslant k} \to A$, *the representation* V_A *obtained from the universal representation over* R *by specialization by* h *satisfies* $c_{V_A,i} = h(c_i)$. *Moreover the* c_i *are contained in the normalization of* $\operatorname{Im}(R_{\leqslant k} \to R_{\leqslant k}[1/p])$.

Proof. The first part follows using an argument similar to that in [Ki 1, 2.3.9].

For the second part we consider (the algebraization of) the universal family of Wach modules $\mathfrak{M}_{\mathscr{W}_{R,\leqslant k}}$ over $\mathscr{W}_{R,\leqslant k}$ (cf. [Ki 1, 2.4.17]), and we set $D_{\mathscr{W}_{R,\leqslant k}} = \mathfrak{M}_{\mathscr{W}_{R,\leqslant k}} / u\mathfrak{M}_{\mathscr{W}_{R,\leqslant k}}$. Then $D_{\mathscr{W}_{R,\leqslant k}}$ is a vector bundle on

$\mathscr{W}_{R,\leqslant k}$, on which $\varphi^{[K_0:\mathbb{Q}_p]}$ acts linearly, and we may form the characteristic polynomial of the endomorphism:

$$P(T) = \det(T - \varphi^{[K_0:\mathbb{Q}_p]} : D_{\mathscr{W}_{R,\leqslant k}}) \in \Gamma(\mathscr{W}_{R,\leqslant k}, \mathcal{O}_{\mathscr{W}_{R,\leqslant k}}).$$

Since Θ_R induces an isomorphism $\mathscr{W}_{R,\leqslant k} \otimes_{\mathbb{Z}_p} \mathbb{Q}_p \xrightarrow{\sim} \operatorname{Spec} R_{\leqslant k}[1/p]$, we may regard the coefficients c_i of $P(T)$ as elements of $R[1/p]$. Since the map Θ_R is proper the c_i are contained in the normalization of $\operatorname{Im}(R_{\leqslant k} \to R_{\leqslant k}[1/p])$.

That the c_i have the property claimed in (2) follows from the fact that in the situation of (3.3), if V is crystalline, and $T \subset V$ is a lattice corresponding to a Wach module $\mathfrak{M}$, then $D_{\mathrm{cris}}(V)$ is canonically isomorphic to $(\mathfrak{M}/u\mathfrak{M})[1/p]$ [Be, Thm. III.4.4]. □

(3.8) We now specialize to the case $K_0 = \mathbb{Q}_p$ and $d = \dim_{\mathbb{F}} V_{\mathbb{F}} = 2$, and we suppose that $k \geq 2$. Let $E/\mathbb{Q}_p$ be a finite extension, and let $a_p \in E$ satisfy $\operatorname{val}(a_p) > 0$, where val denotes the p-adic valuation normalized so that $\operatorname{val}(p) = 1$. We denote by V_{k,a_p} the two dimensional E-representation of $G_{\mathbb{Q}_p}$ which is crystalline with Hodge-Tate weights $0, k-1$ (that is, as a $\mathbb{Q}_p$-representation V_{k,a_p} has $[E:\mathbb{Q}_p]$ Hodge-Tate weights equal to $k-1$ and $[E:\mathbb{Q}_p]$ equal to 0), with determinant χ^{k-1}, and such that the trace of Frobenius on $D_{\mathrm{cris}}(V_{k,a_p}^*)$ is equal to a_p. Here, as usual, V_{k,a_p}^* denotes the E-dual of V_{k,a_p}.

Let $\mathcal{O}$ denote the ring of integers in E, and choose a $G_{\mathbb{Q}_p}$-stable $\mathcal{O}$-lattice in V_{k,a_p}. Reducing this lattice modulo the radical of $\mathcal{O}$, and taking its semi-simplification, we obtain a two dimensional, semi-simple representation $\bar{V}_{k,a_p}$ of $G_{\mathbb{Q}_p}$, over the residue field $\mathbb{F}_E$ of E.

To describe $\bar{V}_{k,a_p}$, for $i \geq 1$ denote by $\mathbb{Q}_{p^i}$ the unramified extension of $\mathbb{Q}_p$ of degree i, and by $\omega_i : G_{\mathbb{Q}_{p^i}} \to \mathbb{F}_{p^i}^\times$ the character which is the fundamental character of level i on inertia, and sends the arithmetic Frobenius element corresponding to $p \in \mathbb{Q}_{p^i}^\times$ to 1. For $\bar{\mathbb{F}}_E$ an algebraic closure of $\mathbb{F}_E$, and $\lambda \in \bar{\mathbb{F}}_E^\times$ write $\mu_\lambda : G_{\mathbb{Q}_p} \to \bar{\mathbb{F}}_E^\times$ for the unramified character which sends an arithmetic Frobenius to λ. For any x in $\mathcal{O}$, we denote by $\overline{x}$ the image of x in $\mathbb{F}_E$.

The following result was conjectured by Breuil [Br 1, Conj. 1.5], and established by Berger-Breuil [BB 1] using the unitary p-adic representations of $\mathrm{GL}_2(\mathbb{Q}_p)$ associated to two dimensional crystalline representations of $G_{\mathbb{Q}_p}$ [BB 2], as well as some explicit calculations with Wach modules, following those of [BLZ].

Theorem (3.9).

(1) *If* $2 \leqslant k \leqslant p+1$, *then* $\bar{V}_{k,a_p} \sim \operatorname{Ind} \omega_2^{k-1}$, *where* Ind *denotes induction from* $G_{\mathbb{Q}_{p^2}}$ *to* $G_{\mathbb{Q}_p}$.

(2) *Suppose that* $k = p + 2$.
- *If* $\mathrm{val}(a_p) < 1$ *then* $\bar{V}_{k,a_p} \sim \mathrm{Ind}\,\omega_2^2$.
- *If* $\mathrm{val}(a_p) \geq 1$ *then* $\bar{V}_{k,a_p} \sim \begin{pmatrix} \mu_\lambda \omega & 0 \\ 0 & \mu_{\lambda^{-1}}\omega \end{pmatrix}$ *with* $\lambda^2 - \overline{\frac{a_p}{p}}\lambda + 1 = 0$.

(3) *Suppose* $p + 3 \leqslant k \leqslant 2p - 1$
- *If* $\mathrm{val}(a_p) < 1$ *then* $\bar{V}_{k,a_p} \sim \mathrm{Ind}\,\omega_2^{k-p}$.
- *If* $\mathrm{val}(a_p) = 1$ *then* $\bar{V}_{k,a_p} \sim \begin{pmatrix} \mu_\lambda \omega^{k-2} & 0 \\ 0 & \mu_{\lambda^{-1}}\omega \end{pmatrix}$ *with* $\lambda = \overline{\frac{(k-1)a_p}{p}}$.
- *If* $\mathrm{val}(a_p) > 1$ *then* $\bar{V}_{k,a_p} \sim \mathrm{Ind}\,(\omega_2^{k-1})$.

(3.10) Suppose now that $V_{\mathbb{F}}$ is as above, and fix $k \geq 2$ as before. We will assume moreover, that $\mathrm{End}_{\mathbb{F}[G_K]} V_{\mathbb{F}} = \mathbb{F}$, since this is the only case in which we will be able to prove a result strong enough to have consequences for modularity.

Lemma (3.11). *There exist a unique quotient* $R_{V_{\mathbb{F}}}^{0,k-1}$ *of* $R_{V_{\mathbb{F}}}$ *with the following properties*

(1) $R_{V_{\mathbb{F}}}^{0,k-1}$ *is p-torsion free, and* $R_{V_{\mathbb{F}}}^{0,k-1}[1/p]$ *is formally smooth over* $W(\mathbb{F})[1/p]$.

(2) *If* $E/W(\mathbb{F})[1/p]$ *is a finite extension,* $x : R_{V_{\mathbb{F}}} \to E$ *a map of* $W(\mathbb{F})[1/p]$*-algebras, and* V_x *the representation of* G_K *obtained by specializing the universal* $R_{V_{\mathbb{F}}}$*-representation by* x*, then* x *factors through* $R_{V_{\mathbb{F}}}^{0,k-1}$ *if and only if* V_x *is crystalline with Hodge-Tate weights* 0 *and* $k-1$.

Proof. Set $R = R_{V_{\mathbb{F}}}$. Then the ring $R_{\leqslant k-1}$ constructed in (3.7) satisfies (1) by (3.7), and its definition as the scheme theoretic image of Θ_R implies that it satisfies a modified version of (2), where one requires only that the Hodge-Tate weights of V_x are in the interval $[0, k-1]$. To see this one uses the description of the Hodge-Tate weights of a crystalline representation in terms of the associated Wach module [Be, III.2].

It is now easy to see, using either Wach lattices or the main result of [Se], that one can construct a quotient $R_{V_{\mathbb{F}}}^{0,k-1}$ with the required properties by taking the closure in $\mathrm{Spec}\, R_{\leqslant k-1}$ of a suitable union of connected components of $\mathrm{Spec}\, R_{\leqslant k-1}[1/p]$. □

(3.12) Next, we fix $E/\mathbb{Q}_p$ a finite extension as above, and a character $\psi : G_{\mathbb{Q}_p} \to \mathcal{O}^\times$ whose restriction to inertia is χ^{k-2}. We denote by $R_{V_{\mathbb{F}}}^{0,k-1,\psi}$ the quotient of $R_{V_{\mathbb{F}}}^{0,k-1} \otimes_{W(\mathbb{F})} \mathcal{O}$ corresponding to deformations with determinant $\psi\chi$. As in (2.10), one sees using (3.11)(1) that $R_{V_{\mathbb{F}}}^{0,k-1,\psi}[1/p]$ is formally smooth over E.

Now suppose, moreover that $2 \leqslant k \leqslant 2p-1$, and that $V_{\mathbb{F}}$ has semi-simplification equivalent to one of the representations listed in (3.9). In other words, we assume that $V_{\mathbb{F}}^{\mathrm{ss}} \sim \bar{V}_{k,a_p}$ with $\mathrm{val}(a_p) > 0$.

Proposition (3.13). *Suppose that* $\mathrm{End}_{\mathbb{F}[G_K]} V_{\mathbb{F}} = \mathbb{F}$. *If* $R_{V_{\mathbb{F}}}^{0,k-1,\psi}$ *is not the zero ring, then*

(1) *The* E*-algebra* $R_{V_{\mathbb{F}}}^{0,k-1,\psi}[1/p]$ *is formally smooth of dimension* 1.
(2) *If* $2 \leqslant k \leqslant 2p-1$, *then* $R_{V_{\mathbb{F}}}^{0,k-1,\psi}$ *is a domain.*

Proof. The first claim follows as in [Ki 1, 2.3.9, 2.3.11].

For the second claim, we may suppose, after twisting that $\psi = \chi^{k-2}$. Moreover, in the case when $V_{\mathbb{F}}$ is ordinary, the list in (3.9) shows that every closed point of $R_{V_{\mathbb{F}}}^{0,k-1}[1/p]$ corresponds to an ordinary representation. Since $\mathrm{End}_{\mathbb{F}[G_K]} V_{\mathbb{F}} = \mathbb{F}$, it is well known that the ordinary deformation ring with fixed determinant is formally smooth of dimension 2. This can be seen by a calculation with Galois cohomology, although an argument analogous to that given below in the non-ordinary case would also work. Hence we may suppose that $V_{\mathbb{F}}$ is non-ordinary, so that $V_{\mathbb{F}}$ has semi-simplification equivalent to one of the representations given in (3.9).

Now suppose that $\bar{V}_{k,a_p} \sim V_{\mathbb{F}}^{\mathrm{ss}}$. If E' denotes the field of definition of V_{k,a_p}, then we claim that there is a unique map $x : R_{V_{\mathbb{F}}}^{0,k-1.\psi}[1/p] \to E'$ such that the induced E'-representation of $G_{\mathbb{Q}_p}$ is isomorphic to V_{k,a_p}. This is clear if $V_{\mathbb{F}}$ is irreducible.

Suppose that $V_{\mathbb{F}}$ is reducible. The uniqueness of such a point follows from the condition $\mathrm{End}_{\mathbb{F}[G_K]} V_{\mathbb{F}} = \mathbb{F}$. For its existence, suppose that $V_{\mathbb{F}}$ is a (necessarily non-trivial) extension of ψ_1 by ψ_2 for some characters ψ_1, ψ_2. Let $L \subset V_{k,a_p}$ be a $G_{\mathbb{Q}_p}$-stable $\mathcal{O}_{E'}$-lattice whose reduction $\bar{L}$ is an extension of ψ_1 by ψ_2. For different choices of L the resulting extensions all lie on an $\mathbb{F}$-line in $\mathrm{Ext}^1_{G_{\mathbb{Q}_p}}(\psi_1, \psi_2)$. In particular, any two such extensions which are non-trivial are isomorphic as $G_{\mathbb{Q}_p}$-representations. Since V_{k,a_p} is irreducible, we can choose L such that $\bar{L}$ is a non-trivial extension. In most cases considered in (3.9) the space of such extensions is one dimensional, so that $\bar{L} \sim V_{\mathbb{F}}$. The only exception is when $k = p+3$, $\mathrm{val}(a_p) = 1$, and $\lambda = \pm 1$. In this case it follows from [BB 1, Thm. 2.2.5] that $\bar{L}$ is always a peu ramifié extension. Hence, $R_{V_{\mathbb{F}}}^{0,k-1,\psi}$ is non-zero only if $V_{\mathbb{F}}$ is peu ramifié, when we again have $V_{\mathbb{F}} \sim \bar{L}$, and that V_{k,a_p} is a deformation of $V_{\mathbb{F}}$.

Now let $a_p \in R_{V_{\mathbb{F}}}^{0,k-1,\psi}[1/p]$ be the image of the function $c_1 \in R_{\leqslant k-1}[1/p]$. Let $\tilde{R}$ denote the normalization of $R_{V_{\mathbb{F}}}^{0,k-1,\psi}$. Then $a_p \in \tilde{R}$, because a_p is a bounded function on the closed points of $R^{0,k-1,\psi}[1/p]$ (this follows for example using Noether normalization). Alternatively, one could use the same argument as in the proof of (3.7). Moreover none of the representations in (3.9) are the reduction of an ordinary representation, so a_p is in fact strictly bounded by 1, and we obtain a map $\mathcal{O}[\![X]\!] \overset{X \mapsto a_p}{\to} \tilde{R}$.

Suppose that $V_{\mathbb{F}} \sim \mathrm{Ind}\,\omega_2^{k-p}$. By (3.9), we see that also $\frac{p}{a_p} \in R_{V_{\mathbb{F}}}^{0,k-1,\psi}[1/p]$ is a function which is strictly bounded by 1. Hence $\frac{p}{a_p} \in \tilde{R}$, and we obtain a map

$$\mathcal{O}[\![X, Y]\!]/(XY - p) \to \tilde{R}. \tag{3.13.1}$$

By (3.9) and what we have seen above, the map obtained from (3.13.1) by inverting p induces a bijection on closed points, and an isomorphism on residue fields between corresponding closed points. In the appendix Gabber proves that any such map between two normal rings which are finite over a power series ring over $\mathcal{O}$ is an isomorphism, and (2) follows.

When we are in one of the other cases given by (3.9) one finds, by a similar argument, that $\tilde{R}$ is isomorphic to a power series ring, so that (2) again follows. For example, in the second case in (3), we have an isomorphism $\mathcal{O}[\![X]\!] \to \tilde{R}$ given by sending X to $\frac{(k-1)a_p}{p} - [\lambda]$. □

Theorem (3.14). *Let F be a totally real field in which p is totally split, and let $\rho : G_{F,S} \to \mathrm{GL}_2(\mathcal{O})$ be a continuous representation. Suppose that for some $2 \leqslant k \leqslant 2p-1$*

1. *ρ is crystalline with Hodge-Tate weights 0 and $k-1$ at each $v|p$.*
2. *There exists a Hilbert modular form over F of parallel weight k and prime to p level, such that $\bar{\rho} \sim \bar{\rho}_f$.*
3. *$\bar{\rho}|_{F(\zeta_p)}$ is absolutely irreducible, and $\bar{\rho}|_{G_{F_v}}$ has only scalar endomorphisms for each $v|p$.*

Then $\rho \sim \rho_g$ for some Hilbert modular eigenform g over F.

Proof. For $v|p$ we set $\bar{R}_v^{\psi} = R_{V_{\mathbb{F}}}^{0,k-1,\psi} \otimes_{R_{V_{\mathbb{F}}}} R_{V_{\mathbb{F}}}^{\square}$, where $V_{\mathbb{F}}$ denotes the underlying $\mathbb{F}$-vector space of $\bar{\rho}|_{G_{F_v}}$, and $\psi = \det \rho|_{G_{F_v}} \cdot \chi^{-1}$. Then (3.13) implies that $\bar{R}_v^{\psi}$ satisfies (2) and (3) of (1.4). That it satisfies (1) follows from an argument analogous to that made in the Barsotti-Tate case in (2.10).

The theorem can now be deduced by combining (1.4) with some level lowering and raising arguments, exactly as for the proof of (2.11). We should remark that the level lowering arguments in [SW] are carried out only for weight two, although the argument in higher weight should be similar, as the authors remark. Thus, the cautious reader may want to admit the theorem as fully proved only in the case when $F = \mathbb{Q}$, for which level lowering is available in greater generality. □

References

[BB 1] L. Berger, C. Breuil, *Sur la réduction des représentations cristallines de dimension 2 en poids moyens*, preprint (2005).

[BB 2] L. Berger, C. Breuil, *Représentations cristallines irréductibles de* $\mathrm{GL}_2(\mathbb{Q}_p)$, preprint (2004).

[Be] L. Berger, *Limites de représentations cristallines*, Compositio Math. **140** (2004), 1473-1498.

[BLZ] L. Berger, H. Li, H.J. Zhu, *Construction of some families of 2-dimensional crystalline representations*, Math. Ann. **329** (2004), 365-377.

[Br 1] C. Breuil, *Sur quelques représemtations modulaires et p-adiques de* $\mathrm{GL}_2(\mathbb{Q}_p)$ *II*, J.Inst. Jussieu **2** (2003), 23-58.

[Br 2] C. Breuil, *Groupes p-divisibles, groupes finis et modules filtrés*, Ann. of Math. **152** (2000), 489-549.

[Br 3] C. Breuil, *Integral p-adic Hodge Theory*, Algebraic Geometry 2000, Azumino, Adv. Studies in Pure Math. 36, 2002, pp. 51-80.

[Br 4] C. Breuil, *Schemas en groupes et corps des normes (unpublished)* (1998), 13 pages.

[Ca] H. Carayol, *Sur les représentations p-adiques associées aux formes modulaires de Hilbert*, Ann. Scient. de l'E.N.S **19** (1986), 409-468.

[CDT] B. Conrad, F. Diamond, R. Taylor, *Modularity of certain potentially Barsotti-Tate Galois representations*, J. Amer. Math. Soc. **12(2)** (1999), 521-567.

[Di] F. Diamond, *The Taylor-Wiles construction and multiplicity one*, Invent. Math. **128** (1997), 379-391.

[DP] P. Deligne, G. Pappas, *Singularités des espaces de modules de Hilbert, en les caractéristiques divisant le discriminant*, Compositio Math. **90** (1994), 59–79.

[DT] F.Diamond, R. Taylor, *Non-optimal levels of mod l modular representations*, Invent. Math. **115** (1994), 253-269.

[Fo] J-M. Fontaine, *Représentations p-adiques des corps locaux*, Grothendieck Festschrift II, Prog. Math. 87, Birkhauser, pp. 249-309, 1991.

[Ja] F. Jarvis, *Correspondences on Shimura curves and Mazur's principle at p.*, Pacific J. Math **213** (2004), 267-280.

[Ki 1] M. Kisin, *Moduli of finite flat group schemes and modularity*, preprint (2004), 74 pages.

[Ki 2] M. Kisin, *Crystalline representations and F-crystals*, Algebraic geometry and number theory. In honour of Vladimir Drinfeld's 50th birthday, Prog. Math. 253, Birkhäuser, pp. 459-496, 2006.

[Ki 3] M. Kisin, *Potentially semi-stable deformation rings*, J.A.M.S, to appear.

[Ma] B. Mazur, *Deforming Galois representations*, Galois groups over $\mathbb{Q}$ (Berkeley, CA, 1987), Math. Sci. Res. Inst. Publ. 16, Springer, New York-Berlin, pp. 395-437, 1989.

[Ne] J. Nekovář, *On p-adic height pairings*, Séminaire de Théorie de Nombres, Paris 1990-91, Progr. Math. 108, Birkhäuser Boston, pp. 127-202, 1993.

[Ram] R. Ramakrishna, *On a variation of Mazur's deformation functor*, Compositio Math. **87** (1993), 269-286.

[Ra] M. Raynaud, *Schémas en groupes de type* $(p, p, \dots, p)$, Bull. Soc. Math. France **102** (1074), 241-280.

[Se] S. Sen, *The analytic variation of p-adic Hodge Structure*, Ann. of Math. **127 (2)** (1988), 647-661.

[SW] C. Skinner, A. Wiles, *Base change and a problem of Serre*, Duke Math. J **107** (2001), 15-25.

[Ta 1] R. Taylor, *On the meromorphic continuation of degree 2 L-functions*, Documenta Math., Extra Volume: John H. Coates' Sixtieth Birthday, 2006, 729-779.
[Ta 2] R. Taylor, *On Galois representations associated to Hilbert modular forms*, Invent. Math. **98** (1989), 265-280.
[Ta 3] R. Taylor, *On Galois representations associated to Hilbert modular forms. II*, Elliptic curves, modular forms, & Fermat's last theorem (Hong Kong, 1993), Ser. Number Theory, I, Internat. Press, Cambridge, MA, 1995., pp. 185-191.
[TW] R. Taylor, A. Wiles, *Ring-theoretic properties of certain Hecke algebras*, Ann. of Math. **141(3)** (1995), 553-572.
[W] A. Wiles, *Modular elliptic curves and Fermat's last theorem*, Ann. of Math. **141(3)** (1995), 443-551.

Appendix by Ofer Gabber

Consider a complete Noetherian local ring V with residue field k and let $\mathcal{C}$ be the category of V-algebras which are finite over some power series ring $V[\![x_1,\dots,x_n]\!]$. Every $R \in \mathrm{Ob}(\mathcal{C})$ is a finite product of complete Noetherian local rings and the maximal ideals of R are the prime ideals whose residue field is a finite extension of k; in what follows $\mathfrak{m}_R$ denotes the Jacobson radical of R and $\mathrm{Max}(R)$ denotes the set of maximal ideals of R.

It follows from ([Ma] Lemma 1 on p. 247) that for a Noetherian local domain R of dimension > 0 the set $P_1 R := \{P \in \mathrm{Spec}\,(R) \mid \dim(R/P) = 1\}$ is dense in $\mathrm{Spec}\,(R)$. In particular for every $R \in \mathrm{Ob}(\mathcal{C})$, every open subscheme U of $\mathrm{Spec}\,(R) - \mathrm{Max}(R)$ is a Jacobson scheme whose set of closed points is $U \cap P_1 R$.

Lemma 0. *Let $f : R \to R'$ be a map in $\mathcal{C}$, $P' \in \mathrm{Spec}\,(R')$ and $P = f^{-1}(P')$. If $\dim(R'/P') \leqslant 1$ then $\dim(R/P) \leqslant \dim(R'/P')$.*

Proof. We may assume $P' = 0$, $P = 0$. If $\dim(R') = 0$ then R' is a finite field extension of k and R is also. Suppose $\dim(R') = 1$ and R is not a field. By the case $\dim(R'/P') = 0$ of the lemma, f is a local homomorphism of local rings. If $0 \neq x \in \mathfrak{m}_R$ then R'/xR' is of dimension 0. Hence $R'/\mathfrak{m}_R R'$ is finite dimensional over $R/\mathfrak{m}_R$. If $g_1, \dots, g_n \in R'$ lift an $R/\mathfrak{m}_R$-basis of $R'/\mathfrak{m}_R R'$ it is easy to see that $R' = \sum_i R g_i$. Thus $\dim(R) = \dim(R')$ by EGA 0_{IV} (16.1.5). □

Theorem. *Let $f : R \to R'$ be a morphism in $\mathcal{C}$ with R' reduced and R normal (i.e. a finite product of normal domains) and U an open subscheme of $\mathrm{Spec}\,(R) - \mathrm{Max}(R)$ such that U is dense in $\mathrm{Spec}\,(R)$, $U' = \mathrm{Spec}\,(f)^{-1}(U)$ is dense in $\mathrm{Spec}\,(R')$, and*

(*) $\mathrm{Spec}\,(f)$ *induces a bijection*

$$\{\textit{closed points of } U'\} \longrightarrow \{\textit{closed points of } U\}$$

and isomorphisms on the residue fields.
Then f is an isomorphism.

Remark 1: Condition $(*)$ is equivalent to the assertion that for every closed point y of U,

$$(U' \times_U y)_{\rm red} \longrightarrow y \text{ is an isomorphism.}$$

Remark 2: Let $J \subset R$ be an ideal with $J\mathcal{O}_U = \mathcal{O}_U$. Consider the blow-ups

$$X = \mathcal{B}l_J(\operatorname{Spec} R) \text{ and } X' = \mathcal{B}l_{JR'}(\operatorname{Spec} R')$$

and write $X_0 = \coprod_{\mathfrak{m}\in \operatorname{Max} R} X \otimes_R (R/\mathfrak{m})$, and define X_0' similarly. Then X' is a closed subscheme of $X \otimes_R R'$, X_0' is a closed subscheme of $X_0 \otimes_{R/\mathfrak{m}_R} (R'/\mathfrak{m}_{R'})$, and $X_0' \to X_0$ is finite. Note that for every $P \in U \cap P_1 R$ the proper transform of Spec (R/P) in X is Spec of some ring between R/P and its normalization. The morphism

$$\varphi : \coprod_{\substack{\xi \\ \text{closed point of } X_0}} \operatorname{Spec}(\widehat{\mathcal{O}}_{X,\xi}) \longrightarrow X$$

becomes an isomorphism over any of those proper transforms. Now let U denote also the preimage of U in X. By Lemma 0, $\varphi^{-1}(U) \to U$ sends closed points to closed points. Thus $\varphi^{-1}(U) \to U$ induces a bijection between the sets of closed points and isomorphisms on their residue fields. Similarly for X', U'.

Let $\xi \in X_0$ be a closed point and $\eta_1, \dots, \eta_n$ its liftings to X_0', $S = \widehat{\mathcal{O}}_{X,\xi}$, $S' = \prod_{i=1}^n \widehat{\mathcal{O}}_{X',\eta_i}$. Then $S \to S'$ equipped with the inverse image of U satisfies the conditions in the theorem except normality. Note that by properties of excellent rings S is reduced and its non-normal locus is contained in $V(J)$. If S^ν is the normalization of S then $S^\nu \to (S' \otimes_S S^\nu)/(J\text{-power-torsion})$ satisfies the conditions of the theorem. Here (J-power-torsion) denotes the ideal of elements annihilated by a power of J, and in our case it is equal to the (I-power-torsion) ideal for every ideal $I \subset R$ such that $V(J) \subset V(I) \subset \operatorname{Spec}(R) - U$, because the schematic denseness of U' and the flatness of Spec $(S') \to X'$ imply that the I-power-torsion ideal of S' is 0.

The theorem will be proved by a flattening technique. Let $\bar{k}$ be an algebraic closure of k.

Lemma 1. *If k_α is a direct system of fields then $\varinjlim_\alpha (k_\alpha[[x_1, \dots, x_n]])$ is Noetherian for all n.*

Proof. Since the class of faithfully flat homomorphisms is closed under filtered direct limits, the map

$$\varinjlim_\alpha (k_\alpha[[x_1, \dots, x_n]]) \to (\varinjlim k_\alpha)[[x_1, \dots, x_n]]$$

is faithfully flat.

Alternatively use EGA $0_{\rm III}$ (10.3.1.3). □

In particular, the rings $P_n = k[\![x_1, \ldots, x_n]\!] \otimes_k \bar{k}$ are Noetherian.

Let $\mathscr{D}$ be the class of $\bar{k}$-algebras which are finite over some P_n. We say that the theorem holds for $B \in \mathscr{D}$ if the theorem holds whenever R is local and, for some V-embedding of the residue field of R into $\bar{k}$, the geometric fiber ring $R' \otimes_R \bar{k}$ is $\bar{k}$-isomorphic to B. Note that for a morphism $f : R \to R'$ in $\mathcal{C}$ with R local, the geometric fiber ring is in $\mathscr{D}$, is independent up to k-isomorphism of the choice of $R \to \bar{k}$, and it is of dimension $\leqslant 0$ if and only if f is finite. Of course, if $B \in \mathscr{D}$ is not isomorphic to $\bar{k}$, "the theorem holds for B" means that B does not arise as a geometric fiber ring under the assumptions of the theorem.

To prove the theorem, if suffices to show that if $B \in \mathscr{D}$ and the theorem holds for B/I for all non-zero ideals I, then the theorem holds for B.

Lemma 2. *The theorem holds when $R \to R'$ is finite.*

Proof. We may assume that R is local. Let $P_1, \ldots, P_n$ be the minimal primes of R'. Since U is Jacobson, the assumption gives that there exists i with $R \to R'/P_i$ injective. Then as $U' \cap \mathrm{Spec}\,(R'/P_i) \to U$ is surjective on closed points we find that all closed points of U' are in $\mathrm{Spec}\,(R'/P_i)$, so $P_i = 0$, $n = 1$.

We have to prove that $\mathrm{Frac}(R) = \mathrm{Frac}(R')$. Let $L/\mathrm{Frac}(R)$ be the maximal separable subextension of $\mathrm{Frac}(R')/\mathrm{Frac}(R)$ and $R'' = R' \cap L$. Then $R \to R''$ satisfies the assumptions of the theorem and it is finite étale of degree $d = [L : \mathrm{Frac}(R)]$ over a dense open $V \subset U$. Taking a closed point of V gives $d = 1$, so it remains to consider the purely inseparable case, which holds by the next lemma. □

Lemma 3. *Let $R \subset R'$ be a finite radicial extension of complete Noetherian local domains of characteristic $p > 0$ and dimension >0, with* $\mathrm{Frac}(R')$ *purely inseparable of degree > 1 over* $\mathrm{Frac}(R)$. *Then for every open dense $U \subset \mathrm{Spec}\,(R)$, there is $P \in U \cap P_1 R$ such that* $\mathrm{Frac}(R/P) \to \mathrm{Frac}(R'/\sqrt{PR'})$ *is of degree > 1.*

Proof. We may assume $R' = R[f^{\frac{1}{p}}]$, with f in the maximal ideal of R, $f \notin \mathrm{Frac}(R)^p$, and that R has a system of parameters $f, g_1, \ldots, g_n$. Let $F = \mathrm{Frac}(R)$. Recall that if $M/L/K$ is a tower of fields and M/L is finite then $\dim_M \Omega^1_{M/K} \geq \dim_L \Omega^1_{L/K}$. Since R is a finite extension of $K[\![x_1, \cdots, x_{n+1}]\!]$ for some field K and $\mathrm{Frac}(K[\![x_1, \cdots, x_{n+1}]\!])$ has $n+1$ linearly independent derivations $\partial/\partial x_i$, we find that $\dim_F \Omega^1_F \geq n+1$. We can find $h_1, \ldots, h_n$ in the maximal ideal of R such that $df, dh_1, \ldots, dh_n$ are linearly independent over F in Ω^1_F. Replacing g_i by $g_i' = g_i^p(1+h_i)$, we see that we can assume $df, dg_1, \ldots, dg_n$ are linearly independent over F.

Let $R'' = R[f^{\frac{1}{p}}, g_1^{\frac{1}{p}}, \ldots, g_n^{\frac{1}{p}}] := R[T_0, \ldots, T_n]/(T_0^p - f, T_1^p - g_1, \ldots, T_n^p - g_n)$ (which is a domain since $f, g_1, \ldots, g_n$ are p-independent over $\mathbb{F}_p$ in F, see proof of [Ma] 26.5).

By ([Ma], Cor. of 30.10) we may shrink U so that R and R'' are regular over U. Choose a coefficient field $K \subset R$, so R is a finite extension of a power series ring $R_0 = K[\![f, g_1, \ldots, g_n]\!]$. There is a non-zero power series $H \in K[\![f, g_1, \ldots, g_n]\!]$ with $\mathrm{Spec}(R[1/H]) \subset U$. One can find $m_1, \ldots, m_n > 0$ such that $H(f, f^{pm_1}, \ldots, f^{pm_n}) \neq 0$. [Take the m_i's such that $(\forall i)\ pm_i > a_0 + pm_1 a_1 + \cdots + pm_{i-1} a_{i-1}$, where $f^{a_0} g_1^{a_1} \ldots g_n^{a_n}$ is the smallest (for the reverse-lexicographic order) monomial which has a non-zero coefficient in H.] Take $P \in U$ which extends $(g_1 - f^{pm_1}, \ldots, g_n - f^{pm_n}) \subset R_0$. The regularity of R''_P implies by EGA 0_{IV} (22.5.4) (which is not correct as stated, but proved when p is in the square of the maximal ideal of A), or by other versions of the Jacobian criterion, that the differentials $df, dg_1, \ldots, dg_n$ are linearly independent in $\Omega := \Omega^1_{R_P} \otimes_{R_P} \kappa(P)$. Use the second fundamental exact sequence

$$0 \to PR_P/P^2R_P \to \Omega \to \Omega^1_{\kappa(P)} \to 0,$$

and note that as R_P is regular of dimension n, PR_P/P^2R_P has $\kappa(P)$-dimension n. Since the g_i are p^{th} powers in $\kappa(P)$ we find that f is not. □

It remains to prove the theorem for $B \in \mathscr{D}$ of dimension > 0 (i.e. to get a contradiction), assuming it holds for proper quotients of B. Suppose we are in the situation of the theorem with R local and B is the geometric fiber ring of $R \to R'$ for some k-embedding of $R/\mathfrak{m}_R$ in $\bar{k}$. Let $\mathscr{F}$ be the set of ideals $I \subset R$ such that R'/IR' is flat over R/I. It follows from the lemma in [Fe] or [GR] 3.4.18 (iii) that $\mathscr{F}$ is closed under finite intersections, hence the set of $I \in \mathscr{F}$ with $\mathfrak{m}_R^n \subset I$ has a smallest element I_n. As $I_n = I_{n+1} + \mathfrak{m}_R^n$, $\varprojlim(I_n/\mathfrak{m}_R^n)$ is an ideal J of R which induces $I_n/\mathfrak{m}_R^n$ on $R/\mathfrak{m}_R^n$, and $J \in \mathscr{F}$ by [Ma] 22.3. It is easily seen to be the smallest element of $\mathscr{F}$. This is an analogue of [RG, I. Thm. 4.1.2] and [HLT, Thm. 1′]. Clearly $J \subset I_1 = \mathfrak{m}_R$.

For every prime ideal $J \subset P \subset R$ with $\dim(R/P) = 1$ we have $\dim(R'/PR') = 1 + \dim(B)$, so $R'/PR' \otimes_{R/P} \mathrm{Frac}(R/P)$ has infinitely many maximal ideals (e.g. by [Ma, Lem. 1, p. 247]). Hence $P \notin U$, so $V(J) \subset \mathrm{Spec}\,(R) - U$. In particular, $J \neq 0$. Let $\underline{K}$ be the kernel of

$$(R'/\mathfrak{m}_R R') \otimes_{R/\mathfrak{m}_R} (J/\mathfrak{m}_R J) \twoheadrightarrow JR'/\mathfrak{m}_R JR'.$$

Note that if J' is an ideal of R with $J^2 \subset J' \subset J$, we have by the local flatness criterion

$$((R'/JR') \otimes_{R/J} (J/J') \xrightarrow{\sim} JR'/J'R') \Leftrightarrow J' \in \mathscr{F}.$$

We apply this to proper subideals of J containing $\mathfrak{m}_R J$. The minimality of J implies that

$(**)$ for every proper subspace $W \subsetneqq J/\mathfrak{m}_R J$,

$$\underline{K} \not\subset (R'/\mathfrak{m}_R R') \otimes_{R/\mathfrak{m}_R} W.$$

In particular, $\underline{K} \neq 0$. Note that $\underline{K}$ is a finitely generated R'-module. Suppose the non-zero elements $\Phi_1, \dots, \Phi_n$ generate $\underline{K}$, and write each Φ_i in a minimal manner as

$$\Phi_i = \sum_{j=1}^{n_i} [f_{ij}] \otimes [g_{ij}], \quad f_{ij} \in R', \; g_{ij} \in J.$$

Thus for each i, the $[f_{ij}] \in R'/\mathfrak{m}_R R'$ are linearly independent over $R/\mathfrak{m}_R$ and the $[g_{ij}] \in J/\mathfrak{m}_R J$ are linearly independent over $R/\mathfrak{m}_R$. By the definition of $\underline{K}$,

$(\forall i)\ \Sigma_j f_{ij} g_{ij} \in \mathfrak{m}_R JR'$, and by $(**)$ the elements $g_{i,j}$ generate J.

Blow up and use notation as in Remark 2. For every $\xi \in X_0$, there exists (i,j) such that $g_{i,j}\mathcal{O}_{X,\xi} = J\mathcal{O}_{X,\xi}$. Then a computation shows that the functions $f_{i,\ell}$ $(1 \leqslant \ell \leqslant n_i)$ restricted to the fiber $X' \times_X \xi$ are linearly dependent over the residue field of ξ. So if ξ is a closed point of X, the geometric fiber ring of $S \to S'$ is a proper quotient of B (for a suitable k-embedding $\kappa(\xi) \to \bar{k}$).

Let X^ν be the normalization of X, U'' the preimage of U' in $X' \times_X X^\nu$ $(U'' \xrightarrow{\sim} U')$, and X'' the schematic closure of U'' in $X' \times_X X^\nu$. The assumption that the theorem holds for proper quotients of B gives that $X'' \to X^\nu$ is a bijection on closed points, and an isomorphism on the completions at these points.

Set $T = V(\mathfrak{m}_R \mathcal{O}_{X^\nu})$ and $T'' = V(\mathfrak{m}_{R'} \mathcal{O}_{X''})$. All the closed points of X^ν (resp. X'') are in T (resp. T''), T'' is a closed subscheme of $T \otimes_{R/\mathfrak{m}_R} (R'/\mathfrak{m}_{R'})$, and T, T'' are of finite type over k. The map $T'' \to T$ is finite, a bijection on closed points, and a closed immersion on the completions at such points. It follows that $T'' \to T$ is a closed immersion and $T''_{\mathrm{red}} \xrightarrow{\sim} T_{\mathrm{red}}$. Let I (resp. I'') be the ideal defining T_{red} in X^ν (resp. T''_{red} in X''). It follows that $I\mathcal{O}_{X''}$ coincides with I'' in the completions of the local rings of X'' at closed points, hence in the local rings.

So $I\mathcal{O}_{X''} = I''$. The set theoretic zero locus of $\mathfrak{m}_R$ on X'' is therefore the set theoretic zero locus of $\mathfrak{m}_{R'}$ on X''. But $X'' \to \operatorname{Spec}(R')$ is proper and an isomorphism over U', so it is surjective, and we get a contradiction, because $\sqrt{\mathfrak{m}_R R'} \neq \mathfrak{m}_{R'}$ as $\dim(B) > 0$. □

This appendix benefitted from many remarks of the referee.

References

[Fe] D. Ferrand, *Descente de la platitude par un homomorphisme fini*, C.R. Acad. Sci. Paris **269** (1969), 946-949.

[GR] O. Gabber, L. Ramero, *Almost ring theory*, LNM **1800**, Springer, 2003.

[HLT] H. Hironaka, M. Lejeune-Jalabert, B. Teissier, *Platificateur local en géométrie analytique*, Astérisque **7 & 8** (1973), 441-463.

[Ma] H. Matsumura, *Commutative ring theory*, Cambridge Univ. Press, 1986.

[RG] M. Raynaud, L. Gruson, *Critères de platitude et de projectivité*, Invent. Math **13** (1971), 1-89.

The Euler system method for CM points on Shimura curves

Jan Nekovář

Université Pierre et Marie Curie (Paris 6)
Institut de Mathématiques de Jussieu
Théorie des Nombres, Case 247
4, place Jussieu
F-75252 Paris cedex 05
FRANCE
nekovar@math.jussieu.fr

Let E be an elliptic curve over $\mathbf{Q}$ of conductor N; let $\pi : X_0(N) \longrightarrow E$ be a modular parametrization sending the cusp ∞ to the origin. If $K = \mathbf{Q}(\sqrt{D})$ is an imaginary quadratic field of discriminant $D < 0$ satisfying the "Heegner condition" of Birch [Bi] ("each prime dividing N splits in $K/\mathbf{Q}$"), a choice of an ideal $\mathcal{N} \subset \mathcal{O}_K$ satisfying $\mathcal{N}\overline{\mathcal{N}} = N\mathcal{O}_K$ and $(\mathcal{N}, \overline{\mathcal{N}}) = (1)$ determines a Heegner point

$$x_1 = [\mathbf{C}/\mathcal{O}_K \longrightarrow \mathbf{C}/\mathcal{N}^{-1}] \in X_0(N)(K[1]),$$

defined over the Hilbert class field $K[1]$ of K. Set $y_1 = \pi(x_1) \in E(K[1])$ and $y = \mathrm{Tr}_{K[1]/K}(y) \in E(K)$. A fundamental result of Kolyvagin [Ko, Thm. A] (explained in a simplified setting in [Gr 1]) states that

(K)
if $y \notin E(K)_{\mathrm{tors}}$, then the abelian groups $E(K)/\mathbf{Z}y$, $Ш(E/K)$ are finite.

More precisely, Kolyvagin showed that $|Ш(E/K)|$ divides a certain multiple of $[E(K) : \mathbf{Z}y]^2$. Kolyvagin's result (K) has been generalized in several directions: Kolyvagin and Logačev [Ko-Lo 1] considered higher-dimensional simple quotients of $J_0(N)$ instead of elliptic curves, as well as [Ko-Lo 2] quotients of Jacobians of some Shimura curves over totally real number fields. Bertolini and Darmon [Be-Da] proved, under some additional assumptions, a "mod p" version of (K) over ring class fields $K[c]$ ($(c, N) = 1$). Tian [Ti] and Tian and Zhang [Ti-Zh] considered a common generalization of [Ko-Lo 2] and [Be-Da]. The more refined "quantitative" results of Kolyvagin were generalized by Bertolini [Be] and Howard [Ho 1,2].

The main result of the present article is the following generalization of (K), which is one of the ingredients in the proof of the parity conjecture for Selmer groups of Hilbert modular forms ([Ne, Ch. 12]).

Theorem. *Let F be a totally real number field, X a Shimura curve over F, and A an F-simple quotient of the Jacobian of X, which corresponds to (the Galois conjugacy class of) a Hilbert modular form f with trivial character (A is an abelian variety of GL_2-type and $\mathrm{End}_F(A)$ is an order in a totally real number field of degree equal to $\dim(A)$). Let x be a CM point on X by a totally imaginary quadratic extension K of F and α a character of the Galois group $\mathrm{Gal}(K(x)/K)$. Assume that A does not acquire CM over any totally imaginary quadratic extension K' of F contained in $K(x)^{\mathrm{Ker}(\alpha)}$ (i.e., f is not a θ-series associated to a Hecke character of K'). If the α-component of the image y of x in A is not torsion, then the α-components of $A(K(x))/\mathrm{End}_F(A)\cdot y$ and $\text{Ш}(A/K(x))$ are finite.*

See Theorem 3.2 for a precise formulation. The proof follows the main lines of the argument of [Be-Da]; the novelty lies in the fact that under our minimalist assumptions there is no "geometric" formula for the action of the complex conjugation on x. Unlike all previous works in this direction, our version of the Euler system argument (in particular, 7.6.4 below) does not require such a formula.

Theorem 3.2 is expected to be related to arithmetic properties of the Rankin-Selberg L-function $L(f_K\otimes\alpha^{-1}, s)$ at the central point $s = 1$ ([Zh 1,2], [Co-Va 1]). The assumption that f has trivial character ensures that $f_K \otimes \alpha^{-1}$ is self-dual in the sense that there is a "symmetric" functional equation relating $L(f_K\otimes\alpha^{-1}, s)$ (which turns out to be equal to $L(f_K \otimes \alpha, s)$ in our case) and $L(f_K \otimes \alpha^{-1}, 2 - s)$. As explained in [Gr 2] (see also [Co-Va 1]), it is possible for $L(f_K \otimes \alpha^{-1}, s)$ to be self-dual even if f has non-trivial character ω, provided that ω is equal to the restriction of α (considered as a character of $\mathbf{A}_K^*/K^*$) to $\mathbf{A}_F^*/F^*$. It is an interesting question whether Theorem 3.2 can be generalized to such a more general self-dual setting.

A substantial part of the present article was written during the author's visit at the COE programme at Nagoya University in spring 2005. The author is grateful to K. Fujiwara for inviting him to Nagoya. He would also like to thank H. Carayol, C. Cornut and M. Dimitrov for useful discussions, and the referee for helpful comments.

1. Quaternionic Shimura curves and their Jacobians

In this section we recall basic properties of Shimura curves associated to quaternion algebras over totally real fields. The standard reference is [Ca 1] (see also [Zh 1,2]), but we follow the notations and sign conventions of [Co-Va 1,2] (with the parameter $\epsilon = 1$). In particular, the reciprocity map of class field theory is normalized by making uniformizers correspond to *geometric* Frobenius elements.

(1.1) Let F be a totally real number field of degree $d = [F : \mathbf{Q}]$. Fix an archimedean prime τ_1 of F and a finite set S_B of non-archimedean primes of F satisfying

$$|S_B| \equiv [F : \mathbf{Q}] - 1 \pmod 2.$$

Let B be the (unique) quaternion algebra over F ramified at the set

$$Ram(B) = \{v|\infty,\, v \neq \tau_1\} \cup S_B.$$

For each real embedding $\tau_j : F \hookrightarrow \mathbf{R}$ $(j = 1, \ldots, d)$ put $B_{\tau_j} = B \otimes_{F,\tau_j} \mathbf{R}$. We fix an isomorphism of $\mathbf{R}$-algebras

$$B_{\tau_1} = B \otimes_{F,\tau_1} \mathbf{R} \xrightarrow{\sim} M_2(\mathbf{R}) \tag{1.1.1}$$

(for $j > 1$, B_{τ_j} is isomorphic to the algebra of Hamilton quaternions, but there is no need to fix a specific isomorphism).

Fix an embedding $\overline{F} \hookrightarrow \mathbf{C}$ extending $\tau_1 : F \hookrightarrow \mathbf{R}$.

(1.2) Let G be the algebraic group over $\mathbf{Q}$ satisfying $G(A) = (B \otimes_{\mathbf{Q}} A)^*$, for every commutative $\mathbf{Q}$-algebra A. Denote by Z the centre of G and by $nr : G(A) \longrightarrow (F \otimes_{\mathbf{Q}} A)^*$ the reduced norm. For $j = 1, \ldots, d$, denote by G_j the algebraic group over $\mathbf{R}$ given by $G_j = G \otimes_{F,\tau_j} \mathbf{R}$; we have $G_{\mathbf{R}} \xrightarrow{\sim} G_1 \times \cdots \times G_d$. We denote, for any abelian group A, $\widehat{A} = A \otimes \widehat{\mathbf{Z}}$.

Quaternionic Shimura curves are associated to the Shimura datum (G, X), where G is as above and X is the $G(\mathbf{R})$-conjugacy class of the morphism

$$\begin{aligned} h_0 : \mathbf{S} = \mathrm{Res}_{\mathbf{C}/\mathbf{R}}(\mathbf{G}_{m,\mathbf{C}}) &\longrightarrow G(\mathbf{R}) = G_1(\mathbf{R}) \times \cdots \times G_d(\mathbf{R}) \\ x + iy &\mapsto \left(\begin{pmatrix} x & y \\ -y & x \end{pmatrix}, 1, \ldots, 1 \right). \end{aligned}$$

In concrete terms, X has a natural complex structure ([Mi 2, p. 320]) and the map

$$\mathrm{Ad}(g)h_0 = gh_0g^{-1} \mapsto g(i) \qquad (g \in G(\mathbf{R}))$$

(where g acts on $\mathbf{C} - \mathbf{R}$ as in 1.3 below) defines a holomorphic isomorphism $X \xrightarrow{\sim} \mathbf{C} - \mathbf{R}$.

(1.3) The Shimura curves in question form a projective system $\{M_H\}$ indexed by open compact subgroups $H \subset G(\mathbf{A}_f) = \widehat{B}^*$. Each curve M_H is smooth over $\mathrm{Spec}(F)$ and each transition map $\mathrm{pr} : M_H \longrightarrow M_{H'}$ (for $H \subset H'$) is finite and flat. The Riemann surface associated to M_H

$$M_H^{\mathrm{an}} = (M_H \otimes_{F,\tau_1} \mathbf{C})(\mathbf{C})$$

is identified with $B^*\backslash(X \times \widehat{B}^*/H)$, where $B^* \subset B^*_{\tau_1}$ acts on $X \xrightarrow{\sim} \mathbf{C} - \mathbf{R}$ via the isomorphism $B^*_{\tau_1} \xrightarrow{\sim} GL_2(\mathbf{R})$ induced by (1.1.1), and the standard action $\begin{pmatrix} a & b \\ c & d \end{pmatrix} \cdot z = \frac{az+b}{cz+d}$ of $GL_2(\mathbf{R})$ on $\mathbf{C}-\mathbf{R}$. We denote by $[z, b] = [z, b]_H$ the point of M_H^{an} represented by a pair $(z, b) \in X \times \widehat{B}^*$. For $H \subset H'$, the transition map $\mathrm{pr} : M_H \longrightarrow M_{H'}$ corresponds to the natural projection

$$\mathrm{pr} : M_H^{\mathrm{an}} \longrightarrow M_{H'}^{\mathrm{an}}, \qquad [z, b]_H \mapsto [z, b]_{H'}.$$

In the "classical case" $F = \mathbf{Q}$, $S_B = \emptyset$ ($\iff B = M_2(\mathbf{Q})$), the curves M_H are the usual modular curves; we denote by

$$M_H^* = M_H \cup \{\mathrm{cusps}\}_H$$

their standard smooth compactifications; they form again a projective system with finite flat transition maps.

If $B \neq M_2(\mathbf{Q})$, then the curves M_H are proper over $\mathrm{Spec}(F)$; we put $M_H^* = M_H$.

(1.4) The group $G(\mathbf{A}_f) = \widehat{B}^*$ acts on the right on the projective systems $\{M_H\}$ and $\{M_H^*\}$. More precisely, the right multiplication by $g \in \widehat{B}^*$ induces an F-isomorphism $[\cdot g] : M_H \xrightarrow{\sim} M_{g^{-1}Hg}$; the corresponding holomorphic isomorphism

$$[\cdot g] : M_H^{\mathrm{an}} = B^*\backslash(X \times \widehat{B}^*/H) \xrightarrow{\sim} M_{g^{-1}Hg}^{\mathrm{an}} = B^*\backslash(X \times \widehat{B}^*/g^{-1}Hg)$$

is given by the formula $[x, b] \mapsto [x, bg]$.

In particular, the centre $\widehat{F}^* = Z(\mathbf{A}_f) \subset \widehat{B}^*$ acts on M_H (and M_H^*) via its finite quotient $\widehat{F}^*/(\widehat{F}^* \cap H) = \widehat{F}^*/(\widehat{\mathcal{O}}_F^* \cap H)$; we denote by $N_H \subset N_H^*$ the corresponding quotient curves, which are again smooth over $\mathrm{Spec}(F)$ [Ka-Ma, p. 508] (the curve M_H (resp., N_H) was denoted by Y (resp., X) in [Zh 1]). The Riemann surface associated to N_H

$$N_H^{\mathrm{an}} = (N_H \otimes_{F,\tau_1} \mathbf{C})\,(\mathbf{C})$$

is identified with $B^*\backslash(X \times \widehat{B}^*/H\widehat{F}^*)$. In particular, N_H (and N_H^*) depend only on the subgroup $H\widehat{F}^* \subset \widehat{B}^*$.

In abstract terms, the projective system $\{N_H\}$ is associated to the Shimura datum $(G/Z, X)$.

Note that, for each open compact subgroup $H \subset \widehat{B}^*$, there exists an $\mathcal{O}_F$-order $R \subset B$ such that $\widehat{R}^* \subset H\widehat{\mathcal{O}}_F^*$. In particular, $\widehat{R}^*\widehat{F}^* \subset H\widehat{F}^*$, hence the curves $N_{\widehat{R}^*}$ (where R runs through all $\mathcal{O}_F$-orders of B) form a co-final subsystem of $\{N_H\}$.

(1.5) Each curve M_H^* (hence also N_H^*) is irreducible ([Ca 1, 1.3], [Co-Va 1, 3.2]), but not necessarily geometrically irreducible. Denote by $F(\mathcal{M}_H)$ (resp., $F(\mathcal{N}_H)$) the algebraic closure of F in the function field of M_H^* (resp., of N_H^*). The reciprocity law [De, 3.9], [Mi 2, II.5.1] (with the sign corrected, [Mi 3, 1.10]) implies that the fields $F(\mathcal{N}_H) \subseteq F(\mathcal{M}_H)$ are abelian over F and that the reciprocity map

$$\mathrm{rec}_F : \widehat{F}^* \longrightarrow \mathrm{Gal}(F^{ab}/F)$$

induces isomorphisms (depending on a choice of embedding $F(\mathcal{M}_H) \hookrightarrow F^{ab}$)

$$F_+^* \backslash \widehat{F}^* / nr(H) \xrightarrow{\sim} \mathrm{Gal}(F(\mathcal{M}_H)/F),$$
$$F_+^* \backslash \widehat{F}^* / \widehat{F}^{*2} nr(H) \xrightarrow{\sim} \mathrm{Gal}(F(\mathcal{N}_H)/F),$$

where $F_+^* = \mathrm{Ker}(F^* \longrightarrow \pi_0((F \otimes_{\mathbf{Q}} \mathbf{R})^*))$ denotes the subgroup of totally positive elements of F^*.

(1.6) Differentials and automorphic forms of "weight 2" ([Co-Va 1, 3.6]) Denote by Ω^{an} the sheaf of holomorphic 1-forms on any given Riemann surface. As in the classical case $B = M_2(\mathbf{Q})$, the space of global holomorphic 1-forms

$$\varinjlim_{H} \Gamma((M_H^*)^{\mathrm{an}}, \Omega^{\mathrm{an}}),$$

equipped with a canonical left action of $\widehat{B}^*$ given by

$$[\cdot g]^* : \Gamma((M_{g^{-1}Hg}^*)^{\mathrm{an}}, \Omega^{\mathrm{an}}) \longrightarrow \Gamma((M_H^*)^{\mathrm{an}}, \Omega^{\mathrm{an}}), \qquad (1.6.1)$$

can be naturally identified with a certain space of automorphic forms on $G(\mathbf{A}) = G(\mathbf{R}) \times \widehat{B}^* = B_{\mathbf{A}}^*$.

More precisely, for each compact open subgroup $H \subset \widehat{B}^*$ we have [1]

$$G(\mathbf{A}) = \coprod_{\alpha \in C} G(\mathbf{Q}) \cdot (G(\mathbf{R})^+ \times \alpha H),$$

where

$$G(\mathbf{R})^+ = GL_2^+(\mathbf{R}) \times G_2(\mathbf{R}) \times \cdots \times G_d(\mathbf{R}),$$
$$G(\mathbf{Q})^+ = G(\mathbf{Q}) \cap G(\mathbf{R})^+ = \mathrm{Ker}\left(B^* \xrightarrow{nr} F^* \longrightarrow F^*/F_+^*\right)$$

---

[1] As the case $B = M_2(\mathbf{Q})$ is well-known, we discuss only the case $B \neq M_2(\mathbf{Q})$, when $M_H^* = M_H$.

and $C \subset \widehat{B}^*$ is a fixed (finite) set of representatives of the double cosets $G(\mathbf{Q})^+\backslash G(\mathbf{A}_f)/H$. Put, for each $\alpha \in C$,

$$\Gamma_\alpha = \alpha H \alpha^{-1} \cap G(\mathbf{Q})^+ \subset G(\mathbf{R})^+, \qquad \overline{\Gamma}_\alpha = \mathrm{Im}\left(\Gamma_\alpha \longrightarrow PGL_2^+(\mathbf{R})\right).$$

Writing $\mathcal{H} = \{z \in \mathbf{C} \mid \mathrm{Im}(z) > 0\}$ for the complex upper half plane, the map

$$\begin{array}{rcl} \coprod_{\alpha \in C} \overline{\Gamma}_\alpha\backslash\mathcal{H} & \xrightarrow{\sim} & M_H^{\mathrm{an}} = (M_H^*)^{\mathrm{an}} \\ \overline{\Gamma}_\alpha \cdot g(i) & \mapsto & [g h_0 g^{-1}, \alpha]_H \end{array} \qquad (g \in GL_2^+(\mathbf{R}))$$

is a holomorphic isomorphism. Using the standard notation

$$j(g,z) = \det(g)^{-1/2}(cz+d), \qquad g = \begin{pmatrix} a & b \\ c & d \end{pmatrix} \in GL_2^+(\mathbf{R}), \qquad z \in \mathcal{H},$$

the formula

$$f_\omega(\gamma(g^+, \alpha h)) = j(g_1^+)^{-2} f_\alpha(g_1^+(i)),$$
$$(\gamma \in G(\mathbf{Q}),\ g^+ = (g_1^+, g_2, \ldots, g_d) \in G(\mathbf{R})^+,\ h \in H)$$

defines a bijection

$$\begin{array}{ccccc} \Gamma((M_H^*)^{\mathrm{an}}, \Omega^{\mathrm{an}}) & \xrightarrow{\sim} & \bigoplus_{\alpha \in C} \Gamma(\overline{\Gamma}_\alpha\backslash\mathcal{H}, \Omega^{\mathrm{an}}) & \xrightarrow{\sim} & \mathcal{S}_2^H \\ \omega & \mapsto & (f_\alpha(z)\,dz)_{\alpha \in C} & \mapsto & (f_\omega : G(\mathbf{A}) \longrightarrow \mathbf{C}), \end{array} \tag{1.6.2}$$

where $\mathcal{S}_2^H$ denotes the space of functions $f : G(\mathbf{A}) \longrightarrow \mathbf{C}$ satisfying

$$f\Big(\gamma\, g\, z_\infty \Big(\begin{pmatrix} \cos(\theta) & -\sin(\theta) \\ \sin(\theta) & \cos(\theta) \end{pmatrix}, g_2, \ldots, g_d \Big) h \Big) = e^{-2i\theta} f(g)$$
$$(\gamma \in G(\mathbf{Q}),\ g \in G(\mathbf{A}),\ z_\infty \in Z(\mathbf{R}),\ \theta \in \mathbf{R},\ g_j \in G_j(\mathbf{R})\ (j > 1),\ h \in H)$$

and such that, for each $g \in G(\mathbf{A})$, the function

$$x + iy \mapsto \frac{1}{y} f\Big(g\Big(\begin{pmatrix} y & x \\ 0 & 1 \end{pmatrix}, 1, \ldots, 1\Big)\Big)$$

is holomorphic on $\mathcal{H}$ (in the case $B = M_2(\mathbf{Q})$ one has to impose an additional cuspidality condition on the function f; this is automatic if $B \neq M_2(\mathbf{Q})$).

Passing to the inductive limit with respect to all open compact subgroups of $\widehat{B}^*$, we obtain a $\widehat{B}^*$-equivariant bijection

$$\varinjlim_{H} \Gamma((M_H^*)^{\mathrm{an}}, \Omega^{\mathrm{an}}) \xrightarrow{\sim} \mathcal{S}_2 := \bigcup_H \mathcal{S}_2^H, \tag{1.6.3}$$

with $\widehat{B}^*$ acting (on the left) on $\mathcal{S}_2$ by right translations: $(b\cdot f)(g) = f(gb)$ (which means that $\mathcal{S}_2^H$ coincides, as the notation suggests, with the space of H-invariant elements of $\mathcal{S}_2$).

Note that the centre $Z(\mathbf{A}_f) = \widehat{F}^*$ acts on $\mathcal{S}_2^H$ via its finite quotient $\widehat{F}^*/(\widehat{F}^* \cap H) = \widehat{F}^*/(\widehat{\mathcal{O}}_F^* \cap H)$, hence (1.6.2) induces a bijection

$$\Gamma((N_H^*)^{\mathrm{an}}, \Omega^{\mathrm{an}}) \xrightarrow{\sim} \mathcal{S}_2^{H\widehat{F}^*}, \tag{1.6.4}$$

where

$$\mathcal{S}_2^{H\widehat{F}^*} = \{f \in \mathcal{S}_2^H \mid (\forall g \in G(\mathbf{A}))\,(\forall z \in Z(\mathbf{A}))\;\; f(gz) = f(g)\}.$$

(1.7) Automorphic representations. The multiplicity one theorem for automorphic forms on $B_{\mathbf{A}}^*$ ([Ge], VI.2) implies that the space $\mathcal{S}_2$ (as an admissible smooth representation of $\widehat{B}^*$) decomposes as a countable direct sum

$$\mathcal{S}_2 \xrightarrow{\sim} \bigoplus_{\pi = \sigma_2 \otimes \pi_f} \pi_f, \tag{1.7.1}$$

where π runs through all irreducible cuspidal (this is automatic if $B \neq M_2(\mathbf{Q})$) unitary automorphic representations of $B_{\mathbf{A}}^*$ whose archimedean component π_∞ is isomorphic to

$$\sigma_2 = \begin{pmatrix} \text{weight 2 holomorphic discrete} \\ \text{series representation of } G_1(\mathbf{R}) \end{pmatrix} \otimes \begin{pmatrix} \text{trivial representation of} \\ G_2(\mathbf{R}) \times \cdots \times G_d(\mathbf{R}) \end{pmatrix}.$$

In particular, (1.6.4) induces an isomorphism of admissible smooth representations of $\widehat{B}^*/\widehat{F}^*$

$$\varinjlim_{H} \Gamma((N_H^*)^{\mathrm{an}}, \Omega^{\mathrm{an}}) \xrightarrow{\sim} \mathcal{S}_2^{\widehat{F}^*} \xrightarrow{\sim} \bigoplus_{\substack{\pi = \sigma_2 \otimes \pi_f \\ \omega_\pi = 1}} \pi_f, \tag{1.7.2}$$

where $\omega_\pi : Z(\mathbf{A}) \longrightarrow \mathbf{C}^*$ denotes the central character of π.

(1.8) The Hecke algebra. For each non-archimedean prime v of F, let dg_v be the (bi-invariant) Haar measure on B_v^* normalized so that $\int_{H_v} dg_v = 1$ for one (hence for every) maximal compact subgroup

$H_v \subset B_v^*$. The product $dg = \prod dg_v$ then defines a bi-invariant Haar measure on $\widehat{B}^*$. We denote by $\mathrm{vol}(X) = \int_X dg$ the corresponding volume. The Hecke algebra

$$\mathcal{H}(\widehat{B}^*) = \mathcal{C}_c^\infty(\widehat{B}^*, \mathbf{C}) = \{\alpha : \widehat{B}^* \longrightarrow \mathbf{C} \mid \text{loc. constant with cpt. support}\}$$

with the convolution product

$$(\alpha * \beta)(g) = \int_{\widehat{B}^*} \alpha(gh^{-1})\beta(h)\, dh$$

is isomorphic to the direct limit (over all open compact subgroups $H \subset \widehat{B}^*$) of the double coset algebras

$$\varinjlim_H \mathbf{C}[H\backslash\widehat{B}^*/H] \xrightarrow{\sim} \mathcal{H}(\widehat{B}^*)$$
$$HbH \mapsto \mathrm{vol}(H)^{-1} 1_{HbH}$$

The product structure on $\mathbf{C}[H\backslash\widehat{B}^*/H]$ is as in [Sh 2, 3.1], [Miy, (2.7.3)] (hence is the *opposite* to that in [Co-Va 1, 3.4, 3.11]).

Any smooth representation (π, V) of $\widehat{B}^*$ gives rise to a left action of $\mathcal{H}(\widehat{B}^*)$ on V, given by the formula

$$\alpha \cdot v = \int_{\widehat{B}^*} \alpha(h)\pi(h)v\, dh \qquad (\alpha \in \mathcal{H}(\widehat{B}^*),\ v \in V).$$

In particular, $\alpha \in \mathcal{H}(\widehat{B}^*)$ acts on $\mathcal{S}_2$ by

$$(\alpha \cdot f)(g) = \int_{\widehat{B}^*} \alpha(h) f(gh)\, dh.$$

Note that, for any open compact subgroups $H, H' \subset \widehat{B}^*$ and $b \in \widehat{B}^*$, the action of $1_{HbH'}$ maps $\mathcal{S}_2^{H'}$ to $\mathcal{S}_2^H$.

(1.9) Hecke correspondences. Let H, H' be open compact subgroups of $\widehat{B}^*$ and $g \in \widehat{B}^*$. The diagram

$$\begin{array}{ccc} M^*_{H \cap gH'g^{-1}} & \xrightarrow{[\cdot g]} & M^*_{g^{-1}Hg \cap H'} \\ \downarrow \mathrm{pr} & & \downarrow \mathrm{pr}' \\ M^*_H & \overset{[HgH']}{\dashrightarrow} & M^*_{H'} \end{array} \qquad (1.9.1)$$

defines a multivalued map (a "Hecke correspondence")

$$[HgH'] : M_H^* \dashrightarrow M_{H'}^*.$$

On non-cuspidal complex points,

$$[HgH'] : M_H^{\mathrm{an}} = B^*\backslash(X \times \widehat{B}^*/H) \dashrightarrow M_{H'}^{\mathrm{an}} = B^*\backslash(X \times \widehat{B}^*/H')$$

is given by the formula

$$[x, b]_H \mapsto \sum_i [x, bg_i]_{H'}, \qquad HgH'F^* = \coprod_i g_i H'F^*. \tag{1.9.2}$$

The degree of $[HgH']$ is the degree of pr, namely $(HF^* : (H \cap gH'g^{-1})F^*)$ (which is equal to $(H : H \cap gH'g^{-1})$, if $\mathcal{O}_F^* \subset H \cap H'$). If $z \in Z(\mathbf{A}_f) = \widehat{F}^*$, then we have

$$[HzH] = [\cdot z] : M_H^* \longrightarrow M_H^*, \qquad [H'zH'] = [\cdot z] : M_{H'}^* \longrightarrow M_{H'}^*,$$
$$[\cdot z] \circ [HgH'] = [HgH'] \circ [\cdot z].$$

For any "reasonable" cohomology theory $H_?^*$, the correspondences $[HgH']$ induce maps

$$[HgH']^* : H_?^q(M_{H'}^*) \longrightarrow H_?^q(M_H^*), \quad [HgH']_* : H_?^q(M_H^*) \longrightarrow H_?^q(M_{H'}^*).$$

As the transpose of the diagram (1.9.1) is given by

$$\begin{array}{ccc} M_{g^{-1}Hg \cap H'}^* & \xrightarrow{[\cdot g^{-1}]} & M_{H \cap gH'g^{-1}}^* \\ \downarrow \mathrm{pr}' & & \downarrow \mathrm{pr} \\ M_{H'}^* & \overset{[H'g^{-1}H]}{\dashrightarrow} & M_H^*, \end{array}$$

we have

$$[HgH']_* = [H'g^{-1}H]^*, \qquad [HgH']^* = [H'g^{-1}H]_* \tag{1.9.3}$$

If we replace H, H' by $H\widehat{F}^*, H'\widehat{F}^*$, then we obtain Hecke correspondences

$$N_H^* \dashrightarrow N_{H'}^*$$

with similar properties.

(1.10) Action on holomorphic differentials. In the situation of (1.9.1), we deduce from (1.9.2) and the $\widehat{B}^*$-equivariance of (1.6.3) that the map

$$[HgH']^* : \Gamma((M^*_{H'})^{\mathrm{an}}, \Omega^{\mathrm{an}}) \longrightarrow \Gamma((M^*_H)^{\mathrm{an}}, \Omega^{\mathrm{an}})$$

corresponds to the action of $1_{HgH'}$, suitably normalized:

$$\begin{aligned} \mathcal{S}_2^{H'} &\longrightarrow \mathcal{S}_2^H \\ f' &\mapsto \mathrm{vol}(H')^{-1}\, 1_{HgH'} \cdot f'. \end{aligned}$$

If $z \in Z(\mathbf{A}_f) = \widehat{F}^*$ and $H = H'$, then

$$[HzH']^* = [\cdot z]^* : \Gamma((M^*_H)^{\mathrm{an}}, \Omega^{\mathrm{an}}) \longrightarrow \Gamma((M^*_H)^{\mathrm{an}}, \Omega^{\mathrm{an}})$$

corresponds to the action of z on $\mathcal{S}_2^H$, which we denote by $[z]$.

(1.11) Adjoints. Up to a scalar multiple, the hermitian scalar product

$$\langle \omega_1, \omega_2 \rangle = \frac{i}{2} \int_{(M^*_H)^{\mathrm{an}}} \omega_1 \wedge \overline{\omega_2} \qquad (\omega_1, \omega_2 \in \Gamma((M^*_H)^{\mathrm{an}}, \Omega^{\mathrm{an}})) \tag{1.11.1}$$

corresponds, via the isomorphism (1.6.2), to the hermitian scalar product

$$\langle f_1, f_2 \rangle = \int_{G(\mathbf{Q})\backslash G(\mathbf{A})/Z_\infty K_\infty^+} f_1(g)\overline{f_2(g)}\, dg \qquad (f_1, f_2 \in \mathcal{S}_2^H),$$

where $Z_\infty K_\infty^+ \subset G(\mathbf{R})^+$ consists of the elements

$$\left(\begin{pmatrix} z & 0 \\ 0 & z \end{pmatrix} \begin{pmatrix} \cos(\theta) & -\sin(\theta) \\ \sin(\theta) & \cos(\theta) \end{pmatrix}, g_2, \ldots, g_d \right) \ (z \in \mathbf{R}^*,\ \theta \in \mathbf{R},\ g_j \in G_j(\mathbf{R}))$$

For $\alpha \in \mathcal{H}(\widehat{B}^*)$ and $f_1, f_2 \in \mathcal{S}_2$, we have

$$\langle \alpha \cdot f_1, f_2 \rangle = \langle f_1, {}^t\alpha \cdot f_2 \rangle, \qquad {}^t\alpha(g) = \overline{\alpha(g^{-1})} \tag{1.11.2}$$

(i.e., ${}^t\alpha$ is the adjoint of α). For example,

$${}^t 1_{HgH'} = 1_{H'g^{-1}H}. \tag{1.11.3}$$

In particular, if $H = H' = H_v H^v$, where $v \notin S_B$, $H_v \subset B_v^*$ is a maximal compact subgroup and $b_v \in B_v^* \subset \widehat{B}^*$, then

$$H_v b_v^{-1} H_v = nr(b_v)^{-1}\, H_v b_v H_v,$$

as H_v is isomorphic to $GL_2(\mathcal{O}_{F,v}) = GL_2(\mathcal{O})$ and

$$GL_2(\mathcal{O}) \begin{pmatrix} t_1^{-1} & 0 \\ 0 & t_2^{-1} \end{pmatrix} GL_2(\mathcal{O}) = (t_1 t_2)^{-1} GL_2(\mathcal{O}) \begin{pmatrix} t_2 & 0 \\ 0 & t_1 \end{pmatrix} GL_2(\mathcal{O}) =$$

$$= (t_1 t_2)^{-1} GL_2(\mathcal{O}) \begin{pmatrix} t_1 & 0 \\ 0 & t_2 \end{pmatrix} GL_2(\mathcal{O}),$$

hence

$${}^t 1_{Hb_vH} = [nr(b_v)^{-1}] \circ 1_{Hb_vH} = 1_{Hb_vH} \circ [nr(b_v)^{-1}] \tag{1.11.4}$$

in this case.

(1.12) Hecke operators and cup products. Fix an open compact subgroup $H \subset \widehat{B}^*$. There exists a finite set $S \supset S_B$ of non-archimedean primes of F such that $H = H_S H^S$, where $H_S \subset \prod_{v \in S} B_v^*$ and

$$H^S = \prod_{v \notin S} H_v = \text{ a maximal compact subgroup of } G(\mathbf{A}_f)^S = {\prod_{v \notin S}}' B_v^*. \tag{1.12.1}$$

More precisely, for each non-archimedean prime $v \notin S$ there exists a unique maximal $\mathcal{O}_{F,v}$-order $R(v) \subset B_v$ such that $H_v = R(v)^*$ (in other words, there exists an isomorphism of groups $B_v^* \xrightarrow{\sim} GL_2(F_v)$ inducing an isomorphism $H_v \xrightarrow{\sim} GL_2(\mathcal{O}_{F,v})$).

For such $v \notin S$, we define the **Hecke correspondence** $T(v)$ as

$$T(v) = [Hb_vH] : \ M_H^* \dashrightarrow M_H^* \ , \tag{1.12.2}$$

for any element $b_v \in R(v) \cap B_v^* \subset B_v^* \subset \widehat{B}^*$ satisfying $\mathrm{ord}_v(nr(b_v)) = 1$. For example, we can fix a uniformizer ϖ_v of $\mathcal{O}_{F,v}$ and take the element b_v which corresponds, under some isomorphism $H_v \xrightarrow{\sim} GL_2(\mathcal{O}_{F,v})$, to the matrix

$$\begin{pmatrix} \varpi_v & 0 \\ 0 & 1 \end{pmatrix}.$$

The **Hecke operator** $T(v)$ is defined as the induced map

$$T(v) := [Hb_vH]^* : \Gamma((M_H^*)^{\mathrm{an}}, \Omega^{\mathrm{an}}) \longrightarrow \Gamma((M_H^*)^{\mathrm{an}}, \Omega^{\mathrm{an}}).$$

According to 1.10, the action of $T(v)$ on $\Gamma((M_H^*)^{\mathrm{an}}, \Omega^{\mathrm{an}}) \xrightarrow{\sim} S_2^H$ is given by

$$T(v) = [Hb_vH]^* \quad \text{corresponds to} \quad \mathrm{vol}(H)^{-1} \, 1_{Hb_vH}.$$

Combining 1.10 with (1.11.4), we see that the adjoint ${}^tT(v)$ of $T(v)$ with respect to the hermitian scalar product (1.11.1) corresponds to

$$\mathrm{vol}(H)^{-1}\,({}^t1_{Hb_vH}) = \mathrm{vol}(H)^{-1}\,1_{Hb_v^{-1}H} = \mathrm{vol}(H)^{-1}\,[\varpi_v^{-1}]\circ 1_{Hb_vH},$$

hence

$${}^tT(v) = [Hb_v^{-1}H]^* = [\varpi_v^{-1}]\circ T(v) = T(v)\circ[\varpi_v^{-1}] \qquad (1.12.3)$$

Fix a prime number ℓ and define

$$\begin{aligned} W &:= H^1((M_H^*)^{\mathrm{an}},\mathbf{Q}), \quad & W_{\mathbf{C}} &:= \Gamma((M_H^*)^{\mathrm{an}},\Omega^{\mathrm{an}}),\\ W_\ell &:= H^1_{et}(M_H^*\otimes_F\overline{F},\mathbf{Q}_\ell), \quad & W_{\mathrm{dR}} &:= H^1_{\mathrm{dR}}(M_H^*/F). \end{aligned}$$

The comparison isomorphisms

$$W\otimes_{\mathbf{Q}}\mathbf{C}\xrightarrow{\sim}W_{\mathbf{C}}\oplus\overline{W_{\mathbf{C}}}, \qquad W\otimes_{\mathbf{Q}}\mathbf{Q}_\ell\xrightarrow{\sim}W_\ell \qquad (1.12.4)$$

are $\widehat{B}^*$-equivariant, hence $\mathbf{Q}[H\backslash\widehat{B}^*/H]$-equivariant (the second comparison isomorphism is defined by using the embedding $\overline{F}\hookrightarrow\mathbf{C}$ that was fixed in 1.1).

Up to a scalar multiple, the hermitian pairing (1.11.1)

$$W_{\mathbf{C}}\otimes_{\mathbf{C}}\overline{W_{\mathbf{C}}}\longrightarrow\mathbf{C}$$

is equal to the $\mathbf{C}$-linear extension of the cup product

$$(\ ,\ ): W\otimes_{\mathbf{Q}}W\xrightarrow{\ \cup\ }H^2((M_H^*)^{\mathrm{an}},\mathbf{Q})\xrightarrow{\ Tr\ }\mathbf{Q}(-1)$$

(note that $W_{\mathbf{C}}$ and $\overline{W_{\mathbf{C}}}$ are isotropic subspaces of $W\otimes_{\mathbf{Q}}\mathbf{C}$ with respect to $(\ ,\ )\otimes_{\mathbf{Q}}\mathbf{C}$). Similarly, $(\ ,\ )\otimes_{\mathbf{Q}}\mathbf{Q}_\ell$ is identified with the étale cup product

$$(\ ,\ )_\ell : W_\ell\otimes_{\mathbf{Q}_\ell}W_\ell\xrightarrow{\ \cup\ }H^2_{et}(M_H^*\otimes_F\overline{F},\mathbf{Q}_\ell)\xrightarrow{\ Tr\ }\mathbf{Q}_\ell(-1).$$

Taking into account (1.12.3), this implies that the adjoint of the operator $T(v) := [Hb_vH]^*$ acting on W (resp., on W_ℓ) with respect to the pairing $(\ ,\ )$ (resp., to $(\ ,\ )_\ell$) is equal to $[\varpi_v^{-1}]\circ T(v) = T(v)\circ[\varpi_v^{-1}]$.

(1.13) Integral models. Integral models of the curves M_H were studied, in an increasing level of generality, by Deligne-Rapoport [De-Ra], Katz-Mazur [Ka-Ma] and Carayol [Ca 1].

Fix a non-archimedean prime $v\notin S_B$ of F; denote by $\mathcal{O}(v)$ the localization of $\mathcal{O}_F$ at v. Assume that $H\subset\widehat{B}^*$ factorizes as $H = H_vH^v$, where H_v (resp., H^v) is an open compact subgroup of B_v^* (resp., of $G(\mathbf{A}_f)^v = \{g \in \widehat{B}^* \mid g_v = 1\}$).

In the classical case $B = M_2(\mathbf{Q})$, the curve M_H is a coarse moduli space of elliptic curves with an H-level structure, for schemes over $\mathrm{Spec}(\mathbf{Q})$. Katz-Mazur [Ka-Ma] extended this moduli problem to schemes over $\mathrm{Spec}(\mathcal{O}(v))$, using the theory of Drinfeld bases. They obtained a regular model $\mathbf{M}_H \longrightarrow \mathrm{Spec}(\mathcal{O}(v))$ of M_H, which they compactified - somewhat artificially - to a regular model $\mathbf{M}_H^*$ of M_H^*, proper over $\mathrm{Spec}(\mathcal{O}(v))$.

Similar techniques apply in the case $F = \mathbf{Q}$, $B \neq M_2(\mathbf{Q})$, when M_H is a coarse moduli space of abelian surfaces with quaternionic multiplication with a suitable level structure [Bu]; one obtains proper regular models $\mathbf{M}_H = \mathbf{M}_H^* \longrightarrow \mathrm{Spec}(\mathcal{O}(v))$ of $M_H = M_H^*$.

If $F \neq \mathbf{Q}$, then M_H has no longer a moduli interpretation, but it can be related to a unitary Shimura curve parametrizing a certain class of abelian varieties of dimension $4[F : \mathbf{Q}]$. Carayol [Ca 1] extended this moduli problem to schemes over $\mathrm{Spec}(\mathcal{O}(v))$ and described, in the case when H^v is *sufficiently small* (this condition depends on H_v), a regular model $\mathbf{M}_H = \mathbf{M}_H^*$ of $M_H = M_H^*$, proper over $\mathrm{Spec}(\mathcal{O}(v))$. If H_v is a maximal open compact subgroup of B_v^*, then $\mathbf{M}_H^*$ is smooth over $\mathrm{Spec}(\mathcal{O}(v))$.

As explained in [Co-Va 2, 3.1.3], Carayol's results [(1)] can be used to construct regular models $\mathbf{M}_H^*$ of M_H^*, proper over $\mathrm{Spec}(\mathcal{O}(v))$, even if H^v is not sufficiently small: fix a sufficiently small open compact normal subgroup H'^v of H^v and put $H' = H_v H'^v \subset \widehat{B}^*$. The right action of the finite group H/H' on $M_{H'}^*$ extends naturally to an $\mathcal{O}(v)$-linear action on $\mathbf{M}_{H'}^*$; one defines $\mathbf{M}_H^*$ as the quotient of $\mathbf{M}_{H'}^*$ by this action of H/H'. The model $\mathbf{M}_H^*$ is regular, proper over $\mathrm{Spec}(\mathcal{O}(v))$, and independent on the choice of H'^v. Taking an additional quotient of $\mathbf{M}_H^*$ by the action of the finite abelian group $\widehat{F}^*/(\widehat{F}^* \cap H)$, we obtain a regular model $\mathbf{N}_H^*$ of N_H^*, proper over $\mathrm{Spec}(\mathcal{O}(v))$. If H_v is a maximal open compact subgroup of B_v^*, then $\mathbf{M}_H^*$ and $\mathbf{N}_H^*$ are again smooth over $\mathrm{Spec}(\mathcal{O}(v))$ (by [Ka-Ma, p. 508]).

(1.14) The Eichler-Shimura congruence relation [Sh 1, Thm 2.2.3], [Oh, Thm. 3.4.3], [Ca 1, 10.3]. In the situation of (1.12.1), fix $v \notin S$ and a prime number ℓ different from the residue characteristic of v. As H_v is a maximal compact subgroup of B_v^*, the model $\mathbf{M}_H^*$ is proper and smooth over $\mathrm{Spec}(\mathcal{O}(v))$, hence the proper and smooth base change theorems yield a canonical isomorphism

$$W_\ell \xrightarrow{\sim} H^1_{et}(\mathbf{M}_H^{*\circ} \otimes_{\kappa(v)} \overline{\kappa(v)}, \mathbf{Q}_\ell) =: W_\ell^\circ,$$

(1) Note that Carayol [Ca 1,2] uses different sign conventions. This is discussed in detail in [Co-Va 2, 3.3.1].

where $\kappa(v)$ is the residue field of v and $\mathbf{M}_H^{*\circ} = \mathbf{M}_H^* \otimes_{\mathcal{O}(v)} \kappa(v)$ the special fibre of $\mathbf{M}_H^*$. In particular, the natural action of $G_F = \mathrm{Gal}(\overline{F}/F)$ on W_ℓ is unramified at v.

As the diagram (1.9.1) for $g = b_v$ from (1.12.2) naturally extends to the integral models, the graph of the Hecke correspondence $T(v)$ (which is a divisor on $M_H^* \times_F M_H^*$) naturally extends to a divisor on $\mathbf{M}_H^* \times_{\mathcal{O}(v)} \mathbf{M}_H^*$; denote by $T(v)^\circ \subset \mathbf{M}_H^{*\circ} \times_{\kappa(v)} \mathbf{M}_H^{*\circ}$ its special fibre. The generalized **Eichler-Shimura congruence relation** states that

$$T(v)^\circ = \Gamma_\varphi + \Gamma_{[\varpi_v \cdot]} \circ {}^t\Gamma_\varphi, \tag{1.14.1}$$

where Γ_φ denotes the graph of the Frobenius morphism $\varphi = \varphi_v : \mathbf{M}_H^{*\circ} \longrightarrow \mathbf{M}_H^{*\circ}$ and ${}^t\Gamma_\varphi$ its transpose. As the contravariant action of Γ_φ (resp., of ${}^t\Gamma_\varphi$) on W_ℓ° coincides with the Galois action of $F = \mathrm{Fr}(v)_{\mathrm{geom}}$ (resp., with the action of ${}^tF = (Nv)F^{-1}$), it follows that

$$(1 - XF)(1 - X[\varpi_v](Nv)F^{-1}) = 1 - XT(v) + X^2(Nv)[\varpi_v] \\ \in \mathrm{End}_{\mathbf{Q}_\ell}(W_\ell^\circ)[X].$$

As $[\varpi_v](Nv)F^{-1}$ is the adjoint of F with respect to the pairing $(\ ,\ )_\ell$, it follows from (1.12.4) and the compatibility of the various pairings discussed in 1.12 that

$$\det(1 - XF \mid W_\ell^\circ) = \det(1 - XT(v) + X^2(Nv)[\varpi_v] \mid W_{\mathbf{C}}). \tag{1.14.2}$$

For the curve N_H^* and its model $\mathbf{N}_H^*$, the congruence relation (1.14.1) simplifies ([Zh 1, Prop. 1.4.10]) to

$$T(v)^\circ = \Gamma_\varphi + {}^t\Gamma_\varphi. \tag{1.14.3}$$

(1.15) The L-function of M_H^*. Combining the discussion from 1.14 (the notations and assumptions of which are still in force) with the decomposition (1.7.1), we obtain

$$W_{\mathbf{C}} \xrightarrow{\sim} \mathcal{S}_2^H \xrightarrow{\sim} \bigoplus_{\pi = \sigma_2 \otimes \pi_f} \pi_f^H.$$

If $\pi = \sigma_2 \otimes \pi_f = \otimes_v' \pi_v$ is an irreducible unitary cuspidal automorphic representation of $B_{\mathbf{A}}^*$ with $\pi_f^H \neq 0$, then $T(v)$ acts on the one-dimensional space $\pi_v^{H_v}$ of spherical vectors (recall that $v \notin S$, by assumption) by a scalar $\lambda_\pi(v) \in \mathbf{C}$. The relation (1.14.2) then gives a formula for the local Euler factor

$$L_v(h^1(M_H^*), s) := \det(1 - (Nv)^{-s}\,\mathrm{Fr}(v)_{\mathrm{geom}} \mid W_\ell^{I_v})^{-1},$$

namely

$$\begin{aligned} L_v(h^1(M_H^*), s) &= \prod_{\pi=\sigma_2\otimes\pi_f} (1 - \lambda_\pi(v)(Nv)^{-s} + \omega_\pi(v)(Nv)^{1-2s})^{-\dim(\pi_f^H)} \\ &= \prod_{\pi=\sigma_2\otimes\pi_f} L_v(\pi, s-1/2)^{\dim(\pi_f^H)}, \end{aligned} \tag{1.15.1}$$

where $\omega_\pi : \mathbf{A}_F^*/F^* \longrightarrow \mathbf{C}^*$ denotes the central character (of finite order) of π. Taking a product of (1.15.1) over all $v \notin S$, we obtain an equality of partial L-functions

$$L^S(h^1(M_H^*), s) = \prod_{\pi=\sigma_2\otimes\pi_f} L^S(\pi, s-1/2)^{\dim(\pi_f^H)}, \tag{1.15.2}$$

where

$$L^S(h^1(M_H^*), s) = \prod_{v\notin S} L_v(h^1(M_H^*), s), \qquad L^S(\pi, s) = \prod_{v\notin S} L_v(\pi, s).$$

Above, the automorphic L-functions are normalized as in [Ja-La] (the centre of symmetry of the functional equation lies at $s = 1/2$).

It is a much deeper fact that the extreme left and right terms in (1.15.1) are equal even for $v \in S$. For $v \in S - S_B$ this is proved in [Ca 2]; for $v \in S_B$ one can reduce to the previous case by switching to a quaternion algebra unramified at v (possibly after a cyclic base change to a suitable totally real number field in which v splits completely).

The Jacquet-Langlands correspondence [Ja-La, §14] associates to each (irreducible, cuspidal) $\pi = \sigma_2 \otimes \pi_f$ with $\pi_f^H \neq 0$ an irreducible cuspidal representation $JL(\pi)$ of $GL_2(\mathbf{A}_F)$, which corresponds to a Hilbert modular newform of weight $(2, \ldots, 2)$ and central character ω_π. This representation is characterized by the property

$$\begin{aligned} &(\forall w \notin Ram(B))\ JL(\pi)_w \xrightarrow{\sim} \pi_w \text{ as representations of } GL_2(F_w) \xrightarrow{\sim} B_w^* \\ &\qquad (\Longrightarrow L_w(JL(\pi), s) = L_w(\pi, s)) \end{aligned}$$

(where w is a prime of F). In particular,

$$L^S(JL(\pi), s) = L^S(\pi, s).$$

In fact, the equality $L_w(JL(\pi), s) = L_w(\pi, s)$ holds also for $w \in Ram(B)$ ([Ja-La], Rmk. before Thm. 14.4).

(1.16) The Jacobian of M_H^* and N_H^*. As in [Co-Va 1, 3.3], we define the Jacobian of M_H^* (resp., N_H^*) to be the abelian variety over F

$$J(M_H^*) := \mathrm{Pic}^\circ(M_H^*/F) = \mathrm{Res}_{F(\mathcal{M}_H)/F}\left(\mathrm{Pic}^\circ(M_H^*/F(\mathcal{M}_H))\right)$$
$$J(N_H^*) := \mathrm{Pic}^\circ(N_H^*/F) = \mathrm{Res}_{F(\mathcal{N}_H)/F}\left(\mathrm{Pic}^\circ(N_H^*/F(\mathcal{N}_H))\right)$$

(recall that $F(\mathcal{M}_H)$ (resp., $F(\mathcal{N}_H)$) denotes the field of constants of M_H^* (resp., of N_H^*)).

As $\mathrm{Pic}^\circ(-/F)$ is both covariant and contravariant for finite flat morphisms of proper smooth curves over $\mathrm{Spec}(F)$, each diagram (1.9.1) induces morphisms of abelian varieties

$$\begin{aligned}[][HgH']_* = [H'g^{-1}H]^* &: J(M_H^*) \longrightarrow J(M_{H'}^*),\\ [HgH']^* = [H'g^{-1}H]_* &: J(M_{H'}^*) \longrightarrow J(M_H^*)\end{aligned} \tag{1.16.1}$$

(and similarly for N_H^*). Each Jacobian $J(M_H^*)$ has a canonical principal polarization. It follows from (1.16.1) that the corresponding Rosati involution ι acts by

$$\begin{aligned}\iota([HgH]_*) &= [HgH]^* = [Hg^{-1}H]_*,\\ \iota([HgH]^*) &= [HgH]_* = [Hg^{-1}H]^*.\end{aligned} \tag{1.16.2}$$

One can consider (1.16.2) as a "motivic" version of (1.11.3).

We let the Hecke correspondences act on the Jacobians *covariantly*. As $[\cdot g] \circ [\cdot g'] = [\cdot g'g]$, the map

$$\begin{aligned}\mathbf{Z}[H\backslash \widehat{B}^*/H]^{\mathrm{op}} &\longrightarrow \mathrm{End}(J(M_H^*))\\ [HgH] &\mapsto [HgH]_*\end{aligned} \tag{1.16.3}$$

is a ring homomorphism (recall that the multiplication on $\mathbf{Z}[H\backslash \widehat{B}^*/H]$ used here is the opposite of the one considered in [Co-Va 1, 3.4, 3.11]).

The space $\Gamma((M_H^*)^{\mathrm{an}}, \Omega^{\mathrm{an}})$ is canonically isomorphic to the cotangent space $(T_0 J(M_H^*)^{\mathrm{an}})^\vee$ of the complex torus $J(M_H^*)^{\mathrm{an}}$ at the origin. For each Hecke correspondence $[HgH'] : M_H^* \dashrightarrow M_{H'}^*$, the induced map

$$[HgH']^* : \Gamma((M_{H'}^*)^{\mathrm{an}}, \Omega^{\mathrm{an}}) \longrightarrow \Gamma((M_H^*)^{\mathrm{an}}, \Omega^{\mathrm{an}})$$

corresponds to the dual of the differential of the morphism

$$[HgH']_* : J(M_H^*) \longrightarrow J(M_{H'}^*)$$

at the origin.

(1.17) The "physical" Hecke algebra. In the situation of (1.12.1), the spherical Hecke algebra

$$\mathbf{T}_H^S := \mathbf{Z}[H^S \backslash G(\mathbf{A}_f)^S / H^S] \subset \mathbf{Z}[H \backslash \widehat{B}^* / H]$$

is commutative; we let it act on the Jacobian $J(M_H^*)$ by the rule (1.16.3), which defines a ring homomorphism

$$\theta : \mathbf{T}_H^S = (\mathbf{T}_H^S)^{\mathrm{op}} \subset \mathbf{Z}[H \backslash \widehat{B}^* / H]^{\mathrm{op}} \longrightarrow \mathrm{End}(J(M_H^*)). \qquad (1.17.1)$$

As a ring, $\mathbf{T}_H^S$ is generated by the double cosets $[Hb_vH]$ and $[H\varpi_v^{\pm 1}H]$ $(v \notin S)$, where b_v is as in 1.12. For each $v \notin S$, the endomorphisms

$$T(v) := \theta([Hb_vH]), \qquad [\varpi_v^{\pm 1}] := \theta([H\varpi_v^{\pm 1}H])$$

of $J(M_H^*)$ induce the eponymous endomorphisms of $\Gamma((M_H^*)^{\mathrm{an}}, \Omega^{\mathrm{an}})$ (which were defined in 1.12)) via the contravariant action on the cotangent space at the origin. In other words, the action of $\mathbf{T}_H^S$ on $\mathcal{S}_2^H \xrightarrow{\sim} \Gamma((M_H^*)^{\mathrm{an}}, \Omega^{\mathrm{an}})$ factors through the morphism θ. Similarly, we let $T(v)$ and $[\varpi_v^{\pm 1}]$ act on the spaces W, W_{dR} and W_ℓ defined in 1.12 by contravariant functoriality, i.e. by $[Hb_vH]^*$ and $[H\varpi_v^{\pm 1}H]^*$.

The commutative ring

$$\mathbf{T} := \theta\left(\mathbf{T}_H^S\right) \subset \mathrm{End}(J(M_H^*))$$

(the "physical" Hecke algebra) is a free $\mathbf{Z}$-module of finite rank, generated (as a ring) by the endomorphisms $T(v)$ and $[\varpi_v^{\pm 1}]$ (for $v \notin S$). It follows from (1.16.2) that

$$(\forall v \notin S) \qquad \iota\left(T(v)\right) = [\varpi_v^{-1}]\, T(v), \qquad \iota\left([\varpi_v^{\pm 1}]\right) = [\varpi_v^{\mp 1}], \quad (1.17.2)$$

hence $\mathbf{T}$ is a commutative subring of $\mathrm{End}(J(M_H^*))$ stable by the Rosati involution ι. The positivity property of this involution implies ([Mu, §21]) that the $\mathbf{Q}$-algebra $\mathbf{T} \otimes \mathbf{Q}$ is isomorphic to a finite product of number fields

$$\mathbf{T} \otimes \mathbf{Q} \xrightarrow{\sim} \prod_{j \in J} L_j, \qquad J = J_1 \cup J_2,$$

where each factor L_j is ι-stable and

$$\begin{aligned} &(\forall j \in J_1) \quad L_j \text{ is totally real and } \iota \text{ acts trivially on } L_j \\ &(\forall j \in J_2) \quad L_j \text{ is a CM field and } \iota \text{ acts on } L_j \text{ by complex conjugation.} \end{aligned} \qquad (1.17.3)$$

For each $j \in J$, we denote by θ_j the ring homomorphism

$$\theta_j : \mathbf{T}_H^S \xrightarrow{\theta} \mathbf{T} \subset \mathbf{T} \otimes \mathbf{Q} \twoheadrightarrow L_j. \tag{1.17.4}$$

The image of θ_j is an order in L_j and $\mathbf{T}$ has finite index in $\prod_j \mathrm{Im}(\theta_j)$. If $\pi = \sigma_2 \otimes \pi_f$ is an irreducible cuspidal unitary representation of $B_{\mathbf{A}}^*$ satisfying $\pi_f^H \neq 0$, then $\mathbf{T}_H^S$ acts on π_f^H through a ring homomorphism $\lambda_\pi : \mathbf{T}_H^S \longrightarrow \mathbf{C}$, which factors as

$$\lambda_\pi : \mathbf{T}_H^S \xrightarrow{\theta_j} L_j \overset{\sigma}{\hookrightarrow} \mathbf{C},$$

for a unique pair $(j, \sigma) \in J \times \mathrm{Hom}(L_j, \mathbf{C})$. The strong multiplicity one theorem for automorphic representations of $B_{\mathbf{A}}^*$ ([Ge], VI.2) implies that

$$\left[\pi = \sigma_2 \otimes \pi_f \neq \pi' = \sigma_2 \otimes \pi'_f,\ \pi_f^H \neq 0 \neq \pi_f'^H \implies \lambda_\pi \neq \lambda_{\pi'}.\right] \tag{1.17.5}$$

The canonical map $J(M_H^*) \longrightarrow J(N_H^*)$ (induced by the projection $M_H^* \longrightarrow N_H^*$) is T_H^S-equivariant and surjective; it follows that $T_H^S \otimes \mathbf{Q}$ acts on $J(N_H^*) \otimes \mathbf{Q}$ (in the category of abelian varieties up to isogeny) by a certain quotient of $\mathbf{T} \otimes \mathbf{Q}$.

Let ℓ be a prime number. For each prime $\mathscr{L} \mid \ell$ of L_j, denote by $V_{\mathscr{L}}(\theta_j)$ the $\mathscr{L}$-adic Galois representation of $G_{F,S}$ associated to $JL(\pi)$ (which is a Hilbert modular form of parallel weight $(2, \ldots, 2)$ over F). This is a two-dimensional, absolutely irreducible ([Tay, Prop. 3.1]) representation of $G_{F,S}$ over $(L_j)_{\mathscr{L}}$, characterized by the property

$$(\forall v \notin S)\ \det(1 - X\,\mathrm{Fr}(v)_{\mathrm{geom}} \mid V_{\mathscr{L}}(\theta_j)) = 1 - \theta_j(T(v))X + \omega_\pi(v)(Nv)X^2.$$

(1.18) Proposition (Decomposition of $J(M_H^*)$ and $J(N_H^*)$ up to isogeny). (i) *For each pair $(j, \sigma) \in J \times \mathrm{Hom}(L_j, \mathbf{C})$ there exists a unique irreducible cuspidal unitary representation $\pi = \sigma_2 \otimes \pi_f$ of $B_{\mathbf{A}}^*$ satisfying $\pi_f^H \neq 0$, $\lambda_\pi = \sigma \circ \theta_j$. We denote $\pi = \pi(\sigma \circ \theta_j)$.*

(ii) *For each prime number ℓ, the $\mathbf{T} \otimes \mathbf{Q}_\ell[G_F]$-module $W_\ell = H^1_{et}(M_H^* \otimes_F \overline{F}, \mathbf{Q}_\ell)$ is isomorphic to*

$$W_\ell \xrightarrow{\sim} \bigoplus_{j \in J} \Big(\bigoplus_{\mathscr{L} \mid \ell} V_{\mathscr{L}}(\theta_j) \Big)^{a_j},$$

where $a_j = \dim_{\mathbf{C}}(\pi(\sigma \circ \theta_j)_f^H) \geq 1$, $\mathscr{L}$ runs through all primes of L_j above ℓ, and $\mathbf{T}$ acts on $V_{\mathscr{L}}(\theta_j)$ via θ_j and the embedding $L_j \hookrightarrow (L_j)_{\mathscr{L}}$.

(iii) *There exists a $\mathbf{T}$-linear isogeny $u : J(M_H^*) \longrightarrow \prod_{j \in J} A_j^{a_j}$ defined over F, where each $a_j \geq 1$ is as in (ii), A_j is an abelian variety over F of dimension $\dim(A_j) = [L_j : \mathbf{Q}]$ satisfying $\mathcal{O}_{L_j} = \mathrm{End}_F(A_j)$ (hence*

A_j is F-simple) on which $\mathbf{T}$ acts via θ_j, $H^1_{et}(A_j \otimes_F \overline{F}, \mathbf{Q}_\ell)$ is isomorphic to $\bigoplus_{\mathscr{L}|\ell} V_{\mathscr{L}}(\theta_j)$ as an $\mathbf{T} \otimes \mathbf{Q}_\ell[G_F]$-module and there is an equality of L-series (Euler factor by Euler factor)

$$L(A_j/F, s) = \prod_{\sigma : L_j \hookrightarrow \mathbf{C}} L(\pi(\sigma \circ \theta_j), s - 1/2).$$

The abelian variety A_j is unique up to an $\mathcal{O}_{L_j}$-linear isogeny.
(iv) *If $j, j' \in J$, $j \neq j'$, then A_j is not isogeneous (over F) to $A_{j'}$.*
(v) *For each $j \in J_1$ (resp., $j \in J_2$) and each polarization of A_j defined over F, the corresponding Rosati involution acts on $L_j = \mathrm{End}_F(A_j) \otimes \mathbf{Q}$ trivially (resp., by the complex conjugation).*
(vi) *There exists a $\mathbf{T}$-linear isogeny $u_1 : J(N_H^*) \longrightarrow \prod_{j \in J_1} A_j^{a_j}$ (defined over F) fitting into a commutative diagram (in which the right vertical arrow is the canonical projection)*

$$\begin{array}{ccc} J(M_H^*) & \xrightarrow{u} & \prod_{j \in J} A_j^{a_j} \\ \downarrow & & \downarrow \\ J(N_H^*) & \xrightarrow{u_1} & \prod_{j \in J_1} A_j^{a_j}. \end{array}$$

Proof. All statements are well-known, but as we are not aware of a good reference, we sketch the argument. In (i), the uniqueness is a consequence of (1.17.5). In order to prove the existence of π, note that $W = H^1(M_H^*, \mathbf{Q})$ is a faithful $\mathbf{T} \otimes \mathbf{Q}$-module, hence it is isomorphic to

$$W \xrightarrow{\sim} \bigoplus_{j \in J} L_j^{\oplus b_j} \qquad (b_j > 0). \tag{1.18.1}$$

Both comparison isomorphisms

$$W \otimes_{\mathbf{Q}} \mathbf{C} \xrightarrow{\sim} W_{\mathbf{C}} \oplus \overline{W_{\mathbf{C}}}, \qquad W_{\mathrm{dR}} \otimes_{F, \tau_1} \mathbf{C} \xrightarrow{\sim} W_{\mathbf{C}}$$

(where $W_{\mathbf{C}} = \Gamma((M_H^*)^{\mathrm{an}}, \Omega^{\mathrm{an}})$ and $W_{\mathrm{dR}} = H^1_{\mathrm{dR}}(M_H^*/F)$) are $\mathbf{T} \otimes \mathbf{C}$-linear. As $\tau_1(F) \subset \mathbf{R}$, the second one implies that the $\mathbf{T} \otimes \mathbf{C}$-modules $W_{\mathbf{C}}$ and $\overline{W_{\mathbf{C}}}$ are isomorphic, which yields an isomorphism of $\mathbf{T} \otimes \mathbf{C}$-modules

$$W_{\mathbf{C}} \xrightarrow{\sim} \bigoplus_{j \in J} (L_j \otimes_{\mathbf{Q}} \mathbf{C})^{\oplus a_j}, \qquad a_j = b_j/2 > 0.$$

The isomorphism

$$L_j \otimes_{\mathbf{Q}} \mathbf{C} \xrightarrow{\sim} \prod_{\sigma : L_j \hookrightarrow \mathbf{C}} \mathbf{C}, \qquad a \otimes z \mapsto (\sigma(a) z)_\sigma$$

implies that each morphism $\sigma \circ \theta_j$ occurs as λ_π in $W_{\mathbf{C}}$, with multiplicity $a_j > 0$. In other words, $\pi(\sigma \circ \theta_j)$ exists and

$$\dim_{\mathbf{C}} \pi(\sigma \circ \theta_j)_f^H = a_j.$$

(ii) As W_ℓ is a semi-simple $\mathbf{Q}_\ell[G_F]$-module ([Fa, Satz 3]), it is sufficient to show that the traces of $\mathrm{Fr}(v)_{\mathrm{geom}}$ ($v \notin S$) acting on both sides coincide, which follows from the $\mathbf{T} \otimes \mathbf{Q}_\ell$-linear comparison isomorphism $W \otimes_{\mathbf{Q}} \mathbf{Q}_\ell \xrightarrow{\sim} W_\ell$ and the formulas (1.14.2) and (1.18.1).

(iii) The existence of a $\mathbf{T}$-linear isogeny $J(M_H^*) \longrightarrow \prod_{j\in J} A_j^{a_j}$ (defined over F) such that $L_j \hookrightarrow \mathrm{End}_F(A_j) \otimes \mathbf{Q}$, $\mathbf{T}$ acts on A_j via θ_j and $H^1_{et}(A_j \otimes_F \overline{F}, \mathbf{Q}_\ell)$ is isomorphic to $\bigoplus_{\mathscr{L}|\ell} V_{\mathscr{L}}(\theta_j)$ follows from (ii) and Faltings' isogeny theorem [Fa, Satz 4] (formerly known as Tate's conjecture). The endomorphism ring $\mathrm{End}_F(A_j)$ contains $\mathrm{Im}(\theta_j)$, which is an order in L_j. Replacing A_j by the F-isogeneous abelian variety $\mathrm{Hom}_{\mathrm{Im}(\theta_j)}(\mathcal{O}_{L_j}, A_j)$ (see [Co, Ch. 10] for a discussion of the formalism of $\mathcal{O}$-transforms), we may assume that $\mathrm{End}_F(A_j)$ contains $\mathcal{O}_{L_j}$.

Fix a prime number ℓ which splits in $L_j/\mathbf{Q}$. If $\mathscr{L}$ and $\mathscr{L}'$ are distinct primes above ℓ in L_j, then the $\mathbf{Q}_\ell[G_F]$-modules $V_{\mathscr{L}}(\theta_j)$ and $V_{\mathscr{L}'}(\theta_j)$ are not isomorphic, as L_j is generated (as a field extension of $\mathbf{Q}$) by the traces $\theta_j(T(v))$ and determinants $\omega_\pi(v)$ of the Frobenius elements $\mathrm{Fr}(v)_{\mathrm{geom}}$ ($v \notin S$) acting on $V_{\mathscr{L}}(\theta_j)$ (as well as on $V_{\mathscr{L}'}(\theta_j)$), and the two embeddings $L_j \hookrightarrow (L_j)_{\mathscr{L}}$, $L_j \hookrightarrow (L_j)_{\mathscr{L}'}$ are distinct. Absolute irreducibility of $V_{\mathscr{L}}(\theta_j)$ yields an inclusion

$$L_j \otimes_{\mathbf{Q}} \mathbf{Q}_\ell \hookrightarrow \mathrm{End}_F(A_j) \otimes_{\mathbf{Q}} \mathbf{Q}_\ell = \mathrm{End}_{\mathbf{Q}_\ell[G_F]}\Big(\bigoplus_{\mathscr{L}|\ell} V_{\mathscr{L}}(\theta_j)\Big) = \bigoplus_{\mathscr{L}|\ell} \mathbf{Q}_\ell, \tag{1.18.2}$$

which must be an equality, by comparing the dimensions of the two flank terms; this proves that $L_j = \mathrm{End}_F(A_j) \otimes \mathbf{Q}$. As $\mathcal{O}_{L_j} \subset \mathrm{End}_F(A_j)$, we must have $\mathrm{End}_F(A_j) = \mathcal{O}_{L_j}$; as its endomorphism ring is an integral domain, the abelian variety A_j is F-simple. The statement about the local L-factor $L_v(A_j/F, s)$ follows from (1.15.2) (resp., the discussion following (1.15.2)) if $v \in S$ (resp., if $v \notin S$), combined with (i) and (ii), since $L_v(h^1(M_H^*), s) = L_v(J(M_H^*), s)$. The uniqueness of A_j (up to an $\mathcal{O}_{L_j}$-linear isogeny) follows from Faltings' isogeny theorem and the equality in (1.18.2).

(iv) The same argument as in the proof of (iii) shows that, if ℓ is a prime number split in both L_j and $L_{j'}$ and $\mathscr{L}, \mathscr{L}'$ are primes above ℓ in $L_j, L_{j'}$, respectively, (not necessarily distinct), then the $\mathbf{Q}_\ell[G_F]$-modules $V_{\mathscr{L}}(\theta_j)$ and $V_{\mathscr{L}'}(\theta_{j'})$ are not isomorphic, hence A_j is not isogeneous to $A_{j'}$ (over F).

(v) As $\mathrm{End}_F(A_j)$ is commutative, the Rosati involution does not depend on the polarization. For the canonical polarization, the statement follows from (1.17.3).

(vi) $\mathbf{T}_H^S \otimes \mathbf{Q}$ acts on $J(N_H^*) \otimes \mathbf{Q}$ through a certain quotient $\mathbf{T}' \otimes \mathbf{Q}$ of $\mathbf{T} \otimes \mathbf{Q}$, which is necessarily isomorphic to $\prod_{j \in J'}$, for some subset $J' \subset J$. As $\Gamma((N_H^*)^{\mathrm{an}}, \Omega^{\mathrm{an}}) \xrightarrow{\sim} \mathcal{S}_2^{H\widehat{F}^*}$ consists precisely of those automorphic forms from $\mathcal{S}_2^H$ which have trivial central character, the formula (1.17.2) implies that the Rosati involution acts trivially on $\mathbf{T}' \otimes \mathbf{Q}$, hence $J' \subset J_1$. The arguments used in the proof of (i), (ii) show that $J(N_H^*)$ is $\mathbf{T}$-linearly isogeneous to $\prod_{j \in J'} A_j^{a'_j}$, where

$$(\forall j \in J')\,(\forall \sigma : L_j \hookrightarrow \mathbf{C}) \qquad a'_j = \dim_{\mathbf{C}} \pi(\sigma \circ \theta_j)_f^{H\widehat{F}^*} > 0.$$

In particular, for each pair $(j, \sigma) \in J' \times \mathrm{Hom}(L_j, \mathbf{C})$, the representation $\pi(\sigma \circ \theta_j)$ satisfies $\pi(\sigma \circ \theta_j)_f^{H\widehat{F}^*} \neq 0$, hence it has trivial central character, which in turn implies that

$$a'_j = \dim_{\mathbf{C}} \pi(\sigma \circ \theta_j)_f^{H\widehat{F}^*} = \dim_{\mathbf{C}} \pi(\sigma \circ \theta_j)_f^{H} = a_j.$$

Conversely, if $j \in J_1$, then the existence of the Weil pairing implies that $G_{F,S}$ acts on $\Lambda^2 V_{\mathscr{L}}(\theta_j)$ by the cyclotomic character ([Cas, Cor. 5.5]; [Ri, Lemma 4.5.1]), hence $\omega_\pi = 1$. It follows that $J' = J_1$.

(1.19) The maps $M_H^* \longrightarrow J(M_H^*)$ and $N_H^* \longrightarrow J(N_H^*)$. In the classical case $B = M_2(\mathbf{Q})$, the class of the divisor (∞) in $\mathrm{Pic}(M_H^*)$ has the following properties: its pull-back to each irreducible component of $M_H^* \otimes_{\mathbf{Q}} \overline{\mathbf{Q}}$ has degree one and there exists an integer $m \geq 1$ such that each Hecke operator $[HbH]_*$ acts on the class of $m(\infty)$ by multiplication by its degree. Indeed, we can take any m that annihilates the cuspidal group $C(M_H^*) \subset \mathrm{Pic}(M_H^*)$, which is defined as the subgroup generated by the divisors whose pull-backs to each irreducible component of $M_H^* \otimes_{\mathbf{Q}} \overline{\mathbf{Q}}$ have degree zero and are supported at the cusps. The group $C(M_H^*)$ is finite by the theorem of Manin-Drinfeld [Dr].

For $B \neq M_2(\mathbf{Q})$, Zhang [Zh 1, Introduction] (see also [Co-Va 1, 3.5]) constructed a certain class $\xi(M_H^*) \in \mathrm{Pic}(M_H^*/F) \otimes \mathbf{Q}$ ("the Hodge class") such that the degree of the pull-back of $\xi(M_H^*)$ to each irreducible component of $M_H^* \otimes_F \overline{F}$ is equal to one. More precisely, $\xi(M_H^*)$ is constructed, for a suitable integer $m \geq 1$, as $\xi(M_H^*) = \widetilde{m\xi(M_H^*)} \otimes 1/m$, where $\widetilde{m\xi(M_H^*)} \in \mathrm{Pic}(M_H^*/F)$ has degree m on each irreducible component of $M_H^* \otimes_F \overline{F}$ and each Hecke operator $[HbH]_*$ acts on it by multiplication by its degree ([Co-Va 1], Remark 3.7).

In general, if

$$M_H^* \otimes_F \overline{F} = \coprod_\alpha \mathcal{C}_\alpha$$

is the decomposition into irreducible components, let $m\delta_\alpha \in \mathrm{Pic}(\mathcal{C}_\alpha)$ be the pull-back to $\mathrm{Pic}(\mathcal{C}_\alpha)$ of the class of $m(\infty)$ (resp., of $m\widetilde{\xi(M_H^*)}$) if $B = M_2(\mathbf{Q})$ (resp., if $B \neq M_2(\mathbf{Q})$). We denote by

$$\iota : M_H^* \longrightarrow J(M_H^*)$$

the morphism (defined over F), which is characterized by the formula

$$(\forall \alpha)\,(\forall P \in \mathcal{C}_\alpha(\overline{F})) \qquad \iota(P) = m(P) - m\delta_\alpha.$$

The same discussion applies to the curves N_H^*; we obtain a morphism

$$\iota : N_H^* \longrightarrow J(N_H^*).$$

2. CM points

In this section we recall basic properties of CM points on the curves M_H and N_H. We follow the notation and conventions of Sect. 1.

(2.1) Let K be a totally imaginary quadratic extension of F such that each prime $v \in S_B$ either ramifies or is inert in K/F. This assumption implies that there exists an F-embedding (= an injective homomorphism of F-algebras) $t : K \hookrightarrow B$; we fix such an embedding and denote by $t_v : K \otimes_F F_v \hookrightarrow B_v$ (resp., by $\widehat{t} : \widehat{K} \hookrightarrow \widehat{B}$) the induced embedding of the completions (resp., of the finite adèles). As in 1.1, we fix an embedding $\overline{K} \hookrightarrow \mathbf{C}$ extending $\tau_1 : F \longrightarrow \mathbf{R}$.

(2.2) Lemma. (i) *There exists a unique point* $z \in \mathbf{C}$ *with* $\mathrm{Im}(z) > 0$, *which is fixed by the action of* $t(K^*) \subset B^* \subset B_{\tau_1}^* \xrightarrow{\sim} GL_2(\mathbf{R})$.
(ii) *We have* $\{\lambda \in B^* \mid \lambda(z) = z\} = t(K^*)$.

Proof. Easy exercise.

(2.3) Definition. *The set of* **CM-points by** K *on the curve* M_H *(resp., on* N_H*) is the set*

$$\begin{aligned} CM(M_H, K) &= \{x = [z, b] \in M_H(\mathbf{C}) \mid b \in \widehat{B}^*\} \\ CM(N_H, K) &= \{x = [z, b] \in N_H(\mathbf{C}) \mid b \in \widehat{B}^*\} \\ &= \text{the image of } CM(M_H, K) \text{ in } N_H(\mathbf{C}) \end{aligned}$$

(these sets do not depend on the choice of t, as two different F-embeddings of K into B are conjugate by an element of B^*, by the Skolem-Noether theorem).

(2.4) The Galois action on CM-points. The reciprocity law [De, 3.9], [Mi 2, II.5.1] (with the sign corrected, [Mi 3, 1.10]) states that $CM(M_H, K) \subset M_H(K^{ab})$ and that the Galois action of $\mathrm{Gal}(K^{ab}/K)$ on $CM(M_H, K)$ is described, via the reciprocity map $\mathrm{rec}_K : \widehat{K}^* \longrightarrow \mathrm{Gal}(K^{ab}/K)$, by the following formula:

$$(\forall a \in \widehat{K}^*) \qquad \mathrm{rec}_K(a)\,[z, b] = [z, \widehat{t}(a)b].$$

(2.5) Proposition. *Denote by $K(x) \subset K^{ab}$ the field of definition over K of a CM point (by K) $x = [z, b]$ on the curve M_H (resp., on N_H). Then the reciprocity map rec_K induces an isomorphism $\mathrm{rec}_K : K^*\backslash\widehat{K}^*/\widehat{t}^{-1}(bHb^{-1}) \xrightarrow{\sim} \mathrm{Gal}(K(x)/K)$ (resp., $\mathrm{rec}_K : K^*\backslash\widehat{K}^*/\widehat{t}^{-1}(bH\widehat{F}^*b^{-1}) \xrightarrow{\sim} \mathrm{Gal}(K(x)/K)$).*

Proof. If $x = [z, b] \in CM(M_H, K)$ and $a \in \widehat{K}^*$, then we deduce from 2.4 and Lemma 2.2(ii) that $\mathrm{rec}_K(a)\,[z, b] = [z, b]$ if and only if

$$\begin{aligned}
&\iff (\exists \lambda \in B^*)\,(\exists h \in H) \qquad (z, \widehat{t}(a)b) = (\lambda(z), \lambda bh) \in (\mathbf{C} - \mathbf{R}) \times \widehat{B}^* \\
&\iff (\exists \mu \in K^*)\,(\exists h \in H) \qquad \widehat{t}(a)b = t(\mu)bh \\
&\iff a \in K^* \cdot \widehat{t}^{-1}(bHb^{-1}).
\end{aligned}$$

The proof for $x \in CM(N_H, K)$ is similar.

(2.6) Ring class fields of K

(2.6.1) From now on, we shall consider only CM points on the curves N_H. For such a point $x = [z, b]$, the formula from Proposition 2.5 can be restated as an isomorphism

$$\mathrm{rec}_K : \widehat{K}^*/K^*\widehat{F}^*Z \xrightarrow{\sim} \mathrm{Gal}(K(x)/K),$$

where $Z = \widehat{t}^{-1}(bH\widehat{\mathcal{O}}_F^*b^{-1}) \subset \widehat{\mathcal{O}}_K^*$ is an open (compact) subgroup of $\widehat{\mathcal{O}}_K^*$ containing $\widehat{\mathcal{O}}_F^*$.

(2.6.2) In the special case when $H = \widehat{R}^*$, for an $\mathcal{O}_F$-order $R \subset B$, then $Z = \widehat{\mathcal{O}}^*$ for some $\mathcal{O}_F$-order $\mathcal{O} \subset K$. Such an order is necessarily of the form $\mathcal{O}_c = \mathcal{O}_F + c\mathcal{O}_K$, where $c \subset \mathcal{O}_F$ is a non-zero ideal of $\mathcal{O}_F$. The corresponding abelian extension $K[c]/K$ satisfying

$$\mathrm{rec}_K : \widehat{K}^*/K^*\widehat{F}^*\widehat{\mathcal{O}}_c^* \xrightarrow{\sim} \mathrm{Gal}(K[c]/K)$$

is called the **ring class field of K of conductor** c (note that Zhang [Zh 2] uses the same terminology, but Cornut and Vatsal [Co-Va 1,2] consider a slightly more general class of extensions).

If $c, c' \subset \mathcal{O}_F$ are non-zero ideals, then we have

$$\widehat{\mathcal{O}}_c^* \cdot \widehat{\mathcal{O}}_{c'}^* = \widehat{\mathcal{O}}_{\gcd(c,c')}^*, \qquad K^*\widehat{F}^*\widehat{\mathcal{O}}_c^* \cap K^*\widehat{F}^*\widehat{\mathcal{O}}_{c'}^* \supseteq K^*\widehat{F}^*\widehat{\mathcal{O}}_{\mathrm{lcm}(c,c')}^*, \tag{2.6.2.1}$$

hence

$$K[c] \cap K[c'] = K[\gcd(c,c')], \qquad K[c]\,K[c'] \subseteq K[\mathrm{lcm}(c,c')].$$

(2.6.3) Each prime of K not dividing $c\mathcal{O}_K$ is unramified in $K[c]/K$. If ℓ is a prime of F which does not divide c and which is inert in K/F, then $\ell\mathcal{O}_K$ splits completely in $K[c]/K$ and each prime above ℓ is totally tamely ramified in $K[c\ell]/K[c]$.

(2.6.4) In general, if $Z \subset \widehat{\mathcal{O}}_K^*$ is an open subgroup containing $\widehat{\mathcal{O}}_F^*$, we denote by $K_Z \subset K^{ab}$ the finite abelian extension of K satisfying

$$\mathrm{rec}_K : \widehat{K}^*/K^*\widehat{F}^*Z \xrightarrow{\sim} \mathrm{Gal}(K_Z/K).$$

It follows from the last paragraph in 1.4 and from Proposition 2.5 that we have

$$\bigcup_Z K_Z = \bigcup_c K[c] =: K[\infty], \qquad CM(N_H, K) \subset N_H(K[\infty]).$$

(2.6.5) The reciprocity map rec_K is compatible with the action of the Galois group $\mathrm{Gal}(K/F) = \{1, \rho\}$ (where ρ denotes the complex conjugation on $\overline{K} \subset \mathbf{C}$) on both sides. As

$$(\forall a \in \widehat{K}^*) \qquad a \cdot \rho(a) \in \widehat{F}^*,$$

it follows that K_Z/F is a Galois extension and the conjugation action of ρ on $\mathrm{Gal}(K_Z/K)$ is given by $\rho g \rho^{-1} = g^{-1}$. In other words,

$$\mathrm{Gal}(K_Z/F) \xrightarrow{\sim} \mathrm{Gal}(K_Z/K) \rtimes \{1, \rho\}$$

is a generalized dihedral group.

(2.7) We shall now describe the Galois group

$$\mathrm{Gal}(K_{Z'}/K_Z) \xrightarrow{\sim} K^*\widehat{F}^*Z/K^*\widehat{F}^*Z', \tag{2.7.1}$$

where $Z' \subset Z \subset \widehat{\mathcal{O}}_K^*$ are two open subgroups containing $\widehat{\mathcal{O}}_F^*$. We use the elementary fact that, whenever A and $B \supset C$ are subgroups of some abelian group, then the obvious inclusion maps give rise to an exact sequence

$$0 \longrightarrow \frac{A\cap B}{A\cap C} \longrightarrow \frac{B}{C} \longrightarrow \frac{AB}{AC} \longrightarrow 0,$$

from which we deduce

$$\begin{aligned}
&\widehat{F}^* \cap \widehat{\mathcal{O}}_K^* = \widehat{\mathcal{O}}_F^* \Longrightarrow \widehat{F}^* \cap Z = \widehat{F}^* \cap Z' = \widehat{\mathcal{O}}_F^* \Longrightarrow Z/Z' \xrightarrow{\sim} \widehat{F}^* Z/\widehat{F}^* Z' \\
&F^* \cap (\mathcal{O}_K^* \cap Z) = F^* \cap (\mathcal{O}_K^* \cap Z') = \mathcal{O}_F^* \Longrightarrow (\mathcal{O}_K^* \cap Z)/(\mathcal{O}_K^* \cap Z') \xrightarrow{\sim} \\
&\xrightarrow{\sim} F^*(\mathcal{O}_K^* \cap Z)/F^*(\mathcal{O}_K^* \cap Z')
\end{aligned} \tag{2.7.2}$$

and, by taking $A = K^*$, $B = \widehat{F}^* Z$, $C = \widehat{F}^* Z'$, an exact sequence

$$0 \longrightarrow \frac{K^* \cap \widehat{F}^* Z}{K^* \cap \widehat{F}^* Z'} \longrightarrow \frac{Z}{Z'} \longrightarrow \mathrm{Gal}(K_{Z'}/K_Z) \longrightarrow 0. \tag{2.7.3}$$

The next step is to determine the abelian group $K^* \cap \widehat{F}^* Z$ (and its counterpart for Z' instead of Z).

(2.8) Proposition. *Put* $G = \mathrm{Gal}(K/F) = \{1, \rho\}$.
(i) $\mathrm{Ker}\left(N_{K/F} : \mathcal{O}_K^* \longrightarrow \mathcal{O}_F^*\right) = (\mathcal{O}_K^*)_{\mathrm{tors}}$.
(ii) *There is an isomorphism* $(K^* \cap \widehat{F}^* \widehat{\mathcal{O}}_K^*)/F^*\mathcal{O}_K^* \xrightarrow{\sim} \mathrm{Ker}(\mathrm{Pic}(\mathcal{O}_F) \longrightarrow \mathrm{Pic}(\mathcal{O}_K))$.
(iii) *([Wa, Thm. 10.3]) The abelian group* $\mathrm{Ker}(\mathrm{Pic}(\mathcal{O}_F) \longrightarrow \mathrm{Pic}(\mathcal{O}_K))$ *naturally injects into* $H^1(G, \mathcal{O}_K^*)$. *The order of the abelian group* $H^1(G, \mathcal{O}_K^*)$ *is equal to* 1 *or* 2.

Proof. (i) As K is a CM field and F its maximal real subfield, the group $\mathrm{Ker}(N_{K/F} : \mathcal{O}_K^* \longrightarrow \mathcal{O}_F^*)$ is finite, hence contained in $(\mathcal{O}_K^*)_{\mathrm{tors}}$. On the other hand, each root of unity $\zeta \in (\mathcal{O}_K^*)_{\mathrm{tors}}$ satisfies $\zeta \cdot \rho(\zeta) = 1$.
(ii) If $\beta \in K^* \cap \widehat{F}^* \widehat{\mathcal{O}}_K^*$, then, for each non-archimedean prime v of F, there exists $\alpha_v \in F_v^*$ such that $\mathrm{ord}_w(\beta) = \mathrm{ord}_w(\alpha_v)$, for all primes w above v in K. Let I be the ideal of $\mathcal{O}_F$ with divisor equal to $\sum_v \mathrm{ord}_v(\alpha_v)\,[v]$; then $I\mathcal{O}_K = (\beta)$. It follows that the map

$$\beta \mapsto \text{ the class of } I\mathcal{O}_K$$

defines a homomorphism

$$(K^* \cap \widehat{F}^* \widehat{\mathcal{O}}_K^*)/F^*\mathcal{O}_K^* \longrightarrow \mathrm{Ker}(\mathrm{Pic}(\mathcal{O}_F) \longrightarrow \mathrm{Pic}(\mathcal{O}_K)),$$

the inverse of which is given by $I \mapsto \beta$, if $I\mathcal{O}_K = (\beta)$.
(iii) The exact sequence

$$0 \longrightarrow P_K \longrightarrow I_K \longrightarrow \mathrm{Pic}(\mathcal{O}_K) \longrightarrow 0$$

(where P_K (resp., I_K) denotes the group of principal (resp., of all) fractional ideals of K) and its counterpart over F give rise to a commutative diagram with exact rows

$$\begin{array}{ccccccccc} 0 & \longrightarrow & P_F & \longrightarrow & I_F & \longrightarrow & \mathrm{Pic}(\mathcal{O}_F) & \longrightarrow & 0 \\ & & \downarrow & & \cap & & \downarrow & & \downarrow \\ 0 & \longrightarrow & P_K^G & \longrightarrow & I_K^G & \longrightarrow & \mathrm{Pic}(\mathcal{O}_K)^G & \longrightarrow & H^1(G, P_K). \end{array}$$

The Snake Lemma then implies that the group $\mathrm{Ker}(\mathrm{Pic}(\mathcal{O}_F) \longrightarrow \mathrm{Pic}(\mathcal{O}_K))$ injects into $\mathrm{Coker}(P_F \longrightarrow P_K^G)$. On the other hand, the cohomology sequence of

$$0 \longrightarrow \mathcal{O}_K^* \longrightarrow K^* \longrightarrow P_K \longrightarrow 0$$

yields an exact sequence

$$0 \longrightarrow \mathcal{O}_F^* \longrightarrow F^* \longrightarrow P_K^G \longrightarrow H^1(G, \mathcal{O}_K^*) \longrightarrow 0,$$

hence $\mathrm{Coker}(P_F \longrightarrow P_K^G) \xrightarrow{\sim} H^1(G, \mathcal{O}_K^*)$. The abelian group $H^1(G, \mathcal{O}_K^*)$ is a quotient of the finite cyclic group $\mathrm{Ker}(N_{K/F} : \mathcal{O}_K^* \longrightarrow \mathcal{O}_F^*) = (\mathcal{O}_K^*)_{\mathrm{tors}}$; on the other hand, it is killed by $|G| = 2$, which implies that $|H^1(G, \mathcal{O}_K^*)| \leq 2$.

(2.9) Proposition. *If $Z' \subset Z$ are open subgroups of $\widehat{\mathcal{O}}_K^*$ containing $\widehat{\mathcal{O}}_F^*$, then the natural inclusions give rise to an exact sequence*

$$0 \longrightarrow \frac{\mathcal{O}_K^* \cap Z}{\mathcal{O}_K^* \cap Z'} \longrightarrow \frac{K^* \cap \widehat{F}^* Z}{K^* \cap \widehat{F}^* Z'} \longrightarrow U_{Z,Z'} \longrightarrow 0,$$

in which $|U_{Z,Z'}| \leq |\mathrm{Ker}(\mathrm{Pic}(\mathcal{O}_F) \longrightarrow \mathrm{Pic}(\mathcal{O}_K))| \leq 2.$

Proof. Injectivity of the first arrow follows from the second isomorphism in (2.7.2) and the equality

$$\begin{aligned} F^*(\mathcal{O}_K^* \cap Z) \cap \widehat{F}^* Z' &= F^*(\mathcal{O}_K^* \cap Z \cap \widehat{F}^* Z') = F^*(\mathcal{O}_K^* \cap Z \cap \widehat{\mathcal{O}}_F^* Z') = \\ &= F^*(\mathcal{O}_K^* \cap Z'). \end{aligned}$$

Define $U_{Z,Z'}$ to be the cokernel of this injective map. Assume first that $\mathrm{Ker}(\mathrm{Pic}(\mathcal{O}_F) \longrightarrow \mathrm{Pic}(\mathcal{O}_K)) = 0$; then $K^* \cap \widehat{F}^* \widehat{\mathcal{O}}_K^* = F^* \mathcal{O}_K^*$, by Proposition 2.8(ii), which implies that

$$\begin{aligned} K^* \cap \widehat{F}^* Z = F^* \mathcal{O}_K^* \cap \widehat{F}^* Z = F^*(\mathcal{O}_K^* \cap \widehat{F}^* Z) &= F^*(\mathcal{O}_K^* \cap \widehat{\mathcal{O}}_F^* Z) \\ &= F^*(\mathcal{O}_K^* \cap Z) \end{aligned}$$

(and similarly for Z'), hence $U_{Z,Z'} = 0$.

If $\mathrm{Ker}(\mathrm{Pic}(\mathcal{O}_F) \longrightarrow \mathrm{Pic}(\mathcal{O}_K)) \neq 0$, then there exists $\beta \in K^*$ such that $K^* \cap \widehat{F}^*\widehat{\mathcal{O}}_K^* = F^*\mathcal{O}_K^* \cup \beta F^*\mathcal{O}_K^*$, by Proposition 2.8(ii)-(iii); thus

$$K^* \cap \widehat{F}^* Z = F^*(\mathcal{O}_K^* \cap Z) \cup F^*(\beta\mathcal{O}_K^* \cap \widehat{F}^* Z).$$

If $u, u' \in \mathcal{O}_K^*$ satisfy $\beta u, \beta u' \in \widehat{F}^* Z$, then $u'/u \in \mathcal{O}_K^* \cap \widehat{F}^* Z = \mathcal{O}_K^* \cap Z$, which shows that $|U_{Z,Z'}| \leq 2$.

(2.10) Proposition. *Let $I_0 \subset \mathcal{O}_K$ be the non-zero ideal $I_0 := \mathrm{lcm}\,\{(u-1) \mid u \in (\mathcal{O}_K^*)_{\mathrm{tors}},\ u \neq 1\}$. If $c \subset \mathcal{O}_F$ is a non-zero ideal such that $c\mathcal{O}_K \nmid I_0$ and $Z \subset \widehat{\mathcal{O}}_c^*$ is a subgroup, then $K^* \cap \widehat{F}^* Z = F^*$ and $\mathcal{O}_K^* \cap Z = \mathcal{O}_F^*$.*

Proof. It is enough to consider the case $Z = \widehat{\mathcal{O}}_c^*$. Assume that $\beta \in K^* \cap \widehat{F}^*\widehat{\mathcal{O}}_c^*$, $\beta = ab$, $a \in \widehat{F}^*$, $b \in \widehat{\mathcal{O}}_c^*$. Then we have

$$u := \beta/\rho(\beta) = b/\rho(b) \in K^* \cap \widehat{\mathcal{O}}_c^* \subset \mathcal{O}_K^*.$$

More precisely, u is contained in

$$\mathrm{Ker}\,(\mathcal{O}_K^* \xrightarrow{N_{K/F}} \mathcal{O}_F^*) \cap \left(\widehat{\mathcal{O}}_c^*\right)^{1-\rho} \subseteq (\mathcal{O}_K^*)_{\mathrm{tors}} \cap \mathrm{Ker}\left(\widehat{\mathcal{O}}_K^* \longrightarrow (\mathcal{O}_K/c\mathcal{O}_K)^*\right);$$

the condition $c\mathcal{O}_K \nmid I_0$ then implies that $u = 1$, hence $\beta \in F^*$. It follows that $K^* \cap \widehat{F}^*\widehat{\mathcal{O}}_c^* = F^*$ and $\mathcal{O}_F^* \subset \mathcal{O}_c^* \subset \mathcal{O}_K^* \cap \widehat{F}^*\widehat{\mathcal{O}}_c^* \subset \mathcal{O}_K^* \cap F^* = \mathcal{O}_F^*$.

(2.11) Corollary. *If $Z' \subset Z \subset \widehat{\mathcal{O}}_K^*$ are open subgroups containing $\widehat{\mathcal{O}}_F^*$ such that $Z \subset \widehat{\mathcal{O}}_c^*$, for some non-zero ideal $c \subset \mathcal{O}_F$ satisfying $c\mathcal{O}_K \nmid I_0$, then the Galois group $\mathrm{Gal}(K_{Z'}/K_Z)$ is isomorphic to Z/Z'.*

Proof. Combine (2.7.3) with Propositions 2.9-10.

3. The main result

We follow the notations defined in Sect. 1-2.

(3.1) Fix the following data:

(3.1.1) An open compact subgroup $H \subset \widehat{B}^*$.

(3.1.2) An F-simple quotient A_j ($j \in J_1$) of $J(N_H^*)$ satisfying $\mathrm{End}_F(A_j) = \mathcal{O}_{L_j}$ and a non-trivial $\mathbf{T}$-linear morphism $J(N_H^*) \longrightarrow A_j$ (defined over F). Denote by ι_j the composite morphism

$$\iota_j : N_H^* \xrightarrow{\iota} J(N_H^*) \longrightarrow A_j.$$

(3.1.3) A totally imaginary quadratic extension K of F such that each prime $v \in S_B$ either ramifies or is inert in K/F.

(3.1.4) An F-embedding $t : K \hookrightarrow B$.

(3.1.5) A CM point $x = [z, b] \in CM(N_H, K)$; denote by $K(x)$ its field of definition over K.

(3.1.6) A character $\alpha : \mathrm{Gal}(K(x)/K) \longrightarrow \mathcal{O}_L^*$, where L is a number field containing the totally real field $L_j = \mathrm{End}_F(A_j) \otimes \mathbf{Q}$.

We denote

$$e_\alpha = \sum_{\sigma \in \mathrm{Gal}(K(x)/K)} \alpha^{-1}(\sigma)\,\sigma \in \mathcal{O}_L[\mathrm{Gal}(K(x)/K)]$$

and, for any $\mathcal{O}_L[\mathrm{Gal}(K(x)/K)]$-module M,

$$M^{(\alpha)} = \{m \in M \mid \forall \sigma \in \mathrm{Gal}(K(x)/K) \quad \sigma(m) = \alpha(\sigma)m\}.$$

The character α factors through an injective character

$$\beta : \mathrm{Gal}(K(\alpha)/K) \hookrightarrow \mathcal{O}_L^*,$$

where $K(\alpha) = K(x)^{\mathrm{Ker}(\alpha)} \subset K(x)$.

(3.2) Theorem. *Assume that we are given the data (3.1.1-6); put $y_j = \iota_j(x) \otimes 1 \in A_j(K(x)) \otimes_{\mathcal{O}_{L_j}} \mathcal{O}_L$. Assume, in addition, that the following condition is satisfied:*

$(\star)$ *The abelian variety A_j does not acquire complex multiplication over any totally imaginary quadratic extension K' of F contained in $K(\alpha)$.*

If $e_\alpha(y_j) \notin \left(A_j(K(x)) \otimes_{\mathcal{O}_{L_j}} \mathcal{O}_L\right)_{\mathrm{tors}}$, then the following abelian groups (more precisely, $\mathcal{O}_L$-modules) are finite:

$$\left(A_j(K(x)) \otimes_{\mathcal{O}_{L_j}} \mathcal{O}_L\right)^{(\alpha)} / \mathcal{O}_L \cdot e_\alpha(y_j), \quad \left(\text{Ш}(A_j/K(x)) \otimes_{\mathcal{O}_{L_j}} \mathcal{O}_L\right)^{(\alpha)}$$

$$\left(A_j(K(x)) \otimes_{\mathcal{O}_{L_j}} \mathcal{O}_L\right)^{(\alpha^{-1})} / \mathcal{O}_L \cdot \rho(e_\alpha(y_j)), \left(\text{Ш}(A_j/K(x)) \otimes_{\mathcal{O}_{L_j}} \mathcal{O}_L\right)^{(\alpha^{-1})}$$

(where the complex conjugation ρ acts on $A_j(K(x)) \otimes_{\mathcal{O}_{L_j}} \mathcal{O}_L$ by $\rho \otimes \mathrm{id}$).

(3.3) Previously known results in this direction ([Ko], [Ko-Lo 1,2], [Be-Da], [Ho 1,2], [Be], [Ti], [Zh 3]) use several additional hypotheses which ensure, among other things, that there is a "geometric" formula relating $\rho(y_j)$ to y_j. Our main observation is that the Euler system argument does not require such a geometric relation.

(3.4) As explained in 6.2.1 below, the condition $(\star)$ can be reformulated as follows: the Hilbert modular forms that occur in the factorization of $L(A_j/F, s)$ do not have CM by K'.

(3.5) It will be convenient to rephrase Theorem 3.2 and its proof in terms of the abelian variety

$$A = A_j \otimes_{\mathcal{O}_{L_j}} \mathcal{O}_L, \qquad \mathcal{O}_L \hookrightarrow \mathrm{End}_F(A)$$

(see [Con, §7] for the general formalism of such tensor products). We have

$$y_j \in A_j(K(x)) \otimes_{\mathcal{O}_{L_j}} \mathcal{O}_L = A(K(x)), \qquad e_\alpha(y_j) = e_\beta(y) \in A(K(\alpha)),$$

where

$$y = \mathrm{Tr}_{K(x)/K(\alpha)}(y_j) \in A(K(\alpha))$$
$$e_\beta = \sum_{\sigma \in \mathrm{Gal}(K(\alpha)/K)} \beta^{-1}(\sigma)\,\sigma \in \mathcal{O}_L[\mathrm{Gal}(K(\alpha)/K)].$$

Fix a maximal ideal $\mathfrak{p} \subset \mathcal{O}_L$ and a sufficiently large integer $M >> 0$ such that the ideal $\mathfrak{p}^M$ is principal. Fixing its generator (which will be, by abuse of language, also denoted by $\mathfrak{p}^M$), the Galois cohomology of

$$0 \longrightarrow A[\mathfrak{p}^M] \longrightarrow A(\overline{F}) \xrightarrow{\mathfrak{p}^M} A(\overline{F}) \longrightarrow 0$$

gives rise to the usual descent sequence

$$0 \longrightarrow A(K(\alpha))/\mathfrak{p}^M \xrightarrow{\delta} S \longrightarrow Ш(A/K(\alpha))[\mathfrak{p}^M] \longrightarrow 0,$$

where we have denoted by $S = Sel(A/K(\alpha), \mathfrak{p}^M)$ the classical Selmer group for the $\mathfrak{p}^M$-descent on A.

In order to prove Theorem 3.2, it will be enough to show, assuming that $e_\beta(y) \notin A(K(\alpha))_{\mathrm{tors}}$, that

$$(\exists C(\mathfrak{p}) \geq 0)\ (\forall M >> 0) \qquad \mathfrak{p}^{C(\mathfrak{p})} \cdot \left(S^{(\beta)}/\mathcal{O}_L \cdot \delta(e_\beta(y))\right) = 0. \tag{3.5.1}$$

$$\text{For all but finitely many } \mathfrak{p}, \quad C(\mathfrak{p}) = 0. \tag{3.5.2}$$

(3.6) Let us briefly outline the proof of (3.5.1), the main steps of which follow [Be-Da]. We denote $\mathcal{O}_\mathfrak{p} = \mathcal{O}_{L,\mathfrak{p}}$ and use temporarily the notation "$X \doteq Y$" (resp., "$x \mathrel{\dot\in} X$") as a shorthand for $\mathfrak{p}^C X = \mathfrak{p}^C Y$ (resp., for $\mathfrak{p}^C x \in X$), where C is a suitable constant, independent of M.

(3.6.1) The Euler system. The first step is to construct an Euler system, by letting various Hecke operators act on the CM point x. Applying Kolyvagin's derivative and e_β, one obtains the derived cohomology classes

$$\kappa_1 = \text{a suitable multiple of } \delta(e_\beta(y)) \in S^{(\beta)},$$
$$\kappa_{\ell_1\cdots\ell_r} \in S^{(\beta)}_{\{\ell_1,\ldots,\ell_r\}} \subset H^1(K(\alpha), A[\mathfrak{p}^M])(\beta),$$

where $\ell_1, \ldots, \ell_r$ are distinct primes of F, inert in K/F, which satisfy "Kolyvagin's condition" modulo a sufficiently high power of $\mathfrak{p}$. The subscript $\ell_1, \ldots, \ell_r$ indicates that the class $\kappa_{\ell_1 \cdots \ell_r}$ satisfies the "Selmer" local conditions only at primes distinct from $\ell_1, \ldots, \ell_r$.

(3.6.2) The Weil pairing. Fix a polarization $\varphi : A_j \longrightarrow \widehat{A}_j$ defined over F; it gives rise to the Weil pairing

$$T_p(A_j) \times T_p(A_j) \longrightarrow \mathbf{Z}_p(1)$$

(where p is the residual characteristic of $\mathfrak{p}$). This extends naturally (see 5.19 below) to an $\mathcal{O}_{\mathfrak{p}}$-bilinear pairing

$$T_{\mathfrak{p}}(A) \times T_{\mathfrak{p}}(A) \longrightarrow \mathcal{O}_{\mathfrak{p}}(1),$$

whose reduction modulo $\mathfrak{p}^M$

$$(\ ,\ )_M : A[\mathfrak{p}^M] \times A[\mathfrak{p}^M] \longrightarrow \mathcal{O}_{\mathfrak{p}}/\mathfrak{p}^M(1)$$

is skew-symmetric, and its kernel is killed by $\deg(\varphi)$.

(3.6.3) The first annihilation relation. For simplicity, assume that $\beta \neq \overline{\beta} := \beta^{-1}$. Let $s \in S^{(\beta)}$. For each Kolyvagin prime ℓ, we have a class $\kappa_\ell \in S^{(\beta)}_{\{\ell\}}$ and its complex conjugate ${}^{\rho}\kappa_\ell \in S^{(\overline{\beta})}_{\{\ell\}}$. Applying the reciprocity law of global class field theory

$$(\forall a \in H^2(K(\alpha), \mathcal{O}_{\mathfrak{p}}/\mathfrak{p}^M(1))) \qquad \sum_v \mathrm{inv}_v(a_v) = 0 \in \mathcal{O}_{\mathfrak{p}}/\mathfrak{p}^M$$

to the cup product $a = s \cup {}^{\rho}\kappa_\ell$, we obtain the relation

$$\sum_v \mathrm{inv}_v(s_v \cup ({}^{\rho}\kappa_\ell)_v) = 0 \in \mathcal{O}_{\mathfrak{p}}/\mathfrak{p}^M, \qquad (3.6.3.1)$$

to which only the primes of $K(\alpha)$ above ℓ contribute. Fixing one such a prime, say, λ, and using the fundamental relation 5.18 between the localizations $(\kappa_1)_\lambda, (\kappa_\ell)_\lambda$, one can rewrite (3.6.3.1) as

$$[K(\alpha) : K]\ (s_\lambda, (\kappa_1)_\lambda)_M = 0 \in \mathcal{O}_{\mathfrak{p}}/\mathfrak{p}^M(1), \qquad (3.6.3.2)$$

where one considers the localizations $s_\lambda, (\kappa_1)_\lambda$ as elements of

$$H^1_{ur}(K(\alpha)_\lambda, A[\mathfrak{p}^M]) \xrightarrow{\sim} A[\mathfrak{p}^M] \quad (\xrightarrow{\sim} (\mathcal{O}_{\mathfrak{p}}/\mathfrak{p}^M)^2),$$

by using evaluation at the Frobenius element $\mathrm{Fr}(\lambda)$.

The assumption $e_\beta(y) \notin A_{\rm tors}$ implies that

$$\langle \kappa_1 \rangle \ \ (= \mathcal{O}_L \cdot \kappa_1) \doteq \mathcal{O}_{\mathfrak{p}}/\mathfrak{p}^M.$$

An application of the Čebotarev density theorem then shows that one can choose a Kolyvagin prime ℓ in such a way that

$$\langle (\kappa_1)_\lambda \rangle \doteq \mathcal{O}_{\mathfrak{p}}/\mathfrak{p}^M$$

(inside $H^1_{ur}(K(\alpha)_\lambda, A[\mathfrak{p}^M]) \xrightarrow{\sim} A[\mathfrak{p}^M]$). As the pairing $(\ ,\ )_M$ on $A[\mathfrak{p}^M]$ is almost symplectic, the annihilation relation (3.6.3.2) yields

$$s_\lambda \mathrel{\dot{\in}} \langle (\kappa_1)_\lambda \rangle^\perp \doteq \langle (\kappa_1)_\lambda \rangle. \tag{3.6.3.3}$$

If we put

$$\widetilde{S}^{(\beta)} = \{ s \in S^{(\beta)} \mid s_\lambda = 0 \},$$

then we deduce from (3.6.3.3) that

$$S^{(\beta)} \doteq \langle \kappa_1 \rangle \oplus \widetilde{S}^{(\beta)}.$$

(3.6.4) The second annihilation relation. Let $s \in \widetilde{S}^{(\beta)}$. In order to complete the proof of (3.5.1), we must show that $s \doteq 0$. If $\ell' \neq \ell$ is another Kolyvagin prime and λ' a prime above ℓ' in $K(\alpha)$, then the reciprocity law applied to the cup product $s \cup {}^\rho\kappa_{\ell\ell'}$

$$\sum_v \mathrm{inv}_v(s_v \cup ({}^\rho\kappa_{\ell\ell'})_v) = 0 \in \mathcal{O}_{\mathfrak{p}}/\mathfrak{p}^M$$

simplifies to

$$[K(\alpha) : K]\ (s_{\lambda'}, (\kappa_\ell)_{\lambda'})_M = 0 \in \mathcal{O}_{\mathfrak{p}}/\mathfrak{p}^M(1). \tag{3.6.4.1}$$

As $(\kappa_1)_\lambda$ (resp., $(\kappa_\ell)_\lambda$) is unramified (resp., totally ramified) at λ, we have

$$\langle (\kappa_1)_\lambda, (\kappa_\ell)_\lambda \rangle \doteq (\mathcal{O}_{\mathfrak{p}}/\mathfrak{p}^M)^2 \subset H^1(K(\alpha)_\lambda, A[\mathfrak{p}^M]),$$

hence

$$\langle \kappa_1, \kappa_\ell \rangle \doteq (\mathcal{O}_{\mathfrak{p}}/\mathfrak{p}^M)^2 \subset S^{(\beta)}_{\{\ell\}}.$$

Another application of the Čebotarev density theorem shows that one can choose ℓ' in such a way that

$$\langle (\kappa_1)_{\lambda'}, (\kappa_\ell)_{\lambda'} \rangle \doteq (\mathcal{O}_{\mathfrak{p}}/\mathfrak{p}^M)^2 = H^1_{ur}(K(\alpha)_{\lambda'}, A[\mathfrak{p}^M]) \xrightarrow{\sim} A[\mathfrak{p}^M] \tag{3.6.4.2}$$

and

$$\langle s_{\lambda'} \rangle \doteq \langle s \rangle. \tag{3.6.4.3}$$

The first annihilation relation (3.6.3.2) applies with ℓ' instead of ℓ, hence

$$s_{\lambda'} \,\dot{\in}\, \langle (\kappa_1)_{\lambda'} \rangle^{\perp}.$$

On the other hand, the second annihilation relation (3.6.4.1) yields

$$s_{\lambda'} \,\dot{\in}\, \langle (\kappa_\ell)_{\lambda'} \rangle^{\perp},$$

thus

$$s_{\lambda'} \,\dot{\in}\, \langle (\kappa_1)_{\lambda'}, (\kappa_\ell)_{\lambda'} \rangle^{\perp} \doteq A[\mathfrak{p}^M]^{\perp} \doteq 0,$$

by (3.6.4.2), which in turns implies that $s \doteq 0$, by (3.6.4.3). This finishes the proof that $\widetilde{S}^{(\beta)} \doteq 0$, hence $S^{(\beta)} \doteq \langle \kappa_1 \rangle = \langle \delta(e_\beta(y)) \rangle$.

4. The Euler system

Assume that we are in the situation of 3.1.

(4.1) According to Proposition 2.5, the field of definition $K(x)$ of x over K is characterized by the isomorphism

$$\mathrm{rec}_K : \widehat{K}^*/K^*\widehat{F}^*Z \xrightarrow{\sim} \mathrm{Gal}(K(x)/K), \qquad Z = \widehat{t}^{-1}(bH\widehat{\mathcal{O}}_F^* b^{-1}).$$

It follows from the first formula in (2.6.2.1) that there exists a smallest non-zero ideal $c(x) \subset \mathcal{O}_F$ (with respect to divisibility) such that $Z \supseteq \widehat{\mathcal{O}}^*_{c(x)}$. Equivalently, $K[c(x)]$ is the smallest ring class field of K (with respect to inclusion) containing $K(x)$.

(4.2) Fix S as in 1.12. The decomposition (1.12.1) implies that we have

$$Z = Z_S Z^S, \qquad Z_S \subset \prod_{v \in S} \mathcal{O}^*_{K,v}, \qquad Z^S = \prod_{v \notin S} Z_v,$$
$$(\forall v \notin S) \quad Z_v = t_v^{-1}(b_v H_v \mathcal{O}^*_{F,v} b_v^{-1}) \subset \mathcal{O}^*_{K,v},$$

where we have denoted, slightly abusively,

$$\mathcal{O}_{K,v} = \mathcal{O}_K \otimes_{\mathcal{O}_F} \mathcal{O}_{F,v} = \bigoplus_{w|v} \mathcal{O}_{K,w}.$$

For each $v \notin S$, we have

$$\mathrm{ord}_v(c(x)) = 0 \iff Z_v = \mathcal{O}^*_{K,v}.$$

(4.3) Definition. *Let $\mathscr{S} = \bigcup_{r\geq 0} \mathscr{S}_r$ be the following set of square-free ideals of $\mathcal{O}_F$: $\mathscr{S}_0 = \{(1)\}$,*

$$\mathscr{S}_1 = \{\ell \in \mathrm{Spec}(\mathcal{O}_F) \mid \ell \text{ is inert in } K/F,\ \ell \notin S,\ \ell \nmid (p)c(x),\ \ell\mathcal{O}_K \nmid I_0\},$$
$$(\forall r > 1) \quad \mathscr{S}_r = \{\ell_1 \cdots \ell_r \mid \ell_j \in \mathscr{S}_1 \text{ distinct}\},$$

where $I_0 \subset \mathcal{O}_K$ is the ideal defined in Proposition 2.10.

(4.4) Definition. *For each $\mathfrak{n} \in \mathscr{S}$, let $h(\mathfrak{n}) \in \widehat{B}^*$ be the following element: $h((1)) = 1$. For each $\ell \in \mathscr{S}_1$, we have $H_\ell = R(\ell)^*$, for a maximal $\mathcal{O}_{F,\ell}$-order $R(\ell) \subset B_\ell$; let $h(\ell) \in R(\ell) \cap B_\ell^* \subset B_\ell^* \subset \widehat{B}^*$ be any element satisfying $\mathrm{ord}_\ell(nr(h(\ell))) = 1$. Finally, for $\mathfrak{n} = \ell_1 \cdots \ell_r$ $(r > 1,\ \ell_j \in \mathscr{S}_1)$, we put $h(\mathfrak{n}) = h(\ell_1)\cdots h(\ell_r) \in \prod_{j=1}^r B_{\ell_j}^* \subset \widehat{B}^*$. Having defined $h(\mathfrak{n})$, we define CM points*

$$(\forall \mathfrak{n} \in \mathscr{S}) \qquad x(\mathfrak{n}) := [z, bh(\mathfrak{n})] \in CM(N_H, K).$$

(4.5) Proposition. *For each $\mathfrak{n} \in \mathscr{S}$, the field of definition $K(x(\mathfrak{n}))$ is contained in $K[c(x)\mathfrak{n}]$ and the reciprocity map induces an isomorphism*

$$\mathrm{rec}_K : \widehat{K}^*/K^*\widehat{F}^*Z(\mathfrak{n}) \xrightarrow{\sim} \mathrm{Gal}(K(x(\mathfrak{n}))/K),$$

where

$$Z(\mathfrak{n}) = \widehat{t}^{-1}(bh(\mathfrak{n})H\widehat{\mathcal{O}}_F^* h(\mathfrak{n})^{-1}b^{-1}) = Z \cap \widehat{\mathcal{O}}_\mathfrak{n}^* = Z_S \prod_{\ell \mid \mathfrak{n}} Z(\mathfrak{n})_\ell \prod_{\ell \notin S(\mathfrak{n})} Z_\ell,$$
$$S(\mathfrak{n}) = S \cup \{\ell \in \mathscr{S}_1,\ \ell \mid \mathfrak{n}\}.$$

The quotient $Z/Z(\mathfrak{n})$ is naturally isomorphic to

$$Z/Z(\mathfrak{n}) \xrightarrow{\sim} \prod_{\ell \mid \mathfrak{n}} Z_\ell/Z(\mathfrak{n})_\ell, \qquad Z_\ell/Z(\mathfrak{n})_\ell \xrightarrow{\sim} \frac{(\mathcal{O}_K/\ell\mathcal{O}_K)^*}{(\mathcal{O}_F/\ell\mathcal{O}_F)^*};$$

in particular, each group $Z_\ell/Z(\mathfrak{n})_\ell$ (for $\ell \in \mathscr{S}_1, \ell \mid \mathfrak{n}$) is cyclic of order $N\ell + 1$.

Proof. The isomorphism comes from Proposition 2.5. As $h(\mathfrak{n}) = 1$ for each $v \nmid \mathfrak{n}$, the abelian group $Z(\mathfrak{n})$ decomposes as required, with $Z(\mathfrak{n})_v = Z_v$ for each $v \nmid S(\mathfrak{n})$ and

$$(\forall \ell \mid \mathfrak{n},\ \ell \in \mathscr{S}_1) \qquad Z(\mathfrak{n})_\ell = t_\ell^{-1}(b_\ell h(\ell)R(\ell)^* h(\ell)^{-1}b_\ell^{-1}).$$

Applying Proposition 4.7(iv) below with $E = F_\ell$, $R = R(\ell) \subset B_\ell \xrightarrow{\sim} M_2(F_\ell)$, $i = \mathrm{Ad}(b_\ell)^{-1} \circ t_\ell : K \otimes_F F_\ell \hookrightarrow B_\ell \xrightarrow{\sim} M_2(F_\ell)$ and $r = 1$ implies that $Z(\mathfrak{n}) = Z \cap \widehat{\mathcal{O}}_\mathfrak{n}^*$ and

$$Z_\ell/Z(\mathfrak{n})_\ell \xrightarrow{\sim} \frac{(\mathcal{O}_K/\ell\mathcal{O}_K)^*}{(\mathcal{O}_F/\ell\mathcal{O}_F)^*}.$$

Finally, it follows from $Z \supseteq \widehat{\mathcal{O}}_{c(x)}^*$ that $Z(\mathfrak{n}) \supseteq \widehat{\mathcal{O}}_{c(x)}^* \cap \widehat{\mathcal{O}}_\mathfrak{n}^* = \widehat{\mathcal{O}}_{c(x)\mathfrak{n}}^*$, hence $K(x(\mathfrak{n})) \subseteq K[c(x)\mathfrak{n}]$.

(4.6) Proposition. *Let v be a non-archimedean prime of F which does not divide $c(x)$.*
(i) *If $\mathfrak{n} \in \mathscr{S}$ and $v \nmid \mathfrak{n}$, then each prime of K above v is unramified in $K(x(\mathfrak{n}))/K$.*
(ii) *If $\mathfrak{n} \in \mathscr{S}$, $v \nmid \mathfrak{n}$ and v is inert in K/F, then $v\mathcal{O}_K$ splits completely in $K(x(\mathfrak{n}))/K$.*
(iii) *If $\mathfrak{n}\ell \in \mathscr{S}_{r+1}$ ($\ell \in \mathscr{S}_1$, $\mathfrak{n} \in \mathscr{S}_r$, $r \geq 0$), then each prime of $K(x(\mathfrak{n}))$ above ℓ is totally tamely ramified in $K(x(\mathfrak{n}\ell))/K(x(\mathfrak{n}))$.*

Proof. (i) As $K(x(\mathfrak{n})) \subseteq K[c(x)\mathfrak{n}]$ and $v \nmid c(x)\mathfrak{n}$, the prime v is unramified in $K(x(\mathfrak{n}))/K$.
(ii) The decomposition group of $v\mathcal{O}_K$ in the extension $K(x(\mathfrak{n}))/K$ is equal to the image of the composite map

$$K_v^* := (K \otimes_F F_v)^* \longrightarrow \widehat{K}^* \longrightarrow \widehat{K}^*/K^*\widehat{F}^*Z(\mathfrak{n}),$$

which factors through $K_v^*/\mathcal{O}_{K,v}^* F_v^* = 0$, as $v\mathcal{O}_K$ is unramified in this extension.
(iii) The inertia group of this extension at any prime λ of $K(x(\mathfrak{n}))$ above ℓ is equal to the image of the composite map

$$f : \mathcal{O}^*_{K(x(\mathfrak{n})),\lambda} \xrightarrow{f_1} \mathcal{O}^*_{K,\ell} \xrightarrow{f_2} \frac{(\mathcal{O}_K/\ell\mathcal{O}_K)^*}{(\mathcal{O}_F/\ell\mathcal{O}_F)^*} = Z(\mathfrak{n}\ell)/Z(\mathfrak{n}) \xrightarrow{f_3}$$
$$\xrightarrow{f_3} \mathrm{Gal}(K(x(\mathfrak{n}\ell))/K(x(\mathfrak{n}))).$$

The arrow f_1 (given by the norm map) is surjective, since λ is unramified in $K(x(\mathfrak{n}))/K$, and the arrows f_2 and f_3 are given by the canonical projections (cf. (2.7.3)). It follows that f is surjective, hence λ is totally ramified; the ramification is tame, as $[K(x(\mathfrak{n}\ell)) : K(x(\mathfrak{n}))]$ divides $N\ell+1$, which is prime to $N\lambda$.

(4.7) Proposition. *Let E be a finite extension of $\mathbf{Q}_p$, E' the unique unramified quadratic extension of E and π a common uniformizer of E and E'. Let $R \subset M_2(E)$ be a maximal $\mathcal{O}_E$-order and $i : E' \hookrightarrow M_2(E)$ an E-embedding. For $r \geq 1$, put $\mathcal{O}^*_{E',r} := \mathrm{Ker}\,(\mathcal{O}^*_{E'} \longrightarrow (\mathcal{O}_{E'}/\pi^r)^*)$.*
(i) *R^* is a maximal compact subgroup of $M_2(E)^* = GL_2(E)$.*
(ii) *$(\forall \gamma \in GL_2(E))\quad i^{-1}(\gamma R^* \gamma^{-1}) \subset \mathcal{O}^*_{E'}$.*
(iii) *If $i^{-1}(R^*) = \mathcal{O}^*_{E'}$ and if $g \in R$ satisfies $\mathrm{ord}_\pi(\det(g)) = 1$, then, for each integer $r \geq 1$, the map of sets*

$$i_r : \mathcal{O}^*_{E'}/\mathcal{O}^*_E\mathcal{O}^*_{E',r} \longrightarrow R^*/(R^* \cap g^r R^* g^{-r})$$

induced by i is bijective.
(iv) *Under the assumptions of (iii), we have, for each $r \geq 1$,*

$$i^{-1}(g^r R^* g^{-r}) = i^{-1}(R^* \cap g^r R^* g^{-r}) = \mathcal{O}^*_E\mathcal{O}^*_{E',r},$$

hence the reduction modulo π^r induces an isomorphism of abelian groups

$$\frac{i^{-1}(R^*)}{i^{-1}(g^r R^* g^{-r})} \xrightarrow{\sim} \frac{\mathcal{O}_{E'}^*}{\mathcal{O}_E^* \mathcal{O}_{E',r}^*} \xrightarrow{\sim} \frac{(\mathcal{O}_{E'}/\pi^r \mathcal{O}_{E'})^*}{(\mathcal{O}_E/\pi^r \mathcal{O}_E)^*}.$$

In particular, reduction modulo π induces an isomorphism

$$\frac{i^{-1}(R^*)}{i^{-1}(gR^*g^{-1})} \xrightarrow{\sim} k_{E'}^*/k_E^*,$$

where k_E and $k_{E'}$ denote the residue fields of E and E', respectively.

Proof. Without loss of generality, we can assume that $R = M_2(\mathcal{O}_E)$ and $g = \begin{pmatrix} 1 & 0 \\ 0 & \pi \end{pmatrix}$. In this case (i) is well-known and (ii) follows from the properness of the map i (which implies that $i^{-1}(\gamma R^* \gamma^{-1})$ is a compact subgroup of E'^*, hence is contained in $\mathcal{O}_{E'}^*$). In order to prove (iii) and (iv), note first that we have, for each $r \geq 0$,

$$\begin{aligned} i^{-1}(g^r R^* g^{-r}) &\overset{(ii)}{=} \mathcal{O}_{E'}^* \cap i^{-1}(g^r R^* g^{-r}) = i^{-1}(R^*) \cap i^{-1}(g^r R^* g^{-r}) = \\ &= i^{-1}(R^* \cap g^r R^* g^{-r}). \end{aligned}$$

On the other hand, the intersection $R_r := R \cap g^r R g^{-r}$ is equal to the Eichler order (of level π^r)

$$R_r = \{ \begin{pmatrix} a & b \\ \pi^r c & d \end{pmatrix} \mid a, b, c, d \in \mathcal{O}_E \} \subset M_2(\mathcal{O}_E) = R,$$

which implies that $i^{-1}(R_r)$ is an $\mathcal{O}_E$-order in $\mathcal{O}_{E'}$, hence $i^{-1}(R_r) = \mathcal{O}_E + \pi^{c(r)} \mathcal{O}_{E'}$, for some $c(r) \geq 0$. As $i^{-1}(R^*) = \mathcal{O}_{E'}^*$, we must have $i^{-1}(R) = \mathcal{O}_{E'}$. For $r \geq 1$, the $\mathcal{O}_E$-module $i^{-1}(R)/i^{-1}(R_r) \xrightarrow{\sim} \mathcal{O}_E/\pi^{c(r)}$ injects into $R/R_r \xrightarrow{\sim} \mathcal{O}_E/\pi^r$, hence $c(r) \leq r$. If $c(r) < r$, then $i(\pi^{r-1}\mathcal{O}_{E'}) \subseteq R_r$, which implies that $i(\mathcal{O}_{E'}) \subseteq R \cap \pi^{1-r} R_r = R_1$. In other words, i induces a k_E-embedding of $k_{E'}$ into the ring $\{ \begin{pmatrix} a & b \\ 0 & d \end{pmatrix} \mid a, b, d \in k_E \}$, which is impossible. This contradiction proves that $c(r) = r$, hence

$$\begin{aligned} i^{-1}(R^* \cap g^r R^* g^{-r}) &= i^{-1}(R_r^*) = i^{-1}(R_r \cap R^*) = (\mathcal{O}_E + \pi^r \mathcal{O}_{E'}) \cap \mathcal{O}_{E'}^* = \\ &= (\mathcal{O}_E + \pi^r \mathcal{O}_{E'})^* = \mathcal{O}_E^* \mathcal{O}_{E',r}^*. \end{aligned}$$

The remaining statements in (iii) and (iv) follow automatically.

(4.8) Proposition (The norm relation). *Let $\mathfrak{n}\ell \in \mathscr{S}_{r+1}$ ($\ell \in \mathscr{S}_1$, $\mathfrak{n} \in \mathscr{S}_r$, $r \geq 0$).*

(i) $K(x(\mathfrak{n}\ell))/K(x(\mathfrak{n}))$ is a cyclic extension of degree $(N\ell+1)/u(r)$, where $u(r) = 1$ for $r > 0$ and $u(0) = (K^* \cap \widehat{F}^* Z : F^*)$ ($= (\mathcal{O}_K^* \cap Z : \mathcal{O}_F^*)$ or $2\,(\mathcal{O}_K^* \cap Z : \mathcal{O}_F^*)$).

(ii) We have an equality of divisors on $N_H^* \otimes_F \overline{F}$

$$T(\ell)\, x(\mathfrak{n}) = u(r) \sum_{\sigma \in \mathrm{Gal}(K(x(\mathfrak{n}\ell)/K(x(\mathfrak{n})))} \sigma(x(\mathfrak{n}\ell)),$$

where $T(\ell)\, x(\mathfrak{n}) = [Hb_\ell H]_*(x(\mathfrak{n}))$ (which is also equal to $[Hb_\ell H]^*(x(\mathfrak{n}))$, as we work with the curve N_H^*).

(iii) The following equality holds in $J(N_H^*)(K(x(\mathfrak{n})))$:

$$T(\ell)\,(\iota(x(\mathfrak{n}))) = u(r)\, \mathrm{Tr}_{K(x(\mathfrak{n}\ell))/K(x(\mathfrak{n}))}(\iota(x(\mathfrak{n}\ell))).$$

Proof. (i) We apply the exact sequence (2.7.3) in the case of the extension $K(x(\mathfrak{n}\ell))/K(x(\mathfrak{n})) = K_{Z(\mathfrak{n}\ell)}/K_{Z(\mathfrak{n})}$. As $\mathfrak{n}\ell\mathcal{O}_K \nmid I_0$, Proposition 2.10 implies that $K^* \cap \widehat{F}^* Z(\mathfrak{n}\ell) = F^*$. If $r > 0$, then $K^* \cap \widehat{F}^* Z(\mathfrak{n}) = F^*$ as well, hence a combination of (2.7.3) with Proposition 4.5 yields an isomorphism

$$\frac{(\mathcal{O}_K/\ell\mathcal{O}_K)^*}{(\mathcal{O}_F/\ell\mathcal{O}_F)^*} \xrightarrow{\ \sim\ } \mathrm{Gal}(K(x(\mathfrak{n}\ell))/K(x(\mathfrak{n}))).$$

If $r = 0$ ($\iff \mathfrak{n} = (1)$), then the same argument shows that the kernel of the natural surjection

$$\frac{(\mathcal{O}_K/\ell\mathcal{O}_K)^*}{(\mathcal{O}_F/\ell\mathcal{O}_F)^*} \longrightarrow \mathrm{Gal}(K(x(\mathfrak{n}\ell))/K(x(\mathfrak{n})))$$

has order equal to $u(0) = (K^* \cap \widehat{F}^* Z : F^*)$, which is in turn equal to $(\mathcal{O}_K^* \cap Z : \mathcal{O}_F^*)$ or $2(\mathcal{O}_K^* \cap Z : \mathcal{O}_F^*)$, thanks to Proposition 2.9.

(ii) By definition, the divisor $T(\ell)\, x(\mathfrak{n})$ contains $x(\mathfrak{n}\ell)$. Proposition 4.7(iii) for $r = 1$ together with the proof of (i) imply that $T(\ell)\, x(\mathfrak{n})$ coincides with the orbit of $x(\mathfrak{n}\ell)$ under the action of $\mathrm{Gal}(K(x(\mathfrak{n}\ell))/K(x(\mathfrak{n})))$, with each point counted with multiplicity $u(r)$.

(iii) This is a consequence of (ii) and the fact that $T(\ell)$ acts on the class of $m(\infty)$ (resp., on $m\widetilde{\xi(N_H^*)}$) by multiplication by $\deg T(\ell) = N\ell + 1$.

(4.9) Proposition (The congruence relation). *Let $\mathfrak{n}\ell \in \mathscr{S}_{r+1}$ ($\ell \in \mathscr{S}_1$, $\mathfrak{n} \in \mathscr{S}_r$, $r \geq 0$). For each prime $\lambda \mid \ell$ above ℓ in $K(\mathfrak{n}\ell)$ the following congruence holds:*

$$x(\mathfrak{n}\ell) \equiv \mathrm{Fr}(\ell)_{\mathrm{arith}}\, x(\mathfrak{n}) \equiv \mathrm{Fr}(\ell)_{\mathrm{geom}}\, x(\mathfrak{n}) \pmod{\lambda}$$

(considered as an equality in $\mathbf{N}_H^*(\kappa(\lambda))$).

Proof. Firstly, $\kappa(\lambda)/\kappa(\ell)$ is a quadratic extension, by Proposition 4.6(ii)-(iii), hence the actions of $\mathrm{Fr}(\ell)_{\mathrm{arith}}$ and $\mathrm{Fr}(\ell)_{\mathrm{geom}}$ on $\kappa(\lambda)$ coincide. Combining this remark with the congruence relation (1.14.3), we deduce that the reduction modulo λ of each point in the support of $T(\ell)\,x(\mathfrak{n})$ (in particular, of $x(\mathfrak{n}\ell)$) is equal to the reduction of $\mathrm{Fr}(\ell)_{\mathrm{arith}}\,x(\mathfrak{n})$.

(4.10) Proposition-Definition. *For each $\mathfrak{n} \in \mathscr{S}$, define the subfield $K(x(\mathfrak{n}))' \subseteq K(x(\mathfrak{n}))$ by*

$$K(x(\mathfrak{n}))' = \begin{cases} K(x), & \mathfrak{n} = (1) \\ K(x(\ell_1)) \cdots K(x(\ell_r)), & \mathfrak{n} = \ell_1 \cdots \ell_r \ (r \geq 1,\ \ell_i \in \mathscr{S}_1) \end{cases}$$

and put $G(\mathfrak{n}) = \mathrm{Gal}(K(x(\mathfrak{n}))'/K(x))$.
(i) *For each $\ell \in \mathscr{S}_1$, the group $G(\ell)$ is cyclic, of order $(N\ell + 1)/u(0)$.*
(ii) *For each $\mathfrak{n} = \ell_1 \cdots \ell_r \in \mathscr{S}_r$, the canonical map $G(\mathfrak{n}) \longrightarrow G(\ell_1) \times \cdots \times G(\ell_r)$ is an isomorphism.*
(iii) *For each $\mathfrak{n} \in \mathscr{S}_r$ $(r \geq 0)$, we have $[K(x(\mathfrak{n})) : K(x(\mathfrak{n}))'] = u(r)u(0)^{r-1}$.*

Proof. (i) This is a special case of Proposition 4.8(i).
(ii) The case $r = 0$ is trivial. For $r \geq 1$ the statement follows from Lemma 4.11 below, applied to $G = Z/Z(\mathfrak{n}) = Z/Z(\ell_1 \cdots \ell_r)$, $G_j = Z(\mathfrak{n}/\ell_j)/Z(\mathfrak{n})$, $\widetilde{G}_j = Z(\ell_j)/Z(\mathfrak{n})$, $A = (K^* \cap \widehat{F}^* Z)/F^*$ (we use (2.7.3) and Proposition 2.10 to view $A \subset G$, and then use Proposition 2.10 again for $\widetilde{G}_i \cap A = 0$).
(iii) The case $r = 0$ is again trivial. For $r \geq 1$ we have

$$[K(x(\mathfrak{n})) : K(x(\mathfrak{n}))'] = |A|^{r-1} = u(0)^{r-1},$$

by the proof of Lemma 4.11.

(4.11) Lemma. *Let $G = G_1 \oplus \cdots \oplus G_r$ $(r \geq 1)$ be a finite abelian group, $A \subset G$ a subgroup and $\pi : G \longrightarrow G/A$ the canonical projection. Assume that, for each $i = 1, \ldots, r$, we have $\widetilde{G}_i \cap A = 0$, where $\widetilde{G}_i = \bigoplus_{j \neq i} G_i \subset G$. Then the canonical homomorphism*

$$\pi(G)/\bigcap_{i=1}^{r} \pi(\widetilde{G}_i) \longrightarrow \bigoplus_{i=1}^{r} \pi(G)/\pi(\widetilde{G}_i)$$

is an isomorphism.

Proof. There is nothing to prove if $r = 1$, so we can assume that $r > 1$. As the homomorphism in question is injective and $|\pi(\widetilde{G}_i)| = |\widetilde{G}_i| = |G|/|G_i|$

(since $\widetilde{G}_i \cap \mathrm{Ker}(\pi) = 0$ by assumption), it is enough to show that

$$\left|\bigcap_{i=1}^{r} \pi(\widetilde{G}_i)\right| \stackrel{?}{=} |\pi(G)| \prod_{i=1}^{r} \frac{|\pi(\widetilde{G}_i)|}{|\pi(G)|} = \frac{|G|}{|A|} \prod_{i=1}^{r} \frac{|A|}{|G_i|} = |A|^{r-1}.$$

Denote, for any subset $I \subset \{1, \ldots, r\}$,

$$G_I = \bigoplus_{i \in I} G_i \subset G, \qquad I^0 = \{1, \ldots, r\} - I, \qquad \widetilde{G}_I = G_{I^0}.$$

Applying the Snake Lemma to the commutative diagrams with exact rows

$$\begin{array}{ccccccccc} 0 & \longrightarrow & \widetilde{G}_i & \longrightarrow & G & \xrightarrow{\pi_i} & G_i & \longrightarrow & 0 \\ & & \downarrow \wr & & \downarrow \pi & & \downarrow & & \\ 0 & \longrightarrow & \pi(\widetilde{G}_i) & \longrightarrow & \pi(G) & \longrightarrow & \pi(G)/\pi(\widetilde{G}_i) & \longrightarrow & 0 \end{array}$$

(where $\pi_i : G \longrightarrow G_i$, $i = 1, \ldots, r$, denotes the natural projection with kernel $\widetilde{G}_i$) and

$$\begin{array}{ccccccccc} 0 & \longrightarrow & \widetilde{G}_{\{1,\ldots,k\}} & \longrightarrow & \widetilde{G}_{\{1,\ldots,k-1\}} & \xrightarrow{\pi_k} & G_k & \longrightarrow & 0 \\ & & \downarrow \alpha_k & & \downarrow \alpha_{k-1} & & \downarrow \pi & & \\ 0 & \longrightarrow & \bigcap_{i=1}^{k} \pi(\widetilde{G}_i) & \longrightarrow & \bigcap_{i=1}^{k-1} \pi(\widetilde{G}_i) & \longrightarrow & \pi(G)/\pi(\widetilde{G}_k) & & \end{array}$$

($k = 2, \ldots, r$), we obtain isomorphisms $A \xrightarrow{\sim} \pi_i(A)$ ($i = 1, \ldots, r$), inclusions

$$\mathrm{Ker}(\alpha_r) \subseteq \mathrm{Ker}(\alpha_{r-1}) \subseteq \cdots \subseteq \mathrm{Ker}(\alpha_1) = 0$$

and exact sequences

$$0 \longrightarrow \pi_k(A) \longrightarrow \mathrm{Coker}(\alpha_k) \longrightarrow \mathrm{Coker}(\alpha_{k-1}) \longrightarrow 0 \qquad (k = 2, \ldots, r),$$

which imply, by induction, the desired equality

$$\left|\bigcap_{i=1}^{r} \pi(\widetilde{G}_i)\right| = |\mathrm{Coker}(\alpha_r)| = \prod_{i=2}^{r} |\pi_i(A)| = |A|^{r-1}.$$

(4.12) Definition of the Euler system. *For each $\mathfrak{n} \in \mathscr{S}_r$ ($r \geq 0$), define the point $y(\mathfrak{n}) \in A(K(x(\mathfrak{n}))') = A_j(K(x(\mathfrak{n}))') \otimes_{\mathcal{O}_{L_j}} \mathcal{O}_L$ by*

$$y(\mathfrak{n}) = \frac{u(0)}{u(r)} \mathrm{Tr}_{K(x(\mathfrak{n}))/K(x(\mathfrak{n}))'}\, \iota(x(\mathfrak{n})) \otimes 1$$

(in particular, $y((1)) = \iota(x) \otimes 1 \in A(K(x))$).

(4.13) Proposition (The Euler system relations). *Let $\mathfrak{n}\ell \in \mathscr{S}_{r+1}$ ($\ell \in \mathscr{S}_1$, $\mathfrak{n} \in \mathscr{S}_r$, $r \geq 0$). Then:*
(i) $\mathrm{Tr}_{K(x(\mathfrak{n}\ell))'/K(x(\mathfrak{n}))'}\, y(\mathfrak{n}\ell) = a_\ell\, y(\mathfrak{n})$, *where* $a_\ell = \theta(T(\ell)) \in \mathcal{O}_{L_j}$ *is the eigenvalue of the Hecke operator* $T(\ell)$ *acting on* A_j.
(ii) *For each prime* $\lambda' \mid \ell$ *above* ℓ *in* $K(x(\mathfrak{n}\ell))'$, *we have*

$$y(\mathfrak{n}\ell) \equiv u(0)\, \mathrm{Fr}(\ell)_{\mathrm{arith}}\, y(\mathfrak{n}) \equiv u(0)\, \mathrm{Fr}(\ell)_{\mathrm{geom}}\, y(\mathfrak{n}) \pmod{\lambda'}.$$

Proof. (i) This follows from Proposition 4.8(iii).

(ii) As $m(\infty)$ (resp., $m\widetilde{\xi(N_H^*)}$) is defined over F, it follows from (the proof of) Proposition 4.9 that we have, for each prime $\lambda \mid \lambda'$ above λ' in $K(x(\mathfrak{n}\ell))$,

$$\iota(x(\mathfrak{n}\ell)) \equiv \mathrm{Fr}(\ell)_{\mathrm{arith}}\, \iota(x(\mathfrak{n})) \equiv \mathrm{Fr}(\ell)_{\mathrm{geom}}\, \iota(x(\mathfrak{n})) \pmod{\lambda},$$

hence (using Proposition 4.10(iii))

$$\begin{aligned} y(\mathfrak{n}\ell) &\equiv \frac{u(0)}{u(r+1)}\, [K(x(\mathfrak{n}\ell)) : K(x(\mathfrak{n}\ell))']\, (\iota(x(\mathfrak{n}\ell)) \otimes 1) \equiv \\ &\equiv u(0)^{r+1}\, \iota(x(\mathfrak{n}\ell)) \otimes 1 \pmod{\lambda} \end{aligned}$$

and, similarly,

$$y(\mathfrak{n}) \equiv u(0)^r\, \iota(x(\mathfrak{n})) \otimes 1 \pmod{\lambda},$$

which proves the desired congruence.

5. The derivative classes

We continue to use the notation of Sect. 1-4.

(5.1) Fix a prime number p, a prime ideal $\mathfrak{p} \subset \mathcal{O}_L$ above p and a prime element $\varpi \in \mathcal{O}_{\mathfrak{p}} = \mathcal{O}_{L,\mathfrak{p}}$. Set $M_0 = \mathrm{ord}_{\mathfrak{p}}(u(0))$.

(5.2) The set $\mathscr{S}_1(M)$ of "Kolyvagin primes"

(5.2.1) Definition. *For each integer $M \geq 1$, define $\mathscr{S}_1(M)$ to be the set of maximal ideals $\ell \subset \mathcal{O}_F$ satisfying the following conditions:*

$$\ell \nmid (p)c(x)d_{K/F}, \qquad \ell\mathcal{O}_K \nmid I_0, \qquad \ell \notin S,$$

and the conjugacy class of the arithmetic Frobenius $\mathrm{Fr}(\ell)_{\mathrm{arith}}$ in the Galois group $\mathrm{Gal}(K(x)(A[\mathfrak{p}^{M+M_0}])/F)$ coincides with the conjugacy class of the complex conjugation ρ. We also define

$$\mathscr{S}_0(M) = \{(1)\}, \quad \mathscr{S}_r(M) = \{\ell_1 \cdots \ell_r \mid \ell_i \in \mathscr{S}_1(M) \text{ distinct}\} \quad (r > 1).$$

(5.2.2) If $\ell \nmid (p)c(x)d_{K/F}$ and $\ell \notin S$, then the field $K(x)(A[\mathfrak{p}^{M+M_0}])$ is unramified over F at ℓ, so it makes sense to consider the conjugacy class of $\mathrm{Fr}(\ell)_{\mathrm{arith}}$.

(5.2.3) By the Čebotarev density theorem, the set $\mathscr{S}_1(M)$ has positive density.

(5.2.4) If $\ell \in \mathscr{S}_1(M)$, then ℓ is inert in K/F. In particular, $\mathscr{S}_1(M) \subseteq \mathscr{S}_1$ (hence $\mathscr{S}_r(M) \subseteq \mathscr{S}_r$ for each $r \geq 0$).

(5.2.5) If $\ell \in \mathscr{S}_1(M)$, then $u(0) \mid (N\ell + 1)$ (by Proposition 4.8(i)) and

$$\begin{aligned} 1 - a_\ell X + (N\ell)X^2 &= \det(1 - \mathrm{Fr}(\ell)_{\mathrm{arith}} X \mid T_\mathfrak{p} A) \equiv \\ &\equiv \det(1 - \rho X \mid T_\mathfrak{p} A) = 1 - X^2 \pmod{\mathfrak{p}^{M+M_0}}, \end{aligned}$$

which implies that the following congruences hold in $\mathcal{O}_L$:

$$a_\ell \equiv 0 \pmod{\mathfrak{p}^{M+M_0}}, \qquad N\ell + 1 \equiv 0 \pmod{u(0)\mathfrak{p}^M}.$$

(5.3) From now on, we fix a sufficiently large integer $M >> 0$ divisible by the order of $\mathfrak{p}$ in the ideal class group of $\mathcal{O}_L$, and a generator of the principal ideal $\mathfrak{p}^M$ (also to be denoted $\mathfrak{p}^M$, by abuse of notation).

(5.4) Definition. *For each $\ell \in \mathscr{S}_1(M)$, fix a generator σ_ℓ of the cyclic group $G(\ell) = \mathrm{Gal}(K(x(\ell))/K(x))$ and define*

$$\mathrm{Tr}_\ell = \sum_{i=0}^{|G(\ell)|-1} \sigma_\ell^i, \qquad D_\ell = \sum_{i=0}^{|G(\ell)|-1} i\sigma_\ell^i \in \mathbf{Z}[G(\ell)];$$

these elements satisfy

$$(\sigma_\ell - 1)D_\ell = |G(\ell)| - \mathrm{Tr}_\ell = (N\ell + 1)/u(0) - \mathrm{Tr}_\ell.$$

For each $\mathfrak{n} = \ell_1 \cdots \ell_r \in \mathscr{S}_r(M)$ $(\ell_i \in \mathscr{S}_1(M))$, *define*

$$D_{\mathfrak{n}} = D_{\ell_1} \cdots D_{\ell_r} \in \mathbf{Z}[G(\ell_1)] \otimes \cdots \otimes \mathbf{Z}[G(\ell_r)] = \mathbf{Z}[G(\mathfrak{n})].$$

(5.5) Lemma. *For each* $\mathfrak{n} \in \mathscr{S}_r(M)$, *the image of the point* $D_{\mathfrak{n}} y(\mathfrak{n}) \in A(K(x(\mathfrak{n}))')$ *in* $A(K(x(\mathfrak{n}))') \otimes \mathcal{O}_{\mathfrak{p}}/\mathfrak{p}^M$ *(which will be denoted by* $(D_{\mathfrak{n}} y(\mathfrak{n})$ $(\mathrm{mod}\, \mathfrak{p}^M)))$ *is contained in* $\left(A(K(x(\mathfrak{n}))') \otimes \mathcal{O}_{\mathfrak{p}}/\mathfrak{p}^M\right)^{G(\mathfrak{n})}$.

Proof. If $\mathfrak{n} = \ell_1 \cdots \ell_r$, then we have, for each $i = 1, \ldots, r$,

$$(\sigma_{\ell_i} - 1)\, D_{\mathfrak{n}}\, y(\mathfrak{n}) =$$
$$= (N\ell_i + 1)/u(0)\, D_{\mathfrak{n}/\ell_i}\, y(\mathfrak{n}) - a_{\ell_i}\, D_{\mathfrak{n}/\ell_i}\, y(\mathfrak{n}/\ell_i) \in \mathfrak{p}^M A(K(x(\mathfrak{n}))').$$

(5.6) As $\mathfrak{p}^M$ is principal, the exact sequence of G_F-modules

$$0 \longrightarrow A[\mathfrak{p}^M] \longrightarrow A(\overline{\mathbf{Q}}) \xrightarrow{\mathfrak{p}^M} A(\overline{\mathbf{Q}}) \longrightarrow 0$$

induces, for each extension K' of F, the cohomology sequence

$$0 \longrightarrow A(K') \otimes \mathcal{O}_{\mathfrak{p}}/\mathfrak{p}^M \xrightarrow{\delta} H^1(K', A[\mathfrak{p}^M]) \longrightarrow H^1(K', A)[\mathfrak{p}^M] \longrightarrow 0.$$

Kolyvagin's **derivative classes** $[c(\mathfrak{n})]$ (which will be constructed in 5.7-9 below) are natural lifts of the classes $\delta(D_{\mathfrak{n}} y(\mathfrak{n})\ (\mathrm{mod}\, \mathfrak{p}^M))$ under the restriction maps [(1)]

$$\mathrm{res} : H^1(K(x), A[\mathfrak{p}^M]) \longrightarrow H^1(K(x(\mathfrak{n}))', A[\mathfrak{p}^M])^{G(\mathfrak{n})} \tag{5.6.1}$$

(of course, there is nothing to do if $\mathfrak{n} = (1)$, in which case one defines $[c((1))] = \delta(y((1))) = \delta(\iota(x) \otimes 1))$.

(5.7) In order to construct the cohomology classes $[c(\mathfrak{n})] \in H^1(K(x),$ $A[\mathfrak{p}^M])$ (where $\mathfrak{n} = \ell_1 \cdots \ell_r \in \mathscr{S}_r$), we fix a point $z(\mathfrak{n}) \in A(\overline{\mathbf{Q}})$ satisfying $\mathfrak{p}^M z(\mathfrak{n}) = D_{\mathfrak{n}} y(\mathfrak{n}) \in A(K(x(\mathfrak{n}))')$. The cohomology class $\delta(D_{\mathfrak{n}} y(\mathfrak{n})$ $(\mathrm{mod}\, \mathfrak{p}^M)) \in H^1(K(x(\mathfrak{n}))', A[\mathfrak{p}^M])$ is then represented by the 1-cocycle

$$(g \mapsto (g-1)z(\mathfrak{n})) \in Z^1(\mathrm{Gal}(\overline{\mathbf{Q}}/K(x(\mathfrak{n}))'), A[\mathfrak{p}^M]).$$

[(1)] As the kernel (resp., cokernel) of the map res is equal (resp., injects) to $H^i(G(\mathfrak{n}), A(K(x(\mathfrak{n}))')[\mathfrak{p}^M])$ for $i = 1$ (resp., $i = 2$) and the torsion submodule $A(K[\infty])_{\mathrm{tors}}$ over $K[\infty] = \bigcup K[c]$ is finite (cf. [Ne-Sch, 2.2]), one sees directly, without any calculation, that such natural lifts exist if we multiply the points $y(\mathfrak{n})$ by the square of any fixed element of $\mathcal{O}_L - \{0\}$ which annihilates $A(K[\infty])_{\mathrm{tors}}$.

Note that the formula $g \mapsto (g-1)z(\mathfrak{n})$ makes sense for $g \in \mathrm{Gal}(\overline{\mathbf{Q}}/K(x))$; it defines a 1-cocycle in $Z^1(\mathrm{Gal}(\overline{\mathbf{Q}}/K(x)), A(\overline{\mathbf{Q}}))$.
Denote by

$$\mathrm{pr} : \mathrm{Gal}(\overline{\mathbf{Q}}/K(x)) \longrightarrow \mathrm{Gal}(K(x(\mathfrak{n}))'/K(x)) = G(\mathfrak{n})$$

the natural surjection. We are going to define a 1-cocycle

$$d'(\mathfrak{n}) \in Z^1(G(\mathfrak{n}), A(K(x(\mathfrak{n}))'))$$

such that the 1-cocycle

$$\begin{aligned} c(\mathfrak{n}) : g \mapsto (\inf(d'(\mathfrak{n})))(g) + (g-1)z(\mathfrak{n}) = \\ = d'(\mathfrak{n})(\mathrm{pr}(g)) + (g-1)z(\mathfrak{n}) \in Z^1(\mathrm{Gal}(\overline{\mathbf{Q}}/K(x), A(\overline{\mathbf{Q}})) \end{aligned}$$

will have values in $A[\mathfrak{p}^M]$. Its cohomology class $[c(\mathfrak{n})] \in H^1(K(x), A[\mathfrak{p}^M])$ will then satisfy

$$\mathrm{res}([c(\mathfrak{n})]) = \delta(D_{\mathfrak{n}} y(\mathfrak{n}) \pmod{\mathfrak{p}^M}).$$

In order to construct $d'(\mathfrak{n})$, we apply the following Lemma to $G = G(\mathfrak{n})$, $G_i = G(\ell_i)$, $\sigma_i = \sigma_{\ell_i}$, $X = A(K(x(\mathfrak{n}))')$.

(5.8) Lemma. *Let $G = G_1 \times \cdots \times G_r$ be a finite product of finite cyclic groups G_i. For each G-module X, the evaluation of 1-cocycles at fixed generators $\sigma_i \in G_i$ induces an isomorphism of abelian groups*

$$Z^1(G, X) \xrightarrow{\sim} \left\{ (x_1, \ldots, x_r) \in X^r \;\middle|\; \begin{array}{l} (\forall i, j)\;\; (\sigma_i - 1)x_j = (\sigma_j - 1)x_i \\ (\forall i)\;\; (1 + \sigma_i + \cdots + \sigma_i^{|G_i|-1})x_i = 0 \end{array} \right\}$$

$$z \mapsto (z(\sigma_1), \ldots, z(\sigma_r)).$$

Proof. Easy exercise.

(5.9) Proposition-Definition. *If $\mathfrak{n} = \ell_1 \cdots \ell_r \in \mathscr{S}_r(M)$ ($r \geq 1$, $\ell_i \in \mathscr{S}_1(M)$), then:*
(i) *The elements*

$$x_i = D_{\mathfrak{n}/\ell_i} \left(\frac{a_{\ell_i}}{\mathfrak{p}^M} y(\mathfrak{n}/\ell_i) - \frac{N\ell_i + 1}{u(0)\mathfrak{p}^M} y(\mathfrak{n}) \right) \in A(K(x(\mathfrak{n}))') \;\; (i = 1, \ldots, r)$$

satisfy

$$(\forall i, j = 1, \ldots, r) \qquad \mathrm{Tr}_{\ell_i} x_i = 0, \qquad (\sigma_{\ell_i} - 1)x_j = (\sigma_{\ell_j} - 1)x_i.$$

(ii) *There exists a unique 1-cocycle* $d'(\mathfrak{n}) \in Z^1(G(\mathfrak{n}), A(K(x(\mathfrak{n}))')$ *satisfying*

$$(\forall i = 1, \dots, r) \qquad d'(\mathfrak{n})(\sigma_{\ell_i}) = x_i.$$

(iii) *The cohomology class of* $d'(\mathfrak{n})$ *lies in* $H^1(G(\mathfrak{n}), A(K(x(\mathfrak{n}))'))[\mathfrak{p}^M]$; *more precisely, we have*

$$(\forall i = 1, \dots, r) \qquad \mathfrak{p}^M d'(\mathfrak{n})(\sigma_{\ell_i}) = -(\sigma_{\ell_i} - 1)D_{\mathfrak{n}} y(\mathfrak{n}).$$

(iv) *Define* $d(\mathfrak{n}) = \inf(d'(\mathfrak{n})) \in Z^1(\mathrm{Gal}(\overline{\mathbf{Q}}/K(x)), A(\overline{\mathbf{Q}}))$, *i.e.* $d(\mathfrak{n})(g) = d'(\mathfrak{n})(\mathrm{pr}(g))$.
(v) *Define* $d((1)) = 0 \in Z^1(\mathrm{Gal}(\overline{\mathbf{Q}}/K(x)), A(\overline{\mathbf{Q}}))$.

Proof. (i) Firstly, the congruences 5.2.5 show that $a_{\ell_i}/\mathfrak{p}^M$ and $(N\ell_i + 1)/u(0)\mathfrak{p}^M$ are contained in $\mathcal{O}_L$, so the definition of x_i makes sense. Secondly, the relation 4.13(i) together with 4.10(i) imply that we have

$$\mathrm{Tr}_{\ell_i} x_i = D_{\mathfrak{n}/\ell_i} \left(\frac{a_{\ell_i}}{\mathfrak{p}^M} |G(\ell_i)| - \frac{N\ell_i + 1}{u(0)\mathfrak{p}^M} a_{\ell_i} \right) y(\mathfrak{n}/\ell_i) = 0$$

and, if $i \neq j$,

$$(\sigma_{\ell_i} - 1)x_j = D_{\mathfrak{n}/\ell_i\ell_j} \left(\frac{N\ell_i + 1}{u(0)} - \mathrm{Tr}_{\ell_i} \right) \left(\frac{a_{\ell_j}}{\mathfrak{p}^M} y(\mathfrak{n}/\ell_j) - \frac{N\ell_j + 1}{u(0)\mathfrak{p}^M} y(\mathfrak{n}) \right) =$$

$$= D_{\mathfrak{n}/\ell_i\ell_j} \Bigg(- \frac{(N\ell_i + 1)(N\ell_j + 1)}{u(0)^2\mathfrak{p}^M} y(\mathfrak{n}) + \frac{(N\ell_i + 1)a_{\ell_j}}{u(0)\mathfrak{p}^M} y(\mathfrak{n}/\ell_j) +$$

$$+ \frac{(N\ell_j + 1)a_{\ell_i}}{u(0)\mathfrak{p}^M} y(\mathfrak{n}/\ell_i) - \frac{a_{\ell_i}a_{\ell_j}}{\mathfrak{p}^M} y(\mathfrak{n}/\ell_i\ell_j) \Bigg) = (\sigma_{\ell_j} - 1)x_i.$$

(ii) This follows from (i) and Lemma 5.8.
(iii) We compute, using again 4.10(i) and 4.13(i), that

$$\mathfrak{p}^M d'(\mathfrak{n})(\sigma_{\ell_i}) = D_{\mathfrak{n}/\ell_i} (a_{\ell_i} y(\mathfrak{n}/\ell_i) - (N\ell_i + 1)/u(0)\, y(\mathfrak{n})) =$$
$$= -D_{\mathfrak{n}/\ell_i}(\sigma_{\ell_i} - 1)D_{\ell_i} y(\mathfrak{n}) = -(\sigma_{\ell_i} - 1)D_{\mathfrak{n}} y(\mathfrak{n}).$$

As $d'(\mathfrak{n})$ is a 1-cocycle, it follows that $\mathfrak{p}^M d'(\mathfrak{n})$ is a coboundary.

(5.10) Proposition-Definition. *Let* $\mathfrak{n} = \ell_1 \cdots \ell_r \in \mathscr{S}_r(M)$ $(r \geq 0$, $\ell_i \in \mathscr{S}_1(M))$. *The 1-cocycle* $c(\mathfrak{n}) \in Z^1(\mathrm{Gal}(\overline{\mathbf{Q}}/K(x)), A(\overline{\mathbf{Q}}))$, *defined by the formula*

$$c(\mathfrak{n}) : g \mapsto d(\mathfrak{n})(g) + (g - 1)z(\mathfrak{n}),$$

has the following properties:
(i) $c(\mathfrak{n}) \in Z^1(\mathrm{Gal}(\overline{\mathbf{Q}}/K(x)), A[\mathfrak{p}^M])$.

(ii) *The cohomology class* $[c(\mathfrak{n})] \in H^1(K(x), A[\mathfrak{p}^M])$ *does not depend on the choice of* $z(\mathfrak{n})$.
(iii) *The image of* $[c(\mathfrak{n})]$ *under the restriction map (5.6.1) is equal to*

$$\mathrm{res}([c(\mathfrak{n})]) = \delta(D_{\mathfrak{n}} y(\mathfrak{n}) \pmod{\mathfrak{p}^M}) \in H^1(K(x(\mathfrak{n}))', A[\mathfrak{p}^M]).$$

(iv) *The image of* $[c(\mathfrak{n})]$ *in* $H^1(K(x), A(\overline{\mathbf{Q}}))$ *is equal to* $[d(\mathfrak{n})]$.

Proof. (i) As $c(\mathfrak{n})$ is a 1-cocycle, it is sufficient to check that $\mathfrak{p}^M c(\mathfrak{n})(g) = 0$, for each element of $\mathrm{Ker(pr)} = \mathrm{Gal}(\overline{\mathbf{Q}}/K(x(\mathfrak{n}))')$ and for each $g \in \mathrm{pr}^{-1}(\sigma_{\ell_i})$ $(i = 1, \ldots, r)$. If $g \in \mathrm{Ker(pr)}$, then

$$\mathfrak{p}^M c(\mathfrak{n})(g) = (g-1) D_{\mathfrak{n}} y(\mathfrak{n}) = 0.$$

If $\mathrm{pr}(g) = \sigma_{\ell_i}$, then we have, by Proposition 5.9(iii),

$$\mathfrak{p}^M c(\mathfrak{n})(g) = -(\sigma_{\ell_i} - 1) D_{\mathfrak{n}} y(\mathfrak{n}) + (g-1) D_{\mathfrak{n}} y(\mathfrak{n}) = 0.$$

(ii) Two choices of $z(\mathfrak{n})$ differ by an element of $A[\mathfrak{p}^M]$.
(iii) For each element $g \in \mathrm{Gal}(\overline{\mathbf{Q}}/K(x(\mathfrak{n}))')$, we have

$$(\mathrm{res}(c(\mathfrak{n})))(g) = (g-1) z(\mathfrak{n}) = \delta(D_{\mathfrak{n}} y(\mathfrak{n}) \pmod{\mathfrak{p}^M})(g).$$

(iv) The cohomology class of $(g-1)z(\mathfrak{n})$ vanishes in $H^1(K(x), A(\overline{\mathbf{Q}}))$.

(5.11) We now investigate the **localizations**

$$[c(\mathfrak{n})_v] \in H^1(K(x)_v, A[\mathfrak{p}^M]),$$
$$[d(\mathfrak{n})_v] \in H^1(K(x)_v, A(\overline{\mathbf{Q}}))[\mathfrak{p}^M] = H^1(K(x)_v, A)[\mathfrak{p}^M]$$

of the cohomology classes $[c(\mathfrak{n})]$ and $[d(\mathfrak{n})]$ at various primes v of $K(x)$.

(5.12) Proposition (Localization outside $\mathfrak{n}$). *Let v be a non-archimedean prime of $K(x)$ lying above a prime v_F of F. Denote by $\widetilde{A}_v$ the special fibre at v of the Néron model of $A \otimes_F K(x)$ over $\mathcal{O}_{K(x)}$, by $\pi_0(\widetilde{A}_v)$ the $\mathcal{O}_L$-module of the connected components of $\widetilde{A}_v$, and define*

$$C_{1,v}(\mathfrak{p}) = \min\{c \geq 0 \mid \mathfrak{p}^c \cdot \pi_0(\widetilde{A}_v)_{\mathfrak{p}} = 0\}.$$

If $v_F \nmid \mathfrak{n}$, then

$$\mathfrak{p}^{C_{1,v}(\mathfrak{p})} \cdot [d(\mathfrak{n})_v] = 0.$$

In particular, if $v_F \nmid \mathfrak{n}$ and $v \notin S$, then $[d(\mathfrak{n})_v] = 0$.

Proof. If $v_F \nmid \mathfrak{n}$, then v is unramified in $K(x(\mathfrak{n}))'/K(x)$, hence the cohomology class $[d(\mathfrak{n})_v]$ is inflated from an element of the cohomology group

$H^1_{ur}(K(x)_v, A)[\mathfrak{p}^M]$, which is isomorphic to $H^1(\kappa(v), \pi_0(\widetilde{A}_v))[\mathfrak{p}^M]$, where $\kappa(v)$ denotes the residue field of v ([Mi 1, I.3.8]). This implies that the annihilator of $\pi_0(\widetilde{A}_v)_{\mathfrak{p}}$ kills $[d(\mathfrak{n})_v]$. If, in addition, $v \notin S$, then A has good reduction at v_F, $A \otimes_F K(x)$ has good reduction at v and $\pi_0(\widetilde{A}_v) = 0$.

(5.13) If $\ell \in \mathscr{S}_1(M)$, fix primes $w \mid v \mid \lambda \mid \ell$ in the fields

$$K(x(\ell)) \supset K(x) \supset K \supset F,$$

respectively (of course, λ is unique and w is uniquely determined by v, by Proposition 4.6). We denote by $\mathrm{Fr}(v) = \mathrm{Fr}(v)_{\mathrm{geom}}$ etc. the various **geometric** Frobenius elements.

The assumption $\ell \in \mathscr{S}_1(M)$ implies that $\mathrm{Fr}(v) = \mathrm{Fr}(\lambda) = \mathrm{Fr}(\ell)^2$ acts trivially on $A[\mathfrak{p}^M]$, hence the evaluation map

$$\begin{array}{ccc} H^1_{ur}(K(x)_v, A[\mathfrak{p}^M]) = \mathrm{Hom}(\langle \mathrm{Fr}(v)\rangle, A(K(x)_v)[\mathfrak{p}^M]) & \xrightarrow{\mathrm{ev}_{\mathrm{Fr}(v)}} & A(K(x)_v)[\mathfrak{p}^M] \\ f & \mapsto & f(\mathrm{Fr}(v)) \end{array}$$

is an isomorphism.

(5.14) Proposition. *If $\ell \in \mathscr{S}_1(M)$ and if $v \mid \ell$ is a prime of $K(x)$ above ℓ, then all maps in the following diagram are isomorphisms and the diagram is commutative.*

$$\begin{array}{ccccc} A(K(x)_v) \otimes_{\mathcal{O}_L} \mathcal{O}_L/\mathfrak{p}^M & \xrightarrow{\delta_v} & H^1_{ur}(K(x)_v, A[\mathfrak{p}^M]) & \xrightarrow{\mathrm{ev}_{\mathrm{Fr}(v)}} & A(K(x)_v)[\mathfrak{p}^M] \\ \downarrow{\scriptstyle \mathrm{red}_v} & & & & \downarrow{\scriptstyle \mathrm{red}_v} \\ \widetilde{A}_v(\kappa(v)) \otimes_{\mathcal{O}_L} \mathcal{O}_L/\mathfrak{p}^M & & \xrightarrow{((N\ell+1) - a_\ell \,\mathrm{Fr}(\ell))/\mathfrak{p}^M} & & \widetilde{A}_v(\kappa(v))[\mathfrak{p}^M] \end{array}$$

Proof. The map δ_v is an isomorphism, as A has good reduction at ℓ and $v \nmid p$. The evaluation map $\mathrm{ev}_{\mathrm{Fr}(v)}$ is an isomorphism, by 5.13. Both vertical maps are isomorphisms, as the kernel of the surjective reduction map $\mathrm{red}_v : A(K(x)_v) \longrightarrow \widetilde{A}_v(\kappa(v))$ is a pro-*char*$(\kappa(v))$-group and $v \nmid p$. It remains to show that the diagram is commutative (which will imply that the bottom horizontal map is also an isomorphism). As the composite map

$$f : A(K(x)_v)[\mathfrak{p}^\infty] \hookrightarrow A(K(x)_v) \twoheadrightarrow A(K(x)_v) \otimes_{\mathcal{O}_L} \mathcal{O}_L/\mathfrak{p}^M$$

is surjective, it is sufficient to compute the value of $\mathrm{ev}_{\mathrm{Fr}(v)} \circ \delta_v \circ f(P)$ for $P \in A(K(x)_v)[\mathfrak{p}^\infty]$. If we fix a point $Q \in A(\overline{K(x)_v})$ such that $\mathfrak{p}^M Q = P$, then

$$\mathrm{ev}_{\mathrm{Fr}(v)} \circ \delta_v \circ f(P) = (\mathrm{Fr}(v) - 1)(Q) = (\mathrm{Fr}(\ell)^2 - 1)(Q) =$$
$$= \frac{a_\ell \,\mathrm{Fr}(\ell) - (N\ell + 1)}{N\ell}(Q) = \frac{a_\ell \,\mathrm{Fr}(\ell) - (N\ell + 1)}{\mathfrak{p}^M N\ell}(P) =$$
$$= \frac{(N\ell + 1) - a_\ell \,\mathrm{Fr}(\ell)}{\mathfrak{p}^M}(P),$$

as

$$N\ell \equiv -1 \pmod{\mathfrak{p}^M}, \qquad Q \in A(K(x)_v)[\mathfrak{p}^\infty]$$

and

$$\mathrm{Fr}(\ell)^{-2} - a_\ell \,\mathrm{Fr}(\ell)^{-1} + N\ell = 0 \quad \text{on} \quad T_\mathfrak{p}(A).$$

(5.15) In the situation of 5.13 we identify the $\mathcal{O}_L/\mathfrak{p}^M$-modules

$$\frac{H^1(K(x)_v, A[\mathfrak{p}^M])}{H^1_{ur}(K(x)_v, A[\mathfrak{p}^M])} = \frac{H^1(K(x)_v, A[\mathfrak{p}^M])}{\mathrm{Im}(\delta_v)} \xrightarrow{\sim} H^1(K(x)_v, A)[\mathfrak{p}^M].$$

The evaluation at the fixed generator $\sigma_\ell \in G(\ell)$ defines an isomorphism

$$\mathrm{ev}_{\sigma_\ell} : \frac{H^1(K(x)_v, A[\mathfrak{p}^M])}{H^1_{ur}(K(x)_v, A[\mathfrak{p}^M])} \xrightarrow{\sim} H^1(K(x)_v^{ur}, A[\mathfrak{p}^M])^{\mathrm{Fr}(v)=1} =$$
$$\mathrm{Hom}(I_v^t, A[\mathfrak{p}^M])^{\mathrm{Fr}(v)=1} = \mathrm{Hom}(I_v^t, A[\mathfrak{p}^M]) \xleftarrow{\sim} \mathrm{Hom}(G(\ell), A(K(x)_v)[\mathfrak{p}^M])$$
$$\xrightarrow{\mathrm{ev}_{\sigma_\ell}} A(K(x)_v)[\mathfrak{p}^M]$$

(where $G(\ell)$ is identified with the quotient $\mathrm{Gal}(K(x(\ell))_w K(x)_v^{ur}/K(x)_v^{ur})$ of the tame inertia group I_v^t at v). We denote by $\Phi_v = \mathrm{ev}_{\sigma_\ell}^{-1} \circ \mathrm{ev}_{\mathrm{Fr}(v)}$ the isomorphism defined by the following commutative diagram:

$$\begin{array}{ccc}
H^1_{ur}(K(x)_v, A[\mathfrak{p}^M]) & \xrightarrow{\Phi_v} & \frac{H^1(K(x)_v, A[\mathfrak{p}^M])}{H^1_{ur}(K(x)_v, A[\mathfrak{p}^M])} \\
\downarrow \mathrm{ev}_{\mathrm{Fr}(v)} & & \downarrow \mathrm{ev}_{\sigma_\ell} \\
A(K(x)_v)[\mathfrak{p}^M] & = & A(K(x)_v)[\mathfrak{p}^M]
\end{array}$$

As $N\ell \equiv -1 \pmod{\mathfrak{p}^M}$, the conjugation action of $\mathrm{Fr}(\ell)$ on $I_v^t \otimes \mathcal{O}_L/\mathfrak{p}^M$ is given by multiplication by -1. In particular,

$$\Phi_v \circ \mathrm{Fr}(\ell) = -\mathrm{Fr}(\ell) \circ \Phi_v. \tag{5.15.1}$$

(5.16) Proposition. *If* $\mathfrak{n}\ell \in \mathscr{S}_{r+1}(M)$ ($\mathfrak{n} \in \mathscr{S}_r(M), \ell \in \mathscr{S}_1(M), r \geq 0$) *and if* $v \mid \ell$ *is a prime of* $K(x)$ *above* ℓ, *then the localization*

$$[d(\mathfrak{n}\ell)_v] \in H^1(K(x)_v, A)[\mathfrak{p}^M] \xleftarrow{\sim} \frac{H^1(K(x)_v, A[\mathfrak{p}^M])}{H^1_{ur}(K(x)_v, A[\mathfrak{p}^M])}$$

is equal to

$$[d(\mathfrak{n}\ell)_v] = -\Phi_v\left(\mathrm{Fr}(\ell)\,[c(\mathfrak{n})_v]\right).$$

Proof. The commutative diagram

$$\begin{array}{ccc}
H^1(K(x(\ell))_w/K(x)_v, A(K(x(\ell))_w))[\mathfrak{p}^M] & \xrightarrow{\mathrm{red}_w} & H^1(G(\ell), \widetilde{A}_v(\kappa(w)))[\mathfrak{p}^M] \\
\downarrow & & \| \\
H^1(K(x(\ell))_w^{ur}/K(x)_v^{ur}, A(\overline{K(x)_v}))[\mathfrak{p}^M] & & \mathrm{Hom}(G(\ell), \widetilde{A}_v(\kappa(v))[\mathfrak{p}^M]) \\
\downarrow & & \uparrow \wr \mathrm{red}_v \\
H^1(K(x)_v^{ur}, A(\overline{K(x)_v}))[\mathfrak{p}^M] & & \mathrm{Hom}(G(\ell), A(K(x)_v)[\mathfrak{p}^M]) \\
\uparrow \wr & & \downarrow \wr \\
H^1(K(x)_v^{ur}, A[\mathfrak{p}^M]) & = & \mathrm{Hom}(I_v^t, A(K(x)_v)[\mathfrak{p}^M])
\end{array}$$

(in which $K(x(\ell))_w^{ur} = K(x(\ell))_w K(x)_v^{ur}$) implies that

$$\begin{aligned}
&\mathrm{red}_v \circ \mathrm{ev}_{\sigma_\ell}([d(\mathfrak{n}\ell)_v]) = \mathrm{red}_v([d'(\mathfrak{n}\ell)_v](\sigma_\ell)) = \\
&= \mathrm{red}_v\left(D_\mathfrak{n}\left(\frac{a_\ell}{\mathfrak{p}^M}\,y(\mathfrak{n}) - \frac{N\ell+1}{u(0)\mathfrak{p}^M}\,y(\mathfrak{n}\ell)\right)\right) \overset{4.12(ii)}{=\!=} \\
&= D_\mathfrak{n}\left(\frac{a_\ell}{\mathfrak{p}^M} - \frac{(N\ell+1)\mathrm{Fr}(\ell)}{\mathfrak{p}^M}\right)\mathrm{red}_v\,y(\mathfrak{n}) = \\
&= \left(\frac{a_\ell\,\mathrm{Fr}(\ell) - (N\ell+1)}{\mathfrak{p}^M}\right)\mathrm{Fr}(\ell)\,\mathrm{red}_v(D_\mathfrak{n}y(\mathfrak{n})) \overset{5.14}{=\!=} \\
&= -\mathrm{Fr}(\ell)\,\mathrm{red}_v\left(\mathrm{ev}_{\mathrm{Fr}(v)} \circ \delta_v(D_\mathfrak{n}y(\mathfrak{n}) \pmod{\mathfrak{p}^M})\right)
\end{aligned}$$

(where we have used the fact that $\mathrm{Fr}(\ell)^2 = \mathrm{Fr}(\lambda) = \mathrm{Fr}(v)$ acts trivially on $\kappa(v)$ and $K(x)_v$). As the reduction map $\mathrm{red}_v : A(K(x)_v)[\mathfrak{p}^M] \longrightarrow \widetilde{A}_v(\kappa(v))[\mathfrak{p}^M]$ is an isomorphism, it follows that

$$[d(\mathfrak{n}\ell)_v] = -\mathrm{ev}_{\sigma_\ell}^{-1} \circ \mathrm{ev}_{\mathrm{Fr}(v)}\left(\mathrm{Fr}(\ell)\,[c(\mathfrak{n})_v]\right) = -\Phi_v\left(\mathrm{Fr}(\ell)\,[c(\mathfrak{n})_v]\right).$$

(5.17) If $\ell \in \mathscr{S}_1(M)$, fix a prime v_α of $K(\alpha)$ above ℓ. As $\ell\mathcal{O}_K$ splits completely in $K(x)/K$ (by Proposition 4.6(ii)), we have, for each prime $v \mid v_\alpha$ of $K(x)$, natural identifications $K(\alpha)_{v_\alpha} = K(x)_v$ and isomorphisms

$$\begin{aligned}
&\mathrm{ev}_{\mathrm{Fr}(v_\alpha)} : H^1_{ur}(K(\alpha)_{v_\alpha}, A[\mathfrak{p}^M]) \xrightarrow{\sim} A(K(\alpha)_{v_\alpha})[\mathfrak{p}^M] \\
&\mathrm{ev}_{\sigma_\ell} : H^1(K(\alpha)_{v_\alpha}, A[\mathfrak{p}^M])/H^1_{ur}(K(\alpha)_{v_\alpha}, A[\mathfrak{p}^M]) \xrightarrow{\sim} A(K(\alpha)_{v_\alpha})[\mathfrak{p}^M] \\
&\Phi_{v_\alpha} = \mathrm{ev}_{\sigma_\ell}^{-1} \circ \mathrm{ev}_{\mathrm{Fr}(v_\alpha)} : H^1_{ur}(K(\alpha)_{v_\alpha}, A[\mathfrak{p}^M]) \xrightarrow{\sim} \\
&\qquad \xrightarrow{\sim} H^1(K(\alpha)_{v_\alpha}, A[\mathfrak{p}^M])/H^1_{ur}(K(\alpha)_{v_\alpha}, A[\mathfrak{p}^M]).
\end{aligned}$$

(5.18) Proposition. *If* $\mathfrak{n}\ell \in \mathscr{S}_{r+1}(M)$ $(\mathfrak{n} \in \mathscr{S}_r(M), \ell \in \mathscr{S}_1(M), r \geq 0)$ *and if* $v_\alpha \mid \ell$ *is a prime of* $K(\alpha)$ *above* ℓ, *then the localization*

$$\left[\left(\mathrm{cor}_{K(x)/K(\alpha)}\, d(\mathfrak{n}\ell)\right)_{v_\alpha}\right] \in H^1(K(\alpha)_{v_\alpha}, A)[\mathfrak{p}^M] \xleftarrow{\sim} \frac{H^1(K(\alpha)_{v_\alpha}, A[\mathfrak{p}^M])}{H^1_{ur}(K(\alpha)_{v_\alpha}, A[\mathfrak{p}^M])}$$

is equal to

$$\left[\left(\mathrm{cor}_{K(x)/K(\alpha)}\, d(\mathfrak{n}\ell)\right)_{v_\alpha}\right] = -\Phi_v\left(\mathrm{Fr}(\ell)\left[\left(\mathrm{cor}_{K(x)/K(\alpha)}\, c(\mathfrak{n})\right)_{v_\alpha}\right]\right).$$

Proof. This follows from Proposition 5.16 and the fact that

$$(\forall X = A, A[\mathfrak{p}^M])\ (\forall c \in H^1(K(x), X)) \qquad \left(\mathrm{cor}_{K(x)/K(\alpha)}(c)\right)_{v_\alpha} = \\ = \sum_{v \mid v_\alpha} \mathrm{cor}_{K(x)_v/K(\alpha)_{v_\alpha}}(c_v) = \sum_{v \mid v_\alpha} c_v.$$

(5.19) The Weil pairings. The Weil pairing

$$(\ ,\ )_{A_j} : T_p(A_j) \times T_p(\widehat{A_j}) \longrightarrow \mathbf{Z}_p(1)$$

is G_F-equivariant and satisfies

$$(\forall f \in \mathrm{End}_F(A_j) = \mathcal{O}_{L_j}) \qquad (f(x), y)_{A_j} = (x, \widehat{f}(y))_{A_j}.$$

Fix a polarization $\varphi : A_j \longrightarrow \widehat{A_j}$ defined over F. As the corresponding Rosati involution acts trivially on $\mathcal{O}_{L_j}$ (by Proposition 1.18(v)), the induced skew-symmetric pairing

$$T_p(A_j) \times T_p(A_j) \longrightarrow \mathbf{Z}_p(1), \qquad x, y \mapsto (x, \varphi(y))_{A_j}$$

factors through $T_p(A_j) \otimes_{\mathcal{O}_{L_j}} T_p(A_j)$ and preserves the decomposition

$$T_p(A_j) = \bigoplus_{P \mid p} T_P(A_j),$$

where P runs through all primes of L_j above p. Denote by

$$(\ ,\ )_{\mathfrak{p}_j} : T_{\mathfrak{p}_j}(A_j) \otimes_{\mathcal{O}_{\mathfrak{p}_j}} T_{\mathfrak{p}_j}(A_j) \longrightarrow \mathbf{Z}_p(1)$$

its $\mathfrak{p}_j$-component, where $\mathfrak{p}_j = \mathfrak{p} \cap \mathcal{O}_{L_j}$ and $\mathcal{O}_{\mathfrak{p}_j} = \mathcal{O}_{L_j, \mathfrak{p}_j}$.
Fix a generator $d \in \mathscr{D}^{-1}_{\mathcal{O}_{\mathfrak{p}_j}/\mathbf{Z}_p}$ of the inverse different; the map

$$\lambda : \mathcal{O}_{\mathfrak{p}_j} \xrightarrow{\sim} \mathrm{Hom}_{\mathbf{Z}_p}(\mathcal{O}_{\mathfrak{p}_j}, \mathbf{Z}_p), \qquad x \mapsto (y \mapsto \mathrm{Tr}_{\mathcal{O}_{\mathfrak{p}_j}/\mathbf{Z}_p}(dxy))$$

is a symmetric isomorphism of $\mathcal{O}_{\mathfrak{p}_j}$-modules. The composition of the maps

$$T_{\mathfrak{p}_j}(A_j) \longrightarrow \mathrm{Hom}_{\mathcal{O}_{\mathfrak{p}_j}}(T_{\mathfrak{p}_j}(A_j), \mathrm{Hom}_{\mathbf{Z}_p}(\mathcal{O}_{\mathfrak{p}_j}, \mathbf{Z}_p(1)))$$
$$x \mapsto (y \mapsto (a \mapsto (x, ay)_{\mathfrak{p}_j}))$$

and

$$\mathrm{Hom}_{\mathcal{O}_{\mathfrak{p}_j}}(T_{\mathfrak{p}_j}(A_j), \mathrm{Hom}_{\mathbf{Z}_p}(\mathcal{O}_{\mathfrak{p}_j}, \mathbf{Z}_p(1))) \xrightarrow{\mathrm{Hom}(\mathrm{id},\lambda^{-1})} \mathrm{Hom}_{\mathcal{O}_{\mathfrak{p}_j}}(T_{\mathfrak{p}_j}(A_j), \mathcal{O}_{\mathfrak{p}_j}(1))$$

defines, by adjunction, a skew-symmetric $\mathcal{O}_{\mathfrak{p}_j}$-bilinear G_F-equivariant pairing

$$(\ ,\ )_j : T_{\mathfrak{p}_j}(A_j) \times T_{\mathfrak{p}_j}(A_j) \longrightarrow \mathcal{O}_{\mathfrak{p}_j}(1)$$

characterized by the property

$$(\forall x, y \in T_{\mathfrak{p}_j}(A_j))(\forall a \in \mathcal{O}_{\mathfrak{p}_j})\ \mathrm{Tr}_{\mathcal{O}_{\mathfrak{p}_j}/\mathbf{Z}_p}(da(x,y)_j) = (x, ay)_{\mathfrak{p}_j} = (ax, y)_{\mathfrak{p}_j}.$$

On tensoring with $\mathcal{O}_{\mathfrak{p}} = \mathcal{O}_{L,\mathfrak{p}}$, we obtain a skew-symmetric $\mathcal{O}_{\mathfrak{p}}$-bilinear G_F-equivariant pairing on $T_{\mathfrak{p}}(A) = T_{\mathfrak{p}_j}(A_j) \otimes_{\mathcal{O}_{\mathfrak{p}_j}} \mathcal{O}_{\mathfrak{p}}$:

$$(\ ,\ ) : T_{\mathfrak{p}}(A) \times T_{\mathfrak{p}}(A) \longrightarrow \mathcal{O}_{\mathfrak{p}}(1)$$
$$x \otimes a,\ y \otimes b \mapsto ab(x,y)_j \qquad (x, y \in T_{\mathfrak{p}_j}(A_j);\ a, b \in \mathcal{O}_{\mathfrak{p}}).$$

The adjoint map $T_{\mathfrak{p}}(A) \longrightarrow \mathrm{Hom}_{\mathcal{O}_{\mathfrak{p}}}(T_{\mathfrak{p}}(A), \mathcal{O}_{\mathfrak{p}}(1))$ is injective and its cokernel is killed by $\deg(\varphi)$. Reducing $(\ ,\ )$ modulo $\mathfrak{p}^M$ yields a skew-symmetric $\mathcal{O}_{\mathfrak{p}}/\mathfrak{p}^M$-bilinear G_F-equivariant pairing

$$(\ ,\ )_M : A[\mathfrak{p}^M] \times A[\mathfrak{p}^M] \longrightarrow \mathcal{O}_{\mathfrak{p}}/\mathfrak{p}^M(1), \qquad (5.19.1)$$

whose kernel is killed by $\deg(\varphi)$. This implies that, for each $\mathcal{O}_{\mathfrak{p}}$-submodule $Y \subset A[\mathfrak{p}^M]$ isomorphic to $\mathcal{O}_{\mathfrak{p}}/\mathfrak{p}^{M-c}$, we have

$$2\deg(\varphi)\,\mathfrak{p}^c Y^{\perp} \subseteq Y. \qquad (5.19.2)$$

(5.20) Proposition. *In the situation of 5.16, let v_α be the prime of $K(\alpha)$ induced by v and $\zeta_v \in K(\alpha)_{v_\alpha} = K(x)_v$ be a root of unity whose image under the reciprocity map $rec_v : K(x)_v^* \longrightarrow G^{ab}_{K(x)_v}$ induces $\sigma_\ell \in \mathrm{Gal}(K(x(\ell))_w/K(x)_v)$. Then we have, for each $x \in H^1_{ur}(K(\alpha)_{v_\alpha}, A[\mathfrak{p}^M])$,*

$$\zeta_v \otimes \mathrm{inv}_{v_\alpha}\Big(x \cup \big(\mathrm{cor}_{K(x)/K(\alpha)}\, c(\mathfrak{n}\ell)\big)_{v_\alpha}\Big) =$$
$$= -\Big(\mathrm{ev}_{\mathrm{Fr}(v_\alpha)}(x), \mathrm{Fr}(\ell)\,\mathrm{ev}_{\mathrm{Fr}(v_\alpha)}\big(\mathrm{cor}_{K(x)/K(\alpha)}\, c(\mathfrak{n})\big)_{v_\alpha}\Big)_M \in \mathcal{O}_{\mathfrak{p}}/\mathfrak{p}^M(1),$$

where $\cup$ denotes the cup product

$$H^1(K(\alpha)_{v_\alpha}, A[\mathfrak{p}^M]) \times H^1(K(\alpha)_{v_\alpha}, A[\mathfrak{p}^M]) \xrightarrow{\cup} H^2(K(\alpha)_{v_\alpha}, \mathcal{O}_{\mathfrak{p}}/\mathfrak{p}^M(1)) \xrightarrow{\mathrm{inv}_{v_\alpha}} \mathcal{O}_{\mathfrak{p}}/\mathfrak{p}^M$$

induced by the pairing $(\ ,\ )_M$.

Proof. Combine Proposition 5.18 with Lemma 5.21 below (using the identification $K(\alpha)_{v_\alpha} = K_\lambda$, where $\lambda = \ell\mathcal{O}_K$).

(5.21) Lemma. *Let $\ell \in \mathscr{S}_1(M)$, $\lambda = \ell O_K$. If $\zeta \in K_\lambda^*$ is a root of unity of order prime to $N\ell$, then we have, for each $x \in H^1_{ur}(K_\lambda, A[\mathfrak{p}^M])$ and $y \in H^1(K_\lambda, A[\mathfrak{p}^M])$,*

$$\zeta \otimes \mathrm{inv}_\lambda(x \cup y) = (x(\mathrm{Fr}(\lambda)), y(\mathrm{rec}_\lambda(\zeta)))_M \in \mathcal{O}_\mathfrak{p}/\mathfrak{p}^M(1),$$

where $\mathrm{rec}_\lambda : K_\lambda^* \longrightarrow G^{ab}_{K_\lambda}$ denotes the reciprocity map (normalized in such a way that the geometric Frobenius element $\mathrm{Fr}(\lambda)$ corresponds to a uniformizer).

Proof. We can assume that ζ is a primitive root of unity of order $n = N\lambda - 1$. Denoting by $\delta : K_\lambda^* \otimes \mathbf{Z}/n\mathbf{Z} \xrightarrow{\sim} H^1(K_\lambda, \mu_n)$ the standard Kummer map, then we have

$$x = \delta\zeta \otimes a \otimes \zeta^{\otimes -1} \in H^1_{ur}(K_\lambda, \mu_n) \otimes A[\mathfrak{p}^M] \otimes \mu_n^{\otimes -1} = H^1_{ur}(K_\lambda, A[\mathfrak{p}^M])$$
$$y = \delta u \otimes b \otimes \zeta^{\otimes -1} \in H^1(K_\lambda, \mu_n) \otimes A[\mathfrak{p}^M] \otimes \mu_n^{\otimes -1} = H^1(K_\lambda, A[\mathfrak{p}^M])$$

for some $a, b \in A[\mathfrak{p}^M]$ and $u \in K_\lambda^*$ (recall that G_{K_λ} acts trivially on $A[\mathfrak{p}^M]$, since $\ell \in \mathscr{S}_1(M)$). As the reciprocity map is normalized by letting the uniformizers correspond to geometric Frobenius elements,

$$x(\mathrm{Fr}(\lambda)) = (\delta\zeta)(\mathrm{Fr}(\lambda)) \otimes a \otimes \zeta^{\otimes -1} = \zeta^{-1} \otimes a \otimes \zeta^{\otimes -1} = -a.$$

For each character $\chi \in \mathrm{Hom}_{\mathrm{cont}}(G_{K_\lambda}, \mathbf{Q}/\mathbf{Z}) = H^1(G_{K_\lambda}, \mathbf{Q}/\mathbf{Z})$, we have ([Se 1, Prop. XI.2])

$$(\forall z \in K_\lambda^*) \qquad \chi(\mathrm{rec}_\lambda(z)) = \mathrm{inv}_\lambda(z \cup \delta\chi) = -\mathrm{inv}_\lambda(\delta z \cup \chi) = \mathrm{inv}_\lambda(\chi \cup \delta z), \tag{5.21.1}$$

where $\delta\chi \in H^2(G_{K_\lambda}, \mathbf{Z})$ denotes the coboundary associated to the exact sequence

$$0 \longrightarrow \mathbf{Z} \longrightarrow \mathbf{Q} \longrightarrow \mathbf{Q}/\mathbf{Z} \longrightarrow 0.$$

Using (5.21.1) with $z = \zeta$ and $z = u$, we obtain

$$y(\mathrm{rec}_\lambda(\zeta)) = (\delta u)(\mathrm{rec}_\lambda(\zeta)) \otimes b \otimes \zeta^{\otimes -1} = -(\delta\zeta)(\mathrm{rec}_\lambda(u)) \otimes b \otimes \zeta^{\otimes -1}$$
$$(= \mathrm{ord}_\lambda(u) b)$$
$$\zeta \otimes \mathrm{inv}_\lambda(x \cup y) = \mathrm{inv}_\lambda(\delta\zeta \cup \delta u)\,(a, b)_M \otimes \zeta^{\otimes -1} = (\delta\zeta)(\mathrm{rec}_\lambda(u))\,(a, b)_M \otimes \zeta^{\otimes -1}$$
$$= (x(\mathrm{Fr}(\lambda)), y(\mathrm{rec}_\lambda(\zeta)))_M \,.$$

6. Linear algebra

In order to simplify the notation, we denote

$$H = K(\alpha), \quad H_M = H(A[\mathfrak{p}^{M+M_0}]), \quad \Delta = \mathrm{Gal}(H/K),$$
$$T = T_\mathfrak{p}(A), \quad V = T_\mathfrak{p}(A) \otimes_{\mathcal{O}_\mathfrak{p}} L_\mathfrak{p},$$
$$U_M = \mathrm{Gal}(H_M/H) \subseteq \mathrm{Aut}_{\mathcal{O}_\mathfrak{p}}(A[\mathfrak{p}^{M+M_0}]) \xrightarrow{\sim} GL_2(\mathcal{O}_\mathfrak{p}/\mathfrak{p}^{M+M_0}).$$

(6.1) The constant $C_2(\mathfrak{p})$

(6.1.1) Proposition. (i) *(Bogomolov [Bo]) The group $Z_p(A/H) := \mathbf{Z}_p^* \cap \mathrm{Im}\left(G_H \longrightarrow \mathrm{Aut}_{\mathbf{Z}_p}(T_p(A)\right)$ is an open subgroup of $\mathbf{Z}_p^*$.*
(ii) *(Serre [Se 3]) There exists a constant $c(A/H)$ such that the inequality $(\mathbf{Z}_p^* : Z_p(A/H)) \leq c(A/H)$ holds for all prime numbers p.*

(6.1.2) Proposition. *Consider the restriction map*

$$\mathrm{res} : H^1(H, A[\mathfrak{p}^M]) \longrightarrow H^1(H_M, A[\mathfrak{p}^M])^{U_M} = \mathrm{Hom}_{U_M}(G_{H_M}^{ab}, A[\mathfrak{p}^M]).$$

There exists an integer $C_2(\mathfrak{p}) \geq 0$ which is equal to zero for all but finitely many $\mathfrak{p}$ and such that $\mathfrak{p}^{C_2(\mathfrak{p})}\,\mathrm{Ker}(\mathrm{res}) = 0$ for all M.

Proof. Proposition 6.1.1(i) implies that there exists $u \in \mathbf{Z}_p^* - \{1\}$ such that

$$(\forall M \geq 0) \qquad u \ (\mathrm{mod}\, \mathfrak{p}^{M+M_0}) \in Z(U_M).$$

It follows from "Sah's Lemma" ([Sa], proof of Prop. 2.7(b)) that $u - 1$ kills $H^1(U_M, A[\mathfrak{p}^M]) = \mathrm{Ker}(\mathrm{res})$, hence $\mathfrak{p}^{C_2(\mathfrak{p})}\,\mathrm{Ker}(\mathrm{res}) = 0$ for $C_2(\mathfrak{p}) = \mathrm{ord}_{\mathfrak{p}}(u-1)$. According to Proposition 6.1.1(ii), we can take $u \not\equiv 1 \pmod p$ (hence $C_2(\mathfrak{p}) = 0$) if $p - 1 > c(A/H)$.

(6.2) The constant $C_3(\mathfrak{p})$

(6.2.1) Proposition. *The following conditions are equivalent.*
(i) *V is an absolutely irreducible representation of G_H.*
(ii) $\mathrm{End}_{L_{\mathfrak{p}}[G_H]}(V) = L_{\mathfrak{p}}$.
(iii) *For every non-trivial character $\eta : \mathrm{Gal}(H/F) \longrightarrow \{\pm 1\}$, the $L_{\mathfrak{p}}[G_F]$-modules V and $V \otimes \eta$ are not isomorphic.*
(iv) *For every totally imaginary quadratic extension K' of F contained in H, the abelian variety A_j does not acquire complex multiplication over K'.*
(v) *For every K' as in (iv), every totally imaginary quadratic extension K_j of L_j, every algebraic Hecke character $\psi : \mathbf{A}_{K'}^* \longrightarrow K_j^*$ and every embedding $\tau : K_j \hookrightarrow \mathbf{C}$, the L-series $L(\psi_\tau, s)$ (where $\psi_\tau : \mathbf{A}_{K'}^*/K'^* \longrightarrow \mathbf{C}^*$ is the idèle class character asociated to ψ at τ) is not equal to any of the L-series $L(\pi(\sigma \circ \theta_j), s - 1/2)$ from Proposition 1.18.*

Proof. If F'/F is any quadratic extension, denote by $\eta_{F'/F} : G_F \longrightarrow \{\pm 1\}$ the associated quadratic character with kernel $\mathrm{Ker}(\eta_{F'/F}) = G_{F'}$.
(i) $\iff$ (ii): Irreducibility of V implies that its restriction to G_H is semi-simple. The equivalence of (i) and (ii) is then well-known ([Cu-Re, Thm. 3.43]).
(ii) $\Longrightarrow$ (iii): According to [Tay], Prop. 3.1 (and using the fact that each complex conjugation acts on $V(f)$ by a matrix with distinct eigenvalues

$\pm 1 \in L_\mathfrak{p}$), V is an absolutely irreducible representation of G_F, hence $\mathrm{End}_{L_\mathfrak{p}[G_F]}(V) = L_\mathfrak{p}$, and both conditions (ii) and (iii) are invariant under finite extensions of the field $L_\mathfrak{p}$. We can assume, therefore, that $L_\mathfrak{p}$ contains all roots of unity of order $[H : K]$. The Frobenius reciprocity yields

$$\begin{aligned}\mathrm{End}_{L_\mathfrak{p}[G_K]}(V) &= \mathrm{Hom}_{L_\mathfrak{p}[G_F]}\left(V, \mathrm{Ind}_{G_K}^{G_F}\left(\mathrm{Res}_{G_F}^{G_K}(V)\right)\right) = \\ = \mathrm{Hom}_{L_\mathfrak{p}[G_F]}\left(V, V \oplus (V \otimes \eta_{K/F})\right) &= L_\mathfrak{p} \oplus \mathrm{Hom}_{L_\mathfrak{p}[G_F]}\left(V, V \otimes \eta_{K/F}\right). \end{aligned} \tag{6.2.1.1}$$

Irreducibility of V implies that any non-trivial element of $\mathrm{Hom}_{L_\mathfrak{p}[G_F]}(V, V \otimes \eta_{K/F})$ is an isomorphism of $L_\mathfrak{p}[G_F]$-modules $V \xrightarrow{\sim} V \otimes \eta_{K/F}$, which proves the equivalence (ii) $\Longleftrightarrow$ (iii) in the case $H = K$. Similarly,

$$\begin{aligned}\mathrm{End}_{L_\mathfrak{p}[G_H]}(V) &= \mathrm{Hom}_{L_\mathfrak{p}[G_K]}\left(V, \mathrm{Ind}_{G_H}^{G_K}\left(\mathrm{Res}_{G_K}^{G_H}(V)\right)\right) = \\ = \mathrm{Hom}_{L_\mathfrak{p}[G_K]}\Big(V, \bigoplus_{\chi \in \widehat{\Delta}} V \otimes \chi\Big) &= \bigoplus_{\chi \in \widehat{\Delta}} \mathrm{Hom}_{L_\mathfrak{p}[G_K]}(V, V \otimes \chi), \end{aligned} \tag{6.2.1.2}$$

where $\widehat{\Delta} = \mathrm{Hom}(\Delta, L_\mathfrak{p}^*)$. Assume that there exists an isomorphism of $L_\mathfrak{p}[G_F]$-modules $V \xrightarrow{\sim} V \otimes \eta$, for a non-trivial character $\eta : \mathrm{Gal}(H/F) \longrightarrow \{\pm 1\}$. If $\eta = \eta_{K/F}$, then (6.2.1.1) implies that $\mathrm{End}_{L_\mathfrak{p}[G_K]}(V) \neq L_\mathfrak{p}$, hence $\mathrm{End}_{L_\mathfrak{p}[G_H]}(V) \neq L_\mathfrak{p}$. If $\eta \neq \eta_{K/F}$, then the character $\chi := \eta|_\Delta \in \widehat{\Delta}$ is non-trivial, hence $\mathrm{End}_{L_\mathfrak{p}[G_H]}(V) \neq L_\mathfrak{p}$, by (6.2.1.2).

(iii) $\Longrightarrow$ (ii): Assume that $\mathrm{End}_{L_\mathfrak{p}[G_H]}(V) \neq L_\mathfrak{p}$. If $\mathrm{End}_{L_\mathfrak{p}[G_K]}(V) \neq L_\mathfrak{p}$, then $V \xrightarrow{\sim} V \otimes \eta_{K/F}$, by (6.2.1.1). If $\mathrm{End}_{L_\mathfrak{p}[G_K]}(V) = L_\mathfrak{p}$, then V is an absolutely irreducible representation of G_K and, thanks to (6.2.1.2), there exists a non-trivial character $\chi \in \widehat{\Delta}$ and a non-trivial morphism of $L_\mathfrak{p}[G_K]$-modules $V \longrightarrow V \otimes \chi$, which is necessarily an isomorphism. As $\det(V) = \det(V)\chi^2$, the values of χ are contained in $\{\pm 1\}$. It follows that χ extends to a character $\chi_F : \mathrm{Gal}(H/F) \longrightarrow \{\pm 1\}$ ($\chi_F \neq 1, \eta_{K/F}$, since $\chi \neq 1$). Applying again the Frobenius reciprocity

$$\begin{aligned} 0 &\neq \mathrm{Hom}_{L_\mathfrak{p}[G_K]}(V \otimes \chi, V) \\ &= \mathrm{Hom}_{L_\mathfrak{p}[G_F]}\left(V \otimes \chi_F, \mathrm{Ind}_{G_K}^{G_F}\left(\mathrm{Res}_{G_F}^{G_K}(V)\right)\right) = \\ &= \mathrm{Hom}_{L_\mathfrak{p}[G_F]}\left(V \otimes \chi_F, V \oplus (V \otimes \eta_{K/F})\right), \end{aligned}$$

we deduce that there exists a character $\eta \in \{\chi_F, \chi_F \eta_{K/F}\}$ (hence $\eta \neq 1$) and a morphism of $L_\mathfrak{p}[G_F]$-modules $V \longrightarrow V \otimes \eta$, which is necessarily an isomorphism.

(ii) $\Longrightarrow$ (iv): If A_j acquires complex multiplication over K' as in (iv), then $\mathrm{Im}\left(L_\mathfrak{p}[G_{K'}] \longrightarrow \mathrm{End}_{L_\mathfrak{p}}(V)\right)$ is a commutative subalgebra of $\mathrm{End}_{L_\mathfrak{p}}(V) \xrightarrow{\sim} M_2(L_\mathfrak{p})$, hence $\mathrm{End}_{L_\mathfrak{p}[G_H]}(V) \supseteq \mathrm{End}_{L_\mathfrak{p}[G_{K'}]}(V) \supsetneq L_\mathfrak{p}$.
(v) $\Longrightarrow$ (iv): This follows from the Main Theorem of Complex Multiplication ([Sh-Ta], [Mi 2, I.5]).
(iv) $\Longrightarrow$ (v): If the L-function $L(\pi(\sigma \circ \theta_j), s - 1/2)$ is given by a Hecke character of K', then $\mathrm{Im}(L_\mathfrak{p}[G_{K'}] \longrightarrow \mathrm{End}_{L_\mathfrak{p}}(V))$ is a commutative subalgebra of $\mathrm{End}_{L_\mathfrak{p}}(V) \xrightarrow{\sim} M_2(L_\mathfrak{p})$. Faltings' isogeny theorem then implies that $\mathrm{End}_{K'}(A_j) \otimes \mathbf{Q}$ contains a commutative subalgebra bigger than L_j, hence A_j acquires CM over K'.
(v) $\Longrightarrow$ (iii): Assume that $\eta : \mathrm{Gal}(H/F) \longrightarrow \{\pm 1\}$ is a non-trivial character such that the $L_\mathfrak{p}[G_F]$-modules V and $V \otimes \eta$ are isomorphic. Then $K' = H^{\mathrm{Ker}(\eta)}$ is a quadratic extension of F and $\mathrm{End}_{L_\mathfrak{p}[G_{K'}]}(V) = L_\mathfrak{p} \oplus L_\mathfrak{p}$, hence V is a semi-simple L-rational abelian representation of $G_{K'}$ with infinite image. This implies, according to [He, Thm. 2] (see also [Se 2, III.2.3, Thm. 2]) that K' is totally imaginary and $G_{K'}$ acts on $V \otimes_{L_\mathfrak{p}} \overline{L}_\mathfrak{p}$ by $\psi_\mathfrak{p} \oplus \psi_\mathfrak{p} \circ \rho$, where $\psi_\mathfrak{p} : G_{K'} \longrightarrow \overline{L}_\mathfrak{p}^*$ is the Galois representation associated to an algebraic Hecke character ψ of K' (cf. [Ne], proof of 12.6.5.2), hence A_j acquires complex multiplication over K'.

(6.2.2) Proposition. *If the equivalent conditions (i)-(v) of Proposition 6.2.1 hold, then there exists an integer $C_3(\mathfrak{p}) \geq 0$, which is equal to zero for all but finitely many $\mathfrak{p}$ and which satisfies*

$$\mathrm{Im}\left(\mathcal{O}_\mathfrak{p}[G_H] \longrightarrow \mathrm{End}_{\mathcal{O}_\mathfrak{p}}(T)\right) \supseteq \mathfrak{p}^{C_3(\mathfrak{p})}\, \mathrm{End}_{\mathcal{O}_\mathfrak{p}}(T).$$

Proof. $\mathrm{Im}\left(\mathcal{O}_\mathfrak{p}[G_H] \longrightarrow \mathrm{End}_{\mathcal{O}_\mathfrak{p}}(T)\right)$ is an $\mathcal{O}_\mathfrak{p}$-lattice in the $L_\mathfrak{p}$-algebra

$$A = \mathrm{Im}\left(L_\mathfrak{p}[G_H] \longrightarrow \mathrm{End}_{L_\mathfrak{p}}(V)\right) \subseteq \mathrm{End}_{L_\mathfrak{p}}(V).$$

As V is a faithful simple A-module satisfying $\mathrm{End}_A(V) = \mathrm{End}_{L_\mathfrak{p}[G_H]}(V) = L_\mathfrak{p}$, a theorem of Burnside ([Cu-Re, Thm 3.32]) implies that $A = \mathrm{End}_{L_\mathfrak{p}}(V)$, which proves the existence of $C_3(\mathfrak{p})$.
It remains to show that, for all but finitely many $\mathfrak{p}$, the map $\mathcal{O}_\mathfrak{p}[G_H] \longrightarrow \mathrm{End}_{\mathcal{O}_\mathfrak{p}}(T)$ is surjective. By the Nakayama Lemma, this amounts to the surjectivity of

$$k[G_H] \longrightarrow \mathrm{End}_k(\overline{T}) \qquad\qquad (k = \mathcal{O}_\mathfrak{p}/\mathfrak{p},\ \overline{T} = T/\mathfrak{p}T),$$

which would follow from absolute irreducibility of $\mathrm{Res}^{G_H}_{G_F}(\overline{T})$. We can assume that $p \nmid [H : F]$ (hence $p \neq 2$) and, after replacing L by a finite extension, that L contains all roots of unity of order $[H : K]$ (hence k does, too). According to [Di, Prop. 3.1], for all but finitely many $\mathfrak{p}$, $\overline{T}$

is an absolutely irreducible $k[G_F]$-module, hence $\mathrm{End}_{k[G_F]}(\overline{T}) = k$. For such $\mathfrak{p}$, $\overline{T}$ is a semi-simple $k[G_H]$-module, by Clifford's Theorem ([Cu-Re, Thm. 11.1(i)]; this applies to an arbitrary group G and its normal subgroup H of finite index), hence

$$\mathrm{Res}^{G_H}_{G_F}(\overline{T}) \text{ is absolutely irreducible} \iff \mathrm{End}_{k[G_H]}(\overline{T}) = k.$$

The same argument as in the proof of 6.2.1 (ii) $\iff$ (iii) shows that

$$\mathrm{End}_{k[G_H]}(\overline{T}) \neq k \iff \left\{ \begin{matrix} \exists \eta : \mathrm{Gal}(H/F) \longrightarrow \{\pm 1\},\ \eta \neq 1 \\ \overline{T} \stackrel{\sim}{\longrightarrow} \overline{T} \otimes \eta \quad \text{as } k[G_F]-\text{modules} \end{matrix} \right\}.$$

If this were true for infinitely many $\mathfrak{p}$'s, we could find a common non-trivial character $\eta : \mathrm{Gal}(H/F) \longrightarrow \{\pm 1\}$ for which the congruences

$$(\forall g \in G_F) \qquad \mathrm{Tr}(g \mid T) \equiv \mathrm{Tr}(g \mid T \otimes \eta) \pmod{\mathfrak{p}}$$

held for infinitely many $\mathfrak{p}$, hence

$$(\forall g \in G_F) \qquad \mathrm{Tr}(g \mid V) = \mathrm{Tr}(g \mid V \otimes \eta),$$

which would imply that $V \stackrel{\sim}{\longrightarrow} V \otimes \eta$ (by irreducibility of V). This contradiction with 6.2.1(iii) shows that $C_3(\mathfrak{p}) = 0$ for all but finitely many $\mathfrak{p}$.

(6.3) Galois groups

(6.3.1) Let $W' \subset H^1(H, A[\mathfrak{p}^M])$ be an $\mathcal{O}_\mathfrak{p}/\mathfrak{p}^M$-submodule of finite type. Denote by $W = \mathrm{res}(W') \subset \mathrm{Hom}_{U_M}(G^{ab}_{H_M}, A[\mathfrak{p}^M])$ its image under the restriction map from Proposition 6.1.2, by $W^\perp \subset G^{ab}_{H_M}$ the annihilator of W, and set $H_M(W) := (H^{ab}_M)^{W^\perp}$. The natural injective map

$$\begin{aligned} G := \mathrm{Gal}(H_M(W)/H) &\hookrightarrow \mathrm{Hom}_{\mathcal{O}_\mathfrak{p}}(W, A[\mathfrak{p}^M]) \\ g &\mapsto (\mathrm{ev}_g : w \mapsto w(g)) \end{aligned}$$

is a morphism of $\mathbf{Z}[U_M]$-modules, where the group $U_M = \mathrm{Gal}(H_M/H) \subset \mathrm{Aut}_{\mathcal{O}_\mathfrak{p}}(A[\mathfrak{p}^{M+M_0}])$ acts trivially on W and by conjugation on G. Denote by

$$X = \mathcal{O}_\mathfrak{p} \cdot G = \mathcal{O}_\mathfrak{p} \cdot \mathrm{Gal}(H_M(W)/H) \subset \mathrm{Hom}_{\mathcal{O}_\mathfrak{p}}(W, A[\mathfrak{p}^M])$$

the $\mathcal{O}_\mathfrak{p}/\mathfrak{p}^M$-submodule of $\mathrm{Hom}_{\mathcal{O}_\mathfrak{p}}(W, A[\mathfrak{p}^M])$ generated by the image of G and by

$$j : X \hookrightarrow \mathrm{Hom}_{\mathcal{O}_\mathfrak{p}}(W, A[\mathfrak{p}^M])$$

the inclusion map. By construction, j is $\mathcal{O}_\mathfrak{p}[U_M]$-linear and the natural map

$$W \longrightarrow \mathrm{Hom}_{\mathcal{O}_\mathfrak{p}}(X, A[\mathfrak{p}^M])$$

is injective.

(6.3.2) Proposition. *If the equivalent conditions of Proposition 6.2.1 are satisfied, then*

$$(\forall M, W') \qquad \mathfrak{p}^{C_3(\mathfrak{p})}\,\mathrm{Coker}(j) = 0.$$

Proof. This is a special case of Proposition 6.4.3 below, for

$$B = \mathrm{Im}\left(L_{\mathfrak{p}}[G_H] \longrightarrow \mathrm{End}_{L_{\mathfrak{p}}}(V)\right) = \mathrm{End}_{L_{\mathfrak{p}}}(V), \qquad D = L_{\mathfrak{p}},$$
$$\Lambda = \mathrm{Im}\left(\mathcal{O}_{\mathfrak{p}}[G_H] \longrightarrow \mathrm{End}_{\mathcal{O}_{\mathfrak{p}}}(T)\right) \supseteq \mathfrak{p}^{C_3(\mathfrak{p})}\,\mathrm{End}_{\mathcal{O}_{\mathfrak{p}}}(T), \qquad R = \mathcal{O}_{\mathfrak{p}}.$$

(6.3.3) Corollary. *If the equivalent conditions of Proposition 6.2.1 are satisfied, then the map*

$$j' : X \xrightarrow{\ j\ } \mathrm{Hom}_{\mathcal{O}_{\mathfrak{p}}}(W, A[\mathfrak{p}^M]) \xrightarrow{\mathrm{Hom(res,id)}} \mathrm{Hom}_{\mathcal{O}_{\mathfrak{p}}}(W', A[\mathfrak{p}^M])$$

satisfies $\mathfrak{p}^{C_2(\mathfrak{p})+C_3(\mathfrak{p})}\,\mathrm{Coker}(j') = 0$.

Proof. It follows from Proposition 6.1.2 that Coker(Hom(res, id)) is killed by $\mathfrak{p}^{C_2(\mathfrak{p})}$.

(6.3.4) Corollary. *If the equivalent conditions of Proposition 6.2.1 are satisfied and if W' is ρ-stable (where $\rho \in G_F = \mathrm{Gal}(\overline{F}/F)$ denotes the complex conjugation with respect to the fixed embedding $\overline{F} \hookrightarrow \mathbf{C}$ extending τ_1), so are W, $H_M(W)$ and X. The maps j, j' are ρ-equivariant and, denoting $(-)^{\pm} = (-)^{\rho=\pm 1}$, we have*

$$2X^+ \subseteq \mathcal{O}_{\mathfrak{p}} G^+ \subseteq X^+$$
$$\mathfrak{p}^{C_2(\mathfrak{p})+C_3(\mathfrak{p})}\,\mathrm{Coker}\left(j' : X^+ \hookrightarrow \mathrm{Hom}_{\mathcal{O}_{\mathfrak{p}}}(W', A[\mathfrak{p}^M])^+\right) = 0.$$

Moreover, the cokernel of the map

$$\mathrm{Hom}_{\mathcal{O}_{\mathfrak{p}}}(W', A[\mathfrak{p}^M])^+ \to \mathrm{Hom}_{\mathcal{O}_{\mathfrak{p}}}((W')^+, A[\mathfrak{p}^M]^+) \oplus \mathrm{Hom}_{\mathcal{O}_{\mathfrak{p}}}((W')^-, A[\mathfrak{p}^M]^-)$$

is killed by 4, and the cokernel of the map

$$\mathcal{O}_{\mathfrak{p}} \cdot 2G^+ \hookrightarrow X^+ \xrightarrow{\ j'\ } \mathrm{Hom}_{\mathcal{O}_{\mathfrak{p}}}((W')^+, A[\mathfrak{p}^M]^+) \oplus \mathrm{Hom}_{\mathcal{O}_{\mathfrak{p}}}((W')^-, A[\mathfrak{p}^M]^-)$$

is killed by $2^4\,\mathfrak{p}^{C_2(\mathfrak{p})+C_3(\mathfrak{p})}$.

(6.4) In this section we prove a general abstract version of Proposition 6.3.2.

(6.4.1) Assume that V is a $\mathbf{Q}_p$-vector space of finite dimension, $B \subset \mathrm{End}_{\mathbf{Q}_p}(V)$ a $\mathbf{Q}_p$-subalgebra, $\Lambda \subset B$ a $\mathbf{Z}_p$-order in B and $T \subset V$ a $\mathbf{Z}_p$-lattice such that $\Lambda T = T$. The $\mathbf{Q}_p$-algebra

$$D = \mathrm{End}_B(V)^{\mathrm{op}}$$

acts on V on the right; its action commutes with the left action of B. The ring

$$R = \{x \in D \,|\, Tx \subset T\}$$

is a $\mathbf{Z}_p$-order in D.

(6.4.2) In the situation of 6.4.1, assume that we are given the following data:

- An integer $N \geq 1$.
- A right $(R/p^N R)$-module $W = W_R$.
- A left $(\Lambda/p^N\Lambda)$-module $X = {}_\Lambda X$.
- A $\mathbf{Z}_p$-bilinear pairing

$$\langle\ ,\ \rangle : X \times W \longrightarrow T/p^N T$$

satisfying

$$\begin{aligned} \langle \lambda x, w\rangle &= \lambda\langle x, w\rangle \\ \langle x, wr\rangle &= \langle x, w\rangle r \end{aligned} \qquad (\forall x \in X,\ \forall w \in W,\ \forall r \in R,\ \forall \lambda \in \Lambda)$$

such that the induced homomorphisms

$$\begin{aligned} i = i_W\ &: W_R \longrightarrow \mathrm{Hom}_\Lambda\left({}_\Lambda X, {}_\Lambda(T/p^N T)_R\right) \\ j = j_X\ &: {}_\Lambda X \longrightarrow \mathrm{Hom}_R\left(W_R, {}_\Lambda(T/p^N T)_R\right) \end{aligned}$$

are both *injective* (i is a morphism of right R-modules, while j is a morphism of left Λ-modules).

(6.4.3) Proposition. *Assume that we are given the data 6.4.2 and that B is a semi-simple $\mathbf{Q}_p$-algebra. Then there exists a maximal $\mathbf{Z}_p$-order $R_{\max} \supset R$ of D; fixing $R_{\max}$, then*

$$\Lambda_{\max} := \{b \in B \,|\, bT \cdot R_{\max} \subset T \cdot R_{\max}\}$$

is a maximal $\mathbf{Z}_p$-order in B containing Λ. Let $c, d \in Z(R_{\max})$ be non-zero central elements in $R_{\max}$ such that $c\Lambda_{\max} \subset \Lambda$ and $dR_{\max} \subset R$ (they always exist). Then

$$cd\,\mathrm{Coker}(j) = 0.$$

Proof. We begin by reducing to the case $R = R_{\max}$, $\Lambda = \Lambda_{\max}$. First of all, $R_{\max}$ exists by [Cu-Re, Thm. 26.5] and $\Lambda_{\max}$ is a maximal $\mathbf{Z}_p$-order in B, by [Cu-Re, Thm. 26.20, Thm. 26.23(iii)]. Consider the maps

$$W_R \xrightarrow{i} \mathrm{Hom}_\Lambda\left({}_\Lambda X, {}_\Lambda(T/p^N T)_R\right) \xrightarrow{\gamma}$$
$$\xrightarrow{\gamma} \mathrm{Hom}_\Lambda\left({}_\Lambda X, {}_{\Lambda_{\max}}(TR_{\max}/p^N TR_{\max})_{R_{\max}}\right);$$

then

$$\widetilde{W}_{R_{\max}} := (\gamma \circ i)(W_R) R_{\max}$$

is a right $R_{\max}$-module. Similarly, consider

$${}_\Lambda X \xrightarrow{j} \mathrm{Hom}_R\left(W_R, {}_\Lambda(T/p^N T)_R\right) \xrightarrow{\delta}$$
$$\xrightarrow{\delta} \mathrm{Hom}_{R_{\max}}\left(\widetilde{W}_{R_{\max}}, {}_{\Lambda_{\max}}(TR_{\max}/p^N TR_{\max})_{R_{\max}}\right)$$

and put

$${}_{\Lambda_{\max}}\widetilde{X} := \Lambda_{\max}(\delta \circ j({}_\Lambda X));$$

this is a left $\Lambda_{\max}$-module and the canonical maps

$$\widetilde{i} : \widetilde{W}_{R_{\max}} \longrightarrow \mathrm{Hom}_{\Lambda_{\max}}\left({}_{\Lambda_{\max}}\widetilde{X}, {}_{\Lambda_{\max}}(TR_{\max}/p^N TR_{\max})_{R_{\max}}\right)$$
$$\widetilde{j} : {}_{\Lambda_{\max}}\widetilde{X} \longrightarrow \mathrm{Hom}_{R_{\max}}\left(\widetilde{W}_{R_{\max}}, {}_{\Lambda_{\max}}(TR_{\max}/p^N TR_{\max})_{R_{\max}}\right)$$

are injective. Assume that the statement has been proved for the maximal orders $R_{\max}, \Lambda_{\max}$ and $c = d = 1$. Then $\widetilde{j}$ is surjective, hence

$$\delta \circ j({}_\Lambda X) \supset c\Lambda_{\max}(\delta \circ j)({}_\Lambda X) = c\,\mathrm{Hom}_{R_{\max}}\left(\widetilde{W}, TR_{\max}/p^N TR_{\max}\right),$$

i.e. c kills $\mathrm{Coker}(\delta \circ j)$. As $(T \cap p^N TR_{\max})d \subset p^N T$, d kills $\mathrm{Ker}(\delta) = \mathrm{Hom}_R(W, (T \cap p^N TR_{\max})/p^N T)$. The exact sequence

$$\mathrm{Ker}(\delta) \longrightarrow \mathrm{Coker}(j) \longrightarrow \mathrm{Coker}(\delta \circ j)$$

then implies that cd kills $\mathrm{Coker}(j)$.

We can assume, therefore, that $R = R_{\max}$ and $\Lambda = \Lambda_{\max}$. The next step is the reduction to the case of a simple $\mathbf{Q}_p$-algebra B. In general there is a finite decomposition $B = \prod e_i B$, where each $e_i B = B_i$ is a simple $\mathbf{Q}_p$-algebra and e_i are the corresponding orthogonal idempotents. Then we have $V = \bigoplus V_i$, $V_i = e_i V$. According to [Cu-Re, Thm. 26.20], $\Lambda = \prod \Lambda_i$, where each $\Lambda_i = e_i \Lambda$ is a maximal $\mathbf{Z}_p$-order in B_i. As $\Lambda T \subset T$, we have $T = \bigoplus T_i$ with $T_i = e_i T$, which implies that $D = \prod D_i$,

$D_i = \mathrm{End}_{B_i}(V_i)^{\mathrm{op}}$ and $R = \prod R_i$, $R_i = \{x \in D_i \mid T_i x \subset T_i\}$. Similarly, $W = \bigoplus e_i W$ and $X = \bigoplus e_i X$.

This implies that we can assume that B is a simple $\mathbf{Q}_p$-algebra, hence $B = M_n(D)$ for a skew-field $D = \mathrm{End}_B(V)^{\mathrm{op}}$ and $V = D^n$. In this case D has a unique maximal order R; every left or right ideal in R is bilateral and is of the form $\varpi_R^m R = R\varpi_R^m$ $(m \geq 0)$, where ϖ_R is a fixed prime element of R ([Cu-Re, Thm. 26.23]).

There is an isomorphism of right R-modules

$$W_R \xrightarrow{\sim} \bigoplus_{i=1}^{k} R/\varpi_R^{n_i} R,$$

where $\varpi_R^{n_i} \mid p^N$ for each i. Using the fact that multiplication by $\varpi_R^{n_i} p^{-N}$ induces an isomorphism between $T \cdot p^N \varpi_R^{-n_i}/p^N T$ and $T/T\varpi_R^{n_i}$, we define a left Λ-module ${}_\Lambda Y$ as the fibre product of the maps

$$T^k \longrightarrow \bigoplus_{i=1}^{k} {}_\Lambda (T/T\varpi_R^{n_i})$$

and

$$\begin{aligned} {}_\Lambda X \xrightarrow{j} \mathrm{Hom}_R\left(W_R, {}_\Lambda(T/p^N)_R\right) \xrightarrow{\sim} \bigoplus_{i=1}^{k} {}_\Lambda \left(T \cdot p^N \varpi_R^{-n_i}/p^N T\right) \xrightarrow{\sim} \\ \xrightarrow{\sim} \bigoplus_{i=1}^{k} {}_\Lambda (T/T\varpi_R^{n_i}). \end{aligned}$$

The injectivity of i_W implies that Y has the following property:

$$(\forall r_1, \ldots, r_k \in R \text{ s.t. } \exists i\; r_i \in R^*)\,(\exists y_1, \ldots, y_k \in Y) \quad \sum_{i=1}^{k} y_i r_i \notin T\varpi_R \tag{$\star_k$}$$

We shall prove, by induction on k, that any Λ-submodule ${}_\Lambda Y \subset {}_\Lambda(T^k)_R$ satisfying $(\star_k)$ is equal to T^k. This will imply that j_X is surjective, as claimed. Let $k = 1$. If ${}_\Lambda Y$ satisfies $(\star_1)$, then $Y \not\subset T\varpi_R$. The lattice T is free over R and, in a suitable R-basis of T, $\Lambda = M_r(R)^{\mathrm{op}}$. As $(Y + T\varpi_R)/T\varpi_R \subset T/T\varpi_R$ is a non-zero $\Lambda/\varpi_R = M_r(R/\varpi_R)$-submodule of the simple Λ/ϖ_R-module $T/T\varpi_R$, we have $Y + T\varpi_R = T$, hence $Y = T$ by the Nakayama Lemma.

Suppose that $k > 1$ and that any ${}_\Lambda Y'$ satisfying $(\star_{k-1})$ is equal to T^{k-1}. If ${}_\Lambda Y \subset {}_\Lambda(T^k)_R$ satisfies $(\star_k)$, then its image under the projection

$\mathrm{pr}_k : T^k \longrightarrow T$ on the k-th factor satisfies $(\star_1)$, hence $\mathrm{pr}_k(Y) = T$. As Λ is a maximal order in a semi-simple $\mathbf{Q}_p$-algebra, it is hereditary, i.e. T is a projective Λ-module [Cu-Re, Thm. 26.12(ii)]. It follows that the (surjective) projection $\mathrm{pr}_k : Y \longrightarrow T$ admits a Λ-linear section $s : T \longrightarrow Y$. For each $i = 1, \ldots, k-1$, the map $T \xrightarrow{s} Y \hookrightarrow T^k \xrightarrow{\mathrm{pr}_k} T$ is an element of $\mathrm{End}_\Lambda(T) = R^{\mathrm{op}}$, hence is given by right multiplication by some $a_i \in R$. It follows that

$$Y = \{(xa_1+y_1, \ldots, xa_{k-1}+y_{k-1}, x) \mid x \in T,\ y_1, \ldots, y_{k-1} \in Y \cap \ker(\mathrm{pr}_k)\}. \tag{6.4.3.1}$$

If $r_1, \ldots, r_{k-1} \in R$ and $(\exists i)$ $r_i \in R^*$, then there exist $y_1, \ldots, y_{k-1} \in Y \cap \ker(\mathrm{pr}_k)$ such that

$$\sum_{i=1}^{k-1} y_i r_i = \sum_{i=1}^{k-1} (xa_i + y_i) r_i - x \sum_{i=1}^{k-1} a_i r_i \notin T\,\varpi_R,$$

since Y satisfies $(\star_k)$. This implies that $Y \cap \ker(\mathrm{pr}_k)$ satisfies $(\star_{k-1})$, hence $Y \cap \ker(\mathrm{pr}_k) = T^{k-1}$ by inductive hypothesis. It follows from (6.4.3.1) that $Y = T^k$, which concludes the proof of the Proposition.

(6.5) Galois groups and Frobenius elements

(6.5.1) In the situation of 6.3.1, assume that W' is ρ-stable. Given an arbitrary element

$$g \in G^+ = \mathrm{Gal}(H_M(W)/H_M)^+,$$

then $h = g^2 = (\rho + 1)g \in 2G^+$. The Čebotarev density theorem implies that there exist infinitely many non-archimedean primes $\mathscr{L}(W)$ of $H_M(W)$ satisfying the following properties:

(6.5.1.1) $\mathscr{L}(W)$ is unramified in $H_M(W)/F$.

(6.5.1.2) $\mathscr{L}(W)$ is prime to $(pu(0))c(x)N(I_0)$ and the prime ℓ of F induced by $\mathscr{L}(W)$ does not lie in S.

(6.5.1.3) $\mathrm{Fr}_{H_M(W)/F}(\mathscr{L}(W)) = \rho g \in \mathrm{Gal}(H_M(W)/F)$.

Denote by ℓ, λ, λ_H and $\mathscr{L}$, respectively, the primes of F, K, H and H_M induced by $\mathscr{L}(W)$.

(6.5.2) Lemma. (i) $\ell \in \mathscr{S}_1(M)$.

(ii) *The prime* $\lambda = \ell\mathcal{O}_K$ *splits completely in* H_M/K.

(iii) $\mathrm{Fr}_{H_M(W)/K}(\mathscr{L}(W)) = (\rho g)^2 = \rho g \rho g = {}^{\rho}g\, g = (\rho + 1)g = g^2 = h \in 2G^+$.

(iv) *The decomposition group of* $\mathscr{L}$ *in* H_M/F *is equal to* $\mathrm{Gal}(H_M/F)_{\mathscr{L}} = \{1, \rho\}$. *In particular,* $\rho(\mathscr{L}) = \mathscr{L}$.

Proof. The statements (i) and (ii) follow from (6.5.1.2) and the equalities $\mathrm{Fr}_{H_M/F}(\mathscr{L}) = \rho g|_{H_M} = \rho$, $\mathrm{Fr}_{H_M/K}(\mathscr{L}) = \mathrm{Fr}_{H_M/F}(\mathscr{L})^2 = \rho^2 = 1$. The statement (iii) is clear and (iv) follows from the fact that $\mathrm{Gal}(H_M/F)_{\mathscr{L}}$ is generated by $\mathrm{Fr}_{H_M/F}(\mathscr{L}) = \rho$.

(6.5.3) In the situation of 6.5.1, the element

$$\mathrm{Fr}(\mathscr{L}) := \mathrm{Fr}_{H_M(W)/H_M}(\mathscr{L}(W)) \in G$$

depends only on $\mathscr{L}$; more precisely, $\mathrm{Fr}(\mathscr{L}) = g^2 = h \in 2G^+$, by Lemma 6.5.2(ii). Its image via the map

$$j' : G^+ \hookrightarrow X^+ \xrightarrow{j} \mathrm{Hom}_{\mathcal{O}_\mathfrak{p}}(W, A[\mathfrak{p}^M])^+ \longrightarrow \mathrm{Hom}_{\mathcal{O}_\mathfrak{p}}(W', A[\mathfrak{p}^M])^+$$

is given by the evaluation map

$$j(\mathrm{Fr}(\mathscr{L}))(w') = \mathrm{res}(w')(\mathrm{Fr}(\mathscr{L})) \qquad (w' \in W') \qquad (6.5.3.1)$$

(6.5.4) The prime $\mathscr{L}$ determines identifications

$$A(H_{\lambda_H})[\mathfrak{p}^M] = A((H_M)_{\mathscr{L}})[\mathfrak{p}^M] = A(H_M)[\mathfrak{p}^M] = A(\overline{F})[\mathfrak{p}^M] \; (=: A[\mathfrak{p}^M]).$$

The image of $W' \subset H^1(H, A[\mathfrak{p}^M])$ under the localization map

$$\mathrm{res}_{\lambda_H} : H^1(H, A[\mathfrak{p}^M]) \longrightarrow H^1(H_{\lambda_H}, A[\mathfrak{p}^M])$$

is contained in $H^1_{ur}(H_{\lambda_H}, A[\mathfrak{p}^M])$ and the composite map

$$W' \xrightarrow{\mathrm{res}_{\lambda_H}} H^1_{ur}(H_{\lambda_H}, A[\mathfrak{p}^M]) \xrightarrow{\mathrm{ev}_{\mathrm{Fr}(\lambda_H)}} A[\mathfrak{p}^M]$$

coincides with the map (6.5.3.1)

$$j(\mathrm{Fr}(\mathscr{L})) : w' \mapsto \mathrm{res}(w')(\mathrm{Fr}(\mathscr{L})).$$

We shall use repeatedly the fact that this map is ρ-equivariant (= is contained in $\mathrm{Hom}_{\mathcal{O}_\mathfrak{p}}(W', A[\mathfrak{p}^M])^+$). In order to stress the dependence on $\mathscr{L}$, we shall write

$$w'_{\mathscr{L}} := \mathrm{res}_{\lambda_H}(w') \qquad (w' \in W').$$

(6.6) Linear and quadratic forms

(6.6.1) Lemma. *Let*

$$f(x) = \sum_{i=1}^{n} f_i\, x_i, \qquad g(x) = \sum_{1 \le i \le j \le n} g_{ij}\, x_i x_j$$

$$\left(f_i, g_{ij} \in \mathcal{O}_{\mathfrak{p}}/\mathfrak{p}^N,\ x = \sum_{i=1}^{n} x_i e_i \in \bigoplus_{i=1}^{n} \mathcal{O}_{\mathfrak{p}}\, e_i = \mathcal{O}_{\mathfrak{p}}^{\oplus n}\right)$$

be a linear and a quadratic form, respectively, in $n \ge 1$ variables, with coefficients in $\mathcal{O}_{\mathfrak{p}}/\mathfrak{p}^N$. Writing $\mathbf{Z}_p^{\oplus n} = \bigoplus_{i=1}^{n} \mathbf{Z}_p\, e_i \subset \mathcal{O}_{\mathfrak{p}}^{\oplus n}$, we have, for each $m \in \{0, \ldots, N\}$:
(i) $f(\mathbf{Z}_p^{\oplus n}) \subseteq \mathfrak{p}^m(\mathcal{O}_{\mathfrak{p}}/\mathfrak{p}^N) \iff (\forall i = 1, \ldots, n)\ \ f_i \in \mathfrak{p}^m(\mathcal{O}_{\mathfrak{p}}/\mathfrak{p}^N) \iff f(\mathcal{O}_{\mathfrak{p}}^{\oplus n}) \subseteq \mathfrak{p}^m(\mathcal{O}_{\mathfrak{p}}/\mathfrak{p}^N)$.
(ii) $g(\mathbf{Z}_p^{\oplus n}) \subseteq \mathfrak{p}^m(\mathcal{O}_{\mathfrak{p}}/\mathfrak{p}^N) \iff (\forall 1 \le i \le j \le n)\ \ g_{ij} \in \mathfrak{p}^m(\mathcal{O}_{\mathfrak{p}}/\mathfrak{p}^N) \iff g(\mathcal{O}_{\mathfrak{p}}^{\oplus n}) \subseteq \mathfrak{p}^m(\mathcal{O}_{\mathfrak{p}}/\mathfrak{p}^N)$.
(iii) If $U \subsetneq \mathbf{Z}_p^{\oplus n}$ is a proper $\mathbf{Z}_p$-submodule and $g(\mathbf{Z}_p^{\oplus n}) \not\subset \mathfrak{p}^m(\mathcal{O}_{\mathfrak{p}}/\mathfrak{p}^N)$ $(1 \le m < N - \mathrm{ord}_{\mathfrak{p}}(2))$, then

$$(\exists x \in \mathbf{Z}_p^{\oplus n},\ x \notin U) \qquad g(x) \notin 2\,\mathfrak{p}^m(\mathcal{O}_{\mathfrak{p}}/\mathfrak{p}^N) = \mathfrak{p}^{m+\mathrm{ord}_{\mathfrak{p}}(2)}(\mathcal{O}_{\mathfrak{p}}/\mathfrak{p}^N).$$

Proof. The statements (i) and (ii) follow from the equalities $f_i = f(e_i)$, $g_{ii} = g(e_i)$, $g_{ij} = g(e_i + e_j) - g(e_i) - g(e_j)$ $(i < j)$.
(iii) There exists an integer $r \ge 1$ and a $\mathbf{Z}_p$-basis $e_1', \ldots, e_n'$ of $\mathbf{Z}_p^{\oplus n}$ such that

$$U \subseteq \bigoplus_{i=1}^{r} p\mathbf{Z}_p\, e_i' \oplus \bigoplus_{j=r+1}^{n} \mathbf{Z}_p\, e_j'.$$

Writing $x = \sum_{i=1}^{n} x_i e_i = \sum_{i=1}^{n} x_i' e_i'$, we have

$$g(x) = \sum_{1 \le i \le j \le n} g_{ij}'\, x_i' x_j',\ g_{ii}' = g(e_i'),\ \ g_{ij}' = g(e_i' + e_j') - g(e_i') - g(e_j')\ (i < j).$$

It follows from (ii) that

$$(\exists k \le l) \qquad g_{kl}' \notin \mathfrak{p}^m(\mathcal{O}_{\mathfrak{p}}/\mathfrak{p}^N).$$

We distinguish three cases:

(1) $k \le l \le r$:

if $k = l$, then $x := e_k' \notin U$ and $g(x) = g_{kk}' \notin \mathfrak{p}^m(\mathcal{O}_{\mathfrak{p}}/\mathfrak{p}^N)$. If $k < l$, then $g(e_k' + e_l') - g(e_k') - g(e_l') \notin \mathfrak{p}^m(\mathcal{O}_{\mathfrak{p}}/\mathfrak{p}^N)$, hence there exists $x \in \{e_k', e_l', e_k' + e_l'\}$ $(\Longrightarrow x \notin U)$ such that $g(x) \notin \mathfrak{p}^m(\mathcal{O}_{\mathfrak{p}}/\mathfrak{p}^N)$.

(2) $k \le r < l$ and $(\forall i \le j \le r) \quad g'_{ij} \in \mathfrak{p}^m(\mathcal{O}_\mathfrak{p}/\mathfrak{p}^N)$:
the congruence

$$(\forall\lambda\in\mathbf{Z}_p)\ g(e'_k+\lambda e'_l) = g'_{kk}+\lambda g'_{kl}+\lambda^2 g'_{ll} \equiv \lambda g'_{kl}+\lambda^2 g'_{ll} \pmod{\mathfrak{p}^m(\mathcal{O}_\mathfrak{p}/\mathfrak{p}^N)}$$

implies that there exists $\lambda \in \{1,2\}$ such that $x := e'_k + \lambda e'_l \notin U$ satisfies $g(x) \notin 2\,\mathfrak{p}^m(\mathcal{O}_\mathfrak{p}/\mathfrak{p}^N)$.
(3) $(\forall i \le r)\ (\forall j \ge i) \qquad g'_{ij} \in \mathfrak{p}^m(\mathcal{O}_\mathfrak{p}/\mathfrak{p}^N)$:
in this case $k > r$ and we have

$$\Big(\forall y \in \bigoplus_{j=r+1}^{n} \mathcal{O}_\mathfrak{p}\, e'_j\Big) \qquad g(e'_1 + y) \equiv g(y) \pmod{\mathfrak{p}^m(\mathcal{O}_\mathfrak{p}/\mathfrak{p}^N)}.$$

If $k = l$, then $x := e'_1 + e'_k \notin U$ satisfies $g(x) \equiv g'_{kk} \not\equiv 0 \pmod{\mathfrak{p}^m(\mathcal{O}_\mathfrak{p}/\mathfrak{p}^N)}$.
If $k < l$, then

$$\begin{aligned} g(e'_1 + e'_k + e'_l) - g(e'_1 + e'_k) - g(e'_1 + e'_l) &\equiv g(e'_k + e'_l) - g(e'_k) - g(e'_l) \equiv \\ &\equiv g'_{kl} \not\equiv 0 \pmod{\mathfrak{p}^m(\mathcal{O}_\mathfrak{p}/\mathfrak{p}^N)}, \end{aligned}$$

hence there exists $x \in \{e'_1 + e'_k + e'_l, e'_1 + e'_k, e'_1 + e'_l\}$ $(\Longrightarrow x \notin U)$ such that $g(x) \notin \mathfrak{p}^m(\mathcal{O}_\mathfrak{p}/\mathfrak{p}^N)$.

(6.6.2) Proposition. *Let $a, b, N \ge 0$ be integers such that $a, 2b + \mathrm{ord}_\mathfrak{p}(2) < N$. Let $Z \subset (\mathcal{O}_\mathfrak{p}/\mathfrak{p}^N)^{\oplus 5}$ be a $\mathbf{Z}_p$-submodule satisfying $\mathcal{O}_\mathfrak{p} \cdot Z \supseteq \mathfrak{p}^a(\mathcal{O}_\mathfrak{p}/\mathfrak{p}^N) \oplus \big(\mathfrak{p}^b(\mathcal{O}_\mathfrak{p}/\mathfrak{p}^N)\big)^{\oplus 4}$. Then there exists an element $z = (z_0, \dots, z_4) \in Z$ $(z_i \in \mathcal{O}_\mathfrak{p}/\mathfrak{p}^N)$ such that*

$$\begin{gathered} \ell_{\mathcal{O}_\mathfrak{p}}\big((\mathcal{O}_\mathfrak{p}/\mathfrak{p}^N)/(\mathcal{O}_\mathfrak{p}/\mathfrak{p}^N)z_0\big) \le a \\ \ell_{\mathcal{O}_\mathfrak{p}}\big((\mathcal{O}_\mathfrak{p}/\mathfrak{p}^N)^{\oplus 2}/(\mathcal{O}_\mathfrak{p}/\mathfrak{p}^N)(z_1, z_2) + (\mathcal{O}_\mathfrak{p}/\mathfrak{p}^N)(z_3, z_4)\big) \le 2b + \mathrm{ord}_\mathfrak{p}(2) \end{gathered}$$

(where $\ell_{\mathcal{O}_\mathfrak{p}}(C)$ denotes the length of a finite $\mathcal{O}_\mathfrak{p}$-module C), hence

$$\begin{gathered} (\mathcal{O}_\mathfrak{p}/\mathfrak{p}^N)z_0 \supseteq \mathfrak{p}^a(\mathcal{O}_\mathfrak{p}/\mathfrak{p}^N), \\ (\mathcal{O}_\mathfrak{p}/\mathfrak{p}^N)(z_1, z_2) + (\mathcal{O}_\mathfrak{p}/\mathfrak{p}^N)(z_3, z_4) \supseteq 2\,\mathfrak{p}^{2b}(\mathcal{O}_\mathfrak{p}/\mathfrak{p}^N)^{\oplus 2}. \end{gathered}$$

Proof. Fix a set of $\mathbf{Z}_p$-generators $z^{(i)} = (z_0^{(i)}, \dots, z_4^{(i)}) \in Z$ $(1 \le i \le n)$ of Z and set

$$f(x) = \sum_{i=1}^{n} x_i z_0^{(i)}, \qquad g(x) = \begin{vmatrix} \sum_{i=1}^n x_i z_1^{(i)} & \sum_{i=1}^n x_i z_2^{(i)} \\ \sum_{i=1}^n x_i z_3^{(i)} & \sum_{i=1}^n x_i z_4^{(i)} \end{vmatrix}$$

$$\Big(x = \sum_{i=1}^{n} x_i\, e_i \in \bigoplus_{i=1}^{n} \mathcal{O}_\mathfrak{p}\, e_i = \mathcal{O}_\mathfrak{p}^{\oplus n}\Big).$$

The assumptions imply that $f(\mathcal{O}_\mathfrak{p}^{\oplus n}) \not\subset \mathfrak{p}^{a+1}(\mathcal{O}_\mathfrak{p}/\mathfrak{p}^N)$, hence

$$U := \{x \in \mathbf{Z}_p^{\oplus n} \mid f(x) \in \mathfrak{p}^{a+1}(\mathcal{O}_\mathfrak{p}/\mathfrak{p}^N)\}$$

is a proper ($U \subsetneq \mathbf{Z}_p^{\oplus n}$) $\mathbf{Z}_p$-submodule of $\mathbf{Z}_p^{\oplus n}$, by Lemma 6.6.1(i). The assumptions also imply that $g(\mathcal{O}_\mathfrak{p}/\mathfrak{p}^N) \not\subset \mathfrak{p}^{2b+1}(\mathcal{O}_\mathfrak{p}/\mathfrak{p}^N)$, hence

$$\Big(\exists x = \sum_{i=1}^{n} x_i e_i \in \mathbf{Z}_p^{\oplus n},\ x \notin U\Big) \qquad g(x) \notin \mathfrak{p}^{2b+1+\mathrm{ord}_\mathfrak{p}(2)}(\mathcal{O}_\mathfrak{p}/\mathfrak{p}^N),$$

by Lemma 6.6.1(iii). The element $z = \sum_{i=1}^n x_i\, z^{(i)} \in Z$ then has the desired properties.

7. The main result

We continue to use the notation from Sect. 3-6.

(7.1) Selmer groups

(7.1.1) Let F'/F be a finite extension and Σ an arbitrary set of primes of F'. We denote

$$S_\Sigma(A/F', \mathfrak{p}^M) := \mathrm{Ker}\Big(H^1(F', A[\mathfrak{p}^M]) \longrightarrow \bigoplus_{v \notin \Sigma} H^1(F'_v, A[\mathfrak{p}^M])/\mathrm{Im}(\delta_v)\Big),$$

where

$$\delta_v : A(F'_v) \otimes \mathcal{O}_\mathfrak{p}/\mathfrak{p}^M \hookrightarrow H^1(F'_v, A[\mathfrak{p}^M])$$

denotes the coboundary map arising from the cohomology exact sequence associated to

$$0 \longrightarrow A[\mathfrak{p}^M] \longrightarrow A(\overline{F}) \xrightarrow{\mathfrak{p}^M} A(\overline{F}) \longrightarrow 0.$$

If $v \nmid p\infty$ and if A has good reduction at the prime of F induced by v, then $\mathrm{Im}(\delta_v) = H^1_{ur}(F'_v, A[\mathfrak{p}^M])$. This implies that, if Σ is finite, so is $S_\Sigma(A/F', \mathfrak{p}^M)$.

(7.1.2) The classical Selmer group

$$S(A/F', \mathfrak{p}^M) := S_\emptyset(A/F', \mathfrak{p}^M)$$

sits in the standard exact sequence

$$0 \longrightarrow A(F') \otimes \mathcal{O}_\mathfrak{p}/\mathfrak{p}^M \xrightarrow{\delta} S(A/F', \mathfrak{p}^M) \longrightarrow \text{Ш}(A/F')[\mathfrak{p}^M] \longrightarrow 0.$$

The inductive limit $\varinjlim_M S(A/F', \mathfrak{p}^M)$ is an $\mathcal{O}_\mathfrak{p}$-module of co-finite type.

(7.1.3) If F''/F' is a finite Galois extension, then the restriction map induces a canonical homomorphism

$$S(A/F', \mathfrak{p}^M) \longrightarrow S(A/F'', \mathfrak{p}^M)^{\mathrm{Gal}(F''/F')},$$

whose kernel and cokernel are killed by $[F'' : F']$.

(7.1.4) The Tate local duality [Mi 1, I.3.4-5] implies that, for each prime v of F', $\mathrm{Im}(\delta_v)$ is an isotropic subspace of $H^1(F'_v, A[\mathfrak{p}^M])$ with respect to the cup product

$$H^1(F'_v, A[\mathfrak{p}^M]) \times H^1(F'_v, A[\mathfrak{p}^M]) \xrightarrow{\cup} H^2(F'_v, \mathcal{O}_\mathfrak{p}/\mathfrak{p}^M(1)) \xrightarrow{\mathrm{inv}_v} \mathcal{O}_\mathfrak{p}/\mathfrak{p}^M$$

induced by the pairing $(\ ,\ )_M$ from (5.19.1).

(7.1.5) If Σ is a finite set of primes of F', $s \in S(A/F', \mathfrak{p}^M)$ and $c \in S_\Sigma(A/F', \mathfrak{p}^M)$, then the reciprocity law

$$(\forall a \in H^2(G_{F'}, \mathcal{O}_\mathfrak{p}/\mathfrak{p}^M(1))) \qquad \sum_v \mathrm{inv}_v(a_v) = 0 \in \mathcal{O}_\mathfrak{p}/\mathfrak{p}^M$$

applied to $a = s \cup c$ yields (thanks to 7.1.4)

$$\sum_{v \in \Sigma} \mathrm{inv}_v(s_v \cup c_v) = 0 \in \mathcal{O}_\mathfrak{p}/\mathfrak{p}^M.$$

(7.2) Annihilation relations

(7.2.1) Kolyvagin's classes. As in Sect. 6, we denote $H = K(\alpha)$. For each $\mathfrak{n} = \ell_1 \cdots \ell_r \in \mathscr{S}_r(M)$ $(r \geq 0,\ \ell_i \in \mathscr{S}_1(M))$, we define

$$\kappa_\mathfrak{n} = \varpi^{C_1(\mathfrak{p})}\, e_\beta \left(\mathrm{cor}_{K(x)/H}(c(\mathfrak{n}))\right) \in H^1(H, A[\mathfrak{p}^M])^{(\beta)},$$

where $C_1(\mathfrak{p}) = \mathrm{max}_v C_{1,v}(\mathfrak{p})$ (the constants $C_{1,v}(\mathfrak{p})$ were defined in 5.2). For $r = 0$, the cohomology class $\kappa_1 := \kappa_{(1)}$ is equal to

$$\kappa_1 = \varpi^{C_1(\mathfrak{p})}\, \delta(e_\beta(y)) \in S(A/H, \mathfrak{p}^M)^{(\beta)},$$

where $y = \mathrm{cor}_{K(x)/H}(y_j) \in A(H)$. For $r \geq 1$, Proposition 5.12 implies that

$$\kappa_\mathfrak{n} \in S_{\{v|\mathfrak{n}\}}(A/H, \mathfrak{p}^M)^{(\beta)}.$$

(7.2.2) Assume that $\mathfrak{n} = \ell_1 \cdots \ell_r \in \mathscr{S}_r(M)$ $(r \geq 1,\ \ell_i \in \mathscr{S}_1(M))$. For each $i \in \{1, \ldots, r\}$, fix a prime $\mathscr{L}_i \mid \ell_i$ of H_M such that $\mathrm{Fr}_{H_M/F}(\mathscr{L}_i) = \rho$ (as an element of $\mathrm{Gal}(H_M/F)$, not only as a conjugacy class) and denote by v_i the prime of H induced by $\mathscr{L}_i$. As in 6.5.4, $\mathscr{L}_i$ determines identifications

$$A(H_{v_i})[\mathfrak{p}^M] = A((H_M)_{\mathscr{L}_i})[\mathfrak{p}^M] = A(\overline{F})[\mathfrak{p}^M] \qquad (=: A[\mathfrak{p}^M]),$$

and the localization map

$$W'_i := S_{\{v|(\mathfrak{n}/\ell_i)\}}(A/H, \mathfrak{p}^M) \xrightarrow{\mathrm{res}_{v_i}} H^1_{ur}(H_{v_i}, A[\mathfrak{p}^M]) \xrightarrow{\mathrm{ev}_{\mathrm{Fr}(v_i)}} A[\mathfrak{p}^M]$$

coincides with the evaluation map $w' \mapsto w'_{\mathscr{L}_i} := (\mathrm{res}(w'))(\mathrm{Fr}(\mathscr{L}_i))$.

(7.2.3) Proposition. *In the situation of 7.2.2, assume that the generators $\sigma_{\ell_i} \in G(\ell_i)$ have been chosen in the following compatible manner: there exists a global root of unity $\zeta \in \mu_{p^\infty}(H_M)$ such that, for each $i = 1, \ldots, r$, the image of σ_{ℓ_i} in $G(\ell_i) \otimes \mathcal{O}_\mathfrak{p}/\mathfrak{p}^M$ coincides with the image of ζ under the composite map*

$$\mu_{p^\infty}(H_M)\otimes\mathcal{O}_\mathfrak{p}/\mathfrak{p}^M \longrightarrow \mu_{p^\infty}((H_M)_{\mathscr{L}_i})\otimes\mathcal{O}_\mathfrak{p}/\mathfrak{p}^M \xleftarrow{\sim} \mu_{p^\infty}(H_{v_i})\otimes\mathcal{O}_\mathfrak{p}/\mathfrak{p}^M \xrightarrow{\mathrm{rec}_{v_i}\otimes\mathrm{id}} G(\ell_i)\otimes\mathcal{O}_\mathfrak{p}/\mathfrak{p}^M$$

(such a compatible choice always exists). Then we have, for each $s \in S(A/H, \mathfrak{p}^M)^{(\beta)}$, an equality

$$[H:K]\sum_{i=1}^{r}\left(s_{\mathscr{L}_i}, (\kappa_{\mathfrak{n}/\ell_i})_{\mathscr{L}_i}\right)_M = 0$$

in $\mu_{p^\infty}(H_M)\otimes\mathcal{O}_\mathfrak{p}/\mathfrak{p}^M = \mathcal{O}_\mathfrak{p}/\mathfrak{p}^M(1)$.

Proof. Applying 7.1.5 to $F' = H$, $\Sigma = \{v \mid \mathfrak{n}\}$, $\rho(s)$ and $c = \kappa_\mathfrak{n}$, we obtain

$$\sum_{i=1}^{r}\sum_{\sigma\in\Delta}\mathrm{inv}_{\sigma(v_i)}\left(\mathrm{res}_{\sigma(v_i)}(\rho(s))\cup\mathrm{res}_{\sigma(v_i)}(\kappa_\mathfrak{n})\right) = 0 \in \mathcal{O}_\mathfrak{p}/\mathfrak{p}^M, \quad (7.2.3.1)$$

where $\Delta = \mathrm{Gal}(H/K)$. As

$$(\forall\sigma\in\Delta)\qquad \sigma(\rho(s)) = \beta^{-1}(\sigma)\rho(s), \qquad \sigma(\kappa_\mathfrak{n}) = \beta(\sigma)\kappa_\mathfrak{n},$$

we deduce from (7.2.3.1) the equality

$$[H:K]\sum_{i=1}^{r}\mathrm{inv}_{v_i}\left(\mathrm{res}_{v_i}(\rho(s))\cup\mathrm{res}_{v_i}(\kappa_\mathfrak{n})\right) = 0 \in \mathcal{O}_\mathfrak{p}/\mathfrak{p}^M. \quad (7.2.3.2)$$

Combined with the formula from Proposition 5.20 (with $\zeta_v = \zeta$), (7.2.3.2) yields

$$[H:K]\sum_{i=1}^{r}\left(\rho(s)_{\mathscr{L}_i}, \mathrm{Fr}(\ell_i)(\kappa_{\mathfrak{n}/\ell_i})_{\mathscr{L}_i}\right)_M = 0 \in \mathcal{O}_\mathfrak{p}/\mathfrak{p}^M(1).$$

Finally, it follows from

$$\rho(s)_{\mathscr{L}_i} = \rho(s_{\mathscr{L}_i}) = \mathrm{Fr}(\ell_i)s_{\mathscr{L}_i} \in A[\mathfrak{p}^M]$$

that

$$\left((\rho(s)_{\mathscr{L}_i}, \mathrm{Fr}(\ell_i)(\kappa_{\mathfrak{n}/\ell_i})_{\mathscr{L}_i}\right)_M = \left(\mathrm{Fr}(\ell_i)s_{\mathscr{L}_i}, \mathrm{Fr}(\ell_i)(\kappa_{\mathfrak{n}/\ell_i})_{\mathscr{L}_i}\right)_M = \\ = \mathrm{Fr}(\ell_i)\left(s_{\mathscr{L}_i}, (\kappa_{\mathfrak{n}/\ell_i})_{\mathscr{L}_i}\right)_M = -\left(s_{\mathscr{L}_i}, (\kappa_{\mathfrak{n}/\ell_i})_{\mathscr{L}_i}\right)_M .$$

(7.2.4) (i) If we do not choose the generators σ_{ℓ_i} compatibly, then there exist $\zeta \in \mu_{p^\infty}(H_M)$ and $u_1, \ldots, u_r \in \mathbf{Z}_p^*$ such that the image of σ_{ℓ_i} in $G(\ell_i) \otimes \mathcal{O}_\mathfrak{p}/\mathfrak{p}^M$ is equal to the image of $\zeta \otimes u_i$ $(i = 1, \ldots, r)$. The statement of Proposition 7.2.3 then becomes

$$[H : K] \sum_{i=1}^{r} \left(s_{\mathscr{L}_i}, (\kappa_{\mathfrak{n}/\ell_i})_{\mathscr{L}_i}\right)_M \otimes u_i = 0 \in \mu_{p^\infty}(H_M) \otimes \mathcal{O}_\mathfrak{p}/\mathfrak{p}^M = \mathcal{O}_\mathfrak{p}/\mathfrak{p}^M(1).$$

(ii) We shall apply Proposition 7.2.3 only in situations when it is known a priori that

$$(\forall i = 2, \ldots, r) \qquad \left(s_{\mathscr{L}_i}, (\kappa_{\mathfrak{n}/\ell_i})_{\mathscr{L}_i}\right)_M = 0.$$

In such a case one does not have to worry about the compatibility of the σ_{ℓ_i}'s.

(7.3) Theorem. *Assume that the condition* ($\star$) *from Theorem 3.2 holds. If* $e_\beta(y) \notin A(H)_{\mathrm{tors}}$, *then there exists an integer* $C(\mathfrak{p}) \geq 0$ *which is equal to zero for all but finitely many* $\mathfrak{p}$ *and such that*

$$(\forall M >> 0) \qquad \mathfrak{p}^{C(\mathfrak{p})}\left(S(A/H, \mathfrak{p}^M)^{(\beta)}/\mathcal{O}_\mathfrak{p} \cdot \kappa_1\right) = 0.$$

(7.4) Thanks to 7.1.2, Theorem 7.3 implies Theorem 3.2 (the statements about the α^{-1}-components follow by applying ρ to the α-components). The proof of Theorem 7.3 will occupy the rest of Sect. 7.

Recall that the constants $C_1(\mathfrak{p}), C_2(\mathfrak{p})$ and $C_3(\mathfrak{p})$ were introduced in 7.2.1, 6.1 and 6.2, respectively. We set

$$C_5(\mathfrak{p}) := \mathrm{ord}_\mathfrak{p}([H : K]), \qquad C_6(\mathfrak{p}) := \mathrm{ord}_\mathfrak{p}(\deg(\varphi)),$$

where $\varphi : A_j \longrightarrow \widehat{A_j}$ is the isogeny from 5.19. If $\beta^2 \neq 1$, then another constant $C_4(\mathfrak{p})$ will be defined in 7.6.1 below.

The assumption $e_\beta(y) \notin A(H)_{\mathrm{tors}}$ implies that the constant

$$C_0(\mathfrak{p}) := \max\{c \in \mathbf{Z}_{\geq 0} \mid e_\beta(y) \in A(H)_{\mathrm{tors}} + \mathfrak{p}^c A(H)\}$$

is defined (and $C_0(\mathfrak{p}) = 0$ for all but finitely many $\mathfrak{p}$).

In order to simplify the notation, we write $C_i = C_i(\mathfrak{p})$ $(i = 0, \ldots, 6)$. We also denote, for each $\mathcal{O}_\mathfrak{p}/\mathfrak{p}^M$-module Y and an element $y \in Y$,

$$\exp(y) := \min\{c \in \mathbf{Z}_{\geq 0} \mid \mathfrak{p}^c y = 0\}, \quad \exp(Y) := \max\{\exp(y) \mid y \in Y\}.$$

Using this notation, we have, for $M >> 0$,

$$\exp(\kappa_1) \geq M - C_0 - C_1. \tag{7.4.1}$$

(7.5) Proof of Theorem 7.3 in the case $\beta^2 = 1$. In this case $S := S(A/H, \mathfrak{p}^M)^{(\beta)}$ is ρ-stable; we denote $S^\pm := S^{\rho=\pm 1}$. As

$$2\kappa_1 = (1+\rho)\kappa_1 + (1-\rho)\kappa_1, \qquad (1 \pm \rho)\kappa_1 \in S^\pm,$$

it follows that

$$(\exists \varepsilon \in \{\pm 1\}) \qquad \exp((1+\varepsilon\rho)\kappa_1) \geq M - C_0 - C_1 - \mathrm{ord}_\mathfrak{p}(2).$$

Fix such an ε; then $x_\varepsilon := (1+\varepsilon\rho)\kappa_1 \in S^\varepsilon$.

(7.5.1) Bounding the exponent of $S^{-\varepsilon}$. Fix $s \in S^{-\varepsilon}$ with maximal $\exp(s)$ and set $x_{-\varepsilon} = s$.

Choosing the first Kolyvagin prime. We apply the discussion from 6.3 and 6.5 to $W' = S$. Set

$$U_{\pm\varepsilon} := \{h \in 2G^+ \mid \exp(j'(h))(x_{\pm\varepsilon}) < \exp(x_{\pm\varepsilon}) - (C_2 + C_3 + 4\,\mathrm{ord}_\mathfrak{p}(2))\}.$$

It follows from Corollary 6.3.4 that $U_{\pm\varepsilon} \subsetneq 2G^+$ are proper subgroups of $2G^+$, hence there exists $h = g^2 \in 2G^+ - (U_\varepsilon \cup U_{-\varepsilon})$. Applying the discussion in 6.5.1 to the element $h = g^2$, choose a prime $\mathscr{L}(W)$ satisfying (6.5.1.1-3). The induced primes $\mathscr{L} \mid \lambda_H \mid \lambda \mid \ell$ of H_M, H, K, F, respectively, satisfy $\ell \in \mathscr{S}_1(M)$ and $\rho(\mathscr{L}) = \mathscr{L}$. The definition of $U_{\pm\varepsilon}$ implies that

$$\exp\left((x_{\pm\varepsilon})_\mathscr{L}\right) \geq \exp\left(x_{\pm\varepsilon}\right) - (C_2 + C_3 + 4\,\mathrm{ord}_\mathfrak{p}(2)),$$

hence

$$\exp\left(((1+\varepsilon\rho)\kappa_1)_\mathscr{L}\right) \geq M - \sum_{i=0}^{3} C_i - 5\,\mathrm{ord}_\mathfrak{p}(2) \tag{7.5.1.1}$$

$$\exp\left(s_\mathscr{L}\right) \geq \exp\left(S^{-\varepsilon}\right) - (C_2 + C_3 + 4\,\mathrm{ord}_\mathfrak{p}(2)). \tag{7.5.1.2}$$

The first annihilation relation. Applying Proposition 7.2.3 with $\mathfrak{n} = \ell$ to s and $\rho(s)$, we obtain (using the ρ-equivariance of the map $w' \mapsto w'_\mathscr{L}$) that

$$[H:K]\left(s_\mathscr{L}, ((1+\varepsilon\rho)\kappa_1)_\mathscr{L}\right)_M = 0 \in \mathcal{O}_\mathfrak{p}/\mathfrak{p}^M(1). \tag{7.5.1.3}$$

Combining (7.5.1.3) with (7.5.1.1) and (5.19.2), we obtain

$$2^6\,\mathfrak{p}^{C_0+C_1+C_2+C_5+C_6}\,s_{\mathscr{L}} \in \mathcal{O}_{\mathfrak{p}}\cdot((1+\varepsilon\rho)\kappa_1)_{\mathscr{L}} \subset A[\mathfrak{p}^M]^{\rho=\varepsilon}.$$

On the other hand, $s_{\mathscr{L}} \in A[\mathfrak{p}^M]^{\rho=-\varepsilon}$ (since $s \in S^{-\varepsilon}$) and $2(A[\mathfrak{p}^M]^{\rho=\varepsilon} \cap A[\mathfrak{p}^M]^{\rho=-\varepsilon}) = 0$, hence

$$2^7\,\mathfrak{p}^{C_0+C_1+C_2+C_5+C_6}\,s_{\mathscr{L}} = 0.$$

Combined with (7.5.1.2), we obtain

$$\mathfrak{p}^{M_1}\,S^{-\varepsilon} = 0, \qquad M_1 = C_0 + C_1 + 2C_2 + 2C_3 + C_5 + C_6 + 11\,\mathrm{ord}_{\mathfrak{p}}(2). \tag{7.5.1.4}$$

In particular, $\mathfrak{p}^{M_1}(1-\varepsilon\rho)\kappa_1 = 0$, hence $2\,\mathfrak{p}^{M_1}\kappa_1 = \mathfrak{p}^{M_1}(1+\varepsilon\rho)\kappa_1$.

(7.5.2) Bounding the exponent of $S^{\varepsilon}/\mathcal{O}_{\mathfrak{p}}\cdot(1+\varepsilon\rho)\kappa_1$. Denote

$$\widetilde{S} = \mathrm{Ker}\left(\mathrm{res}_{\lambda_H} : S \longrightarrow H^1(H_{\lambda_H}, A[\mathfrak{p}^M])\right)$$

and fix $\widetilde{s} \in \widetilde{S}^{\varepsilon} = \widetilde{S}^{\rho=\varepsilon}$ with maximal $\exp(\widetilde{s})$. It follows from Proposition 5.18 combined with (5.15.1) that

$$\exp((1-\varepsilon\rho)\kappa_\ell) \geq \exp\left(f\left(((1-\varepsilon\rho)\kappa_\ell)_{\mathscr{L}}\right)\right) = \exp\left(((1+\varepsilon\rho)\kappa_1)_{\mathscr{L}}\right) \geq$$
$$\geq M - \sum_{i=0}^{3} C_i - 5\,\mathrm{ord}_{\mathfrak{p}}(2),$$

where f denotes the map

$$f : H^1(H_{\lambda_H}, A[\mathfrak{p}^M]) \longrightarrow H^1(H_{\lambda_H}, A[\mathfrak{p}^M])/H^1_{ur}(H_{\lambda_H}, A[\mathfrak{p}^M])$$

(cf. 7.6.4 below).

Choosing the second Kolyvagin prime. We apply the discussion from 6.3 and 6.5 to the submodule $W' = S_{\{v|\ell\}}(A/H, \mathfrak{p}^M)^{(\beta)}$. Set $x'_{\varepsilon} = \widetilde{s} \in (W')^{\varepsilon}$ and $x'_{-\varepsilon} = (1-\varepsilon\rho)\kappa_\ell \in (W')^{-\varepsilon}$. The argument as in 7.5.1 shows that there exists a prime $\ell' \in \mathscr{S}_1(M)$ ($\ell' \neq \ell$) and a prime $\mathscr{L}' \mid \ell'$ of H_M such that $\rho(\mathscr{L}') = \mathscr{L}'$ and

$$\exp\left(((1-\varepsilon\rho)\kappa_\ell)_{\mathscr{L}'}\right) \geq M - \sum_{i=0}^{3} C_i - 5\,\mathrm{ord}_{\mathfrak{p}}(2) - (C_2 + C_3 + 4\,\mathrm{ord}_{\mathfrak{p}}(2)), \tag{7.5.2.1}$$

$$\exp\left(\widetilde{s}_{\mathscr{L}'}\right) \geq \exp\left(\widetilde{S}^{\varepsilon}\right) - (C_2 + C_3 + 4\,\mathrm{ord}_{\mathfrak{p}}(2)). \tag{7.5.2.2}$$

The second annihilation relation. Applying Proposition 7.2.3 with $\mathfrak{n} = \ell\ell'$ to $\widetilde{s}$ and $\rho(\widetilde{s})$, we obtain (using the ρ-equivariance of the map $w' \mapsto w'_{\mathscr{L}'}$ and the assumption $\widetilde{s}_{\mathscr{L}} = 0$) that

$$[H:K]\,(\widetilde{s}_{\mathscr{L}'}, ((1-\varepsilon\rho)\kappa_\ell)_{\mathscr{L}'})_M = 0 \in \mathcal{O}_\mathfrak{p}/\mathfrak{p}^M(1). \qquad (7.5.2.3)$$

Combining (7.5.2.3) with (7.5.2.1) and (5.19.2), we deduce that

$$2^{10}\,\mathfrak{p}^{C_0+C_1+2C_2+2C_3+C_5+C_6}\widetilde{s}_{\mathscr{L}'} \in \mathcal{O}_\mathfrak{p}\cdot((1-\varepsilon\rho)\kappa_\ell)_{\mathscr{L}'} \subset A[\mathfrak{p}^M]^{\rho=-\varepsilon}.$$

On the other hand, $\widetilde{s} \in \widetilde{S}^\varepsilon$ implies that $\widetilde{s}_{\mathscr{L}'} \subset A[\mathfrak{p}^M]^{\rho=\varepsilon}$, hence

$$2^{11}\,\mathfrak{p}^{C_0+C_1+2C_2+2C_3+C_5+C_6}\widetilde{s}_{\mathscr{L}'} = 0.$$

Applying (7.5.2.2), we obtain

$$2^{15}\,\mathfrak{p}^{C_0+C_1+3C_2+3C_3+C_5+C_6}\,\widetilde{S}^\varepsilon = 0. \qquad (7.5.2.4)$$

(7.5.3) Bounding the exponent of $S/\mathcal{O}_\mathfrak{p}\,\kappa_1$. As $2H^1_{ur}(H_{\lambda_H}, A[\mathfrak{p}^M])^\varepsilon$ is a cyclic $\mathcal{O}_\mathfrak{p}$-module, it follows from (7.5.1.1) that

$$2^5\,\mathfrak{p}^{C_0+C_1+C_2+C_3}\left(S^\varepsilon/\left(\widetilde{S}^\varepsilon + \mathcal{O}_\mathfrak{p}\cdot(1+\varepsilon\rho)\kappa_1\right)\right) = 0.$$

Putting this together with (7.5.1.4) and (7.5.2.4), we deduce

$$2^{21}\,\mathfrak{p}^{2C_0+2C_1+4C_2+4C_3+C_5+C_6}\,(S/\mathcal{O}_\mathfrak{p}\,\kappa_1) = 0,$$

which finishes the proof of Theorem 7.3 in the case $\beta^2 = 1$.

(7.6) Proof of Theorem 7.3 in the case $\beta^2 \neq 1$. We write $\overline{\beta} := \beta^{-1} \neq \beta$, $S := S(A/H, \mathfrak{p}^M)$ (for $M >> 0$) and, as in 7.5, $C_i := C_i(\mathfrak{p})$ $(i = 0, \dots, 6)$.

(7.6.1) Lemma-Definition. *Define*

$$C_4(\mathfrak{p}) = \begin{cases} 0, & p \nmid \text{order of } \beta^2 \\ \mathrm{ord}_\mathfrak{p}(p)/(p-1), & p \mid \text{order of } \beta^2. \end{cases}$$

(i) $C_4(\mathfrak{p}) \in \mathbf{Z}_{\geq 0}$.
(ii) $C_4(\mathfrak{p}) \leq C_5(\mathfrak{p})/(p-1)$.
(iii) *For any $\mathcal{O}_\mathfrak{p}[\Delta]$-module Y (where $\Delta = \mathrm{Gal}(H/K)$), $\mathfrak{p}^{C_4(\mathfrak{p})}(Y^{(\beta)} \cap Y^{(\overline{\beta})}) = 0$.*

Proof. (i) If p divides the order of β^2, then $L_\mathfrak{p}$ contains the values of β^2, hence $L_\mathfrak{p} \supset \mathbf{Q}_p(\mu_p)$.

(ii) This is clear.
(iii) $(\exists \sigma \in \Delta)$ $\beta(\sigma) \neq \overline{\beta}(\sigma)$; then $Y^{(\beta)} \cap Y^{(\overline{\beta})}$ is killed by $\beta^2(\sigma) - 1 \neq 0$, but

$$\mathrm{ord}_{\mathfrak{p}}(\beta^2(\sigma) - 1) = \begin{cases} 0, & \beta^2(\sigma) \notin \mu_{p^\infty} \\ \leq \mathrm{ord}_{\mathfrak{p}}(p)/(p-1), & \beta^2(\sigma) \in \mu_{p^\infty}. \end{cases}$$

(7.6.2) Definition. (i) *Fix* $t_\pm \in T^\pm = T^{\rho=\pm1}$ *such that* $T^\pm = \mathcal{O}_{\mathfrak{p}}\, t_\pm$ *and let* $t_{\pm,M}$ *be the image of* $t_\pm$ *in* $(T/\mathfrak{p}^M T)^\pm = A[\mathfrak{p}^M]^\pm$.
(ii) $\exp(t_{\pm,M}) \geq M - \mathrm{ord}_{\mathfrak{p}}(2)$; *fix* $u_\pm \in (\mathcal{O}_{\mathfrak{p}}/\mathfrak{p}^M) t_{\pm,M}$ *with* $\exp(u_\pm) = M - \mathrm{ord}_{\mathfrak{p}}(2)$.
(iii) *Let* W' *be as in 6.3.1; let* $f \in \mathrm{Hom}_{\mathcal{O}_{\mathfrak{p}}}(W', A[\mathfrak{p}^M])^+$. *Then* $2f((W')^\pm) \subset (\mathcal{O}_{\mathfrak{p}}/\mathfrak{p}^{M-\mathrm{ord}_{\mathfrak{p}}(2)}) u_\pm$; *we define the morphisms*

$$f_\pm (W')^\pm \longrightarrow \mathcal{O}_{\mathfrak{p}}/\mathfrak{p}^{M-\mathrm{ord}_{\mathfrak{p}}(2)}, \quad (\forall w \in (W')^\pm) \quad f_\pm(w) u_\pm = 2f(w).$$

(iv) *For each* $w \in W'$, $2w = (1+\rho)w + (1-\rho)w$, *hence*

$$4f(w) = f_+((1+\rho)w)u_+ + f_-((1-\rho)w)u_-.$$

(7.6.3) Choosing the first Kolyvagin prime. We have

$$\kappa_1 \in S^{(\beta)}, \qquad \rho(\kappa_1) \in S^{(\overline{\beta})}, \qquad \exp(\kappa_1) = \exp(\rho(\kappa_1)) \geq M - C_0 - C_1.$$

As $\mathfrak{p}^{C_4}\left(S^{(\beta)} \cap S^{(\overline{\beta})}\right) = 0$, it follows that the elements $(1 \pm \rho)\kappa_1 \in S^\pm = S^{\rho=\pm1}$ satisfy

$$\exp((1 \pm \rho)\kappa_1) \geq M - C_0 - C_1 - C_4 - \mathrm{ord}_{\mathfrak{p}}(2). \tag{7.6.3.1}$$

Applying Corollary 6.3.4 to $W' = S$, we obtain, as in 7.5.1, primes $\ell \in \mathscr{S}_1(M)$ and $\mathscr{L} \mid \ell$ in H_M such that $\rho(\mathscr{L}) = \mathscr{L}$ and

$$\exp\left(((1 \pm \rho)\kappa_1))_{\mathscr{L}}\right) \geq \exp((1 \pm \rho)\kappa_1) - (C_2 + C_3 + 4\,\mathrm{ord}_{\mathfrak{p}}(2)) \geq M - M_2, \tag{7.6.3.2}$$

where

$$M_2 = \sum_{i=0}^{4} C_i + 5\,\mathrm{ord}_{\mathfrak{p}}(2).$$

As $2\kappa_1 = (1+\rho)\kappa_1 + (1-\rho)\kappa_1$ and $2(A[\mathfrak{p}^M]^+ \cap A[\mathfrak{p}^M]^-) = 0$, it follows that

$$\exp\left((\kappa_1)_{\mathscr{L}}\right), \exp\left((\rho(\kappa_1))_{\mathscr{L}}\right) \geq M - M_2 - 2\,\mathrm{ord}_{\mathfrak{p}}(2). \tag{7.6.3.3}$$

The first annihilation relation. Let $s \in S^{(\beta)}$. Applying Proposition 7.2.3 with $\mathfrak{n} = \ell$, we obtain

$$[H : K]\,(s_{\mathscr{L}}, (\kappa_1)_{\mathscr{L}})_M = 0 \in \mathcal{O}_{\mathfrak{p}}/\mathfrak{p}^M(1),$$

hence, using (5.19.2) and (7.6.3.3),

$$2^3\,\mathfrak{p}^{M_2+C_5+C_6}\,s_{\mathscr{L}} = 2^8\,\mathfrak{p}^{\sum_{i=0}^{6} C_i}\,s_{\mathscr{L}} \in \mathcal{O}_{\mathfrak{p}}\,(\kappa_1)_{\mathscr{L}}. \tag{7.6.3.4}$$

Set

$$\widetilde{S} = \operatorname{Ker}\Big(S \xrightarrow{(\mathrm{res}_v)} \bigoplus_{v|\ell} H^1(H_v, A[\mathfrak{p}^M])\Big);$$

then

$$\widetilde{S}^{(\beta)} = \operatorname{Ker}\left(\mathrm{res}_{\lambda_H} : S^{(\beta)} \longrightarrow H^1(H_{\lambda_H}, A[\mathfrak{p}^M])\right),$$

and the inclusion (7.6.3.4) can be reformulated as follows:

$$\begin{aligned} &2^3\,\mathfrak{p}^{M_2+C_5+C_6}\left(S^{(\beta)}/\left(\widetilde{S}^{(\beta)} + \mathcal{O}_{\mathfrak{p}}\,\kappa_1\right)\right) = \\ &= 2^8\,\mathfrak{p}^{\sum_{i=0}^{6} C_i}\left(S^{(\beta)}/\left(\widetilde{S}^{(\beta)} + \mathcal{O}_{\mathfrak{p}}\,\kappa_1\right)\right) = 0. \end{aligned} \tag{7.6.3.5}$$

(7.6.4) Choosing the second Kolyvagin prime. In the exact sequence of $\mathcal{O}_{\mathfrak{p}}/\mathfrak{p}^M$-modules

$$0 \longrightarrow H^1_{ur}(H_{\lambda_H}, A[\mathfrak{p}^M]) \longrightarrow H^1(H_{\lambda_H}, A[\mathfrak{p}^M]) \xrightarrow{f} \frac{H^1(H_{\lambda_H}, A[\mathfrak{p}^M])}{H^1_{ur}(H_{\lambda_H}, A[\mathfrak{p}^M])} \longrightarrow 0,$$

both flank terms are isomorphic to $A[\mathfrak{p}^M] \xrightarrow{\sim} (\mathcal{O}_{\mathfrak{p}}/\mathfrak{p}^M)^{\oplus 2}$, hence the sequence splits and the middle term is isomorphic, as an $\mathcal{O}_{\mathfrak{p}}/\mathfrak{p}^M$-module, to $(\mathcal{O}_{\mathfrak{p}}/\mathfrak{p}^M)^{\oplus 4}$. We know that

$$\begin{gathered} (\rho(\kappa_1))_{\mathscr{L}} = \rho((\kappa_1)_{\mathscr{L}}) \in H^1_{ur}(H_{\lambda_H}, A[\mathfrak{p}^M]) \xrightarrow{\sim} A[\mathfrak{p}^M], \\ ((1 \pm \rho)\kappa_1)_{\mathscr{L}} \in A[\mathfrak{p}^M]^{\pm} \end{gathered}$$

and

$$\exp\left(((1 \pm \rho)\kappa_1)_{\mathscr{L}}\right) \geq M - M_2.$$

On the other hand, (5.15.1) and Proposition 5.18 imply that

$$f\left(((1 \pm \rho)\kappa_\ell)_{\mathscr{L}}\right) = -\Phi_{\lambda_H}\left(((1 \mp \rho)\kappa_1)_{\mathscr{L}}\right),$$

hence

$$\exp\left(f\left(((1 \pm \rho)\kappa_\ell)_{\mathscr{L}}\right)\right) \geq M - M_2. \tag{7.6.4.1}$$

Set

$$W' = S_{\{v|\ell\}}(A/H, \mathfrak{p}^M), \qquad \widetilde{W}' = \mathrm{Ker}\Big(W' \xrightarrow{(\mathrm{res}_v)} \bigoplus_{v|\ell} H^1(H_v, A[\mathfrak{p}^M])\Big);$$

then

$$(\widetilde{W}')^{(\beta)} = \widetilde{S}^{(\beta)}, \qquad (\widetilde{W}')^{(\overline{\beta})} = \widetilde{S}^{(\overline{\beta})}, \qquad \kappa_1, \kappa_\ell \in (W')^{(\beta)},$$
$$\rho(\kappa_1), \rho(\kappa_\ell) \in (W')^{(\overline{\beta})}.$$

Denote by $U \subseteq W'$ the $\mathcal{O}_\mathfrak{p}/\mathfrak{p}^M$-submodule generated by $(1+\rho)\kappa_1$, $(1-\rho)\kappa_1, (1+\rho)\kappa_\ell$ and $(1-\rho)\kappa_\ell$. It follows from (7.6.3.2) and (7.6.4.1) that

$$\mathrm{res}_{\lambda_H}(U) \supseteq \mathfrak{p}^{M_2}\, H^1(H_{\lambda_H}, A[\mathfrak{p}^M]), \tag{7.6.4.2}$$

hence

$$\mathfrak{p}^{M_2}(U \cap \widetilde{W}') = 0 \tag{7.6.4.3}$$

and U contains an $\mathcal{O}_\mathfrak{p}/\mathfrak{p}^M$-submodule isomorphic to $(\mathfrak{p}^{M_2}(\mathcal{O}_\mathfrak{p}/\mathfrak{p}^M))^{\oplus 4}$. Fix an element $\widetilde{s} \in (\widetilde{W}')^{(\beta)} = \widetilde{S}^{(\beta)}$ with maximal $\exp(\widetilde{s})$; then the argument used in the proof of (7.6.3.1) shows that

$$\exp((1+\rho)\widetilde{s}) \geq \exp(\widetilde{s}) - C_4 - \mathrm{ord}_\mathfrak{p}(2). \tag{7.6.4.4}$$

Define a homomorphism of $\mathcal{O}_\mathfrak{p}/\mathfrak{p}^M$-modules

$$z : \mathrm{Hom}_{\mathcal{O}_\mathfrak{p}}(W', A[\mathfrak{p}^M])^+ \longrightarrow \big(\mathcal{O}_\mathfrak{p}/\mathfrak{p}^N\big)^{\oplus 5} \qquad (N = M - \mathrm{ord}_\mathfrak{p}(2))$$

by the formula

$$z(f) = (f_+((1+\rho)\widetilde{s}),\ f_+((1+\rho)\kappa_1), f_-((1-\rho)\kappa_1),$$
$$f_+((1+\rho)\kappa_\ell),\ f_-((1-\rho)\kappa_\ell)))$$

and set

$$Z = (z \circ j')(2G^+) = (z \circ j')(2\,\mathrm{Gal}(H_M(W)/H_M)^+),$$

which is a $\mathbf{Z}_p$-submodule of $(\mathcal{O}_\mathfrak{p}/\mathfrak{p}^N)^{\oplus 5}$. It follows from (7.6.3.2), (7.6.4.1), (7.6.4.3) and (7.6.4.4) that

$$\mathrm{Im}(z) \supseteq 2\,\mathfrak{p}^{a'}(\mathcal{O}_\mathfrak{p}/\mathfrak{p}^N) \oplus (2\,\mathfrak{p}^{M_2}(\mathcal{O}_\mathfrak{p}/\mathfrak{p}^N))^{\oplus 4},$$

where

$$N - a' \geq \exp(\widetilde{s}) - (M_2 + C_4 + \mathrm{ord}_\mathfrak{p}(2)).$$

Corollary 6.3.4 implies that

$$\mathcal{O}_{\mathfrak{p}} \cdot Z \supseteq \mathfrak{p}^a(\mathcal{O}_{\mathfrak{p}}/\mathfrak{p}^N) \oplus \mathfrak{p}^b(\mathcal{O}_{\mathfrak{p}}/\mathfrak{p}^N))^{\oplus 4},$$

where

$$\begin{aligned} N - a &\geq \exp(\widetilde{s}) - (M_2 + C_2 + C_3 + C_4 + 6\,\mathrm{ord}_{\mathfrak{p}}(2)),\\ b &= M_2 + C_2 + C_3 + 5\,\mathrm{ord}_{\mathfrak{p}}(2). \end{aligned}$$

According to Proposition 6.6.2, there exists an element $h = g^2 \in 2G^+$ such that the corresponding vector

$$(z_0, \ldots, z_4) = (z \circ j')(h) \in (\mathcal{O}_{\mathfrak{p}}/\mathfrak{p}^N)^{\oplus 5}$$

satisfies

$$z_0 \notin \mathfrak{p}^{a+1}(\mathcal{O}_{\mathfrak{p}}/\mathfrak{p}^N), \qquad \begin{vmatrix} z_1 & z_2 \\ z_3 & z_4 \end{vmatrix} \notin 2\,\mathfrak{p}^{2b+1}(\mathcal{O}_{\mathfrak{p}}/\mathfrak{p}^N). \tag{7.6.4.5}$$

Applying the discussion from 6.5.1, we choose $\mathcal{L}'(W)$ satisfying (6.5.1.1-3) and not dividing ℓ. We denote by $\mathcal{L}', \lambda'_H, \ell' \in \mathcal{S}_1$ ($\ell' \neq \ell$) the induced primes of H_M, H and F, respectively. By construction, we have

$$\begin{aligned} 2\,((1+\rho)\widetilde{s})_{\mathcal{L}'} &= z_0 u_+,\\ 2\,((1+\rho)\kappa_1)_{\mathcal{L}'} = z_1 u_+, &\quad 2\,((1-\rho)\kappa_1)_{\mathcal{L}'} = z_2 u_-,\\ 2\,((1+\rho)\kappa_\ell)_{\mathcal{L}'} = z_3 u_+, &\quad 2\,((1-\rho)\kappa_\ell)_{\mathcal{L}'} = z_4 u_-, \end{aligned}$$

hence

$$4(\kappa_1)_{\mathcal{L}'} = z_1 u_+ + z_2 u_-, \qquad 4(\kappa_\ell)_{\mathcal{L}'} = z_3 u_+ + z_4 u_-.$$

It follows from (7.6.4.5) that

$$\begin{aligned} \exp\left(((1+\rho)\widetilde{s})_{\mathcal{L}'}\right) \geq M - a &\geq \exp(\widetilde{s}) - (M_2 + C_2 + C_3 + C_4 + 5\,\mathrm{ord}_{\mathfrak{p}}(2))\\ &= \exp(\widetilde{s}) - (b + C_4) \end{aligned} \tag{7.6.4.6}$$

and

$$\mathrm{res}_{\lambda'_H}(\mathcal{O}_{\mathfrak{p}}\,\kappa_1 + \mathcal{O}_{\mathfrak{p}}\,\kappa_\ell) \supseteq \mathfrak{p}^{2b} H^1_{ur}(H_{\lambda'_H}, A[\mathfrak{p}^M]) \xrightarrow{\sim} \mathfrak{p}^{2b} A[\mathfrak{p}^M]. \tag{7.6.4.7}$$

The second annihilation relation. Applying Proposition 7.2.3 with $\mathfrak{n} = \ell\ell'$ to the above element $s = \widetilde{s} \in \widetilde{S}^{(\beta)}$, we obtain, as in (7.5.2.3),

$$[H:K]\,(\widetilde{s}_{\mathcal{L}'}, (\kappa_\ell)_{\mathcal{L}'})_M = 0 \in \mathcal{O}_{\mathfrak{p}}/\mathfrak{p}^M(1).$$

Applying Proposition 7.2.3 with $\mathfrak{n} = \ell'$ to $s = \widetilde{s}$ (i.e., the first annihilation relation with ℓ' instead of ℓ), we obtain

$$[H : K]\,(\widetilde{s}_{\mathscr{L}'}, (\kappa_1)_{\mathscr{L}'})_M = 0 \in \mathcal{O}_{\mathfrak{p}}/\mathfrak{p}^M(1).$$

Combining these two relations with (7.6.4.7), we deduce that the element $\mathfrak{p}^{2b+C_5}\,\widetilde{s}_{\mathscr{L}'} \in A[\mathfrak{p}^M]$ lies in the kernel of the pairing $(\ ,\ )_M$, hence

$$\mathfrak{p}^{2b+C_5+C_6}\,\widetilde{s}_{\mathscr{L}'} = 0.$$

It follows from (7.6.4.6) that

$$\begin{aligned} \exp\left(\widetilde{S}^{(\beta)}\right) &= \exp(\widetilde{s}) \le \exp\left(((1+\rho)\widetilde{s})_{\mathscr{L}'}\right) + (b + C_4) \\ &\le (2b + C_5 + C_6) + (b + C_4) \end{aligned}$$
$$= 3b + C_4 + C_5 + C_6 = 3C_0 + 3C_1 + 6C_2 + 6C_3 + 4C_4 + C_5 + C_6 + 30\,\mathrm{ord}_{\mathfrak{p}}(2).$$

Combined with (7.6.3.5), this relation yields

$$2^{38}\,\mathfrak{p}^{4C_0+4C_1+7C_2+7C_3+5C_4+2C_5+2C_6}\left(S^{(\beta)}/\mathcal{O}_{\mathfrak{p}}\,\kappa_1\right) = 0,$$

which concludes the proof of Theorem 7.3 (hence also the proof of Theorem 3.2) in the case $\beta^2 \neq 1$.

References

[Be] M. Bertolini, *Selmer groups and Heegner points in anticyclotomic* $\mathbf{Z}_p$*-extensions*, Compositio Math. **99** (1995), 153–182.

[Be-Da] M. Bertolini, H. Darmon, *Kolyvagin's descent and Mordell-Weil groups over ring class fields*, J. Reine Angew. Math. **412** (1990), 63–74.

[Bi] B. Birch, *Heegner points of elliptic curves*, In: Symposia Mathematica, Vol. XV (Convegno di Strutture in Corpi Algebrici, INDAM, Rome, 1973), pp. 441–445. Academic Press, London, 1975.

[Bo] F.A. Bogomolov, *Sur l'algébricité des représentations l-adiques*, C. R. Acad. Sci. Paris Sér. A-B **290** (1980), no. 15, A701–A703.

[Bu] K. Buzzard, *Integral models of certain Shimura curves*, Duke Math. J. **87** (1997), 591–612.

[Ca 1] H. Carayol, *Sur la mauvaise réduction des courbes de Shimura*, Compositio Math. **59** (1986), 151–230.

[Ca 2] H. Carayol, *Sur les représentations ℓ-adiques attachées aux formes modulaires de Hilbert*, Ann. Sci. E.N.S. **19** (1986), 409–469.

[Cas] W. Casselman, *On abelian varieties with many endomorphisms and a conjecture of Shimura's*, Invent. Math. **12** (1971), 225–236.

[Con] B. Conrad, *Gross-Zagier revisited*, in: Heegner points and Rankin L-series, (H. Darmon, S.-W. Zhang, eds.), MSRI Publ. **49**, Cambridge Univ. Press, Cambridge, 2004, pp. 67–163.

[Co] C. Cornut, *Réduction de Familles de Points CMs*, Thesis, Strasbourg, 2000.

[Co-Va 1] C. Cornut, V. Vatsal, *Nontriviality of Rankin-Selberg L-functions and CM points*, this volume, pp. 121-186.

[Co-Va 2] C. Cornut, V. Vatsal, *CM points and quaternion algebras*, Documenta Math. **10** (2005), 263-309.

[Cu-Re] C.W. Curtis, I. Reiner, *Methods of Representation Theory, Vol. I*, Wiley, New York, 1990.

[De] P. Deligne, *Travaux de Shimura*, Sém. Bourbaki, exp. 389, 1970/71, Lect. Notes in Math. **244**, Springer, Berlin, 1971, pp. 123–165.

[De-Ra] P. Deligne, M. Rapoport, *Les schémas de modules de courbes elliptiques*, in: Modular functions of one variable II (Antwerp, 1972), Lect. Notes in Math. **349**, Springer, Berlin, 1973, pp. 143–316.

[Di] M. Dimitrov, *Galois representations modulo p and cohomology of Hilbert modular varieties*, Ann. Sci. E.N.S. **38** (2005), 505–551.

[Dr] V.G. Drinfeld, *Two theorems on modular curves* (Russian), Funkcional. Anal. i Priložen. **7** (1973), 83–84. English translation: Functional Anal. Appl. **7** (1973), 155–156.

[Fa] G. Faltings, *Endlichkeitssätze für abelsche Varietäten über Zahlkörpern*, Invent. Math. **73** (1983), 349–366. Erratum: Invent. Math. **75** (1984), 381.

[Ge] S. Gelbart, *Lectures on the Arthur-Selberg trace formula*, Univ. Lect. Series **9**, Amer. Math. Soc., Providence, 1996.

[Gr 1] B.H. Gross, *Kolyvagin's work on modular elliptic curves*, in: L-functions and arithmetic (Durham, 1989; J. Coates, M.J. Taylor, eds.), London Math. Soc. Lect. Note Ser. **153**, Cambridge Univ. Press, Cambridge, 1991, pp. 235–256.

[Gr 2] B.H. Gross, *Heegner points and representation theory*, in: Heegner points and Rankin L-series, (H. Darmon, S.-W. Zhang, eds.), MSRI Publ. **49**, Cambridge Univ. Press, Cambridge, 2004, pp. 37–65.

[He] G. Henniart, *Représentations l-adiques abéliennes*, in: Séminaire de Théorie des Nombres de Paris 1980/81, Progress in Math. **22**, (M.-J. Bertin, ed.), Birkhäuser, Boston, 1982, pp. 107–126.

[Ho 1] B. Howard, *The Heegner point Kolyvagin system*, Compos. Math. **140** (2004), 1439–1472.

[Ho 2] B. Howard, *Iwasawa Theory of Heegner points on abelian varieties of GL_2-type*, Duke Math. J. **124** (2004), 1–45.

[Ja-La] H. Jacquet, R. Langlands, *Automorphic forms on $GL(2)$*, Lect. Notes in Math. **114**, Springer, Berlin-New York, 1970.

[Ka-Ma] N. Katz, B. Mazur, *Arithmetic moduli of elliptic curves*, Annals of Mathematics Studies **108**, Princeton Univ. Press, Princeton, 1985.

[Ko] V. A. Kolyvagin, *Euler systems*, in: The Grothendieck Festschrift II, Progress in Mathematics **87**, Birkhäuser, Boston, Basel, Berlin, 1990, pp. 435–483.

[Ko-Lo 1] V.A. Kolyvagin, D.Yu. Logachev, *Finiteness of the Shafarevich-Tate group and the group of rational points for some modular abelian varieties* (Russian), Algebra i Analiz **1** (1989), 171–196. English translation: Leningrad Math. J. **1** (1990), 1229–1253.

[Ko-Lo 2] V.A. Kolyvagin, D.Yu. Logachev, *Finiteness of* Ш *over totally real fields* (Russian), Izv. Akad. Nauk. SSSR, Ser. Math. **55** (1991), 851–876. English translation: Math. USSR-Izv. **39** (1992), 829–853.

[Mi 1] J.S. Milne, *Arithmetic duality theorems*, Persp. in Math. **1**, Academic Press, Boston, 1986.

[Mi 2] J.S. Milne, *Canonical models of (mixed) Shimura varieties and automorphic vector bundles*, in: Automorphic forms, Shimura varieties, and L-functions, Vol. I (Ann Arbor, MI, 1988), Perspect. Math. **10**, Academic Press, Boston, 1990, pp. 283–414,

[Mi 3] J.S. Milne, *The points on a Shimura variety modulo a prime of good reduction*, in: The zeta functions of Picard modular surfaces, (R. Langlands, D. Ramakrishnan, eds.), Univ. Montreal, Montreal, 1992, pp. 151–253.

[Miy] T. Miyake, *Modular forms*, Springer, Berlin, 1989.

[Mu] D. Mumford, *Abelian varieties*, Oxford Univ. Press, London, 1970.

[Ne] J. Nekovář, *Selmer complexes*, to appear in Astérisque. Available at `http://www.math.jussieu.fr/~nekovar/pu/`

[Ne-Sch] J. Nekovář, N. Schappacher, *On the asymptotic behaviour of Heegner points*, Turkish J. of Math. **23** (1999), 549–556.

[Oh] M. Ohta, *On l-adic representations attached to automorphic forms*, Japan. J. Math. **8** (1982), 1–47.

[Ri] K. Ribet, *Galois action on division points of Abelian varieties with real multiplications*, Amer. J. Math. **98** (1976), 751–804.

[Sa] C.-H. Sah, *Automorphisms of finite groups*, J. Algebra **10** (1968), 47–68.

[Se 1] J.-P. Serre, *Corps locaux*, Hermann, Paris, 1962.

[Se 2] J.-P. Serre, *Abelian l-adic representations and elliptic curves*, Benjamin, New York-Amsterdam, 1968.

[Se 3] J.-P. Serre, *Lettre à Marie-France Vignéras du 10/2/1986*, Collected Papers IV, 38–55.

[Sh 1] G. Shimura, *On canonical models of arithmetic quotients of bounded symmetric domains*, Ann. of Math. **91** (1970), 144–222.

[Sh 2] G. Shimura, *Introduction to the Arithmetic Theory of Automorphic Functions*, Princeton Univ. Press, Princeton, 1971.

[Sh-Ta] G. Shimura, Y. Taniyama, *Complex multiplication of abelian varieties and its applications to number theory*, Publ. Math. Soc. Japan **6**, Tokyo, 1961.

[Tay] R. Taylor, *On Galois representations associated to Hilbert modular forms II*, in: Elliptic Curves, Modular Forms and Fermat's Last Theorem, (J. Coates, S.T. Yau, eds.), International Press, Boston, 1995, pp. 185–191.

[Ti] Y. Tian, *Euler systems of CM points on Shimura curves*, Thesis, Columbia University, 2003.

[Ti-Zh] Y. Tian, S.-W. Zhang, in preparation.

[Wa] L.C. Washington, *Introduction to cyclotomic fields*, Graduate Texts in Mathematics **83**, Second ed., Springer, New York, 1997.

[Zh 1] S.-W. Zhang, *Heights of Heegner points on Shimura curves*, Ann. of Math. (2) **153** (2001), 27–147.

[Zh 2] S.-W. Zhang, *Gross-Zagier formula for GL_2*, Asian J. Math. **5** (2001), 183–290.

Représentations irréductibles de GL(2,F) modulo p

Marie-France Vignéras
Institut de Mathematiques de Jussieu
175 rue du Chevaleret
Paris 75013
France
vigneras@math.jussieu.fr

Résumé. This is a report on the classification of irreducible representations of $GL(2,F)$ over $\overline{\mathbf{F}}_p$ when F is a local field of finite residual field contained in the algebraically closed field $\overline{\mathbf{F}}_p$ of characteristic p.

1 Toutes les représentations des groupes seront lisses, i.e. chaque vecteur est fixe par un sous-groupe ouvert.

Soit p un nombre premier, F un corps local complet pour une valuation discrète de corps résiduel fini $\mathbf{F}_q$ de caractéristique p ayant q éléments, $\overline{F}$ un clôture séparable de F et $\overline{\mathbf{F}}_p$ une clôture algébrique de $\mathbf{F}_q$. Soit $n \geq 1$ un entier. Pour un nombre premier $\ell \neq p$, la correspondance semisimple de Langlands modulo ℓ, est une bijection "compatible avec la réduction modulo ℓ" entre les classes d'isomorphisme des représentations irréductibles de dimension n du groupe de Gal$(\overline{F}/F)$ sur $\overline{\mathbf{F}}_\ell$ et les classes d'isomorphisme des représentations irréductibles supercuspidales de $GL(n,F)$ sur $\overline{\mathbf{F}}_\ell$, étendue de façon à inclure les représentations semi-simples de dimension n de Gal$(\overline{F}/F)$, et toutes les représentations irréductibles de $GL(n,F)$ sur $\overline{\mathbf{F}}_\ell$ [V2].

On s'intéresse au cas $\ell = p$. On connait bien les représentations irréductibles de dimension finie de Gal$(\overline{F}/F)$ sur $\overline{\mathbf{F}}_p$, mais quelles sont les représentations irréductibles de $GL(n,F)$ sur $\overline{\mathbf{F}}_p$? La réponse est connue uniquement pour le groupe $GL(2,\mathbf{Q}_p)$; c'est un problème ouvert pour $n \geq 3$ ou pour $F \neq \mathbf{Q}_p$.

2 Les représentations irréductibles de dimension n du groupe de Galois Gal$(\overline{F}/F)$ sur $\overline{\mathbf{F}}_p$ se classent facilement [V3] 1.14 page 423.

2.1 Lorsque $n = 1$, l'isomorphisme de réciprocité du corps de classes, qui envoie une uniformisante p_F de F sur un Frobenius géométrique Frob$_F$, identifie les caractères (représentations de dimension 1) de F^* et de Gal$(\overline{F}/F)$. Nous utiliserons systématiquement cette identification. Pour une extension finie F' de F contenue dans $\overline{F}$, la restriction du côté galoisien correspond à la norme $F'^* \to F^*$.

2.2 Les représentations irréductibles de Gal$(\overline{F}/F)$ sur $\overline{\mathbf{F}}_p$ de dimension $n \geq 2$ sont induites par les caractères réguliers sur F du groupe multiplicatif de l'unique extension F_n de F non ramifiée de degré n sur F contenue dans $\overline{F}$. Un caractère de F_n^* est régulier sur F si ses n conjugués par le groupe de Galois $\mathrm{Gal}(F_n/F)$ sont distincts. Les représentations $\rho(\chi), \rho(\chi')$ de $\mathrm{Gal}(\overline{F}/F)$ induites de deux caractères χ, χ' de F_n^* réguliers sur F sont isomorphes si et seulement si χ, χ' sont conjugués par le groupe de Galois $\mathrm{Gal}(F_n/F)$. Le déterminant de $\rho(\chi)$ est la restriction de χ à F^*.

2.3 On note O_F l'anneau des entiers de F; un caractère $O_F^* \to \overline{\mathbf{F}}_p^*$ s'identifie à un caractère de $\mathbf{F}_q^*$. L'uniformisante p_F de F est aussi une uniformisante de l'extension non ramifiée F_n. Le caractère ω_n de F_n^* tel que $\omega_n(p_F) = 1$ et dont la restriction à $O_{F_n}^*$ s'identifie au plongement naturel $\iota_n : \mathbf{F}_{q^n}^* \to \overline{\mathbf{F}}_p^*$, est appelé un caractère de Serre; il est régulier. Pour $\lambda \in \overline{\mathbf{F}}_p^*$, on note $\mu_{n,\lambda}$ le caractère non ramifié de F_n^* tel que $\mu_{n,\lambda}(p_F) = \lambda$. On supprime l'indice $n = 1$ pour F^*. Le composé de la norme $F_n^* \to F^*$ et du caractère ω, resp. μ_λ, est le caractère $\omega_n^{1+q+\ldots+q^{n-1}}$, resp. μ_{n,λ^n}.

Les caractères de F_n^* sont $\mu_{n,\lambda}\, \omega_n^a$ pour un unique couple $(\lambda, a) \in \overline{\mathbf{F}}_p^* \times \{1, \ldots, q^n - 1\}$.

2.4 *Les représentations irréductibles de dimension $n \geq 1$ du groupe de Galois $\mathrm{Gal}(\overline{F}/F)$ sur $\overline{\mathbf{F}}_p$ sont*

$$\rho_n(a,\lambda) = \mu_\lambda \otimes \mathrm{ind}_{\mathrm{Gal}(\overline{F}/F_n)}^{\mathrm{Gal}(\overline{F}/F)} \omega_n^a = \mathrm{ind}_{\mathrm{Gal}(\overline{F}/F_n)}^{\mathrm{Gal}(\overline{F}/F)} (\mu_{n,\lambda^n}\, \omega_n^a)$$

pour les entiers $a \in \mathbf{Z}/(q^n - 1)\mathbf{Z}$ tels que $a, qa, \ldots, q^{n-1}a$ sont distincts. Les isomorphismes sont les suivants

$$\rho_n(a,\lambda) \simeq \rho_n(aq^i, \zeta\lambda)$$

pour les entiers $1 \leq i \leq n-1$ et $\zeta \in \overline{\mathbf{F}}_p^*$ avec $\zeta^n = 1$.

Le déterminant de $\rho_n(a,\lambda)$ est $\omega^{a'}\mu_{\lambda^n}$ où $a' \in \mathbf{Z}/(q-1)\mathbf{Z}$ est l'image de a. Le nombre de représentations irréductibles avec $\det \mathrm{Frob}_F$ fixé est fini, égal au nombre de polynômes irréductibles unitaires de degré n dans $\mathbf{F}_q[X]$ [V1] 3.1 (10). Lorsque $n = 2$, ce nombre est $q(q-1)/2$.

2.5 Lorsque $F = \mathbf{Q}_p$, les représentations irréductibles de dimension 2 du groupe de Galois $\mathrm{Gal}(\overline{\mathbf{Q}}_p/\mathbf{Q}_p)$ sur $\overline{\mathbf{F}}_p$ sont

$$\sigma(r,\chi) = \chi \otimes \mathrm{ind}_{\mathrm{Gal}(\overline{F}/F_2)}^{\mathrm{Gal}(\overline{F}/F)} \omega_2^{r+1}$$

pour les entiers $r \in \{0, \ldots, p-1\}$ et les caractères $\chi : \mathbf{Q}_p^* \to \overline{\mathbf{F}}_p^*$. On a $\sigma(r,\chi) = \rho_2(a,\lambda)$ avec $\chi = \mu_\lambda \omega^b, a = (p+1)b + r + 1$. Le déterminant de $\sigma(r,\chi)$ est $\chi^2\omega^{r+1}$. Les isomorphismes sont les suivants

$$\sigma(r,\chi) \simeq \sigma(p-1-r,\chi) \simeq \sigma(r,\chi\mu_{-1}) \simeq \sigma(p-1-r,\chi\mu_{-1}).$$

3 Les représentations irréductibles de $GL(2, \mathbf{Q}_p)$ sur $\overline{\mathbf{F}}_p$ avec un caractère central sont classées [BL2] [Br]. On dispose d'une liste pour $GL(2, F)$ lorsque $F \neq \mathbf{Q}_p$ [V0], [Pa], probablement non complète.

3.1 Les représentations irréductibles de $GL(2, \mathbf{F}_q)$ sur $\overline{\mathbf{F}}_p$ sont $(\chi \circ \det) \otimes \mathrm{Sym}_{\underline{r}} \overline{\mathbf{F}}_p^2$ pour un unique couple (r, χ) avec $0 \leq r \leq q-1$ et un caractère $\chi : \mathbf{F}_q^* \to \overline{\mathbf{F}}_p^*$; on peut aussi remplacer χ par l'unique entier $1 \leq a \leq q-1$ tel que $\chi(?) = ?^a$. On a le développement p-adique $r = r_1 + pr_2 + \ldots + p^{f-1} r_f$ avec $0 \leq r_i \leq p-1$, $q = p^f$, et

$$\mathrm{Sym}_{\underline{r}} \overline{\mathbf{F}}_p^2 = \otimes_{i=1}^f \mathrm{Sym}^{r_i} \overline{\mathbf{F}}_p^2 \circ \mathrm{Fr}^{i-1}$$

où

$$\mathrm{Fr}\begin{pmatrix} a & b \\ c & d \end{pmatrix} = \begin{pmatrix} a^p & b^p \\ c^p & d^p \end{pmatrix}$$

et $\mathrm{Sym}^r \overline{\mathbf{F}}_p^2$ est la représentation de $GL(2, \mathbf{F}_q)$ sur les polynômes homogènes de degré $r \geq 0$ dans $\overline{\mathbf{F}}_p[X, Y]$ vérifiant

$$\begin{pmatrix} a & b \\ c & d \end{pmatrix} X^i Y^j = (aX + cY)^i (bX + cY)^j \quad (i, j \geq 0, i + j = r).$$

La représentation triviale et la représentation spéciale (ou de Steinberg) sont $\mathrm{Sym}_{\underline{0}} \overline{\mathbf{F}}_p^2$ et $\mathrm{Sym}_{\underline{q-1}} \overline{\mathbf{F}}_p^2$.

La représentation irréductible $\mathrm{Sym}_{\underline{r}} \overline{\mathbf{F}}_p^2$ s'identifie à une représentation irréductible de $GL(2, O_F)$, ou à une représentation de $K_o = GL(2, O_F) p_F^{\mathbf{Z}}$ triviale sur p_F. On lui associe par induction compacte une représentation lisse de $GL(2, F)$

$$E(\underline{r}) = \mathrm{ind}_{K_o}^G \mathrm{Sym}_{\underline{r}} \overline{\mathbf{F}}_p^2.$$

3.2 *Les représentations irréductibles de* $G = GL(2, F)$ *sur* $\overline{\mathbf{F}}_p$ *sont:*
(i) Les caractères $\chi \circ \det$ *pour les caractères* $\chi : F^* \to \overline{\mathbf{F}}_p^*$.
(ii) Les séries principales $\mathrm{ind}_B^G(\chi_1 \otimes \chi_2)$ *induites par le caractère*

$$(\chi_1 \otimes \chi_2)\begin{pmatrix} a & b \\ 0 & d \end{pmatrix} = \chi_1(a)\chi_2(d)$$

du sous-groupe triangulaire supérieur B*, pour les caractères distincts* $\chi_1, \chi_2 : F^* \to \overline{\mathbf{F}}_p^*$, $\chi_1 \neq \chi_2$.

(iii) Les séries spéciales (appelées aussi de Steinberg) $\mathrm{Sp} \otimes (\chi \circ \det)$*, pour les caractères* $\chi : F^* \to \overline{\mathbf{F}}_p^*$*, où* Sp *est le quotient de la représentation induite* $\mathrm{ind}_B^G \mathrm{id}_{\overline{\mathbf{F}}_p}$ *du caractère trivial de* B*, par le caractère trivial de* G.

(iv) Les représentations irréductibles supersingulières.

Il n'y a pas d'isomorphisme entre ces représentations.

Les représentations supersingulières sont les représentations supercuspidales (non sous-quotient d'une représentation induite de B). La représentation spéciale ne se plonge pas dans une représentation induite parabolique; elle est cuspidale (l'espace de ses coinvariants par le radical unipotent N de B est nul) sans être supercuspidale.

Barthel et Livné ([BL2] prop.8) montrent que $\mathrm{End}_{\overline{\mathbf{F}}_p G}(\mathrm{ind}_{K_o}^G \mathrm{id}) \simeq \overline{\mathbf{F}}_p[T]$ où T correspond à la double classe de $h = \begin{pmatrix} p_F & 0 \\ 0 & 1 \end{pmatrix}$; les algèbres $\mathrm{End}_{\overline{\mathbf{F}}_p G} E(\underline{r})$ sont canoniquement isomorphes. Ils ont appelé supersingulières les représentations irréductibles supercuspidales ayant un caractère central, et démontré que ce sont les quotients irréductibles de

$$V(\underline{r}, \chi) = (\chi \circ \det) \otimes \frac{E(\underline{r})}{TE(\underline{r})}$$

pour les caractères $\chi : F^* \to \overline{\mathbf{F}}_p^*$ et les entiers $0 \leq r \leq q-1$; le caractère central de $V(\underline{r}, \chi)$ est $\chi^2 \omega^r$. La représentation de G

$$V(\underline{r}, \lambda, \chi) = (\chi \circ \det) \otimes \frac{E(\underline{r})}{(T-\lambda)E(\underline{r})} \qquad (\lambda \in \overline{\mathbf{F}}_p^*)$$

est isomorphe à

i) la série principale irréductible $(\chi \circ \det) \otimes \mathrm{ind}_B^G(\mu_{\lambda^{-1}} \otimes \mu_\lambda \omega^r)$ si $\mu_{\lambda^{-1}} \neq \mu_\lambda \omega^r$, i.e. si $\lambda \neq \pm 1$ ou $\underline{r} \neq (0, \ldots, 0), (p-1, \ldots, p-1)$,

ii) la représentation $(\chi \circ \det) \otimes \mathrm{ind}_B^G \mu_\lambda$ de longueur 2 non scindée, contenant $\chi\mu_\lambda \circ \det$ et de quotient $(\chi\mu_\lambda \circ \det) \otimes \mathrm{Sp}$ lorsque $\lambda = \pm 1$ et $\underline{r} = (p-1, \ldots, p-1)$,

iii) une représentation de longueur 2 non scindée contenant $(\chi\mu_\lambda \circ \det) \otimes \mathrm{Sp}$ et de quotient $\chi\mu_\lambda \circ \det$ lorsque $\lambda = \pm 1$ et $r = (0, \ldots, 0)$.

3.3 Dans le cas particulier mais important $F = \mathbf{Q}_p$, Breuil [Br] 4.1.1, 4.1.4, a montré que les représentations $V(r, \chi)$ sont irréductibles.

Les représentations irréductibles supersingulières de $GL(2, \mathbf{Q}_p)$ sont les $V(r, \chi)$ pour $0 \leq r \leq p-1$ et χ un caractère $F^* \to \overline{\mathbf{F}}_p^*$; les isomorphismes sont :

$$V(r, \chi) \simeq V(p-1-r, \chi\omega^r) \simeq V(r, \chi\mu_1) \simeq V(p-1-r, \chi\omega^r \mu_{-1}).$$

3.4 Breuil [Br] 4.2.4 en déduit une bijection unique "compatible avec la réduction modulo p" de la correspondance donnée par la "cohomologie étale des courbes modulaires"

$$\sigma(r, \chi) \leftrightarrow V(r, \chi)$$

entre les classes d'isomorphisme des représentations irréductibles de dimension 2 du groupe de Galois $\mathrm{Gal}(\overline{\mathbf{Q}}_p/\mathbf{Q}_p)$ sur $\overline{\mathbf{F}}_p$ (2.5) et les classes d'isomorphisme des représentations irréductibles supersingulières de $GL(2,\mathbf{Q}_p)$ sur $\overline{\mathbf{F}}_p$ (3.3), qu'il étend de façon à inclure les représentations semi-simples de dimension 2 de $\mathrm{Gal}(\overline{\mathbf{Q}}_p/\mathbf{Q}_p)$,

$$\chi\otimes(\omega^{r+1}\mu_\lambda\oplus\mu_{\lambda^{-1}})$$
$$\leftrightarrow (\chi\circ\det)\otimes[\mathrm{ind}_B^G(\mu_{\lambda^{-1}}\otimes\omega^r\mu_\lambda)\oplus\mathrm{ind}_B^G((\omega^r\mu_\lambda)\omega\otimes\mu_{\lambda^{-1}}\omega^{-1})]^{ss},$$

notant $?^{ss}$ la semi-simplifiée d'une représentation ? de longueur finie. Soit $r'\in\{0,\ldots,p-2\}$ congru à $p-3-r$ modulo $p-1$; le membre de droite est aussi (3.2):

$$V(r,\lambda,\chi)^{ss}\oplus V(r',\lambda^{-1},\omega^{r+1}\chi)^{ss}.$$

Le déterminant $\chi^2\omega^{r+1}$ de la représentation galoisienne ne coincide pas par l'isomorphisme de la théorie du corps de classes avec le caractère central $\chi^2\omega^r$ de la représentation de $GL(2,\mathbf{Q}_p)$.

3.5 Une représentation irréductible avec un caractère central de $GL(2,\mathbf{Q}_p)$ est caractérisée par sa restriction au sous-groupe triangulaire B, qui est irréductible sauf pour une série principale où elle est de longueur 2; ceci est démontré par Berger [Be] lorsque $F=\mathbf{Q}_p$ en utilisant les représentations de $B(\mathbf{Q}_p)$ construites par Colmez avec les (ϕ,Γ)-modules de Fontaine; une preuve non galoisienne est donnée dans [Vc] (voir 3.6) pour tout F, mais uniquement pour les séries principales et la Steinberg.

Toute représentation irréductible W de $\mathrm{Gal}(\overline{\mathbf{Q}}_p/\mathbf{Q}_p)$ *de dimension finie sur $\overline{\mathbf{F}}_p$, définit une représentation irréductible de dimension infinie Ω_W de $B(\mathbf{Q}_p)$ sur $\overline{\mathbf{F}}_p$. Deux représentations W non isomorphes donnent des représentations Ω_W non isomorphes.*

Les représentations irréductibles contenues dans la série principale ou la série spéciale (resp. dans les supersingulières) de $GL(2,\mathbf{Q}_p)$ sont les Ω_W pour $\dim W=1$ (resp. $\dim W=2$).

Remarque. Notons P le sous-groupe mirabolique formé des matrices de seconde ligne $(0,1)$ dans B. Il est isomorphe au produit semi-direct du groupe additif G_a et du groupe multiplicatif G_m (agissant naturellement sur G_a). On identifie G_a au radical unipotent de P ou de B.

Sur un corps algébriquement clos de caractéristique différente de p, le groupe mirabolique $P(F)$ a une unique représentation irréductible de dimension infinie τ; les autres sont des caractères. La restriction à $P(F)$ d'une représentation irréductible π de dimension infinie de $GL(2,F)$ contient τ et π/τ est de longueur $2,1,0$ selon que π est de la série principale, spéciale, ou est supercuspidale [V4].

3.6 On peut décomposer les séries principales avec les arguments suivants [Vc]. La représentation naturelle ρ de $P(F)$ sur l'espace des fonctions localement constantes à support compact sur F et à valeurs dans $\overline{\mathbf{F}}_p$ est irréductible; ceci utilise que l'algèbre de groupe complétée d'un pro-p-groupe sur corps fini de caractéristique p est locale. Joint au fait que les F-coinvariants de ρ sont nuls car un pro-p-groupe n'a pas de mesure de Haar à valeurs dans $\overline{\mathbf{F}}_p$, on obtient que la restriction à $B(F)$ de $\operatorname{ind}_B^G \chi_1 \otimes \chi_2$ est de longueur 2, de quotient le caractère $\chi_1 \otimes \chi_2$, et l'image de $\operatorname{ind}_B^G \chi_1 \otimes \chi_2$ par le foncteur de Jacquet (les $U(F)$-coinvariants) est $\chi = \chi_1 \otimes \chi_2$. Cette démonstration n'utilisant pas l'arbre de $PGL(2)$, est généralisable à $GL(n)$.

4 Généralités.

4.1 On dit que V est admissible si l'espace V^K des vecteurs de V fixes par K est de dimension finie pour tout sous-groupe ouvert compact K de G. Lorsque $C = \overline{\mathbf{F}}_p$, il suffit que ce soit vrai pour un seul pro-p-sous-groupe ouvert $\mathcal{P}$ de G. Voici la preuve simple et astucieuse due à Paskunas. La restriction de V à $\mathcal{P}$ se plonge dans $(\dim_{\overline{\mathbf{F}}_p} V^{\mathcal{P}}) \operatorname{Inj} 1_{\overline{\mathbf{F}}_p}$, où $\operatorname{Inj} 1_{\overline{\mathbf{F}}_p}$ est l'enveloppe injective de la représentation triviale de $\mathcal{P}$; pour tout sous-groupe ouvert distingué K de $\mathcal{P}$, on a $(\operatorname{Inj} 1_{\overline{\mathbf{F}}_p})^K = \overline{\mathbf{F}}_p[\mathcal{P}/K]$, donc la dimension de V^K est finie.

Toutes les représentations irréductibles de $GL(2,F)$ sur $\overline{\mathbf{F}}_p$ connues sont admissibles, ont un caractère central et sont définies sur un corps fini.

4.2 Tout pro-p-groupe agissant sur un $\overline{\mathbf{F}}_p$-espace vectoriel non nul a un vecteur non nul invariant. Ceci implique qu'une représentation admissible non nulle V de G sur $\overline{\mathbf{F}}_p$, contient une sous-représentation irréductible W.

4.3 Soit W une représentation lisse sur un corps commutatif C de dimension finie d'un sous-groupe ouvert K de G. On lui associe la représentation lisse $\operatorname{ind}_K^G W$ de G, par induction compacte. L'algèbre de Hecke de (K, W) dans G,

$$H(G, K, W) = \operatorname{End}_{CG} \operatorname{ind}_K^G W$$

s'identifie à l'algèbre de convolution des fonctions $f : G \to \operatorname{End}_C W$ de support une union finie de doubles classes de G modulo K, satisfaisant $f(kgk') = k \circ f(g) \circ k'$ pour $g \in G$ et $k, k' \in K$; la condition sur $F = f(g) \in \operatorname{End}_C W$ est $F \circ k = gkg^{-1} \circ F$ pour tout $k \in K \cap g^{-1}Kg$. On associe à V le $H(G, K, W)$-module à droite

$$\operatorname{Hom}_{CG}(\operatorname{ind}_K^G W, V) \simeq \operatorname{Hom}_{CK}(W, V),$$

par adjonction.

4.4 Les algèbres de Hecke fournissent un critère bien utile d'irréductibilité [V0] Criterium 4.5 page 344. Si $\mathcal{P}$ est un pro-p-sous-groupe ouvert, une représentation V de G sur $\overline{\mathbf{F}}_p$ engendrée par $\pi^{\mathcal{P}}$ est irréductible, si $\pi^{\mathcal{P}}$ est un $H(G, \mathcal{P}, \mathrm{id}_{\overline{\mathbf{F}}_p})$-module simple.

4.5 Nous donnons maintenant deux propriétés générales et techniques utilisées dans la démonstration par Barthel et Livné qu'une représentation irréductible V de $GL(2, F)$ sur $\overline{\mathbf{F}}_p$ ayant un caractère central est quotient d'un $V(r, \lambda, \chi)$ (3.2), que nous expliquerons en (6.3).

(i) Soit V irréductible tel que $\mathrm{Hom}_{CK}(W, V)$ contienne un sous-$H(G, K, W)$-module de dimension finie sur C (vrai si V est admissible); alors il existe un $H(G, K, W)$-module simple à droite M tel que V est quotient de

$$M \otimes_{H(G,K,W)} \mathrm{ind}_K^G W.$$

(ii) Soit un quotient $j : W \to W'$ de la représentation W de K; l'application $\mathrm{ind}_K^G(j) : \mathrm{ind}_K^G W \to \mathrm{ind}_K^G W'$ est surjective et induit une application injective

$$? \circ \mathrm{ind}_K^G(j) : \mathrm{Hom}_{CG}(\mathrm{ind}_K^G W', V) \to \mathrm{Hom}_{CG}(\mathrm{ind}_K^G W, V)$$

pour toute représentation V de G. Soit M' un sous-espace de Hom_{CG} $(\mathrm{ind}_K^G W', V)$ dont l'image par $? \circ \mathrm{ind}_K^G(j)$ dans $\mathrm{Hom}_{CG}(\mathrm{ind}_K^G W, V)$ est $H(G, K, W)$-stable. Si tout morphisme $h' \in H(G, K, W')$ se relève en un morphisme $h \in H(G, K, W)$ tel que $h' \circ \mathrm{ind}_K^G(j) = \mathrm{ind}_K^G(j) \circ h$, alors M' est $H(G, K, W')$-stable.

La condition sur les algèbres de Hecke signifie que pour chaque g dans un système de représentants des doubles classes $K \backslash G / K$, tout morphisme $F' \in \mathrm{End}_C W'$ vérifiant $F' \circ f = gkg^{-1} \circ F'$ pour tout $k \in K \cap g^{-1} K g$ se relève en un morphisme $F \in \mathrm{End}_{CK} \mathrm{ind}_K^G W$ vérifiant la même relation tel que $F' \circ j = j \circ F$.

5 Algèbre de Hecke du "pro-p-Iwahori" $I(1)$.

5.1 L'image inverse dans $K = GL(2, O_F)$ de $B(\mathbf{F}_q)$ par la réduction $K \to GL(2, \mathbf{F}_q)$ est le groupe d'Iwahori I; celle du groupe strictement triangulaire supérieur est le pro$-p$-sous-groupe $I(1)$; c'est un pro-p-Sylow de I. La $\mathbf{Z}$-algèbre de Hecke du pro-p-Iwahori $I(1)$ [V1]

$$H(2, q) = \mathrm{End}_{\mathbf{Z}G} \mathbf{Z}[I(1) \backslash GL(2, F)]$$

ne dépend que de $(2, q)$. La $\mathbf{Z}$-algèbre $H(2, q)$ a une grosse sous-algèbre commutative de type fini avec une action naturelle de S_2, style Bernstein, dont les S_2-invariants forment le centre de $H(2, q)$. L'algèbre $H(2, q)$ est un module de type fini sur son centre. Le centre de $H(2, q)$ est une $\mathbf{Z}$-algèbre de type fini.

5.2 Un $H(2,q) \otimes_{\mathbf{Z}} \overline{\mathbf{F}}_p$-module simple (à droite ou à gauche) a un caractère central, est de dimension finie, et est défini sur un corps fini.

C'est l'analogue pour l'algèbre de Hecke du pro-p-Iwahori de "une sous-représentation d'une représentation de type fini de G sur $\overline{\mathbf{F}}_p$ est de type fini" et de "une représentation irréductible de G sur $\overline{\mathbf{F}}_p$ est admissible définie sur un corps fini" (questions ouvertes).

Cela résulte de:

5.3 *Soit C un corps commutatif parfait, Z une C-algèbre commutative de type fini, H un anneau qui est un Z-module de type fini. Alors un H-module (à droite ou à gauche) simple est de dimension finie sur C.*

Preuve (implicite dans [V4] I.7.11): Un H-module M simple (à droite par exemple) étant un Z-module de type fini, admet un quotient Z-simple. Si C est algébriquement clos, un Z-module simple est de dimension 1 sur C; il existe donc un morphisme $\chi : Z \to C$ tel que $M(\chi) = \{mz - \chi(z)m, \ (m \in M, z \in Z)\}$ est distinct de M et $M(\chi)$ est stable par A. Comme M est simple, $M(\chi) = 0$ i.e. Z agit sur M par χ; ceci implique que M est de dimension finie sur C. Si C est parfait, il existe une extension finie galoisienne C'/C et un morphisme $\chi : Z \to C'$ tel que $M \otimes_C C'$ est distinct de $(M \otimes_C C')(\chi)$. Le $H \otimes_C C'$-module $M \otimes_C C'$ est une somme directe finie de modules simples, conjugués par $\mathrm{Gal}(C'/C)$. L'un d'entre eux N est distinct de $N(\chi)$, donc $N(\chi) = 0$ et N est de dimension finie sur C'. La dimension de $M \otimes_C C'$ sur C' est donc de dimension finie; c'est aussi celle de M sur C.

5.4 Définition d'un $H(2,q) \otimes_{\mathbf{Z}} \overline{\mathbf{F}}_p$-module simple (à droite ou à gauche) supersingulier [V1]. Un caractère du centre de $H(2,q)$ à valeurs dans $\overline{\mathbf{F}}_p$ a nécessairement beaucoup de "zéros". Lorsqu'il y a des zéros supplémentaires, le caractère est dit singulier (non singulier est appelé régulier). Le pire cas plein de zéros est appelé supersingulier. La terminologie s'étend à un $H(2,q) \otimes_{\mathbf{Z}} \overline{\mathbf{F}}_p$-module simple via son caractère central (5.3). Le nombre de $H(2,q) \otimes_{\mathbf{Z}} \overline{\mathbf{F}}_p$-modules simples supersinguliers de dimension 2 connus avec une action de p_F fixée, est exactement le nombre de représentations continues irréductibles de $\mathrm{Gal}(\overline{F}/F)$ de dimension 2 avec le déterminant de Frob_F fixé.

6 Nous expliquons le rôle du foncteur $\pi \to \pi^{I(1)}$ des $I(1)$-invariants dans les démonstrations de (3.2), (3.3) et nous donnons des propriétés de ce foncteur [V0], [O].

6.1 Le cas du groupe fini $GL(2,\mathbf{F}_q)$ [BL2] [Pa]. Un p-groupe de Sylow est le groupe des matrices strictement triangulaires supérieures $U(\mathbf{F}_q)$; le foncteur $\rho \to \rho^{U(\mathbf{F}_q)}$ définit une bijection des représentations irréductibles de $GL(2,\mathbf{F}_q)$ sur $\overline{\mathbf{F}}_p$ sur les modules simples de la $\overline{\mathbf{F}}_p$-algèbre de Hecke de $U(\mathbf{F}_q)$.

6.2 Soit ρ une représentation irréductible de $GL(2,\mathbf{F}_q)$; le poids de ρ est le caractère η de $T(F)$ sur les invariants $\rho^{U(\mathbf{F}_q)}$ (qui est de dimension 1); le copoids de ρ est le caractère η' de $T(F)$ sur les coinvariants $\rho_{U(\mathbf{F}_q)}$ (qui est de dimension 1); par adjonction, ρ est un quotient de $\mathrm{ind}_{B(\mathbf{F}_q)}^{GL(n,\mathbf{F}_q)}\eta$ et une sous-representation de $\mathrm{ind}_{B(\mathbf{F}_q)}^{GL(2,\mathbf{F}_q)}\eta'$. Comme la contragrédiente de $\mathrm{ind}_{B(\mathbf{F}_q)}^{GL(n,\mathbf{F}_q)}\eta$ est $\mathrm{ind}_{B(\mathbf{F}_q)}^{GL(n,\mathbf{F}_q)}\eta^{-1}$, les poids et copoids de la contragrédiente de ρ sont $(\eta')^{-1}$ et η^{-1}. Explicitement, la représentation irréductible $\det^a\otimes \mathrm{Sym}_{\underline{r}}\overline{\mathbf{F}}_p^2$ pour $1\le a\le q-1, 0\le r\le q-1$ a pour contragrédiente $\det^{-r-a}\otimes\mathrm{Sym}_{\underline{r}}\overline{\mathbf{F}}_p^2$, et pour poids

$$\mathrm{diag}(1,?)\to ?^a,\quad \mathrm{diag}(?,?^{-1})\to ?^r;$$

le poids est le caractère $\mathrm{diag}(x,z)\to x^cz^d$, les deux entiers $1\le c,d\le q-1$ étant liés aux deux entiers a,r par les congruences $d\equiv a \pmod{q-1}$, $c\equiv a+r \pmod{q-1}$. Les représentations irréductibles non isomorphes

$$\det{}^a\otimes\mathrm{Sym}_{\underline{r}}\overline{\mathbf{F}}_p^2,\quad \det{}^{a+r}\otimes\mathrm{Sym}_{\underline{q-1-r}}\overline{\mathbf{F}}_p^2,$$

ont des poids conjugués par S_2; le poids est fixe par S_2 si $r=0$ ou $r=q-1$.

Le nombre de caractères de $T(\mathbf{F}_q)$ modulo l'action naturelle du groupe symétrique S_2, i.e. de représentations semi-simples de dimension 2 de $\mathbf{F}_q^*$ sur $\overline{\mathbf{F}}_p$, est $q(q-1)/2$ comme en (2.4).

6.3 Nous expliquons comment l'isomorphisme [BL2] prop.8:

$$H(G,Kp_F^{\mathbf{Z}},\mathrm{id}_{\overline{\mathbf{F}}_p})\simeq\overline{\mathbf{F}}_p[T],$$

le résultat crucial [BL1] prop.15, [BL2] prop.18: si $E\ne 0$ est une sous-représentation de $E(\underline{r})$, alors $E^{I(1)}$ est de codimension finie dans $E(\underline{r})^{I(1)}$, et (4.5) impliquent qu'une représentation irréductible V de $G=GL(2,F)$ sur $\overline{\mathbf{F}}_p$ ayant un caractère central est quotient d'un $V(r,\lambda,\chi)$ (3.2).

Le premier groupe de congruence $K(1)$ des matrices congrues modulo p_F à l'identité dans $K=GL(2,O_F)$, étant un pro-p-groupe, V contient une représentation irréductible de K triviale sur $K(1)$. On en déduit qu'il existe $\underline{r},\chi$ comme en (3.2) tel que V est quotient de $E(\underline{r},\chi)=(\chi\circ\det)\otimes E(\underline{r})$. On se ramène à χ trivial par torsion. Il existe un G-morphisme non nul

$$E(\underline{r})\to\mathrm{ind}_B^G\chi$$

pour tout caractère $\chi: B\to\overline{\mathbf{F}}_p^*$, trivial sur $\mathrm{diag}(p_F,p_F)$ et de restriction à $T(O_F)$ le copoids η' de $\mathrm{Sym}_{\underline{r}}$ (6.1), par adjonction [V4] I.5.7 et l'isomorphisme $(\mathrm{ind}_B^G \chi)^{K(1)} \simeq \mathrm{ind}_{B(\mathbf{F}_q)}^{GL(2,\mathbf{F}_q)} \eta'$. Donc $E(\underline{r})$ est réductible; le résultat crucial implique que l'image de $E(\underline{r})^{I(1)}$ dans $V^{I(1)}$ est un sous-$H(G, I(1), \mathrm{id}_{\overline{\mathbf{F}}_p})$-module de dimension finie. Les algèbres de Hecke $H(G, I(1), \mathrm{id})$ et $H(G, K_o, \mathrm{ind}_{I(1)}^{K_o} \mathrm{id})$ sont isomorphes et l'on a une surjection canonique $\mathrm{ind}_{I(1)}^{K_o} \mathrm{id} \to \mathrm{Sym}_{\underline{r}} \overline{\mathbf{F}}_p^2$. Les isomorphismes $\overline{\mathbf{F}}_p[T] \simeq H(G, Kp_F^{\mathbf{Z}}, \mathrm{id}_{\overline{\mathbf{F}}_p}) \simeq H(G, Kp_F^{\mathbf{Z}}, \mathrm{Sym}_{\underline{r}})$ et (4.5) impliquent alors que V est quotient d'un $V(r, \lambda)$.

6.4 Structure de $H(2,q) \otimes_{\mathbf{Z}} \overline{\mathbf{F}}_p$ et modules simples à droite [V0].

L'algèbre $H(2,q) \otimes_{\mathbf{Z}} \overline{\mathbf{F}}_p$ est une somme directe $\oplus_{\eta_1 \oplus \eta_2} H(G, I, \eta)$, paramétré par les représentations semi-simples $\eta_1 \oplus \eta_2$ de dimension 2 de $\mathbf{F}_q^*$ sur $\overline{\mathbf{F}}_p$ (les "poids modulo S_2" (6.1)), où $\eta = \eta_1 \otimes \eta_2$ si $\eta_1 = \eta_2$ et $\eta = (\eta_1 \otimes \eta_2) \oplus (\eta_2 \otimes \eta_1)$ si $\eta_1 \neq \eta_2$. Le nombre de facteurs est $q(q-1)/2$ comme en (2.4). Les modules simples sont génériquement déterminés par leur caractère central; pour chaque facteur $H(G, I, \eta)$, chaque $(a, z) \in \overline{\mathbf{F}}_p \times \overline{\mathbf{F}}_p^*$ détermine un ou deux modules simples "jumeaux"; l'uniformisante p_F agit par multiplication par z.

a) Pour $\eta_1 = \eta_2$, le facteur $H(G, I, \eta)$ est isomorphe à la $\overline{\mathbf{F}}_p$-algèbre de Hecke du groupe d'Iwahori I; elle est engendrée par les fonctions S, T, égales à 1 sur $\begin{pmatrix} 0 & 1 \\ 1 & 0 \end{pmatrix}, \begin{pmatrix} 0 & 1 \\ p_F & 0 \end{pmatrix}$ respectivement, de support une double classe modulo I, vérifiant les relations $T^2 S = S T^2, S^2 = -S$; le centre est engendré par $ST + TS + T, T^{\pm 2}$. Les modules à droite simples sont génériquement les modules $M_2(a, z)$ de dimension 2, déterminés par leur caractère central $(ST + TS + T, T^2) \to (a, z)$, où T, S agissent respectivement par

$$\begin{pmatrix} 0 & z \\ 1 & 0 \end{pmatrix}, \quad \begin{pmatrix} -1 & a \\ 0 & 0 \end{pmatrix}.$$

Les modules à droite simples sont

(i) $M_2(a, z)$ si $z \neq a^2$,

(ii) les caractères $M_1(a, -1), M_1(a, 0)$ tels que $T \to a \in \overline{\mathbf{F}}_p^*$ et $S \to \varepsilon \in \{-1, 0\}$ sont respectivement contenus et quotients du module $M_2(a, a^2)$; ce sont des "jumeaux".

b) Pour $\eta_1 \neq \eta_2$, le facteur $H(G, I, \eta)$ est isomorphe à $M(2, R)$ où R est la $\overline{\mathbf{F}}_p$-algèbre commutative engendrée par $Z^{\pm 1}, X, Y$ vérifiant $XY = 0$. En effet, $\begin{pmatrix} 0 & 1 \\ p_F & 0 \end{pmatrix}$ normalise I et permute $\eta_1 \otimes \eta_2$ et $\eta_2 \otimes \eta_1$, donc induit un isomorphisme $\mathrm{ind}_I^G(\eta_1 \otimes \eta_2) \simeq \mathrm{ind}_I^G(\eta_2 \otimes \eta_1)$. L'isomorphisme $R \simeq H(G, I, \eta_1 \otimes \eta_2)$ envoie $Z^{\pm 1}, X, Y$ sur les fonctions supportées sur une doubles classe modulo I et égales à 1 sur $p_F^{\pm 1}, \begin{pmatrix} p_F & 0 \\ 0 & 1 \end{pmatrix}, \begin{pmatrix} 1 & 0 \\ 0 & p_F \end{pmatrix}$, respectivement. L'automorphisme de R déduit de l'isomorphisme

$H(G, I, \eta_1 \otimes \eta_2) \simeq H(G, I, \eta_2 \otimes \eta_1)$ induit par $\begin{pmatrix} 0 & 1 \\ p_F & 0 \end{pmatrix}$ fixe Z et permute X, Y.

Les modules à droite simples $M_2(xa, ya, z)$ de $M(2, R)$ sont de dimension 2, déterminés par leur caractère central $X \to xa, Y \to ya, Z \to z$, pour tout $a \in \overline{\mathbf{F}}_p, z \in \overline{\mathbf{F}}_p^*$ et $(x, y) = (1, 0)$ ou $(0, 1)$.

Si $a \neq 0$, les modules simples $M_2(a, 0, z), M_2(0, a, z)$ sont "jumeaux"; ils n'ont pas le même caractère central. Si $a = 0$, on note $M_2(0, z) = M_2(0, 0, z)$.

L'isomorphisme $R \simeq H(G, I, \eta_2 \otimes \eta_1)$ aurait permuté les jumeaux. Comme les caractères η_1, η_2 jouent des rôles symétriques, il faut regrouper les "jumeaux" si l'on veut rester canonique.

On déduit de a), b):

Les modules simples supersinguliers de $H(2, q) \otimes_{\mathbf{Z}} \overline{\mathbf{F}}_p$ sont les modules $M_2(0, z)$ pour chacun des $q(q-1)/2$ facteurs de $H(2, q) \otimes_{\mathbf{Z}} \overline{\mathbf{F}}_p$, où $z \in \overline{\mathbf{F}}_p^*$ est l'action de p_F. On notera $M_{\eta,z}$ le module simple supersingulier $H(2, q) \otimes_{\mathbf{Z}} \overline{\mathbf{F}}_p$ de poids η modulo S_2 où p_F agit par z.

6.5 On donne les $I(1)$-invariants des représentations $\mathrm{ind}_B^G \chi, \mathrm{Sp}, V(\underline{r}, \chi)$ de G sur $\overline{\mathbf{F}}_p$ introduites en (3.2), en utilisant la classification des $H(2, q) \otimes_{\mathbf{Z}} \overline{\mathbf{F}}_p$-modules simples (6.4).

a) Soit $\chi : B \to \overline{\mathbf{F}}_p^*$ un caractère de B défini par un caractère $\chi_1 \otimes \chi_2$ du groupe diagonal. On pose

$$a^{-1} = \chi_1(p_F), \quad z^{-1} = \chi_1(p_F)\chi_2(p_F), \quad \eta_? = \chi_?|_{O_F^*}.$$

Le $H(2, q) \otimes_{\mathbf{Z}} \overline{\mathbf{F}}_p$-module à droite $(\mathrm{ind}_B^G \chi)^{I(1)}$ est de dimension 2, s'identifie à un module du facteur associé à $\eta_1 \oplus \eta_2$ égal à $M_2(a, z)$ si $\eta_1 = \eta_2$ et à $M_2(a, 0, z)$ si $\eta_1 \neq \eta_2$ pour le choix de l'ordre (η_1, η_2). On a

$$M_2(a, 0, z) \oplus M_2(0, a, z) = (\mathrm{ind}_B^G \chi_1 \otimes \chi_2)^{I(1)} \oplus (\mathrm{ind}_B^G \chi_2 \otimes \chi_1)^{I(1)}$$

lorsque $\eta_1 \neq \eta_2$.

b) On a $1^I = M_1(1, 0)$ et $\mathrm{Sp}^I = M_1(1, -1)$ [BL1] lemma 26.

c) L'irréductibilité des représentations $V(r) = E(r)/TE(r)$ de $GL(2, \mathbf{Q}_p)$, avec $0 \leq r \leq p-1$, se montre ainsi. La représentation $V(r)$ étant engendrée par $V(r)^{I(1)}$, il suffit de montrer que $V(r)^{I(1)}$ est un $H(2, p)$-module simple. Breuil montre [Br] 4.1.2, 4.1.3:

$$V(0) \simeq V(p-1),$$

puis pour $0 \leq r \leq p-2$, que l'image par l'application $E(r)^{I(1)} \to V(r)^{I(1)}$ des fonctions $A, B \in E(r)^{I(1)}$ concentrées sur une classe de

$GL(2,O_F)p_F^{\mathbf{Z}}$, avec

$$A(\mathrm{id}) = X^r, \quad B\begin{pmatrix} 1 & 0 \\ 0 & p_F^{-1} \end{pmatrix} = Y^r,$$

forment une base de $V(r)^{I(1)}$ [Br1 3.2.4, 4.1.4]; il est facile d'en déduire que le $H(2,\mathbf{Q}_p)\otimes_{\mathbf{Z}}\overline{\mathbf{F}}_p$-module $V(r)^{I(1)}$ est le module simple $M_2(0,1)$ du facteur correspondant à $\omega^r \oplus \mathrm{id}$.

On tord par le caractère $\chi = \omega^a \mu_\lambda$ avec $0 \leq a \leq p-2, \lambda \in \overline{\mathbf{F}}_p^*$, et l'on voit que le $H(2,\mathbf{Q}_p)\otimes_{\mathbf{Z}}\overline{\mathbf{F}}_p$-module $V(r,\chi)^{I(1)}$ est le module $M_2(0,\lambda^2)$ du facteur correspondant à $\omega^{a+r}\oplus\omega^a$.

6.6 La décomposition des induites paraboliques $\mathrm{ind}_B^G \chi$ (3.2), l'irréductibilité des $V(r)$ et les isomorphismes entre les $V(r,\chi)$ si $F = \mathbf{Q}_p$ (3.3), se déduisent de (6.3), (6.5), en appliquant le critère d'irréductibilité (4.5). On obtient aussi les propriétés suivantes du foncteur des $I(1)$-invariants:

(i) Le foncteur $\pi \to \pi^{I(1)}$ définit une bijection des représentations irréductibles non supercuspidales de $GL(2,F)$ sur $\overline{\mathbf{F}}_p$, sur les $H(2,q)\otimes_{\mathbf{Z}}\overline{\mathbf{F}}_p$-modules simples non supersinguliers.

(ii) Le foncteur $\pi \to \pi^{I(1)}$ définit une bijection des représentations irréductibles de $GL(2,\mathbf{Q}_p)$ sur $\overline{\mathbf{F}}_p$ ayant un caractère central sur les $H(2,p)\otimes_{\mathbf{Z}}\overline{\mathbf{F}}_p$-modules simples.

(iii) Toute représentation irréductible de $GL(2,\mathbf{Q}_p)$ sur $\overline{\mathbf{F}}_p$ où p_F opère trivialement est admissible.

6.7 Pour $G = GL(2,F)/p_F^{\mathbf{Z}}$ et $p \neq 2$, Rachel Ollivier [O] a montré que la bijection (6.6 (iii)) provient d'une équivalence de catégories:

(i) Le module universel $\overline{\mathbf{F}}_p[I\backslash G]$ d'un Iwahori I est projectif sur l'algèbre de ses $\overline{\mathbf{F}}_p[G]$-endomorphismes.

Le module universel $\overline{\mathbf{F}}_p[I(1)\backslash G]$ d'un pro-p-Iwahori $I(1)$ est plat sur l'algèbre $H(2,q)\otimes_{\mathbf{Z}}\overline{\mathbf{F}}_p$ de ses $\overline{\mathbf{F}}_p[G]$-endomorphismes, si et seulement si $q = p$.

(ii) Lorsque $F = \mathbf{Q}_p$ et $p \neq 2$, la catégorie des représentations π de G sur $\overline{\mathbf{F}}_p$ engendrées par $\pi^{I(1)}$ est abélienne et équivalente à celle des $H(2,p)\otimes_{\mathbf{Z}}\overline{\mathbf{F}}_p$-modules à droite sur lesquels p agit trivialement, par le foncteur des $I(1)$-invariants $?^{I(1)}$ d'inverse $?\otimes_{H(2,p)\otimes_{\mathbf{Z}}\overline{\mathbf{F}}_p}\overline{\mathbf{F}}_p[I(1)\backslash G]$.

(ii) est faux lorsque $q \neq p$ ou lorsque $p \neq 2$ et $F = \mathbf{F}_p((t))$ est le corps des séries de Laurent en la variable t à coefficients dans $\mathbf{F}_p$, car l'espace des $I(1)$-invariants de $M\otimes_{H(2,p)\otimes_{\mathbf{Z}}\overline{\mathbf{F}}_p}\overline{\mathbf{F}}_p[I(1)\backslash G]$ est de dimension infinie lorsque M est supersingulier.

7 Nous expliquons la construction par Paskunas [Pa] d'une représentation irréductible admissible ayant un caractère central de $G = GL(2,F)$ sur $\overline{\mathbf{F}}_p$, tel que les $I(1)$-invariants du socle de sa restriction à $K = GL(2, O_F)$ est un $H(2,q) \otimes_{\mathbf{Z}} \overline{\mathbf{F}}_p$-module simple supersingulier quelconque (6.4). Une telle représentation est supersingulière (3.2), par (6.6)(i).

7.1 La construction part du principe que se donner une action de G sur un groupe abélien V est équivalent à se donner une action sur V de $K_o = p_F^{\mathbf{Z}} K$, et une action de $K_1 = p_F^{\mathbf{Z}} I \cup p_F^{\mathbf{Z}} I \begin{pmatrix} 0 & 1 \\ p_F & 0 \end{pmatrix}$ qui coincident sur $K_o \cap K_1 = p_F^{\mathbf{Z}} I$, où I le sous-groupe d'Iwahori supérieur. Ceci se démontre en utilisant l'arbre X de $PGL(2,F)$. Les groupes K_o, K_1 sont les stabilisateurs dans $GL(2,F)$ d'un sommet et d'une arête contenant ce sommet; les actions compatibles de K_o et de K_1 sur V définissent un système de coefficients $\mathcal{V}$ sur X qui est G-equivariant, d'homologie $H_o(X, \mathcal{V})$ est isomorphe à V; la représentation de G sur $H_o(X, \mathcal{V})$ prolonge les actions de K_o et de K_1 [Pa] 5.3.5.

7.2 Soit $M_{\eta,z}$ le $H(2,q) \otimes_{\mathbf{Z}} \overline{\mathbf{F}}_p$-module à droite simple supersingulier, de poids η modulo S_2 où p_F agit par z (6.3). Soit ρ_η la somme directe des deux représentations irréductibles de $GL(2, \mathbf{F}_q)$ de poids η modulo S_2 (6.2), vue comme une représentation de K triviale sur $K(1)$; on note $\rho_{\eta,z}$ l'action étendue à K_o en faisant agir p_F par z; les $H(2,q) \otimes_{\mathbf{Z}} \overline{\mathbf{F}}_p$-modules $M_{\eta,z}$ et $\rho_{\eta,z}^{I(1)}$ sont isomorphes. Soit $\mathrm{Inj}\,\rho_\eta$ l'enveloppe injective de la représentation ρ_η de K; on note $\mathrm{Inj}\,\rho_{\eta,z}$ l'action étendue à K_o en faisant agir p_F par z. Le point crucial est [Pa] 6.4 page 76:

(i) $\mathrm{Inj}\,\rho_{\eta,z}$ *est munie d'une action de* K_1 *compatible avec celle de* K_o*, donc d'une action de* G *(7.1), telle que*

(ii) l'inclusion $\rho_{\eta,z}^{I(1)} \subset (\mathrm{Inj}\,\rho_{\eta,z})^{I(1)}$ *est* $H(2,q) \otimes_{\mathbf{Z}} \overline{\mathbf{F}}_p$*-équivariante.*

La représentation de G recherchée est la représentation $\pi_{\eta,z}$ engendrée par $\rho_{\eta,z}$ dans la représentation $\mathrm{Inj}\,\rho_{\eta,z}$ de G (i). Elle est irréductible par l'argument suivant. Si π' est une sous-représentation non nulle de $\pi_{\eta,z}$, le socle de $\pi'|_K$ est non nul et contenu dans le socle $\rho_{\eta,z}$ de $\mathrm{Inj}\,\rho_{\eta,z}|_K$; la simplicité de $\rho_{\eta,z}^{I(1)}$ comme $H(2,q) \otimes_{\mathbf{Z}} \overline{\mathbf{F}}_p$-module déduite de (ii), implique $\rho_{\eta,z}^{I(1)} \subset (\pi')^{I(1)}$; comme $\rho_{\eta,z}$ et $\pi_{\eta,z}$ sont engendrés par $\rho_{\eta,z}^{I(1)}$ on déduit $\pi' = \pi_{\eta,z}$ donc $\pi_{\eta,z}$ est irréductible.

Les $I(1)$-invariants du socle de $\pi_{\eta,z}|_K$ est isomorphe au $H(2,q) \otimes_{\mathbf{Z}} \overline{\mathbf{F}}_p$-module simple supersingulier $M_{\eta,z}$; on ne sait pas si $\pi_{\eta,z}^{I(1)}$ est isomorphe à $M_{\eta,z}$. La représentation $\mathrm{Inj}\,\rho_{\eta,z}$ de G est admissible (4.2), donc $\pi_{\eta,z}$ est admissible.

7.3 Paskunas [Pa] 6.2 construit un autre système de coefficients G-équivariant $\mathcal{V}_{\eta,z}$ sur l'arbre tel que les $I(1)$-invariants de tout quotient irréductible de la représentation de G sur $H_o(X, \mathcal{V}_{\eta,z})$ contiennent $M_{\eta,z}$; il est associé à une action de K_1 sur $\rho_\eta^{I(1)}$ telle que l'inclusion

$\rho_\eta^{I(1)} \to \rho_\eta$ soit $K_o \cap K_1$-équivariante; la représentation de K_1 est isomorphe à $\mathrm{ind}_{Ip_F^{\mathbf{Z}}}^{K_1} \eta_z$, où η est relevé en un caractère η_z de $Ip_F^{\mathbf{Z}}$ sur lequel p_F agit par z.

Toute représentation irréductible π de G sur $\overline{\mathbf{F}}_p$ telle que $\pi^{I(1)} = M_{\eta,z}$, est quotient de $H_o(X, \mathcal{V}_{\eta,z})$. Si $p = q$ ou si le poids η est fixe par S_2, l'inclusion $M_{\eta,z} \subset \pi^{I(1)}$ suffit pour que π soit quotient de $H_o(X, \mathcal{V}_{\eta,z})$ [Pa] Cor.6.8, 6.10.

8 Nous montrons que la partie lisse de la contragrédiente d'une $\overline{\mathbf{F}}_p$-représentation irréductible lisse de $GL(2,F)$ ayant un caractère central et de dimension infinie, est nulle.

8.1 *Une forme linéaire lisse sur* $\mathrm{ind}_B^G \chi$ *est nulle, pour tout caractère* $\chi : B \to \overline{\mathbf{F}}_p^*$.

Preuve. Pour tout entier $n \geq 1$, on note $K(n)$ le n-ième sous-groupe de congruence de $GL(2, O_F)$.

Le caractère χ est trivial sur le pro-p-groupe $gK(n)g^{-1} \cap B$, aussi pour tout $g \in G$, il existe une fonction $f_{g,n} : G \to \overline{\mathbf{F}}_p$ de support $BgK(n)$ égale à $\chi(b)$ sur $bgK(n)$ pour tout $b \in B$. Comme $K(n)$ normalise $K(n+1)$, on a $f_{gh,n+1} = h^{-1} f_{g,n+1}$ pour tout $h \in K(n)$ et

$$f_{g,n} = \sum_h f_{gh,n+1} \quad \text{pour} \quad h \in (g^{-1}Bg \cap K(n))K(n+1) \backslash K(n).$$

Une forme linéaire lisse L sur $\mathrm{ind}_B^G \chi$ est fixée par un groupe de congruence assez petit. Il existe un entier $r \geq 1$ tel que $L(h^{-1} f_{g,n}) = L(f_{g,n})$ pour tout $h \in K(n)$ et pour tout $n \geq r$ et $g \in G$. On en déduit $L(f_{g,n}) = \sum_h L(h^{-1} f_{g,n+1}) = 0$, car les pro-$p$-groupes $K(n)$ et $(g^{-1}Bg \cap K(n))K(n+1)$ sont toujours distincts. Donc $L = 0$.

8.2 [L1] [L2] *Une forme linéaire lisse sur* $E(0)/TE(0)$ (3.2) *est nulle.*

Preuve. Soit $K_o = p_F^{\mathbf{Z}} GL(2, O_F)$. L'ensemble X_o des sommets de l'arbre de $PGL(2,F)$ est en bijection avec $GL(2,F)/K_o$. Une forme linéaire L sur $E(0)/TE(0)$ s'identifie à une fonction $f : X_o \to \overline{\mathbf{F}}_p$ de somme nulle sur les voisins de chaque sommet.

On note x_o le sommet fixe par K_o et $C(?)$ l'ensemble des sommets à distance ? de x_o, pour tout entier $? \geq 1$; le sous-groupe de congruence $K(?)$ fixe chaque sommet de $C(?)$ et agit transitivement sur les sommets de $C(?+1)$ se projetant sur le même sommet de $C(?)$; tout sommet $x \in C(?)$ est voisin d'un unique sommet $x_- \in C(?-1)$ (avec $C(0) = x_o$) et de q sommets $x_1, \ldots, x_q \in C(?+1)$.

Si L est lisse, il existe un entier $r \geq 1$ tel que L est fixe par $K(r)$; alors pour tout $x \in C(n+1)$, on a $L(x_-) + qL(x_1) = L(x_-) = 0$; donc L est nulle sur $C(n)$ pour $n \geq r$. Donc le support de L est fini. Le même argument montre que si L est nulle sur $C(? + 1)$, alors L est nulle sur $C(?)$ pour tout $? \geq 0$, car pour $y \in C(?)$ il existe $x \in C(? + 1)$ avec $x_- = y$. Donc $L = 0$.

8.3 *Il n'y a pas de forme linéaire lisse non nulle sur une représentation irréductible de dimension infinie de $GL(2, F)$ sur $\overline{\mathbf{F}}_p$ ayant un caractère central.*

Preuve. Pour une série principale ou une représentation spéciale Sp par (8.1) et (3.2). Pour une supersingulière par (3.2), (8.2) et sa généralisation que nous admettons: pour $0 \leq r \leq q-1$, une forme linéaire lisse sur $E(\underline{r})/TE(\underline{r})$ (3.2) est nulle (je ne l'ai pas vérifié si $r \neq 0$, mais Ron Livné dit l'avoir fait).

8.4 Soit π une représentation irréductible de $GL(2, F)$ sur $\overline{\mathbf{F}}_p$ de caractère central ω_π. Notons

$$\pi^* = \pi \otimes (\omega_\pi^{-1} \circ \det).$$

Avec les notations de (3.2), on a

$$\mathrm{Sp}^* = \mathrm{Sp}, \quad (\mathrm{ind}_B^G(\chi_1 \otimes \chi_2))^* = \mathrm{ind}_B^G(\chi_2^{-1} \otimes \chi_1^{-1}).$$

Lorsque $F = \mathbf{Q}_p$, on a $V(r, \chi)^* = V(r, \omega^{-r}\chi^{-1})$.

Pour π une représentation irréductible de $GL(2, F)$ sur $\mathbf{C}$, la contragrédiente de π est isomorphe à π^* [Bu] 4.2.2.

8.5 Lorsque $F = \mathbf{Q}_p$, le dual de Cartier d'une représentation σ de $\mathrm{Gal}(\overline{\mathbf{Q}}_p/\mathbf{Q}_p)$ est son dual usuel tordu par le caractère ω (2.4). *La correspondance $\sigma \leftrightarrow \pi$ de Breuil (3.4) envoie le dual de Cartier de σ sur π^* et le déterminant de σ sur le caractère $\omega_\pi\omega$ produit du caractère central de π par ω.*

9 Remarques finales. Pour $F \neq \mathbf{Q}_p$, on s'attend à ce qu'il existe d'autres $\overline{\mathbf{F}}_p$-représentations irréductibles supersingulières de $GL(2, F)$ que celles construites par Paskunas. Nous avons essayé de dégager les principes généraux des preuves de [BL], [Br], [Pa], dans le but d'une généralisation éventuelle. Certains résultats présentés ici sont déja étendus à $GL(3)$ [O], ou à $GL(n)$ ou même à un groupe réductif général.

Bibliographie modulo p

Barthel Laure, Livné Ron,

[BL1] Modular representations of GL_2 of a local field: the ordinary, unramified case. J. Number Theory 55 (1995), 1-27.

[BL2] Irreducible modular representations of GL_2 of a local field. Duke Math. J. 75 (1994), 261-292.

[Be] Berger Laurent, Représentations modulaires de $GL_2(\mathbf{Q}_p)$ et représentations galoisiennes de dimension 2. Preprint 2005.
[Br] Breuil Christophe, Sur quelques représentations modulaires et p-adiques de $GL_2(\mathbf{Q}_p)$. I, Compositio Math. 138 (2003), 165-168.
[L1] Livné Ron, lettre à Joseph Bernstein Nov 1996.
[L2] Livné Ron, lettre 31 oct 2000.
[O] Ollivier Rachel, Modules sur l'algèbre de Hecke du pro-p-Iwahori de $GL_n(F)$ en caractéristique p. Thèse 2005.
[Pa] Paskunas Vytautas, Coefficient systems and supersingular representations of $GL_2(F)$. Mémoires de la S.M.F. 99 (2004).

Vignéras Marie-France,

[V0] Representations modulo p of the p-adic group $GL(2,F)$. Compositio Math. 140 (2004) 333-358.
[V1] On a numerical Langlands correspondence modulo p with the pro-p-Iwahori Hecke ring. Mathematische Annalen 331 (2005), 523-556. Erratum: 333 (2005), 699-701.
[Vc] Représentations lisses irréductibles de $GL(2,F)$. Notes de cours. http://www.math.jussieu.fr/ vigneras/cours2MP22.pdf

Bibliographie modulo $\ell \neq p$

[V2] Correspondance de Langlands semi-simple pour $GL_n(F)$ modulo $\ell \neq p$, Invent. Math. 144 (2001), 177-223.
[V3] A propos d'une conjecture de Langlands modulaire. Dans "Finite Reductive Groups: Related Structures and Representations. Marc Cabanes, Editor. Birkhauser PM 141, 1997, 415-452.
[V4] Représentations ℓ-modulaires d'un groupe réductif p-adique avec $\ell \neq p$. Birkhäuser PM137 (1996).
[V5] Induced representations of reductive p-adic groups in characteristic $l \neq p$. Selecta Mathematica New Series 4 (1998) 549-623.

Bibliographie sur $\mathbf{C}$

[Bu] Bump Daniel, Automorphic forms and representations. Cambridge Studies in Advances Mathematics 55 (1997).

Printed in the United States
by Baker & Taylor Publisher Services